Nelson

BIOLOGY

Nelson
BIOLOGY

BOB RITTER
*University of Alberta/Austin O'Brien
High School
Edmonton, Alberta*

RICHARD F. COOMBS
*Science Coordinator
Roman Catholic School Board
for St. John's
St. John's, Newfoundland*

DR. R. BRUCE DRYSDALE
*Coordinator of Instruction:
Science/Health
Red Deer Public School District #104
Red Deer, Alberta*

DR. GRANT A. GARDNER
*Associate Professor
Department of Biology
Memorial University of Newfoundland
St. John's, Newfoundland*

DAVE T. LUNN
*Science Department Head
Henry Wise Wood High School
Calgary, Alberta*

Nelson Canada

© Nelson Canada,
A Division of Thomson Canada Limited, 1993

Published in 1993 by
©Nelson Canada,
A Division of Thomson Canada Limited
1120 Birchmount Road, Scarborough, Ontario, M1K 5G4

Care has been taken to trace ownership of copyright material contained in this publication. The publisher will gladly receive any information that will rectify any reference or credit line in subsequent editions.

∞

This book is printed on acid-free paper, approved under Environment Canada's "Environmental Choice Program." The choice of paper reflects Nelson Canada's goal of using, within the publishing process, the available resources, technology, and suppliers that are as environment friendly as possible.

All student investigations in this textbook have been designed to be as safe as possible, and have been reviewed by professionals specifically for that purpose. As well, appropriate warnings concerning potential safety hazards are included where applicable to particular investigations. However, responsibility for safety remains with the student, the classroom teacher, the school principal, and the school board.

ISBN 0-17-603860-4

Canadian Cataloguing in Publication Data

Main entry under title:

Nelson biology

Includes index.
ISBN 0-17-603860-4

1. Biology. I. Ritter, Robert John, 1950-

QH308.7.N45 1992   574   C92-093829-9

Cover photo: The cover shows a stand of yellow cedars in British Columbia. The inset shows a cougar, a common inhabitant of western Canada. *Tim Fitzharris;* **6**: Jim Brandenburg/Minden Pictures; **7**: Left, Illustrated by Dave Mazierski; Right, Dr. Bryan Eyden/Science Photo Library; **8**: Left, CNRI/Science Photo Library; Right, Illustrated by Dave Mazierski; **9**: Left, Illustrated by Dave Mazierski; Right, Monfred Kage/Peter Arnold Inc.; **10**: Left, Joseph Drivas/The Image Bank; Right, Illustrated by Dave Mazierski; **11**: Left, Illustrated by Dave Mazierski; Right, Howard Sochurek/Masterfile; **12**: Science Source/Photo Researchers; **13**: Top left, Dr. E.R. Degginger; Bottom left, Illustrated by Dave Mazierski; Right, P. Wallick/Miller Comstock; **14**: Mark Tomalty/Masterfile; **Unit 1** Main photo: This spectacular photograph is a compilation of thousands of separate images from the Tiros-N series of meteorological satellites; Sidebar: A glacier stream; **Unit 2** Main photo: Giant trees in a rain forest on Vancouver Island; Sidebar: Microscopic image of B-lymphocyte cells; **Unit 3** Main photo: A capillary (green) densely packed with red blood cells; Sidebar: The beginning of a blood clot. The slim threads are fibrinogen fibers; **Unit 4** Main photo: Skydivers join together to form a complex display while falling toward the earth; Sidebar: A laser is used to correct an eye disorder; **Unit 5** Main photo: Birth of a baby; Sidebar: Sunrise over the Atlantic Ocean; **Unit 6** Main photo: Stickleback eggs close to hatching; Sidebar: Seeds of a bean plant; **Unit 7** Main photo: A school of tropical fish; Sidebar: Penguins on the Antarctic continent.

Sponsoring Editor:	*Lynn Fisher*	**Photo Research:**	*Sandra Mark, Debbie Lonergan,*
Project Manager:	*Ruta Demery*		*Ann Ludbrook, Leonard Lessin*
Project Coordinator:	*Jennifer Dewey*	**Photo Editor:**	*Stephen Cowie*
Senior Supervising Editor:	*Susan Green*	**Design and Art Direction:**	*John Robb*
Copy Editor:	*James Leahy*	**Assistant Designer:**	*Liz Nyman*

Printed and Bound in the United States of America

10   (QPK)   03 02

C O N T E N T S

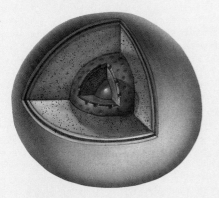

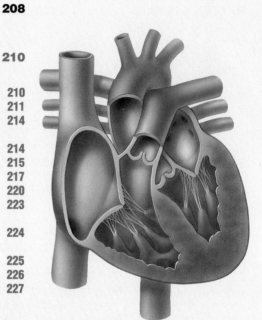

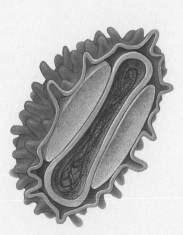

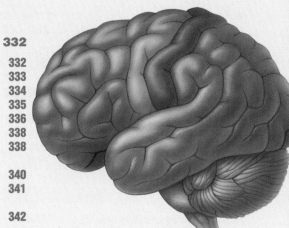

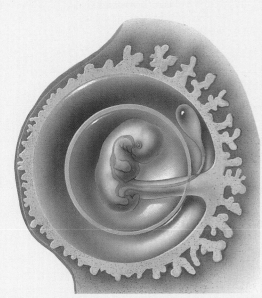

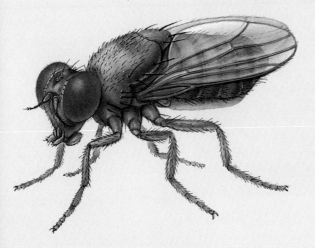

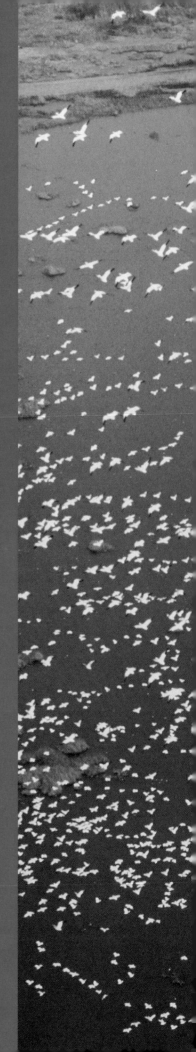

A number of science-related social issues are introduced throughout this textbook. At the end of each chapter, you are given an opportunity to debate one of these issues with other members of your class. An understanding of the issues requires an understanding of scientific principles, and the potentials and limitations of technological applications. Background information is provided in each Social Issue section as well as throughout the chapter. However, it is recommended that you do further research before the decision-making and debating begins. Collect a number of articles that present different viewpoints on the issue. In addition, you may want to speak with scientists, economists, or other professionals on the topic as you are formulating an opinion.

Read the point-counterpoint arguments presented in the debates. Note that these are not designed to provide a summary, nor are they meant to be comprehensive arguments. Rather, the arguments outline the complexity of the issues and act as a springboard for further research. Once you have completed your research, take an opportunity to reflect on your findings and discuss your viewpoint with others both inside and outside the classroom.

If you are not accustomed to viewing science and technology from a social context, you may expect science to provide you with unequivocal answers.

It is easy to fall into the trap of reducing any controversial issue to a "good-versus-bad" argument. The key to developing an understanding of why experts occasionally arrive at different conclusions is grounded in an understanding of the nature of science. Science does not provide absolute truths but presents an approach for interpreting nature. While research answers some questions, it raises many others.

When preparing for your debate it is also important to be aware of the limitations of scientific and technological problem-solving. Different social, moral, religious, and economic considerations must be recognized and valued. Controversy arises when people with competing world views draw divergent conclusions about the impact of science and technology on society. The debate format will enable you to recognize the difference between rejecting an idea and rejecting the person with the idea. The uniqueness of Canadian society stems, in part, from a plurality of beliefs and values.

The authors hope that these conflict-resolution scenarios will provide you with an opportunity to think critically, to reason, to argue logically, to devise answers that are supported by evidence, to reflect on your thinking, and to listen to others. The group debates will encourage you to share your concerns and will foster an appreciation of divergent world views.

Almost all human action involves a degree of risk to some component of the biosphere. When science is applied through technology, human influences are amplified. Evaluation of the technology requires careful analysis of its risks and benefits, as well as its short- and long-term consequences.

Although short-term consequences often provide empirical evidence, long-term consequences tend to be more difficult to predict. For example, the pesticide DDT proved very effective in killing malaria-carrying mosquitoes and in significantly reducing the incidence of malaria. The economic and social benefits were so conspicuous that DDT was hailed as a miracle chemical. Reasonably inexpensive to produce, DDT was used to control a variety of other insect pests.

Unfortunately, the long-term effects were not so favorable. It was discovered that DDT accumulates in fatty tissues and its concentrations increase as it passes through the food chain. DDT hampers egg formation in a number of birds, the peregrine falcon being the most familiar victim. DDT has also been found in the breast milk of mammals, including humans, and has been linked with birth defects. Even though DDT has not been used in Canada for many years, the problems associated with its long-term effects remain. There is no defined time limit when evaluating the risks of technology.

System Dependability

The difficulty in calculating the long-term risks of established technologies is increased when scientists attempt to determine how new technologies will affect the biosphere. Engineers and systems analysts construct models to identify things that may go wrong. However, their calculations of a system's reliability are based on probabilities. They must also consider both the reliability of the technology and the reliability of the people who control the technology. The following formula is used to calculate system reliability, or dependability:

System reliability = technological reliability × human reliability

As an example, consider the risks involved in the use of nuclear energy. Suppose a nuclear plant that is 98% reliable has a poorly trained staff that is only 60% reliable. Its system dependability, then, would be (0.98 × 0.60) or 58.8%. The Chernobyl disaster occurred when poorly trained individuals decided to disarm safety devices during a test of the plant. The near disaster at Three Mile Island was also attributed to human error.

Desirability of a Technology

To determine the desirability of a technology, scientists develop scenarios that examine cause-and-effect relationships. Desirability is expressed by means of the following formula:

$$\text{Desirability} = \frac{\text{social benefits}}{\text{social risks}}$$

One of the greatest difficulties in determining desirability involves attaching values to the risks and benefits. Not all risks and benefits have the same value. For example, does the satisfaction that smoking brings some individuals outweigh the health risks? The diagram below shows a risk-benefit analysis for medical X rays, coal-power plants, and smoking.

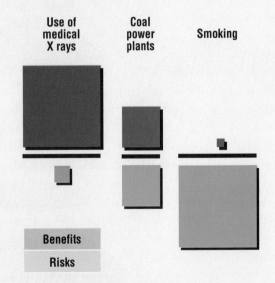

The desirability of a technology or product is determined by calculating social benefits and risks.

*C*oncept maps are used to link ideas. Although a variety of concept maps are presented throughout this textbook, the most meaningful ones are those you construct yourself. Research indicates that students who use concept maps as a study tool are often more successful on exams. Concept maps help you connect the facts you have learned in a way that is personally meaningful.

Your concept maps may be somewhat sparse at first—it is often difficult to identify the relationships between things you have studied. However, if you persist and go back to your concept maps before a test or exam, you will find that they provide an excellent summary as well as a strategy for understanding the material covered.

You may find the following points useful when using concept maps.

• Construct a concept map after every lesson. Because concept maps for most lessons do not need to be detailed, this process should only take you a few min-

utes. Check to see that all concepts presented in the lesson have been included in the concept map. If one seems to be missing, ask the teacher where it fits in.

• After every chapter, review your lesson concept maps and make a detailed chapter concept map linking the concepts learned in each lesson. You may want to look at concept maps made by other people in your class. Have they included any relationships that you might have missed?

• To prepare for unit tests, look for links between chapter concept maps. To accomplish this major review, you will need several sheets of paper. You may wish to use string or thread to make the interconnections.

The following concept map shows some of the relationships that exist in an ecosystem. How many relationships can you identify?

Relationships in Ecosystems

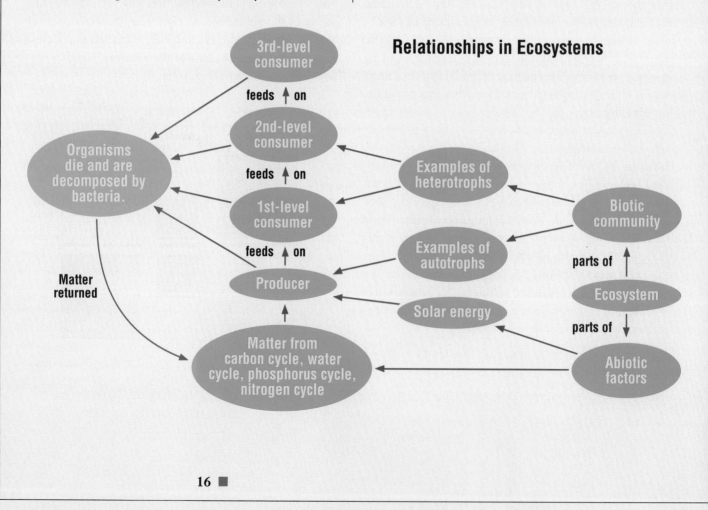

*T*he laboratory is a dynamic and exciting setting for learning biology; however, it is also a setting that can be potentially dangerous. The laboratory exercises in this textbook have been designed to minimize any hazards. However, to prevent placing yourself or other students in a harmful situation, it is important that you follow all safety procedures.

Safety Equipment

Your instructor will acquaint you with the safety equipment found in your laboratory. Locate and familiarize yourself with the use of the following safety items:

- Eyewash bottle or eye bath   • Chemical shower
- Fire extinguisher   • Fire blanket   • First-aid kit

Safety Symbols

The following symbols are used throughout the textbook to alert you to the use of specific safety equipment and procedures. You should recognize these symbols and know what they represent.

 Safety goggles: You must wear eye protection during any lab that involves flames, chemicals, or the possibility of broken glass or other small particles. Since safety goggles are shatter resistant and protect the side of the eyes, you must wear them even if you wear glasses.

 Laboratory apron: Wear a laboratory apron when handling any materials that could damage your skin or clothing.

 Wash your hands: Always wash your hands and forearms thoroughly with soap after any experiment.

General Safety Rules

1 Before you begin a lab, clear all unnecessary items from your work area and remove them to a designated place. Ensure that your hair, clothing, or jewelry will not interfere with your work. Maintain a clean and tidy work area during the laboratory. **2** Read all the directions for an experiment carefully several times before you begin your laboratory work. Before beginning any step, make sure that you understand what to do. Review any safety procedures that are specific to the exercise. Follow the steps exactly as they are written or modified by your instructor. Do not experiment on your own unless your instructor has approved your procedures and you have received permission to proceed. **3** Report all injuries, accidents, and spills, no matter how minor, to your instructor immediately. **4** Notify your instructor of any piece of equipment that is broken or does not work. Do not attempt to fix any broken or defective equipment on your own. **5** Before using any glassware, inspect it for chips or cracks. If it is chipped or cracked, do not use it. If glassware breaks, notify your instructor and dispose of it in the proper waste container using a safe procedure. Never force glass tubing into a rubber stopper; use gloves and a turning motion and a lubricant to help you. When heating a test tube, always point the open end away from yourself and others and move the tube back and forth through the flame. Use a clamp or tongs when handling glassware that has been heated. Before putting glassware away, clean it thoroughly as instructed by your teacher. **6** When you smell something, it should be done by a wafting of the fumes. Do not put your nose directly over the substance. **7** Do not taste or touch any material unless you are told to do so by your teacher. Do not eat or chew gum in the laboratory. **8** Caution must be exercised when dissecting an organism. **9** Disconnect electrical equipment by removing plugs from sockets and not pulling on the cord. Also remember that it is hazardous to have water or wet hands near an electrical source. **10** In microbiology labs, it is standard practice to assume that all microorganisms are potentially hazardous. **11** Follow carefully all directions given regarding handling, sterilization, and disposal of cultures and associated materials. **12** Dispose of any unused or waste materials, specimens, chemicals, etc., as instructed by your teacher. **13** When you have completed your laboratory exercise, clean up your work area, and return all equipment to its proper place and wash your hands thoroughly before leaving.

*L*ife in the Biosphere

■

U N I T

Equilibrium in the Biosphere

IMPORTANCE OF EQUILIBRIUM IN THE BIOSPHERE

One can only imagine how the Apollo astronauts felt when they first set foot on the moon and saw the spectacle of the living earth rising above the lifeless lunar rock. This contrast of life versus desolation must have made them appreciate the uniqueness of living organisms and the narrow range of physical conditions—known only on the earth—within which life can exist.

Think of our planet as a spaceship. Travelling around the sun in a slightly elliptical orbit, the earth carries with it the only forms of life known in the universe. It is a closed system. There is no outside source for life-sustaining raw materials or an interplanetary garbage dump to store wastes. Life is totally dependent on solar energy and the matter available aboard the spaceship earth.

J.E. Lovelock, a British scientist, compares the earth to a living body. The metaphor is referred to as the *Gaia* (pronounced "gay-ah") *hypothesis*, named after the Greek goddess of earth. Although a controversial idea in the

Figure 1.1

Earthrise. A blue and white earth rises over the horizon of the moon.

scientific community, it serves to emphasize that all living things interact with each other and with the nonliving components of our planet. In much the same way that the brain requires oxygen and nutrients from the circulatory system to function properly, each component of the earth's environment must be in a state of balance or equilibrium with every other component. What affects one part affects all parts. The expression "dynamic equilibrium" is used to describe any system in which changes are continuously occurring but whose components have the ability to adjust to these changes without disturbing the entire system.

As the 21st century approaches, ecologists have evidence to suggest that the earth is facing a crisis in which its dynamic equilibrium is being upset. However, scientists have not reached a consensus about the magnitude of the predicament or what can be done. The problems appear to be the result of the activities of a single dominant form of life: humans.

Humans have the ability to understand natural processes and act on this knowledge. By studying some of the well-established principles of ecology, you, as a member of the earth's community, can become a knowledgeable decision-maker. The decisions you make will, in part, determine the future direction of life on this small and fragile planet.

THE BIOSPHERE

There are three basic structural zones of the earth: the lithosphere (land), the hydrosphere (water), and the atmosphere (air). All are visible in Figure 1.1. However, at this distance, no signs of life can be seen. Living organisms are found in all three zones, an area referred to as the **biosphere,** whose limits extend from the ocean depths through the lower atmosphere. Most organisms are confined to a narrow band where the atmosphere meets the surface of the land and water. The regions of the planet that are not within the biosphere, such as the upper atmosphere or the earth's core, are also important because they affect living organisms.

Life forms are referred to as the **biotic,** or living, component. Chemical and geological factors, such as rocks and minerals, and physical factors, such as temperature and weather, are referred to as the **abiotic,** or nonliving, component. It is the interactions within and between the biotic and abiotic components that the ecologist endeavors to understand and explain.

When biologists investigate how a complex organism functions, they must study its various levels of organization. Moving from the simple to the more complex, biologists study individual cells, then tissues, organs and organ systems, and finally the integrated, functioning body. Ecologists investigating the biosphere proceed in a similar manner. Table 1.1 compares the levels of organization of an individual organism with those of the biosphere. By examining its individual parts, ecologists are able to bring together the various data and provide a picture of how the biosphere operates as an integrated unit.

*The **biosphere** is the narrow zone around the earth that harbors life.*

***Biotic** components are the biological or living components of the biosphere.*

***Abiotic** components are the nonliving components of the biosphere. They include chemical and physical factors.*

Table 1.1 Organizational Levels of an Organism and the Biosphere

Organism	Biosphere
Cell	Organism
Tissue	Population
Organ	Community
Organ system	Ecosystem
Complete organism	Biosphere

Ecological studies begin at the organism level. Investigations are designed to determine how the individual interacts with its biotic and abiotic environment. However, an organism does not live in isolation. It tends to group with others of the same species into **populations.** A population influences and is influenced by its immediate environment. When more than one population lives in an area, a **community** of organisms is established.

An **ecosystem,** the functional unit of the biosphere, has both biotic and abiotic components. The physical and chemical environment, as well as the community of organisms interact with each other in an ecosystem.

Table 1.2 Gases of the Atmosphere

Name of gas	Approximate percentage
Nitrogen	78.08
Oxygen	20.95
Argon	0.93
Carbon dioxide	0.03
Trace gases (e.g., ozone, methane)	±0.01
Water vapor	variable (0–4)

EARTH'S VITAL ATMOSPHERE

The **atmosphere** encircles the earth like a spherical skin. It is held to the earth by gravitational attraction. The atmosphere is composed of gases and contains a variety of solid particles such as pollen, dust, spores, bacteria, and viruses. With the exception of water vapor, the percentage composition of the other atmospheric gases is relatively fixed. The amounts of water vapor vary, depending on the temperature and the availability of water.

The atmosphere is described in terms of zones moving upward from the earth's surface to a height of some 900 km,

Figure 1.3

Graph showing temperature versus altitude for the zones of the atmosphere.

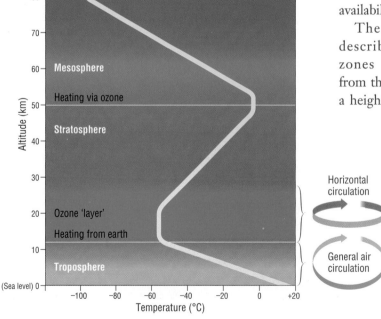

where it melds into outer space. The lowest level, the **troposphere,** has a thickness ranging from about 15 km at the equator to 8 km at the poles. This zone contains 80% of the atmosphere's mass along with most of its water vapor and dust. In the higher regions of the troposphere, the temperature drops steadily to about –60°C. Gigantic convection currents arise due to the difference in temperature between the air at the earth's surface and the air in the upper troposphere. These convection currents are the source of winds, turbulence, and weather.

Above the troposphere is the **stratosphere,** which extends to a height of approximately 50 km. The weather in this zone is stable. Here oxygen (O_2) actively absorbs **ultraviolet radiation** from the sun, forming **ozone** (O_3). Ozone is a bluish gas that is spread so thinly that it is barely noticeable. It is possible to catch a strong whiff of this sharp, clean-smelling gas after a thunderstorm. Lightning can split molecules of oxygen, which then combine to form ozone. Because ozone is unstable and reacts quickly with many other chemicals, its smell quickly dissipates after a storm.

Ozone can also form from air pollutants such as gasoline vapors and emissions from automobiles. High levels of ozone have been known to collect over large cities on hot summer days. Although the complete story is not known, high ozone concentrations at lower levels of the atmosphere can be harmful. Prolonged exposure to high concentrations of ozone will damage plants and cause respiratory problems.

The **mesosphere** contains little in the way of atmospheric gases—only traces of water vapor. At its upper limits, temperatures of –113°C are common. The mesosphere represents the limit of what might be called the true atmosphere.

Figure 1.4

The northern lights are caused by charged subatomic particles from the sun interacting with atoms in the upper atmosphere.

*The **troposphere** is the lowest region of the earth's atmosphere and extends upward about 12 km. Most weather occurs here.*

*The **stratosphere** is the region of the earth's atmosphere found above the troposphere. The ozone layer is found in this region.*

***Ultraviolet radiation** is the electromagnetic radiation from the sun and can cause burning of the skin (sunburn) and cellular mutations.*

***Ozone** is an inorganic molecule. A layer of ozone found in the stratosphere helps to screen out ultraviolet radiation.*

*The **mesosphere** is the region of the atmosphere found between the stratosphere and the upper atmosphere.*

*The **ionosphere** is a region of the upper atmosphere consisting of layers of ionized gases that produce the northern lights and reflect radio waves.*

*The **magnetosphere** is a region found above the outer atmosphere consisting of magnetic bands caused by the earth's magnetic field.*

The region above the mesosphere is referred to as the upper atmosphere. Of direct significance to life in this region is the **ionosphere,** which absorbs most of the deadly X rays and gamma rays produced by solar radiation. The aurora borealis, or northern lights, is produced in this zone.

Just beyond the limits of the atmosphere is a region sometimes referred to as the **magnetosphere.** Created by the earth's magnetic field, this layer can deflect most of the charged and potentially lethal particles that are emitted by the sun or that come from interstellar space. The Van Allen belts, discovered in 1958, are found at a height of some 950 km. They trap many of these particles in their doughnut-shaped bands.

HOLES IN THE OZONE LAYER

Without the ozone layer the earth would be bathed in ultraviolet radiation from the sun. The damage this form of radiation causes is so severe it is doubtful that plants or animals could survive its effects.

1979

1981

1987

1989

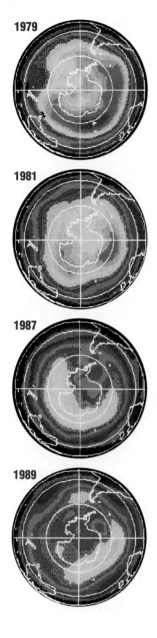

*An **ozone hole** is a region in the ozone layer in which the ozone levels have been considerably reduced and the layer has become very thin.*

1990

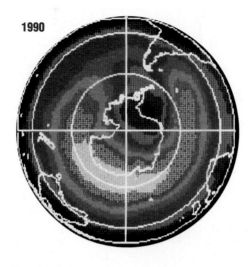

Figure 1.5

Pictures of the antarctic showing levels of ozone from 1979 to 1990. High ozone values are shown in orange/red, while low values are purple. In 1990 the black area inside the purple shows a hole in the ozone layer.

Fortunately, oxygen in the upper atmosphere is converted to ozone, which then screens out ultraviolet (UV) radiation. As a result, nearly 99% of the ultraviolet radiation striking the atmosphere never reaches the earth's surface.

Recent studies indicate that the ozone layer is slowly being depleted. The identification of **ozone holes** above Antarctica and the Canadian Arctic and the concurrent increase in skin cancers and eye problems associated with ultraviolet radiation are raising much concern among ecologists and the general public. Although a number of reasons for this depletion have been suggested, most of the attention has been focused on the release of chlorofluorocarbons (CFCs) into the atmosphere. These odorless compounds have been widely used as propellants in aerosol cans, as coolants in air conditioners, refrigerators, and freezers, and are also the waste products in the

manufacturing of some foam plastics. When the relatively inert CFC compounds diffuse into the atmosphere they are broken down by ultraviolet radiation. This process releases reactive chlorine molecules, each of which is capable of reacting with and breaking down thousands of ozone molecules.

Some scientists have speculated that holes in the ozone layer may have been developing since the beginning of time. They have suggested that we were simply unaware of the holes until recent technological advances allowed us to measure accurately the ozone concentrations in the upper atmosphere.

The rate at which the ozone layer is decreasing has alarmed the world scientific community. Predictions that as much as 60% of the ozone layer will be destroyed by the middle of the next century are not uncommon. Many countries have begun to take action on the problem by either banning or reducing freon-based propellants in aerosol cans, and by modifying the methods of producing plastic foam that produce chlorofluorocarbons. Canada has been heavily involved in international negotiations with countries that produce and release ozone-depleting materials. The Montreal Protocol, an international treaty, was signed in 1987 by over 31 members of the United Nations. Although a total ban on ozone-depleting products has not yet been attained, several governments have set up timetables for cutting emissions by specific percentages over a period of time. The first steps have been taken to correct the ozone problem. Whether the treaty will have a major effect or simply slow down a potentially disastrous situation is not yet known.

FRONTIERS OF TECHNOLOGY: PROBING THE UPPER ATMOSPHERE

The aurora borealis, or northern lights, is commonly observed in Canada and other northern countries. Since development in the Canadian North began, a variety of electronic communications problems, including total radio blackouts, have been associated with the aurora. This prompted Canadian scientists to study the upper atmosphere. By the 1930s Canada had taken a leading role in upper-atmosphere research.

In the late 1950s, Canada's National Research Council (NRC) began to use balloons to probe the atmosphere. These balloons carry instrumentation to altitudes of up to 40 km. The largest has a volume of 570 000 m³ and carries an instrument package of 1600 kg.

The NRC also supported the development of the Black Brant family of rockets and the Alouette 1515 scientific satellites. The largest, Black Brant X, is a three-stage rocket that can carry a 200 kg payload to an altitude of 700 km. The launch site for these rockets was located at the Churchill Research Range in northern Manitoba. Its three launch pads were used extensively in the 1970s and early 1980s, handling 20 m rockets, each weighing 5000 kg. Significant information on the magnetosphere and ionosphere was obtained. Both the balloon and rocket-based experiments have resulted in a new understanding of the properties of the northern lights and the transfer of solar energy down to the lower atmosphere. Additional information has been provided on the ozone layer, the electromagnetic fields surrounding the earth, cosmic radiation, magnetic storms, and even acid rain.

Through their association with industry, the U.S. National Aeronautics and Space Administration (NASA), the European Space Agency, and joint projects with scientists from many other countries, Canadian scientists are playing a major role in the development of highly sophisticated recording instruments and satellites, keeping Canada in the forefront of atmospheric research.

a)

b)

■ REVIEW QUESTIONS ?

1 In what ways is the earth like a spaceship?
2 What is meant by a closed system?
3 What are the abiotic and biotic components of the biosphere?
4 In what way does a community differ from an ecosystem?
5 Name the levels of organization in the biosphere.
6 Name the specific gases found in the atmosphere. What other materials are found there?
7 State where the ozone layer is located in the atmosphere and explain why it is important.
8 In what ways do the ionosphere and the magnetosphere protect living organisms?
9 What is meant by an ozone hole?

Figure 1.6

(a) A launch of balloons with experimental payloads as part of an arctic ozone hole research project. (b) A rocket leaves the launch pad carrying *Alouette I* spacecraft into orbit.

SOLAR ENERGY AND GLOBAL TEMPERATURES

Because of its unique position in the solar system, earth is the only planet that meets the delicate requirements for sustaining life. At an average distance of 14.5×10^7 km from the sun, the earth receives its primary source of energy in the form of solar radiation. Solar radiation interacts with the oceans, the atmosphere, and the continental masses to help maintain the narrow range of temperatures that can support life forms. Solar radiation is a combination of many forms of energy at different wavelengths, including light, heat, and ultraviolet radiation. Although heat from the earth's core and the tidal effect between the earth, the moon, and the sun do provide some energy to the biosphere, 99% of the earth's energy comes from solar radiation. Table 1.3 shows what happens to solar energy as it strikes the earth.

Much of the incoming radiation is reflected back out into space by the atmosphere and the earth's surface. The remaining portion is absorbed by the earth, converted to long-wave infrared wavelengths, and re-emitted as heat. Clouds and gases in the atmosphere trap surface heat radiation, reflecting it back to the earth and maintaining a global temperature nearly 33°C warmer than would be possible without an atmosphere.

Not all of the earth's surface is heated equally. Rays of sunlight striking the earth perpendicular to its surface transmit a specific amount of energy per unit area. As shown in Figure 1.8, the same rays hitting the surface at an angle spread the same amount of energy over a larger area. Therefore, each unit area receives less energy with a corresponding temperature drop.

Figure 1.7

Electromagnetic spectrum chart.

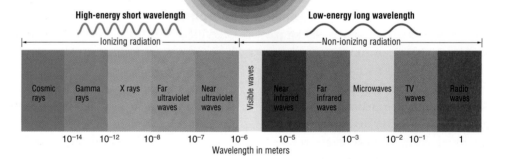

High-energy short wavelength   Low-energy long wavelength

Ionizing radiation   Non-ionizing radiation

| Cosmic rays | Gamma rays | X rays | Far ultraviolet waves | Near ultraviolet waves | Visible waves | Near infrared waves | Far infrared waves | Microwaves | TV waves | Radio waves |

10^{-14} 10^{-12} 10^{-8} 10^{-7} 10^{-6} 10^{-5} 10^{-3} 10^{-2} 10^{-1} 1

Wavelength in meters

Table 1.3 Distribution of Solar Energy

Incoming solar radiation	100%
Reflected by the atmosphere	30%
Absorbed by the atmosphere	45%
Used to drive the water cycle	23%
Used to drive winds and current	<1%
Used for photosynthesis	0.02%

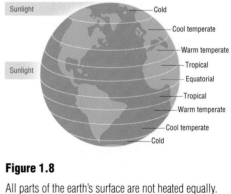

Figure 1.8

All parts of the earth's surface are not heated equally.

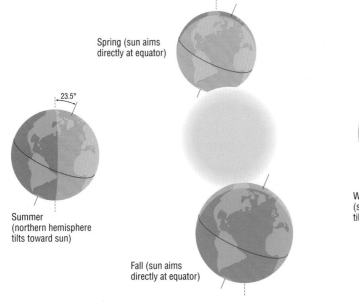

Spring (sun aims
directly at equator)

23.5°

Summer
(northern hemisphere
tilts toward sun)

Fall (sun aims
directly at equator)

Winter
(southern hemisphere
tilts toward sun)

Figure 1.9

Annual variation in the amount of incoming solar radiation.

If the earth's axis were perpendicular to the plane of its orbit around the sun, all areas of the earth would receive light during its 24-hour rotation and the amount of radiation would be the same at equal latitudes. Temperatures would not only become cooler, moving from the equator toward the poles, but the overall climate of any one region would be similar throughout the year; seasons would not exist. However, because the earth's axis is tilted at an angle of 23.5° from the perpendicular, the planet experiences seasons. Seasons are fairly well defined, particularly in the higher latitudes, and are reversed in the northern and southern hemispheres. As seen in Figure 1.9, there is an unequal distribution of incoming solar radiation, with the result that each of the polar regions experiences periods of continuous 24-hour daylight or darkness at some time during the year.

THE ALBEDO EFFECT

The term **albedo** is used to describe the extent to which a material can reflect sunlight. The higher the albedo, the greater the ability to reflect sunlight. Applying this principle to the solar radiation striking the earth, the higher the overall albedo of the earth, the less energy will be absorbed and available for maintaining the earth's global temperature. For example, the albedo of snow cover is extremely high. The presence of snow is a contributing factor to the low temperatures experienced during winter. Snow also delays warming in the spring, even though there is more solar radiation available per unit area. Lighter-colored areas of the earth's surface caused by the chemical composition of rock, the presence of very light-colored sand, or deforestation have a similar effect.

Water vapor added to the atmosphere causes more extensive cloud cover. The high albedo of cloud then causes more of the incoming radiant energy to be reflected directly back into space. An increase of dust in the atmosphere will produce the same effect. This phenomenon has been observed following volcanic activity. The cooling associated with a "nuclear winter," the result of atomic warfare or volcanic ash, increases the albedo of the atmosphere.

Figure 1.10

A column of ash, hot gases, and pulverized rock shoot out of Mt. St. Helens, Washington, during the second eruption in 1980. The increase of dust in the atmosphere produces a cooling effect.

Albedo *is a term used to describe the extent to which a surface can reflect light that strikes it. An albedo of 0.08 means that 8% of the light is reflected.*

LABORATORY

INVESTIGATING THE ALBEDO EFFECT

Objective

To examine the ability of selected colors and surface conditions to reflect light.

Materials

light source
2 ring stands
extension clamp
photocell
voltmeter
colored card stock
 (including black and white)

dissecting pan
sand
gravel
soil
water
snow and/or ice

Procedure

1 Attach the light source to the top of one of the ring stands. Aim the light down at the bench surface.

2 Attach the photocell unit to the second stand so that it is higher than the light source. Its active surface must face down and have a clear path to the surface of the bench. Attach the voltmeter to the photocell unit so that the photocell, wires, and meter do not interfere with the light source.

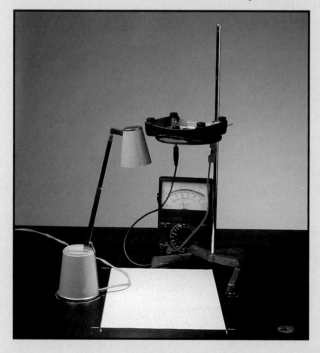

3 Place the sheet of white card stock directly under the light source.

4 Turn on the light source and measure the voltage produced by the photocell. Record the voltage.

   a) Why is it necessary to keep the photocell, wire, and voltmeter out of the direct line of light?

   b) Why does the voltmeter register an electrical change?

5 Repeat the experiment, using the other colored sheets. Remember to control all your variables. Record your data.

   c) What variable must be controlled?

   d) Why are black and white card stock colors used in this experiment?

6 Prepare a table similar to the one below.

Color of surface	Reflected light (in volts)

   e) List the colored surfaces in order from least to greatest reflected light.

7 Set the dissecting pan directly under the light source.

8 Record the reflective value of the pan when empty. Repeat, using a sand surface, a gravel surface, soil, water, and snow and ice.

9 Prepare a table similar to the one in part 6.

   f) Use your data to make comments on the albedo effect of the materials used.

Laboratory Application Questions

1 Design an experiment to measure the effect of color or surface conditions on the absorption of heat. State your hypothesis and the predictions resulting from it.

2 Perform the investigation (approval of instructor required) and draw conclusions based on the collected data.

3 What variable were you unable to control in part 8 of the laboratory procedure? How could you redesign these steps to account for the variable?

4 In order to melt a pile of snow more quickly, a researcher sprayed dye-colored water on it. Which color was probably selected and why? ■

RESEARCH
IN CANADA

Dr. Jill Oakes was born in British Columbia and received her Ph.D. from the University of Manitoba. Her research in northern Alaska, Canada, and Greenland has focused on the Inuit peoples, especially the way in which their clothing has helped them withstand the extreme climatic conditions of the arctic environment. With the arrival of northern European, American Alaskan, and southern Canadian influences into the arctic, the various Inuit societies have undergone significant changes. Most Inuit have lost the skills required to produce arctic clothing and many have switched to synthetic materials.

Dr. Oakes has lived with Inuit from two different regions of the arctic and has learned how to make Inuit clothing from the few individuals who still retain the skills. Travelling across the arctic, she has carefully recorded the details of how each item is constructed. Dr. Oakes' investigations led to the discovery of a unique type of parka made by the Inuit of the Belcher Islands in Hudson Bay. Lacking caribou from which to obtain the customary skins used in arctic clothing, the islanders had to rely on the skins and attached feathers of eider ducks. Dr. Oakes is continuing her work at the University of Alberta. She has obtained her commercial flying licence and is in the process of constructing a biplane.

Inuit Clothing

DR. JILL OAKES

BIOGEOCHEMICAL CYCLES

Table 1.4 The Four Groups of Organic Compounds

Group	Primary Function
Carbohydrates	cell energy
Lipids	energy storage
Proteins	cell structure
Nucleic acids	heredity material

If energy or matter were removed from the biosphere, life would cease to exist. Energy is required to drive both the biotic and abiotic components. All living things depend on abiotic materials to build their body structures and to serve as raw materials during energy-requiring metabolic reactions.

The materials used in building the geological structure of the earth or the bodies of living organisms are limited to those atoms and molecules that make up the planet. In effect, no alternative source of matter is available. Therefore, to maintain the biosphere, matter must be recycled. The elements carbon, hydrogen, oxygen, and nitrogen are the main constituents of the four basic organic compounds found in living organisms: carbohydrates, lipids, proteins, and nucleic acids. When an organism dies and decays, its complex organic molecules are broken down into their basic constituents and once again become available to the biosphere. It is quite possible that atoms such as carbon, hydrogen, oxygen, and nitrogen, which are basic components of your body, could have been part of another organism in the past, possibly a dinosaur. The same atoms will eventually pass from you into the abiotic environment and could once again become part of the biotic world at some future time. This use and reuse of materials of the earth is referred to as a **biogeochemical cycle.**

One biochemical process fundamental to life is **photosynthesis.** Photosynthetic organisms combine atmospheric carbon dioxide with water to produce oxygen gas and glucose, a simple sugar molecule. **Chlorophyll** must be present, along with certain enzymes and sunlight (solar radiant energy) for photosynthesis to occur. The balanced chemical equation below (simplified) illustrates the photosynthetic process.

$$\underset{\text{carbon}}{6CO_2} + \underset{\text{water}}{6H_2O} + \text{solar energy} \xrightarrow{\text{enzymes}} \underset{\text{glucose}}{C_6H_{12}O_6} + \underset{\text{oxygen}}{6O_2}$$

chlorophyll

Solar energy is used to bond carbon, hydrogen, and oxygen atoms into glucose, which is thus made available to other organisms as their primary energy source. Glucose is one of a number of organic molecules produced by photosynthetic

*A **biogeochemical cycle** is the complex, cyclical transfer of nutrients from the environment to an organism and back to the environment.*

***Photosynthesis** is the process by which plants and some bacteria use chlorophyll, a green pigment, to trap sunlight energy. The energy is used to synthesize carbohydrates.*

***Chlorophyll** is the pigment that makes plants green. Chlorophyll traps sunlight energy for photosynthesis.*

Figure 1.11

A simplified pathway of matter as it cycles between living organisms and the nonliving environment.

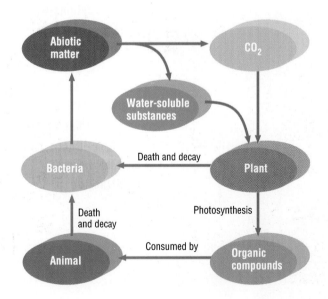

Unit One:
Life in the Biosphere

organisms. Other molecules, including complex carbohydrates, lipids, proteins, and nucleic acids, are also required by all living organisms. Animals and other non-photosynthetic organisms obtain their organic molecules from the photosynthesizers and, in turn, pass these molecules to other animals. A few primitive organisms are capable of generating organic molecules by means of a process called **chemosynthesis.**

THE HYDROLOGIC CYCLE

Water, a requirement for all living organisms, plays a critical role in maintaining a relatively stable global heat balance. Water is the solvent in which most metabolic reactions take place. It is the major component of a cell's cytoplasm. Many organisms live within water's stable environment, while others depend on water to carry dissolved nutrients to their cells and organs. Table 1.5 lists several ways in which organisms depend on water.

The volume of water in the biosphere in its many forms remains fairly constant. However, the specific amount in any one phase can vary considerably; water is continuously entering and leaving living systems.

Table 1.5 Importance of Water to Organisms

Absorbs and releases heat energy.

Is the medium in which metabolic reactions take place in organisms.

Is an excellent solvent.

Composes about 60% of a cell's mass.

Supplies the hydrogen to organisms during the metabolism of key organic molecules and oxygen atoms for atmospheric oxygen production during photosynthesis.

Is a reactant in some metabolic activities and a product in others.

The pathway of water through the biosphere is called the **hydrologic** or **water cycle.** This cycle is shown in Figure 1.12. Water reaching the earth's surface as precipitation (rain, snow, sleet, hail, or any combination of these forms) can enter a number of pathways. It may remain on the surface as standing water (lakes, swamps, sloughs) or form rivers and streams, which terminate in the oceans, where the bulk of the earth's water reserves are held. Some precipitation seeps into the soil and subsurface

Chemosynthesis *occurs when some organisms use energy from chemicals rather than sunlight to produce glucose.*

The **hydrologic cycle,** *or* **water cycle,** *is the movement of water through the environment—from the atmosphere to earth and back.*

Figure 1.12

Simplified diagram of the water cycle.

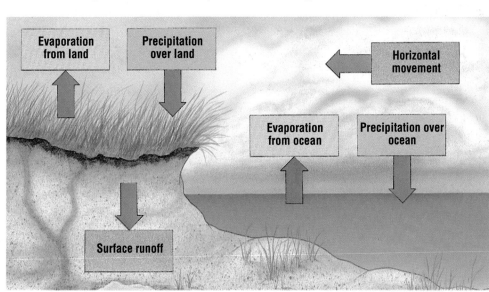

crustal rock to form groundwater. If the rock is permeable, some of the ground-water may seep to the surface, forming individual springs or adding water to existing lakes and streams. The movement of water through rock is slow but measurable. When groundwater is removed from an area during large-scale deep-well drilling, the rock can slowly be depleted of its water. This water may take centuries to be replaced.

By absorbing heat energy from the sun, some of the surface water evaporates as water vapor. It rises upward into the atmosphere until it reaches a point where the temperature is low enough for the water vapor to condense into tiny droplets of liquid water. These droplets are so light that they remain suspended in the atmosphere as clouds, each supported by rising air currents and winds. When conditions are right (i.e., a temperature drop), the droplets come together to form larger drops or sometimes ice crystals. Once the mass of the droplet or ice crystal can no longer be supported by air currents, precipitation occurs. The cycle repeats itself endlessly.

Living organisms also play a vital role in the water cycle. Water enters all organisms through osmosis and is utilized in various ways during their metabolic activities. It would appear then that living things tend to remove water from the environment, thus interfering with the cycle. However, through such processes as **cellular respiration** and the decay of dead organisms, water is released back into the atmosphere. Plants, particularly broadleaf trees and shrubs, play a major role in water recycling through the process of **transpiration.** Where there has been extensive removal of forests by logging or burning, there is less water in the atmosphere and noticeable climatic changes can occur. Surface runoff patterns become disturbed and the water-

holding capability of the soil may be reduced. This helps explain why the destruction of Brazilian rain forests only provides temporary usable land for agriculture.

Acid Deposition and the Water Cycle

Technology is often described as a doubled-edged sword, cutting through problems with one edge, while scarring the environment with the other. Nowhere is this more evident than with the technologies that contribute to acid deposition. Coal-burning plants, metal smelters, and oil refineries provide energy for the industrial world, but at the same time produce oxides of sulfur and nitrogen, listed among the most dangerous of air pollutants.

When fossil fuels and metal ores containing sulfur undergo combustion, the sulfur is released in the form of sulfur dioxide (SO_2), a poisonous gas. Combustion in automobiles, fossil-fuel-burning power plants, and the processing of nitrogen fertilizers produce various nitrous oxides (NO_x). Sulfur and nitrous oxides often enter the atmosphere and combine with water droplets to form acids. On entering the water cycle, the acids return to the surface of the earth in the form of snow or rain. The term *acid rain* is used to describe the movement of sulfur- and nitrogen-containing acids from the atmosphere to the land and water. Acid rain 40 times more acidic than normal rain has been recorded. Such acid precipitation, some as acidic as lemon juice, reacts with marble, metal, mortar, rubber, and even plastics. The impact of acid rain on ecosystems has been well documented. Acid precipitation kills fish, soil bacteria, and both aquatic and terrestrial plants. However, the devastation is rarely uniform in all areas. Some regions are more sensitive to acid rain than others. Alkaline soils neutralize

Cellular respiration *is the process by which living things convert the chemical energy in sugars into the energy used to fuel cellular activities.*

Transpiration *is the loss of water through the leaves of a plant.*

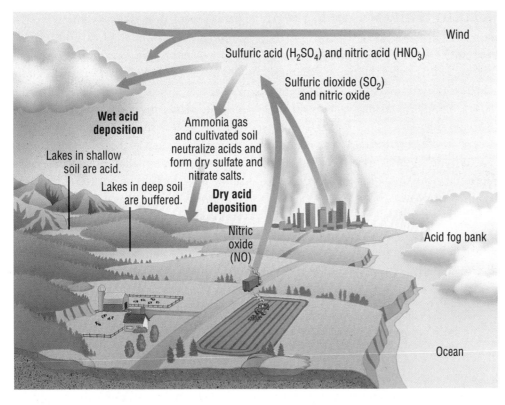

Wet acid deposition

Lakes in shallow soil are acid.

Lakes in deep soil are buffered.

Ammonia gas and cultivated soil neutralize acids and form dry sulfate and nitrate salts.

Dry acid deposition

Nitric oxide (NO)

Sulfuric acid (H$_2$SO$_4$) and nitric acid (HNO$_3$)

Sulfuric dioxide (SO$_2$) and nitric oxide

Wind

Acid fog bank

Ocean

Figure 1.13

Wet and dry acid deposition.

Figure 1.14

Devastation of forest areas caused by long-term exposure to acid rain. Conifers are particularly susceptible to acid rain. Scarring caused by acids makes the trees vulnerable to a variety of infections.

acids, minimizing the impact of the corrosive acids before runoff carries them to streams and lakes. Unfortunately, soils in much of southeastern Canada have a thin layer of rich soil on top of a solid granite base. Granite offers little in the way of buffering agents that would neutralize the acids.

The sulfur and nitrous oxides released from smokestacks do not always enter the water cycle in the atmosphere. Depending on weather conditions, particles of sulfur and nitrogen compounds may remain airborne and settle out. These dry pollutants then form acids when they combine with moisture. The dew from a lawn, the surface of a lake, or the water inside your respiratory tract are but a few of the potential sites where such acids can form.

One early solution to the problem of acid deposition was to build taller smokestacks. The average height of smokestacks in the 1950s was about 100 m. By the 1980s it was 224 m, with some exceeding 300 m. It was reasoned that if the pollutants could be expelled farther into the upper atmosphere, fewer would fall on any specific area. However, such thinking turned a local problem into an international one. Acid deposition remained suspended in the air and was often carried hundreds of kilometers from the site of the pollution across international borders and to areas less able to buffer the acids.

Technology offers some solutions to the problems that have been created by oxides of sulfur and nitrogen. Scrubbers have been placed in smokestacks to remove harmful emissions, and lime has been added to lakes in an attempt to neutralize acids that have descended from the atmosphere. However, both of the solutions are expensive.

The prospect of improving smelters is equally difficult. Mining companies are already battling to remain operational and compete in a world market. Would tougher legislation result in higher levels of unemployment?

THE CARBON CYCLE

Carbon is the key element in all organic compounds. In the abiotic environment, carbon is part of the carbon dioxide (CO_2) molecule. Some carbon dioxide exists in the atmosphere, but most is dissolved in the oceans. Each year about 50 to 70 billion tonnes of carbon enter the biotic environment through the photosynthetic process.

Some of the organic carbon is released back to the atmosphere as carbon dioxide through cellular respiration, as illustrated in the simplified equation below:

$$C_6H_{12}O_6 + 6O_2 \xrightarrow{\text{enzymes}} 6H_2O + 6CO_2$$

glucose   oxygen   water   carbon dioxide

Because photosynthesis and cellular respiration are complementary processes, this phase is often called the *carbon-oxygen cycle*. Most of the carbon that remains as part of living organisms is returned to the atmosphere or water as carbon dioxide when wastes and the bodies of dead organisms decay. Under certain conditions the decay process is delayed and the organic matter may be converted into fossil fuels such as coal, petroleum, and natural gas. This carbon is then unavailable to the cycle unless natural or human events allow these fuels to be burned. The burning process releases carbon dioxide to the atmosphere.

In aquatic systems, particularly the oceans, inorganic carbon can be incorporated into carbonate compounds, which form the shells and other hard structures of many organisms. If the carbonates become part of sedimentary rock, the carbon can be trapped for millions of years until geological conditions bring it back to the surface. Natural heat produced by volcanic activity can break down carbonate-containing rocks such as limestone, releasing carbon dioxide.

Figure 1.15

Simplified diagram of the carbon cycle.

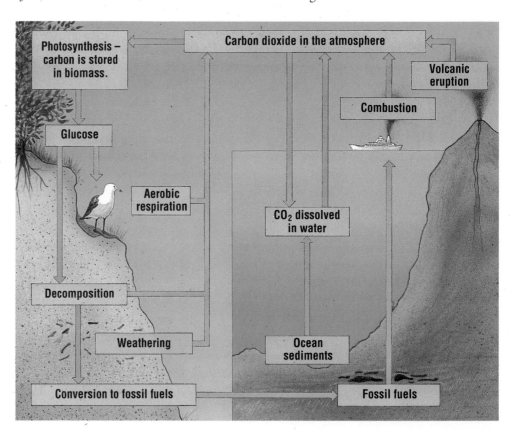

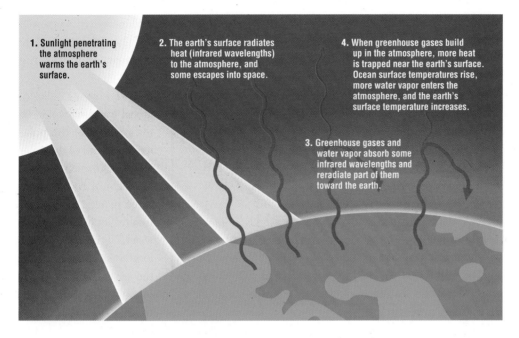

1. Sunlight penetrating the atmosphere warms the earth's surface.

2. The earth's surface radiates heat (infrared wavelengths) to the atmosphere, and some escapes into space.

4. When greenhouse gases build up in the atmosphere, more heat is trapped near the earth's surface. Ocean surface temperatures rise, more water vapor enters the atmosphere, and the earth's surface temperature increases.

3. Greenhouse gases and water vapor absorb some infrared wavelengths and reradiate part of them toward the earth.

Figure 1.16

The greenhouse effect.

Humans have modified the global carbon cycle through the accelerated use of fossil fuels and by the burning of forests. These processes release carbon dioxide at a rate well above that of natural cycling. In addition, the destruction of vegetation reduces the capacity for photosynthesis and thus decreases the volume of carbon dioxide removed from the atmosphere. Most carbon dioxide released to the air becomes dissolved in the oceans, but the oceans can only hold so much. The level of carbon dioxide in the atmosphere is definitely rising and with it comes the possibility of "greenhouse" global warming.

The Greenhouse Effect and the Carbon Cycle

The term "greenhouse effect," first coined in the 1930s, is used to compare the heat-blocking action of atmospheric gases to the glass that covers a greenhouse. When the sun shines on a glass-covered greenhouse, the shorter wavelengths of light enter through the glass, strike objects within the greenhouse, and are converted to radiation of a longer wavelength. The glass acts as a shield, preventing the longer infrared wavelengths from escaping. Have you ever noticed that on a sunny winter day the temperature inside an unheated car can increase to such an extent that it can become uncomfortable? This trapping of thermal energy is caused by the greenhouse effect.

Many of the constituents of the atmosphere, such as carbon dioxide, produce the same greenhouse effect but on a global scale. Wavelengths in the visible spectrum strike the earth's surface and are reflected in the form of longer wavelengths. Atmospheric gases trap the longer infrared radiation in a blanket of heat.

Table 1.6 Greenhouse Gases

Name of gas	Chemical formula	Source
Carbon dioxide	CO_2	Burning of fossil fuels
Methane	CH_4	Bacteria decay activity and natural feed lot
Nitrous oxide	N_2O	Fertilizers and animal wastes
Chlorofluorocarbons (CFCs)		Plastic foams, refrigerator coolants

Since the beginning of the industrial revolution, and particularly during the latter part of the 20th century, human activities have increased the concentration of greenhouse gases in the atmosphere. Since the middle of the 19th century, carbon dioxide levels in the atmosphere have increased by about 25%. Worldwide estimates indicate that nearly five billion tonnes are now being added to the atmosphere each year. There is no sign that these percentages will soon decrease. Although Canada produces only slightly more than two percent of the world's total carbon dioxide output, it is a world leader on a per capita basis.

*The **tundra** is the most northerly major life zone, and is characterized by low precipitation, low temperatures, permafrost, and a lack of trees.*

*The **boreal forest** is a worldwide forest region found in the upper latitudes that is characterized by coniferous trees.*

Figure 1.17

Temperature change.

climate changes and potential warming. However, many of the models to date support the prediction of increasing world temperatures.

One obvious effect of global warming would be the melting of the polar ice caps. If all of the polar ice were to melt, the sea level could rise by over three meters. This could totally flood most of the world's major cities, located as they are along the sea coasts. The various climate patterns of the world would also undergo change. Some regions of the earth would receive an increase in rainfall, while others would experience a decrease. These factors would have an impact on agriculture and food production and alter natural vegetation patterns.

The first regions to feel the effect would be the north—the **tundra** and the **boreal forest.** Winter temperatures over Hudson Bay, the Davis Strait, and the Arctic Ocean might rise by as much as 11°C, making these bodies of water ice free in the winter. Precipitation patterns would also be disrupted. Since ocean water can modify the climate of land, the tundra conditions of the far north could disappear. This may seem potentially beneficial since it would increase the area of land suitable for agriculture. Whether the vegetation could tolerate such a sudden transition or whether soil could form normally or rapidly enough are some of the questions that remain unanswered. The southern prairies, with a predicted annual increase

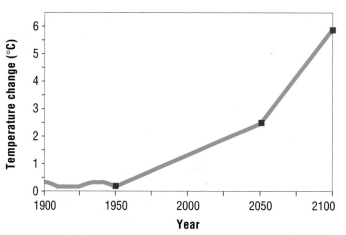

Because of this massive increase in carbon dioxide levels, some scientists have predicted an increase in global temperatures of 2 to 5°C by the end of the next century. This may not sound like much of a change, but on a global scale it is enormous. Mathematical computer models have been developed using available data on warming and climate to predict future events. At best these models can only show a trend, since they cannot account for all of the variables involved in both

Unit One:
Life in the Biosphere

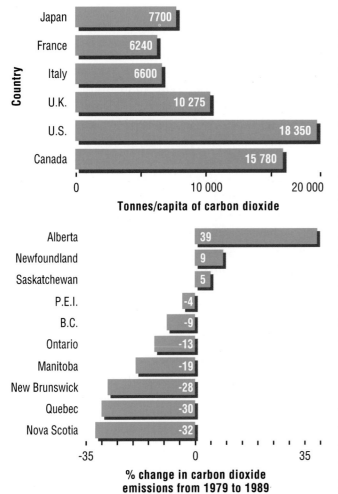

Figure 1.18

Carbon dioxide emissions in 1987.

Figure 1.19

Percentage change in carbon dioxide emissions.

to remove carbon dioxide from the atmosphere. When vast areas of forest are burned to clear areas for agriculture, carbon dioxide is added to the atmosphere as well as dust and ash. These solid particles raise the albedo of the atmosphere, lowering solar radiation input and compensating for any potential greenhouse warming.

There are still many unanswered questions concerning the overall effects of global warming. The latest information seems to indicate that temperatures *are* on the increase and that the amount of carbon dioxide entering the atmosphere is well above natural levels. It will take time and long negotiations at an international level to deal with the problem. On spaceship earth everyone is involved.

of 4 to 6°C, could become desert, even if precipitation also increased. In much of northern Canada, Europe, and Asia, the cold and acid conditions in bogs and muskegs cause dead organic matter to decompose very slowly. Any rise in temperature would speed up decay and release methane gas in abnormally high quantities. Since methane is a greenhouse gas, it would enhance the rise in temperatures.

Some experts contend that there is no reason for alarm. Excess carbon dioxide, they say, can normally dissolve in surface water. With 66% of the earth covered by water there may be room in this "sink" for much more carbon dioxide. Extra carbon dioxide also increases plant growth. Since carbon dioxide is used in photosynthesis, this process would tend

■ REVIEW QUESTIONS ■ ?

10 List the factors responsible for determining how much heat from solar radiation is retained by the earth.

11 What is the effect of the tilt of the earth's axis on the absorption of heat at the earth's surface?

12 How does water help in maintaining global temperatures?

13 The albedo of a rocky surface is 0.6 (60%). What does this mean in terms of heat absorption?

14 What is the greenhouse effect?

15 Describe some of the possible effects of global warming.

THE NITROGEN CYCLE

Without nitrogen, proteins and nucleic acids cannot be produced. When you consider that nitrogen gas (N_2) composes nearly 79% of the earth's atmosphere, easy access to nitrogen would not seem to be a problem for living organisms. Unfortunately, this is not the case. Nitrogen gas, a very stable molecule, reacts chemically only under unique conditions. As a result, nitrogen is most often utilized by organisms in the form of the nitrate ion (NO_3^-) found in nitrate salts. Even then, only those organisms capable of making their own food can use nitrates to generate proteins, nucleic acids, and other nitrogen compounds.

The **nitrogen cycle** is very complex. In this simplified description, you will examine two ways in which atmospheric nitrogen can be converted into the nitrates, the usable form of nitrogen. The process is called **nitrogen fixation.**

Lightning causes nitrogen gas to react with oxygen in the air to produce nitrates. Nitrates dissolve in the rain or surface waters, enter the soil, and then move into plants through their roots. The plants can then produce a variety of proteins for their own use. If consumed by animals, the plant proteins are reorganized to provide the type of protein molecules required by animals. Plants are the only nitrogen source for animals.

All organisms produce wastes and eventually die. In bacterial decay, proteins are slowly broken down into ammonia compounds, then to nitrites, and finally back to nitrates. The nitrates can then re-enter the cycle.

At various stages in the decay process, **denitrifying bacteria** can break down nitrites and nitrates, releasing nitrogen gas back into the atmosphere. This ensures the balance between soil nitrates and nitrites, and atmospheric nitrogen. Older lawns often contain many of these microbes. The fact that denitrifying bacteria respire best under **anaerobic** conditions may help explain why people often aerate their lawns in early spring. By exposing these denitrifying bacteria to oxygen, the breakdown of nitrates to free nitrogen is reduced—nitrates and nitrites remain in the soil, and can be used by

Figure 1.20

Nitrogen (N_2) must be converted into a form that plants can use to make proteins.

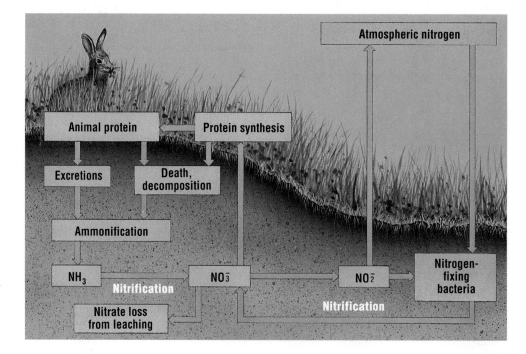

plants to make proteins. This also explains why older lawns lack a rich green color. Chlorophyll, the green pigment of plants, is a protein and therefore requires sufficient levels of nitrates to maintain the green color.

The denitrification process is enhanced when the soil is very acid or waterlogged. Bogs, for example, are deficient in nitrogen. Unique plants, such as sundews and pitcher plants, which are native to bogs, obtain their nitrogen by digesting trapped insects. In an interesting reversal of roles, these plants obtain their protein from animals.

An alternate pathway for the production of nitrates is through **nitrogen-fixing bacteria.** These bacteria grow in nodules on the roots of legumes such as clover or alfalfa. They convert atmospheric nitrogen to nitrates. The bacteria provide these plants with a built-in supply of usable nitrogen, while the plant supplies the nitrogen-fixing bacteria with needed sugars. There is usually much more nitrate produced than required. The excess nitrates move into the soil, providing an alternate source of nitrogen for other plants. The traditional agricultural practice of rotating crops capitalizes on bacterial nitrogen fixation. Plants that are fertilized with manure and/or other decaying matter also take advantage of the nitrogen cycle. Throughout this cycle the mass of nitrogen gas removed from the atmosphere is balanced by the mass being returned.

Have you ever had to pick clover from your lawn by hand? There is an easier way to get rid of it. Because clover contains microbes capable of fixing nitrogen, it has a special advantage over other plants that cannot fix their own nitrogen. This means that clover grows well on lawns that lack nitrogen. Healthy lawns, which are rich in nitrogen, rarely allow clover to establish itself.

THE PHOSPHORUS CYCLE

Phosphorus is a key element in cell membranes, the energy storage molecules, and the calcium phosphate of mammalian bone. (You can learn more about energy storage molecules in the chapter Energy within the Cell.) Environmental phosphorus tends to recycle in two ways: a long-term cycle involving rocks of the earth's crust and a shorter cycle involving living organisms. Phosphorus is usually found in the form of phosphate ions, combined with a variety of elements, as part of the continental rock. Phosphates are soluble and dissolve in water as part of the hydrologic cycle. During this phase phosphates can be absorbed by photosynthetic organisms and passed through the food chains. Erosion carries phosphates from the land to streams and rivers, and then finally to the oceans. The ocean phosphates form sediments that, through geological activities, may be thrust upward and once again become part of the land surface. The overall process can take millions of years.

Figure 1.21

Nitrogen-fixing bacteria grow in root nodules of plants in the legume family.

Figure 1.22

Phosphate cycles through both long and short cycles.

Nitrogen-fixing bacteria *convert atmospheric nitrogen to nitrogen compounds such as ammonia and nitrate.*

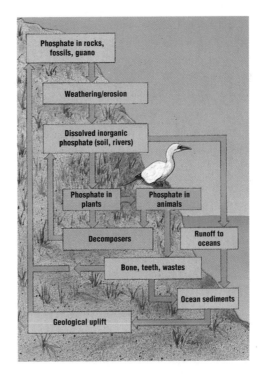

Most of the phosphates used in photosynthesis enter the shorter cycle. This phase is able to meet the high demand for phosphates by all organisms through rapid recycling. Photosynthesis is required for usable phosphates to enter the food chain. Plants are eaten by animals, which, in turn, are eaten by other animals. When organisms decompose, wastes and dead tissue are broken down and the phosphates are released. The phosphates are now available to enter the photosynthesis process again as soluble phosphate ions. During this phase phosphates can be absorbed by photosynthetic organisms and passed through the food chains.

Agriculture and the Nitrogen and Phosphorus Cycles

As crops are harvested, valuable nitrogen and phosphates are removed. The interruption of the nitrogen and phosphate cycles soon depletes the soil. Soil fertilizers restore required nutrients and increase food production. Some estimates suggest that fertilizers containing nitrogen and phosphates can almost double cereal-crop yields. However, fertilizers must be used responsibly. As anyone who has overfertilized a lawn has discovered, more is not necessarily better. Soil bacteria convert ammonia or urea fertilizers into nitrates, but the nitrates in the soil may generate nitric acids. Depending on the amount of buffer, a typical application of between 14 and 23 kg of nitrogen fertilizer per year over 10 years can produce a soil that is 10 times more acidic. This can have devastating effects on food production. Most prairie soils have a neutral pH near 7.0. Should the pH drop to 6.0, some sensitive crops, such as alfalfa and barley, are affected. The effect on soils near the Great Lakes is even more pronounced, as these soils have considerably less

Figure 1.23
Spring runoff carries nitrogen and phosphate fertilizers to streams, promoting aquatic plant growth.

buffering capacity and, therefore, are even more vulnerable to acids. A drop to pH 5.0 will affect almost all commercial crops. The combination of fertilizers and acid deposition only accelerates the destruction of crops.

The accumulation of nitrogen and phosphate fertilizers produces yet another environmental problem. As spring runoff carries decaying plant matter and fertilizer-rich soil to lakes and streams, aquatic plant growth is promoted. Once the aquatic plants die, decomposing bacteria use valuable oxygen from the water to complete the carbon, nitrogen, and phosphate cycles. Because the decomposers flourish in an environment with such an abundant food source, lake oxygen levels drop quickly and fish often begin to die. Unfortunately, this makes the problem worse as decomposers begin to recycle the matter from the dead fish, enabling even more bacteria to flourish and oxygen levels to be reduced even further.

■ REVIEW QUESTIONS ?

16 What is a biogeochemical cycle? Why are these cycles important to living organisms?

17 In terms of biogeochemical cycles, why is photosynthesis critical to the biosphere?

18 Describe the role of plants in the water cycle.

19 What do nitrogen-fixing bacteria and lightning have in common?

20 Why is carbon critical to the biosphere?

21 Where is most of the biosphere's carbon dioxide stored after it is released into the atmosphere?

22 Why are photosynthesis and cellular respiration often called the biotic phase of the carbon cycle?

LABORATORY

RECYCLING PAPER

Objective

To recycle paper.

Materials

newspapers
large screen
100 mL graduated cylinder
beaker tongs
goggles
blotting paper (or paper towel)

10% starch solution
rolling pin
250 mL beaker
lab apron
hot plate

Procedure

1 Measure 150 mL of the starch solution and pour it into the 250 mL beaker.
Using a hot plate, heat the starch solution to boiling.
2 Tear a sheet of newspaper into very small pieces and place it in the boiling starch solution. Remove the beaker from the hot plate and allow the mixture to soak for 10 min.
3 Place the screen over a large dish and pour the solution on the screen. Allow the fluids to collect in the dish.

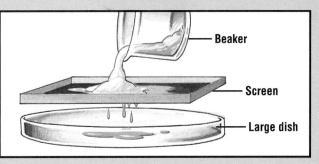

4 Smooth the pulp evenly over the screen and then place the blotter on top of the screen.
5 Place the screen inside the fold of a number of newspapers and roll with a rolling pin.

Laboratory Application Questions

1 Why was the starch solution used?
2 Why was the starch solution boiled?
3 Propose a method for recycling paper on a larger scale. ■

IMPORTANCE OF RECYCLING

The renewed environmental awareness of the 1990s has prompted widespread interest in recycling. Nearly 50% of household wastes can be recycled. It takes 95% less energy to reuse an aluminum can than to make a new one. One estimate suggests that if Canadians recycled all of their aluminum food and soft-drink cans, enough metal could be saved to make 145 000 cars. Recycling a tonne of paper saves 11 to 17 trees and uses 50% less energy.

Besides the economic advantages of recycling, there are many environmental benefits as well. Landfill sites take up valuable agricultural land and cause the accumulation of potentially dangerous household wastes. The example most frequently mentioned is disposable diapers. Composed of fibers and plastic, these diapers take several hundred years to decompose. It has been estimated that two million disposable diapers are used in Canada every day and that each child produces one tonne of soiled diapers each year.

SOCIAL ISSUE:
The Greenhouse Effect

Scientists predict that, if global warming trends continue, the average temperature on earth may increase by as much as 5°C by the end of the next century. Global warming could result in changes in weather patterns and growing seasons, the melting of the polar ice cap, rising sea levels, and the flooding of coastal regions.

Statement:

Action must be taken to prevent greenhouse warming.

Point

- The level of greenhouse gases, particularly CO_2, has been on the rise continuously over the past 100 years. This suggests that global warming is more than a distinct possibility and should be prevented.
- The Canadian climate will warm up sufficiently to push the treeline and prairie further north, quickly altering our present ecosystems.
- Computer and mathematical models all suggest that warming is a reality and that within 100 years significant changes to temperatures and ocean levels will take place.

Research the issue.
Reflect on your findings.
Discuss the various viewpoints with others.
Prepare for the class debate.

Counterpoint

- Carbon dioxide is highly soluble in water. Therefore, the oceans will absorb any excess CO_2, and global warming will not be a problem.
- If the climate warms up, it will free areas in Canada from the extremes of winter and put more land to use.
- There are so many variables, including the warming and cooling that have occurred in the recent past, that the "models" cannot predict anything accurately.

CHAPTER HIGHLIGHTS

- Ecology is the study of the interaction between living (biotic) organisms and their nonliving (abiotic) environment.
- Often compared to a spaceship, the earth supports the only known life forms in existence.
- Living organisms are found in a limited region of the earth known as the biosphere.
- The atmosphere is organized in zones. It protects living organisms from excessive radiation from the sun and other sources in space.
- The protective ozone layer appears to be breaking down, resulting in an increase in the amount of harmful ultraviolet radiation reaching the earth.

- Some gases in the atmosphere, such as carbon dioxide, are responsible for trapping heat as it radiates to outer space. This is called the greenhouse effect.
- Greenhouse gases are accumulating in the atmosphere at an ever-increasing rate. Scientists predict that this will cause global warming.
- The tilt of the earth's axis, the physical properties of water, and the albedo of the atmosphere and the earth's surface contribute to the global distribution of heat by causing climate and weather.
- The cycling of matter between the abiotic and biotic environments is called biogeochemical cycling.
- Photosynthesis is the process by which abiotic matter is converted into the organic molecules.

- The hydrologic, or water, cycle describes water balance by tracing the phase changes of water in the abiotic environment and its role in living organisms.
- Nitrogen atoms, required for the manufacture of proteins and nucleic acids, are derived from and eventually returned to the atmosphere in a process called the nitrogen cycle.

- Carbon is the basic atom on which life is based. Incorporated into organisms through photosynthesis, carbon maintains a global balance as it is exchanged between organisms and the environment in the carbon cycle.

APPLYING THE CONCEPTS

1 Without a greenhouse effect, life could not exist. Discuss this statement in terms of the present concern with global warming.
2 The albedo of the planet Venus is very high. At the same time, the atmosphere of the planet has an exceptionally high concentration of greenhouse gases. How might these two factors affect the surface temperature of Venus?
3 List several tactics that would help reduce acid depositions. Predict areas of opposition and suggest reasons for the opposition.
4 Hypothesize how radiation from the Chernobyl nuclear power plant may have entered the water cycle.

5 Explain how chlorofluorocarbons (CFCs) can contribute to the depletion of the ozone layer.
6 Suggest a strategy for ridding your lawn of clover. Provide a scientific explanation for the strategy.
7 A town that obtains its water supply from deep-well drilling discovers toxic wastes in the water. Blame is placed on a toxic waste dump over 600 km from the town. The waste had been dumped into an abandoned mine shaft.
   a) How could you prove that the toxic wastes had come from the dump site?
   b) Assuming the dump site was the source of the pollution, explain how it could have contaminated the town's water supply.

CRITICAL-THINKING QUESTIONS

1 A plowed field is adjacent to the fairway of a golf course. During the winter, equal depths of snow covered the field and the fairway. Assuming that both fields are level and there is no disturbance to the snow pack, explain why the plowed field loses most of its snow before the fairway even begins to lose its cover.
2 To provide land for agriculture, the tropical rain forests of the Amazon are being destroyed by fire at an alarming rate. Describe any possible effects this clearing process may have.

3 A farmer's field became waterlogged during a very wet spring. This delayed planting for about five weeks. Planting was followed by excellent weather for the remainder of the growing season, yet the crop growth was poor. A local expert suggested that the reason for this poor growth lay in the nitrogen cycle. Study the nitrogen cycle carefully. Then write a report explaining to the farmer why the crop failed.
4 Provide evidence to support Lovelock's Gaia hypothesis.

ENRICHMENT ACTIVITIES

1 With parental approval develop and implement a plan that would reduce the amount of garbage your household produces each week. Record the results.

2 Suggested reading: the September 1989 special issue of *Scientific American*, "Managing the Earth," provides an excellent resource.

Energy and Ecosystems

IMPORTANCE OF ENERGY IN THE ECOSYSTEM

Coniferous forests, deserts, and freshwater lakes, despite their tremendous diversity, share many similarities. Each region functions as a system. Ecological systems, or **ecosystems,** are smaller regions within the biosphere. Within each ecosystem biotic and abiotic components are interdependent.

The size and boundaries of each ecosystem are determined by the ecologists who study them. An ecosystem with a distinct boundary and size, such as a pond, makes the investigative work of the ecologist much simpler. Grassland or woodland ecosystems often have less distinct boundaries. Many organisms move into and out of ecosystems during daily activities. The scale and complexity of an ecosystem varies, depending not only on the organisms found there, but on abiotic factors such as climate and local geology. By studying a variety of ecosystems and combining the various data gathered, scientists are able to piece together a picture of the biosphere as a whole. Two details are clear: (1) there is a cycling of matter between the biotic and abiotic compo-

*An **ecosystem** is a community and its physical and chemical environment.*

Figure 2.1

In any ecosystem, photosynthesis and the feeding activities of non-photosynthetic organisms are responsible for integrating energy into the biosphere.

nents of an ecosystem; (2) energy drives the ecosystem.

Functioning organisms and ecosystems both require energy to operate. You have already seen how solar energy drives many abiotic activities of the biosphere such as weather and biogeochemical cycles. However, unlike matter, energy cannot be continually recycled. Consider what would happen if energy were not added to an ecosystem. In this chapter you will examine how energy, in the form of chemical bond energy, is integrated into the biosphere through photosynthesis and the feeding activities of non-photosynthetic organisms.

AUTOTROPHIC AND HETEROTROPHIC NUTRITION

Autotrophs are "self-feeders." Most autotrophs require an input of energy from the sun. As you read in the previous chapter, photosynthesis is the process in which solar energy is converted into a form of energy that is usable by organisms. Without photosynthesis, little energy would move from the abiotic environment to living things. Lakes might warm up, but warm water does not supply the energy that organisms living in a lake need for growth and development. Solar energy must be converted into chemical energy before it can be used by living things. Photosynthesis provides chemical energy storage.

Photosynthesis also provides the basic organic molecules required by the entire living ecosystem. For this reason, autotrophs are referred to as **producers.** Those organisms incapable of photosynthesis are referred to as **consumers,** or **heterotrophs.** Consumers must feed directly on autotrophs or other heterotrophs for both their chemical energy

and their basic organic molecules. Growing plants take up carbon dioxide, water, nitrogen, and phosphate, which are continuously exchanged between producers and consumers through biogeochemical cycles. Carbon dioxide, water, nitrogen, and phosphate are essential for the synthesis of carbohydrates, proteins, lipids, and nucleic acids—the organic materials found in all living things. The energy obtained from photosynthesis drives the biogeochemical cycles.

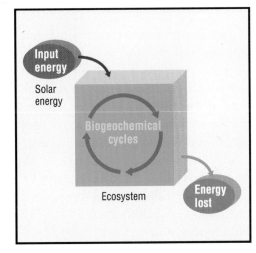

Figure 2.2

Energy from the sun drives the cycling of matter in ecosystems. Note that ecosystems are "open systems," which means that energy can move in or out of the system.

Autotrophs *are organisms capable of obtaining their energy from the physical environment and building their required organic molecules.*

Producers *are autotrophic organisms.*

Consumers *are heterotrophic organisms.*

Heterotrophs *are organisms that obtain food and energy from autotrophs or other heterotrophs; they are unable to synthesize organic food molecules from inorganic molecules.*

As mentioned earlier, ecosystems rarely have rigid boundaries. A deer may leave a woodland ecosystem and temporarily become part of a grassland ecosystem. The dragonfly spends its larval life in a freshwater ecosystem, but will live its adult term in a woodland or grassland ecosystem. Ecosystems are often referred to as "open systems." The fact that energy and some matter can both enter and leave the ecosystem means that ecosystems are not self-sustaining. It is important to account for the energy lost from an ecosystem. When the deer moves from the woodlands to grasslands, energy is lost from the woodland ecosystem. However, not all of the energy from the deer will remain within the grassland ecosystem. Most of the energy of metabolism is lost as heat. Table 2.1 compares the incoming solar radiation and the amount of energy actually used in photosynthesis.

Table 2.1 Energy Used in Photosynthesis

Item	Energy /m2/year
Solar radiation	7 106 000 kJ
Solar radiation reflected from vegetation	7 019 014 kJ
Solar radiation used in photosynthesis	86 986 kJ

Not all organisms depend on the sun. A few ecosystems, such as those found in dark caves, must depend on another source for producers. Through the process of **chemosynthesis,** a limited number of bacteria and blue-green algae are able to obtain energy from the breakdown of inorganic substances such as sulfur. As such, they are true autotrophs. The nitrogen-fixing bacteria described in the previous chapter also have this ability. However, these chemosynthetic organisms are relatively minor producers in the structure of most ecosystems. Because of their dominant position in ecosystems, this text will deal only with photosynthetic autotrophs as producer organisms.

DEFINING PLANTS AND ANIMALS

The most familiar photosynthetic organisms are terrestrial plants. The most familiar heterotrophs are animals. The biological meaning of plant and animal is determined by the classification system used to group organisms into their basic kingdoms. Organisms have been classified in groups ranging from a basic two-kingdom system (plants and animals) to groups describing as many as eight or more kingdoms. In systems that use more than two kingdoms, many organisms that have been traditionally classified as plants or animals now belong to other kingdoms. For example, in the five-kingdom system, photosynthetic organisms are found in the Kingdoms Monera and Protista as well as the Plant Kingdom itself. This chapter refers to autotrophic (producer) organisms as plants and heterotrophic (consumer) organisms as animals.

Figure 2.3

A grasshopper is snared in a Venus flytrap, a carnivorous plant.

Figure 2.4

Representative organisms from the five kingdoms.

Chemosynthesis *is the formation of carbohydrates from energy resulting from the breakdown of inorganic substances rather than from light.*

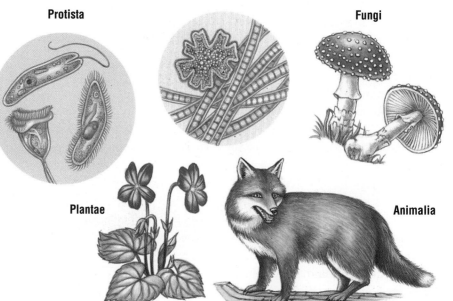

DECOMPOSERS

The breakdown of complex organic molecules found in the wastes and bodies of all organisms is the work of a special group of heterotrophs referred to as **decomposers,** or microconsumers. They consist mainly of bacteria and fungi. During decomposition activities decomposers obtain energy through respiration from the breakdown of tissues and waste materials. They are also referred to as *saprophytes* when their food source is limited to dead tissue.

The role of decomposers in the recycling of matter is so important that they are usually treated as a separate group of organisms.

TROPHIC LEVELS AND THE FOOD CHAIN

The term **trophic level** is used to locate the position or level of an organism during its energy-seeking activities. Plants are said to belong to the first trophic (energy) level since the chemical energy they both store and utilize is one step (level) from the original solar energy they trapped. When one organism eats another, energy is transferred. Since heterotrophs obtain their energy from autotrophs, they occupy higher trophic levels. To follow this energy transfer in terms of trophic levels it is helpful to examine a simple feeding sequence, or **food chain.** The arrows indicate the direction in which the energy is being transferred.

Plant tissue ⟶ White-tailed deer ⟶ Wolf
(Producer)   (Consumer)   (Consumer)

The plant tissue consumed by a white-tailed deer is at the first trophic level. The deer receives its energy and organic nutrients from the vegetation, thus making it a

heterotroph. Since it receives its energy two steps from the original source (sunlight), it is at the second trophic level. By the same reasoning, the wolf, also a heterotroph, is a member of the third trophic level.

Other terms are used to describe the feeding habits of animals. **Herbivores** feed on plants, while **carnivores** attack and eat herbivores. Carnivores may also kill and eat other carnivores. The terms *predator* and *prey* describe the hunter and victim relationship. Animals that eat both plants and animals are referred to as **omnivores.**

Figure 2.5

The elephant and the larva are both examples of herbivores. Tigers are carnivores and the grizzly bear is an omnivore.

Decomposers *are generally bacteria and fungi that break down the remains or wastes of other organisms in the process of obtaining their organic nutrients.*

Trophic level *refers to the number of energy transfers an organism is from the original solar energy entering an ecosystem.*

Food chains *illustrate a step-by-step sequence of who eats whom in the biosphere.*

A **carnivore** *is an animal that eats other animals in order to obtain food.*

An **omnivore** *is an organism that eats both animals and plants.*

A **herbivore** *is an animal that obtains its food exclusively from plant tissue.*

Consumers are categorized according to their specific trophic level in a food chain. Herbivores are considered to be *primary consumers* since they obtain their energy/nutrients from plants. A carnivore directly feeding on a primary consumer is a *secondary consumer*. A carnivore that eats a secondary consumer is referred to as a *tertiary consumer*. The final carnivore in all food chains is called a *top carnivore*. In the previous example, the deer is a primary consumer because it feeds on vegetation. The wolf is both a secondary consumer and top carnivore since it obtains its energy and organic molecules from the deer and is not fed on by other carnivores.

There are two basic types of natural food chains. **Grazer food chains** have characteristics similar to the plant–deer–wolf chain described previously. They include any herbivore that grazes on or consumes plants (producers) and is, in turn, usually eaten by a carnivore. In a **decomposer,** or **detritus, food chain** the food (energy/nutrient) sources for the decomposer organisms come from both plant and animal wastes as well as their remains. Detritus is a term used to describe any organic waste from an organism. Dead leaves and intestinal wastes from animals are examples of detritus.

Grazer food chains *originate with plants that are consumed by herbivores (grazers).*

Detritus *is any organic waste from animals and plants.*

Decomposer food chains, *usually bacteria and fungi, consume wastes and dead tissue from organisms.*

Figure 2.6

Example of a grazer food chain occurring in the water.

Phytoplankton
(plants)

Zooplankton
(animals)

Osprey

Larger fish

Small fish

Detritus feeders include such animals as ravens, vultures, carrion beetles, earthworms, and the larvae of a variety of flesh flies, commonly called *maggots*. Larger detritus feeders that consume animal remains are often called *scavengers*. The wastes and dead bodies of the detritus feeders are, in turn, acted upon by fungi and bacteria. Decomposer food chains are critical in the recycling of chemical substances such as nitrogen and carbon back to the abiotic environment.

Table 2.2 A Simple Grazing Food Chain

Organism	Category	Trophic Level
Aquatic plants	Producer	First
Krill	Primary consumer	Second
Cod fish	Secondary consumer	Third
Seal	Tertiary consumer	Fourth
Polar bear	Top carnivore (quaternary consumer)	Fifth

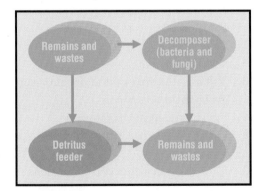

Figure 2.7

A generalized decomposer food chain.

FOOD WEBS

Consider what would happen if deer depended exclusively on a single type of plant tissue such as aspen poplar buds for food. How would a caterpillar infestation affect the deer? What effect would this have on the wolves? Although the interdependence of predator and prey serves to emphasize the reliance of all living things on each other, such dramatic cause-and-effect relationships do not normally occur in natural ecosystems. Deer feed not only on poplar buds but on the bark, stems, and buds of a variety of trees and shrubs, as well as certain grasses. The wolf includes in its diet many different animals such as mice, ground-nesting birds, beaver, and muskrat. In reality, each individual organism in an ecosystem is involved in a number of food chains. Many interlock, forming feeding relationships called **food webs.**

Most food webs are quite complex. Extensive studies on the feeding habits of individual organisms are required before a food web in an ecosystem can be described. It is often easier to examine feeding interactions in parts of the biosphere where climatic conditions are more extreme, such as the arctic or desert areas. Since fewer organisms have been able to adapt to these regions, there is a greater chance of unraveling the food webs and discovering the principles that may govern their structure. This knowledge can then be applied to more complex ecosystems.

The most stable ecosystems have complex and well-developed food webs. A reduction in numbers or complete removal of one of its organisms may have a minimal effect on the overall web. However, where natural conditions limit the number of kinds of organisms, the webs become simplified into structures more like food chains. This is particularly true in the high arctic. A limited number of organisms means that relationships are well defined. In these situations the loss of any one member will have a profound effect on the remaining organisms. The

*A **food web** is a series of interlocking food chains representing the transfer of energy through various trophic levels in an ecosystem.*

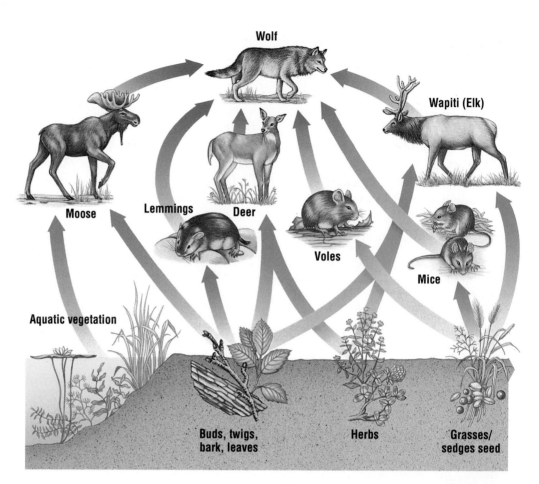

Wolf

Wapiti (Elk)

Moose Lemmings Deer Voles Mice

Aquatic vegetation

Buds, twigs, bark, leaves Herbs Grasses/ sedges seed

Figure 2.8

A simplified food web with only the wolf as the top carnivore. The organisms in a food web are positioned such that the producers (first trophic level) appear on the bottom and each succeeding row represents a higher trophic level.

arctic ecosystem, with its limited number of organisms, can be compared to a small company. In much the same way as a small company may have to suspend operations if one or more employees is sick, the arctic ecosystem depends on each member of the food web for survival. Food webs in warmer climates, with their greater number of organisms, tend to be less fragile. Returning to the analogy of the companies, it would be unlikely that a large corporation would be severely affected by the loss of an employee for even a few days. Without care and consideration, human interference, in terms of pollution and habitat destruction, can easily upset any fragile ecosystem.

REVIEW QUESTIONS ?

1 Why are producer organisms called autotrophs?

2 How does a heterotroph differ from an autotroph?

3 What is the role of the decomposer group in an ecosystem?

4 Explain the term "trophic level."

5 Distinguish between a food chain and a food web.

6 How does an omnivore differ from a herbivore and a carnivore? Give an example of each.

7 What type of food would be consumed by a secondary consumer? Explain your answer.

8 What is meant by the term "top carnivore"?

LABORATORY

INVESTIGATING A MICROECOSYSTEM

Objective

To investigate relationships between the abiotic and biotic environments in a microecosystem.

Materials

hand lens or dissecting microscope
plastic bags and labels
spade
penknife
tweezers
small jars with screw caps

Procedure

1 Locate rotting logs in various stages of decay. (Each lab group may be assigned a different log.) Using the chart below, assign qualitative descriptions indicating the stage of decomposition.

A Rating:	Log is firm. Bark remains on log even when the log is stepped on.
B Rating:	Log is less firm. Bark easily pulls away when pressure is applied, but log maintains shape.
C Rating:	Log feels spongy. The log breaks when pressure is applied. Wood begins to crumble.
D Rating:	Log is in the last stages of decomposition. Log shows evidence of decomposition even before any pressure is applied.

a) Give a rating for the log that your group is studying.

2 Observe the outer surface of the log for signs that animals have burrowed into or out of the log. Are you able to predict a general grouping—for example, insects or small mammals?
b) Record your observations.

3 Examine the outer side of the bark for bracket fungi, small mushrooms, puff balls, mosses, and lichens.
c) Draw a diagram of the outer surface of the log indicating the position of the organisms you identified.

4 Using your penknife, remove a small section of the bark and look for insect larvae, centipedes, millipedes, ants, and wood-boring beetles. You may wish to take a small sample of the bark back to the laboratory for later study.
d) Record your observations.

5 Roll the log over and feel the underside of the bark.
e) Compare the moisture on top of and underneath the log.

6 Look for white or yellow threads from fungi.
f) Record your observations.

7 Use the data collected by another group to compare at least two different logs at different stages of decay. If time permits, you may want to examine samples taken from the other group's log.

Laboratory Application Questions

1 Compare the abiotic factors such as sunlight, moisture, wind, and temperature that influence life on the top and bottom of the log. Use a chart to record your observations.

2 Categorize the members of the biotic community as autotrophs, heterotrophs (carnivores or herbivores), or decomposers. Distinguish between those found on the top and the bottom of the log. Use a table to record your observations.

Biotic community	Top of log	Bottom of log
1 Autotrophs		
2 Heterotrophs		
a) Carnivores		
b) Herbivores		
3 Decomposers		

3 Using concept mapping techniques, construct a food web showing the manner in which energy and matter cycle through an ecosystem. You may present a food web of organisms found either on the top or the bottom of the log.

4 Select two logs at different stages of decomposition and combine the class data. Which log has the greater number of fungi? Explain your observations.

5 Provide a hypothesis that explains why bracket fungi are most often found on top of the log, while the yellow and white threads from molds, another type of fungus, are found under the log.

6 In what ways are the microecosystems under a log and inside a running shoe similar? ■

THE LAWS OF THERMODYNAMICS

Food chains and webs provide a model for studying the movement of energy and nutrients in ecosystems. The amount of energy flowing through the biosphere obeys basic scientific principles known as the laws of thermodynamics. Thermodynamics refers to the use of heat as a convenient measurement of chemical energy in any reaction. The **first law of thermodynamics** has traditionally been called the *law of conservation of energy*. In its simplified form it states that, although energy can be changed (transformed) from one form to another, it cannot be created or destroyed. Therefore, the energy input must equal the energy output. The efficiency of energy transfer in a system can be calculated by dividing the usable energy output by the total energy input. The **second law of thermodynamics** states that, during any energy change, some of the energy is converted into unusable forms, particularly heat. Often referred to as waste energy, these forms are not capable of doing useful work within the system. Each time energy is transformed, some energy is lost from the system.

Photosynthesis transforms solar energy into chemical bond energy of organic molecules in plant tissue. In turn, plant tissue is eaten by a consumer. According to the first law, the consumer should now contain an amount of energy equivalent to the original solar energy. The second law, reminds us that there is a loss of energy, usually in the form of heat, during

*The **first law of thermodynamics** states that energy can be changed in form but cannot be created or destroyed. It is often referred to as the law of conservation of energy.*

*The **second law of thermodynamics** states that, during an energy transformation, some of the energy produced, usually in the form of heat, is lost from the system.*

Figure 2.9

Only 16.2% of the energy stored in chemical bonds by the producer is transferred to the primary consumer. More than 63% of the energy stored in plants is used to meet their own needs. Tissue growth, cell repair, and the transportation of nutrients all require energy. During decomposition, bacteria and fungi use 20.4% of the energy found in the producers.

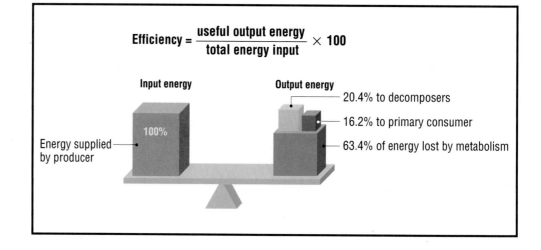

$$Efficiency = \frac{useful\ output\ energy}{total\ energy\ input} \times 100$$

Input energy

Output energy

Energy supplied by producer

100%

20.4% to decomposers

16.2% to primary consumer

63.4% of energy lost by metabolism

all energy transformations. The key transformation reactions in living organisms occur in cellular respiration. On this basis alone, the higher the position of an organism in the trophic levels, the less usable energy there is available for it. If you follow the energy flow in the simple food chain, vegetation–deer–wolf, you can see that there are further factors that reduce the available energy moving from one trophic level to the next. For example, a deer grazing on vegetation only eats a portion of the vegetation. Not all of this food is digested by the animal and eliminated in its wastes (feces). A portion of the potential energy in the remaining food is used to fuel cellular respiration. In turn, respiration and other metabolic reactions produce considerable heat energy. While some is necessary to maintain the animal's body temperature, most is lost to the surrounding air. The remainder of the organic tissue consumed is used to build the deer's own tissues. As a result, less than 20% of the original energy in the plant transferred to the deer becomes available for the wolf. By not consuming such parts of the deer as its bones, hooves, skin, and fur, the wolf uses only a portion of the potential energy stored in the total deer tissue. The wolf normally occupies the top position in the food chain. In general, the overall loss of energy at each step sets a limit on the number of trophic levels in a food chain to about five.

Table 2.3 Laws of Thermodynamics

Law	Concept
First law	Energy input = energy output
Second law	Energy input ➔ desired energy + waste energy

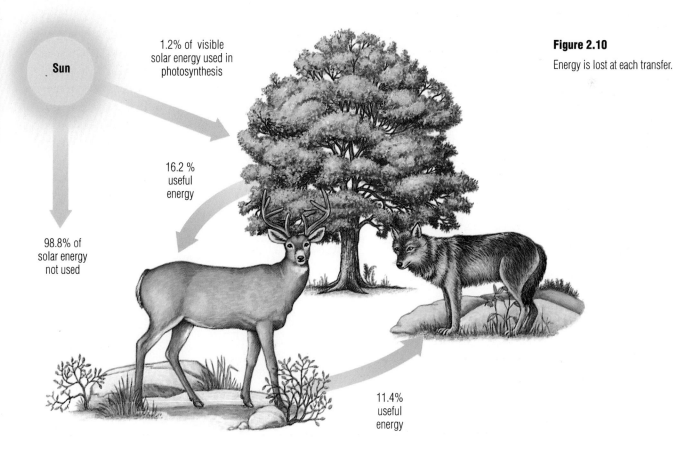

1.2% of visible solar energy used in photosynthesis

Sun

16.2 % useful energy

98.8% of solar energy not used

11.4% useful energy

Figure 2.10

Energy is lost at each transfer.

LABORATORY
MEASURING ENERGY LOSS DURING METABOLIC ACTIVITIES

Objective

To determine whether energy loss, in terms of heat production, occurs during the metabolic activities of germinating pea seeds.

Materials

Note: *To initiate germination, the pea seeds must be soaked in water for 24 h prior to the experiment.*

3 thermos bottles   60 dry pea seeds
3 thermometers   60 germinating pea seeds
cotton wool (absorbent cotton)

Procedure

1 Obtain a thermometer, a thermos bottle, and some cotton wool.
2 Wrap a sufficient amount of cotton around the thermometer at the 10°C marking so that the cotton will both plug the opening of the thermos and hold the thermometer in an upright position. The temperature graduations from 10°C and above must be clearly visible.
3 Carefully plug the mouth of the thermos with the cotton wool/thermometer so that the bottle is airtight.

The plug must be inserted firmly.

4 Label the thermos as the "control."
   a) What is the function of a control in an experiment?
5 Place 60 dry pea seeds into the second thermos bottle. Insert the thermometer and cotton plug as in steps 2 and 3 above.
6 Label the thermos "dry seeds."
7 Place 60 water-soaked pea seeds in the third thermos bottle. Repeat the procedures for inserting the thermometer and cotton plug.
8 Label the thermos "germinating seeds."
   b) What is meant by germination?
   c) Why was it necessary to soak the seeds in water to initiate germination?
9 Place the three thermos bottles in a location where the temperature is not likely to change significantly.

10 Create a table for recording data. Record the temperature within each bottle. Record the precise time when the temperatures were recorded. Calculate and record the time elapsed.

Reading number	Time of day	Time elapsed (hrs)	Control temp. (°C)	Dry seed temp. (°C)	Wet seed temp. (°C)

11 Predict what might happen to the temperature in each thermos over three days. Write these predictions in your lab notebook.
12 Record the temperature within each thermos at least three times per day. Record the time when each reading was made, then calculate the "time elapsed" since the first measurement was taken.
13 Continue recording temperatures and times for at least three consecutive days.
14 Draw a graph plotting temperature against the time elapsed for each thermos. All three curves should appear on the same graph.
   d) Which is the manipulated variable? the responding variable? Explain.
   e) What would a temperature change in the control imply about the temperatures in the other two bottles?

Laboratory Application Questions

1 Account for any temperature changes during the experiment.
2 How did your predicted results compare with the actual temperatures? Comment on any similarities and/or differences.
3 Why were dry seeds used in one thermos?
4 What do the results of the experiment indicate about the energy available to a potential consumer if it selects either the dry or the germinating seeds? Assume each seed has the same amount of bond energy at the start of the experiment.
5 Since all the seeds are alive, explain the temperature difference between the dry and the moistened seeds.
6 In a similar experiment, germination did not occur in the moistened seeds. Nevertheless, there was a distinct rise in the temperature in the thermos. How could you account for this observation? ■

RESEARCH IN CANADA

Janet Edmonds, a researcher with the wildlife division of Alberta Forestry Lands and Wildlife, conducted research on the mountain caribou of west central Alberta during the early 1980s. Her work focused on the interactions of the caribou and their relationships with plants and animals in the region. An understanding of the delicate balance of producers and consumers was especially important because of intensified forestry activities in the area. Many ecologists believed that the mountain caribou would be vulnerable to the environmental changes brought on by logging.

Edmonds' research project extended over four winters, beginning in 1980. Much of the data were collected from the air. Twenty-one mountain caribou were captured and fitted with radio transmitters so that they could be tracked from the air. On the ground, vegetation and snow sampling were done to determine how caribou use their winter habitat to find food and avoid deep snow.

The project sought answers to a number of questions: After logging was completed, would sufficient food sources remain to support the herd? Would logging cuts make it easier for predators to find the caribou? Would a herd that had little to eat become easy prey? Are fluctuations in population a response to the changing environment or just part of a natural cycle?

Important recommendations for managing the mountain caribou herds were put forth, including special timber harvest guidelines for caribou habitats, total protection of some winter ranges, wildlife sanctuaries along major roads, predator management to reduce the high rates of adult mortality, as well as an extensive public education progam.

Managing Mountain Caribou Herds

JANET EDMONDS

ECOLOGICAL PYRAMIDS

Graphs called pyramids can be used to represent energy flow in food chains and webs. Three basic types of pyramids are in general use, all based on the idea that, due to energy loss, fewer animals can be supported at each additional trophic level in a food chain. The base of the pyramid always indicates a fixed amount of energy and the remainder of the pyramid indicates what happens to this energy.

A **pyramid of numbers** can be drawn by counting the number of plants (producers) living in an area that are required to support a number of herbivores and, in turn, higher-order carnivores. When these numbers are represented on a vertical graph, the graph takes on the general shape of a pyramid. The areas of each bar represent relative numbers. An example is shown in Figure 2.11. Biologists have found that there are many exceptions to this pyramid due to the physical size of the members of a food chain. For example, many aphids (tiny insects) can be found feeding off a single plant (see Figure 2.12). In this case the bottom layer of the pyramid of numbers might be inverted.

A **pyramid of biomass** provides a more accurate representation of energy, although it has its limitations. In these pyramids the mass of the dry tissue in the plants and animals is calculated and graphed in a manner similar to the numbers pyramid. Since biomass is an easily measured product of energy use, it illustrates the decrease of energy in terms of relative mass more accurately than numbers of organisms. It is usual to obtain all the plant tissue from a measured area and calculate its biomass. The biomass of the herbivores that are supported by that amount of plant tissue is then measured. Carnivore biomass is determined in a similar fashion.

The most accurate representation of energy loss can be seen in a **pyramid of energy**. The chemical bonds of the

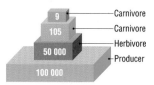

Figure 2.11

A theoretical pyramid of numbers. The numbers are relative and do not represent a specific food chain.

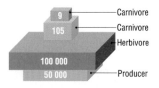

Figure 2.12

An inverted pyramid of numbers. Because the aphids are much smaller than the plant, a single plant can provide food for many aphids.

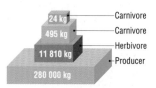

Figure 2.13

A theoretical pyramid of biomass.

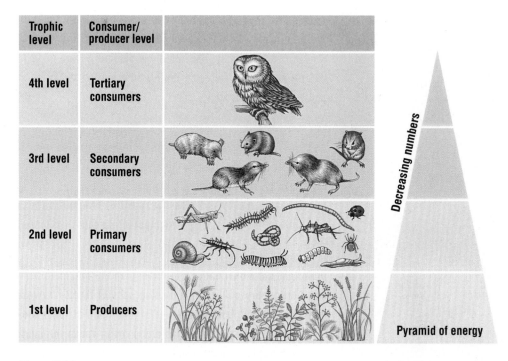

Figure 2.14

In actual studies the structure of a pyramid is not a perfect geometrical shape.

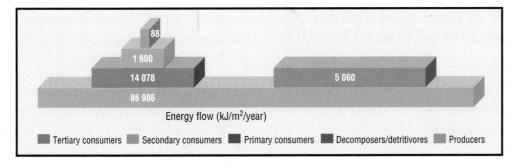

Figure 2.15

A pyramid of energy flow during one year. The numbers are approximate.

organic molecules that make up tissue contain vast amounts of stored energy. When organic matter is burned, this energy is released as thermal energy and is measured using a technique called *calorimetry*. Thus, the amount of energy at each trophic level can be calculated in terms of thermal energy and the relative energy at each level can be compared.

So far we have only examined a grazing food chain. If we combine it with the activity of the decomposers, a more com-plicated but more realistic pattern can be described, as shown in Figure 2.16.

Regardless of the pyramid used to illustrate a food chain or web, each shows the same end result. The energy available to maintain a food chain inevitably runs out, unless the original energy, sunlight, is continuously fed into the system. To obtain the most energy it is best to be a primary consumer. This has real implica-tions for humans as the world population continues to rise dramatically.

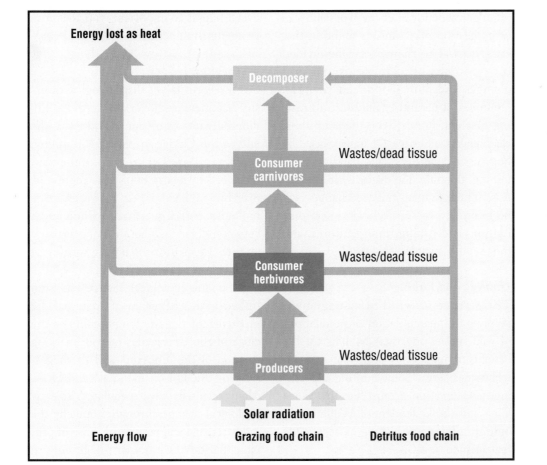

Figure 2.16

An integrated food chain including both grazer and detritus components.

*A **pyramid of numbers** is an energy pyramid based on the numbers of organisms at each trophic level.*

*A **pyramid of biomass** is an energy pyramid based on the dry mass of tissue of the organism at each trophic level.*

*A **pyramid of energy** is a pyramid drawn on the basis of the energy (as heat) produced at each trophic level.*

INTERFERING WITH ECOLOGICAL PYRAMIDS

Most ecosystems can adapt to small changes. For example, in temperate ecosystems variations between the seasons are characterized by fluctuations in populations. We would expect to find different food webs in a forest in winter and summer. Not surprisingly, the biomass of producers in spring, summer, fall, and winter is different. For most plants, winter has the lowest amount of energy available to sustain life. With less biomass at the level of producer, fewer consumers can be supported. The winter migrations of birds and the hibernation of some mammals help reduce energy demands from the producers. Even larger predators such as bears, which do not hibernate, slow their activity and live off stored body fat.

Large-scale changes in an ecosystem often permanently alter the types and size of population of organisms found in that ecosystem. The changes can be divided into two major categories: natural changes and human-induced changes. Examples of natural changes are retreating glaciers, floods, volcanic eruptions, earthquakes, and fires. Changes caused by humans include fires, land clearing, strip mining, flooding due to dams, and pollution. Three human-induced changes will be examined in the following sections: hunting and fishing, monocultures, and the use of pesticides.

Hunting and Fishing

The science of wildlife management involves the manipulation of populations of wild species and their habitats for the benefits of humans. Wildlife management techniques permit some hunting and fishing. Conservation groups like the Sierra Club and the Defenders of Wildlife recognize hunting and fishing as acceptable management tools.

Despite protests from some conservation groups, it would appear that controlled hunting or fishing has its place in the natural order. Aboriginal plains cultural groups hunted the bison for centuries without depleting the herds. However, with the introduction of Spanish horses the aboriginal groups increased their ability to kill larger numbers of the animals, and the ecology of the plains bison herds were impacted in a dramatic fashion.

A confrontation between technology and nature is unfolding in Canadian coastal waters. Improved factory ships, larger nets, and an increased number of boats have dramatically increased the harvest of marine fish. As a result, prized fish such as cod, halibut, and salmon have been drastically reduced. The pursuit of short-term economic gain at the expense of long-term economic collapse from overfishing is an important issue that governments must address.

Monocultures

Fossil records tell us that biological diversity has increased over time. About 150 different families of animals existed at the end of the Cambrian period 500 million years ago. Since then the number has increased to nearly 800. This represents over two million species. However, many biologists will argue that this number is very conservative. There may be as many as 6 to 15 million different species of organisms now living on earth.

Dramatic ecological change has been linked with extinction of species. The most recent era of change came at the close of the Cretaceous period, marking the end of the dinosaurs and the beginning of the age of mammals and birds. Many scientists have speculated that the ecological changes brought about by the industrial age will far exceed those of previous periods. Are humans beginning to

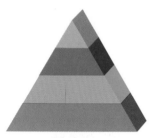

Summer

Winter

Tertiary consumers
Secondary consumers
Primary consumers
Producer

Figure 2.17

Decreased radiant energy during the winter months changes the number of producers and hence the energy made available to consumers. Winter ecosystems have less biomass.

Figure 2.19

Mist rises from a forest after a heavy rain in Ecuador. This demonstrates the rapid recycling of rainwater back to the atmosphere that is characteristic of rain forests.

reduce the number of species on earth? If so, how will decreased biodiversity affect food webs?

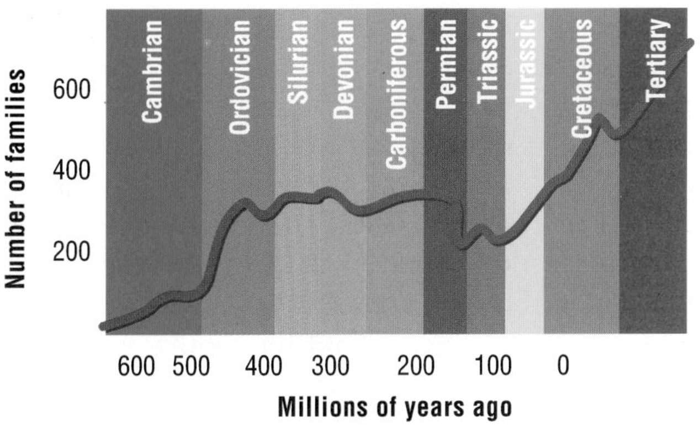

Number of families

600

400

200

Cambrian | Ordovician | Silurian | Devonian | Carboniferous | Permian | Triassic | Jurassic | Cretaceous | Tertiary

600 500 400 300 200 100 0

Millions of years ago

Figure 2.18

The graph shows a trend toward greater biodiversity.

Historically, humans have used about 700 different species of plants. According to the noted biologist Edward Wilson, today we rely heavily on about 20 species—wheat, rice, cotton, barley, and corn being the most important. A great deal of interest in agriculture has been directed at propagating plants for use as food, but at what expense? A great number of wild plants are equally important. For example, the rosy periwinkle, *Obignya*

phalerata, a plant native to Madagascar, produces two important chemicals. The chemicals are particularly necessary medicines for anyone with Hodgkin's disease, a form of leukemia, or cancer of the white blood cells. However, many useful wild plants have already been destroyed to grow food crops, especially in the tropical rain forests.

What makes this situation even worse is that the rich fertile soil of the tropical rain forests is not well suited for **monocultures** of cereal grains such as wheat and barley. These soils require the renewal of decomposed matter to maintain adequate levels of nitrogen and phosphorus. Nitrogen and phosphorus cycles should not be disrupted in the delicate rain forests. A few seasons after planting, the soil will no longer support the growth of crops. What makes the situation even more critical is that the greatest biodiversity exists in the tropical rain forests. Many species have yet to be classified, let alone investigated for possible

*A **monoculture** involves growing a single species of plant to the exclusion of others.*

medicines. Converting a natural ecosystem to an artificial ecosystem has many implications. Should we drain a marsh to plant rice? Should we irrigate a desert to plant wheat or cotton? Most important, we must consider the people who live in these areas. Many grow the cereal grains to feed their families or sustain an income. Are these people left with a choice?

Pesticides

Nowhere is the impact of human interference with food webs more dramatic than in the use of pesticides. In her book, *Silent Spring*, the American author Rachel Carson gave many examples of how pesticides have become incorporated into food chains. Pesticides are designed to reduce the populations of unwanted organisms, both plant and animal. One estimate suggests that as much as 30% of the annual crop in Canada is lost to nuisance pests. The pests include weeds, rusts and molds, insects, birds, and small mammals. The cost to consumers is staggering. For example, in 1954 three million tonnes of wheat were destroyed by stem rust. Another pest, the anopheles mosquito, carries a parasite responsible for malaria. With a simple mosquito bite, the parasite can enter the human circulatory system, causing fever and possible death. As late as 1955, malaria affected more than 200 million people worldwide.

During World War II a host of modern insecticides were developed to protect troops fighting in the tropical jungles of Asia and the Pacific. The most important was dichloro diphenyl trichloroethane, or DDT. In the 1950s the World Health Organization began using DDT and related compounds to control insect pests.

Figure 2.20

The peregrine falcon has become a victim of the use of pesticides. DDT interferes with the formation of the shell of the falcon's egg.

> **BIOLOGY CLIP**
> Although associated with tropical climates, malaria was once a serious problem in Canada. A malaria outbreak occurred during the building of Ontario's Rideau Canal in the 1830s.

The following account of an island north of Borneo dramatically illustrates the effects of DDT and similar compounds not only on malaria control but also on an area's food web. Prior to 1955 about 90% of the island's population was infected with malaria. When disease-carrying mosquitoes were sprayed with Dieldrin, a DDT-related compound, malaria was nearly eliminated, but other effects soon were noted. Other insects began to disappear. Soon the lizards that fed on the insects began to disappear. Despite these changes, no alarm was sounded. Most of the inhabitants were not particularly worried about the disappearance of a few insects and lizards. However, residents took notice when their cats, which had fed on dead lizards, began to get sick and die. Without cats the rat population soon increased. Fearing an outbreak of the plague and other diseases linked with rats, the World Health Organization brought other cats onto the island.

The disruption to the food web became even more evident when caterpillar populations began to increase. Apparently, the Dieldrin affected wasps and other predators of the caterpillar, but had little effect on the caterpillars themselves. With the natural predators eliminated, the caterpillar populations increased dramatically. Eventually, the caterpillars, searching for new food sources, devastated food crops and even began eating the leaves that were used to thatch roofs. Although the ecosystem eventually stabilized, the example illustrates how a change in one part of a food web can affect a number of interrelationships.

This type of scenario has been repeated in many parts of the world. In Canada the peregrine falcon has become a victim of insecticides. Like the cats on the island north of Borneo the peregrine falcon occupies the uppermost level of the ecological pyramid. Toxins such as DDT and Dieldrin accumulate in the fatty tissues of all consumers. The problem intensifies because the concentrations of toxins become magnified as you move up the food chain. Although relatively low levels of toxins are consumed with the prey, toxins accumulate in predators like the falcon. Because the toxins are soluble in fat but not in water, they are not released with most other waste products. When low levels of the toxin are taken in with each organism consumed, levels in the predator begin to rise. At each stage of the food chain the concentration becomes greater. The higher the trophic level, the greater the concentration of toxins. The process is referred to as **biological amplification.**

The irony of insecticides is that although they were developed to rid the world of harmful insects, they have had a much greater effect on humans. Like other predators at the top rung of the food pyramid, humans are subject to biological amplification. During the 1950s and 1960s, fat-soluble DDT turned up in breast milk and was passed from mothers to their babies. DDT levels were especially high in humans who lived in areas where DDT was used for spraying crops. Even those who ate crops from these areas or consumed the animals that fed on these crops were exposed to DDT. The

> **BIOLOGY CLIP**
> Invented by a graduate student in chemistry in 1873, DDT was almost forgotten until 1939, when the Swiss entomologist, Paul Mueller, rediscovered it. In 1948 Mueller was awarded the Nobel Prize for the discovery of the insect-killing properties of DDT. Since that time nearly two million tonnes of DDT have been used worldwide.

fact that DDT was banned for use in Canada and the United States during the late 1970s has not totally eliminated the problem. Migratory birds like the mallard duck, Canada goose, and peregrine falcon winter in Central America and Mexico, where DDT is still used. A similar scenario exists for the migrating fish of the Atlantic and Pacific Oceans.

Chemical factories, many of which are owned by stockholders from countries that ban DDT and similar chemicals, continue to make these pesticides available for use by the poorer nations of the world. Despite all of the insidious effects of these toxins, the chlorine pesticides still have one alluring feature: they are incredibly cheap to manufacture.

Biological amplification *refers to the buildup of toxic chemicals in organisms as tissues containing the chemical move through the food chain.*

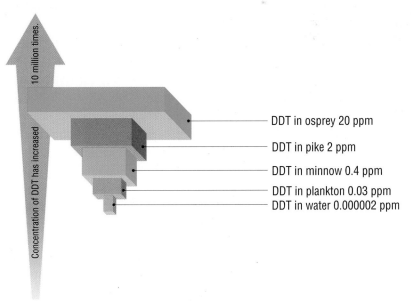

Figure 2.21

Biological amplification. The concentration of DDT increases as you move up the food chain. The greater the number of trophic levels, the greater the amplification. (ppm = parts per million)

DDT in osprey 20 ppm
DDT in pike 2 ppm
DDT in minnow 0.4 ppm
DDT in plankton 0.03 ppm
DDT in water 0.000002 ppm

10 million times.

Concentration of DDT has increased

9 State the first law of thermodynamics.

10 Describe the second law of thermodynamics in your own words.

11 Why is less than 20% of available chemical energy transferred from one organism to another during feeding?

12 Why do energy pyramids have their specific shape?

13 What would be the best source of energy for an omnivore: the plant or animal tissue it feeds on? Explain.

FRONTIERS OF TECHNOLOGY: ARTIFICIAL ECOSYSTEMS

Artificial ecosystems allow ecologists to study food webs while controlling different variables. Biosphere II, the largest of the artificial environments, spans nearly one hectare in the Arizona desert. Many smaller-scale artificial ecosystems exist in Canada, yet the scientific community remains divided on their usefulness. Science writer and broadcaster David Suzuki, for one, suggests that these projects are not worth the money spent on them. By controlling variables, we may get information about interactions within the artificial environment, but how much does that tell us about natural ecosystems? Although scientists have long recognized the value of using models to describe chemical structures or physiological systems, they are divided on the usefulness of models for studying the environment. Can ecosystems devised by humans over a period of a few years approximate natural ecosystems that have evolved over millions of years? Although the artificial ecosystems allow scientists to control variables, can any of their findings be transferred to natural ecosystems?

Research into another type of artificial ecosystem, the indoor environment, has emerged as an important area of study. Many Canadians spend a great deal of time indoors. Downtown office complexes are linked with one another by a series of underground tunnels. Even malls enable shoppers to travel from store to store without going outside. Light, temperature, and moisture are artificially controlled. The scientific community does not agree on the amount of natural light required for a healthy environment. Research shows that workers who receive little natural light are adversely affected.

Pollution has traditionally been seen as a problem affecting outside environments, but studies indicate that indoor environments may pose even greater risks. One research team found that indoor air quality is often much worse than outdoor air. The research team found noxious indoor vapors such as carbon monoxide, nitric oxide, and nitrogen dioxide sometimes exceed the maximum allowable levels for outside air.

Indoor ecologists point out the importance of cleaning air conditioners and heat-exchange systems to prevent the circulation of dust and pollen in buildings. The problems created by office crowding have also been addressed. Have you ever noticed how difficult it can be to breathe if you are working in a crowded classroom? Planting trees and ferns in office buildings and shopping malls to supply more oxygen, and designing buildings with large open areas and skylights are just a few of the newer strategies that attempt to create more livable indoor environments.

SOCIAL ISSUE:
Economics and the Environment

Canada's 1990 Green Plan calls for "responsible sustained development." It recognizes a complex interaction of interests when it calls for global action to ensure that all human activities, including economies, be balanced with environmental protection.

Statement:

Protecting the environment should have priority over economic interests.

Point

- Unless environmental legislation has teeth, it will fail to bring about action. Government initiatives such as Canada's Green Plan of 1990 should make clearer statements about the quality of the air, soil, and water, and ensure that they are put into effect.
- Each country has a responsibility to act for the good of the environment without waiting for other countries to join in. We cannot continue to pass the buck; time is running out. Some sacrifices in terms of jobs and prices are necessary, and Canadians should be willing to make them.
- Improving the environment is expensive. However, less than two weeks' worth of global military spending would pay for a proposed UN plan to minimize clean-water problems in third world countries.

Counterpoint

- Many people blame politicians for short-term goals, but the political system reinforces short-term planning— elections occur every 4 to 5 years.
- Canada should not enact environmental legislation alone. Strict environmental controls could lead to a drop in living standards. Jobs could be lost if companies relocated in developing countries where there are less stringent standards.
- Governments, many of whom have huge debts, cannot afford the enormous amounts of money required to clean up the environment. One estimate states that $774 billion would be needed to reverse negative global trends in soil erosion, to implement reforestation, and to develop renewable energy.

Research the issues.
Reflect on your findings.
Discuss the various viewpoints with others.
Prepare for the class debate.

CHAPTER HIGHLIGHTS

- The energy required for all living organisms originates in solar radiation, which is converted to chemical energy during photosynthesis and stored in the chemical bonds of organic molecules such as glucose.
- Energy is transferred from organism to organism during the feeding process. Therefore, organisms exist in different trophic levels, depending on the number of steps in a food chain they are away from the original source of energy, sunlight.
- Plants belong to the first trophic level and are called producers. Through photosynthesis, they manufacture chemical energy. All animals are called consumers because they must feed on other organisms to obtain energy.
- Primary consumers, called herbivores, eat plant tissues. They are at the second trophic level. Other animals may be carnivores or omnivores, depending on their specific diet.

- A food chain describes a particular feeding (energy) pathway. A food web is a pattern of natural energy flow in which a large number of food chains interlock in an ecosystem.
- Complex food webs are an indication of a stable ecosystem.
- Grazer food chains start with plants that are consumed by herbivores. Decomposer food chains involve the breakdown of wastes and dead tissues from organisms.
- Energy flow in ecosystems must obey the laws of thermodynamics. Therefore, as energy (food) is transferred from one trophic level to the next, over 80% of the original energy is lost.
- For an ecosystem to exist there must be a continuous energy input in the form of sunlight. Energy flows through an ecosystem and is eventually lost.
- Energy flow can be shown graphically in the form of pyramids of numbers, biomass, and energy.

APPLYING THE CONCEPTS

1 In underground caves, where there is permanent darkness, a variety of organisms exist. In terms of energy flow, explain how this can be possible.

2 Based on what you have learned about energy pyramids, comment on the practice of cutting down rain forests to grow grain for cattle.

3 Design complex food webs for a tundra ecosystem and a middle-latitude woodland ecosystem. Reference books can be used to determine the other members of the food web.
   a) Which ecosystem has the greatest biomass? Provide your reasons.
   b) Which ecosystem has the greatest number of organisms? Provide your reasons.
   c) Which ecosystem has the greatest energy requirement? Provide your reasons.
   d) Comparing the tundra and middle-latitude woodland ecosystems, indicate which is more susceptible to environmental pollution. Explain your answer.

4 By law, the cutting of forests must be followed by replanting. Why do some environmentalists object to monoculture replanting programs?

5 Assume that the plant material in a plant–deer–wolf food chain contains a toxic material. Why would the wolf's tissue contain a higher concentration of the toxin than the plant tissue?

6 Provide examples of the two laws of thermodynamics in terms of some common, everyday events.

7 Of the three basic energy pyramids, which best illustrates energy transfer in a food chain? Explain.

8 Provide examples of how technological innovations have altered ecosystems.

9 Assuming an 80% loss of energy across each trophic level, state how much energy would remain at the fourth trophic level if photosynthesis makes available 100 000 kJ of potential energy. Show your reasoning. Construct a properly labelled pyramid to represent this situation. Could a fifth-level organism be added to the chain? Explain.

CRITICAL-THINKING QUESTIONS

1 Assume that a ski resort is proposed in a valley near your favorite vacationing spot. What type of environmental assessment should be done before the ski resort is built? In providing an answer, pick an actual location you are familiar with and give specific examples of studies that you would like to see carried out.

2 Atmospheric warming may cause drought in some parts of the world. Illustrate the impact of drought by drawing an energy pyramid before and after the drought. Explain why the two pyramids are different.

3 Insect-eating plants such as the sundew are commonly found in bogs all across the country. Although referred to as "carnivorous" plants, they are still considered to be members of the first trophic level. Is this the proper trophic level to assign to these plants? Research information on carnivorous plants, then state the trophic level you think is most appropriate. Present the reasoning behind your choice.

4 Some ecologists have stated that to maximize the food available for the earth's exploding human population, we must change our trophic level position. What is the probable reasoning behind this statement? Could any potential biological problems occur if this switch were actually made?

5 A team of biologists notes that the population of white-tail deer is decreasing in a particular area. Hunting pressures have remained stable and the biomass of producers in the area has not changed. What other factors might the biologists consider to determine why the deer population is decreasing?

ENRICHMENT ACTIVITIES

1 Obtain a copy of Canada's Green Plan from Environment Canada.

2 Many businesses have made an attempt to protect Canada's environment. Prepare a visual display explaining some of the initiatives taken by industry.

3 Identify an environmental issue and do a risk-analysis assessment. Show your calculations, identify your assumptions, and justify your conclusions.

CHAPTER 3

Aquatic and Snow Ecosystems

IMPORTANCE OF WATER IN ECOSYSTEMS

Water is the world's greatest natural resource, yet most people undervalue its worth. Gold, diamonds, and emeralds are prized because they are so rare, yet none has the value of water, a chemical so common that it is found almost everywhere you look.

Approximately 70% of your body is composed of water. Water is so vital that most people will die within 48 hours if they cannot replenish their water reserves. By comparison, most people can survive for a week or more without food. No living organism is less than 50% water, and some creatures are as much as 97% water.

Water covers more than two-thirds of our planet, yet only a small portion of it is actually drinkable. Ninety-seven percent of the world's water is salt water. Although unfit for human consumption, the oceans' great reserves of salt water are of tremendous value to all living things. The oceans control the weather patterns on our planet. They also act as a huge reservoir that provides fresh water by evaporation. By far the greatest portion of fresh water is stored as snow and ice,

Figure 3.1

As terrestrial beings we think of the earth in terms of land. However, the vast majority of the earth's surface is covered with water.

a phenomenon that may not be as surprising to Canadians as to some other people. A much smaller portion of the world's fresh water is readily accessible for industrial and human use. It has been estimated that one two-hundred-thousandth (0.0005%) of the world's available fresh water is sufficient to sustain the entire human population of the earth.

In one of nature's unusual twists, water, a chemical used to extinguish fires, is composed of two highly combustible gases: oxygen and hydrogen. The water molecule has properties that are quite unlike other molecules. This unique molecule attracts many other molecules, especially other water molecules. Have you ever noticed how small water droplets draw toward each other when they touch? The attraction is so strong that the outside edge of a water droplet actually forms a type of skin.

Another unique property of water can be observed when water is cooled. Most matter contracts as it is cooled, but water expands. When a bottle of water is frozen, the expanding ice breaks the glass.

Water is such a good solvent that it is rarely found in its pure form. Minerals, vitamins, and sugars found in your body are dissolved in water and carried by the blood to all your tissues. The fact that fish can exist in lakes and rivers depends on the presence of oxygen dissolved in the water. Water also has the ability to store heat. The heat capacity of water not only makes it an ideal medium for chemical reactions, but also serves to moderate our climate.

In this chapter you will have the option of studying one of two ecosystems that are essentially water ecosystems. Should you begin studying ecology in spring or fall you might study ecosystems in lakes or ponds. By studying these freshwater ecosystems you will be able to explore specific interactions between the physical environment and the biotic community. Should you begin studying ecology during the winter months you might study snow ecology. Contrary to popular belief, terrestrial ecosystems are active during the winter, even though much of the activity remains hidden beneath a blanket of snow.

BIOMES OF CANADA

Worldwide belts of climate have resulted in fairly distinct vegetation patterns that show little change over the years. These general vegetation zones are called

Table 3.1 Biomes of Canada

Name	Abiotic factors	Biotic community
Tundra	Arctic circumpolar.	Caribou
	Extremely short growing season.	Rapid-flowering plants
	Permafrost layer immediately below surface of soil will not thaw.	Moss and lichens Ptarmigan
	Low temperatures.	Polar bears
	Precipitation (10–12 cm/year).	Seal
Boreal forest	Immediately south of tundra.	Coniferous forest
	Precipitation (35–40 cm/year).	Moose
	Change in weather patterns.	Goldenrod (plant)
	Soil contains water and is acidic.	Caribou
Mixed woodland forest	Generally south of boreal forest.	Decidous trees
	Increased sunlight strikes forest floor during the spring and fall.	Moose Weasels
	Precipitation (100 cm/year).	Many shrubs
	Rich fertile soil.	Woodpeckers
Grassland	Same latitude as mixed woodland.	Hawks
	Less precipitation than woodland.	Rattlesnakes
	Soils hold less water.	Plains bison

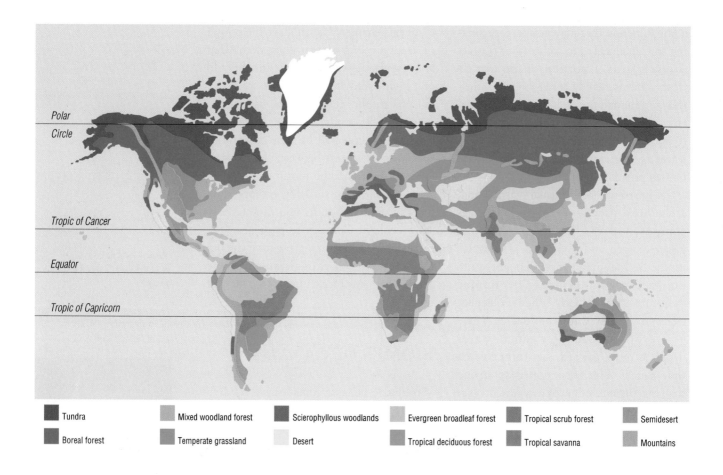

■ Tundra	■ Mixed woodland forest	■ Scierophyllous woodlands	■ Evergreen broadleaf forest	■ Tropical scrub forest	■ Semidesert
■ Boreal forest	■ Temperate grassland	■ Desert	■ Tropical deciduous forest	■ Tropical savanna	■ Mountains

Figure 3.2

A map of the world showing the generalized location of the major biomes.

Biomes *are large-scale ecosystems such as tundra, boreal forest, or grassland.*

biomes. Biomes support characteristic types of terrestrial organisms. When the biotic and abiotic components are considered, biomes are in effect worldwide ecosystems. Aquatic and terrestrial ecosystems are the basic units of the biosphere. Knowledge of how they function individually and integrate with each other allows us to more fully understand the biosphere as a whole.

BIOLOGY CLIP
The tundra and boreal forest biomes dominate. They cover over 80% of the land surface of Canada. Tundra and boreal forest biomes are referred to as circumpolar because they form a band around the northern half of the earth. Scandinavia and Russia also have tundra and boreal forest biomes.

AQUATIC COMMUNITIES

Water-based communities are distinguished by a number of factors. Most water is found in the oceans, where marine communities are located. The ocean waters surround the continents and are significant along coastlines, where tides and wave action continuously change the local environment. Water also accumulates on land surfaces in lakes, ponds, bogs, sloughs, and other types of wetlands. Running-water ecosystems include permanent and temporary, slow- and fast-running streams. Due to the variety of freshwater environments, many specialized communities, varying in size, uniqueness, and complexity, have developed. Freshwater and marine organisms, although basically similar, have developed physiological variations that allow them to cope with the presence or lack of dissolved salts in their water surroundings.

There are many variations in the aquatic environments described above. Of these, the **estuary** may be the most important in terms of biological production. In an estuary, fresh water from rivers and streams mixes with the salt water of the seas, producing a **brackish** mixture that is neither fresh nor as salty as ocean water.

Many inland waters are landlocked and thus become salty to some degree, particularly if the evaporation rate is high. Although we usually associate the salt in the ocean with sodium chloride (table salt), many other salts are also present. In inland "salt lakes" these different salts may predominate, depending on the local geology (soluble minerals), drainage, and groundwater flow. Many of the shallow lakes in western Canada also contain various concentrations and types of salts that affect their use by animal life. Often water is collected in temporary sites, ranging in size from puddles to fairly large but extremely shallow lakes. Local precipitation, melting of snow and ice, and evaporation determine how long the water may remain. Whether the water is permanent or temporary, fresh or salty, a variety of aquatic ecosystems can develop.

CHARACTERISTICS OF STANDING WATER

Lakes and ponds are the most obvious examples of inland standing waters. Ecologists usually distinguish ponds from lakes by the penetration of sunlight into the water. By this criterion, light reaches the bottom of a pond but not that of a lake. Lakes, therefore, have a region in which photosynthesis cannot occur.

Regardless of whether it is a lake or pond, if a body of water is to contain a varied and thriving community, it must contain sufficient nutrients to support a broad base of producer organisms. Enough oxygen must also be present to satisfy the demands of all organisms, the higher-order consumers in particular.

LAKE AND POND SUCCESSION

Some standing water is only temporary. A heavy snow accumulation over winter or heavy spring rains may produce a number of ponds that will survive for all or most of a summer. The progressive change in the composition of plants and animals in the pond throughout the summer is referred to as **succession.** Similarly, some lakes go through a number of stages

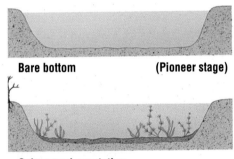

Bare bottom　　　　(Pioneer stage)

Submerged vegetation

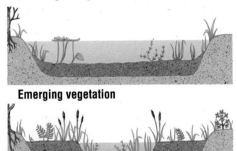

Emerging vegetation

Temporary pond and meadow

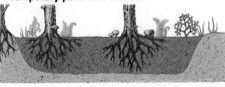

Black spruce and tamarack　　(Climax stage)

*An **estuary** is the place where rivers enter the ocean.*

***Brackish** refers to a mixture of fresh and salt water.*

***Succession** refers to the changes in plant and animal populations between colonization and the final community.*

Figure 3.3

Possible pond succession pattern in the boreal forest.

Eutrophication *is the filling in of a lake by organic matter and silt.*

Climax vegetation *refers to the long-enduring steady-state plant community.*

Oligotrophic *lakes are cold and deep, and have only begun the process of eutrophication.*

Eutrophic *lakes are shallow and warm, and are rapidly becoming filled in.*

over a period of many years until they end up as dry land. Matter is continuously being added to the ecosystem by various forms of erosion, mainly from wind and running water. Dead animal and plant materials fall to the bottom and undergo various levels of decomposition and accumulation. Over time, the lake slowly fills. The more shallow parts, usually along the shoreline, are the first to show signs of aging. This process is called **eutrophication.** The rate of eutrophication depends on both natural and human factors.

As a lake becomes shallower, sunlight is able to reach more of the lake bottom. In turn, this increases water temperature and plant growth. Increased productivity from photosynthesis causes even more matter and nutrients to accumulate at the bottom. Eventually, the lake becomes a shallow pond almost completely covered by submerged and emergent plant species. You might think that the greater number of producers would make the lake more productive, but this is not so. Plants will eventually die, increasing the amount of detritus found in the lake. Bacteria decompose the plants and in doing so use valuable oxygen from the water.

Oxygen deprivation, the result of decomposer action and warmer water temperatures, reduces the number of fish species. Fish such as trout and salmon, which are particularly active and therefore have a higher metabolic rate, are especially sensitive to lower levels of oxygen. In contrast, plankton, insect larvae, worms, leeches, and mollusks have a much lower metabolic rate, and thrive in this oxygen-depleted environment. Slug worms, a close relative to terrestrial earthworms, are often used as indicators to identify lakes with lower levels of oxygen.

The plankton and invertebrates provide a food supply for waterfowl, shore birds, and animals living along the shore-line. Both the emergent vegetation and the shoreline plants and trees provide a home and nesting site for many birds. The red-winged blackbird is a dominant species of this region. Thousands of hatching insects are preyed upon by bats, fly-catching birds, and amphibians such as frogs. As the emergent and submerged plants advance into the lake, dry land slowly forms along the edge and the process of terrestrial succession begins. When complete, the only suggestion that the area was once a lake is the flatland and the special type of climax community that usually develops on it. Because of the drainage, a fair amount of water is still retained in the soils. Two boreal forest trees, the larch (tamarack) and black spruce, are adapted to moisture and thus become the main species of the **climax vegetation.**

Oligotrophic lakes are typically deep and cold. Lake Baykal in Russia, the world's largest lake, and Lake Superior are prime examples of oligotrophic lakes. Nutrient levels are low, limiting producer organism activity. With limited numbers of organisms present, the water is usually very clear and the bottom visible through great depths. These coldwater lakes change little over time and are particularly valuable for sport fishing. Conversely, **eutrophic** lakes are generally shallow and warmer, and have an excellent supply of vital nutrients. Many species of photosynthetic organisms take advantage of these favorable abiotic conditions.

In general, it would appear that, during its life, a lake slowly changes from oligotrophic to eutrophic—eventually filling in and becoming dry land. Humans sometimes accelerate eutrophication of lakes by adding nutrient-rich substances such as human wastes, fertilizers (in the runoff from agricultural land), and other household and industrial products.

1   Why is a water environment less subject to change than a terrestrial ecosystem?

2   Distinguish between freshwater and marine communities.

3   What is the major difference between a lake and a pond?

4   Where would you expect to find brackish water?

5   What is meant by temporary standing water?

6   Compare an oligotrophic lake with a eutrophic lake.

7   What are some major causes of eutrophication?

THE STRUCTURE OF LAKES AND PONDS

The diagram in Figure 3.4 represents a cross section of a typical lake. As with most generalizations, there are variations and exceptions. The **littoral zone** represents the area extending outward from the lake margins, where the lake is shallow enough for rooted aquatic plants (emergent vegetation) such as bullrushes and water lilies to take hold. In this shallow region, rooted plants may also thrive.

Beyond this region is the open water area of the lake, or **limnetic zone.** It extends downward to a depth at which there is insufficient light for photosynthesis. The most common form of organism within the limnetic zone is plankton, which includes both photosynthetic and non-photosynthetic microscopic organisms. Plankton becomes the food for higher orders of consumers in the various limnetic food chains and webs. The bottom of any body of water is called the **benthos** and contains a variety of living forms called benthic organisms. Where light strikes the benthos, photosynthetic organisms can thrive.

The region beneath the limnetic zone is called the **profundal zone.** In the profundal zone of most lakes, the only source of nutrients is the decaying matter or detritus that falls from the limnetic zone. Without sunlight, photosynthesis will not occur in the profundal zone. (A few lakes have a significant number of chemosynthetic bacteria that act as producers for the food chain.) In eutrophic lakes decaying matter is slowly broken down by bacteria or consumed by bottom-dwelling invertebrates and fish, which are called *benthic detritus feeders*. However, the completion of the biogeochemical cycles in this part of the lake changes the ecosystem. Bacteria use oxygen to decompose detritus, thereby reducing oxygen levels. Without sunlight and plants in this area to replenish oxygen, the oxygen levels can be reduced to such an extent that few fish species can survive. Invertebrate populations are confined to those tolerant of a reduced dissolved oxygen concentration.

*The **littoral zone** is the edge around a lake or pond where the water is shallow enough to permit the growth of aquatic vegetation.*

*The **limnetic zone** is the open water area of a lake.*

*The **benthos** is the bottom of any body of water.*

*The **profundal zone** is the region of a lake to which light cannot penetrate.*

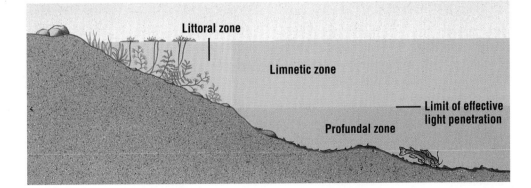

Figure 3.4

A cross section of a typical lake showing lake zones.

Littoral zone

Limnetic zone

Limit of effective light penetration

Profundal zone

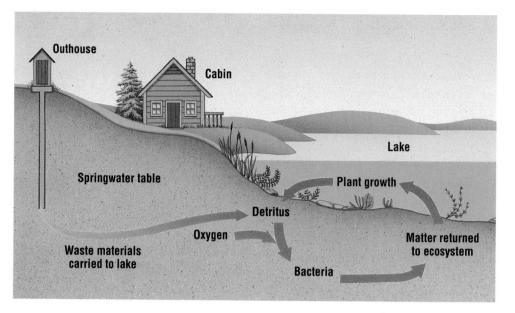

Figure 3.5

Human wastes are broken down by bacteria, which function as decomposers in the lake's ecosystem.

In the figure: Outhouse, Cabin, Lake, Springwater table, Plant growth, Detritus, Oxygen, Waste materials carried to lake, Bacteria, Matter returned to ecosystem.

Problems are created when additional detritus is introduced into a lake ecosystem, especially from the seepage of sewage into lakes from cottage outhouses. Decomposition of human wastes by oxygen-consuming bacteria creates difficulties for fish. The greater the amount of decaying matter introduced into the lake, the greater will be the population of decomposer bacteria. Unfortunately, the bacteria and fish both use oxygen. Not surprisingly, the more complicated life forms, the fish, have greater oxygen requirements and will die if oxygen levels drop too low. Moreover, the death of fish adds even more detritus to the ecosystem, promoting increased numbers of bacteria. In turn, this causes oxygen levels to drop even more. To make matters worse, human wastes act much like fertilizers by introducing additional nitrogen and phosphates into the ecosystem. The phosphates and nitrogen compounds, in turn, promote plant growth, which will eventually die and be decomposed. Each time matter is returned to the ecosystem, oxygen levels are further reduced. Can you imagine what would happen to the profundal zone of a lake if oxygen levels continued to drop? Fortunately, there are

ways in which oxygen and other dissolved materials can be redistributed.

Seasonal Variations

All lakes in Canada are directly influenced by the changing seasons. During the winters the lakes are covered with ice and snow. This prevents atmospheric oxygen from dissolving in the water. At the same time, however, the temperature of the water under the ice is at or slightly above 0°C. Temperatures measured from the lake bottom to the ice surface drop from 4°C to 0°C, regardless of the air temperatures above the ice. If the ice is wind-blown and transparent, light can penetrate into the water to support photosynthesis and other biological activity. However, if winter is unusually long or the ice freezes to a greater thickness than normal, problems can occur. The dissolved oxygen in the water may be insufficient to support organisms requiring higher levels of oxygen. Fish are particularly sensitive to dissolved oxygen concentrations. The end result can be a massive die-off of some fish species. In many of the shallow arctic lakes ice may form right to the bottom, virtually eliminating most life forms.

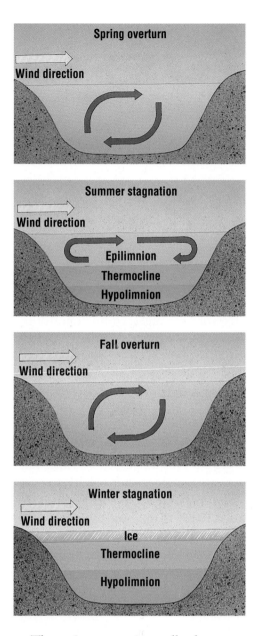

Spring overturn

Wind direction

Summer stagnation

Wind direction

Epilimnion

Thermocline

Hypolimnion

Fall overturn

Wind direction

Winter stagnation

Wind direction

Ice

Thermocline

Hypolimnion

In most lakes the summer season results in the development of layers of water at different temperatures. Anyone swimming in a lake during the summer has experienced the warm upper layer; when your feet slowly descend, you feel the colder region. Since water above 4°C is less dense, it forms a layer that rests on and does not mix with the cooler water beneath. These layers are called the **epilimnion** and **hypolimnion** respectively. Between the two layers is found the **thermocline,** a narrow zone where the temperature drops from warm to cold. The lack of mixing blocks the movement of oxygen into the hypolimnion. Organisms in this zone must rely on oxygen reserves from the spring overturn. The epilimnion has a different oxygen problem. The ability of water to hold dissolved gases is inversely proportional to the temperature of the water. Therefore, the warmer the water, the lower the amounts of dissolved oxygen that can be held. If a shallow lake experiences a prolonged warming there may be sufficient oxygen depletion to cause the death of many species of aquatic organisms.

Figure 3.6

Spring and fall overturn carry oxygen to the bottoms of many deep lakes.

*The **epilimnion** is the upper layer of water in a lake; it heats up in the summer.*

*The **hypolimnion** is the lower layer of a lake; it maintains a constant low temperature.*

*The **thermocline** is the zone separating the epilimnion from the hypolimnion.*

The spring season is usually characterized by winds, rain, and storms. The now ice-free lake surface begins to dissolve oxygen from the air and to warm up. Winds and storms, by stirring up the water, increase the surface area and the rate of dissolved oxygen accumulation. As the temperature of the surface water approaches 4°C, it slowly sinks through the less dense water beneath it, carrying its precious supply of oxygen to all depths of the lake. This overall mixing process is called the *spring overturn.*

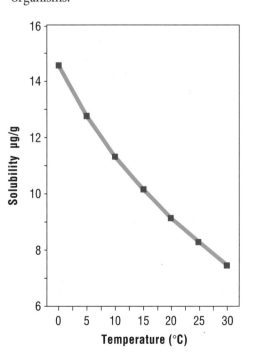

Figure 3.7

Solubility of oxygen in water for atmospheric air at standard pressure.

In the autumn, with dropping air temperatures, there is a reversal of spring conditions. This *fall overturn* renews oxygen levels at all depths by the breakdown of the thermal layers. In the arctic many of the lakes experience only one turnover because of the higher temperatures at the peak of summer. Some lakes may not even thaw. Canada's arctic lakes and ponds are unique and must be considered distinct from the lake ecosystems familiar to most people.

Every lake has its own special conditions of temperature and oxygen levels. The fish are adapted to a specific range of oxygen levels. For example, trout, which require high concentrations of oxygen, can be found at all levels of a lake in spring and autumn. During summer they are confined to the deeper, cooler regions, where more oxygen is available. Lakes with generally warmer water support varieties of fish such as perch, which are able to live in waters with less oxygen.

Table 3.2 Oxygen Concentrations (μg/g) during Summer in an Oligotrophic and Eutrophic Lake

Region	Oxygen Concentration (μg/g)	
	Oligotrophic	**Eutrophic**
Epilimnion area 1	8.3	7.8
Epilimnion area 2	7.9	7.9
Thermocline	7.5	7.0
Hypolimnion area 1	7.0	5.2
Hypolimnion area 2	6.9	2.0
Hypolimnion area 3	6.8	0.2

Oxygen levels are also influenced by the rate of decay of organic matter or detritus on the lake bottom. When organisms die, there is an increase in the number of decomposer organisms and a subsequent decrease in oxygen levels. The situation escalates when natural conditions or pollutants cause the nutrient levels in the lake to increase. Consider the effects of large-scale logging, in which trees are cut and pushed into the water for easy transport. Although the trees are eventually collected by the logging company, much of the bark and many of the smaller branches remain. This detritus provides nutrients, which, in turn, cause a rapid growth of plant species, particularly algae. This process is called a *bloom*. Blooms are followed by the death of the plants and rapid decomposition, with the inevitable reduction of oxygen and the subsequent death of many animal species. The lake also fills in more rapidly, thus increasing its eutrophication rate. The shallower the lake, the greater the warming and the lower the oxygen levels. To make matters worse, the cutting of trees increases soil erosion. During the spring runoff more decaying plant material is carried to the lakes, thereby increasing levels of nutrients even further. The increased nutrients stimulate algal growth, and the problem spirals.

Figure 3.8

Water samples to be tested for oxygen concentrations are collected for various regions of a lake.

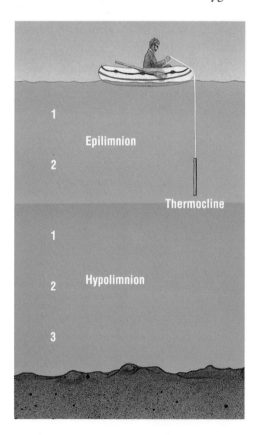

Epilimnion

Thermocline

Hypolimnion

Summerlike conditions can be maintained throughout the year because of thermal pollution. Water removed from a lake by industries and used as a coolant is sometimes returned warm enough to prevent freezing over. Many organisms thus continue to thrive through the winter, in turn producing more detritus with its resultant decomposition. Once again an oxygen-depleted hypolimnion may result, with all of its implications. Additional organic matter in the benthic zone causes more rapid eutrophication.

Canadian Lakes

The most productive part of a lake is the littoral zone. Here algae and other green plants take advantage of the sunlight for their photosynthetic activities. The depth of a lake and the extent of the littoral zone are influenced by the type of lake bed in which the water is found. In central Canada into the western Northwest Territories, lakes were typically formed in basins carved in the granite rock of the Canadian Shield during the most recent glaciation. These lakes are usually deep and, due to their granite base, limited in the dissolved natural nutrients (minerals). Most therefore are oligotrophic. Although lakes in the Atlantic region are somewhat similar, the underlying rock is more varied, since it originated from ancient mountains. In contrast, lakes in the prairies are found in depressions formed in thick glacial deposits. This varied base is richer in soluble nutrients, making the ecosystems of these lakes more productive. These lakes also tend to be fairly shallow and collect sediments more rapidly than the Shield lakes. They are usually classified as eutrophic lakes.

The largest lakes run in a curve from Great Bear Lake in the Northwest Territories to the Great Lakes. They lie on the boundary between the glacial deposits to the south and west and the Shield rock to the north and east. They are considered to be oligotrophic due to their depth and low water temperatures. The major exception is Lake Erie, which is fairly shallow. Agricultural and industrial pollutants plus human wastes have caused rapid eutrophication of Lake Erie, resulting in noticeable changes in the population and species of organisms. International cooperation between Canada and the United States appears to have reduced the process to a considerable extent.

▮ REVIEW QUESTIONS ▮ ?

8 Distinguish between the littoral, limnetic, and profundal zones of a lake.

9 What is meant by the term plankton?

10 What is the effect of the fall overturn on a lake?

11 What is unique about arctic lakes?

12 What is the relationship between dissolved oxygen and the temperature of a lake?

13 Why does a thermocline form in the summer?

14 Define algal bloom. What are its effects on a lake?

15 Outline the steps of pond succession through to a mature forest.

16 Explain how waste runoff from a cattle ranch can cause eutrophication.

Figure 3.9

Thermal pollution from industries can produce summerlike conditions in a lake throughout the year.

LABORATORY

ACID RAIN AND AQUATIC ECOSYSTEMS

Objective

To design an experiment to investigate how acid rain will affect an aquatic ecosystem.

Background Information

Instructions will be given for building an acid-rain generator. Suggestions will be provided for an aquatic ecosystem. The experimental design is up to you. Approval should be obtained from the instructor before beginning.

Materials (suggested)

acid-rain generator	glass tubing
2 125-mL flasks	1 M NaHSO$_4$
1 M HCl	aquarium pump
2 rubber (2-hole) stoppers	lab apron
for flasks	rubber tubing
safety goggles	aquatic ecosystem
field guide for lake and	dissecting microscope
pond organisms	2 L pop bottle
graduated cylinder	pH paper

CAUTION: HCl is very corrosive. Avoid skin and eye contact. Wash all splashes off your skin and clothing thoroughly. If you get any chemical in your eyes, rinse for at least 15 min and inform your teacher.

Procedure

Experimental Design

1 Formulate a hypothesis. Most hypotheses are presented in an "if ... then" format (i.e., "If acids are added to an aquatic environment, then ...").
   a) State your hypothesis.
   b) Identify dependent and independent variables.
2 Working in small groups, devise a procedure that allows you to determine the effect of acid rain on an aquatic ecosystem. Remember to accept all opinions of group members and work to a consensus. Attempt to provide a measurement for all data collected.

c) Outline your group's procedure. Your teacher should approve the procedure before you begin.
3 Prepare data tables. Remember: attempt to quantify all possible recordings.
   d) Present your data. Graphs and data tables should be considered.

Constructing an Acid-rain Generator

4 Using a graduated cylinder, add 100 mL of distilled water to flask #2.
5 Arrange the apparatus as shown in the diagram. Connect the rubber tubing.

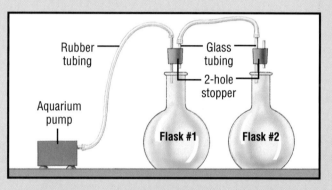

6 Using a graduated cylinder, add 50 mL of NaHSO$_4$ to flask #1. Rinse the graduated cylinder and then follow the same procedure by adding 50 mL of HCl to flask #1.
7 Place stoppers in the flasks and begin your experiment. (It is recommended that you allow the generator to run for at least 20 min.)

Aquatic Ecosystems

8 Fill a 2 L plastic bottle with lake or pond water. Decide whether or not you will aerate the water.
9 Survey the organisms found in the pond water. You may wish to develop a technique for determining the number of organisms.

Aquatic ecosystem

Laboratory Application Questions

1 List the conclusions that you have drawn from the laboratory data.
2 Provide reasons for each of your conclusions.
3 Identify modifications or changes to the procedure that you would incorporate should you wish to carry out the investigation again. ■

FRONTIERS OF TECHNOLOGY: ECHO SOUNDING

One of the most important research tools used for investigating the aquatic environment is the echo sounder, or *sonar* device (sonar is an acronym for *so*und *nav*igation and *r*anging). Originally designed to detect submarines in World War II, echo sounding is used by researchers to map the depth and the bottom contours of bodies of water. As well, it is sensitive enough to record the presence and location of fish and large water-dwelling mammals, enabling the researcher to follow their activities and movement underwater.

As the name suggests, the echo sounder works on the principle of the echo. Sound is transmitted into the water and the reflected signals are detected by a receiver. The distance of an object is determined by the speed of that particular sound in the water. The results are often displayed on a TV screen.

Modern sonar devices use a variety of sound frequencies, both audible and inaudible. They are also able, to a certain extent, to correct for the variables that affect the speed of sound in water, such as water temperature, density, and salinity. Layers of water having different properties can give an inaccurate reading. However, without echo sounding, knowledge of the physical structure of bodies of water and its effects on the distribution of life forms would be severely limited.

Branches of biology such as marine fisheries management depend on a knowledge of the precise location and boundaries of areas such as Georges Bank, an area with large stocks of fish, located at the edge of the Atlantic continental shelf. In areas of extensive deep water such as the Great Lakes and the oceans, echo sounding has provided vital information on aquatic ecosystems and water circulation patterns.

WINTER IN NORTHERN ECOSYSTEMS

Periods of snow and subzero temperatures characterize ecosystems found in regions that experience a distinct winter season. Much of Canada is included in the arctic or subarctic region of North America.

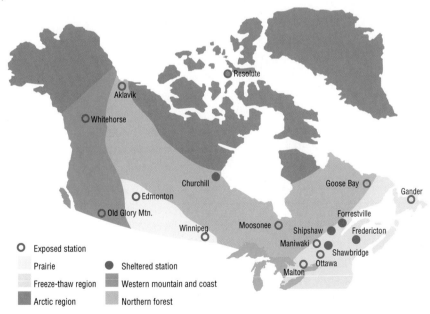

Figure 3.10

Major snow regions of Canada.

Various groups of northern native peoples are so dependent on snow conditions for their survival in winter that they have developed an extensive vocabulary to describe the properties of snow. The Kobuk Valley Inuit of Alaska have devised more than 17 terms to describe different kinds of snow. These terms have now been made part of the language of snow ecology. Snow terminology also uses words from other native groups and from the languages of the countries surrounding the North Pole (i.e., the Russian word *taiga* refers to an area of coniferous forest).

Figure 3.11

Snowshoes can be constructed in different shapes and sizes to suit varying types of snow cover.

Table 3.3 Some Common Inuit Snow Terms

Term	Pronunciation	Description
Qali	Kall-ee	Snow that collects on trees, i.e., on the branches of spruce and other conifers.
Qamaniq	Com-an-nique	The depression in the snow found around the base of trees, i.e. the snow shadow beneath most conifers.
Anniu	An-nee-you	Falling snow.
Api	Aye-pee	Snow on the ground.
Siqoq	See-cok	Drifting (blowing) snow.
Upsik	Up-sik	Wind-beaten snow.
Pukak	Pew-cak	A crystalline snow layer found at the ground–snow interface.
Siqoqtoaq	See-cok-tow-ak	Sun-crusted snow.
Kimoaqruk	Kee-mow-ak-rook	Drift.
Kaioglaq	Kay—oh-glak	Sharply etched wind-eroded snow surface.

Snow characteristics have influenced the mode and form of transport used by northern natives. For example, different shapes and sizes of snowshoes were constructed based on the type of snow cover found in a specific region. The Inuit sled (komatik) is suitable for travel on the hard-packed snow of the tundra, but useless in the soft snow of the forests to the south. The toboggan was developed for travel in softer snow.

ADAPTATIONS FOR WINTER

For the organisms living in northern ecosystems, winter is the most critical time of the year. These organisms must have some way of dealing with limiting factors of snow, cold, and the reduction in or lack of a food supply, to survive. The problem they face was best described by the Russian ecologist, A.N. Formozov, when he stated, "Snow cover for many species is the most important element of environmental resistance, and the struggle against this particular element is almost beyond the species' ability." Yet even when snow and cold persist for most of the year, many organisms can and do survive. It is the extremes in a winter environment that have the most impact on an organism.

Five key conditions—snow, cold, radiation (of heat), energy (food), and wind,—determine whether an organism can survive in a winter ecosystem. These five factors determine successful adaptation. Adaptations can be anatomical, physiological, or behavioral.

RESPONSE TO SNOW

The classification of animals based on the impact of snow on their behavior was suggested by Formozov in 1946 (see Table 3.4). The Russian word for snow, *chion*, prefixes each of the terms. These terms will be used to describe a number of situations.

Table 3.4 Classification of Animals Based on Their Ecological Relationships with Snow

Classification	Description	Examples
Chionophobes	Animals that avoid snow and winter conditions. Do not inhabit snowy regions.	Pronghorns, many terrestrial birds.
Chioneuphores	Animals that are able to withstand winter conditions.	Deer, voles, small mammals.
Chionophiles	Animals that have adapted to snow. The range of the animal is limited to regions of long, cold winters.	Snowshoe hare, polar bear, caribou, musk-ox.

There are many ways in which animals can adjust to winter conditions. **Migration** is the answer for some. Entire populations of some animals move to regions with a more favorable climate and a plentiful food supply. This is most typical of bird species, some of which migrate thousands of kilometers each year. Many of the waterfowl that breed in the arctic spend the winter along the coast of the Gulf of Mexico.

Some migrations may simply involve a shift in population from an area of one set of winter conditions to that of another. The most spectacular of these migrations is made by the barren-land caribou, which migrate thousands of kilometers. They winter in the boreal forest, which is protected from severe winds, and where lichens, a preferred food, are easily obtained. In the summer the caribou move out into the tundra to reproduce. Elk in the Rocky Mountains usually spend their summers at high altitudes, then, in the autumn, move down into the valleys where there are fewer temperature extremes and greater supplies of food. Deer may simply move from one snowy place to another depending on how easy it is to obtain food. Prior to European settlement, the enormous herds of bison used to migrate the length of the prairies to find available food.

Figure 3.13

A migrating caribou herd numbering 10 000.

Figure 3.12

Animals can be classified according to the impact of snow on their behavior.

Migration *is the movement of organisms between two distant geographic regions.*

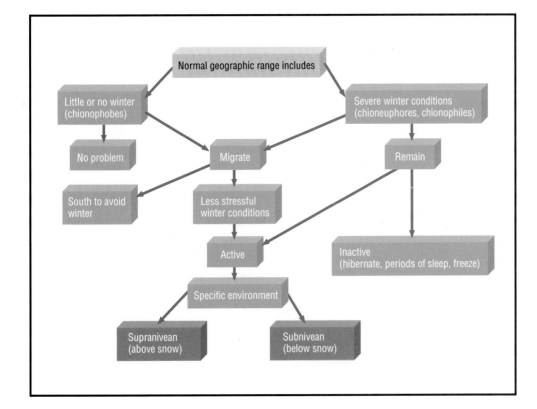

Figure 3.14

Gray squirrel.

Figure 3.15

Moth pupa in cutaway cocoon.

Chionophobes (*snow haters*) *are animals that avoid snow-covered regions.*

Hibernation *is a dormant (sleep) state in which body temperature and functions are reduced well below normal.*

Supercooling *occurs when a water solution is chilled well below the point at which the solution crystallizes spontaneously into ice.*

Chionophobes

The pronghorn of the southern prairies are nearly unable to cope with snow cover. They rarely attempt to dig through the snow to reach their food and must move to where the food is actually visible. If snow cover persists for an unusual length of time in their range, massive starvation and death can result. Because pronghorn have short legs, walking long distances is a major problem and so they are often forced to use roads for travel. Vehicle traffic then kills many of these unique animals. Their range therefore is limited to regions with little or no snow. Animals that avoid snowy regions are called **chionophobes** (literally, "snow haters").

A number of animals **hibernate** during the most severe part of the winter. Skunks, chipmunks, ground squirrels, and marmots (groundhogs) enter a period of low metabolic activity (low body temperature, slow heart and breathing rates) that may extend for six months or more. The arctic ground squirrel (Sic-Sic) hibernates for as long as 10 months. These animals enter their hibernation periods at various times before winter. Most of their nesting areas are in burrows below ground, which help protect them from the winter temperatures. Some will come out of hibernation for brief periods during winter to feed on food stores that were built up during the summer. While "sleeping," they rely on stored fat, particularly a high-energy form called brown adipose (fat) tissue, to maintain their body functions.

Although bears enter their dens by early winter, they are not true hibernators. They may sleep for short periods but

with no appreciable change in heartbeat, breathing rate, or body temperature. Some bears have been known to burst from their caves within a few seconds of being disturbed, even in midwinter. In pleasant weather they may roam the areas near their dens. In the Mountain National Parks their activity has often forced the closure of cross-country ski trails. Polar bears are exceptionally well adapted to winter and are active the year round.

A number of insects utilize their life cycle to survive the winter. In the fall they lay their eggs, which then remain dormant until spring conditions trigger their hatching. The adult insects usually do not survive the cold. A few insects overwinter in the pupa stage.

When ice forms, the normal result is increased tissue destruction due to the growth of ice crystals within the cytoplasm of the cells. However, some organisms have been known to survive the cold of winter by freezing solid. Animals that can survive this freezing process include many insects of the arctic and subarctic regions, and animals such as barnacles and mussels of the tidal zone of the northern oceans. Apparently these species can produce an antifreeze mixture of proteins and glycerol that is effective in lowering their freezing point.

The **supercooling** of water also plays a role in organism survival. In this process water can be gradually cooled down to as low as $-41°C$ before it changes to ice. The larvae of *Bracon cephi*, an insect parasite of the Canadian wheat-stem sawfly, can survive a temperature of $-47°C$. It produces an antifreeze—a concentrated glycerol solution—that lowers

> **BIOLOGY CLIP**
> In the high arctic, a limited number of plant species freeze during their flowering period and remain frozen till the next summer, when the flowering process continues where it left off.

its freezing point to about –17°C. In addition, through supercooling, the water molecules in the tissues produce ice crystals so slowly that an additional 30°C drop can be reached before tissue damage results.

Recent research has discovered a number of amphibians and reptiles that freeze solid when air and soil temperatures drop below freezing, then thaw in spring and continue on with their lives. Some can even survive several freeze-thaw sequences during the winter. Forest floor frogs, such as the spring peeper, the wood frog, and the striped chorus frog, regularly survive freezing. Young painted turtles, which hatch in late summer, remain in their nests and freeze solid when the ground temperatures fall below –3°C.

Chioneuphores

Animals whose ranges extend into regions where winter snow and cold temperatures are the dominant climatic conditions are called **chioneuphores.** They are capable of surviving in snowy regions, but may also be found in other regions less affected by winter. By remaining in the winter environment they are faced with a reduction in their food supply. Because cold temperatures cause the animals to lose heat more rapidly, they require additional food in order to maintain their normal body temperatures. Energy-intensive activities, such as wading through snow, put additional stress on animals. Some waders, such as the fox, coyote, and wolf, step in their own footprints in the soft snow to reduce the energy requirements of travel.

Animals referred to as *waders*, such as deer, struggle to walk through snow. The deer usually herd together where there is food and protection, trampling down the snow and forming "yards." Moose, on the other hand, are able to walk through deep snow using a gait that involves stepping upward, lifting their hooves above the snow, then "post-holing" their way about. At snow depths greater than 60 cm, however, moose must start wading through the snow, using up precious energy.

Since snow will cover food sources found at the ground level, many of the larger-hoofed animals have the ability and instinct to dig for food, forming craters in the snow. This activity may also expose food sources for other animals. Grouse often depend on food uncovered by deer. However, if the wind has hard-packed the snow or if sleet has formed a thick crust, animals may be unable to dig craters and therefore can starve. The effort required to walk through these snow conditions demands increased energy. The hard-packed surface snow can cut legs, causing blood loss and placing additional stress on the animals.

Animals such as the elk may be forced to change their diet in winter. Often the plants they consume in summer are no longer growing or are covered by deep snow and become unavailable. Elk switch from summer **grazing** to winter **browsing,** feeding on buds, twigs, and the bark of trees and shrubs. Bite marks leave characteristic scars on the trunks of aspen poplar trees where elk have stripped bark in times of low food supplies. Elk, deer, and other animals chew a cud of vegetation prior to digesting the material. Bacteria in the digestive system help the digestive process and act on specific foods. However, the bacteria must change with the diet. This is why, when hay and other foods are brought to deer herds that are trapped by a snowstorm, the starving animals continue to die off—they no longer have "summer-type" bacteria capable of breaking down the food.

The lynx, snowshoe hare, ptarmigan, and ruffed grouse are called *floaters.* They have developed adaptations that enable them to walk or "float" on the snow sur-

Figure 3.16
In winter, snow covers the food sources of deer.

Figure 3.17
Elk winter diet. Bite marks leave scars on aspen trees.

Chioneuphores *are species of animals that can withstand winters with snow and cold temperatures.*

Grazing *is feeding on grass or grasslike vegetation.*

Browsing *is feeding on leaves, twigs, buds, bark, and similar vegetation.*

a)

b)

c)

Figure 3.18

(a) Ruffed grouse.
(b) Willow ptarmigan. (c)Lynx.
(d) Snowshoe or varying hare.

Chionophiles *(snow lovers) are animals whose ranges lie within regions of long, cold winters.*

face. In each case, the feet enlarge in the fall and develop excess hair or feathers. This spreads the overall mass of the animal over a larger foot area, sufficient to keep them from sinking in the snow. If the snow is too soft to support them, hares make trails that they and other animals follow regularly, keeping the energy demands of travel to a minimum.

d)

The ruffed grouse of the boreal forest has a unique method of surviving the intense cold nights of winter. It will fly directly into the soft snow and wiggle its wings and legs to create a tunnel. Snow then fills the tunnel and completely cov-

ers the animal. This behavior allows the birds to take advantage of the warmer temperature in the snow pack. In addition, the food it has eaten, usually frozen buds from trees and shrubs, requires less body heat to thaw and to be digested.

Winter winds pose additional problems for animals. Air moving over the surface of the skin removes heat at a rapid rate. The higher the wind speed, the greater the loss of heat from the skin. The windchill factor, combined with low temperatures, puts extra stress on organisms. Animal herding is one way of reducing the effects of the winds.

Snow can be beneficial to animals that live on its surface. The snowshoe (varying) hare is an excellent example. The winter diet of these animals consists of the buds and branch tips of a variety of shrubs. If the hare population is high, it takes little time for the most nutritious buds to be eaten. Remember, however, that the snowshoe hare is a floater. Each snowfall increases the height of the snow pack. As a result, the hares have access to the upper regions of the shrubs, where a new supply of food awaits them. Eating terminal buds can also trigger lateral branch growth the following summer, providing an even greater food source for the next winter. In addition, the weight of the snow on shrubs and trees can lower the branches, allowing animals to feed.

Chionophiles

Chionophiles are animals whose ranges are usually limited to climatic regions dominated by long, cold winters. Characteristic species include the polar bear, lemmings, musk-ox, barren-ground caribou, arctic fox, arctic hare, and a variety of small rodents and insectivorous mammals. Also referred to as "snow lovers," the chionophiles are so well adapted to winter conditions that they have limited success in southern locations.

Physiological and Anatomical Adaptations

Most northern mammals have some adaptations or behaviors that reduce the effects of the intense cold and temperature extremes of the arctic and subarctic regions. Any protruding body parts are often smaller in size to prevent heat loss. The arctic fox, for example, has tiny, rounded, fur-covered ears compared with the red fox of the southern boreal forest. Another adaptation is a coat of fur that can insulate the body. As indicated by the graph below, the thicker the fur, the greater is the conservation of body heat. Animal molting (shedding) produces a winter fur cover that usually includes inner and outer hair layers. Some hair is specialized for insulation. In musk-ox the hair grows to such an extent that it nearly covers the animal to the ground, thus protecting its legs from the winds and cold.

BIOLOGY CLIP

Many marathon swimmers and some individuals who live in severely cold climates share a common practice: both apply animal fats to the outer surfaces of their bodies. The fat acts as an insulation, reducing heat loss.

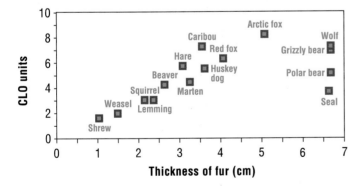

Thickness of fur (cm)

Coat color has always been considered a form of camouflage. For example, the snowshoe (varying) hare has a white coat in winter and brown fur in the summer. Recent studies of a variety of white-coated northern animals have indicated that the white hairs have superior insulation properties to the dark ones. This confers a distinct advantage to those animals that change their coat color to white in the winter.

It is not common practice for humans to walk about in the snow without covering on their feet or legs. A person who did so would suffer frostbite and would rapidly lose sufficient body heat to develop hypothermia. This is because very cold temperatures affect the extremities more quickly than the main core of the body. Frostbite usually hits the smaller projections of the body (nose, ears, toes, fingers), which have large surface areas compared with the volume of tissue they enclose. Although blood is pumped from the body core to the extremities to provide more heat, there is a net heat loss and the cooled blood returns to the core, reducing the core temperature. This can eventually lead to hypothermia and death. In nature, animals in winter must face this problem all the time.

Many mammals and birds have unprotected extremities and therefore face a similar situation in cold conditions. First described in 1957, a unique variation in the vessels of the circulatory system called the *rete mirabile*, or "wonderful net," is found in the extremities of these animals.

Today this type of circulatory adaptation is most often called a **counter-current exchange** system. In this adaptation, the arteries carrying warm blood to the extremities are adjacent to the veins that bring the blood back to the

Figure 3.19

A "CLO" unit equals the amount of insulation provided by the clothing a human usually wears at room temperature.

Counter-current exchange *for the circulatory system is an anatomical variation designed to reduce heat loss.*

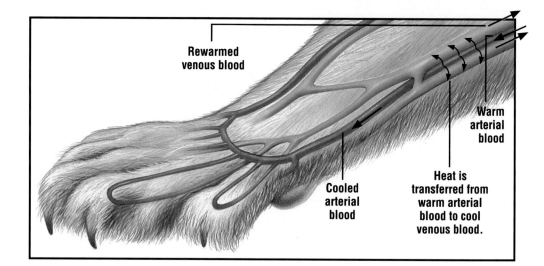

Figure 3.20

Counter-current heat exchange system. Warm arterial blood transfers heat to cold blood returning from the extremities.

heart. The warm arterial blood transfers some of its heat to the blood in the veins and cools down in the process. Thus, cooled blood reaches the extremities, and heat loss from appendages is considerably reduced, allowing them to continue functioning. The blood returning to the heart is warmed up by the blood in the arteries and does not affect the animal's core temperature.

Many organisms increase their activity when their temperature drops. This activity will generate additional heat but also requires extra energy. Shivering, for example, is a significant factor in maintaining body temperature, although it depletes energy reserves during a time of limited food.

Adipose, or fat tissue, is an excellent source of stored energy. A special form called brown fat is associated with animals living in cold conditions. It produces a greater amount of heat than regular fats. Brown fat is formed when food is easily available, and is then used to meet the additional heat requirements of animals active in the winter environment. It is also a storehouse of energy for many animals that hibernate. In cold conditions, fat tissue tends to become hard and brittle. It would therefore be expected that the legs and feet of mammals that normally contain fat would become stiff and make walking difficult. However, the fat found in the feet and lower leg bones of the caribou and other hoofed mammals (including domestic cattle) is usually soft and does not freeze. Extracts of this fat, called neat's-foot oil, are often used to keep leather flexible in cold weather.

Figure 3.21

The snowshoe hare loses less heat to the environment by huddling under the lower branches of a spruce tree.

Animals without large fat stores use insulation provided by the environment. The lynx and snowshoe hare spend time under cover of the lower branches of conifer trees or the branches of any trees or shrubs that have been weighed down by snow. Some of the heat lost by their bodies is reflected back by the overhanging branches, helping to keep the animals warmer. Other animals simply become more active hunters and eat to produce heat. Arctic foxes associate themselves with polar bears, attempting to scavenge the dead seals killed by the bears.

Northern mammals tend to be larger than southern species. This lowers the surface area of the animal where heat is lost, compared with the overall volume of the animal that retains heat. In smaller mammals, the area-to-volume ratio becomes less and body heat is lost more rapidly. In some mammals heat loss cannot be balanced by food intake. These animals must exhibit some form of energy-saving behavior or die of hypothermia. This might mean remaining inactive in an insulated nesting site (e.g., the red squirrel of the boreal forest) for the duration of an intensive cold period and feeding on stored food. Our smallest mammals, those smaller than a squirrel (e.g., mice, voles, shrews, and lemmings), must move under the snow for protection from heat loss. There they remain somewhat active all winter long. These mammals are referred to as **subnivean** ("beneath the snow") animals and are completely dependent on snow cover for their survival. The less active subnivean animals such as mice supply their nests with food before winter, then spend much of their time in small groups huddling together and sharing their body heat. Others, such as shrews and voles, remain active throughout winter. To understand why, it is necessary to examine some characteristics of snow and snow cover.

SNOW ECOLOGY: ABIOTIC FACTORS

When snow falls on the ground it affects everything it touches and, in turn, is modified by everything it touches. The properties of old snow are different from those of fresh snow. The changing characteristics of snow can mean life or death to a large number of winter residents.

In the summer the ground absorbs heat from solar radiation. This heat is slowly lost back to the atmosphere as the year progresses. In addition, geothermal energy produced in the earth's core is also being radiated outward through the soil. During a snowfall the well-developed snowflakes interlock as they land on the ground and trap air between them. Air is a poor conductor of heat, making it an excellent insulator. This slows down heat loss from the soil beneath. The snow therefore becomes an insulating blanket. Snow is most effective as an insulator when it is light and fluffy. Fluffy snow, much like down, contains many small air spaces. This type of snow is most often found in areas such as forests, where maximum wind protection is provided.

The insulation provided by snow can best be illustrated by the following demonstration. Place pieces of metal, wood, and styrofoam in a refrigerator overnight. Remove each of the objects the following day and hold each in the palm of your hand for one minute. Which one feels coldest? Even though all three have the same temperature, it should come as no surprise that the metal object feels coldest. That is because the material is denser; fewer air spaces are found between the molecules, and heat is conducted at a faster rate. *Conduction* occurs best when molecules are touching. Heat flows from your hand to the metal object very quickly. By comparison, the wood has many air spaces

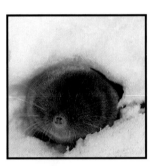

Figure 3.22

A shorttail shrew takes advantage of the insulating properties of snow.

Subnivean *means "beneath a snow cover."*

left from dead cells. Air is a poor conductor of heat. Most molecules in the air are not touching. Styrofoam has even more air spaces than does the wood, making it a poorer conductor and better insulator. To emphasize this concept, place the block of wood in water and refrigerate overnight. Water will enter many of the air spaces in the wood. The denser wood is a better conductor of heat, making it a poorer insulator. That helps explain why wet wood feels colder than dry wood, or why woolen socks lose their insulating value when they become wet.

As the snow pack increases in depth, more of the stored heat in the ground is trapped. The heat warms up the air, which then slowly moves upward in the snow. This results in a snow pack of different temperatures, from warmest at the snow-ground interface to coldest at the snow surface.

The heat also causes sublimation to occur at the tips of the snow crystals at the snow-ground interface. This destroys the original structure of snowflakes. The water vapor that moves upward is recrystallized on the cooler snow at upper levels. This process results in a change in the form of the upper snow crystals. These crystal changes are called *destructive* and *constructive metamorphism* respectively. The bottom crystals tend to change from their usual six-sided structure to pieces of ice the shape of ball bearings. By determining the average time required for the snow crystal to change shape, it is possible to determine the general age of an undisturbed snow pack.

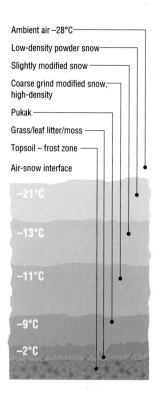

Figure 3.23

The effect of snow cover in a cold-weather environment.

Figure 3.24

(a) Changes in snow crystal form in the snow pack. (b) The growth of snow crystals in the snow pack due to heat from the earth. (c) The destructive metamorphism of a stellar snow crystal. The numerals give the age of the snow crystal in days.

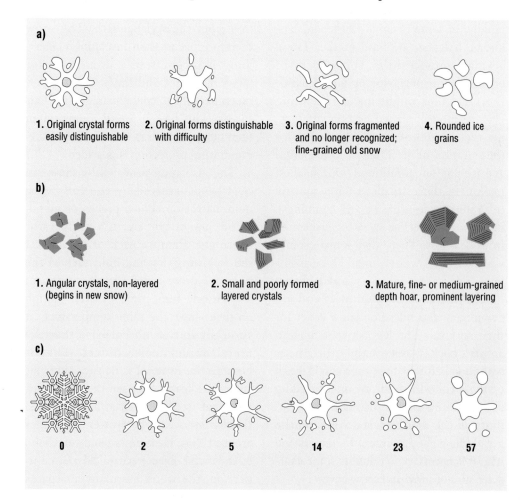

86

On a mountain slope the development of these ball-shaped crystals makes the snow unstable and can lead eventually to the mass movement of the snow pack in the form of an *avalanche*. Studying a core sample of the snow can provide advance warning of avalanches.

The upper surface of the snow is acted on by heat from the sun, rain, sleet, and the movement of the snow by wind. These factors also change the original structure of the snow crystals. Since the different crystal structures at the boundary of each snowfall are usually quite visible, it is often easy to determine the number of snowfalls that have accumulated. In a cross section of a snow pack the number of observed snow layers or strata is an indicator of the number of snowfalls. As well, the age of the snow in each stratum can be estimated.

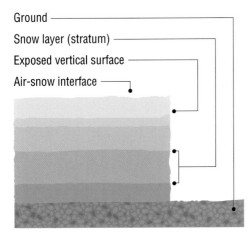

Figure 3.25
Cross section through a snow pack outlining the snow strata.

Crystals of snow at the base of the snow pack sublimate, causing a layer of open space containing tiny ice crystals to develop. This layer is called **pukak.** As more of the ice crystals sublimate, the pukak layer becomes a partially empty gap between the soil below and the snow pack above. Subnivean mammals and insects use this space to move about in search of food. The environment of the pukak layer is dark, moist, and quiet. When the snow pack reaches a depth of about 30 cm, the temperature of the pukak stabilizes and remains at one or two degrees below freezing regardless of the air temperature. By the end of winter the pukak layer can be quite extensive. It will usually not form where the snow is very thin or where ice has formed a layer against the ground.

The most favorable subnivean environment occurs where the snow covers areas of heavy leaf litter, moss, and/or grasses, all of which have additional insulation properties of their own. A subnivean space is formed after the first snowfall as the low-lying vegetation keeps the snow from contact with the ground. This space can be used immediately by organisms without waiting for pukak to develop. Occupied grass nest sites under the snow have actually reached temperatures as high as 18°C, even though the air above the snow was –17°C.

Figure 3.26
Pukak. Water vapor moves up from warm to colder regions of snow forming loosely-arranged, brittle crystals at the base of the snowpack.

Pukak *is a layer of open space containing ice crystals at the base of a snow pack .*

SNOW ECOLOGY: THE BIOTIC COMMUNITY

Many factors make life under the snow advantageous for small mammals, insects, and other life forms. Snow offers some protection from larger predators such as the coyote. There is a fairly constant high humidity, which reduces evaporation and its resultant additional heat loss. Most significant is the fact that snow is an excellent insulator and can protect organisms beneath it from the lower air temperatures. The deeper the snow cover, the warmer the subnivean temperature. If an animal wants to reach the surface it simply digs upward, producing a ventilator shaft.

When snow first falls, particularly in the boreal forest regions, it is light, fluffy, and has a low density. This light snow has nearly the insulation value of eiderdown. As a general rule, wherever the average snow depth is from 10 to 15 cm, the subnivean temperature is usually warm enough to provide some protection for the organisms beneath it.

Although snow cover usually provides protection from predators, some carnivores use the subnivean environment to their advantage to hunt for prey. The weasel actively hunts voles and locates them by following their ventilator shafts down to their trails and tunnels. Where the snow is not too deep, coyotes and foxes will remain motionless on the surface, listening for sounds of small mammal activity. The subnivean mammals are often vocal, giving off squeaks as they travel about. They also cause a tinkling sound as they collide with and break off the tiny crystals of ice in the pukak layer.

Moving to different locations and listening to sounds, the coyote processes this information and is able to locate the small mammal. Leaping into the air and planting its forelegs forcibly into the snow, the coyote can break through to the pukak layer, pinning its prey to the ground. Biologists marvel at the mathematical calculations that the coyote must instinctively perform to locate its prey.

Snow cover also prevents the ground from freezing too deeply in areas where there is no permafrost. In the subarctic the frost line usually reaches a depth of 1 to 1.5 m. When it thaws in spring, vegetation can begin to grow. In years of low snowfall or in places where the snow has blown away, the depth of the frost increases. This often delays the growth of some vegetation in the spring.

The density of the snow cover is a significant factor in maintaining a constant subnivean environment. Any increase in snow density reduces its insulating properties. The snow density will increase if it is wet or if the wind blows the snow about, smashing the crystals into small granules that pack tightly together to form a layer with a density approaching that of ice. Time also allows the snow to settle and pack.

Figure 3.27

The relationship between the insulation properties and relative density of snow cover.

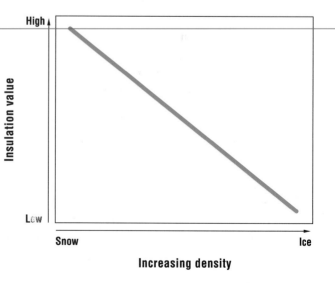

Figure 3.28

Qaminiq.

Figure 3.29

Qali.

In open spaces in the forest or on the prairies, the winds are able to break down the upper snow surface sufficiently to form a hard, thick crust that is able to support animals, including humans. This snow is quite dense and loses much of its ability to insulate. A crust can block access to the surface for small mammals. As well, it can prevent grazers such as caribou and deer from getting down to their food supply. If the wind blows sufficient snow from an area, the reduced snow cover may no longer act as protection, and subnivean organisms must move or perish. In the tundra, a suitable subnivean environment is found only where snow has been deposited behind windbreaks in depressions in this otherwise wind-swept environment. Here it provides protection for lemmings, ptarmigan, and various forms of plant life.

Trees also have adapted to cold climates. Conifers generally have resins within their vascular tissues that act as antifreeze. Supercooling is also a help. If the temperature drops rapidly, however, water in the tissues can expand to such an extent that a tree may split from top to bottom, producing a sound similar to that of a gunshot. Deciduous trees usually transport most fluids to the root system, below the frost line, thus reducing the chance of freezing and tissue destruction.

Coniferous trees of the boreal forest, such as the white spruce, have adapted to **qali,** the snow that sticks to its branches. In these trees, the branches radiate outward and down from a main trunk. If too much snow builds up, it can slide off with minimum damage to the tree. If heavy snow falls on your head it can cause death. In conifers the orientation of the branches results in the formation of a snow shadow, or **qamaniq,** beneath each tree. Little subnivean protection is found in these depressions, where the snow is not deep enough to insulate the ground. Therefore, small mammals are rarely found here.

If hit by an early or late snowstorm, deciduous trees cannot handle the weight of snow on their branches, particularly if leaves are still present. Extensive tree damage can occur, opening up areas in the boreal forest to secondary succession.

Qali *is snow that collects on the branches of trees.*

Qamaniq *(snow shadow) is the depression in the snow found around the base of trees, particularly conifers.*

THE SUBNIVEAN ECOSYSTEM

Many people believe that insects and small mammals are dormant during the winter. The truth is that some continue to forage for food during the winter—they just are not as visible as in the summer. Insects and small mammals can be found scurrying through the pukak layer. Protected by a blanket of insulating snow cover, these animals must adapt to a very different and often changing ecosystem.

As winter progresses, the upper layers of snow become more dense. Not only are they less able to insulate, they also tend to reduce the air exchange between the subnivean space and the air above. An accumulation of carbon dioxide resulting from bacterial decomposition begins to occur beneath the snow, especially in low areas or depressions. Small mammals must tunnel to the surface to breathe fresh air. The ventilation shafts they create help to improve air exchange. If the temperature is mild, the animals may spend short periods on the surface. However, predators such as owls are aware of this activity and often position themselves near tunnel openings.

There are three critical periods for subnivean mammals. The first occurs prior to the appearance of permanent snow in early winter. At this time of the year, daytime air temperatures are gradually dropping and nighttime temperatures are usually below freezing. Only an animal's heat-producing metabolic processes, its coat, and the insulating effects of nests

Hypothermia is a condition in which an animal's body core temperature drops to a level that will eventually cause death.

Figure 3.30

Now rarely used by the Inuit, the snow igloo is made from blocks of densely packed snow.

BIOLOGY CLIP
Snow fleas are small insects that hatch under the snow in late winter. They work their way upward and roam about the snow surface, particularly on sunny days. The snow may actually look black as millions of the insects gather where it is warm. If an animal makes a footprint in the snow in a sunny location, the snow fleas will gather on the side of the depression facing the sun, producing a "reverse shadow."

and dead grass and leaves on the ground prevent **hypothermia** and death. If temperatures continue to drop without snow cover, there is a high mortality rate among the small mammal population. With a normal pattern of snowfall, a depth of as little as 6 cm will provide the subnivean environment with some insulation from the colder ambient air temperature above.

The second critical period for subnivean animals occurs with the January, or midwinter, thaw. This is a common weather pattern in the eastern half of North America. It is initiated by periods of rain or warm weather, which melt the snow to some extent, increasing its density while reducing the depth of the snow pack. The snow is less able to provide protection from the below-freezing air temperatures that usually follow the thaw. Sleet can form a layer of ice that may prevent the normal exchange of oxygen and carbon dioxide as well as block an animal's access to the surface. Where the snow is melted away during a thaw or blown away by high winds, insulation becomes nonexistent. Bare patches in the snow indicate zones of low temperatures, and the animals keep away from them. This additional stress placed on the subnivean animals may be more than they can handle.

A final challenge to subnivean life occurs during the spring thaw. The density of the remaining snow has usually increased to the point where it has lost most of its insulation properties. It is now often warmer above the snow than beneath it, particularly in the daytime, and the animals may move above or below depending on local conditions. As the snow disappears and the air begins to warm up, the animals can usually survive the daily temperature fluctuation. However, if the loss of snow cover is followed by a period of very cold weather, hypothermia can once again become a problem, and a late winterkill may result.

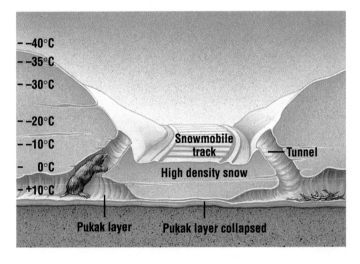

- --40°C
- --35°C
- --30°C
- --20°C
- --10°C
- 0°C
- +10°C

Snowmobile track

High density snow

Tunnel

Pukak layer   **Pukak layer collapsed**

Vehicles such as snowmobiles may compress snow to such an extent that the pukak layer is destroyed. Any animals that would normally burrow to a feeding or nesting site on the opposite side of the trail now face a barrier. Voles and shrews often dig upward to the surface of the trail, cross over, then dig back down to the pukak layer. However, this practice presents many hazards for the small mammals forced to the surface. Because these dark-coated animals are highly visible when crossing the path, the trails provide an excellent hunting ground for the predaceous snowy owl. Additionally, when animals leave the warm confines of the pukak layer, some, like the shrew, are so small that they may freeze to death before they can get across the track. This provides additional food for scavengers such as the raven.

The Inuit and other northern native peoples have taken advantage of snow's insulation properties to construct snow shelters. The Inuit are famous for the now rarely used snow igloo made from the densely-packed snow typical of the tundra regions. This snow is hard enough to be cut into rigid blocks. When appropriately fitted together, the blocks provide considerable insulation even though the snow density is quite high and its ability to insulate reduced. An igloo also protects individuals from the wind-chill characteristic of the tundra.

Snow in the boreal forest regions that make up most of Canada tends to be softer and cannot be used to construct igloos. The Chipewyans and the Athapaskans developed a unique snow shelter called a quinzhee, which allows them to use the excellent insulation properties of soft snow. With this type of shelter they have been able to create an environment similar to that of subnivean plants and animals. With one or two people inside the quinzhee, a combination of the heat rising from the ground plus body heat can raise the inside temperature five to ten degrees above freezing. It is not unusual to find inside temperatures of 10°C with outside temperatures of -30°C or lower.

Figure 3.31

Snowmobile trails and subnivean mammal movement.

Figure 3.32

The quinzhee, another form of snow shelter, takes advantage of the insulation properties of soft snow.

■ REVIEW QUESTIONS ■ ?

17 Provide examples of anatomical, physiological, and behavioral adaptations to a winter ecosystem.

18 Define: chionophobes, chioneuphores, and chionophiles.

19 Explain why snow acts as an insulator.

20 Why does a pukak layer form?

RESEARCH IN CANADA

Dr. William O. Pruitt, Jr., is a professor in the zoology department at the University of Manitoba. He has made unique and outstanding contributions to our knowledge of the physical characteristics of snow and the impact of winter conditions on living organisms. His research findings have been published in major North American and international journals. Dr. Pruitt has also contributed articles to such magazines as *Nature Canada* and *Scientific American.* His many investigations on how snow characteristics influence the behavior and life-style of both large and small mammal populations have set benchmarks for other researchers.

Winter Environment

Educating Canadians about the winter environment has been one of Dr. Pruitt's major goals. As he has stated, "In the literature of the sciences that ought to be most concerned, there is even yet little to suggest that snow is a major element in the environment of life." The number of books on snow and winter that are now appearing in schools and bookstores is, in part, due to his efforts.

Dr. Pruitt is an outstanding teacher. He has helped to produce films on winter ecology studies for classroom use and personally conducts many of his classes where they are most effective—outdoors in the middle of winter.

DR. WILLIAM O. PRUITT, JR.

LABORATORY

ABIOTIC CHARACTERISTICS OF SNOW

Objective

To examine the effect of snow cover on temperatures beneath the snow.

Materials

3 thermometers (reinforced)
shovel
black cardboard
meterstick
stopwatch
hand lens
winter gloves, hat,
    adequate warm clothing

Procedure

Note: This investigation is to be conducted outdoors. Be prepared for severe winter conditions. Air temperatures of −10°C or lower are recommended for this activity.

1 Select a location where the snow is fairly deep.
2 Dig a snow pit down to the ground level. Be sure the ground is uncovered.
   a) Describe the snow in terms of appearance, hardness, and the size of the snow crystals at different depths.
3 Cut away one side of the pit to form a vertical surface.

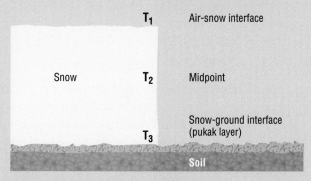

T_1 Air–snow interface

Snow T_2 Midpoint

T_3 Snow–ground interface (pukak layer)

Soil

Refer to the diagram for assistance.
   b) Note whether a pukak layer is visible.
4 Quickly insert one thermometer lengthwise into the vertical surface at the midpoint between the upper surface and the ground (see the diagram). Simultaneously insert a second thermometer under the snow at the ground level. Leave the thermometers in place for two minutes.
5 Remove each thermometer and immediately record the two temperatures.

   c) Why is it necessary to wait two minutes before removing the thermometers?
   d) Why must the temperatures be read immediately after the thermometers are removed from the snow?
6 Place the third thermometer on the upper surface of the snow. The thermometer must be shaded. Leave it in place for two minutes, then record the temperature.
7 Record your observations in a table similar to the one below.

	Site 1	Site 2
Snow depth		
Snow surface temperature		
Midpoint temperature		
Snow–ground temperature		

8 Measure and record the snow depth along the vertical face.
   e) Is there any evidence of snow strata? If so, record the thickness of each layer and its position in the snow pack. A diagram may be helpful.
9 Select a second site where the snow is significantly thinner but has similar characteristics to the snow at site 1.
10 Dig a second snow pit and repeat procedures 3 to 8. Record all your data.
11 Place snow crystals collected from different strata on black cardboard and examine them with the hand lens.
   f) Referring to the diagrams of snow crystals provided earlier in the chapter, determine the age of the snow.

Laboratory Application Questions

1 Is there any indication that snow has any effect on the temperature within and beneath it? Explain, using data from your investigation.
2 Discuss the results you obtained from site 1 and site 2. Are any trends observable? Use your graph and tables of data to support your conclusions.
3 Would ground-level conditions be any different if there were no snow cover? Explain.
4 Would the effects have been any different if the air temperature had been above 0°C? How could you confirm your opinion?
5 Survival experts often suggest that if you are stranded in a snowstorm you should dig a hole in a snowdrift and crawl in. Comment on this idea, using data from your observations. ◼

SOCIAL ISSUE:
Northern Development

The James Bay Project, a huge development on the east coast of James Bay, has greatly increased Quebec's production of hydroelectric power. However, the project has generated much controversy because of its effects on native groups and the environment. Entire villages have been relocated and rivers reduced to streams; forests have been destroyed by fire and large wilderness areas flooded.

The Canadian North is rich in untapped natural resources. Northern megaprojects would boost the Canadian economy and provide much-needed jobs. But are the short-term profits for developers worth the potential long-term harm done to the environment? Can the delicate northern ecosystems survive such disruptions?

Statement:

All northern megaprojects should be halted in order to protect delicate arctic and subarctic ecosystems.

Point

- A simple tire track across the snow can spell death for many small mammals. Depressions in the soft summer soils become puddles, which act as breeding grounds for mosquitoes. Work camps often attract bears and other wildlife unaccustomed to humans. Low temperatures slow decomposition of garbage and other wastes.

> **Research the issue.**
> **Reflect on your findings.**
> **Discuss the various viewpoints with others.**
> **Prepare for the class debate.**

Counterpoint

- Northern communities need economic growth. Canada's North provides a wealth of crude oil and minerals. This untapped resource may well be the key to Canada's future. Development must be responsible, but waiting one to two years for an environmental assessment will kill most major projects.

CHAPTER HIGHLIGHTS

- Biomes are global ecosystems recognized by their distinctive vegetation and controlled by climate and other abiotic factors.
- The major abiotic factors determining the vegetation within an ecosystem are precipitation and temperature. Large bodies of water modify the temperature of any adjoining land mass.
- Light can penetrate a lake only to a specific depth. The bottom is in darkness. Ponds are shallow enough to allow light to reach the bottom.
- The amount of dissolved oxygen in a lake or pond depends on surface area, water movement, water temperature, and the rate of organic decomposition. Oxygen depletion in standing water can result in a

die-off of oxygen-tolerant species such as fish. Winter conditions can alter the lake or pond ecosystem.
- The process of eutrophication initiates a sequence of changes (succession) in lakes and ponds, which can undergo succession from open, oligotrophic water to mature forest.
- The major ecosystems in Canada include the tundra, boreal forest, west-coast temperate rain forest, western cordilleran (mountain) region, and the eastern deciduous forest.
- Animals have various ways of coping with winter conditions: by migrating, hibernating, or freezing during winter or by remaining active in snowy regions through physical or behavioral adaptations.

- Chionophobes avoid winter if possible; chion-euphores can live in snowy regions; chionophiles must live in regions dominated by long, cold winters.

- To conserve heat, the most northerly mammals tend to be large (reduced surface area–to–volume ratio), are usually covered with a dense winter fur, and have small projecting body structures (e.g., ears).

APPLYING THE CONCEPTS

1 Describe how a water-treatment plant slows down eutrophication.
2 Following a late spring, the ice cover on a lake lasts an extra four weeks past normal breakup. Describe the possible effects on the lake ecosystem.
3 Explain why certain species of animals are used as markers for determining the oxygen levels of lakes. (For example, many leeches are found along the bottoms of lakes with low levels of oxygen; stonefly larvae are indicators of higher levels of oxygen.)
4 How does an increased number of decomposers affect the life of a lake or pond?
5 Explain why eastern lakes are more seriously affected by acid rain than their western counterparts.
6 After a wind, many of the projections break from snow crystals. Explain how winds affect the insulating value of snow.

7 Does your paper carrier walk across your lawn? Explain why the grass often appears dead along the paper carrier's path, even in early spring. Why does the grass next to the path appear much healthier?
8 A conservationist feeds grain to a herd of starving deer during a particularly difficult winter. Discuss the implications of such practices.
9 Considering the low levels of solar radiation available for heating in the subarctic winter, would a change of coat color from dark to white really be an advantage to an animal such as the snowshoe (varying) hare? Explain.
10 Outline how activities such as cross-country skiing and snowshoeing can affect snow ecosystems.

CRITICAL-THINKING QUESTIONS

1 Discuss the various roles that detritus can play in aquatic ecosystems. Why have some cottage towns insisted that their residents equip cottages with sewage-holding tanks?
2 If humans stopped interfering with lake ecosystems, would eutrophication be stopped? Explain.
3 Thermal pollution has significant effects on Canadian lakes and ponds and the communities associated with them.

a) Why are Canadian standing waters particularly affected?
b) Comment on specific effects of thermal pollution.
4 An early autumn storm dumps 30 cm of heavy, wet snow on a section of the boreal forest prior to leaf fall. What effects, both positive and negative, might the storm have on this section of the forest?

ENRICHMENT ACTIVITIES

1 The barren-land caribou population has dropped significantly over the last 80 to 90 years. Research this subspecies of caribou and discuss the reasons for this negative population growth, the status of the population today, and ways in which the herds can be protected from destruction.
2 Build a small model of a quinzhee. Record the temperature inside and outside of the quinzhee.

*A*daptation and Change

IMPORTANCE OF ADAPTATION

No two organisms are exactly alike—not even identical twins! This observation demonstrates an important feature of all life: diversity. Equally important, but often less apparent, are the similarities shared by all living things. All plants and animals, regardless of size, shape, or level of complexity, share certain characteristics. These include requirements for energy, basic cell structure and function, and **adaptation** to a particular habitat.

All living organisms are "adapted" in the sense that their appearance, behavior, structure, and mode of life make them well-suited to survive in a particular environment. Fish are not generally adapted for flying, and forest birds are not commonly found swimming in the ocean depths. Similarly, pine trees and elms are dryland inhabitants, whereas moss plants and ferns are found in water-laden or damp habitats. In other words, organisms have their own special adaptations and cannot survive or thrive in the habitats of others.

The theory of evolution attempts to explain why living organisms, so similar in their biochemistry and

Figure 4.1

This photograph does not show a twig but the caterpillar of the peppered moth.

Adaptation *is an inherited trait or set of traits that improve the chances of survival and reproduction of organisms.*

Figure 4.2

What similarities or unifying features are most apparent in these organisms? How do they differ? How is each adapted to its habitat?

molecular biology, are so different in form and function. In its simplest form, **evolution** is the process by which populations of living things change over a series of generations. It is important to note that evolution does not refer to individual change or development. Although you have changed greatly since childhood, it would not be correct to refer to your changes as evolution. Evolutionary biologists study changes that occur within a population of organisms.

EVIDENCE FOR EVOLUTION

Evidence for evolution comes from many lines of investigation. Some evidence comes from direct observation, while some is more indirect. Included in these investigations are studies of rocks and fossils, comparative anatomy and embryology, geographic distribution, physiology and biochemistry, plant and animal breeding, and genetics. It is important to keep in mind that the evidence for evolution is distinct from the theories that attempt to explain the mechanisms by which evolution might have occurred.

While it is impossible at this time to explore all the available evidence, an examination of selected examples provides insight into the importance of evolution to the diversity and unity of life on this planet.

DIRECT EVIDENCE: FOSSILS

Paleontology, the study of fossils, occupies a central position in the study of evolution. The remains, impressions, and traces of organisms from past geological ages provide scientists with a record of past life. Fossils offer direct evidence of the pathways taken by living organisms in their evolutionary history, or *phylogeny*.

Many fossils represent species that have become extinct. Others represent organisms that have undergone very little change over long periods of time, resulting in what are termed "living fossils." However, fossils alone do not provide absolute proof of evolution. Scientists recognize that a complete set of fossils of all previous existing life forms will never be uncovered. The approximately 250 000 known fossil species represent only a fraction of the estimated 4.5 million species alive today. Many organisms are consumed or decomposed after death, leaving no evidence of their existence. But the fossils that are known can be used to construct convincing arguments for the process of evolution.

How Are Fossils Formed?

The hard parts of organisms, like teeth, shells, and bones, resist the action of weathering for long periods, particularly in dry environments. This explains why dinosaur fossils have remained intact for over 100 million years in the desert sands of Mongolia. In other cases, the soft tissues of extinct organisms have been preserved. For example, the bodies of extinct mammoths have been discovered almost entirely intact. They are believed to have been preserved by a "deep-freeze" process that occurred over 10 000 years ago.

In many cases the impressions or imprints of plants and animals, tracks made in soft mud, and the fecal material of animals, have been preserved. Other organisms such as insects may become entrapped and embedded in amber (a hardened gum given off by trees), leaving fossils that show structural detail. Fossils of the sabre-toothed tiger have been retrieved from old tar pits, along with those of other organisms such as mammoths, horses, birds, and camels.

a)

b)

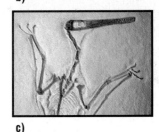
c)

Figure 4.3

A sample of fossils.
(a) Hominid fossil skull.
(b) Fossilized fish.
(c) *Pterodactylus* fossil.

Evolution *refers to the cumulative changes in characteristics of populations of organisms in successive generations.*

When the cell cavities and intracellular spaces of the skeletal material of animals or plants are replaced with mineral matter such as silica, calcite, or iron compounds, they are said to be *petrified*. The petrification process explains the presence of "petrified forests." These forests have literally turned to stone.

Most fossils are found in sedimentary rock, which is formed when soil and rock particles are laid down, usually at the bottom of lakes and oceans. Gradually, newer layers form over older ones, and the layers compress and harden. Over long periods of time, these layers can be uplifted by geological processes and appear on the earth's surface.

Geological Time Scale

Because of the gradual formation of sedimentary rock and the steady accumulation of the remains of organisms, the deepest sedimentary layers are usually the oldest. While some layers may be missing (due to erosion or nondeposition), deformed, or even reversed in some cases, the overall pattern of deposition is usually the same.

During the early work on evolution, investigators had no way to estimate the actual age of the earth. However, geologists and paleontologists did determine a relative time scale for dating rocks and fossils. This scale, which consists of a series of major and minor divisions, remains in use today (see Figure 4.4).

The divisions correspond to major evolutionary events and in this sense may be considered as much a biological as a geological time scale. The names of the largest divisions, called *eras*, are based on the Greek root *zoe*, meaning "life." They include the *Paleozoic* (ancient life), *Mesozoic* (middle life), and *Cenozoic* (modern life) eras. Eras are subdivided into *periods*. The most recent era, the Cenozoic, is further divided into shorter intervals called *epochs*. Together, the geological time scale and the fossil record provide a "calendar of events" that assists scientists in understanding the great diversity and unity of life on this planet.

Figure 4.4

Geological time scale.

Era	Period	Epoch	Age (millions of years)
Cenozoic	Quarternary	Recent	0.01
		Pleistocene	2.5
	Tertiary	Pliocene	7
		Miocene	25
		Oligocene	38
		Eocene	54
		Paleocene	65
Mesozoic	Cretaceous	Late	100
		Early	135
	Jurassic		195
	Triassic		240
Paleozoic	Permian		285
	Carboniferous		375
	Devonian		420
	Silurian		450
	Ordovician		520
	Cambrian		570
Proterozoic	Oxygen (O_2) abundant		2000
			2500
Archean	Oldest fossils known		3500
	Oldest dated rocks		3800
	Approximate origin of the earth		4600

Unit One:
Life in the Biosphere

Dating the Geological Past

It takes approximately 1000 years to form 30 cm of sedimentary rock. Therefore, when the thickness of a rock layer (and its location above and below other rock layers) is determined, geologists can estimate its age.

Recently, scientists have been able to calculate the ages of rocks and fossils more accurately, using radioactive dating. This method involves measurements of the decay of radioactive isotopes such as potassium-40, which becomes argon-40; uranium-238, which changes to lead-206; and carbon-14, which becomes nitrogen-14 (see Figure 4.5).

Radioactive isotope	Half-life (years)	Stable product	Useful range (years)
Rubidium 87	49 billion	Strontium 87	>100 million
Thorium 232	14 billion	Lead 208	>200 million
Uranium 238	4.5 billion	Lead 206	>100 million
Uranium 235	704 million	Lead 207	>100 million
Potassium 40	1.25 billion	Argon 40	>100 000
Carbon 14	5730	Nitrogen 14	0–60 000

Although radioactive dating has proved to be an accurate timekeeper, it is not perfect. Potassium and uranium are used to date materials older than 50 000 to 60 000 years. Radioactive carbon dating, also called the carbon-14 method, is used for measuring rocks and fossils younger than about 50 000 years.

Scientists have learned to link sedimentary rocks and their fossils. This allows them to establish a worldwide inventory of former life forms and their environments. Radioactive dating enables scientists to estimate actual ages of fossils, which are then used to interpret changes over time. Collectively, this information has been useful in determining both the age of the earth and the time that the first life forms appeared. Scientists now believe that the earth was formed about 4.5 billion years ago and that life first appeared about 3.5 billion years ago.

The geological record contained in rocks is not perfect. Gaps in the record occur when a layer does not contain fossils of certain kinds of organisms found both in older and in more recent layers. The organisms in the more recent layers are frequently quite different and more advanced. These gaps, or so-called "missing links," have encouraged much debate about the pattern of evolution. Ongoing investigations and discoveries continue to fill in these gaps. This has led many scientists to believe that the gaps represent incomplete evidence, rather than exceptions to the pattern of evolution.

The human life span is so short in relation to the earth's history that it is difficult to visualize the enormous time represented by the fossil record. The fossil record is not complete in any one loca-

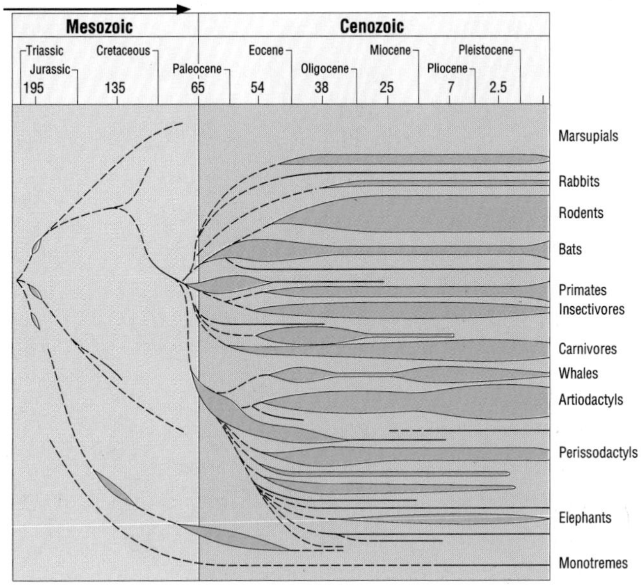

Figure 4.5

Radioactive dating.

Figure 4.6

Phylogenetic relationships and geological range of principal orders of mammals.

tion but is compiled from rock layers in many locations around the globe. Evidence and research are not based on guesswork but on careful observations, comparisons of rock layers and fossils, and quantitative measurement and analysis of age. The oldest parts of the fossil record contain fossils of very simple organisms. Fossils of more recent origin represent more complex organisms. If the time difference between two groups of fossils is great, the differences between the groups are usually great.

By piecing together fossil evidence according to age and similarity of structure, scientists have been able to study patterns of relationships among organisms. These patterns are frequently referred to as "trees of life" or phylogenetic trees. As shown in Figure 4.6, the oldest and simplest life forms constitute the base or trunk of the tree, and the more recent forms make up the branches.

INDIRECT EVIDENCE: LIVING ORGANISMS

Direct evidence provided through fossils is not the only evidence that supports the theory of evolution. Other evidence is readily observable in living organisms, which, like fossils, show the links between existing forms and their ancestors.

Embryology

Embryology, the study of organisms in the early stages of development, offers valuable insight into the process of evolution. During the late 1800s, scientists noted a striking similarity between the embryos of different species (see Figure 4.7). At the time, a German embryologist, K.E. von Baer, wrote that because he had not labelled the two similar embryos he had in his possession, he was unable to identify whether they were the embryos of lizards, birds, or mammals. At a later date, biolo-

Figure 4.7

Embryological development of a salamander, chicken, pig, and human.

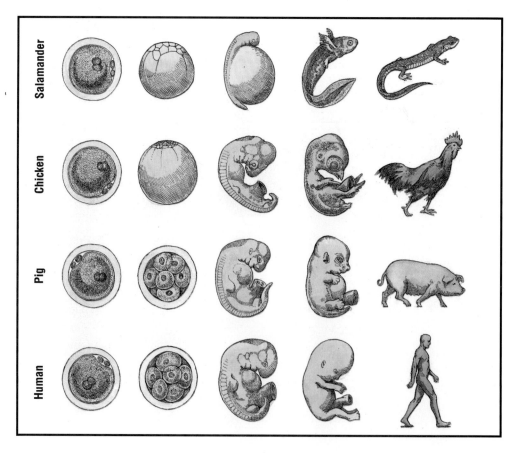

gists suggested that the similarity of the embryos was due to their evolution from a common ancestor. This does not mean that birds necessarily evolved from lizards, or mammals from birds, but rather that the young forms of these organisms resemble the young of related species.

Around the same time, another German biologist, E.H. Haeckel, advanced the theory of recapitulation, more commonly expressed as "ontogeny recapitulates (repeats) phylogeny." In other words, every organism repeats its evolutionary development in its own embryology. The theory is applicable only in a very broad sense. Scientists believe that many of the structures in an embryo are similar to those found in common ancestors.

Homologous and Analogous Structures

When the anatomies of various organisms are studied and compared, the suggestion that organisms with similar structures evolved from a common ancestor becomes increasingly obvious. For example, the flipper of a seal, the leg of a pig, the wing of a bat, and the human arm all have the same basic structure and the same pattern of early growth. These **homologous structures** in some cases serve different functions. However, they are sufficiently similar to suggest that they have the same evolutionary origin. Many other examples of homologies can be found in living organisms. For example, the eustachian tube, which leads from the middle ear to the mouth of humans is homologous to one of the gill slits of fish, and the middle-ear bones of humans are homologous to certain jawbones of fish. On the other hand, **analogous structures,** such as the wing of a bird and that of a butterfly, have similar functions but are quite different anatomically. They are good indicators that these organisms did not evolve from a common ancestor.

Figure 4.8

A seal's flippers and a bat's wings are homologous structures. A bird's and butterfly's wings are examples of analogous structures.

Physiological Evidence

Just as anatomical evidence reveals structural similarities among organisms, physiological studies (studies of the function of any part of an organism) are another basis for establishing relationships. Research has shown that the wastes excreted from the kidneys of birds and reptiles have the same chemical makeup. Perhaps a more striking case is that certain hormones from sheep and pigs can be injected into humans if required. For example, the insulin used to treat diabetes is obtained from cow or pig insulin. (In the chapter Protein Synthesis you can read about a technological breakthrough that uses genetically engineered bacteria to produce human insulin.)

Homologous structures *have similar origin but different uses in different species. The front flipper of a dolphin and the forelimb of a dog are homologous structures.*

Analogous structures *are similar in function and appearance but not in origin. The wing of an insect and the wing of a bird are analogous structures.*

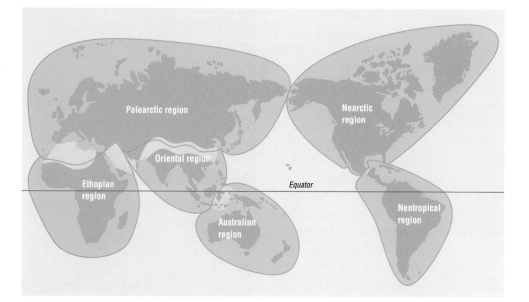

Figure 4.9

Biogeographical regions on earth. Barriers such as deserts, high mountain ranges, and deep water channels that separate the regions are shown in yellow.

Pangaea *was a large supercontinent that existed approximately 225 million years ago.*

Convergent evolution *refers to the development of similar forms from unrelated species due to adaptation to similar environments.*

Biochemical Evidence

The biochemical details shared by all organisms are further evidence to suggest that all life descended from a distant, common ancestor—the first cell. A variety of chemical analysis techniques have been used to show that all organisms share similar DNA molecules and certain proteins, such as cytochrome C. Furthermore, analysis and comparison of proteins and DNA from different organisms indicate that similar organisms also have similar chemical structures. When species of different orders are compared, the difference in their DNA structures is greater than when species within the same genus are compared.

BIOGEOGRAPHY

Biogeography is the study of the geographical distribution of plants and animals. One of the earliest books on biogeography was Alfred Wallace's *The Geographic Distribution of Animals*, written in 1876. This book established six biogeographical regions, which are still used today to describe the broad distribution of living organisms.

Each biogeographical region contains plants and animals that are unique to that region. Fossils found in one region are often found nowhere else. Furthermore, the regions are separated from each other by impassable barriers that influence where these organisms can live (see Figure 4.9). Climate and other environmental factors play a large role in affecting distribution patterns over "short" periods of time (e.g., thousands of years).

Over much longer periods of time (e.g., millions of years), changes in the positions of the continental land masses themselves are important. Changes in the position of continental land masses can be explained by the theory of *continental drift.* In the mid- to late-1960s, scientists such as Canadian geologist and geophysicist Dr. John (Tuzo) Wilson developed a hypothesis to explain the drifting of continents. The hypothesis, called *plate tectonics*, explains how the earth changed from a single supercontinent (approximately 225 million years ago), called **Pangaea,** to the continents known today, by the process of continental drift (see Figure 4.10). Changes are still occurring in the present land masses, even as you read this text.

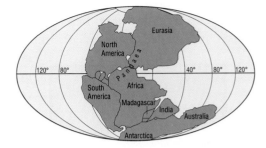

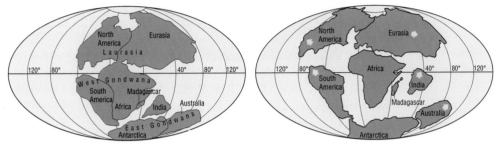

Figure 4.10

The earth changed from a single supercontinent to the continents known today.

a)

b)

Figure 4.11

(a) A kangaroo is an example of a marsupial (pouched mammal). (b) A monkey is a placental mammal.

The distribution of marsupials in Africa, Australia, and South America can be explained by the continental drift that followed the breakup of Pangaea some 150 million years ago. Before the emergence of the Central American landbridge during this period, the animals of South America included a wide array of marsupials. The American Virginia opossum is one of the rare descendants from that period. The marsupials in what is now Australia subsequently became isolated and occupied areas that, on most other continents, are occupied by placental mammals. This development of similar forms in geographically different areas in response to similar environments is an example of **convergent evolution.**

Differences between the distribution of amphibians and reptiles, and that of mammals can be explained by biogeography. Amphibians and reptiles, which arose during the time of the supercontinent, are widely distributed on practically all continents. Mammals, which did not appear until after the breakup of Pangaea, are limited in their distribution. In fact, today, each type of mammal tends to be unique to a specific continent.

Exceptions, such as the presence of moose in both Europe and North America, can usually be explained by landbridges. Landbridges are narrow strips of land that connected certain continents. Most likely, these "bridges" existed until recent times and permitted the migration of animals between certain continents.

Recently, more and more biogeographers have turned their attention to a discipline called *island biogeography* to study the geographical distribution of organisms. Islands have fewer species than their mainland counterparts. Thus they serve as an exciting natural laboratory for biogeographical study.

One obvious question in the study of island biogeography is "How do organisms get to islands?" For animals that can fly or "hitch a ride," the answer is simple. But islands, like their mainland counterparts, house a broad range of living organisms, many of which had to cross water barriers to get to the islands. The usual methods by which animals reached the islands included swimming, crossing ice and landbridges, and "rafting" on vegetation and ice floes.

Those modes of island immigration explain the presence or scarcity of organisms in various parts of Canada. Mice have been observed swimming distances of almost one kilometer between islands in the frigid waters of Labrador. Mammologists at the University of British Columbia have proposed that rafts of timber formed by landslides on the rugged B.C. coast might have transported animals to islands. Ice rafts and ice bridges provide another means of distribution. As recently as the spring of 1991, polar bears have been seen on the northeastern coast of Newfoundland. The polar bears, along with arctic fox, arctic hare, ermine, and wolf are presumed to have arrived there via ice floes. The strong currents in the 19-km-wide Strait of Belle Isle, which separates Newfoundland from southern Labrador (and mainland North America), prevent the formation of ice bridges. What structures constructed by humans might contribute to the present-day distributions of animals?

Further evidence of the importance of natural modes of immigration is the uniform absence of certain species from islands, particularly *oceanic islands* such as Newfoundland. Oceanic islands are islands that have always been isolated from the mainland. Animals such as the porcupine, gray squirrel, and raccoon, and many hibernating animals such as the woodchuck and the eastern chipmunk, shun the cold water of the North Atlantic and are absent from islands of North America.

In contrast to oceanic islands, *landbridge islands* are characterized by a wealth of species from the past. Landbridge

> **BIOLOGY CLIP**
> The width of the Atlantic Ocean is increasing by about 2 cm each year, while the width of the Pacific Ocean is decreasing.

islands were once connected to a continent. Before rising sea levels or crustal movements, these islands contained as many species as their mainland (continental) counterparts. Cape Breton Island and Prince Edward Island were once connected to Nova Scotia before glaciation. Today, mainland Nova Scotia has 38 kinds of terrestial mammals, Cape Breton supports 24, while Prince Edward Island has 31 species. In contrast, Newfoundland, an oceanic island, has only 12 native terrestrial mammals and lacks many of the smaller mammals common to the rest of Canada.

While islands are often regarded as specialized, even unique, environments, biologists find them very valuable in the study of biogeography. They provide important information on the evolution and distribution of organisms throughout the world.

■ REVIEW QUESTIONS ■ ?

1 What does "adaptation" mean in biological terms?

2 Why is evolution such an important scientific concept?

3 What is a fossil? How are fossils formed?

4 How do homologous and analogous structures differ? Give examples.

5 Why is the fossil record considered to be incomplete?

6 Why is biochemical evidence of evolution considered to be indirect?

7 What does the name "Pangaea" refer to?

8 Why are there greater numbers of species of organisms on landbridge islands than on oceanic islands?

EARLY BELIEFS

Evolution, like most great human concepts, is not a recent idea. Moreover, this principle was not always accepted by scientists as it is today. Before the 18th century, humans relied heavily on what they heard or were told in order to form opinions about the natural world. It was widely believed that living things were "fixed" and that they existed much as they did when they first appeared on earth. Although some early Greek philosophers may have held views that foreshadowed the principle of evolution, none could answer the question of how evolution occurred.

During the second half of the 18th century, a number of scholars began to speculate and speak out on the issue of evolution. Pierre-Louis de Maupertuis, a French mathematician, was one the earliest scholars to propose a rudimentary explanation of evolution. In 1751 he wrote that the multiplication of species was a result of accidental recombination of elementary particles (probably genes in modern biological terms), leading to offspring that differed from their ancestral forms. Around the same time, Georges-Louis Leclerc de Buffon, a leading naturalist, proposed two hypotheses to explain the world distribution of plants and animals known at the time. He struggled to answer the question of how organisms could have been created at the same time in the same place when so many organisms—e.g., kangaroos and cactus plants—were restricted to certain parts of the world.

> **BIOLOGY CLIP**
> Up to the late 18th and early 19th centuries it was generally accepted that creation was completed on Sunday, October 23, 4004 B.C. This date had been calculated by the Irish prelate James Ussher (1581–1656). He counted the generations, using the "begats" from the book of Genesis up to the birth of Christ.

One of Buffon's hypotheses suggested that the "creation of species" did not occur in one single place but, rather, that there may have been a number of "centers of creation." Buffon also hypothesized that species were not created in a perfect state, but underwent modifications over time. He later wondered whether certain species may have descended from a common ancestor. Buffon even went as far as to suggest that all animals could be related. Much of the evidence to support his hypotheses came from early fossil discoveries.

Buffon's unorthodox viewpoint was raised again in 1760 by the Swedish botanist Carl Linneaus (founder of biological nomenclature), and again in 1794 by the grandfather of Charles Darwin, Erasmus Darwin, a physician and naturalist. Darwin's interest was sparked largely by changes he observed in animal development—for example, when a tadpole "changes" into a frog—and in the changes seen in plants and animals as a result of cultivation and domestication.

In the early 1800s, a student of Buffon, Jean-Baptiste de Lamarck, presented the first theory that recognized the possibility of evolution. Lamarck believed that organisms have an imaginary "force" or "desire" to change themselves for the better. He felt that organisms had the ability to produce new parts to satisfy these needs and to become better adjusted to their environment. Lamarck further reasoned that the use and disuse of certain structures could be passed on to the offspring. He believed that organs that were used regularly became stronger and more highly developed to perform their function; organs that were not used simply withered away. Today the term "Lamarckism" is used to describe this concept of inheritance of acquired characteristics.

Although Lamarck's hypothesis was rejected by his contemporaries and is today regarded as invalid, nevertheless he did offer an explanation for the mechanism of evolution. The foundation for his theory that species change over time and that environment is a factor in that change helped pave the way for succeeding theories. His main contribution was to show that evolution is adaptive, and that the diversity of life is the result of adaptation. Even today, any theory on the mechanism of evolution must explain adaptation.

Almost half a century passed from the time Lamarck proposed the concept of evolution until the English naturalist Charles R. Darwin published *On the Origin of Species by Means of Natural Selection* in 1859. Darwin's publication had immediate repercussions throughout Western civilization and became an instant bestseller. The book sparked great controversy and changed human thinking forever.

Views that supported the idea of evolution were beginning to surface in other parts of Europe. A British scientist, Alfred Russell Wallace, who had been working in the East Indies, sent Darwin a scientific paper in 1858 describing his work. In his paper, Wallace presented many of the same ideas on evolution that Darwin had worked a lifetime to formulate and support. In the same year, the work of the two men was presented at the Linnaean Society. Wallace insisted that Darwin receive credit for the theory, and finally, in 1859, Darwin's *On the Origin of Species* was published. It would take almost another century before the British statistician, Sir Ronald Fisher, and others would integrate the information from many disciplines into the modern theory of evolution, which combines both genetics and Darwin's theory of natural selection to explain evolution.

Figure 4.12

Charles Darwin. Alfred Wallace.

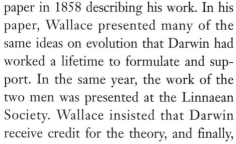

Figure 4.13

On December 27, 1831, HMS *Beagle,* with Charles Darwin aboard, set sail for a five-year trip around the world.

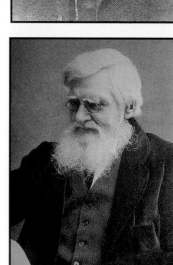

CHARLES DARWIN (1809–1882)

As a young boy, Charles Darwin was extremely curious about the natural world. During his college years, he studied medicine and, later, theology. At the age of 22, he was elated by an invitation to join the crew of the HMS *Beagle* as the ship's chief naturalist. As a naturalist, Darwin was expected to make field observations and collections of plants and animals throughout the voyage. In 1831, the *Beagle* embarked on a five-year mapping and collecting expedition to South America and the South Sea islands.

Darwin took every opportunity to study the diverse forms of life in the countries and islands visited during the voyage. He was especially surprised to find so many unusual species along the west coast of South America. These included the armadillo (Figure 4.14), giant turtles (the Galápagos Islands derive their name from the Spanish word for turtle), and numerous other unique plant and animal species. When he observed living armadillos and the fossils of ancient armadillo-like creatures (glyptodonts) in the same location, Darwin began to wonder why the armadillo had survived while the glyptodont had not. He would later conclude that one form had evolved from the other.

Grasping, probing bill
Eats insects

Chisels through tree bark to find insects.
Uses a tool (a cactus spine or small twig) to probe insects.

Large crushing bill
Eats seeds

Parrotlike bill
Eats fruit

Figure 4.15

The variations among the beak shapes of Darwin's finches as they relate to feeding habits.

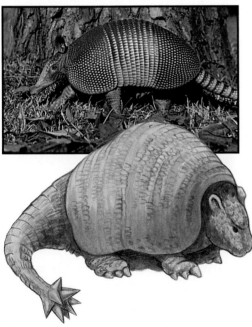

Figure 4.14

The existence of living armadillos and fossils of armadillo-like animals in the same location led Darwin to conclude that one form had evolved from the other.

While exploring the Galápagos Islands, located approximately 1000 km west of Ecuador, Darwin noted slight variations among similar species of organisms from island to island. In particular, he observed 14 species of finch that were similar to each other in many ways and similar to a species of finch found on the mainland. The notable difference in the finches lay in the shape of their beaks. It appeared to Darwin that the different beak shapes were adaptations for eating a certain kind of food characteristic of the various geographic locations (Figure 4.15). He assumed that these different species had evolved from a single ancestral species that probably originated on the mainland.

On his return home, Darwin began sorting out his collections and observations and started to formulate ideas about evolution. He firmly believed that it would be useless to argue for the occurrence of evolution until he was able to explain how the process, in particular adaptation, took place. While he had worked out the process of natural selection in 1838, it took another 21 years before his work was published and finally acknowledged.

Table 4.1 Characteristics of Scientific Theories

- Theories try to explain observed events in a manner that shows cause-and-effect relationships.
- Theories are developed through the imagination of scientists, and are often based on experimental results.
- Theories allow scientists to make predictions about future events.
- Theories are based on several assumptions, many of which are also theories.
- Theories will be changed if they can no longer explain observed events.

THEORY OF NATURAL SELECTION

Darwin's theory of evolution by natural selection can be divided into five distinct ideas:

- Overproduction
- Struggle for existence (competition)
- Variation
- Survival of the fittest (natural selection)
- Origin of new species by inheritance of successful variations

By briefly examining the main ideas, it is possible to understand the basis of Darwin's theory of evolution by natural selection.

Overproduction. In simple terms, overproduction means that the number of offspring produced by a species is greater than the number that can survive, reproduce, and live to maturity. For example, if the millions of eggs laid annually by a female codfish were to survive to adulthood, the oceans would be filled solid with codfish in a few short years. However, this does not happen, because only a few survive and reproduce.

Darwin's concept of overproduction was influenced by Thomas Malthus's *Essay on Population*, written in 1798. The essay pointed out that while populations of organisms increase geometrically, the available food supply increases only by arithmetic progression. The resulting gap between the two explains the idea of overproduction.

Struggle for existence (competition). Because of overproduction, organisms of the same species, as well as those of different species, must compete for limited resources such as food, water, and a place to live.

Variation. Differences among traits occur among members of the same species. Therefore, no two individuals are exactly alike. Darwin believed that these variations (including acquired variations) are passed on to the next generation.

Survival of the fittest (natural selection). Those individuals in a species with traits that give them an advantage (those that are well adapted to their environment) are better able to compete, survive, and reproduce. All others die off without leaving offspring. Since nature selects the organisms that survive, the process is called natural selection.

Origin of new species (speciation). Over numerous generations, new species arise by the accumulation of inherited variations. When a type is produced that is significantly different from the original, it becomes a new species.

Objective

To examine how changes in the peppered moth are an example of a workable adaptation.

Background Information

Although uncertain about the mechanisms involved, Darwin felt that natural selection was the basis of evolutionary change. Today, scientists know that while the effect of natural selection may be subtle and complex, the basic concept is simple. Every population has a broad range of inheritable variations. Some characteristics make an organism better adapted to survive and reproduce, while others impede adaptation. Organisms that are best adapted will be most successful at reproducing, and relatively more of their offspring will enter the population. Over time, an increasing proportion of the population will have the adaptive trait.

The Case Study

In 19th-century England, collecting butterflies and moths was a popular hobby. Records indicate that one species, the peppered moth (*Biston betularia*), was especially common during the first half of the century. It had a whitish appearance with minute dots and speckles, from which it got its common name. The pigmentation gave it a remarkable resemblance to the light-colored lichens on the tree barks on which it rested. Because it was difficult to see, it was protected from predatory birds.

b)

Whitish and melanic forms of white and black peppered moths (a) on light-colored bark, and (b) on polluted background.

The first *melanic*, or black, specimen was recorded near Manchester, England, around 1848. The difference between the black and white forms is due to a single mutation. The factor that produces black forms is dominant to the factor that produces the normal, whitish appearance. Although melanic forms were quite rare at the time, the mutation is known to occur quite frequently. There is little doubt that melanic individuals appeared before 1848, but they would have been at a distinct disadvantage. Their dark color would have made them easy prey for insect-eating birds.

The industrial revolution, which reached its peak in Manchester in the mid-1800s, introduced an unpredictable change to the environment of the peppered moth. Industrial fumes, containing excessive amounts of sulfur dioxide, destroyed almost all the lichens in the region. The tree bark became coated with coal dust, soot, and other dirt, and the adaptive advantage of the peppered moth was quickly diminished. In fact, the melanic individuals now had a selective advantage as they were more difficult for predators to see against the background of soot-stained tree bark. Eventually, this selection pressure caused the population to change from one dominated by light-colored individuals to one dominated by melanic individuals.

From what has been presented so far, answer the following questions:

a) What might have caused the appearance of the first melanic form? Explain.

b) If the environment caused the selection pressure for change, what was the actual selecting agent in this case?

a)

Repeated studies and experiments have been undertaken during this century to determine the extent of the selection pressure on the peppered moth in both polluted and nonpolluted locations. In one experiment, recordings have been made from direct observations of large numbers of both melanic and normal forms of *Biston betularia*. The procedure involved the release of both types of moths, followed by tabulations based on recapture rate. The results (shown in the table below) have enabled scientists to calculate the selection pressure against melanic forms as well as that against normal whitish forms.

Location	No. released		No. recaptured		% recaptured	
	M	N	M	N	M	N
Non-polluted	473	496	(30)	62	6.3	12.5
Polluted	447	137	(130)	(18)	27.5	13.0

Adapted from H.B.D. Kettlewell, *Annual Review of Entomology 6* (1961): pp. 245—62. Note: M = melanic, N = normal; numbers in parentheses have been rounded to the next highest whole number.

c) What might have happened to the moths that were not recaptured?

d) How can you account for the differences in the recapture numbers for polluted and nonpolluted sites?

e) What generalization do the results suggest about environmental selection for the two forms of moth?

f) Explain why the melanic form is more abundant today than in the early part of the 19th century.

Case-Study Application Questions

1 Even a population that is 98% melanic retains the factor for light color in some of its members. What would happen if the environmental conditions were again reversed?

2 Explain the following statements as they apply to this case study:

- Evolution and adaptation need not always involve long periods of time.
- While the change was quick, it was actually quite small.
- Evolution and adaptation usually occur by means of small changes. ■

TYPES OF ADAPTATION

Living organisms are unique in their ability to adapt to changing environments. The accumulation of characteristics that improve a species' ability to survive and reproduce is called adaptation, and occurs over long time periods. The mechanism for adaptation, as noted in Darwin's theory, is natural selection. Characteristics that make an organism better suited, or "fitted," to its environment enhance its opportunity to reproduce. Hence, those characteristics are more likely to be passed on to successive generations than are characteristics that make an organism less suited to its environment and less likely to reproduce.

The natural world is full of examples of adaptations. This section takes a brief look at the three broad categories of adaptation: structural, physiological, and behavioral.

Figure 4.16

The leaves of the pitcher plant (above and right) are shaped like jars, an adaptation that permits the plant to feed on insects.

110 ■

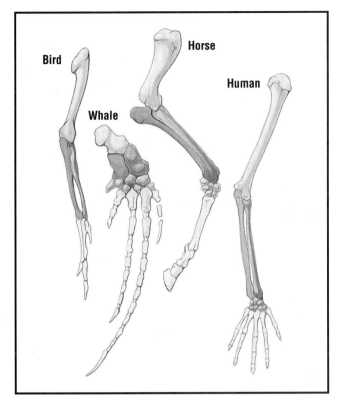

Bird

Horse

Whale

Human

for flying. In contrast to humans and bats, the shorter forelimb and short, flat fingers of seals are adapted to support the flipper, which is used for swimming. The survival value of forelimb adaptations is further demonstrated in most other terrestrial animals, including amphibians, reptiles, and birds.

A number of plants have become adapted to feeding on animals. Perhaps the best known are the Venus flytrap and the pitcher plant. Their leaves have become adapted in different ways to capture prey, usually insects. The leaves of the pitcher plant, for example, are shaped like jars—once the insects enter the "jar," escape is practically impossible. Although these plants photosynthesize, they also require the nutrients that are obtained from the digested remains of their prey. This adaptation likely arose because these plants usually live in bogs and wetlands where the soil is poor in nitrogen-containing substances.

Figure 4.17

The forelimbs of these mammals are adapted to carry out different functions.

*A **pheromone** is a chemical substance produced by an organism that serves as a stimulus to another organism of the same species.*

Structural Adaptations

Most organisms show clear structural adaptations to their environment. Whether these adaptations involve modification to the pentadactyl limb (five digits) in mammals such as humans, seals, and bats, or modification of pitcher plant leaves to capture insects, they represent the most obvious of the three categories of adaptation. But how do these adaptations help the organisms survive and function?

Mammals, including humans, have the same basic limb design. However, in each case the forelimbs have become adapted to do a specific job. In humans, for example, the forelimb (with its five digits) has become particularly well-adapted for grasping and holding things. This feature is important in the human environment for accomplishing everything from eating and caring for the young to providing shelter and developing new technology. The fingers of bats are greatly lengthened to support the wing, which is used

Physiological Adaptations

The example of the pitcher plant illustrates the fact that one type of adaptation frequently depends on other types. While the structural adaptation of the leaf parts is quite apparent, the physiological adaptation that enables the plant to produce chemicals for the digestion of its prey is less obvious, though by no means less crucial for the organism's ability to survive and reproduce.

The production of chemicals such as **pheromones**—chemicals secreted by organisms to influence the behavior of other organisms—also has a physiological basis. In insects, these chemicals can have a number of functions, including serving as sexual attractants and alarm signals. As sexual attractants, the chemicals have reproductive value, whereas the latter adaptation clearly suggests the importance of pheromones as a survival mechanism.

A number of physiological adaptations involve specialized **enzymes** that control body functions such as temperature, respiration, digestion, circulation and blood clotting, and muscle and nerve coordination. The secretion of venom by snakes and the production of toxins by certain plants and animals are further examples of physiological adaptations.

Enzymes are special protein structures that regulate chemical reactions.

Behavioral Adaptations

Behavioral adaptation is a key factor in keeping organisms alive and enabling them to reproduce. Some behavioral adaptations are better known than others. For example, the southward migration of the Canada goose, the hibernation of certain mammals, and the storage of nuts by squirrels are all behaviors involving reactions to the environment. These actions help the organisms adapt to their surroundings.

Have you ever seen the "broken wing" behavior of a bird, which is designed to draw the enemy away from the nest site? Although this behavioral response may signal death for the parent, it increases the likelihood of survival for the offspring.

Another familiar behavioral adaptation can be noticed in cats and dogs. Originally cats were adapted to live in forests, where living or hunting in groups had no advantage. In contrast, dogs in earlier times lived in a grassland environment, where they adapted to hunt cooperatively. This evolutionary history helps explain why cats today tend to be solitary animals while dogs enjoy human companionship.

Most organisms, both plants and animals, respond to stimuli, and may show interesting adaptations in the way they behave. For example, protozoans such as amoebas and hydras respond quickly to touch, and react by moving away. They also react to temperature changes and chemical substances in the water. Depending on whether the experience is stressful or not, the organisms may move away from or toward the stimulus. Similarly, earthworms will react to touch, light, and chemicals in the soil. It is their reaction to light that explains why they burrow into the soil and only come out at night.

Plants also exhibit behavioral adaptations, but instead of being controlled by nerves, as is the case with animals, plants are controlled by chemicals called hormones. Many people are familiar with the behavior of plants that "bend" toward light. The term *tropism* is used to describe the orientation of plants according to some stimulus. In the case of light the orientation is called phototropism.

■ REVIEW QUESTIONS   ?

9 How does the modern view of evolution differ from earlier beliefs?

10 What adaptation in the Galápagos finches made the greatest impression on Darwin?

11 What is the significance of the fourth step in Darwin's theory of natural selection?

12 What contribution did Lamarck make to our understanding of the mechanism of evolution?

13 How did Buffon's theory influence Darwin's thinking about evolution?

14 What are the three general types of adaptation?

15 What adaptation did the peppered moth make to the pollution around Manchester in the mid 1800s?

FRONTIERS OF TECHNOLOGY: USING COMPUTERS TO ASSESS ENVIRONMENTAL IMPACT

Much scientific knowledge comes from laboratory experiments in which variables that might affect the results are controlled. This approach has yielded valuable information about many of the components of complex systems. But imagine trying to test in a laboratory all the factors that affect the "health" of a salt marsh, or attempting to enclose a major ecological region to investigate the environmental impact of pollution, acid rain, or ozone depletion.

Supercomputers, the fastest of all computers, provide an approach to scientific investigation that is radically different from traditional observation and experimentation. The combination of computer-enhanced photographic images obtained from spacecraft, satellites, and aircraft, plus improvements in the modelling and simulation capabilities of computers, is playing a vital role in allowing researchers to assess environmental impact on local and global scales.

Supercomputers permit researchers to gather data and test theories on detailed simulations, or models, of physical reality. This modelling provides a framework for comprehending and predicting how nature acts. The supercomputer creates a "virtual reality," in which users participate directly in the world created by the computer. Numerical experiments can be performed and then used with laboratory data to predict what is likely to happen in "real life" situations that cannot be tested directly.

One example of the use of computer simulations to study environmental impact is a project entitled "Designer Wetlands" developed by Ken Pittman at the Visual Environment Laboratory at the North Carolina State University. The program, which puts together general scientific information such as topography, weather, geology, and wildlife, is used to simulate a wetland ecosystem. The computer can actually take the observer on a visual tour of the "designed" wetlands.

The implications of programs of this type are far-reaching. Researchers can quickly see if a given design is feasible and determine its application to wetlands that may be in trouble in the real world.

Figure 4.18

The CRAY X-MP/48 supercomputer consists of four central processors and has a capacity of eight million bits of shared memory.

While supercomputers have allowed us to create and manipulate complex mathematical models, it is important to remember that they are just that—models. They are no better than the set of mathematical relationships on which they are based, and they can never (at least not in the foreseeable future) recreate all of the factors that exist in a real system. They are one more tool in the scientist's repertoire, to be used with caution and care. The best use of such models might well be to direct us toward the best questions to ask, and to help us design experiments that can test the predictions of our models.

CASE STUDY
RESISTANCE TO DDT

Objective

To investigate the connection between pesticides and natural selection.

Background Information

Environmental change introduced by human activities provides good evidence for natural selection. Examples of such activities include the extensive use of drugs and antibiotics to treat certain kinds of pathogens, and the use of pesticides to control diseases such as malaria and yellow fever. Bacteria and insects, like all living organisms, demonstrate variability. Variability results from mutations and chromosomal rearrangements, which are present in a population at very low levels at all times. This variability enables natural selection to take place.

The application of pesticides to target populations favors strains that are resistant to the pesticides. Normally, such strains are uncommon in a population; however, once the environment is changed, the resistant forms have the better chance of survival. The resistant forms will then be reproduced in larger numbers in succeeding generations. Thus, they are spread rapidly throughout the population. As resistance develops, the frequency of application and/or the dosage is increased to maintain levels of control. Increased application and dosages of chemicals further increase the size of the resistant populations. Thus, chemicals that once controlled a pest population are no longer effective.

The Case Study

DDT is a chemical pesticide that has had enormous success in the battle against insect pests. Since it was first used in World War II, to get rid of head lice, it has been effective in the fight against controlling mosquito and other insect populations known to cause diseases such as malaria and yellow fever. Because of its potentially harmful accumulation in the food chain and the possible danger to humans, regulations banning or restricting the use of DDT were implemented by a number of countries in the early 1970s. However, DDT continued to be used in many less developed countries.

Some scientists believe that the resistance displayed by the pests did as much to spell the demise of DDT as did regulations directed at curtailing its use. After 1950, when pesticides came into widespread use, the number of pest species, including insects, that have developed resistant strains has increased dramatically.

Examine the following illustration to determine the effect of repeated sprayings or applications of DDT on a hypothetical population.

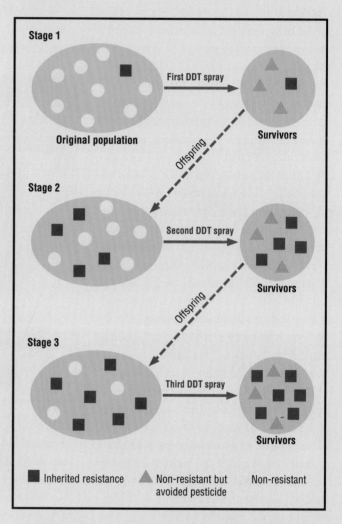

a) Where did the resistant trait come from in the original population?

b) What happens to the nonresistant trait over the three stages of the population? Why?

c) What happens to the resistant trait over the course of evolution in this hypothetical situation? Why?

d) Explain the difference in the selection pressure (the trait natural selection is favoring) in stages 1 and 3.

e) Is this a case of "evolution in action"? Explain.

A complicating factor in the use of pesticides such as DDT is that some pests develop multiple resistance and cross-resistance. In cross-resistance, the organism develops a resistance to one compound and then achieves resistance to others, usually in the same chemical group. For example, DDT-resistant houseflies tolerate higher levels of a closely related chemical that is also used to control this common household pest. Multiple resistance is much more serious and has wide-ranging consequences. In this type of resistance, pests develop a tolerance for many classes of different compounds. This effect is common in pests that have a strong resistance to DDT. Insects with this factor, called *kdr* (knock-down resistance), are preadapted to synthetic chemicals that are chemically quite different from DDT.

An understanding of the effects of multiple resistance can be illustrated by the World Health Organization's (WHO) 1955 global education program to control the anopheles mosquito (carriers of the malarial parasite). The WHO strategy included a two-stage offensive: (1) spraying DDT and other pesticides inside dwellings where anopheles mosquitoes frequently obtain blood meals from humans, and (2) treating infected individuals with anti-plasmodium drugs to destroy the blood parasites that cause malaria.

For about 15 years, the program showed remarkable success. For example, within a 10-year period in India the annual incidence of malaria was reduced from 100 million cases to 50 000, with a corresponding reduction from three million to 25 in Sri Lanka. By 1970, efforts to control the mosquitoes that carried plasmodia were failing. By 1980, 51 of the known 60 species of anopheles mosquitoes were capable of transmitting the malarial parasite. During the same time as the WHO mosquito spray programs, many countries had implemented extensive agricultural programs. Crops such as cotton, rice, and tobacco were sprayed with pesticides, including many of the same ones used in the malaria-eradication program.

f) What explanation can be given for the early success of the malarial spray program and its eventual failure in later years?

g) What combination of factors most likely contributed to a strong selective force for developing resistant strains in the mosquito population?

h) Because the anopheles mosquito has developed resistance to DDT, what might be the long-term effectiveness of other known synthetic chemicals in controlling malaria? Explain.

Case-Study Application Questions

1 The resistance of mosquitoes to DDT is a good example of natural selection. Explain how this case study supports Darwin's observation that
   a) hereditary variations exist among species
   b) natural selection acts on organisms with these variations.

2 Discuss the following statement: "Natural selection as the cause of evolution has been neither proved nor disproved." ∎

RESEARCH IN CANADA

Dr. Ford Doolittle's work at Dalhousie University in Halifax has focused on applying the recently developed methods of molecular genetics and "genetic engineering" to questions about evolution that have troubled biologists for decades.

In the early 1970s, Lynn Margulis had proposed that chloroplasts and mitochondria—the organelles of photosynthesis and respiration in higher cells—were once free-living bacteria that had become trapped inside the cytoplasm of some larger host cell. Doolittle's laboratory was one of the first to use a molecular approach to provide evidence that at least one type of organelle, chloroplast, was indeed the descendant of ancient bacteria.

Molecular Genetics and Evolution

Recently, Doolittle's laboratory has focused on archaebacteria. Carl Woese, at the University of Illinois, had shown that these bacteria are as different from the more familiar "eubacteria" like *Escherichia coli* as eubacteria are from humans or plants, and likely branched off from the trunk of the tree of life over three billion years ago. Archaebacteria live in extreme environments such as the bottoms of swamps, salt evaporation ponds, hot springs, and deep-ocean geothermal vents called "black smokers"; some even grow best at temperatures above that of boiling water. Doolittle's laboratory has worked toward developing tools for the genetic engineering of archaebacteria, so that we can both exploit and understand their remarkable and ancient adaptations.

DR. FORD DOOLITTLE

CAREER INVESTIGATION CAREER CAREER

With the population of spaceship earth exceeding 6 billion people, our survival depends more and more on an intelligent, well-planned management of the world's resources. A growing number of careers require an awareness of environmental issues. The demand for people with knowledge and skills in environmental science is projected to grow well into the next century.

Politician/civil servant

Voters expect their politicians and civil servants to take strong actions to protect the environment. Government officials have to make daily decisions on a wide variety of environmental issues, from water quality and waste disposal sites to the preservation of wilderness areas.

- Identify a career that requires a knowledge of environmental issues.
- Investigate and list the features that appeal to you about this career. Make another list of features that you find less attractive about this career.
- Which high-school subjects are required for this career? Is a postsecondary degree required?
- Survey the newspapers in your area for job opportunities in this career.

Architect

An architect must take environmental factors into account when designing a structure. For example, a high-rise building in an earthquake zone must be designed to move with possible earth tremors. In planning any project, architects, developers, and city planners must consider such factors as wind tunnels and access to sunlight for neighboring buildings. A building design must also consider the people who will be working in it. Workers do not want to work in stuffy, airtight boxes.

Retailer

In the 1990s, consumers are demanding "environmentally friendly" products such as phosphate-free detergents or cosmetics that are not tested on animals. Retailers must address consumers' concerns about wasteful packaging by providing products in recyclable containers.

Forestry worker

Canada's vast forests are a valuable, renewable resource that must be handled with care. Forestry workers must understand the impact that the harvesting of trees has on forests and wildlife. The forestry industry also must plan effective reforestation policies so that our forests can be used and enjoyed by future generations.

SOCIAL ISSUE:
Pesticides and Evolution

The development of chemical pesticides has led to the evolution of more resistant forms of organisms. Although the controlled and prudent use of pesticides can improve agricultural yields, there remain several concerns about their long-term effects in the ecological chain, including water contamination, which also kills fish; the poisoning of birds and other non-target organisms; as well as possible harm to humans.

Statement:

There should be a worldwide ban on the use of pesticides in agriculture.

Point

- Since the introduction of DDT in 1939, hundreds of insect species have developed resistant strains. As a result, crop damage from insects is now greater than it was before the introduction of chemicals. Some people are concerned that we will run out of new and effective pesticides before the pests run out of resistant strains.

- Nonchemical practices such as crop rotation, burning or plowing of plant refuse, and proper timing for planting were used before chemical pesticides were introduced. These methods worked adequately and kept a healthy balance of nutrients in the soil.

Counterpoint

- Resistance does not always lead to the evolution of new species. Consider the cotton boll weevil and the coddling moth. In both cases, after more than 30 years of exposure to pesticides, there is no evidence that the effectiveness of the pesticide has been reduced or that the organisms have evolved any resistance.

- Alternative methods of pest control are expensive and out of reach for those who will be affected most: underdeveloped and developing nations. An end to pesticide use would mean a decreased quality of life for all.

Research the issue.
Reflect on your findings.
Discuss the various viewpoints with others.
Prepare for the class debate.

CHAPTER HIGHLIGHTS

- Evolution is a major theme in biology. It attempts to explain how living organisms change through time. Earlier beliefs held that organisms were "fixed" and did not change.
- Adaptation refers to the ways in which whole organisms or their individual parts are suited to carrying out the processes of life. Adaptations improve an organism's chances of survival and reproduction.

- Indirect and direct evidence can be used to support the theory of evolution. Some evidence (e.g., that from fossils) is direct, while other evidence (e.g., that from living organisms) is indirect.
- Darwin's theory of evolution by natural selection helped convince the scientific and intellectual communities that evolution was possible.

- Biogeography is the study of the distribution and dispersal of plants and animals.

- Scientific laws are descriptions based on observed events. Theories are attempted explanations based on observations and tested by hypotheses.

APPLYING THE CONCEPTS

1 In what way is the geological time scale both a biological calendar and a framework that describes major geological events?
2 Contrast the views on evolution held by modern scientists with those of earlier proponents.
3 How is the theory of continental drift useful in the study of evolution?
4 How did Darwin use the ideas of Lamarck and Malthus in formulating his theory of evolution?
5 What does the peppered moth case illustrate with respect to the rate of evolutionary change?

6 Using examples, demonstrate from an evolutionary viewpoint that the survival of species is more critical than the survival of specific individuals.
7 How could supercomputers and computer simulation play a useful role in predicting how plants and animals might adapt to a changing environment? Give examples.
8 Give an example of how environmental impact caused by human activity can alter the natural course of evolution.
9 Explain why organisms with a high reproductive capacity can adapt more readily than many organisms that have a lower reproductive capacity.

CRITICAL-THINKING QUESTIONS

1 The wombat, a marsupial mammal found in Australia, and the rabbit, a placental mammal that was brought into Australia by explorers, have many of the same requirements. A biologist studying both animals suggests that rabbits will eventually take over the ranges occupied by wombats because they reproduce at a faster rate. What evidence would you want to gather before accepting this conclusion?
2 Many of Darwin's ideas about evolution came from his observations on the Galápagos Islands. Today, scientists recognize that islands are literally "laboratories of evolution." Discuss the validity of this viewpoint.

3 Scientists suggest that theories provide tentative explanations of the natural world and should never be interpreted as absolute fact. Provide an example that indicates how biological explanations of the natural world would change if a Lamarckian explanation of evolution were endorsed by the scientific community rather than the Darwinian explanation of evolution by natural selection.
4 Adaptations are so numerous that they embrace practically every structure or occurrence in biology. Even when organs have fallen into disuse they often become adapted to a second purpose. Investigate the present function of the pelvic girdle of snakes and the three hinge bones of reptile jaws.

ENRICHMENT ACTIVITIES

1 Read "In the Beginning" by Wallace Raven in *Discover* (October 1990) and write a book report on the research of paleobiologist J. William Schopf.
2 Research the social differences between two designers of the theory of evolution: Darwin and Wallace.

3 How do the creation stories of the North American aboriginals refute the theory that aboriginals migrated to the North American continent?

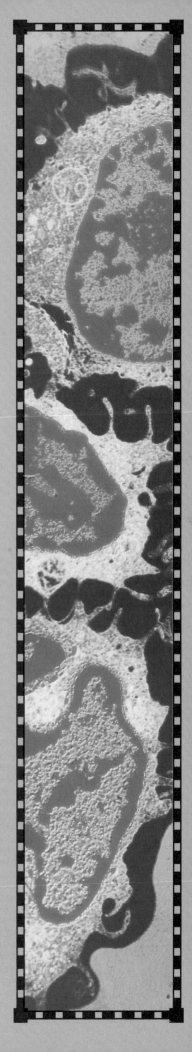

Exchange of Energy and Matter in Cells

■

U N I T 2

Development of the Cell Theory

Figure 5.1

A head louse clings to a human hair. The head louse inhabits the hair of the head, gluing its eggs to the individual hairs.

IMPORTANCE OF THE CELL THEORY

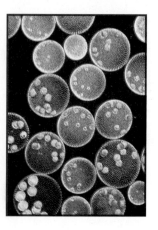

The cell is the basis of life. It has been estimated that the human body is made up of about one hundred trillion cells. Different cells in the body are specialized to perform various tasks. Muscle cells, for example, are capable of rapid contraction. If they did not function properly, movement would not be possible. Nerve cells transmit electrochemical messages. If they did not work properly, a person would not be aware of his or her environment, or be able to respond to changes in it. Vision, hearing, taste, smell, and touch all depend on nerves. The transport of oxygen, defense against disease, and communication all depend on specialized body cells. Cells are not exclusive to humans; they are also found in plants, bread mold, and pond scum. In fact, all life forms are composed of at least one cell.

Most of the cells that you encounter every day are invisible to the naked eye. The idea that you are not alone becomes evident if you examine the cells that co-exist with you. A small, rod-shaped bacterium, called *Escherichia coli*, lives in your gut. This microbe, or its descendants, supplies you with important vitamins and will likely be with you until the end of your life. A survey of your skin may reveal different types of fungi. Athlete's foot is an example of a fungus. Your skin also harbors a sphere-shaped bacterium called *Staphylococcus epidermis*. You might also see tiny mites in your eyebrows, plant spores and pollen that have found their way into your lungs, and a host of microbes and microscopic animal eggs that enter your body with the food you consume. Your body is a walking ecosystem.

Cells outside your body also play an important role in your life. The yogurt that you ate for lunch is a living community of bacteria. A bacterium called *Lactobacillus bulgaricus* causes the milk in the yogurt to sour, while a second bacterium, *Streptococcus lactis*, enhances the taste. The processing of cheese, beer, and wine depends on cells. Even the tanning of leather could not be accomplished if it were not for cells. Is it any wonder that an introduction to biology can begin with a study of the cell?

In spite of their varied size, shape, and appearance, cells have several things in common. All cells digest nutrients, excrete wastes, synthesize needed chemicals, and reproduce. Cells define both life and death.

ABIOGENESIS

Biology, like other forms of science, progresses by observation. Unfortunately, observations of nature are often flawed by interpretations that attempt to explain what was observed. This is due, in part, to the fact that the observer is actually part of nature. Therefore, the observer is subject to many of the same factors as the objects being observed. In an attempt to interpret natural events, scientists often propose explanations. An explanation is also called a **hypothesis.** The hypotheses proposed by early scientists were almost never tested by experiment. Often one unsubstantiated hypothesis became the basis for another, and scientists moved further and further from the truth.

Early scientists noticed that ponds dried up during a long period of drought and that no living fish were found in the mud. When rain began to fall in the spring and the pond filled with water, observers noticed that the pond was teeming with frogs and fish. Some concluded that the frogs and fish must have fallen to earth during the rainstorm. Incredible as this explanation may seem now, it seemed logical to many people in earlier times. The fact that nobody had ever been hit by a frog or a fish during a rainstorm did not seem to have occurred to anyone!

Aristotle, the great philosopher of ancient Greece, did not accept the hypothesis of fish and frogs falling from the sky. He proposed that fish and frogs came from the mud. He also believed that flies came from rotting meat, because he had always observed flies on decayed meat. Aristotle was so persuasive in his arguments that scientists accepted his theory of **abiogenesis** for nearly 2000 years. Abiogenesis is the theory that proposes that nonliving things can be transformed into living things spontaneously. The theory is sometimes referred to as "spontaneous generation." A mere 300 years ago, a Belgian doctor, Jean van Helmont, concluded that mice could be created from grains of wheat and a dirty shirt. Van Helmont had placed grains of wheat and a dirty shirt in a container, and within 21 days mice appeared. According

*A **hypothesis** is a possible solution to a problem or an explanation of an observed phenomenon.*

Abiogenesis *is a theory that states that nonliving things can be transformed into living things.*

wheat and a dirty shirt in a container, and within 21 days mice appeared. According to van Helmont, the sweat in the shirt caused the wheat to ferment. The fermenting wheat bubbled and was eventually transformed into mice.

In 1668, Francesco Redi, an Italian physician, conducted an experiment to test the hypothesis that rotting meat can be transformed into flies. Prior to Redi's work, science was based on logical analysis rather than on experimentation. Redi placed bits of snake, eel, fish, and veal in four different jars. He repeated the same steps in four other jars, but sealed the second set of jars. The open set of jars was designated the **experimental** group, while the closed set was designated the **control** group. What do you think happened next? After a period of time, Redi noticed that maggots were crawling all over the meat in the open jars. Apparently, flies had been attracted to the meat and began laying eggs on the food supply. The eggs hatched into maggots, which began feeding on the meat. The maggots then became flies and the cycle continued. Redi concluded that flies come from other flies, not from rotting meat. However, Redi's critics replied that the sealed jars were different from the control set, because no fresh air circulated around the meat. Air, claimed the critics, is the "active ingredient" that causes spontaneous generation. Fresh air must circulate around the meat in order for flies to appear.

Once again, Redi turned to experimentation for his answer. He repeated the experiment, but this time placed fine, meshed wire over the opening of the experimental set of jars. As Redi had predicted, flies were not found inside the experimental jars, despite the fact that air circulated around the meat. Once again, Redi proved that rotting meat cannot be transformed into flies.

Experimental variables are designed to test a hypothesis. Experimental groups test a single variable at a time.

Controls are standards used to verify a scientific experiment. Controls are often conducted as parallel experiments.

A WINDOW ON THE INVISIBLE WORLD

Inventions in one field of science often contribute to advancements in others. Three thousand years ago, when the artisans of Egypt and Mesopotamia first combined silicon dioxide, boric oxide, and aluminum oxide, they produced a supercooled liquid that would forever change the world. The miracle substance, glass, became the material of lenses. In the early 1200s a famous scientist named Roger Bacon described how crystal lenses might help improve the vision of the elderly. By the late 1200s, an Italian scientist, Salvino degli Amati, provided the world with spectacles. Lenses were now fashioned by craftsmen, and a new branch of physics, called optics, began to explain the movement of light.

The two greatest inventions of the 1600s, the telescope and the microscope, changed the way in which people understood and explained the universe. The telescope allowed scientists to see a far greater universe than was previously known, while the microscope opened the doors to a hidden world. The boundaries of the universe were no longer limited by what people could see with the naked eye.

No one scientist is responsible for the development of the microscope. Like many other inventions, the development of the microscope was an ongoing process that involved technological advances in glass making and lens polishing, along with refinements to existing models. One of the first great contributors to the development of the microscope was the Dutch naturalist, Anton van Leeuwenhoek (1632–1723). A draper by profession, van Leeuwenhoek became an expert lens maker, and proved to be a remarkably observant scientist. Although simple by today's standards, his microscope was the first to reveal microorganisms and human blood cells. Van Leeuwenhoek's techniques, however, remain shrouded in mystery. Modern scientists are still trying to explain how he was able to record and describe the invisible world so accurately. A private man, van Leeuwenhoek did not sell or display his most effective microscopes. It was only after his death, when his daughter made over 200 of his finest microscopes available to the scientific community, that his true genius was recognized.

Early microscopes were named after two great astronomers, Galileo and Kepler, as an acknowledgment of their pioneering work with telescope lenses. While the telescope was designed to magnify distant objects—to bring large objects near—the microscope took advantage of a similar construction to magnify near, but tiny objects. In fact, the microscope is so similar to the telescope that it is often described as bringing small objects near. Although this may seem an odd way to describe magnification under a microscope, consider microscopic objects as being so distant from the human eye that they are invisible. The microscope brings tiny objects closer to the eye, making them appear larger. To focus on a human blood cell, you would have to place it within one millimeter of your eye. However, the human eye cannot focus on objects that close. The microscope brings the image close to the eye through a series of lenses.

Magnification and Resolution

The most important feature of a microscope is not its magnification, but its ability to distinguish fine detail. This is referred to as *resolving power*. The resolving power of a human eye is about 0.1 mm. This means that the eye can distinguish one dot from another as long as the dots are separated by a distance of at least 0.1 mm.

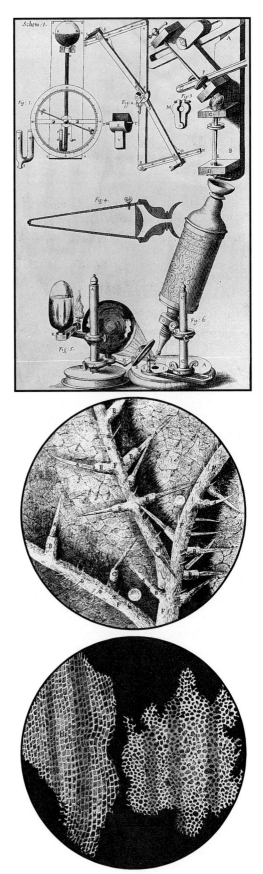

Figure 5.2

The impact of the microscope was so great that most scholars of the 17th century proudly displayed their microscopes in full view. In 1637, the famous mathematician and philosopher, René Descartes, published drawings of specimens he had observed under a microscope. Like many new inventions, the microscope at first was greeted with suspicion. Descartes was alarmed when his fellow townspeople referred to a prepared slide of a flea as "a devil shut up in a glass."

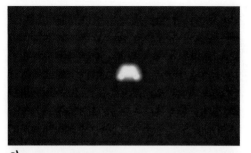

a)

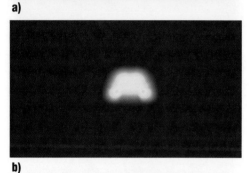

b)

c)

Figure 5.3

Demonstration of magnification with resolution. (a) Car headlights as they appear at a distance. (b) Magnification without resolution. The eye and camera fail to detect a space between the headlights. (c) Magnification with resolution. Each headlight appears as a distinct object.

Figure 5.4

The compound light microscope uses a system of two lenses.

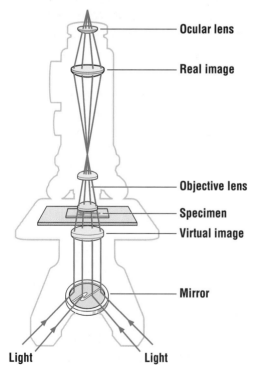

Ocular lens

Real image

Objective lens

Specimen

Virtual image

Mirror

Light   Light

The microscope employs two lenses that bring the specimen closer to the eye. The image is magnified once by the *ocular lens*, which is near the observer's eye, and again by an *objective lens*, which rests above the specimen. Should the objective lens provide a magnification of 4× (four times) and the ocular lens an additional 10× magnification, the image would be enlarged 40×. Likewise, an objective of 40×, along with a 10× ocular magnification, would increase the size of the image 400×.

Microscope Safety

1 Always carry the microscope with two hands. One hand is placed on the arm of the microscope, while the other supports the base. Always carry the microscope in the upright position so that it will not slip.

2 Use only lens paper to clean the lenses. Coarse paper will scratch the lens. Never allow your fingers to touch the lenses.

3 Never allow the lens to touch the cover slip of a slide or wet mount stains. Corrosive chemicals may destroy the lens.

4 Never attempt to repair your microscope. Always notify your teacher when problems arise. Microscopes require special repairs to ensure proper functioning.

5 Use stage clips or a mechanical stage to secure the slide.

6 Before returning your microscope to the storage area, remove any slides from the stage and rotate the nosepiece to the lowest-power objective. The low-power lens is farthest from the stage and, therefore, is less likely to be damaged. Slides left on the stage may damage lenses.

7 Remove the electrical cord from the socket by the plug. Do not pull on the cord.

Unit Two:
Exchange of Energy and Matter in Cells

LABORATORY

AN INTRODUCTION TO THE MICROSCOPE

Objective

To investigate the features and basic operations of the compound microscope.

Materials

compound microscope	cover slips
newspaper picture	medicine dropper
scissors	colored thread
lens paper	ruler
glass slides	small beaker (50 or 100 mL)

Procedure

Part 1: Parts of the Microscope

1 Locate the parts of the microscope shown in the diagram below.

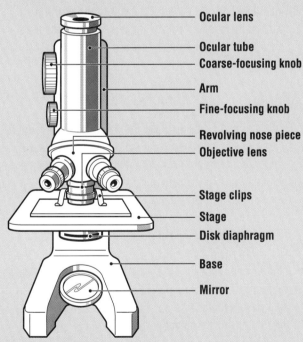

- Ocular lens
- Ocular tube
- Coarse-focusing knob
- Arm
- Fine-focusing knob
- Revolving nose piece
- Objective lens
- Stage clips
- Stage
- Disk diaphragm
- Base
- Mirror

Part 2: Resolving Power

2 Cut a 1 cm square from a newspaper picture.
   a) Describe its appearance with the naked eye.
3 Obtain a glass slide, a cover slip, and a medicine dropper.

4 Make a temporary wet mount of the newspaper picture, as shown in the diagram. Add the cover slip.

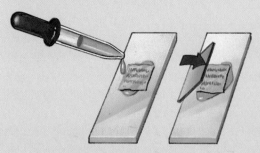

5 Clean the ocular and objective lenses with the lens paper. Adjust the nosepiece to the lowest-power objective lens. Make sure that the lens clicks into place.
6 Place the slide on the stage with the picture centered over the hole. Use the clips to hold the slide in position.
7 Rotate the coarse-adjustment knob forward to lower the low-power objective lens. If your microscope has an automatic stop, the lens will stop at a specific distance above the slide. If the microscope does not have a stop, do not let the objective lens get closer than 0.5 cm from the slide.
8 Using one eye, look through the ocular lens. (If you wear glasses, focusing with or without them on should not matter.) It is easier on your eyes if you keep both eyes open. You should be able to see a circular field of view.
9 Slowly rotate the coarse-adjustment focus backwards until the image is clear.
   b) Describe the appearance of the picture under low-power magnification.

Part 3: Determining Focal Distance

10 Cut two different-colored threads and place them on a glass slide.
11 Make a temporary dry mount by placing one of the threads over the other in the form of an X. Cover the threads with a cover slip.

12 Place the slide on the stage so that the crossed threads are in the center over the hole. Use the clips to hold the slide in position.

13 Turn on the substage light. If your microscope has a mirror, make sure that it is adjusted to reflect the light through the objective lens.

14 Rotate the coarse-adjustment knob forward so that the low-power objective approaches the slide. Do not allow the objective to touch the slide.

15 View the crossed threads by slowly rotating the coarse-focus knob. At this point you may wish to adjust the diaphragm for better lighting.

    c) Are both threads in focus?

16 Use the fine adjustment to focus on the top thread.

    d) Measure and record the distance (in mm) between the bottom of the objective lens and the top of the cover slip.

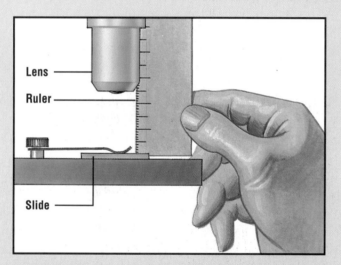

Lens —
Ruler —
Slide —

17 Rotate the revolving nosepiece to the medium-power objective. The objectives in most microscopes are *parfocal*, which means that once the low-power objective is in focus, the higher-power lenses are in focus. Usually some minor adjustment is required for sharp focusing. If your microscope is not parfocal, check with your teacher for special instructions.

18 Use the fine adjustment to focus on the upper thread. As a matter of procedure, always bring the image into focus with the low-power objective first and then proceed to use higher-power objectives. The coarse adjustment should only be used for low-power focusing.

    e) Measure and record the distance between the cover slip and the objective lens.

19 Rotate the nosepiece to the high-power objective and fine focus.

    f) Measure and record the distance between the cover slip and the objective lens.

    g) As you move from the low- to higher-power objectives, describe the change in light intensity.

    h) Which objective is the best for showing the detail of the threads?

    i) Under which objective is the bottom thread clearest when the top thread is in focus? (You may wish to re-examine the threads with each of the magnifications.)

Part 4: The Field of View
The circle of light seen through the microscope is called the *field of view*. It represents the observed area.

20 Switch the nosepiece to the low-power objective and examine the length of thread seen.

21 Repeat the procedure for the medium- and high-power objectives.

    j) Compare the length of thread seen under each objective.

22 Clean the slide and cover slip and return them to their appropriate location. Rotate the nosepiece to the low-power objective and return the microscope to the storage area.

Laboratory Application Questions

1 Explain why microscopes are stored with the low-power objective lens in position.

2 *Astigmatism* is a common disorder in which the lens of the eye has an asymmetrical shape. Most people have symmetrical lenses—the top half is identical in shape to the bottom half. Explain why individuals who have astigmatism may experience difficulties distinguishing fine detail with the naked eye.

3 Explain why resolving power decreases as the thickness of the objective lenses increases.

4 Why should the coarse-adjustment focus not be used with a high-power objective lens?

5 The microscope invented by van Leeuwenhoek consisted of a single lens. What advantages do compound microscopes have over single-lens microscopes? ■

ABIOGENESIS AND MICROBES

Even as the invention and refinement of the light microscope revolutionized the study of biology, scientists continued to make mistakes as they sought to interpret their observations. Such was the case of the English biologist John Needham (1713–1781) when he set out to re-examine the theory of abiogenesis. Needham observed that meat broth left unsealed soon changed color and gave off a putrid smell. Mold and bacteria were found growing in the rich nutrient, but it was unclear where these microbes came from. Unlike the early supporters of abiogenesis who used logical analysis, John Needham tested his hypothesis through experimentation. Experimentation had become an essential component of science.

Needham boiled flasks containing nutrient meat broth in loosely sealed flasks for a few minutes in order to kill the microbes. The solutions appeared clear after boiling. The flasks were then left for a few days and the murky contents were examined under a microscope. The broth was teeming with microorganisms. Could this mean that the broth had spontaneously created microbes? Needham rushed to retest the experiment, using different nutrient solutions. Despite the boiling, the microbes reappeared a few days later. Needham concluded that the microbes had come from nonliving things in the nutrient broth.

Needham's conclusions sent many scientists down the wrong pathway. Let us re-examine his experiment to understand why. One of the difficulties arose from the fact that the flasks were not sealed properly—the tiny microbes could have entered the flasks after boiling. Another difficulty resulted from the design of his experiment. The fact that the flasks appeared clear immediately after boiling did not mean that all the microorganisms were destroyed. If only a few of the tiny microbes had survived, they would be able to multiply to millions within a few days. Needham did not check the flasks for microbes immediately after boiling. Even if he had checked the flasks, it is unlikely that he would have found any of the remaining microbes. Each drop of the nutrient would have to be examined, and such an examination might even infect the flask.

Needham's conclusions were upheld for nearly 25 years. Lazzaro Spallanzani (1729–1799) repeated Needham's experiment, but boiled the flasks longer. Spallanzani also took special care to seal the flasks completely. No microorganisms were found; abiogenesis did not occur. Needham's supporters were cautious about Spallanzani's experiments. They suggested that because Spallanzani had completely sealed the jars, the active principle had been destroyed. You will recall that the active principle objection had been used to oppose the work of Francesco Redi about 100 years earlier. Others claimed that the boiling had destroyed the nutrients. Although Spallanzani did not believe that an active principle existed, he was unable to overcome the objections that centered on the active-principle hypothesis.

The final blow to the theory of abiogenesis was delivered by the great French scientist Louis Pasteur (1822–1895). In 1864, Pasteur had a glassworker develop a special swan-necked flask. Broth was placed in the flask and subsequently boiled to destroy the microbes. Air passed from the flask during boiling. Fresh air entered the flask as the flask cooled. However, the microbes were trapped in the curve of the flask and were not carried into the broth from the surrounding air. Because the broth remained clear, Pasteur predicted that microbes were not present.

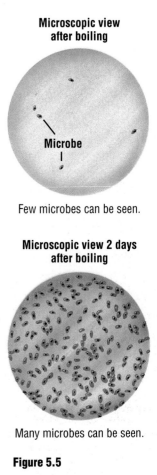

Microscopic view after boiling

Microbe

Few microbes can be seen.

Microscopic view 2 days after boiling

Many microbes can be seen.

Figure 5.5

Supplied with enough nutrients, microbes reproduce quickly.

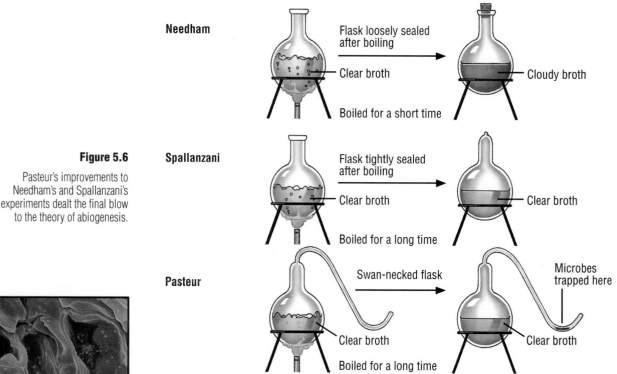

Needham

Flask loosely sealed
after boiling

Clear broth

Boiled for a short time

Cloudy broth

Figure 5.6

Pasteur's improvements to Needham's and Spallanzani's experiments dealt the final blow to the theory of abiogenesis.

Spallanzani

Flask tightly sealed
after boiling

Clear broth

Boiled for a long time

Clear broth

Pasteur

Swan-necked flask

Microbes
trapped here

Clear broth

Boiled for a long time

Clear broth

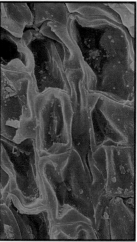

Figure 5.7

Cork cells seen through a microscope. The visible objects are the cell walls that surrounded the living cells.

A microscopic examination of the nutrient broth confirmed his prediction. Microbes could not be created from nonliving broth. As a finale, Pasteur tipped the broth in one of the flasks, allowing it to run into the curve of the swan-necked flask. As Pasteur had predicted, the broth became contaminated by the microorganisms trapped there. In a few days, the flask became cloudy.

EMERGENCE OF THE CELL THEORY

No one scientist developed the cell theory. Cells were probably first described in 1665, when the English scientist, Robert Hooke, noticed many repeating honeycomb-shaped structures while viewing a thin slice of cork under his primitive microscope. In his book, *Micrographia*, Hooke used the word "cell" to describe these structures. However, cork, the spongy tissue from cork trees, has few liv-

ing cells. What Hooke observed were the rigid cell walls that surrounded the once-living plant cells.

A few years later, Anton van Leeuwenhoek observed living blood cells, bacteria, and single-cell organisms in a drop of water. As microscopes improved, more structures were described. Around 1820, Robert Brown described the appearance of a tiny sphere in plant cells. He called the structure the *nucleus* (plural: "nuclei"). Nuclei were soon discovered in animal cells. A zoologist, Theodor Schwann, and a botanist, Mathias Schleiden, concluded that plant and animal tissues are composed of cells. Schwann and Schleiden prepared the foundations for the modern *cell theory*. The modern cell theory states:

- All living things are composed of cells. The cell is the basic living unit of organization.
- All cells arise from pre-existing cells. Cells do not come from nonliving things.

1 What is a scientific hypothesis?
2 Why do scientists test hypotheses by experimentation?
3 What observations led to the theory of abiogenesis?
4 How did Francesco Redi's experiment challenge the theory of abiogenesis?
5 Using Redi's experiment, differentiate between experimental variables and controls.
6 Who was one of the first contributors to the development of the microscope?
7 How did the work of John Needham cause a resurgence of the theory of abiogenesis?
8 Explain how Pasteur refuted the theory of abiogenesis.
9 Why were Robert Hooke's discoveries important to the development of the cell theory?
10 What are the two components of the cell theory?

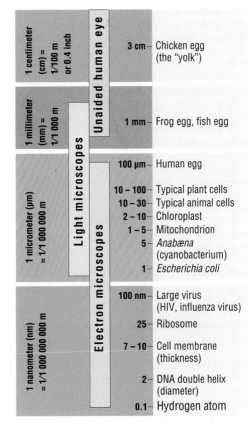

$1 \text{ m} = 10^2 \text{ cm} = 10^3 \text{ mm} = 10^6 \text{ µm} = 10^9 \text{ nm}$

Figure 5.8

Comparison of cell sizes.

OVERVIEW OF CELL STRUCTURE

Cells vary in size, shape, and function. Although there is no one common cell, all plant and animal cells are organized along a similar plan. Near the center of all cells is the nucleus, which is surrounded by fluid cytoplasm. The entire cell is covered by a membrane envelope, the cell membrane. Both the nucleus and the cytoplasm are collectively referred to as the **protoplasm.** The word protoplasm is a nonspecific term that refers to all substances within the cell.

The **cytoplasm** is the region of the protoplasm outside of the nucleus and the location where nutrients are absorbed, transported, and processed. During the processing of nutrients, waste products accumulate. The cytoplasm stores the

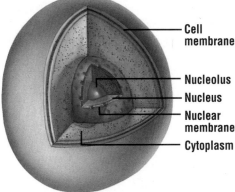

Figure 5.9

Basic cell structures of an animal cell.

wastes until proper disposal can be carried out.

The *cell membrane* is the outermost edge of the cell. Composed of a double layer of lipid or fat molecules and embedded proteins, the cell membrane provides the cell with a connection to the external environment. The membrane holds the contents of the cell in place and regulates the movement of molecules into and out

Protoplasm *refers to all the material within a cell. The protoplasm is composed of the nucleus and the cytoplasm.*

Cytoplasm *is the area of the protoplasm outside of the nucleus.*

of the cell. The cell membrane also contains receptor sites, which serve as docks for the entry of molecules that affect cell activity.

The **nucleus** is the cell's control center. Inside the nucleus the instructions for life are encoded within a molecule called DNA. DNA carries heredity information. In most higher life forms, the nucleus is enclosed by a nuclear membrane. Cells that have a true nuclear membrane are called **eukaryotic cells.** Plant and animal cells are eukaryotic cells. The nucleus of a eukaryotic cell appears darker than the watery cytoplasm when viewed under the microscope. Cells that lack a nuclear membrane are referred to as **prokaryotic cells.** Bacteria and blue-green algae are prokaryotic cells. Prokaryotic cells are the oldest known forms of life. The hereditary material of these cells is not contained in a nucleus, but rather is spread throughout the cytoplasm.

*The **nucleus** is the control center for the cell and contains hereditary information.*

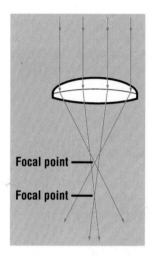

Focal point ——

Focal point ——

Figure 5.10

An image through a thick lens appears blurry.

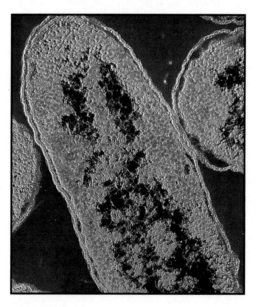

Figure 5.11

A prokaryotic cell, such as the bacterium *Escherichia coli*, lacks a nuclear membrane.

The hereditary or genetic material in the nucleus is organized into threadlike structures called **chromosomes.** Each chromosome contains a number of different characteristics, or genes. Genes are units of instruction that determine the specific traits of an individual. Some cells contain a **nucleolus,** a dark-stained, spherical structure within the nucleus. Although the entire function of the nucleolus is not known, scientists believe that it is involved with the synthesis of proteins in the cytoplasm.

THE SECOND TECHNOLOGICAL REVOLUTION

Oil-Immersion Lenses

As light rays move through the lenses of a microscope, the different wavelengths form various focal points. Because the image produced has many points of focus, it appears blurred. In order for things to appear larger, thicker lenses are required. But as lenses become thicker, light rays are bent even more, and the image becomes blurry. All microscopes sacrifice clarity, or resolution, in order to increase magnification.

The dilemma of increasing magnification while preserving resolution has plagued scientists for many years. One of the major breakthroughs in microscopy came when a drop of oil was placed on a slide, and the large, high-power objective was lowered into the oil. The denser oil limited the scattering of light and provided a clearer image. Oil-immersion lenses are still used today, especially for studying bacteria.

But oil-immersion lenses have their limitations. Even the most sophisticated techniques limit the light microscope to $2000\times$ magnification. For very tiny viruses or the details within a human cell to be seen, greater magnification is required. The electron microscope provides the answer.

Electron Microscope

A very crude electron microscope was invented in Germany in 1932. Although it provided 400× magnification, the image was grainy. In 1937, at the University of Toronto, James Hillier and Albert Prebus unveiled an electron microscope with 7000× magnification. Today, transmission electron microscopes are capable of 2 000 000× magnification.

The electron microscope uses electrons instead of light. Electrons are tiny subatomic particles that orbit around the nucleus of an atom. First described by J. J. Thompson in 1897, electrons move about in waves in a manner similar to visible light. However, the wavelength of the negatively charged electron is approximately 100 000 times shorter than that of visible light. Because this shorter wavelength is scattered less than longer wavelengths, a sharper image results.

The electron microscope looks much like a very large, upside-down light microscope. The object to be viewed is encased in thin plastic and placed on a stage. Electrons are beamed through the specimen and an image is projected on a screen. Because the electrons pass through the object, this microscope is often called a *transmission electron microscope*.

One of the greatest disadvantages of the electron microscope is that electrons are easily deflected or absorbed by other molecules. For this reason, the microscope's transmission chamber is equipped with a vacuum pump that removes air molecules. Because electrons pass more easily through single layers of cells, very thin sections of cells must be used. A thick specimen would absorb all the electrons and produce a blackened image. The cell sections are mounted in plastic, which means that only dead cells can be observed. Thus the transmission electron microscope is best suited for examining the structures within a nonfunctioning cell.

Scanning Electron Microscope

The scanning electron microscope provides even greater scope for scientific investigation. Made popular by physicist Victor Crewe in about 1970, this instrument produces a three-dimensional image. Electrons are passed through a series of magnetic lenses to a fine point. This fine point of electrons scans the surface of the specimen. The electrons are then reflected and magnified onto a TV screen where they produce an image. Specimens are often coated with a thin layer of gold to produce a sharper image.

Eukaryotic cells are cells that have a true nucleus. The nuclear membrane surrounds a well-defined nucleus.

Prokaryotic cells are primitive cells that do not have a true nucleus or a nuclear membrane.

Chromosomes are threadlike structures of DNA that contain genes.

The nucleolus is a small spherical structure located inside the nucleus.

Figure 5.13

Scanning electron micrograph of the surface of a spider mite.

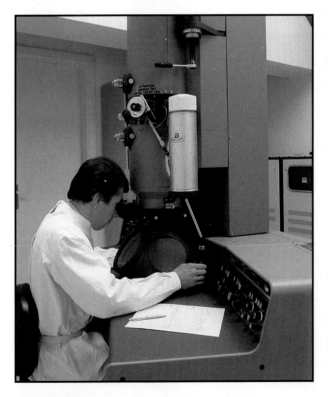

Figure 5.12

The transmission electron microscope permits scientists to view cell structures at extremely high magnification.

RESEARCH IN CANADA

James Hillier was born in Brantford, Ontario, in 1915 and trained as a physicist at the University of Toronto. Along with his colleague, Albert Prebus, Hillier has had a tremendous impact on the advancement of cell biology. Credited with designing the first truly functional transmission electron microscope, Hillier and Prebus permitted biologists to peer into a previously unseen world.

Dr. Thomas Chang, a scientist at McGill University, invented artificial cells in 1957 when he was still an undergraduate student. Artificial cells, which function in much the same way as natural cells, offer many opportunities for replacing the functions of biological cells.

Continuing research by Dr. Chang and other scientists around the world has resulted in artificial blood soon ready for human use. Artificial cells have also been tested for the treatment of diabetes and liver failure. Other types of artificial cells are being tested for the treatment of hereditary diseases and metabolic disorders. Researchers are also studying cell membranes of artificial cells to gather information about drug delivery systems. The possibilities of artificial cell technology seem almost limitless.

DR. JAMES HILLIER
Electron Microscope

DR. THOMAS CHANG
Artificial Cells

THE CYTOPLASMIC ORGANELLES

A cell can be compared to a factory. Like factories, new cell structures are erected and damaged structures are repaired. Nutrient molecules are absorbed and fashioned into molecules essential for life. The cytoplasm is the area of the cell in which work is accomplished, while the nucleus provides the directions. The working area of the cell has specialized structures called organelles ("small organs"), which are visible when viewed under an electron microscope. The following sections discuss the organelles that play a major role in the proper functioning of the cell "factory."

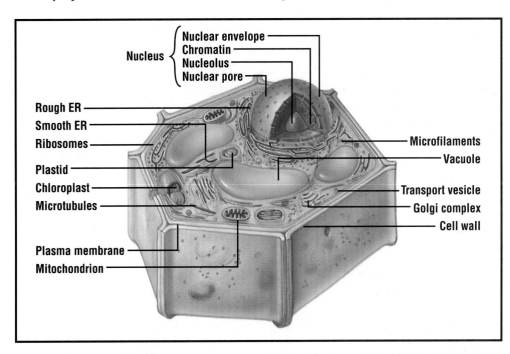

Figure 5.14

Plant cell and organelles.

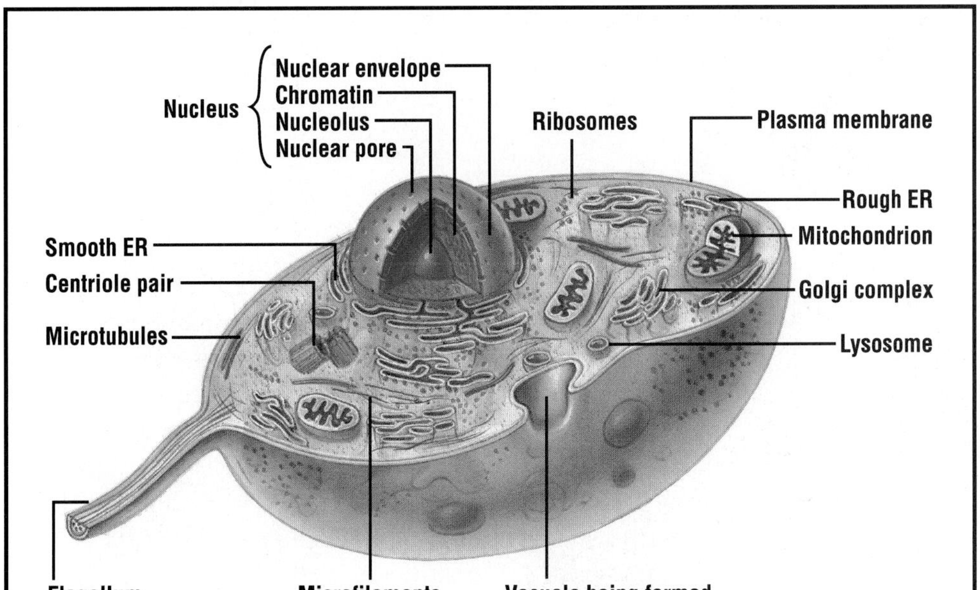

Figure 5.15

Animal cell and organelles.

Mitochondria

Tiny oval-shaped organelles called mitochondria (singular: "mitochondrion") are often referred to as the "power plant" of the cell. They provide the body with needed energy in a process called cell **respiration.** In this process, sugar molecules are combined with oxygen to form carbon dioxide and water. As the products are formed, energy is released.

BIOLOGY CLIP
Although most pictures show mitochondria as oval structures, scientists know that the shape can change quickly. Mitochondria swell and shrink in response to certain hormones or drugs.

$$C_6H_{12}O_6 + 6O_2 \longrightarrow 6CO_2 + 6H_2O + energy\ (ATP)$$
sugar   oxygen   carbon   water
                    dioxide

It is important to note that energy is not made in the mitochondria. Chemical bonds within the sugar molecule are broken and chemical energy is converted

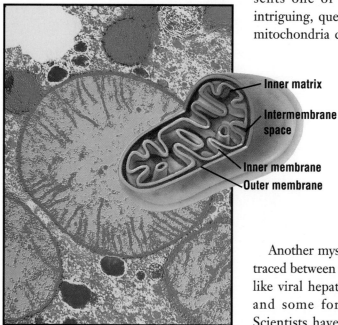

Figure 5.16

Electron micrograph of the mitochondria. Mitochondria are the largest of the cytoplasmic organelles.

Inner matrix
Intermembrane space
Inner membrane
Outer membrane

into other forms of energy. The energy permits muscle contraction, the synthesis of new molecules, and the transport of certain molecules within the cell. Energy is essential to life.

The mitochondria have two separate membranes: a smooth outer membrane and a folding inner membrane. The inner membrane consists of fingerlike projections called *cristae*. Special proteins called **enzymes** are located on the cristae. These enzymes assist in the breakdown of the sugar molecules in the mitochondria. Approximately 36% of the energy from sugar molecules is converted into a molecule called *adenosine triphosphate (ATP)*, a chemical storage compound. Most of the remaining 64% is converted into thermal energy. This means that animals that maintain a constant body temperature use a great number of sugar molecules to keep warm.

The origin of the mitochondria presents one of the most baffling, but intriguing, questions for biologists. The mitochondria contain their own hereditary material. However, the DNA of the mitochondria is not like that found in other eukaryotic cells. The DNA is much closer to bacterial DNA. Could this mean that the mitochondria were once separate organisms that invaded eukaryotic cells?

Another mysterious connection can be traced between mitochondria and diseases like viral hepatitis, obstructive jaundice, and some forms of muscle disease. Scientists have found large amounts of nutrients stockpiled within the mitochondria of afflicted cells. Scientists believe that the nutrients are waiting to be processed and that the enzymes required for this processing may be missing.

Ribosomes

Ribosomes are the organelles in which *proteins* are synthesized. Proteins are the molecules that make up cell structure. Cell growth and reproduction require the constant synthesis of protein. The nucleus provides the information for the type of protein needed. The chief building blocks of proteins, *amino acids*, are fused together by enzymes within the ribosome. There are 20 different amino acids. A change in position of a single amino acid can create a different protein.

Measuring just 20 nm (nanometers) in length, ribosomes are among the smallest organelles found in the cytoplasm. (There are one million nanometers in a single millimeter.) Yet, despite their minute size, ribosomes make up a great portion of the cytoplasm. It has been estimated that one-quarter of the cell mass of *Escherichia coli* are ribosomes. The large number of ribosomes permits the simultaneous construction of many proteins within a single cell.

Endoplasmic Reticulum

A series of canals carry materials throughout the cytoplasm. The canals, composed of parallel membranes, are referred to as *endoplasmic reticulum* or *ER*. The membranes can appear either rough or smooth when viewed under the electron microscope. The rough endoplasmic reticulum (RER) has many ribosomes attached to it. These ribosomes synthesize proteins. Rough endoplasmic reticulum is especially prevalent in cells that specialize in secreting proteins. For example, RER is highly developed in cells of the pancreas that secrete digestive enzymes. Smooth endoplasmic reticulum (SER) is free of ribosomes and is the area in which fats or lipids are synthesized. Smooth endoplasmic reticulum is prevalent in cells of developing seeds and animal cells that secrete steroid hormones.

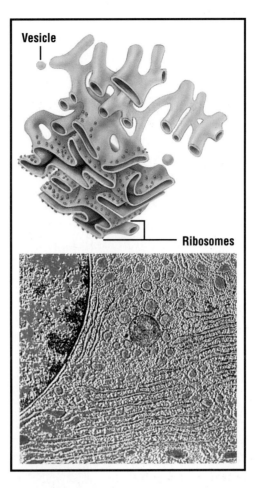

Vesicle

Ribosomes

Golgi Apparatus

The **Golgi apparatus** was first described by the Italian physician Camillo Golgi in 1898. Golgi had stained cells from a barn owl and found a new cytoplasmic structure. Half a century later, electron microscopy confirmed Golgi's observations. The structure appears like a stack of pancakes. The pancake-like structures are actually membranous sacs piled on top of each other.

Protein molecules from the rough endoplasmic reticulum are stored within the Golgi apparatus. The packed protein membranes move toward the cell membrane. Once the Golgi apparatus fuses to the outer membrane, small packets, called *vesicles*, are released. This process, described as **exocytosis,** is the means by which large molecules such as hormones and enzymes are released from cells.

Figure 5.17

Diagram and electron micrograph of rough endoplasmic reticulum. The micrograph shows the endoplasmic reticulum as yellow and green parallel, linear structures.

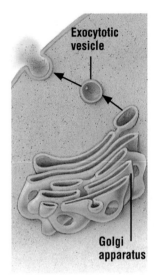

Exocytotic vesicle

Golgi apparatus

Figure 5.18

Large molecules are transported out of a cell by means of a vesicle.

*The **Golgi apparatus** is a protein-packaging organelle.*

Exocytosis *is the passage of large molecules through the cell membrane to the outside of the cell.*

Lysosomes

Lysosomes, formed by the Golgi apparatus, are saclike structures that contain digestive enzymes. Slightly smaller than the mitochondria, the lysosomes break down large molecules and cell parts within the cytoplasm. Food particles that are brought into the cell are broken down into the smaller molecules, which can then be used by the cell.

Lysosomes also play an important role in the human body's defense mechanism by destroying harmful substances that find their way into the cell. In the case of white blood cells that engulf invading bacteria, the lysosomes release their digestive enzymes, destroying both the bacterium and the white blood cell. The fluid and protein fragments that remain after the cells have been destroyed make up a substance called *pus*. The enzymes released from the lysosomes also destroy damaged or worn-out cells.

More than 30 different hereditary diseases have been linked to defective digestive enzymes in the lysosome. Tay-Sachs disease, for example, results from an enzyme deficiency that causes waste materials to accumulate. The buildup of waste products can cause brain damage.

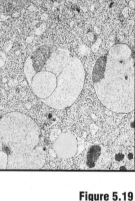

Figure 5.19

Acting as "suicide sacs," lysosomes release enzymes that destroy damaged or worn-out cells.

Plastids *are organelles that function as factories for the production of sugars or as storehouses for starch and some pigments.*

Chloroplasts *specialize in photosynthesis, and contain the green pigment chlorophyll found in plant cells.*

proteins, the microtubules are found in structures called *cilia* and *flagella*. The cilia are tiny hairlike structures that aid in movement. The cells that line the windpipe, for example, use cilia to sweep foreign materials from the lungs. The flagella are longer whiplike tails that propel some cells forward. Sperm cells, for example, use flagella to move.

SPECIAL STRUCTURES OF PLANTS

The organelles discussed to this point are found in both plant and animal cells. However, plant cells differ from animal cells in that they can produce their own food. Specialized organelles, called **plastids,** are associated with the production and storage of food in plant cells. In addition, a large part of the cytoplasm of plant cells is composed of a fluid-filled space, called a *vacuole*. The vacuole serves as a storage space for sugars, minerals, and proteins. The vacuole also increases the size, and hence the surface area, of the cell, thereby increasing the rate of absorption of minerals necessary for plant nutrition.

> **BIOLOGY CLIP**
> During the metamorphosis of insects, the membranes of the lysosomes become permeable, releasing digestive enzymes. The discovery of the reason behind this process may give scientists a better understanding of arthritis. The swelling and pain associated with arthritis have been linked to the seepage of lysosome enzymes into the cytoplasm. Drugs like cortisone, which reduce swelling, are known to strengthen lysosomal membranes.

Microfilaments and Microtubules

Microfilaments are pipelike structures that help provide shape and movement for the cells. Muscle cells have many microfilaments. Microtubules are tiny threadlike fibers that transport materials throughout the cytoplasm. Composed of

Plastids

Plastids are chemical factories and storehouses for food and color pigments. **Chloroplasts** are plastids that contain the green pigment chlorophyll. They specialize in *photosynthesis*, a process by which plants combine carbon dioxide from the air with water from the roots in the presence of light, producing sugar and releasing oxygen.

Another type of plastid, called a **chromoplast,** stores the orange and yellow pigments found in numerous plants. Colorless plastids, called **amyloplasts,** are storehouses for starch. Potato tubers and seeds contain many amyloplasts.

Cell Walls

Most plant cells are surrounded by a nonliving cell wall. Cell walls are composed of cellulose. Their main function is to protect and support plant cells. Some plants have a single cell wall, referred to as the *primary cell wall*, but others also have a *secondary cell wall*, which provides the cell with extra strength and support. The petals of a flower are composed of thin primary cell walls, as are cherries, strawberries, and lettuce. Particularly rigid secondary cell walls can be found in trees, even after the plant cells have died. In fact, most of a tree trunk is made up of hollow cell walls.

Plants cells are organized in regular patterns. The layer between the cell walls is referred to as the *middle lamella* (plural: "lamellae"). The middle lamella contains a sticky fluid, called *pectin*, that helps to hold the cells together. Have you ever noticed the sweet gooey material that forms on top of an apple pie after baking? This material, called pectin, is released during cooking, and forms a jellylike substance as it cools.

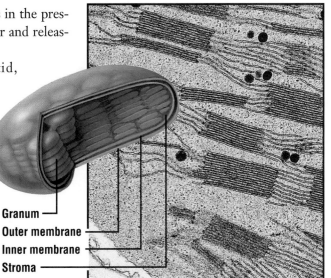

Granum
Outer membrane
Inner membrane
Stroma

Table 5.1 Summary of Cell Components

Cell structure	Prokaryotic	Eukaryotic	
		Plant	*Animal*
Cell wall	yes	yes	no
Cell membrane	yes	yes	yes
Nucleus	no	yes	yes
DNA	yes	yes	yes
Mitochondria	no	yes	yes
Ribosomes	yes	yes	yes
Lysosomes	no	yes	yes
Golgi bodies	no	yes	yes
Endoplasmic reticulum	no	yes	yes
Plastids	some species	yes	no

■ REVIEW QUESTIONS   ?

11 How do prokaryotic cells differ from eukaryotic cells?

12 Identify the two structures of protoplasm and give a generalized function of each.

13 What are chromosomes and genes?

14 Discuss the contributions of the Canadian scientist James Hillier to the study of cell biology.

15 List the cytoplasmic organelles found in animal cells and state the function of each.

16 What are chloroplasts?

17 Why would you expect to find more amyloplasts in a potato tuber (root) than in a potato leaf?

Figure 5.20

Electron micrograph and diagram showing the general internal structure of a chloroplast.

Chromoplasts *store orange and yellow pigments.*

Amyloplasts *are colorless plastids that store starch.*

Primary cell wall
Secondary cell wall

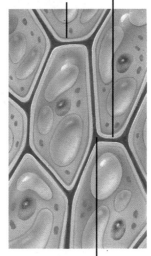

Middle lamella

Figure 5.21

The middle lamella of plant cells contains pectin.

LABORATORY

PLANT AND ANIMAL CELLS

Objectives

To identify the components and compare the structures of plant and animal cells.

Materials

compound microscope	water
lens paper	cover slip
onion	iodine stain
scalpel	paper towels
forceps	prepared slide of human
medicine dropper	cheek epithelium
glass slide	50 mL beaker

Procedure

1 Use lens paper to clean the ocular and objective lenses of the microscope.

2 Obtain a 2 cm² section of a scale from an onion bulb.

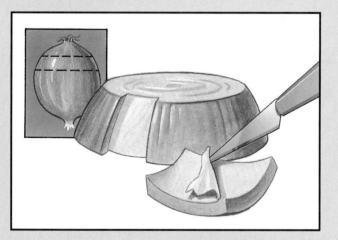

3 Using the scalpel, remove a single layer from the inner side of the scale leaf of the onion. If the extracted tissue is not transparent to light, try again.

4 Using forceps, place the tissue on a glass slide and add two drops of water. Holding the cover slip with your thumb and forefinger, touch it to the surface of the slide at a 45° angle. Gently lower the cover slip, allowing the air to escape. If air bubbles are present, gently tap the slide with the eraser end of a pencil to remove them.

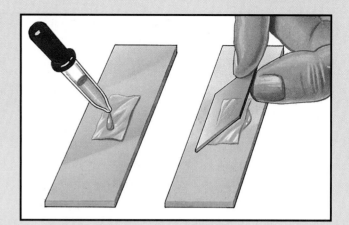

5 Focus the cells under low-power magnification. Identify a group of cells that you wish to study, and move the slide so that the cells are in the center of the field of view.

6 Rotate the nosepiece of the microscope to the medium-power objective and view the cells. Using your fine-adjustment focus, bring the cells into clear view.

  a) Draw and describe what you see.

7 Slowly decrease light intensity by adjusting the diaphragm of the microscope.

  b) Which light intensity reveals the greatest detail?

8 Remove the slide from the microscope and place a drop of iodine on one edge of the cover slip. Touch the opposite edge of the cover slip with the edge of a paper towel. This will draw the stain onto the slide.

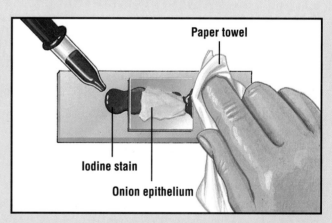

Paper towel

Iodine stain

Onion epithelium

9 Select a section of the tissue and view the cells under medium-power magnification.

  c) Why was iodine used?

10 Rotate the revolving nosepiece to high-power magnification. Using the fine adjustment, focus on a group of cells.

  d) Draw a four-cell grouping and label as many cell structures as you can see.

  e) Estimate the diameter of one cell.

11 Rotate the nosepiece back to low-power magnification and remove the slide containing the plant cells. Place the prepared slide of animal tissue epithelium on the stage of the microscope. Using the coarse-adjustment focus, locate a group of animal cells.

  f) How does the arrangement of plant and animal cells differ?

12 Locate a number of cells under medium power and focus using the fine-adjustment focus.

  g) Draw three different cells and label those cell structures that are visible.

  h) Estimate the diameter of the cells.

Laboratory Application Questions

1 What function is served by the cells of the epithelial tissue of plants and animals?

2 In what ways do the onion cells differ from those of the animal epithelium?

3 Explain why the cells of the onion bulb do not appear to have any chloroplasts.

4 Why are cells of the onion and animal epithelium classified as eukaryotic cells? ■

FRONTIERS OF TECHNOLOGY: THE THIRD REVOLUTION

The electron microscope opened a new window onto the cell, allowing scientists to observe things they had only been able to imagine before. As we discussed earlier, however, the transmission electron microscope has its limits. Because cell structures must be fixed in plastic to be viewed through this microscope, scientists are limited to observing nonliving things. A clearer understanding of a living cell requires different technologies.

Cell fractionation provides information about the thousands of chemical reactions that occur simultaneously in a cell. In this technique, cells are ground into fragments and placed in a test tube, which is then spun in a centrifuge. The centrifuge is a machine that rotates at high speeds to produce a force of gravity hundreds of thousands of times greater than normal. Heavy structures are driven to the bottom of the test tube, while lighter objects remain nearer the surface. Because each layer contains different cell

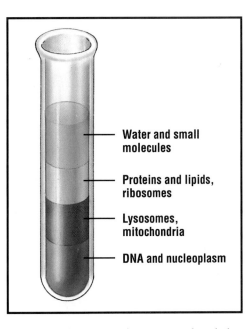

Water and small molecules

Proteins and lipids, ribosomes

Lysosomes, mitochondria

DNA and nucleoplasm

Figure 5.22

Cell fractionation.

parts, the layers can be separated and the chemical reactions in each layer studied.

In another technique, living cells are treated with **radioisotopes.** The radioactivity emitted from the radioisotopes can be traced with special equipment. The radioactive chemicals can then be followed through chemical reactions. Radioisotopes permit scientists to follow specific chemical reactions in specific organelles.

Cell fractionation *is the process by which cell fragments are separated by centrifugation.*

Radioisotopes *are unstable chemicals that emit bursts of energy as they break down.*

THE LIVING CELL MEMBRANE

The cell membrane separates the protoplasm from the nonliving environment. However, the cell membrane is much more than a plastic envelope that holds the cytoplasmic organelles in place. It regulates what enters and leaves the cell. If the cell membrane is pierced with a pin, some cytoplasm will ooze out, but the puncture will soon be sealed. Cell membranes are living structures.

In order to survive, cells must extract nutrients from the nonliving environment. As nutrients are processed, wastes build up inside the cell. This time, the nonliving environment acts as a waste disposal site. Without it, the accumulation of poisonous wastes within a cell would soon cause cell death.

The fact that the cell membrane is involved in the transportation of materials into and out of the cell raises some questions: How does the cell know which materials must be absorbed and which must be excreted? How does the cell select molecules?

To understand the movement of molecular traffic across the cell membrane, the structure of the living membrane must be examined. The membrane appears as two layers of lipids. Each lipid molecule found in the cell membrane can be represented by a head and two tails. The head is hydrophilic, or "water loving," and is soluble in water. The tails are hydrophobic, or "water hating," and are not soluble in water. The water-soluble ends face the outer environment and the inner cell components. The lipid molecules are mobile, contributing to the movement within the cell membrane.

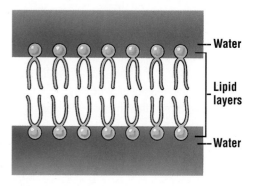

A variety of different protein molecules are embedded within the two layers of lipid. Most of these protein molecules carry a special sugar molecule and are thus referred to as **glycoproteins.** The sugar molecules provide the cell with a special signature. Like written signatures, the sugar molecules vary between different organisms, and even between individuals within the same species. These unique sugar proteins help distinguish a type A red blood cell from a type B red blood cell. Your immune system identifies foreign invaders by recognizing their unique structure on the cell membrane. This helps explain why transplanted organs are often rejected by recipients. Some of the lipid molecules also contain a specialized sugar called a **glycolipid.**

The protein molecules serve different functions. Some act as gatekeepers, opening and closing paths through the cell membrane. For example, specialized protein gatekeepers, located in nerve cells,

---

Glycoproteins *are compounds consisting of specialized sugar molecules attached to the proteins of the cell membrane. Many distinctive sugar molecules act as signatures that identify specialized cells.*

Glycolipids *are compounds consisting of specialized sugar molecules attached to lipids.*

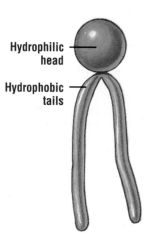

Hydrophilic head

Hydrophobic tails

Lipid molecule

BIOLOGY CLIP

Some scientists have speculated that many hormone disorders are due not to inadequate hormone production but to low numbers of receptor sites on the membranes of target cells. An excess number of receptor sites could also cause severe problems. Hypertension, or high blood pressure, may be related to an excess number of receptor sites for stress-related hormones.

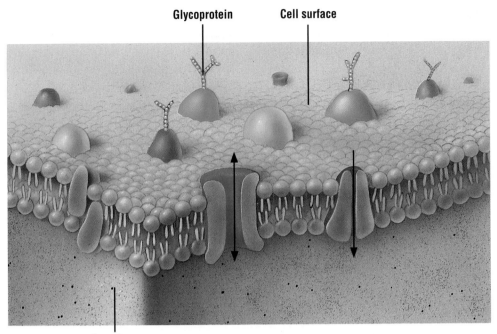

Glycoprotein   **Cell surface**

Cytoplasm

Figure 5.23

A cell membrane is formed by a double layer of lipid molecules in which protein molecules are embedded.

allow potassium ions to move through the pores when the membrane is at rest. However, when the nerve is excited, a sodium ion gateway is opened. This is what causes the nerve impulse to fire.

Other proteins act as receptor sites for hormones. Hormones are chemical messengers that allow cells to communicate with one another. A third type of protein is involved in transport, using cell energy to pick up needed materials and moving them into or out of the cell. You will read more about transport proteins later in this chapter.

PASSIVE TRANSPORT

The cell membrane acts as a barrier. However, cells must take in food and eliminate wastes in order to maintain a constant internal environment. The movement of materials across a cell membrane without the expenditure of cell energy is called *passive transport*. The following section discusses two types of passive transport: diffusion and osmosis.

Diffusion

Molecules move about randomly and collide. This random movement is referred to as **Brownian motion.** Molecules move in all directions with equal frequency, bouncing off each other when they collide. This causes molecules concentrated in one area to spread outward. **Diffusion** can be explained by the movement of molecules from an area of high concentration to an area of lower concentration. (Concentration can be described as the number of molecules per volume.) You can experience diffusion simply by opening a perfume bottle. The molecules inside the perfume bottle move into the surrounding air. However, if you drop ink into a glass of water and watch the color of the water change, you will note that diffusion is not confined to gases. Water molecules collide with the molecules of ink, spreading them apart. The water molecules alone are not responsible for diffusion, however. The ink molecules are also colliding with other ink molecules. In addition, the ink molecules strike the water molecules.

Brownian motion *refers to the random movement of molecules.*

Diffusion *is the movement of molecules from an area of higher concentration to an area of lower concentration.*

Molecular collision causes diffusion. Can you predict how an increase in water temperature would affect the rate of diffusion? The faster the molecules move, the more often they collide and the faster they will move apart. In other words, diffusion rates increase with temperature. Pressure also speeds up molecular motion. Molecules are bunched close together if the pressure is high. Molecules in high-pressure areas will collide more frequently. Molecules in areas of lower pressure are spread out and collide less frequently. Molecules move by diffusion from areas of high pressure to areas of low pressure.

To summarize, diffusion is affected by concentration, temperature, and pressure and will occur until molecules are equally distributed within an area. At this point, **equilibrium** is reached. Although individual molecules will continue to move, there is no net movement from one area to another.

Oxygen and carbon dioxide move across cell membranes by diffusion. Oxygen diffuses from the blood, an area of high concentration, into the cell, an area of low concentration. Because oxygen is continuously consumed within a cell, the concentration of oxygen tends to be low. Carbon dioxide, by contrast, accumulates within a cell and diffuses from the cell to the blood.

Osmosis
Consider the system shown in the diagram below. Osmosis is the diffusion of

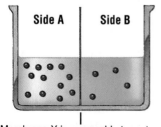
Membrane X is permeable to water, but not to protein.

water through a **selectively permeable membrane.** Membrane X is permeable to water, but impermeable to the larger protein molecules. Protein cannot diffuse from side A into side B. Does this mean that the system will never reach equilibrium?

The concentration of protein in side A is greater than in side B. Did you notice that the volume of fluid in side A is equal to that in side B? Which side do you think has the greater concentration of water? There are fewer protein molecules on side B, but many more water molecules. The spaces between protein molecules are filled with water molecules. Therefore, the fewer the number of protein molecules, the greater is the amount of space for water molecules.

Remember that osmosis is actually a diffusion of water. Water will diffuse from side B, the area of high water concentration, to side A, the area of lower water concentration. Like other molecules, water follows the diffusion gradient. The diagram below shows what happens when water moves from side B into side A.

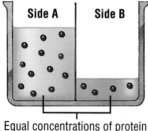
Equal concentrations of protein and water in side A and side B.

As water moves from side B into side A by osmosis, the protein in side A becomes less concentrated. As water leaves side B, the protein in side B becomes more concentrated. Eventually, the concentrations of protein and water in sides A and B will become equal. Once the system reaches equilibrium, water molecules continue to move between the two sides; however, the number of molecules gained from side A

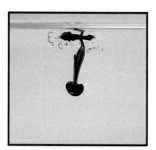

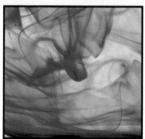

Figure 5.24
Ink diffusing in water.

Equilibrium *is a condition in which all acting influences are balanced, resulting in a stable condition.*

Selectively permeable membranes *allow some molecules to pass through the membrane, but prevent other molecules from penetrating the barrier.*

equals the number of molecules gained from side B. There is no net movement of water at equilibrium.

The movement of water into and out of living cells is vital to life processes. Ideally, cells are bathed in **isotonic solutions**—solutions in which the **solute** concentration outside the cell is equal to that inside the cell. In isotonic solutions, the water movement into a cell is balanced by the water movement out of a cell.

Cells that are placed in **hypotonic solutions** (solutions that have a lower concentration of solute relative to the solute concentration inside the cell) are not at equilibrium. The concentration of water in a hypotonic solution is greater than that found inside a cell. Fewer solutes means that there is more room for water molecules. The water molecules move from the area of high water concentration—the area outside the cell—into the area of lower water concentration—the area inside of the cell. The cell expands as water moves in.

Some animals are able to expel the excess water absorbed from freshwater environments where solute concentrations are very low. The single-cell paramecium uses a contractile vacuole to expel water from the cytoplasm. Any failure of the contractile vacuole would be disastrous for the cell. Water would continue to diffuse into the cell, eventually causing it to explode.

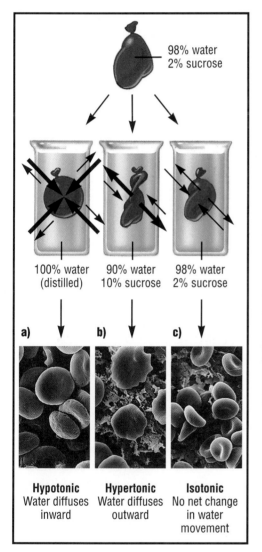

a) Red blood cells in hypotonic solution. (b) Red blood cells in hypertonic solution. (c) Red blood cells in isotonic solution.

Figure 5.25

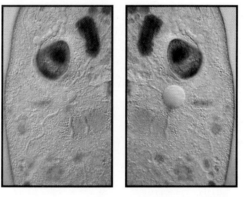

On the other hand, cells placed in **hypertonic solutions** tend to shrink. Hypertonic solutions have greater concentrations of solutes outside of the cell than inside the cell. The solutes outside the cell create an osmotic force and draw water from the cell.

Have you ever noticed that the salt sprinkled on sidewalks during winter kills the surrounding grass in spring? Salt creates a hypertonic solution that draws

Isotonic solutions *are solutions in which the concentration of solute molecules outside the cell is equal to the concentration of solute molecules inside the cell.*

Solutes *are molecules that are dissolved in water. Salt and sugars are common solutes.*

Hypotonic solutions *are solutions in which the concentration of solutes outside the cell is lower than that found inside the cell.*

Figure 5.26

A contractile vacuole expels water from a paramecium, thereby permitting it to exist in a hypotonic environment.

Hypertonic solutions *are solutions in which the concentration of solutes outside the cell is greater than that found inside the cell.*

water from the cells of the grass. As water leaves the plant cells, they wilt. Water pressure, referred to as **turgor pressure,** pushes the cytoplasm of the plant cell against the nonliving cell wall. Turgor pressure is the reason plants are rigid. The principle of turgor pressure also explains why plants die if they are exposed to too much fertilizer and why vegetables are sprayed with water in your local grocery store.

Figure 5.27

As the plant loses turgor pressure, it begins to wilt.

Facilitated Diffusion

Protein carrier molecules, located in the cell membrane, can aid in passive transport. Although the precise action of these carriers is not well understood, scientists believe that the protein carriers speed up the movement of molecules already moving across the cell membrane. Glucose

diffuses into red blood cells hundreds of times faster than other sugar molecules that have similar properties. Why would one sugar molecule diffuse faster than another? Scientists have proposed facilitated diffusion. The carrier proteins must be specialized to aid the diffusion of glucose molecules, but not other sugars.

ACTIVE TRANSPORT

The concentration of sodium ions outside a nerve cell is much greater than the concentration of sodium ions inside. Potassium ions, by contrast, are more concentrated inside the nerve cell. How do the nerve cells maintain different concentrations when the cell membrane is permeable to both sodium and potassium ions? Root cells from plants absorb minerals and ions from the surrounding soil despite the fact that these ions are more concentrated inside the root cell. How do the root cells transport minerals against the concentration gradient? Bacteria in milk have a higher concentration of milk sugar than that found in the surrounding milk, despite the fact that the bacteria consume milk sugar for energy. How do the bacteria get so much sugar inside their cells?

The answer to these questions lies in the fact that some materials are transported against the concentration gradient. In such cases, energy must be used. The movement of materials from an area of low concentration to an area of higher concentration is referred to as **active transport.** It has been estimated that while you sleep between 30 and 40% of your total energy budget is used for active transport. Without active transport, your kidneys would fail to reabsorb needed materials, your muscles would not contract, and your nerves would not carry impulses.

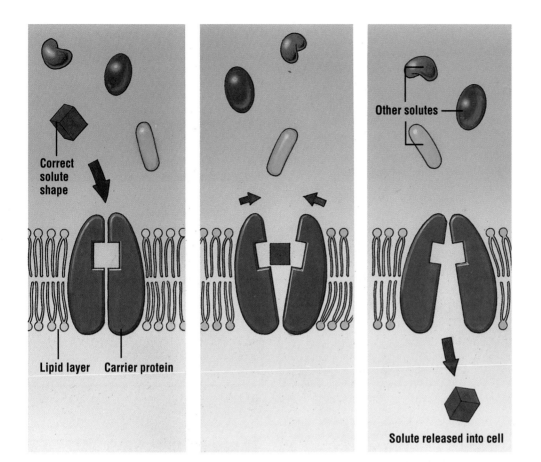

Figure 5.28
Model of active transport.

Carrier protein molecules suspended within the cell membrane receive an energy boost that permits them to aid active transport. The energized proteins capture specific solute molecules and move them either into or out of the cell. Although the exact mechanism is not completely understood, scientists have devised models to help explain how transport molecules actively transport solutes against the concentration gradient. Figure 5.28 shows how the specific geometry of the protein molecule is well suited for trapping and then transporting solute molecules that have complementary shapes. Once the solute gains access to the binding site of the carrier protein, energy is released and the carrier molecule binds with another solute molecule. After the energy is consumed, the binding site loosens, and the solute molecule is released. Although the model presented shows the delivery of the solute into the cytoplasm, it is important to note that some carrier proteins actively pump materials out of cells.

Endocytosis and Exocytosis

Cells take in smaller solutes by means of transport carrier molecules, but how do cells absorb larger molecules? Some molecules, many of which are essential to life, do not fit between the pores of the cell membrane. The cell must expend energy to transport these larger substances.

Endocytosis is the process by which cells engulf large particles by extending their cytoplasm around the particle. As two cell membranes come together, the ingested particle is trapped within a pouch, or vacuole, inside the cytoplasm. Enzymes from the lysosomes are then often used to digest the large molecules absorbed by endocytosis.

Active transport *involves the use of cell energy to move materials across a cell membrane against the concentration gradient.*

Endocytosis *is the process by which particles too large to pass through cell membranes are transported within a cell.*

Figure 5.29

Electron micrographs illustrate the process of phagocytosis. (a) Membrane begins to fold around molecules. (b) Membrane traps molecules. (c) Cell membranes come together and vacuole is formed. (d) Large molecules are digested within the vacuole.

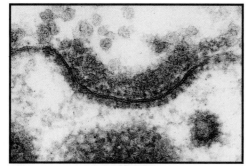

a)

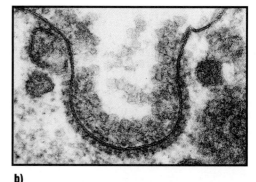

b)

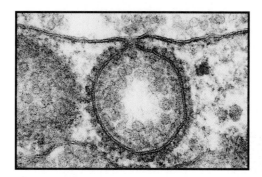

c)

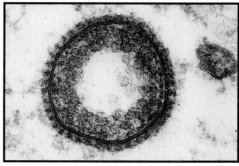

d)

Pinocytosis is a form of endocytosis in which liquid droplets are engulfed by cells.

Phagocytosis is a form of endocytosis in which solid particles are absorbed by cells.

There are two types of endocytosis. In **pinocytosis,** cells absorb liquid droplets by engulfing them. Cells of your small intestine engulf fat droplets by pinocytosis. **Phagocytosis** is the process by which cells engulf solid particles. Some white blood cells are often referred to as phagocytes (cell eaters) because they consume invading microbes by engulfing them. The microbe, trapped in the vacuole, is digested when the vacuole fuses with lysosomes.

Exocytosis is the process by which large molecules held within the cell are transported to the external environment. Waste materials are often released by exocytosis. Useful materials, like transmitter chemicals emitted from nerve cells, are also released by exocytosis. The Golgi apparatus holds the secretions inside the fluid-filled membranes. Small vesicles break off and move toward the cell membrane. The vesicles fuse with the cell membrane and the fluids are released.

■ REVIEW QUESTIONS ?

18 How have cell fractionation techniques and the use of radioisotopes helped advance the knowledge of cell biology?

19 In what ways does a cell membrane differ from a cell wall?

20 Describe the structure of a cell membrane.

21 List three factors that alter diffusion rates.

22 Define isotonic, hypertonic, and hypotonic solutions.

23 How does facilitated diffusion differ from normal diffusion and osmosis?

24 Why does grass wilt if it is overfertilized?

25 How does passive transport differ from active transport?

SOCIAL ISSUE:
Limits to Cell Technology

Dr. John Roder of Queen's University in Kingston has created a type of immortal human cell that churns out antibodies to fight cancer, leprosy, and tetanus.

*Dr. Gerald Price, from McGill University, has fused conventional white blood cells with cancer cells. The resulting cells, called **hybridomas,** produce huge amounts of antibodies for defense against disease.*

Other scientists are looking to lysosomes to unravel the mysteries of aging and disease. Lysosomes, the digestive packets held within the cells, release enzymes that destroy cells. Researchers believe that the inflammation and pain associated with arthritis may be caused by the leakage of enzymes from the lysosomes in white blood cells. This hypothesis is supported by the fact that a drug known to reduce swelling strengthens the lysosome membrane.

A Halifax hospital may soon begin transplanting brain tissue from aborted fetuses into patients with Parkinson's disease. This hereditary disease affects the nervous system, and is characterized by uncontrollable tremors. People with Parkinson's disease would be helped immensely by this new procedure.

Statement:

Cell research is progressing so quickly that many of the long-range effects of the research have not been considered. A private body, representing community values, should be established to oversee the technological applications of cell biology.

Point

- The general public has a right to know how scientific research will affect their lives. Research on lysosomes can forestall aging. Can you imagine the consequences of increasing life expectancy by 50%?
- Scientists have developed skills related to research. They are not trained to evaluate the moral, social, and economic consequences of their work.

Counterpoint

- Regulatory bodies have already been established by scientists who carefully monitor their own work. Greater interference by nonprofessionals would only slow the rate at which science progresses.
- There has been a long history of public monitoring of cell research. In 1978, the town council of Cambridge, Massachusetts, closed a cell biology lab at Harvard University. In the mid-1980s, a group in California worked to prevent the release of genetically engineered bacteria into the environment.

Research the issues.
Reflect on your findings.
Discuss the various viewpoints with others.
Prepare for the class debate.

CHAPTER HIGHLIGHTS

- A scientific hypothesis is an explanation based on the observation and interpretation of events in nature.
- A scientific hypothesis must be tested by experimentation.
- The theory of abiogenesis attempted to explain the origin of living things from a nonliving world.
- Francesco Redi used controlled experiments to refute the theory of abiogenesis.
- Pasteur and Spallanzani disproved the theory of spontaneous generation of microbes.
- The invention of the light microscope led to the discovery of cells and the development of the cell theory.
- The protoplasm of cells is composed of a nucleus and cytoplasm.
- Primitive prokaryotic cells do not contain a nuclear membrane or many of the organelles found within the more advanced eukaryotic cells.
- The transmission and scanning electron microscopes have made the study of cytoplasmic organelles possible.
- Cell fractioning techniques and the use of radioisotope tracers have enabled cell biologists to study the chemical reactions that occur within the cytoplasmic organelles.
- Plant cells contain cell walls and plastids.
- The cell membrane acts as a boundary between the cell and its external environment. The cell membrane regulates the movement of molecular traffic into and out of the cell.
- Materials are transported across cell membranes by either active or passive transport.
- Diffusion, osmosis, and facilitated diffusion are forms of passive transport in which molecules permeable through the cell membrane move from an area of higher concentration to an area of lower concentration.
- Materials can move against the concentration gradient if energy is used. The process of active transport requires the use of carrier molecules.
- Endocytosis and exocytosis are processes by which cells take in or release molecules too large to pass through the pores in cell membranes.

APPLYING THE CONCEPTS

1 Biology, like other sciences, progresses by observation. Unfortunately, the observations of nature are often flawed by interpretation. Provide two examples of faulty conclusions that supported the theory of abiogenesis and explain the source of the error.

2 Explain why John Needham came to an incorrect conclusion about abiogenesis. What was his error?

3 Those who supported the theory of abiogenesis used the critical factor of fresh air to refute Spallanzani's experiments. Explain why Spallanzani was unable to overcome the challenge from those who supported the theory of abiogenesis. How did Pasteur finally disprove the theory of abiogenesis?

4 A student suggests that flour sealed in a jar has been transformed into flour beetles after it has been sitting for six weeks. Provide a probable explanation for the flour beetles and then design an experiment to test your hypothesis.

5 A cell is viewed under low-power magnification. When the revolving nosepiece is turned to high-power magnification, the object appears to disappear, despite many attempts to refocus the slide. Provide a possible explanation for the disappearance of the cell.

6 By comparing a bee's body mass to its wing span, a physicist once calculated that bees should not be able to fly. Cell biologists have found that the muscles that control the wing of the bee have an incredible number of mitochondria. Indicate why this finding may help explain why bees can fly.

7 Explain why stomach cells have a large number of ribosomes and Golgi apparatus.

8 Hormones are the body's chemical messengers. Protein hormones, such as insulin, must attach themselves to a receptor site on the cell membrane; fat-soluble steroid hormones, such as sex hor-

mones, pass directly into the cell. Explain why steroid hormones pass directly into the cytoplasm of a cell.

9 Identify some limitations of the light microscope and explain why the transmission and scanning electron microscopes have had major impacts on the study of cell biology.

10 A marathon runner collapses after running on a hot day. Although the runner consumed water along the route, analysis shows that many of the runner's red blood cells had burst. Why did the red blood cells burst? (Hint: On hot days many runners consume drinks that contain sugar, salt, and water.)

CRITICAL-THINKING QUESTIONS

1 The statement "We are not alone" is often used in science-fiction movies. Explain how this statement applies to the cells within your body.

2 Early scientists often proposed hypotheses but did not test them by experimentation. Instead, they used logic to translate the hypotheses into an explanation or theory. Using the theory of abiogenesis, explain why this method of inquiry is susceptible to errors.

3 Why was Francesco Redi's experiment considered to be a significant turning point for the way in which scientific experiments were performed?

4 Many people argue that technology follows science. In many cases, this may be true; however, technology can assume a leading role. Using the development of the lens as an example, explain how technology changed the manner in which humans perceived themselves and their relationship to their environment.

5 Some unrelated diseases may have an interesting link: the mitochondria. Disorders of the liver (i.e., viral hepatitis, obstructive jaundice, and cirrhosis) and some muscle disorders (i.e., muscular dystrophy) are characterized by abnormally shaped mitochondria. Large amounts of unprocessed chemicals appear to accumulate in the mitochondria. Indicate why this link may be important. Does it provide any clues to the cause of the diseases? How might scientists use this information to develop a cure?

ENRICHMENT ACTIVITIES

1 Suggested reading:
- Chollar, Susan. "The Poison Eaters." *Discover*, April 1990, p. 76. A description of how microbial cells are used to solve the toxic-waste problem.
- Cook-Varsat, Alice, and Harvey F. Lodish. "How Receptors Bring Proteins and Particles into Cells." *Scientific American*, May 1984, p. 52. A technical presentation of transport systems.
- Kiester, Edward. "A Bug in the System." *Discover*, February 1991, p. 70. An investigation of how the DNA from mitochondria has been linked with hereditary disease.
- Murray, Mary. "Life on the Move." *Discover*, March 1991, p. 72. A description of microfilaments and movement.

2 Many industries and businesses use scientific research in their advertising to legitimize their products. Locate a magazine or TV advertisement that attempts to use scientific data to support a product. Does the ad present any data? Comment on the presentation of research.

3 Scientists at the Pasteur Institute in Paris were reported to be the first to grow HIV (the virus that causes AIDS) in tissue culture. Researchers from the United States indicated that they had also grown the virus, but insisted that they were first. Conduct research into the discovery of HIV.

Chemistry of Life

IMPORTANCE OF CELL CHEMISTRY

Defining the difference between living and nonliving things has presented an array of problems for scientists. During the 19th century, biologists such as Pasteur and Spallanzani performed experiments to dispel the belief that nonliving materials could be transformed into living organisms. Today there is little argument that living organisms only arise from other living organisms. Ironically, living things *are* composed of nonliving chemicals. Proteins, carbohydrates, lipids, and nucleic acids are often categorized as the chemicals of life despite the fact that none of them are capable of life by themselves.

Another group of scientists, called *vitalists*, believed that nonliving things were distinctly different from living things. Some even believed that life had a vital force that was neither chemical nor physical. They believed that living things had their own type of substance that was not found in nonliving things.

Scientific investigations have shown that the same principles of chemistry apply in both the physical world

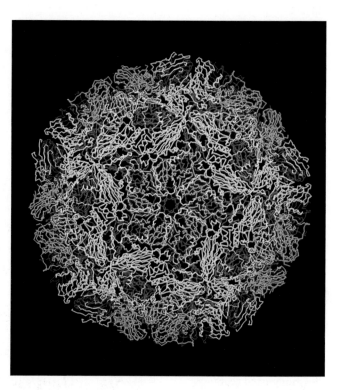

Figure 6.1

Computer graphics representation of the structure of a virus. Precise positions have been assigned to over 300 000 atoms that comprise the spherical coat that encloses the genetic material.

and the living world. An understanding of the chemistry of life comes in part from an understanding of how chemical reactions are regulated within cells. Chemicals from your surroundings are altered within your cells. Bonds are broken and new bonds are formed in a continuous cycling of matter. The absorption of nutrients, the synthesis of vital chemicals, and the excretion of wastes are vital to life. Consider the oxygen you breathe. Mitochondria use oxygen to break the bonds in nutrient sugar molecules. The breakdown of the sugar yields energy, carbon dioxide, and water. Plants and some bacteria utilize the water and the carbon dioxide released from animal cells to make sugars and to release oxygen. An oxygen molecule that you have absorbed may once have been part of a potato, a giant maple tree, or it may be the same oxygen molecule once used by Julius Caesar!

It has been estimated that over 200 000 chemical reactions occur in the cells of your body. The host of chemical reactions that take place within cells are referred to as **metabolism.** Metabolic reactions can be classified as either **catabolic**—reactions in which large chemical complexes are broken down into smaller components—or **anabolic**—reactions in which complex chemicals are built from smaller components.

ORGANIZATION OF MATTER

Matter can be described as the material of which things are made. Matter occupies space and has mass, and can exist in the form of a solid, liquid, or gas. Matter is composed of tiny particles called **molecules.** In solids, the molecules are packed tightly together and are held in a relatively fixed position. The molecules vibrate, creating small spaces between themselves. In a liquid, molecules move more freely, and the spaces between them are greater. Molecules in gases can have very large spaces between them. If energy is added to liquid water molecules by heating, the movement of the molecules intensifies, and molecules are liberated from the surface of the liquid. Liquid water is then converted into vapor.

The search for the composition of matter began with the ancient Greeks. Around 450 B.C., the Greek philosopher Empedocles speculated that matter was composed of four **elements:** air, water, fire, and earth. The four elements were in turn produced by combinations of four properties: coldness, wetness, hotness, and dryness. Water was thought to be a combination of coldness and wetness, while air was said to be a combination of wetness and hotness. By 400 B.C., another Greek, Democritus, provided a theory that was to serve as the basis for the development of the modern scientific theory of matter. He believed that matter was composed of many small subunits that were indivisible. These particles were called **atoms.** The word atom comes from the Greek word *atomos*, meaning "indivisible." Democritus proposed that all matter is composed of essentially the same materials, but that atoms come in different sizes and shapes. He reasoned that a cube of lead would have different particles than a sphere of lead.

Today we know that atoms are composed of protons, neutrons, and electrons. Protons carry a positive charge, electrons have a negative charge, and neutrons, as the name suggests, carry no charge at all. The larger protons and neutrons are located at the center of the atom, called the *nucleus.* The much smaller electrons orbit around the nucleus in an energy level, or *electron cloud.* Protons and neutrons make up about 99.9% of the total

Metabolism *is the sum of all chemical reactions that occur within the cells.*

Catabolism *refers to reactions in which complex chemical structures are broken down into simpler molecules.*

Anabolism *refers to chemical reactions in which simple chemical substances are combined to form complex chemical structures.*

Molecules *are units of matter. They can be composed of one or more different atoms.*

Elements *are pure substances that cannot be broken down into simpler substances. There are 109 different elements.*

Atoms *are the smallest particles of matter and are composed of smaller subatomic particles: neutrons, protons, and electrons.*

mass of an atom, yet only a very small part of the total volume. Most of the volume of an atom is taken up by the space in which electrons orbit. To get an idea of the relative dimensions of an atom, consider a bee in the center of the baseball field at Toronto's SkyDome. The bee would represent the size of the nucleus of the atom, while the rest of the stadium would represent the part of the atom in which the electrons are found.

Isotopes *have the same number of protons but different numbers of neutrons.*

Figure 6.2

The nucleus of an atom can be compared to a bee in the center of the SkyDome.

Figure 6.3

Hydrogen, helium, and lithium atoms.

Figure 6.4

Three forms of hydrogen.

Table 6.1 The Subatomic Particles of Atoms

Particle	Relative mass	Relative charge
Electron	1/1836	−1
Proton	1	+1
Neutron	1	0

The number of protons and neutrons in an atom determines its *atomic mass.* Hydrogen, with one proton and no neutrons, has an atomic mass of 1. Helium, with two protons and two neutrons, has an atomic mass of 4. Figure 6.3 shows three different atoms. The proton number, or *atomic number,* defines the type of element. Any element with two protons is helium and only helium atoms have two protons. Similarly, any atom that has three protons must be lithium and only lithium atoms have three protons. Each chemical element is made up of a different type of atom.

Each element is represented by a symbol, as shown in Figure 6.3. Lithium is represented by the symbol Li, hydrogen by H, and helium by He. Did you notice that the number of protons is always equal to the number of electrons? The reason is that atoms are not charged. The number of neutrons, however, can vary even within the same element. Figure 6.4 shows three different forms of hydrogen that exist because of differences in atomic mass. Protium, deuterium, and tritium all have one proton and therefore are forms of hydrogen. Each form has a different number of neutrons. Deuterium, with an atomic mass of 2, has one proton and one neutron; tritium, with an atomic mass of 3, has one proton and two neutrons. Deuterium and tritium are often described as heavy hydrogen. Atoms that have the same atomic number, but different atomic mass, are called **isotopes.** Protium, the most common type of hydrogen, comprises 99.985% of naturally occurring hydrogen. A few atoms of the deuterium isotope can be expected to be found in nature. That is why atomic masses are not whole numbers. Atomic mass is calculated by averaging the most commonly occurring form of the element with all of its less common isotopes. The atomic mass of hydrogen is actually 1.0079.

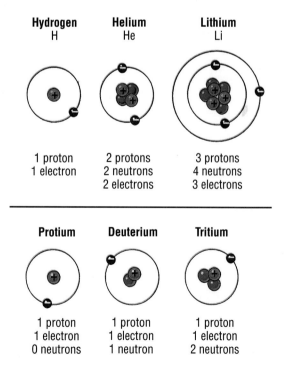

Most elements have at least one isotope. Tin has 11 different isotopes. Many isotopes are not stable and disintegrate spontaneously. These isotopes give off radiation as they decay. During decay, subatomic particles escape from the nucleus, and bursts of energy are emitted. For example, radioactive uranium is converted into lead after it decays. Not surprisingly, these isotopes are referred to as *radioactive isotopes*.

A device called a *Geiger counter* can detect the radiation released during radioactive decay. This permits scientists to follow the movement of radioactive molecules, making them ideal as tracers. For example, by adding a radioactive isotope of chromium to a red blood cell, scientists can follow the movement of the blood cell through the organs of the body. The tracer even allows scientists to monitor the life span of the cells.

BIOLOGY CLIP

How fast do electrons move? One estimate places their speed at about 13 000 km/h. An electron moving at this incredible speed would orbit the nucleus billions of times every second.

Chemical Bonding

The electron holds the key to understanding why some chemicals combine with others. If you look again at the diagram of hydrogen, helium, and lithium in Figure 6.3, you will notice that a maximum of two electrons fill the first electron cloud, or energy level. The third electron of lithium is found in an outer energy level. Table 6.2 presents some of the common elements that you will study in this unit.

Why do some elements react quickly, while others tend not to take part in any chemical reactions? To find an answer you must look to the group of elements called the *noble gases*. Helium, neon, and argon are examples of noble gases. The noble gases do not react with other elements. The stability of this group of elements can be explained by the fact that they contain a full complement of electrons in their outermost energy

Table 6.2 Common Elements

Element	Symbol	Atomic number	Electron number	Number of electrons in each energy level		
				1	*2*	*3*
Hydrogen	H	1	1	1	0	0
Helium	He	2	2	2	0	0
Lithium	Li	3	3	2	1	0
Beryllium	Be	4	4	2	2	0
Boron	B	5	5	2	3	0
Carbon	C	6	6	2	4	0
Nitrogen	N	7	7	2	5	0
Oxygen	O	8	8	2	6	0
Fluorine	F	9	9	2	7	0
Neon	Ne	10	10	2	8	0
Argon	Ar	18	18	2	8	8

level. Helium has two electrons in the outermost energy level, while neon and argon both have eight electrons in their outermost energy levels. Scientists discovered that elements without a full complement of electrons in the outermost energy level lose or gain electrons until the outermost energy level is full. Consider the options for the atoms in Figure 6.5.

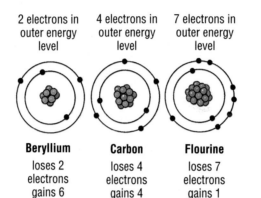

2 electrons in outer energy level	4 electrons in outer energy level	7 electrons in outer energy level

Beryllium	**Carbon**	**Flourine**
loses 2 electrons gains 6 electrons	loses 4 electrons gains 4 electrons	loses 7 electrons gains 1 electron

Figure 6.5

Elements lose or gain electrons from their outer energy level.

Beryllium can become stable by either losing two electrons or gaining six electrons. Which situation do you think is more likely? It is much easier to give up two electrons to another atom than it is to gain six electrons. Notice that once the beryllium gives up two electrons from its outermost energy level, the inner level acts as the outermost shell. This energy level carries a full complement of electrons. Beryllium is considered to be an *electron donor*. Carbon has two options: it can either lose four electrons or gain four electrons. Carbon is neither a strong electron donor nor a strong electron acceptor. Fluorine can either lose seven electrons or gain one electron. The fact that it is more likely to gain one electron makes fluorine a strong *electron acceptor*.

Atoms that have gained or lost electrons have an imbalance of charged particles. Atoms that donate electrons have an excess of positive charge. These atoms become positive **ions**. Atoms that accept electrons have an excess of negative charge. These atoms become negative ions. Electron acceptors will accept electrons from strong electron donors. In the process, both atoms become stable.

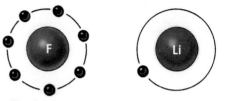

The fluorine atom has seven electrons in its outermost energy level, while the lithium atom has only one electron in its outermost level.

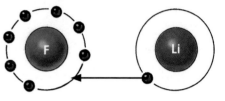

The lithium atom donates an electron to the fluorine atoms.

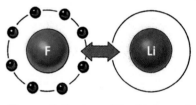

The negatively charged flouride ion and positively charged lithium ion are drawn to each other by charge attraction.

Figure 6.6

Fluorine and lithium combine to form the compound lithium fluoride.

Chemical compounds are formed when two or more elements are joined. For example, lithium and fluorine combine to form the compound lithium fluoride, as shown in Figure 6.6. Lithium fluoride is classified as an ionic compound because it was formed from a metal (lithium) and a nonmetal (fluorine). The bond that joins the lithium and fluoride ions is referred to as an **ionic bond.** Ionic bonds are formed when electrons are transferred from strong electron donors to strong electron acceptors.

Ions *are atoms that have either lost electrons or gained electrons to become positively or negatively charged.*

Chemical compounds *are formed when two or more elements are joined by chemical bonds.*

Ionic bonds *are formed when electrons are transferred between two atoms.*

Unit Two:
Exchange of Energy and Matter in Cells

Another type of chemical bond is formed when electrons are shared rather than transferred. This type of bond is referred to as a **covalent bond.** Carbon, for example, which can either gain or lose four electrons, is capable of forming covalent bonds with other atoms.

Polar Molecules

Although electrons are shared between two atoms in a covalent bond, the sharing may not be equal. Water molecules are created by covalent bonds that join one oxygen and two hydrogen atoms. The electrons are drawn closer toward the oxygen atom, creating a region of negative charge near the oxygen end of the compound and a region of positive charge near the hydrogen end of the molecule. Despite the fact that the entire molecule is balanced in the number of positive and negative charges, the molecule has a positive pole and a negative pole. It is for this reason that water is referred to as a **polar molecule.**

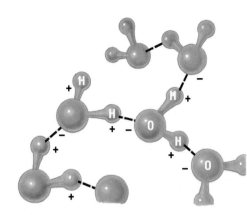

which has a single proton. Attraction between opposing charges of different polar molecules creates a special **hydrogen bond.** Hydrogen bonds pull polar molecules together.

Hydrogen bonding accounts for the strength of fibers in wood. It also helps explain some of the physical characteristics of water. Water boils at $100°C$ and freezes at $0°C$. By comparison, sulfur dioxide, a molecule of similar size, boils at $62°C$ and freezes at $-83°C$. The higher boiling point and melting point of water can be explained by the hydrogen bonds that help hold groups of water molecules together. These hydrogen bonds must be broken before water molecules can escape into the air. This requires energy in the form of heat. Molecules such as sulfur dioxide and carbon dioxide do not have hydrogen bonds and, consequently, require less energy to boil.

Covalent bonds *are formed when electrons are shared between two or more atoms.*

Polar molecules *are molecules that have positive and negative ends.*

Hydrogen bonds *are formed between a hydrogen proton and the negative end of another molecule.*

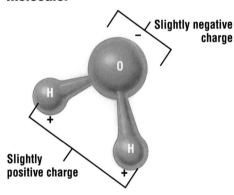

Slightly negative charge

Slightly positive charge

Figure 6.7
The electrons are pulled closer toward the oxygen end of the molecule. The single proton of the hydrogen causes a positively charged end near the hydrogen.

The negative end of a water molecule repels the negative end of another water molecule, but attracts its positive end

ACIDS AND BASES

Chemical reactions occur within the water of the cells and tissues. As mentioned before, water exists as two atoms of hydrogen attached to an atom of oxygen. However, a small number of water molecules dissociate into two separate ions: a positive hydrogen ion and a negative hydroxide ion. The equation below shows the dissociation of water into ions:

$$H_2O \longrightarrow H^+ + OH^-$$

water   hydrogen   hydroxide
          ion   ion

The number of hydrogen ions is equal to the number of hydroxide ions in water. A solution in which the number of hydrogen ions equals the number of hydroxide ions is called a *neutral* solution. Solutions in which the concentration of hydrogen ions is greater than the number of hydroxide ions are called **acids.** For example, stomach juices contain high amounts of hydrochloric acid (HCl). Hydrochloric acid dissociates into positive hydrogen ions and negative chloride ions. Note that the addition of HCl to water would bring about an imbalance between hydrogen and hydroxide ions. The reaction below shows the dissociation of HCl in water:

$$HCl \longrightarrow H^+ + Cl^-$$

Bases are formed when the concentration of hydroxide ions is greater than the concentration of hydrogen ions. Sodium hydroxide, a base, dissociates into hydroxide and sodium ions. Should sodium hydroxide be added to water, the number of hydroxide ions would be greater than the number of hydrogen ions. The reaction below shows the dissociation of sodium hydroxide:

$$NaOH \longrightarrow Na^+ + OH^-$$

The **pH scale** measures the concentration of acid and base solutions. The concentration of hydrogen ions in solution is used to measure the pH. The most useful part of the scale ranges from 1, indicating a very strong acid, to 14, indicating a very strong base. Water has a pH of 7, indicating a neutral solution— the

concentration of hydrogen ions is equal to the concentration of hydroxide ions. The concentration of both ions would be expressed as 1×10^{-7} mol/L. The pH is calculated from the negative logarithm of the hydrogen ion concentration. A pH of 6, therefore, represents a hydrogen ion concentration of 1×10^{-6} mol/L. A solution that has a pH of 6 has a hydrogen ion concentration 10 times greater than that of a solution with a pH of 7.

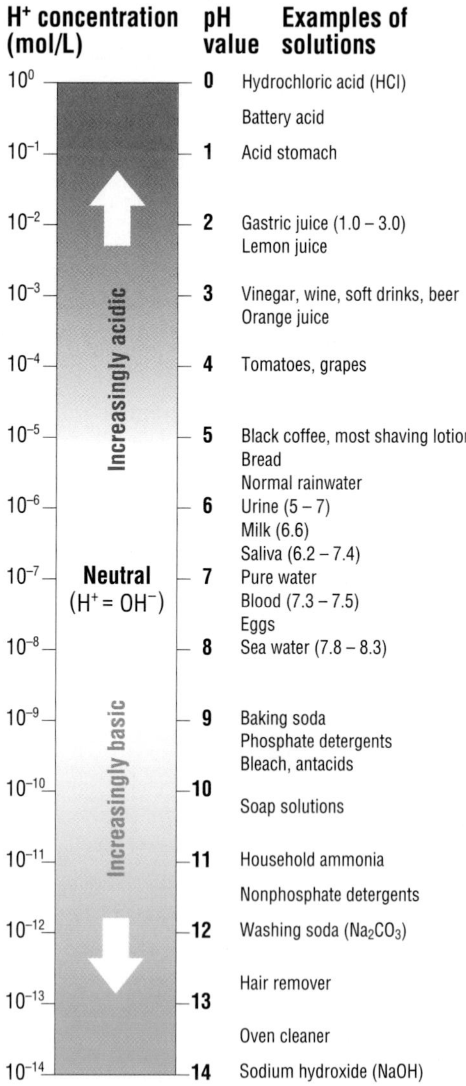

H$^+$ concentration (mol/L)	pH value	Examples of solutions
10^0	0	Hydrochloric acid (HCl)
		Battery acid
10^{-1}	1	Acid stomach
10^{-2}	2	Gastric juice (1.0 – 3.0) Lemon juice
10^{-3}	3	Vinegar, wine, soft drinks, beer Orange juice
10^{-4}	4	Tomatoes, grapes
10^{-5}	5	Black coffee, most shaving lotion Bread Normal rainwater
10^{-6}	6	Urine (5 – 7) Milk (6.6) Saliva (6.2 – 7.4)
10^{-7}	7 Neutral (H$^+$ = OH$^-$)	Pure water Blood (7.3 – 7.5)
10^{-8}	8	Eggs Sea water (7.8 – 8.3)
10^{-9}	9	Baking soda Phosphate detergents Bleach, antacids
10^{-10}	10	Soap solutions
10^{-11}	11	Household ammonia Nonphosphate detergents
10^{-12}	12	Washing soda (Na$_2$CO$_3$)
10^{-13}	13	Hair remover Oven cleaner
10^{-14}	14	Sodium hydroxide (NaOH)

Increasingly acidic / Neutral / Increasingly basic

Figure 6.9

Each increase of 1 in the pH scale actually represents a tenfold decrease in the concentration of hydrogen ions.

Acids *are substances that release hydrogen ions in solution.*

Bases *are substances that release hydroxide ions in solution.*

The **pH scale** *is used to measure the concentrations of acids and bases.*

Neutralization *occurs when the pH is brought to 7, i.e., the [H⁺] = [OH⁻].*

Bases can be **neutralized** by acids. This means that the excess number of hydroxide ions can be eliminated by the addition of more hydrogen ions. Once the concentration of hydrogen ions is equal to that of the hydroxide ions, the base becomes neutral. Similarly, acids can be neutralized by bases. However, if an excess amount of base is added to the acid solution, the concentration of hydroxide ions will exceed the number of hydrogen ions, and the acid will be converted into a base. The following chemical reaction shows how hydrochloric acid can be neutralized by the addition of sodium hydroxide:

$$HCl + NaOH \longrightarrow \underset{\substack{salt \\ (neutral\ pH)}}{NaCl} + \underset{water}{HOH}$$

BUFFERS

Different cells work at various levels of acidity. Chemical reactions in the cells of the stomach work best at a pH between 2.5 and 3.0, while chemical reactions in cells of the small intestine operate best at a pH of 8. A sharp drop in pH could cause stomach or intestinal ulcers, while an increase beyond the normal range would slow digestion and cause a variety of problems. The pH must be maintained within a rather narrow range. But how can the body maintain a constant pH when the foods you eat contain such varying pH levels? The maintenance of a constant pH is accomplished by a system of **buffer** pairs found in every living cell of the body. One of the buffer pairs absorbs excess amounts of acid, while the other buffer pair absorbs any excess bases. This prevents large changes in pH.

It is important not to confuse buffering with neutralization. Buffers only absorb excess acids or bases. They are not designed to bring the pH to 7 but, rather,

to maintain a constant pH. The pH of the blood, for example, must be maintained between 7.35 and 7.45; a lowering of the blood pH to 7 would cause death.

■ REVIEW QUESTIONS ■ ?

1 Define the following terms: atom, element, molecule, compound, and ion.
2 Element X has 6 protons, 8 neutrons, and 6 electrons. Predict the atomic mass of element X.
3 How does carbon-14 ($^{6}_{14}C$) differ from carbon-12 ($^{6}_{12}C$)?
4 Differentiate between an ionic bond and a covalent bond.
5 What are polar molecules?
6 How does hydrogen bonding help explain the different boiling and freezing points of water and carbon dioxide?
7 How do acids differ from bases?
8 How is the buffering of an acid different from its neutralization?

ORGANIC CHEMISTRY

The element carbon provides the structural framework of all living things. The fact that carbon is neither a strong electron acceptor nor a strong electron donor enables it to form covalent bonds with a great many elements. The importance of carbon is underscored by the fact that over two million different carbon compounds have been identified. By comparison, only a half-million compounds without carbon have been identified. Compounds containing carbon are classified as **organic molecules.** Compounds without carbon are classified as *inorganic.*

Organic molecules can be broken down within cells for energy or linked

Buffers *absorb excess acid or base, thereby preventing any significant fluctuation in pH values.*

Organic molecules *are compounds that contain carbon.*

together to form large macromolecules essential for life. Table 6.3 summarizes some of the organic compounds that you will study in this chapter.

Table 6.3 Some Important Classes of Organic Compounds

Class	Some characteristic functional groups		Occur in
Hydrocarbons	methyl (— CH$_3$)	H | — C — H | H	fats, oils, waxes
Alcohols	hydroxyl	— OH	sugars
Carbonyls	aldehyde (— CHO)	O || — C — H	sugars
	ketone ($>$C = O)	O || — C — C — C	sugars
	carboxyl (— COOH)	O || — C — OH	sugars, fats, amino acids
Amines	amino (— N, — NH, — NH$_2$)	H | — N — H	amino acids, proteins
Phosphate compounds	phosphate (—Ⓟ))	O || — P — O | O	energy carriers (such as ATP)

CARBOHYDRATES

Carbohydrates are the body's most important source of energy. However, the human body is not able to make these vital chemicals by itself. You rely on plants as your source of carbohydrates. Using the energy provided by the sun, plants combine carbon dioxide and water to synthesize carbohydrates by photosynthesis.

Carbohydrates are either single sugar units or **polymers** of many sugar units. Single sugar units contain carbon, hydrogen, and oxygen in a 1:2:1 ratio. For example, *triose* sugars have the molecular formula $C_3H_6O_3$, and *hexose* sugars have the molecular formula $C_6H_{12}O_6$. The

word *triose* refers to the fact that the sugars have a three-carbon chain (the prefix *tri-* means three). Hexose sugars contain six-carbon-chain sugars (the prefix *hex-* means six). Many of the most important sugars contain either three-, five-, or six-chain sugars. Those that contain more than five carbons are often arranged in a ring form.

Common sugars like glucose, found in human blood; fructose, a plant sugar commonly found in fruits; and deoxyribose, a sugar component of the large DNA molecule, can be identified as sugars by the *-ose* suffix. Even the large molecule cellulose, which makes up plant cell walls, is a carbohydrate.

Carbohydrates can also be classified according to the number of sugar units they contain. **Monosaccharides** are the simplest of sugars, containing single sugar units. Glucose and fructose are two of the more common monosaccharides. Figure 6.10 shows that glucose and fructose are **isomers**—that is, both molecules have the same molecular formula, $C_6H_{12}O_6$, but different structural arrangements. The different chemical properties of fructose and glucose can be explained by their different structural arrangements. Fructose is much sweeter than glucose and is often used by food manufacturers to sweeten their products. Both glucose and fructose rotate between the straight-chain form and the ring structure, shown in Figure 6.10.

The combination of two monosaccharides forms a **disaccharide.** Sucrose (white table sugar) is a disaccharide formed from glucose and fructose. Sucrose is extracted from plants such as sugar cane and sugar beet. Maltose (malt sugar) is a disaccharide formed from two glucose units. Maltose is commonly found in the seeds of germinating plants. Lactose (milk sugar) is composed of glucose and galactose units. All disaccharides

Polymers *are molecules composed of from three to several million subunits. Many polymers contain repeating subunits.*

Monosaccharides *are single sugar units. All monosaccharides include carbon, hydrogen, and oxygen in a ratio of 1:2:1.*

Isomers *are chemicals that have the same chemical formula but a different arrangement of molecules.*

Disaccharides *are formed by the joining of two monosaccharide subunits.*

Dehydrolysis synthesis *is the process by which larger molecules are formed by the removal of water from two smaller molecules.*

are formed by a process called **dehydrolysis synthesis,** in which a water molecule is extracted from the two monosaccharide sugar molecules.

Polysaccharides are carbohydrates formed by the union of many monosaccharide subunits. **Starch,** for example, is a plant polysaccharide that is composed of multiple subunits of glucose. Plants store energy as starch. Sugars formed in the leaves of potatoes, for example, are linked together into double sugar molecules and transported to the roots for storage. Once in the roots, the double sugar units are linked to form the larger starch molecules. Starches can exist in two different forms: *amylose* and *amylopectin.* The amylose molecules contain up to 1000 or more glucose units with the first carbon of a glucose molecule linked to the fourth carbon in the next molecule. The starch molecules tend to bend in the shape of a helix, or a coil. The amylopectins contain between 1000 and 6000 glucose subunits and have short branching chains of between 24 and 36 glucose units extending from the main branch.

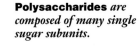

Polysaccharides *are composed of many single sugar subunits.*

Starch *is a plant storage carbohydrate.*

Figure 6.10

Glucose and fructose are isomers. Glucose is classified as an aldehyde sugar because the carbonyl group (purple) is located on a terminal, or end, carbon group. Fructose is classified as a ketone sugar because its carbonyl group (green) is located on a nonterminal, or middle, carbon molecule.

Figure 6.11

The formation of a disaccharide by dehydrolysis synthesis. Note: Sucrose can only exist in the ring form.

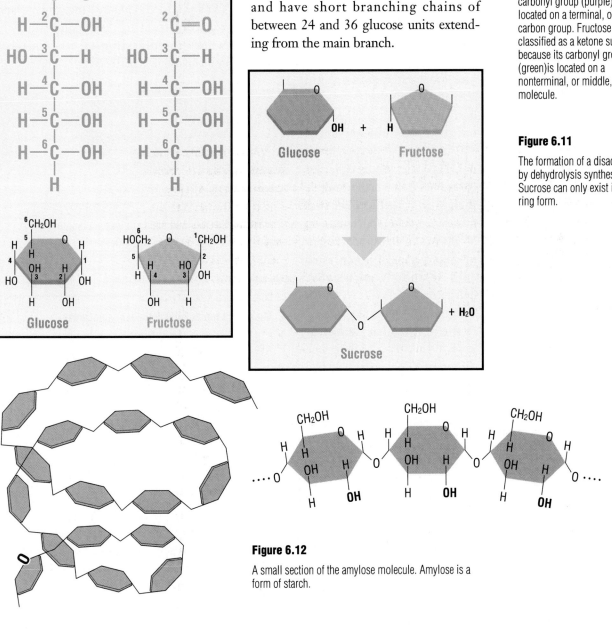

Figure 6.12

A small section of the amylose molecule. Amylose is a form of starch.

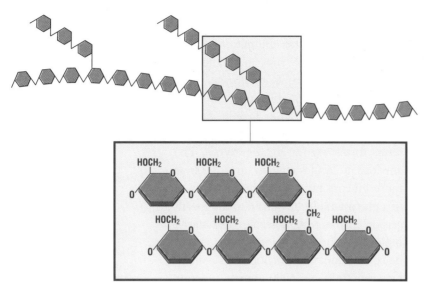

Its unique bonding arrangement gives cellulose distinctive properties. Unlike starch, cellulose cannot be digested by humans. Often referred to as *fiber* or *roughage*, cellulose holds water in the large intestine, thereby aiding the elimination of wastes. Only a few microbes contain enzymes that enable them to break down cellulose. Even cows cannot digest plant cell walls. Tiny microbes housed in the cow's first stomach, the rumen, cleave the bonds between the adjoining sugar molecules. Once the cellulose has been digested, the cow regurgitates the food, chews it once again, and then diverts it to the second stomach, where the digestion of other macromolecules takes place.

Figure 6.13

The branched structure of the glycogen molecule is similar to that of some forms of starch.

Figure 6.14

Cellulose is composed of many glucose subunits. Hydrogen bonds bring together multiple chains of cellulose.

Glycogen *is the form of carbohydrate storage in animals. Glycogen is often called the animal starch.*

Cellulose *is a plant polysaccharide that makes up plant cell walls.*

Animals store carbohydrates in the form of a polysaccharide called **glycogen.** The structure of glycogen resembles that of the amylopectin starch molecule, except that its branching structures contain only 16 to 24 glucose units. The excess sugars carried by the blood are linked together to form glycogen, which is then stored in the liver and muscles. As the concentration of glucose in the blood begins to drop, glycogen is converted back to individual glucose units.

Plant cell walls are made up of the polysaccharide **cellulose.** Over 50% of all organic carbon in the biosphere is tied up as cellulose. Cellulose molecules, like starch and glycogen, are composed of many glucose subunits. However, the bonding of the linking oxygen atoms differs between starch and cellulose; cellulose tends not to form coiled structures. The many layers of cellulose are attracted to one another by hydrogen bonds between the —OH groups.

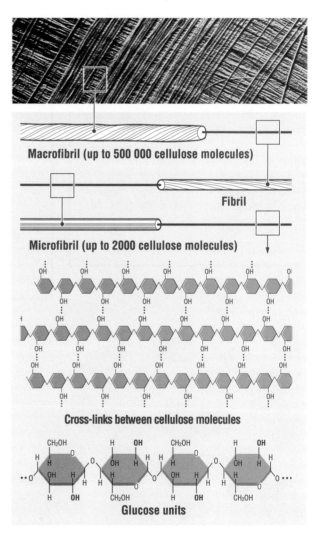

Macrofibril (up to 500 000 cellulose molecules)

Fibril

Microfibril (up to 2000 cellulose molecules)

Cross-links between cellulose molecules

Glucose units

LABORATORY
IDENTIFICATION OF CARBOHYDRATES

Objective

To identify reducing sugars qualitatively and quantitatively.

Background Information

Benedict's solution identifies reducing sugars, and iodine solution identifies starches. Iodine turns blue-black in the presence of starches. The Cu^{2+} ions in the Benedict's solution are converted to Cu^+ ions if they react with a reducing sugar. Not all sugars are reducing sugars. All monosaccharides are reducing sugars, but some disaccharides will not react with Benedict's solution.

The chart summarizes the quantitative results obtained when a reducing sugar reacts with Benedict's solution:

Color of Benedict's solution	Approximate % of sugar
Blue	negative
Light green	0.5% – 1.0%
Green to yellow	1.0% – 1.5%
Orange	1.5% – 2.0%
Red to red brown	+ 2.0%

Materials

safety goggles	5% maltose solution
lab apron	5% starch solution
test-tube brushes	8 test tubes
detergent	test-tube rack
400 mL beaker	Benedict's solution
hot plate	wax pencil
thermometer	test-tube clamp
10 mL graduated cylinder	5 medicine droppers
distilled H_2O	depression plates
5% fructose solution	iodine solution
5% sucrose solution	solutions X, Y, and Z
5% glucose (dextrose) solution	

> **CAUTION:** The chemicals used are toxic and irritant. Avoid skin and eye contact. Wash all splashes off your skin and clothing thoroughly. If you get any chemical in your eyes, rinse for at least 15 min and inform your teacher.

Procedure

Before you begin:
- Make sure that all the glassware is clean and well rinsed.
- Note the location of the eyewash station.

Part I : Reducing Sugars

1 Prepare a water bath by heating 300 mL of tap water in a 400 mL beaker. Heat the water until it reaches approximately 80°C. (Use the thermometer to monitor the temperature.)

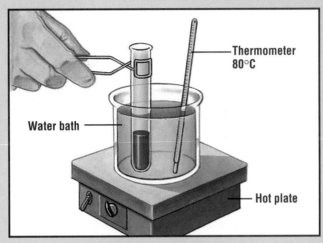

2 Using a 10 mL graduated cylinder, measure 3 mL each of distilled water, fructose, glucose, maltose, sucrose, and starch solutions. Pour each solution into a separate test tube. Clean and rinse the graduated cylinder after each solution. Add 1 mL of Benedict's solution to each of the test tubes and label using the wax pencil.

   a) Why should the graduated cylinder be cleaned and rinsed after the measurement of each solution?

3 Using a test-tube clamp, place each of the test tubes in the hot water bath. Observe for six minutes.

   b) Record any color changes in a chart like the one below.

Solution	Initial color	Final color	% sugar
Water	Blue		
Glucose			
Fructose			
Maltose			
Sucrose			
Starch			

4 Using a medicine dropper, place a drop of water on a depression plate and add a drop of iodine.
  c) Record the color of the solution.

5 Repeat the procedure, this time using drops of starch, glucose, maltose, and sucrose instead of water.
  d) Record the color of the solutions. Which solutions indicate a positive test?

Part III: Checking Unknown Solutions

6 Test the unknown solutions for reducing sugars and starches. Design your own table, showing both qualitative and quantitative data.
  e) Record your data.

Laboratory Application Questions

1 Which test tube served as a control in the test for reducing sugars and starches?

2 What laboratory data suggest that not all sugars are reducing sugars?

3 A student decides to sabotage the laboratory results of his classmates and places a sugar cube into solution Z. Explain the effect of dissolving a sugar cube in the solution.

4 A drop of iodine accidently falls on a piece of paper. Predict the color change, if any, and provide an explanation for your prediction. ■

LIPIDS

Triglycerides are lipids composed of glycerol and three fatty acids.

Fats are animal lipids composed of glycerol and saturated fatty acids.

Oils are plant lipids composed of glycerol and unsaturated fatty acids.

Phospholipids are the major components of cell membranes in plants and animals. Phospholipids have a phosphate molecule attached to the glycerol backbone, making the molecule polar.

Lipids are nonpolar compounds that are insoluble in polar solvents like water. You may have noticed how fat floats on the surface of water while you are doing dishes. Many lipids are composed of two structural units: glycerol and fatty acids. Like complex carbohydrates, glycerol and fatty acids can be combined by dehydrolysis synthesis. An important function of lipids is the storage of energy. Glycogen supplies are limited in most animals. Once glycogen stores have been built up, excess carbohydrates are converted into fat. This helps explain why the eating of carbohydrates can cause an increase in fat storage. Other lipids serve as key components in cell membranes, act as cushions for delicate organs of the body, serve as carriers for vitamins A, D, E, and K, and act as the raw materials for the synthesis of hormones and other important chemicals. A layer of lipids at the base of the skin insulates you against the cold. The thicker the layer of fat, the better the insulation. Taking their cue from whales, seals, and other marine mammals,

marathon swimmers often coat their bodies with a layer of fat just before entering cold water.

Triglycerides are formed by the union of glycerol and three fatty acids. Triglycerides that come from animals are often called **fats.** Most of the fatty acids in animal fats are said to be *saturated.* This means that only single covalent bonds exist between the carbon atoms. Because the single covalent bonds tend to be stable, animal fats are difficult to break down. Animal fats are usually solid or semi-solid at room temperature. Plant triglycerides that are liquid are called **oils.** The fatty-acid components of plants have some double bonds between carbon atoms. Plant oils are often described as *polyunsaturated.* The unsaturated double bonds are somewhat reactive, and, therefore, plant oils are more easily broken down than animal fats.

A second group of lipids, called **phospholipids,** have a phosphate molecule attached to the glycerol backbone of the molecule. The negatively charged phosphate replaces one of the fatty acids, providing a polar end to the lipid. The polar end of phospholipids is soluble in water, while the nonpolar end is insoluble. These special properties make phospholipids well suited for cell membranes.

Waxes make up a third group of lipids. In waxes, long-chain fatty acids are joined to long-chain alcohols or to carbon rings. These long, stable molecules are insoluble in water, making them well suited as a waterproof coating for plant leaves or animal feathers and fur.

Fats and Diet

Almost everyone likes the taste of french fries. Unfortunately, when you eat french fries, you are also eating the fat they were cooked in. Although fats are a required part of your diet, problems arise when you consume too much fat. Doctors recommend that no more than 30% of total energy intake should be in the form of fats.

Fats are concentrated packages of energy containing more than twice as much energy as an equivalent mass of carbohydrate or protein. By eating 100 g of fat, you take in about 3780 kJ of energy. (The kilojoule, kJ, is a unit used to measure food energy.) By comparison, 100 g of carbohydrates or protein yield 1680 kJ of energy. When energy input or consumption exceeds energy output, the result is a weight gain. If you were to drink a 350 mL chocolate milkshake containing about 2200 kJ of energy, you would have to jog for about one hour to burn off the energy taken in.

Heart disease has been associated with diets high in saturated fats. Recall that the single bonds between the carbon molecules make the fats stable. The stable fats tend to remain intact inside the cells of the body much longer than more reactive macromolecules. High-fat diets and obesity have also been linked to certain types of cancer, such as breast cancer, cancer of the colon, and prostate cancer. Obesity has also been linked to high blood pressure and adult diabetes. According to one report, over 80% of people with adult-onset diabetes are overweight.

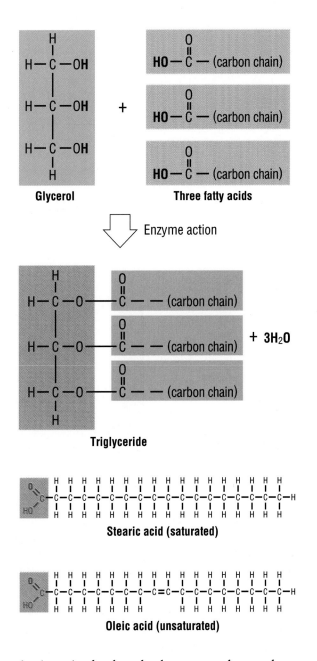

Triglyceride

Stearic acid (saturated)

Oleic acid (unsaturated)

Figure 6.15

Triglycerides are formed by the union of glycerol and three fatty acids. Note the removal of water in the synthesis. The terms monoglyceride or diglyceride are used to describe the joining of glycerol with one or two fatty acids respectively.

Figure 6.16

Saturated fats do not have double bonds between carbon atoms. Therefore, these compounds tend to be stable.

Waxes *are long-chain lipids that are insoluble in water.*

The Cholesterol Controversy

Heart disease, the number-one killer of North Americans, can be caused by the accumulation of cholesterol in the blood vessels. Scientific research about cholesterol has changed direction in recent years. Lipid-rich foods, such as fish and olive oil, were once thought to raise blood cholesterol levels. Currently, most scientists believe that these foods may actually reduce blood cholesterol levels. Similarly, alcohol, in moderate consumption may contribute to a decrease in blood cholesterol levels. Added to this confusion is the fact that genes play a major role in determining cholesterol levels. Research indicates that people with a certain genetic makeup are predisposed to **atherosclerosis.** Individuals who are not susceptible to atherosclerosis can tolerate higher levels of cholesterol in their blood.

Not all cholesterol is bad. Cholesterol is found naturally in cell membranes and acts as the raw material for the synthesis of certain hormones. The difference between males and females would be less pronounced if it were not for cholesterol—sex hormones are made from cholesterol.

All of the cholesterol that the body needs can be obtained from the fats consumed. The cells of the body package cholesterol in a water-soluble protein in order to transport it in the blood. Scientists have identified two very important types of cholesterol packaging: low-density lipoproteins and high-density lipoproteins. Low-density lipoproteins, or LDLs, are considered to be "bad" cholesterol. About 70% of cholesterol intake is in the form of LDLs. High levels of LDLs have been associated with the clogging of arteries—LDL particles bind to receptor sites on cell membranes and are removed from the blood (principally by the liver). However, as the levels of LDLs increase and exceed the number of receptor sites, excess LDL-cholesterol begins to form deposits on the walls of arteries. The accumulation of cholesterol and other lipids on the artery walls is known as *plaque*. Unfortunately, plaque restricts blood flow to the heart and brain and can lead to heart attack or stroke.

The second type of cholesterol—high-density lipoproteins, or HDLs—is often called "good" cholesterol. The HDLs carry bad cholesterol back to the liver, which begins breaking it down. The HDLs lower blood cholesterol. Most researchers now believe that the balance between LDL and HDL is critical in assessing the risks of cardiovascular disease. A desirable level of HDL is 35 mg/100 mL of blood or higher. Some researchers believe that exercise increases the level of HDLs. Strong evidence also supports the theory that fiber or cellulose in the diet helps reduce cholesterol. It is believed that fiber binds to cholesterol in the gastrointestinal tract. However, it should be pointed out that fiber does not affect everyone the same way.

The controversy over the ideal diet continues. The most important thing to keep in mind is maintaining a balance between food intake and energy needs.

BIOLOGY CLIP

Have you ever gone on a diet only to discover that after days of reducing your food intake, you have actually gained weight? This can occur because as fat is used up by the body, it is replaced by water. Since water weighs more than fat, you may experience a temporary gain in weight. Eventually, the water will leave the tissues and your body weight will decrease.

Atherosclerosis is a disorder of the blood vessels characterized by the accumulation of cholesterol and other fats along the inside lining.

RESEARCH IN CANADA

A research group consisting of Drs. Norman Boyd, Leslie Levin, Barbara Heartwell, and Dianne Goettler is conducting a study to determine whether breast cancer can be reduced by a decrease in dietary fat and an increase in dietary carbohydrate.

Strong evidence suggests that the development of breast cancer is influenced by environmental factors. The incidence of breast cancer around the world varies, being five times greater in North America and Western Europe than in Japan and most of Asia. The Ontario research group believes that the differences are not genetic, since Asians who have emigrated to high-risk countries have displayed a similar predisposition to breast cancer. According to the research group, the differences in breast cancer rates can be linked to the consumption of dietary fats. Evidence gained from animal research supports their hypothesis.

The research group will follow 9500 high-risk patients at centers in Canada and the United States for 10 years to determine if low-fat diets reduce the incidence of breast cancer.

A Low-Fat Diet to Prevent Breast Cancer

DR. LESLIE LEVIN
DIANNE GOETTLER

PROTEINS

Proteins are the structural components of cells. Cytoplasmic organelles like the mitochondria and ribosomes are composed largely of protein. The predominant part of muscles, nerves, skin, and hair is protein. *Antibodies* are specialized proteins that help the body defend itself against disease; *enzymes* are proteins that speed chemical reactions. Proteins are essential for the building, repair, and maintenance of cell structure. Like lipids and carbohydrates, proteins are composed of carbon, hydrogen, and oxygen. However, proteins contain nitrogen and, most often, sulfur molecules as well. Like sugars and lipids, proteins can supply energy for the tissues, although energy production is not their main function.

The diversity among people and among different species can, in part, be explained by proteins. A limited number of carbohydrates and lipids are found in all living things, but the array of proteins is almost infinite. Proteins are composed of 20 different amino acid building blocks. With a change in position of a single amino acid, the structure of a protein can be altered. The diagram below shows a generalized plan for an amino acid. The —NH$_2$ group is referred to as the amino group and the —COOH group as the acid group. The R group can represent a number of different structures.

Small protein molecules are composed of as few as eight amino acids, while large molecules often contain thousands of amino acids. The order and number of amino acids determine the type of protein. Fish protein is distinctively different from cow protein and human protein. Did you ever hear anyone say, "I eat so much fish, I'm going to turn into one"? By consuming fish protein, you *are* taking in the structural characteristic of fish; however, fish protein does not remain as fish protein once it is digested. Following digestion and absorption, the individual amino acids are carried in the blood to the cells of your body. Your cells reassemble the amino acids from the fish to make human proteins. The sequencing of amino acids is regulated by the genes located on your chromosomes. Without the digestion and restructuring of amino acids into human proteins, your diet would change your appearance. By eating a fish, you would begin to take on the structural appearance of a fish. The implications of consuming poultry, beef, or soya bean protein would be equally frightening! Figure 6.17 shows how amino acids are joined. As in carbohydrate and lipid synthesis, a water molecule is removed during the synthesis of protein. The covalent bond that forms between the acid group of one amino acid and the amino group of the adjoining amino acid is called a **peptide bond.** For this reason, proteins are often referred to as **polypeptides.**

The importance of proteins in the diet cannot be underestimated. Although the body is capable of making many of the amino acids, there are eight amino acids that the body cannot synthesize. These eight amino acids, called *essential amino acids,* must be obtained from your foods. The lack of any one of the essential amino acids will lead to specific protein deficiencies and disease.

Peptide bonds *join amino acids.*

Polypeptides *are proteins. They are composed of amino acids joined by peptide bonds.*

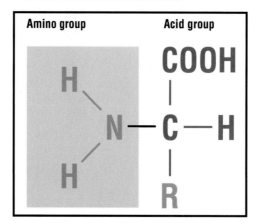

Amino group   Acid group

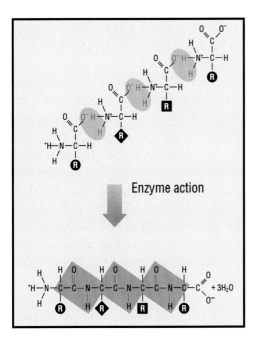

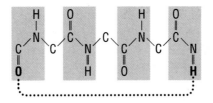

Figure 6.17

Dehydrolysis synthesis of amino acids to form a protein.

Types of Protein

Proteins can be classified into four main groups according to the arrangement of the amino acids. *Primary proteins* have amino acids organized in linear arrangements. Figure 6.18 shows cow insulin, an example of a primary protein. Identified by the British chemist Frederick Sanger in 1953, cow insulin is composed of two chains of amino acids, linked by sulfur bonds. The sulfur bonds are called *disulfide bridges*.

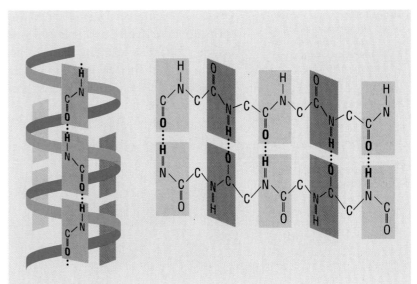

Figure 6.19

Hydrogen bonding pulls long chains into coiled or sheetlike chains.

Secondary proteins are arranged in coils. Hydrogen bonding from the negative end of the acid groups and the positive end of the amino groups pull what would have been long chains into helical coils. Figure 6.19 shows a secondary protein.

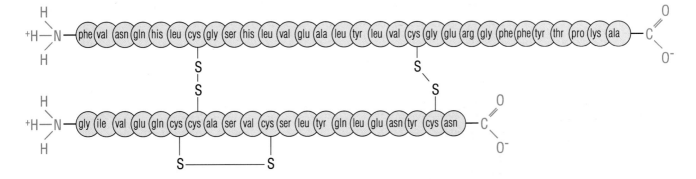

Figure 6.18

Cow insulin is a primary protein.

Figure 6.20

Three-dimensional
representation of the tertiary
protein of a hemoglobin
molecule. The tertiary
polypeptide chain is shown in
red. The oxygen will bond to the
iron-containing pigment (heme
group).

Tertiary proteins result from R group interactions. A single polypeptide chain of hemoglobin, the iron-containing pigment found in red blood cells, is a tertiary protein. The normal coil structure, characteristic of secondary proteins, has been altered by the R groups (see Figure 6.20). *Quaternary proteins* are large globular proteins formed from the interactions between different protein chains. The four protein chains of the hemoglobin molecule interact to form quaternary proteins (see Figure 6.21).

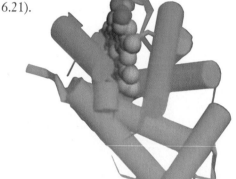

DENATURATION AND COAGULATION

Exposing a protein to excess heat, radiation, or a change in pH will alter its shape. Physical or chemical factors that disrupt bonds, thereby causing changes in the configuration of the protein, are said to **denature** the protein. The protein may uncoil or assume a new shape. The result is a change in the protein's physical properties as well as its biological activities. Once the physical or chemical factor is removed, the protein may assume its original shape.

A permanent change in protein shape is referred to as **coagulation.** The boiling of an egg, for example, causes the shape of proteins to be altered. The proteins in the egg are said to have coagulated, because no matter how much cooling takes place, the proteins of the egg will never assume their original shape.

Figure 6.21

Three-dimensional diagram of
four protein chains found in a
hemoglobin molecule. The
interaction between the protein
chains causes the formation of a
quaternary protein. The special
shape of the hemoglobin permits
the transport of oxygen.

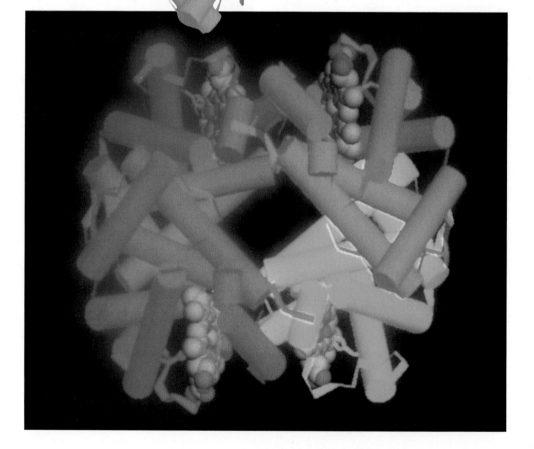

LABORATORY

IDENTIFICATION OF LIPIDS AND PROTEINS

Objective

To identify lipids and proteins.

Background Information

Proteins can be identified by means of the biuret test. The biuret reagent reacts with the peptide bonds that join amino acids together, producing color changes from blue, indicating no protein, to pink (+), to violet (++), and to purple (+++). The (+) sign indicates the relative amounts of peptide bonds. Lipids can be identified by a Sudan IV solution, which is soluble in nonpolar solvents. Lipids turn from a pink to a red color, but polar compounds will not assume the pink color of the Sudan IV solution. Lipids can also be identified by unglazed brown paper. Because lipids allow the transmission of light through the brown paper, the test is often called the *translucence test*.

Materials

safety goggles	8 test-tube stoppers
lab apron	5 medicine droppers
test-tube rack	5% glucose (dextrose)
test-tube brush	solution
8 test tubes	5% gelatin solution
wax pencil	liquid soap
10 mL graduated cylinder	egg albumin
distilled water	liquid detergent
vegetable oil	simulated blood plasma
Sudan IV	solutions X, Y, and Z
unglazed brown paper	biuret reagent

CAUTION: Avoid skin and eye contact. Wash all splashes off your skin and clothing thoroughly. If you get any chemical in your eyes, rinse for at least 15 min and inform your teacher.

Procedure

Part I: Sudan IV Lipid Test

1 Obtain two test tubes. Label one C, for control, and the other T, for lipid test. Add 3 mL of water to test tube C. Add 3 mL of vegetable oil to test tube T.

2 Add 6 drops of Sudan IV solution to each test tube. Place stoppers in the test tubes and shake them vigorously for two minutes.

   a) Record the color of the mixtures.

Part II: Translucence Lipid Test

3 Label a 10-cm square piece of unglazed paper C, for control. Place a drop of water on the paper. Label a second piece of paper T for lipid test. Place a drop of vegetable oil on the paper labelled T.

4 Wave papers C and T in the air until the water from paper C has evaporated. Hold both papers to the light and observe.

   b) Record whether or not the papers appear translucent.

Part III: Biuret Test for Proteins

5 Using a wax pencil, label two test tubes C and T. Add 2 ml of water to test tube C. Add 2 mL of gelatin to test tube T. Add 2 mL of biuret reagent to each of the test tubes, then tap the test tubes with your fingers to mix the contents.

   c) Record any color changes.

Part IV: Testing for the Presence of Protein and Lipid

6 Use the appropriate solution to test the following: 5% glucose(dextrose), 5% gelatin, liquid soap, egg albumin, liquid detergent, simulated blood serum, solution X, solution Y, solution Z.

   d) Record your results.

Laboratory Application Questions

1 Why were controls used for the experiments?

2 A student heats a test tube containing a large amount of protein and notices a color change in the test tube. Explain why heating causes a color change.

3 Explain the advantage of using two separate tests for fats.

4 Would you expect to find starches and sugars in the blood plasma? Explain your answer. ■

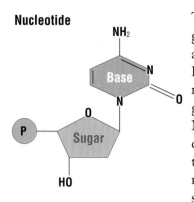

Nucleotide

Nucleotides *are the functional units of nucleic acids.*

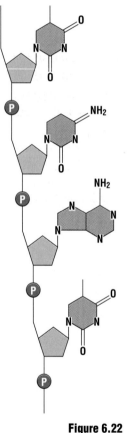

Figure 6.22

Nitrogen bases, shown in purple, are either single-ring molecules (pyrimidines) or double-ring molecules (purines).

NUCLEIC ACIDS

The hereditary material found within the genes of chromosomes comprises yet another type of organic compound. Deoxyribonucleic acid, or DNA, as it is more commonly known, belongs to a group of compounds called nucleic acids. Nucleic acids are nitrogen-containing compounds and are composed of functional units called **nucleotides.** Each nucleotide is composed of a five-carbon sugar unit, a phosphate molecule, and a nitrogen base. DNA nucleotides contain the five-carbon sugar deoxyribose, while RNA nucleotides are made up of the five-carbon sugar ribose .

Nucleic acids are either double-stranded (DNA) or single-stranded (RNA) molecules whose nucleotides are linked by sugar and phosphate molecules. DNA and RNA each have four of the five nitrogen bases attached to their sugar-phosphate backbones. (See Figure 6.22.) Adenine and guanine are purines; cytosine, thymine and uracil are pyrimidines. DNA contains thymine; RNA contains uracil. The sequencing of the nitrogen bases determines the genetic code.

■ REVIEW QUESTIONS ■ ?

9 Differentiate between organic and inorganic molecules. Provide an example of each.

10 Name three classes of carbohydrates.

11 What are the two structural components of triglycerides?

12 In what ways do plant and animal fats differ?

13 Differentiate between monoglycerides, diglycerides, and triglycerides.

14 What is cholesterol, and why has it become so important to health-conscious consumers?

15 State the function of HDLs and LDLs.

16 In what ways do proteins and nucleic acids differ from carbohydrates and lipids?

17 Why do so many different proteins exist?

18 Using examples of carbohydrates, lipids, and proteins, explain the process of dehydrolysis synthesis.

19 Define denaturation and coagulation.

20 What are nucleotides?

ANABOLIC STEROIDS

Steroids are lipids produced by the cells of the body. Many steroids act as chemical messengers, regulating cell function. Anabolic steroids are tissue-building messengers. Most anabolic steroids are either synthetic or natural versions of the male sex hormone, testosterone. Male sex hormones in general are known as *androgens*.

At puberty, the level of testosterone increases dramatically in males, triggering muscle development and other secondary sex characteristics. Androgens from the blood attach themselves to a binding site on muscle cell membranes, causing the muscle cells to grow and develop. Women can increase muscle development by injecting male sex hormones. Although they can grow bigger and stronger, many experience side effects associated with male characteristics.

Males do not escape without side effects either. Males who rely on large doses of testosterone cannot utilize all of the testosterone. Consequently, the excess testosterone is gradually converted into estrogen, the female sex hormone. Although every male contains some estrogen, few contain the same levels as women. Breast enlargement, a shrinking of the testes, and liver and kidney dysfunctions have been associated with steroid use.

SOCIAL ISSUE:
Athletes and Performance-Enhancing Drugs

In 1988, Canadian sprinter Ben Johnson was stripped of his gold medal at the summer Olympics after a drug test indicated anabolic steroids were in his system. At the subsequent Dubin Inquiry, held in Toronto, it was revealed that steroid use was commonplace among the world's top athletes.

Anabolic steroids, stimulants, and painkillers have improved performance levels of athletes. In many respects, today's athletes are a product of better chemistry. However, the use of performance-enhancing drugs raises some important issues. Current studies show that prolonged use of steroids can result in certain side effects, such as the accumulation of plaque inside the blood vessels. On a wider scale, it could be contended that the use of drugs has corrupted athletic competition and that only those countries with the best technology can win medals.

Statement:

Major athletic competitions such as the Olympics have become meaningless because of some athletes' use of performance-enhancing drugs.

Point

- The use of anabolic steroids, stimulants, and painkillers has improved performance levels of athletes. In many respects, these athletes are a product of better chemistry. How can other athletes afford to compete fairly?
- Improvements in drug testing will only lead to the use of better masking agents that disguise the drugs, or to the use of new synthetics that are more difficult to detect. A new peptide hormone, used in place of conventional anabolic steroids, has surfaced already. The peptide hormones are very difficult to detect.

Counterpoint

- The fact that some athletes have misused drugs should not take away from the nature of athletic competition. Improved drug testing at the Olympic games and other major sporting events will curtail the use of performance-enhancing drugs.
- Any chemical can be detected if enough care is exercised during the testing. However, testing must be combined with an education program. If performance-enhancing drugs are not accepted, their use will soon decrease.

Research the issue.
Reflect on your findings.
Discuss the various viewpoints with others.
Prepare for the class debate.

CHAPTER HIGHLIGHTS

- Metabolism is the sum of all chemical reactions that occur within cells. Metabolism can be divided into two main groups: catabolism, which involves the breakdown of macromolecules into smaller sub-units, and anabolism, which involves the synthesis of large macromolecules from their component sub-units.
- Molecules are the units of matter. Molecules are composed of one or more atoms.
- Elements are pure substances that cannot be broken apart into simpler substances. There are 109 elements, each with a specific number of protons.
- Chemical compounds are created by the joining of two or more elements.
- Ionic bonds are formed by the transfer of electrons from an electron donor to an electron acceptor. Ions of opposing charge attract.
- Covalent bonds form when electrons are shared by two or more atoms.
- Acids are solutions that release hydrogen ions in solution. Bases are substances that combine with hydrogen ions. The concentration of an acid or base can be measured by the pH scale.
- Organic compounds contain carbon; inorganic compounds do not contain carbon. Living things are made up of organic compounds.
- Carbohydrates are complex molecules that contain hydrogen, carbon, and oxygen. Carbohydrates are the preferred source of energy for cells. Cells break down carbohydrates readily.
- Lipids are organic compounds formed from glycerol and fatty acids. Lipids are energy-storage compounds.
- Proteins are large molecules constructed of many amino acids. Proteins are the structural components of cells.
- Nucleic acids are composed of carbon, oxygen, hydrogen, nitrogen, and phosphorus. Nucleic acids are the molecules of heredity.
- Steroids are lipids formed from carbon rings. Cholesterol and sex hormones are steroids.

APPLYING THE CONCEPTS

1. Explain why lithium is more likely to react with fluorine than with beryllium or sodium. Use the information provided in the chart below.

Element	Symbol	Atomic number	Electron number	Energy level		
				1	2	3
Helium	He	2	2	2	0	0
Lithium	Li	3	3	2	1	0
Beryllium	Be	4	4	2	2	0
Fluorine	F	9	9	2	7	0
Sodium	Na	11	11	2	8	1

2. Use proteins to provide examples of catabolism and anabolism.

3. Explain why marathon runners consume large quantities of carbohydrates a few days prior to a big race.

4. Indicate some of the symptoms of individuals who are deficient in carbohydrates, proteins, and lipids, respectively.

5. Why is cellulose, or fiber, considered to be an important part of the diet?

6. Why are cows able to digest plant matter more effectively than humans?

7. Margarine is processed by attaching hydrogen atoms to unsaturated double bonds of plant oils. The oil becomes solid or semi-solid. Have you ever noticed that some margarines are stored in plastic tubs, while others are stored in wax paper? Compare the two types of margarine. Which would you recommend?

8. Why are phospholipids well suited for cell membranes?

9. Explain why many physicians suggest that the ratio of HDL-cholesterol to LDL-cholesterol is more significant than the cholesterol level.

10. List some of the side effects of prolonged anabolic steroid use.

CRITICAL-THINKING QUESTIONS

1 A student believes that the sugar inside a diet chocolate bar is actually sucrose because a test with Benedict's solution yields negative results. How would you go about testing whether or not the sugar present is a nonreducing sugar?

2 Three different digestive fluids are placed in test tubes. The fluid placed in test tube #1 was extracted from the mouth. The fluids in test tubes #2 and #3 were extracted from what was believed to be the stomach. Five milliliters of olive oil are placed in each of the test tubes, along with a pH indicator. The initial color of the solutions is red, indicating the presence of a slightly basic solution. The solution in test tube #3 turns clear after 10 min, but all of the other test tubes remain red. State the conclusions that you would draw from the experiment and support each of the conclusions with the data pro-

vided. (Hint: Consider which substance is digested. What are the structural components?)

3 High cholesterol levels and high risk of atherosclerosis are in part related to diet and in part determined by genetics. The LDL-cholesterol receptor sites located on cell membranes are controlled by genetics. Explain how the number of receptor sites may cause a predisposition to atherosclerosis.

4 Why do some athletes feel compelled to take anabolic steroids?

5 See the diagrams below. Three experimental designs are presented to determine the best pH for the hydrolysis of starch to smaller glucose units. Specified quantities of starch and 2 mL of an enzyme capable of initiating starch breakdown are added to each of the test tubes. Which experimental design would you choose? Justify your answer.

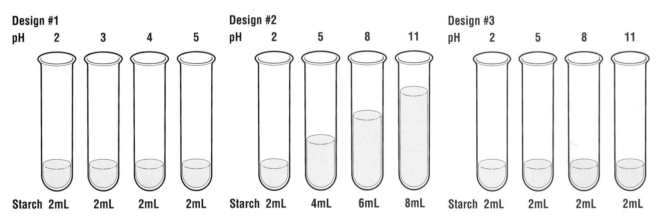

Design #1
pH   2   3   4   5
Starch 2mL 2mL 2mL 2mL

Design #2
pH   2   5   8   11
Starch 2mL 4mL 6mL 8mL

Design #3
pH   2   5   8   11
Starch 2mL 2mL 2mL 2mL

ENRICHMENT ACTIVITIES

Suggested reading:

- Beckett, Arnold. "Philosophy, Chemistry and the Athlete." *New Scientist*, August 1984, pp. 18–19.
- Blumenthal, Daniel. "Do You Know Your Cholesterol Level?" *FDA Consumer*, U.S. Food and Drug Administration, March 1989, pp. 24–27.
- Franklin, D. "Steroids Heft Heart Risk in Iron Pumpers." *Science News* 126 (21 July 1984): p. 38.
- Fritz, Sandy. "Drugs and Olympic Athletes." *Scholastic Science World* 40(16): pp. 7–14 (13 April, 1984).
- Harland, Susan. "Biotechnology: Emerging and Expanding Opportunities for the Food Industry." *Nutrition Today* (22)3: p. 21 (July/August).
- Harper, Alfred. "Killer French Fries." *Sciences*, Jan/Feb 1988, pp. 21–27.
- Monmaney, Terrence, and Karen Springen. "The Cholesterol Connection." *Newsweek*, 8 February 1988, pp. 56–58.
- Pertutz, Max. "The Birth of Protein Engineering." *New Scientist*, June 1985, pp. 12–16.
- Sperryn, Perry. "Drugged and Victorious: Doping and Sport." *New Scientist*, August 1984, pp. 16–18.

Energy within the Cell

IMPORTANCE OF CELL ENERGY

Energy can be defined as the ability to do work. For living things, the definition of work must be considered in a broad sense. Cell reproduction, the synthesis of cytoplasmic organelles, the repair of damaged cell membranes, movement, and active transport are just a few of the things that can be considered work.

Energy is found everywhere you look. Light, sound, and electricity are all forms of energy. The food that you ate for breakfast this morning provides the energy to move your limbs, maintain a constant body temperature, manufacture new chemicals, and maintain the cells of your body. This energy comes from the breakdown of macromolecules into their component parts.

There are two ways in which organisms can acquire food. *Autotrophic* organisms are organisms that are capable of making their own food. Plants use the energy from sunlight to make carbohydrates through a process called **photosynthesis.** In this process, energy from the sun converts low-energy compounds (carbon diox-

Photosynthesis *is the process by which plants and some bacteria use chlorophyll, a green pigment, to trap sunlight energy. The energy is used to synthesize carbohydrates.*

Figure 7.1

The sun is the source of life on earth. Sunlight provides the energy that powers the process of photosynthesis.

ide and water) into high-energy compounds (carbohydrates). *Heterotrophic* organisms are not capable of making their own food so they must take in food that is already made. All animals and fungi as well as many bacteria and protists are heterotrophs. Their food comes either directly from plants or from other animals that have eaten plants. Heterotrophs break chemical bonds in the large food molecules, thereby returning carbon dioxide and water to the environment and releasing energy. The energy-generating process is referred to as **cellular respiration.** (It should be noted that plants also carry out respiration to provide cell energy.)

Energy Systems

Energy is most often investigated within **energy systems.** Energy systems involve energy input, energy transformations, and energy output. Energy conversion or transformation occurs within cells.

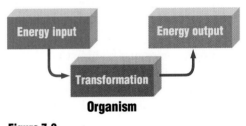

Figure 7.2

An energy system.

Unlike matter, energy does not cycle between autotrophs and heterotrophs. When living things die, their chemicals are returned to the soil. Decomposers ensure the cycling of matter. However, not all of an organism's energy is returned to the soil. The energy an organism uses for warmth and movement can never be recaptured and used by another organism. This energy is lost from the ecosystem. Fortunately, energy in the form of sunlight is constantly added, preventing the energy system from expiring.

LAWS OF THERMODYNAMICS

The study of energy relationships is called thermodynamics. The **first law of thermodynamics** states that energy cannot be created or destroyed, but energy can change forms. The amount of energy within a closed system remains constant. In other words, the same amount of energy was present at the beginning of the universe and will continue to be present until the end of time. The first law of thermodynamics has two essential parts. The first part indicates that energy cannot be destroyed. The energy held within a plant is not destroyed when it is eaten by a heterotroph; it merely changes form. The second part of the law indicates that energy may take a number of different forms. For example, plants use light energy to make sugar. In turn, the sugar is converted into starch for storage. During the process, light energy is converted into chemical energy.

The **second law of thermodynamics** states that all conversions of energy produce heat, which is not useful energy. Heat, which is a by-product of energy conversion, can be described as thermal exhaust, or waste energy. The fact that your muscles get warm during strenuous exercise attests to the fact that not all of the energy you generate is used for muscle contraction; some is wasted as heat.

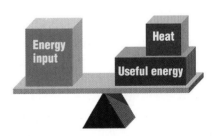

The amount of energy unavailable to do work is referred to as **entropy.** First described in 1850 by Rudolf Clausius, a

Figure 7.3

The second law of thermodynamics.

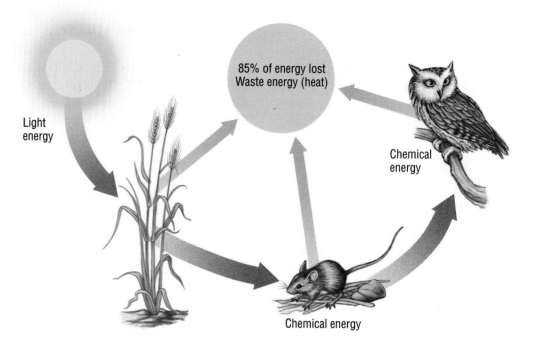

85% of energy lost
Waste energy (heat)

Light
energy

Chemical
energy

Chemical energy

Figure 7.4

Much of the light energy
reaching the plant is transformed
into non-useful heat.

*Entropy is a measure of
non-usable energy within
a system.*

Exergonic reactions
release energy.

Activation energy *is the
energy required to initiate
a chemical reaction.*

German physicist, entropy is most often associated with non-useful heat. Clausius described entropy as the "running down of the universe" or "the heat death of the universe." Entropy reflects the disorder of a system. The greater the disorder, the greater the entropy.

As stated previously, plants use energy from sunlight to make sugars, another form of energy. However, not all the light energy that reaches the plant is transformed into chemical energy. Some of the light energy warms the soil and the air surrounding the plant. In turn, not all of the chemical energy held within the plant is transferred to the animal. The mouse in Figure 7.4 uses energy in moving toward the plant and in chewing. The mouse also uses a significant amount of food energy just to maintain a constant body temperature. The owl will eat the mouse, but again not all of the chemical energy can be converted into useful energy. The transfer of energy is incomplete. As heat escapes from the mouse and the owl, energy is lost from the system. Where does the energy go? The energy may be lost to the owl and the mouse, but it is not destroyed. The heat from the owl warms the air surrounding the owl. Eventually, the energy is radiated into space.

METABOLIC REACTIONS

A comparison of input and output energy during transformations can be applied to the chemical reactions that occur in cells. Reactions that release energy are referred to as **exergonic reactions.** These reactions show a net loss of energy because the products of the reaction have less energy than the reactants. Some chemical energy may be lost as heat, as light, or as sound. The burning of wood is an example of an exergonic reaction. Once lit, the wood continues to release heat and light energy, eventually ending in a lower energy form. However, it is important to note that wood is not capable of spontaneous combustion. Energy is required to initiate the reaction. This energy is called the **activation energy.** In the case of a wood fire, a lit match provides activation energy.

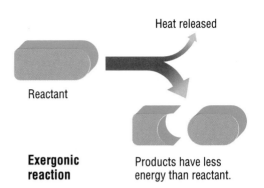

Heat released

Reactant

Exergonic reaction

Products have less energy than reactant.

A second type of chemical reaction, referred to as **endergonic,** requires the constant addition of energy. In endergonic reactions, the products have more energy than the reactants. It is important to note that energy is not created within an endergonic system; rather, energy is added from an external source. Photosynthesis provides an excellent example of an endergonic reaction. Without the addition of sunlight, the synthesis of carbohydrates would not be possible.

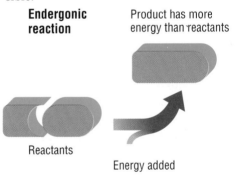

Endergonic reaction

Product has more energy than reactants

Reactants

Energy added

■ REVIEW QUESTIONS ■ ?

1 Provide two examples of energy transformation within your body.

2 Differentiate between heterotrophs and autotrophs.

3 In what ways are photosynthesis and respiration complementary processes?

4 State the first two laws of thermodynamics.

5 Define entropy.

6 Provide an example of a reaction that requires activation energy.

ENZYMES AND METABOLIC REACTIONS

Molecules are in constant motion. Even molecules within solids vibrate within fixed positions. Although chemical reactions sometimes occur when molecules collide, most reactions do not occur spontaneously. Adding thermal energy to a system increases the system's kinetic energy. This means that the molecules move faster, increasing the number of collisions. As the number of collisions increases, the probability of a reaction taking place increases. However, heating cells is dangerous—thermal energy could destroy the cell.

Chemical reactions must proceed at relatively low temperatures within cells. **Catalysts** are chemicals that control the speed of chemical reactions without altering the products formed by the reaction. This means that the catalyst remains unchanged after the chemical reaction. It is for this reason that catalysts can be used again and again. Reactions that occur within living organisms are regulated by protein catalysts called **enzymes.** Enzymes permit low-temperature reactions by reducing the reaction's activation energy. In much the same way as you would use less energy to pass through a tunnel than to scale a mountain, chemical reactions proceed faster when enzymes provide alternate pathways for chemical reactions. Figure 7.7 compares two exergonic reactions—Reaction I occurs with an enzyme, while Reaction II occurs without an enzyme. Because the enzyme-catalyzed reaction requires less activation energy, this reaction proceeds at a faster rate than the reaction that does not have an enzyme.

Figure 7.5
Exergonic reaction.

Endergonic *reactions require the continuous addition of energy. Low-energy reactants are converted into high-energy products.*

Figure 7.6
Endergonic reaction.

Catalysts *are chemicals that regulate the rate of chemical reactions without themselves being altered.*

Enzymes *are special protein catalysts that permit chemical reactions within your body to proceed at low temperatures.*

Figure 7.7

The effect of an enzyme on
activation energy.

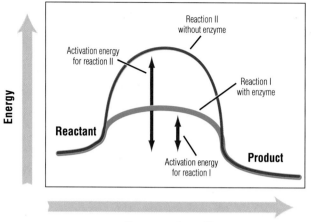

The molecules on which the enzyme
works are called the **substrate.** Each
substrate molecule combines with a specific enzyme. The substrate molecules are
changed during the reaction, and a product is formed. It has been estimated that
about 200 000 different chemical reactions occur within the cells of your body.
Each reaction uses a specific enzyme to
catalyze it.

Enzymes are identified by the suffix
-ase, which is added to the name of the
reaction that the enzyme controls. For
example, the enzyme that controls the
hydrolysis of sucrose into its two component parts—glucose and fructose—is
called sucrase. A protease enzyme will
cause the breakdown of protein, while
urease is related to the substrate urea.

Enzyme Models

Enzymes increase the probability of reactions by bringing substrate molecules
together. Enzymes have folded surfaces
that act as molds for trapping the correct
substrate molecules, aligning them to
cause the desired chemical reaction.
Having large molecules collide is not
enough—the molecules must collide at
the appropriate bond sites if the reaction
is to proceed.

The **active site** of the enzyme is the
area that joins with the substrate mole-

cules. Each enzyme has
a specially-shaped active
site that provides a "dock"
for specific substrate
molecules. This long-standing model, called the
"lock-and-key model," was
first proposed by Emil
Fischer in 1890. The temporary joining of the
enzyme with the substrate
molecule forms the
enzyme-substrate complex.
A modified theory, called
the "induced-fit model" replaced the
lock-and-key model in 1973. The
induced-fit model suggests that the actual
shape of the active site is altered slightly
when the substrate molecules are trapped,
making the fit between enzyme and substrate even tighter during the formation
of the enzyme-substrate complex.

Some enzymes require **cofactors** or
coenzymes to help them bind to substrate molecules. Cofactors are inorganic
molecules such as iron, zinc, and potassium, as well as copper-containing compounds. Coenzymes are organic
molecules that are synthesized from vitamins. Coenzymes and some cofactors
may work with more than one enzyme.

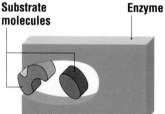

**Substrate
molecules**

Enzyme

Substrate molecules do not fit
in the active site of the enzyme.

Coenzyme

The coenzyme alters the active
site of the enzyme. The enzyme
can bind with the substrate.

Substrate

**Enzyme-
substrate
complex**

Product

Figure 7.8

Note how the active site of the
enzyme changes slightly once
the substrate is trapped. The
enzyme returns to its original
shape after the product is
formed.

Figure 7.9

Model of an enzyme and
coenzyme.

Factors That Affect Enzyme Reactions

It has been estimated that a single enzyme can catalyze between 100 reactions and 30 million reactions every minute. Why do some reactions occur much faster than others? In order to compare reaction rates, you must examine the different factors that affect enzymes. The graph in Figure 7.10 indicates how temperature affects enzyme-catalyzed reactions. The fact that reaction rates increase as temperatures increase should not be surprising. As energy is added, the molecules begin to move faster. The faster the molecules move, the greater the number of collisions. Subsequently, the greater number of collisions cause a greater number of products to be formed. But why do reaction rates peak at about 37°C and then drop even though the molecules are moving faster and colliding more often?

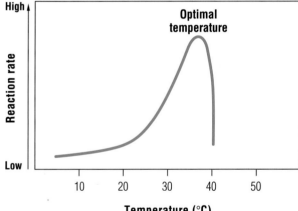

Temperature (°C)

To answer these questions, recall some important facts about enzymes. The fact that enzymes are proteins is particularly significant, because at high temperatures proteins change shape or are denatured. Any change in enzyme shape will have an effect on the formation of the enzyme-substrate complex. The greater the change to the active site

of the enzyme, the less effective the enzyme. Once the enzyme is denatured, the active site is so severely altered that the substrate can no longer bind with the enzyme. The reaction is no longer catalyzed by the enzyme and therefore proceeds at a much slower rate.

The effect of temperature on enzymes helps explain why high fevers can be so dangerous. The relationship between temperature and enzyme-catalyzed reactions also indicates some of the advantages of being a *homeotherm*, an animal that maintains a constant body temperature. Mammals, birds, and other homeotherms keep their bodies at optimal temperatures for reactions.

The graph in Figure 7.11 indicates that enzymes function best within certain pH ranges. The enzyme pepsin, shown in blue, operates best in an acidic pH. Not surprisingly, this enzyme is found in the stomach, an area of low pH. The enzyme shown in gray operates in mildly acidic and mildly basic solutions; however, it is most effective at a neutral or near neutral pH. The third enzyme, trypsin, shown in green, is most effective in a basic medium. Not surprisingly, trypsin is found in the small intestine, an area that is generally about pH 9.

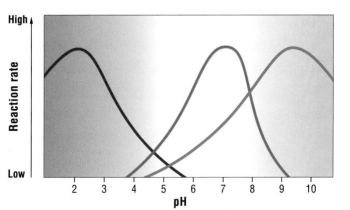

pH

Substrate *molecules attach to enzymes.*

The **active site** *of an enzyme is the area of the protein that combines with the substrate.*

Cofactors *are inorganic molecules that help enzymes combine with substrate molecules.*

Coenzymes *are organic molecules synthesized from vitamins that help enzymes combine with substrate molecules.*

Figure 7.10

Effect of temperature on enzyme reaction.

Figure 7.11

The enzymes indicated by the blue, gray, and green lines are most effective at a pH of 2.5, 7.0, and 9.0 respectively.

To understand why pH affects enzyme activity, you must look at the molecular structure of the protein molecule. Remember that the folds in the protein molecule are created by hydrogen bonds between negatively charged acid groups and positively charged amino groups. The addition of positively charged H^+ ions, characteristic of an acidic solution, or the introduction of negatively charged OH^- ions, characteristic of a basic solution, will affect the hydrogen bonds. Thus, the three-dimensional shape of an enzyme is altered by a change in pH. When the folds in the protein are changed, the active site of the enzyme is transformed, altering the reaction.

Enzyme activity can also be affected by the concentration of substrate molecules. For chemical reactions, the greater the number of substrate molecules, the greater the number of collisions, and the greater the rate of the reaction. Up to a point, enzyme-catalyzed reactions behave in the same manner. The reaction rate shown in Figure 7.12 begins to level off at point X because there is a limit to the amount of enzyme available. Substrate molecules cannot join with the active site of an enzyme until it is free. Once the number of substrate molecules exceeds the number of enzyme molecules, the reaction rate begins to level off. Therefore, the excess substrate molecules will not gain access to the active site of an enzyme.

*A **competitive inhibitor** has a shape complementary to a specific enzyme, thereby permitting it access to the active site of the enzyme. Inhibitors block chemical reactions.*

Inhibitor molecules can affect enzyme reactions. Often referred to as **competitive inhibitors,** these molecules have shapes very similar to that of the substrate. The inhibitors actually compete with the substrate molecules for the active sites of the enzymes. As long as the inhibitors remain joined to the enzyme it is prevented from functioning properly.

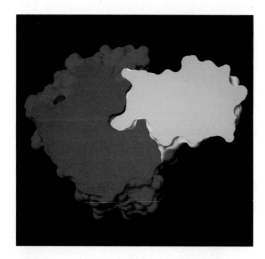

Figure 7.13

Computer graphic of the enzyme trypsin (red) and an inhibitor (green) bound to the active site. The active site of the enzyme is tied up by the inhibitor.

Figure 7.13 shows an important function of inhibitors in the body. Trypsin is a protein-digesting enzyme that is stored in the cells of the pancreas. An interesting problem arises when a protein-digesting enzyme must be stored in cells that are essentially protein in structure. The protein-digesting enzyme must be prevented from digesting the cell. An inhibitor ties up the active site, thereby preventing the stored trypsin from digesting the cell. The inhibitor is removed once the enzyme is needed. The premature activation of the enzyme produces pancreatitis, an inflammation of the pancreas.

Figure 7.12

The rate of a metabolic reaction is affected by the concentration of substrate molecules and the concentration of enzymes.

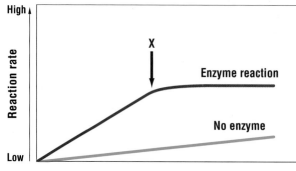

Many poisons and drugs work as competitive inhibitors. Carbon monoxide competes with oxygen for the active site on the hemoglobin molecule. Unfortunately, carbon monoxide fits into the hemoglobin about 200 times faster than does oxygen. Once carbon monoxide attaches itself, it is rarely dislodged. Cyanide, a lethal poison, works in much the same manner. Cyanide attaches itself to an enzyme in the mitochondria, preventing the breakdown of sugars for energy. Without the energy needed for active transport, protein synthesis, the transport of nutrients, and the elimination of wastes, a cell dies almost instantly.

All drugs are, in effect, poisons. Their harmfulness or usefulness depends on which chemical reactions they inhibit. Both cyanide and penicillin, for example, kill invading bacteria. Penicillin, however, blocks an enzyme essential for the synthesis of protective walls in bacteria, but has no effect on human enzymes.

Regulation of Enzyme Activity

Metabolic pathways are orderly sequences of chemical reactions, with enzymes regulating each step of the reaction. Consider the following example of a metabolic pathway. Testosterone is a male sex hormone synthesized from cholesterol or other steroids. The hormone, which is produced in larger quantities from puberty onward, is responsible for the development of secondary male sex characteristics. The maturation of the larynx, the lowering of the voice, the development of body hair, and the stimulation of the sex drive are a few of the effects of this hormone.

Can you imagine what would happen if all of the steroids in the body were converted into testosterone? The regulation of chemicals produced by metabolic pathways is essential. The production of chemicals within a cell is regulated by the need for those chemicals. As the product from a series of chemical reactions begins to accumulate within a cell, the product interferes with one of the enzymes in a process known as **feedback inhibition.** The interference slows the reaction rate, preventing the accumulation of final products, and thereby opening other side-reaction pathways.

Feedback inhibition *is the inhibition of an enzyme in a metabolic pathway by the final product of that pathway.*

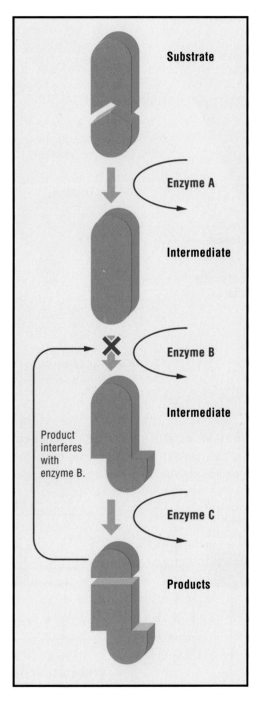

Figure 7.14

Typical metabolic pathway.

The final product of the metabolic pathway interferes with the enzyme by combining with its regulatory site. The binding of the final product with the regulatory site of the enzyme alters the active site, and thus prevents the union of the enzyme and substrate.

combines at the regulatory site of one of the enzymes, **precursor activity** occurs. During precursor activity, the combination of the substrate and enzyme actually improves the fit of the enzyme-substrate complex. This speeds up the formation of final products. Both feedback inhibition and precursor activity involve the binding of a molecule with the regulatory site of the enzyme. Both processes are called **allosteric activity.** The binding of the final product with the regulatory site of the enzyme will change the enzyme's active site, thereby inhibiting subsequent reactions. The binding of one of the initial reactants with the regulatory site will help mold the active site of the enzyme, improving the fit between substrate and enzyme.

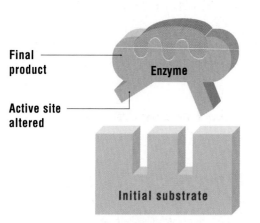

Figure 7.15

The final product binds at the regulatory site of the enzyme and alters the active site. Feedback inhibition occurs because the enzyme will no longer combine with its substrate molecule.

Labels: Final product, Regulatory site, Enzyme, Initial substrate

Precursor activity *is the activation of the last enzyme in a metabolic pathway by the initial reactant.*

Allosteric *activity is the change in the protein enzyme caused by the binding of a molecule to the regulatory site of the enzyme.*

Labels: Final product, Enzyme, Active site altered, Initial substrate

REVIEW QUESTIONS ?

7 Explain the importance of enzymes in metabolic reactions.
8 How do enzymes increase the rate of reactions?
9 List and explain four factors that affect the rate of chemical reactions.
10 How do cofactors and coenzymes work?
11 What are competitive inhibitors?
12 What is allosteric activity?
13 How are metabolic pathways regulated by the accumulation of the final products of the reaction?

Regulatory sites are not just used to turn off metabolic pathways. A build-up of the initial substrate can turn on enzyme activity. If the substrate molecule

Figure 7.16

Precursor activities turn metabolic pathways "on," while feedback inhibition activities turn metabolic pathways "off."

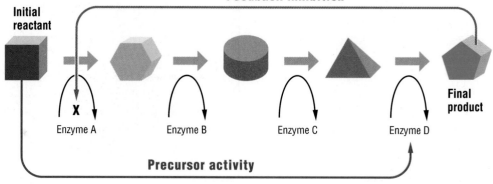

Feedback inhibition

Initial reactant — X — Enzyme A — Enzyme B — Enzyme C — Enzyme D — Final product

Precursor activity

Unit Two:
Exchange of Energy and Matter in Cells

RESEARCH IN CANADA

Scientists use chemical models to simulate interactions between drugs and target molecules. As you read earlier in the chapter, a drug must have the correct geometry to gain access to the appropriate target site. In a process described as the "lock-and-key" fit, the drug attaches to the target site, but avoids other non-target sites, thereby reducing unwanted side effects. Three-dimensional computer graphics and a chemical analysis technique called X-ray crystallography allow researchers to construct the chemical models of both the target and the drug, reducing the element of chance in drug research.

Chemical Models

Dr. Penelope Codding of the University of Calgary has modeled the site where Valium binds to its target molecule. Valium is a widely used antidepressant that has several unwanted side effects. The information that Dr. Codding has gained about the target site may enable researchers to design new drugs with fewer side effects. Dr. Codding is presently investigating new drugs for the control of epilepsy.

DR. PENELOPE CODDING

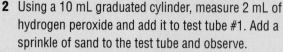

LABORATORY

ENZYMES AND H₂O₂

Objective

To identify factors that affect the rate of enzyme-catalyzed reactions.

Background Information

Organisms that live in oxygen-rich environments need the catalase enzyme. The catalase enzyme breaks down hydrogen peroxide (H_2O_2), a toxin that forms readily from H_2O and O_2. The reaction below describes the effect of the catalase enzyme.

The formation of hydrogen peroxide:

$$2H_2O + O_2 \longrightarrow 2H_2O_2 \text{ (hydrogen peroxide)}$$

The effect of catalase enzyme:

$$2H_2O_2 \xrightarrow{\text{catalase enzyme}} 2H_2O + O_2$$

Materials

safety goggles	fine sand
lab apron	scalpel
6 test tubes	potato
wax marker	chicken liver (fresh)
3% hydrogen peroxide	stirring rod
tweezers or forceps	mortar and pestle
10 mL graduated cylinder	

CAUTION: Hydrogen peroxide is a strong irritant. Avoid skin and eye contact. Wash all splashes off your skin and clothing thoroughly. If you get any chemical in your eyes, rinse for at least 15 min and inform your teacher.

Procedure:

Part I: Identifying the Enzyme

1 Label three clean test tubes #1, #2, and #3.

2 Using a 10 mL graduated cylinder, measure 2 mL of hydrogen peroxide and add it to test tube #1. Add a sprinkle of sand to the test tube and observe.

3 Add 2 mL of H_2O_2 to test tubes #2 and #3. Using the scalpel, remove a slice of potato approximately the size of a raisin and add it to test tube #2. Observe the reaction. Repeat the procedure once again, but this time add a piece of liver to test tube #3. Observe the reaction.

4 Compare the reaction rates of the three test tubes. Use 0 to indicate little or no reaction, 1 to indicate slow, 2 to indicate moderate, 3 for fast, and 4 for very fast.

   a) Record your results in a table similar to the following.

Test-tube number	Reaction rate	Product of reaction
1		
2		
3		

Part II: Factors that Affect Reaction Rates

5 Divide the hydrogen peroxide used in test tube #3 into two clean test tubes. Label one of the test tubes #4 and the other #5. Using tweezers or forceps, remove the liver from test tube #3 and divide it equally into test tubes #4 and #5. Add a second piece of liver to test tube #4 and observe. Add 1 mL of new hydrogen peroxide to test tube #5 and observe.

   b) Record your results.

6 Using a scalpel, cut another section of liver the size of a raisin. Add sand to a mortar and grind the liver into smaller pieces with the pestle. Remove the liver and place it in a clean test tube labelled #6. Add 2 mL of H_2O_2 and compare the reaction rate of the liver in test tube #6 with the uncrushed liver in test tube #3.

   c) Record your results.

Laboratory Application Questions

1 In part I , which test tube served as the control?

2 Account for the different reaction rates between the liver and potato.

3 Explain the different reaction rates in test tubes #4 and #5.

4 Why did the crushed liver in test tube #6 react differently from the uncrushed liver in test tube #3?

5 Predict what would happen if the liver in test tube #3 were boiled before adding the H_2O_2. Give the reasons for your prediction. ■

ENERGY STORAGE AND TRANSFORMATION

Energy storage and conversion processes are critical for sustaining life. Imagine the difficulty you would have storing solar energy in a box. If you opened the box one week later, would you be able to put the solar energy to use?

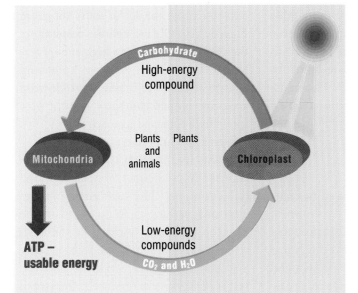

In this chapter you will study two important types of energy storage and transformation: photosynthesis and cellular respiration. During photosynthesis, solar energy is converted to chemical bond energy within carbohydrates. During cellular respiration, carbohydrates are converted into a storage compound called **adenosine triphosphate,** or **ATP**. ATP is the usable form of chemical energy. Before cells can use chemical energy, carbohydrates and other organic compounds must be converted into ATP. ATP provides the energy required for the synthesis of needed chemicals within cells, the active transport of materials across cell membranes, and the contraction of muscle fibers.

Adenosine triphosphate is composed of a five-carbon sugar (ribose), a nitrogen

base (adenine), and three phosphate molecules. A diagram of the ATP molecule is shown in Figure 7.18.

ATP can be thought of as a gold coin that cells use to pay for work. In much the same way as you draw money from your bank account, special enzymes can extract a phosphate molecule from ATP whenever energy is required. The new product is called *adenosine diphosphate*, or *ADP*, because it has only two phosphate molecules. ADP has less energy than the ATP molecule. The ADP molecule can be thought of as a silver coin. Although the silver coin is capable of paying for energy demands, it has less value than the ATP gold coin. Occasionally, a second phosphate bond can be extracted and ADP is converted into *adenosine monophosphate*, or *AMP*. The third phosphate group will not be removed. To return to the money analogy, once the silver coins have been spent, no additional energy can be purchased. The bank account is empty. The reaction in Figure 7.18 summarizes how ATP provides energy for a cell.

ATP *is a compound that stores chemical energy.*

Figure 7.17

Two important organelles for energy transformation. The chloroplasts are the sites of photosynthesis and the mitochondria are the sites of cellular respiration.

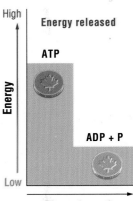

$$ATP \rightarrow ADP + P + energy$$

Figure 7.18

ATP can be thought of as a gold coin and ADP as a silver coin.

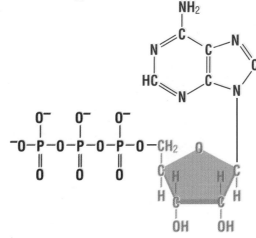

Figure 7.19

The structural formula for adenosine triphosphate. The ribose sugar is shown in green, the adenine molecule in red, and the three phosphate molecules in purple.

Figure 7.20

(a) The combination of oxygen and hydrogen releases energy as a burst of heat. (b) The step-by-step release of energy by electron transport systems enables cells to convert much of the energy into ATP.

In the same way as you must deposit more money into your bank account to continue withdrawing, ATP supplies must be constantly replenished. Energy must be added to refill ATP supplies, but where does the energy come from? Exergonic chemical reactions supply needed energy for the synthesis of ATP. A tremendous amount of potential energy is stored in chemical bonds. The conversion of high-energy molecules into lower-energy molecules releases energy, which, in turn, can be used to attach high-energy phosphate bonds to ADP, thereby making ATP. The addition of a phosphate molecule is called **phosphorylation.**

Regulating Energy Input for ATP Production

An electrical spark can be used to combine hydrogen and oxygen. As shown in Figure 7.20, such a combination produces water. Because water contains much less energy than the reactants (hydrogen and oxygen), energy is released. Unfortunately, this rapid release of energy is not ideally suited for cells. In many situations, cells could be damaged by the uncontrolled release of heat. The heat may actually distort proteins and damage the cell. However, a step-by-step release of energy allows cells time to transform and store much of the energy.

Energy production in cells takes place within the **electron transport system.** In many ways, the electron transport system resembles the set of stairs shown in Figure 7.20. During cellular respiration, the electrons move from high energy levels to lower energy levels, releasing energy at each step. The final electron acceptor, oxygen, combines with the proton of hydrogen and its electron to form a low-energy compound, water.

Plants use electron transport systems to make high-energy compounds during photosynthesis. The solar energy excites electrons, pushing them up the stairs (the electron transport system). As the electrons climb each step of the pathway, they gain potential energy. Once the electrons return to lower energy levels, the stored energy is released.

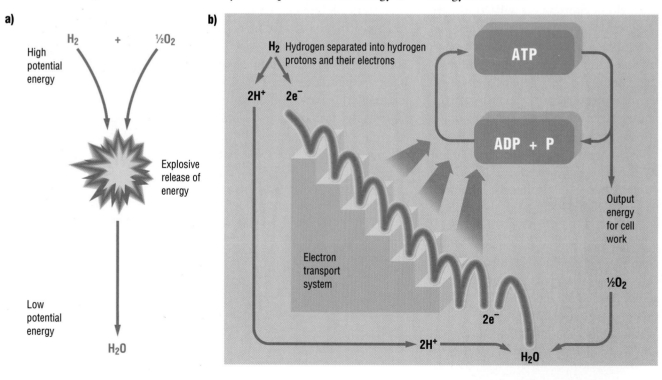

The stairway analogy may help you picture how a ball moves from one level to another, but why do electrons move along transfer systems? Elements such as lithium, calcium, and sodium are strong electron donors. This means that they tend to lose electrons to other elements in an attempt to become stable. Elements such as fluorine, chlorine, and oxygen are strong electron acceptors. By accepting electrons, these elements fill their outermost energy levels, thereby becoming stable. The process in which elements lose electrons in order to become stable is called **oxidation.** The process in which elements accept electrons is called **reduction.** When organic molecules are oxidized, the hydrogen nuclei and their electrons (H^+ and e^-) are lost. Organic molecules are reduced when they gain hydrogen nuclei and their electron.

The transfer of electrons from one reactive atom to another produces more stable ions or compounds. The fact that the products have less energy than the reactants indicates that energy is released during the oxidation reaction. This energy can be used to make ATP. The diagram in Figure 7.21 shows how the energy from an oxidation-reduction reaction is used to attach phosphates to ADP. The product, ATP, is a high-energy compound.

Each time electrons are transferred in oxidation-reduction reactions, energy is made available for the cell to make ATP. Electron transport systems shuttle electrons from one molecule to another. Electron and proton carriers like **NAD⁺** (nicotinamide adenine dinucleotide), an important coenzyme for cellular respiration, and **NADP⁺** (nicotinamide adenine dinucleotide phosphate), an important coenzyme in photosynthesis, act as strong electron acceptors. The electron acceptors strip a hydrogen proton and its electron from a number of organic compounds. Once the coenzyme picks up a hydrogen proton and its electron oxidation occurs, ATP is formed. Eventually, the coenzyme loses electrons to an even stronger electron acceptor. Each time the electron is transferred in the electron transport system, energy is released.

PHOTOSYNTHESIS AND ENERGY

The extent to which our planet relies on plants can be demonstrated by a study of photosynthesis. The word is a composite of *photo*, meaning "light," and *synthesis* meaning "to make or build." In photosynthesis, carbon dioxide from the air and water from the soil are combined to form glucose, a carbohydrate. Light energy is used to convert the low-energy reactants into a complex, high-energy product. Although animals rely on glucose to fuel the cells of their bodies, they cannot produce it themselves—they must rely on plants for their food. The chemical reaction for photosynthesis is shown in Figure 7.22 on the following page.

Oxidation *occurs when an atom or molecule loses electrons.*

Reduction *occurs when an atom or molecule gains electrons.*

NAD⁺ *is a strong electron acceptor important for electron transport systems in cellular respiration. NADH is the reduced form of NAD⁺.*

NADP⁺ *is a strong electron acceptor important for electron transport systems in photosynthesis. NADPH is the reduced form of NADP⁺.*

Figure 7.21

The energy released from the oxidation-reduction reaction is used to attach a free phosphate to ADP to make ATP. Note that ATP is a high-energy compound.

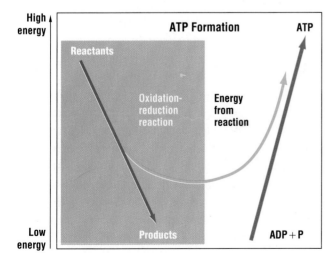

$$6CO_2 + 12H_2O + \text{energy (sunlight)} \longrightarrow C_6H_{12}O_6 + 6O_2 + 6H_2O$$

carbon dioxide + water + sunlight $\longrightarrow$ glucose + oxygen + water

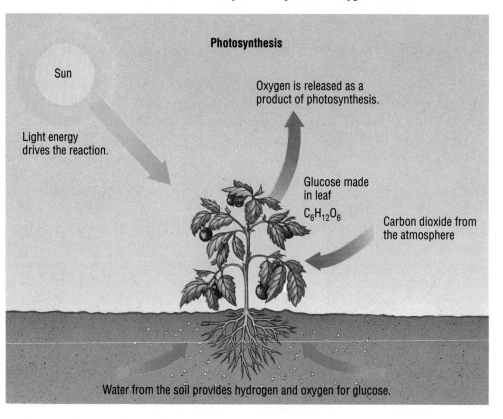

Photosynthesis

Sun

Light energy drives the reaction.

Oxygen is released as a product of photosynthesis.

Glucose made in leaf $C_6H_{12}O_6$

Carbon dioxide from the atmosphere

Water from the soil provides hydrogen and oxygen for glucose.

Figure 7.22

Plants utilize light energy to form glucose from water and carbon dioxide.

Chlorophyll *is the pigment that makes plants green. Chlorophyll traps sunlight energy for photosynthesis.*

Chloroplasts *are cytoplasmic organelles that contain chlorophyll.*

Thylakoid membranes *are specialized cell membranes found in chloroplasts.*

Grana *are green disks stacked together. The disks are part of the thylakoid membrane.*

Stroma *is the gel-like substance containing proteins that surrounds the grana.*

During photosynthesis, plants convert light energy into chemical energy. However, the conversion is not a single-step process. Photosynthesis can be broken down into two components: the light-dependent reaction and the carbon-fixation reaction. During the light-dependent reaction, light energy is converted into chemical energy and temporarily stored as ATP. Light energy is used to split the water molecule into its two component parts: hydrogen and oxygen. The oxygen is released into the atmosphere. The hydrogen temporarily combines with the coenzyme $NADP^+$. During the oxidation-reduction reaction, the $NADP^+$ accepts a hydrogen proton and its electron to become NADPH, the reduced form of $NADP^+$. As in all oxidation-reduction reactions, energy is released. This energy is then used to form ATP.

During the carbon-fixation phase,

hydrogen atoms extracted from water molecules are joined with carbon dioxide molecules to make glucose. The energy required to make glucose comes from the NADPH and ATP made during the light reaction.

Chlorophyll

More than 2000 years ago, Aristotle discovered that plants need sunlight. Deprived of sunlight, a plant turns yellow and begins to wilt. If the plant is put back in the sunlight, it turns green once again. **Chlorophyll** is the pigment that makes plants green. The green pigment is located in special structures within the cytoplasm, called **chloroplasts.** The chlorophyll within the chloroplasts traps sunlight energy for photosynthesis.

Photosynthesis takes place within specialized membranes of the chloroplasts, called **thylakoid membranes.** When

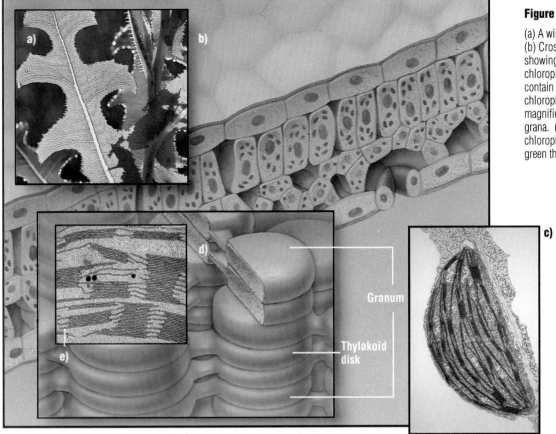

Figure 7.23

(a) A wild prickly lettuce leaf. (b) Cross section of a leaf showing the cells that contain chloroplasts. Chloroplasts contain the green pigment chlorophyll. (c) Chloroplast magnified. (d) Diagram showing grana. (e) Grana magnified. The chlorophyll is located in the green thylakoid disks.

Figure 7.23

(a) A wild prickly lettuce leaf. (b) Cross section of a leaf showing the cells that contain chloroplasts. Chloroplasts contain the green pigment chlorophyll. (c) Chloroplast magnified. (d) Diagram showing grana. (e) Grana magnified. The chlorophyll is located in the green thylakoid disks.

examined under the transmission electron microscope, the chloroplasts appear to be composed of stacks of small green disks, arranged in regular arrays, called **grana.** The grana are surrounded by a gel-like substance called **stroma.**

White light is composed of all wavelengths of visible light. The green color of a plant is produced because the chlorophyll reflects green wavelengths of light. However, the chlorophyll absorbs blue and red wavelengths. Other pigments, like the carotenoids, absorb violet and blue wavelengths but reflect yellow. The different pigments absorb the wavelengths of light that provide the right amount of energy to the electrons within them. The trapped energy excites the electrons, boosting them to a higher energy level.

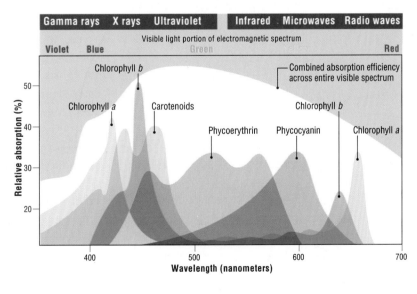

Figure 7.24

Plants use only visible light for photosynthesis. Two types of chlorophyll—chlorophyll *a* and chlorophyll *b*—reflect green light, but tend to absorb wavelengths from the red and blue parts of the electromagnetic spectrum.

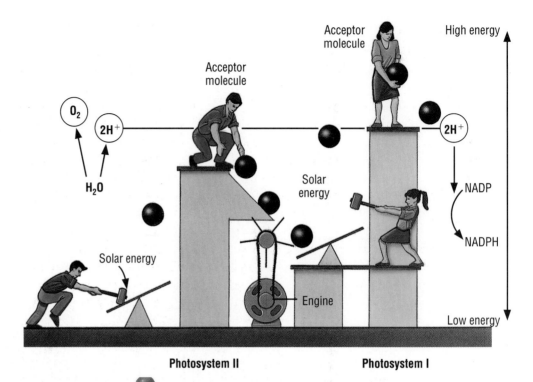

Figure 7.25

A mechanical analogy for photosynthesis. Solar energy, represented by the hammer, is used to split water molecules. Electrons are pushed to higher energy levels in photosystem II. The electron acceptor passes the electrons down a transport system to photosystem I. As the electron moves down the transport system, the energy released can be used to perform work, represented by the engine (an ATP generator). The ATP generator pushes H^+ protons up the energy gradient. Once again, solar energy is used to raise the energy level of the electron. The electron is held by an acceptor molecule in photosystem I in much the same way as it was held in photosystem II. When the electrons move to lower energy levels, the energy released can be used to drive the linkage of two electrons and H^+ to $NADP^+$. Reduced NADPH is formed from photolysis.

Figure 7.26

Summary of light-dependent reaction and carbon-fixation reaction.

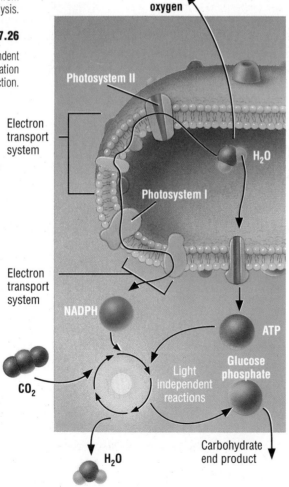

Light-Dependent Reactions

Light-absorbing pigments are not randomly scattered in the thylakoid membranes of the chloroplast; they are organized into clusters, called **photosystems.** The light energy captured by the photosystem is used to push electrons to higher energy levels. In much the same way as walking upstairs requires the expenditure of energy, the excitation of electrons also requires energy. Each time an electron moves up to a higher step, it is said to have greater energy. However, the energized chlorophyll is not stable. Once the electron returns to the lower energy level, the energy is released. Electrons move back down the energy hill in a step-by-step transport system. The energy released through this process is used to form ATP.

Not all of the light energy that strikes chlorophyll is used to make ATP. Some of the energy is used to split water molecules. Water is composed of hydrogen protons, electrons, and oxygen.

When the water molecule is split, its electrons are transferred through the two photosystems and the two electron transport systems. In photosystem II, a chlorophyll molecule excites the electrons removed from the water molecule. Electrons are delivered to the electron acceptor, which, in turn, delivers them to photosystem I. In turn, the electrons are excited by a second chlorophyll molecule in photosystem I.

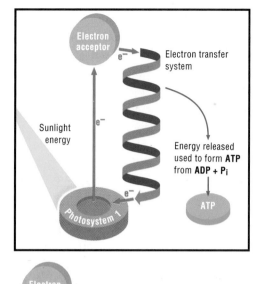

Figure 7.27

Cyclic pathway transfers electrons to make ATP. Sunlight energy excites electrons in the chlorophyll. The electron moves toward a molecule that acts as an electron acceptor. The electrons are driven uphill by light. When electrons move back to lower energy levels, energy is released. The energy released can be used to make ATP.

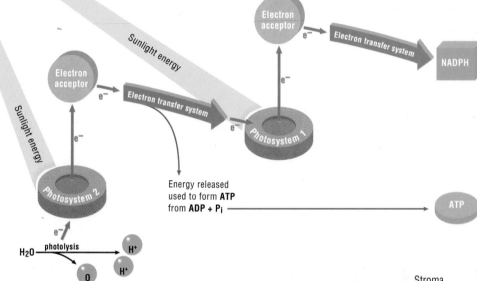

Figure 7.28

Noncyclic pathway, yielding ATP and NADPH. Electrons obtained from photolysis of water are transported through two photosystems.

Photolysis *refers to the splitting of water by means of light energy.*

Photosystems *are light-trapping units composed of pigments within the thylakoid membranes.*

Consider the products of **photolysis:** hydrogen protons, their electrons, and oxygen. The oxygen produced from the splitting of the water is released into the atmosphere. The oxygen that you are consuming right now was once part of a water molecule. The electrons replace those electrons in the chlorophyll that captured light energy has pushed to higher energy levels. The hydrogen path is more difficult to follow. A coenzyme NADP, found within the chloroplast, acts as a hydrogen acceptor for the hydrogen protons released by the splitting of water. NADP is converted into NADPH. The

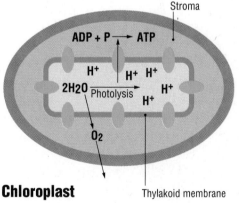

Chloroplast

Figure 7.29

The H$^+$ released from photolysis are transported to the compartment surrounded by the thylakoid membrane. H$^+$ begin to accumulate in the compartment, thereby increasing the amount of positive charge. As the concentration and electrical gradients begin to build, H$^+$ move through special protein channels in the thylakoid membrane. The H$^+$ flow drives enzymes, which convert ADP into ATP.

NADPH moves through the thylakoid membrane to the stroma of the chloroplasts. Here the hydrogen is released, to be used during the carbon-fixation reaction to form glucose.

Carbon-Fixation Reactions

During the second phase of photosynthesis, carbon dioxide molecules are linked into chains. Carbon is said to be fixed. The hydrogen molecule extracted from the NADPH of the light reaction is linked to the carbon to form carbohydrates ($C_nH_{2n}O_n$).

Energy for the carbon-fixation reaction comes from a supply of ATP produced during the light-dependent reaction. A complex cycle of enzyme reactions links carbon atoms, producing carbohydrates and water. Named after its co-discoverers, Melvin Calvin and Andrew Benson, the Calvin-Benson cycle occurs in the stroma of the chloroplasts. During this cycle, a five-carbon sugar called ribulose biphosphate, or RuBP, acts as a carbon dioxide acceptor. Carbon dioxide and RuBP create an unstable six-carbon sugar. This sugar immediately divides into two three-carbon sugars called phosphoglyceric acid, or PGA. The PGA uses ATP as a source of energy to remove hydrogen from NADPH and to form phosphoglyceraldehyde (PGAL) and water. The NADPH and ATP are supplied from the light-dependent reaction.

In turn, PGAL can be used as chemical energy, or two molecules of PGAL can unite to form glucose. Glucose is a stable six-carbon sugar. The glucose acts as the base from which more complicated carbohydrates such as sucrose, starch, and cellulose can be formed.

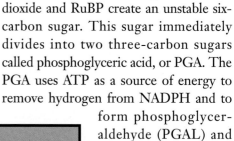

BIOLOGY CLIP

Some 40 to 60% of the world's photosynthesis occurs in the oceans. It is estimated that microscopic ocean plants produce one hundred billion tons of oxygen each year.

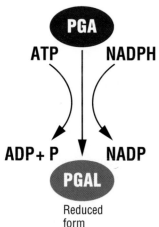

Figure 7.30

The reaction that reduces PGA to PGAL uses NADPH and ATP formed in the thylakoid membrane during the light-dependent reactions. The energy stored in PGAL is used to facilitate later reactions.

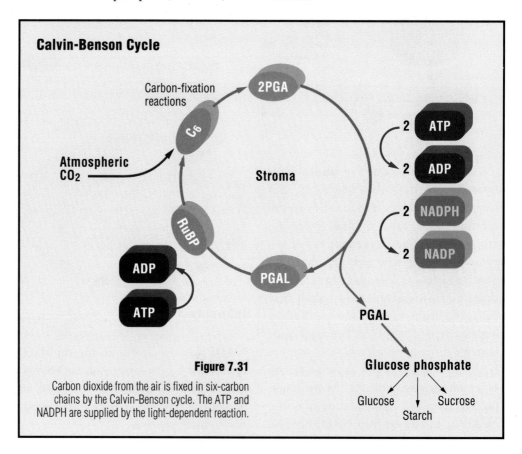

Calvin-Benson Cycle

Figure 7.31

Carbon dioxide from the air is fixed in six-carbon chains by the Calvin-Benson cycle. The ATP and NADPH are supplied by the light-dependent reaction.

In many ways, PGAL can be considered the most important product of photosynthesis. PGAL serves three purposes. First, most of the PGAL is used for energy to fuel the carbon-fixation reaction. Second, some of the PGAL can be converted to glucose for energy storage. Third, a portion of the PGAL molecules is used to replenish RuBP. The build-up of additional supplies of RuBP can assure the continued fixation of carbon dioxide and the continuation of the Calvin-Benson cycle. A summary of the light-independent reaction is provided in Figure 7.31.

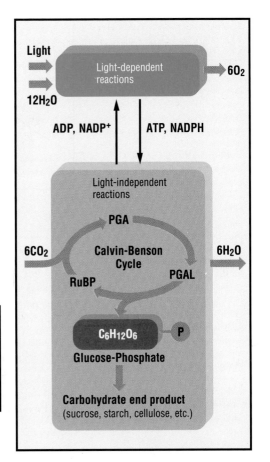

Figure 7.32

Photosynthesis consists of two reactions; the light-dependent reaction that produces ATP and NADPH and the light-independent reaction that produces sugar phosphates such as glucose.

> **B I O L O G Y   C L I P**
> Each year five billion people on earth appropriate more than 40% of the organic matter produced by photosynthesis on land.

Summary of Photosynthesis

Photosynthesis is the process by which plants and some bacteria use chlorophyll, a green pigment, to trap sunlight energy. During photosynthesis, inorganic compounds are converted into organic compounds. The reaction can be summarized by the following equation:

$$12H_2O + 6CO_2 + \text{sunlight} \rightarrow C_6H_{12}O_6 + 6H_2O + 6O_2 \text{ energy}$$

Light energy is used to split water molecules. The energized chlorophyll forms ATP and attaches hydrogen atoms to NADP. During the light-independent phase of the reaction, carbon dioxide from the air is combined with a five-carbon sugar called RuBP to form PGA. The PGA needs ATP from the light reaction to remove hydrogen from the NADPH and form PGAL. PGAL can be converted into glucose or additional RuBP molecules.

■ REVIEW QUESTIONS ■ ?

14 What is ATP?

15 Explain why oxidation reactions are often associated with the creation of ATP.

16 Differentiate between oxidation and reduction.

17 Explain why electron transport systems are important in living things.

18 Explain the meaning of the word photosynthesis.

19 Summarize the events of the light-dependent reaction.

20 In what part of the chloroplast does the light-dependent reaction take place?

21 What function does chlorophyll serve in the light-dependent reaction?

22 What is the Calvin-Benson cycle? Where does it occur?

LABORATORY

PHOTOSYNTHESIS

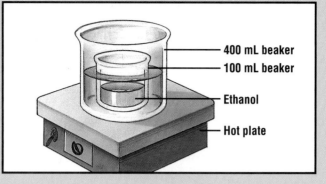

400 mL beaker
100 mL beaker
Ethanol
Hot plate

Objective

To demonstrate the importance of sunlight during photosynthesis.

Materials

safety goggles
lab apron
10 mL graduated cylinder
distilled water
5% starch solution
2 test tubes
iodine solution
medicine dropper

400 mL beaker
beaker tongs
hot plate
denatured ethanol
variegated coleus plant
2 100-mL beakers
forceps
petri dish

CAUTION: The chemicals used are toxic and irritant. Avoid skin and eye contact. Wash all splashes off your skin and clothing thoroughly. If you get any chemical in your eyes, rinse for at least 15 min and inform your teacher. Ethanol is flammable. Do not use near open flames.

Procedure

Part I: Testing for Starch

1 Using a 10 mL graduated cylinder, add 3 mL of distilled water and 3 mL starch solution to two separate test tubes. Add four drops of iodine to each test tube and observe.
   a) Record the color of each test tube.

Part II: Sunlight and Photosynthesis

2 Add 200 mL of water to a 400 mL beaker and, using beaker tongs, place the beaker on a hot plate.
3 Add 40 mL of ethanol to a 100 mL beaker. Using beaker tongs, place the 100 mL beaker inside the 400 mL beaker. Turn on the hot plate and bring the water to a boil.

CAUTION: Always use a water bath.

4 Remove the leaf of a coleus plant that has been exposed to direct light for at least 72 h.
   b) Sketch the leaf, and label areas of red and green pigment.
5 Using forceps, place the leaf in boiling water for three seconds and then immerse it in the boiling alcohol for approximately five minutes, or until the pigment disappears completely from the leaf.
6 Once again, immerse the leaf in boiling water for approximately three seconds.
   c) Describe the appearance of the leaf.
7 Place the leaf in an open petri dish and carefully spread out the leaf. Add 20 to 30 drops of iodine, completely covering the leaf.
   d) Describe the appearance of the leaf.
   e) Sketch the leaf, and label the areas that contain starch.
8 Repeat the same procedure for a plant that has been kept in total darkness for 72 h.
   f) Describe the appearance of the leaf.
   g) Sketch the leaf, labelling the areas that contain starch.

9 Ask your teacher to dispose of the alcohol.

Laboratory Application Questions

1 Why was iodine used in the experiment?
2 From the data that you have collected, support the fact that light is required for photosynthesis.
3 Explain why coleus was used rather than a plant that contains only green pigments.
4 Predict how the results would have been affected if the plant exposed to light had been covered by a transparent green bag. State the reasons for your prediction.
5 Design an experiment to determine which visible colors provide the optimal energy for photosynthesis.
6 Why was a Bunsen burner not used as a heat source for the ethanol? ■

CELLULAR RESPIRATION IN PLANTS AND ANIMALS

Cellular respiration includes all the chemical reactions that provide energy for life. Carbohydrates, most notably in the form of glucose, are the most usable source of energy. Only after glucose supplies have been depleted do cells turn to other fuels. For animals, glycogen, a storage carbohydrate composed of many glucose units, breaks down, releasing single glucose units into the blood in an attempt to maintain blood glucose levels. Once glycogen supplies from the liver and muscles are depleted, fat becomes the preferred energy source. Proteins, the organic compounds of cell structure, are used as a final resort once fat supplies have been exhausted. The utilization of proteins for energy means that the cell begins breaking down its own structures in order to obtain energy. Plants use starch as an energy storage compound in much the same manner as animals use glycogen. When needed, starch can be broken down to maltose, a disaccharide sugar. Unlike starch, maltose is soluble. Eventually, the disaccharide can be broken down into monosaccharide units.

During respiration, chemical bonds are broken. Chemical energy is stored in these bonds. It has been estimated that, at best, only 36% of the energy is used to make ATP; the remaining 64% is released as heat. The thermal energy helps maintain a constant body temperature in mammals and birds. ATP, as mentioned before, is a form of stored energy that is readily accessed for active transport, the synthesis of needed chemicals, the contraction of muscle fibers, and other energy-requiring functions.

Oxidation and Phosphorylation

As the glucose molecule is oxidized, energy is released. Hydrogen and its electron move from a weak electron acceptor to a stronger electron acceptor. NAD (nicotinamide adenine dinucleotide) is a typical electron acceptor in cells, accepting hydrogen and its electron from activated carbohydrates. Electrons can move from one electron acceptor to progressively stronger acceptors. In much the same way as you descend a flight of stairs, the electron moves from one acceptor to another, losing energy at each step. Electrons found on the top step are able to provide the greatest amount of energy for ATP formation. Electrons found on the lower steps are less able to provide energy for ATP formation.

During oxidation, high-energy compounds like glucose are converted into low-energy compounds, such as carbon dioxide and water. Each time a hydrogen and its electron move from glucose, energy is released. Not surprisingly, the greater the number of hydrogen atoms extracted, the greater the amount of energy released. The release of energy

Cellular respiration *is the oxidation of glucose to produce ATP.*

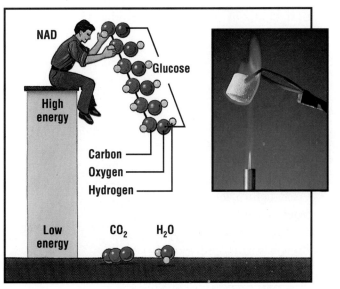

Figure 7.33

NAD acts as electron acceptor. High-energy glucose donates hydrogen and its electron to NAD. The final products— carbon dioxide and water—have less energy. The inset shows glucose burning. Oxygen removes hydrogen and its electron from glucose. A high-energy compound, glucose, is converted to low-energy compounds, carbon dioxide and water.

during an oxidation reaction might lead you to conclude that energy is created; however, as you read earlier, energy cannot be created—the total amount of energy remains constant within the system. The energy released during the chemical reactions comes from the chemical bonds of the high-energy molecules. Thus, the products of oxidation contain less energy than the reactants. The fact that the products have less energy can be easily proved. Place glucose in one test tube and light the glucose. The glucose burns. The substitution of glucose with carbon dioxide produces different results. Carbon dioxide does not burn. Oxygen from the atmosphere is capable of pulling electrons from glucose but is incapable of oxidizing carbon dioxide. Carbon dioxide has already been oxidized.

The energy released during the oxidation of glucose comes from the formation of new low-energy products. The chemical energy of glucose can be converted into heat or other forms of chemical energy storage such as ATP. However, glucose and other sugars will not release hydrogen and its electrons without first being activated. Consider the following scenario. The lid from a sugar bowl is removed exposing the sugar to oxygen, a strong electron acceptor. Does the plentiful supply of sugar become oxidized? The answer is no. The prospect of a bowl of sugar bursting into flames every time the top is removed would make our morning cups of coffee extremely dangerous! Sugar molecules must be activated before they can react.

The match provides the activation energy needed to initiate the

burning of sugar. However, activating sugar in this manner is unacceptable for your cells.

ATP is used to activate glucose by a process known as *phosphorylation*. During phosphorylation, a high-energy phosphate bond is linked with a glucose molecule. As in other chemical reactions, activation energy must be reached before the oxidation of glucose begins. Energy must be used before greater energy can be released.

ANAEROBIC RESPIRATION

Anaerobic respiration takes place in the absence of oxygen. In animal cells, glucose can be partially broken down to lactic acid, which contains less energy than glucose; however, many hydrogen atoms remain attached to the carbon atoms. If oxygen is available, even more energy can be released as lactic acid is oxidized. Another type of anaerobic respiration, *fermentation*, occurs in yeast. The products of fermentation are carbon dioxide and alcohol. During both alcoholic fermentation and lactic acid anaerobic respiration a limited number of chemical bonds are broken. A total of two ATP are produced by each process.

Anaerobic respiration *takes place in the absence of oxygen.*

Figure 7.34

Higher-energy glucose is oxidized to lower-energy products. Alcohol and lactic acid have less potential energy than glucose.

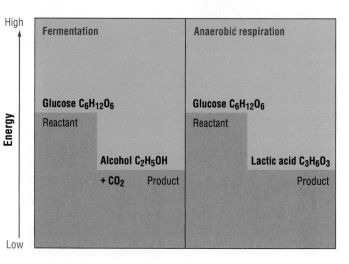

Glycolysis is the first step in cellular respiration. High-energy phosphate molecules from the cytoplasm of the cells are used to activate the six-carbon sugar, glucose. The activated glucose is converted into two three-carbon PGAL molecules. A coenzyme, NAD, accepts a hydrogen proton and its electron from each of the PGAL molecules, and is converted to NADH. The oxidized PGAL becomes phosphoglyceric acid, or PGA. The energy released by this oxidation reaction can now be used to form ATP. Other enzymes now remove the high-energy phosphates loaned to the activated glucose molecules, and PGA is converted into pyruvic acid. Pyruvic acid plays an important part in both anaerobic respiration and fermentation. Being an even stronger electron acceptor than NAD, pyruvic acid pulls hydrogen and its electron away from NADH. In turn, the NADH releases hydrogen and its electron to pyruvic acid, thereby restoring NAD. Pyruvic acid is said to be the terminal, or final, electron acceptor in anaerobic respiration. Figure 7.35 below summarizes the events of glycolysis in animals.

The final step shown in Figure 7.35 indicates that pyruvic acid is converted into lactic acid. However, lactic acid is not normally a product of fermentation. Yeast cells contain enzymes that extract carbon dioxide before lactic acid can be formed. The end-products of fermentation, therefore, are alcohol and carbon dioxide.

Lactic acid will accumulate in muscles during strenuous exercise if sufficient amounts of oxygen are not delivered to the tissues. Have you ever felt a sharp pain in your side while running? This pain may have been caused by a build-up of lactic acid. If you continue to run, the pain often intensifies. Anaerobic respiration does not supply enough energy to meet the demands that you are placing on your body, and the depletion of ATP reserves causes fatigue. The pain is merely providing a warning message.

Fermentation occurs in the cytoplasm of yeast cells. Like anaerobic lactic acid respiration, fermentation yields two ATP. However, special enzymes extract carbon dioxide from PGA, thereby creating different products. Bread dough rises when carbon dioxide gases are released during fermentation. The bubbles released in

Glycolysis is the process in which ATP is formed by the conversion of glucose into pyruvic acid.

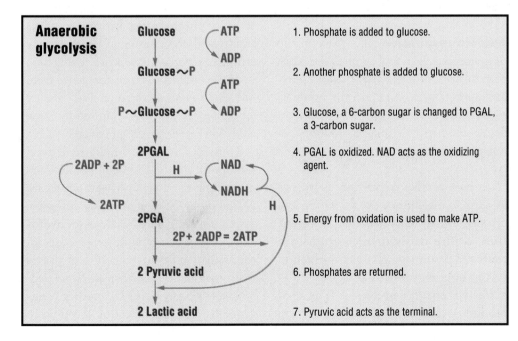

Anaerobic glycolysis

1. Phosphate is added to glucose.
2. Another phosphate is added to glucose.
3. Glucose, a 6-carbon sugar is changed to PGAL, a 3-carbon sugar.
4. PGAL is oxidized. NAD acts as the oxidizing agent.
5. Energy from oxidation is used to make ATP.
6. Phosphates are returned.
7. Pyruvic acid acts as the terminal.

Figure 7.35

Anaerobic glycolysis is the first step in cellular respiration.

$$C_6H_{12}O_6 + 6O_2 \longrightarrow 6CO_2 + 6H_2O$$
glucose oxygen   carbon   water
dioxide

typical energy yield = 36 ATP

Figure 7.36

Vast energy supplies come from ATP, which is synthesized during aerobic respiration; however, when energy demands exceed the oxygen supply, some ATP is formed by anaerobic respiration. The product of anaerobic respiration is lactic acid. It is now known that some athletes can tolerate higher than average levels of lactic acid, enabling them to continue energy production longer.

Figure 7.37

Yeast fermentation produces alcohol and carbon dioxide.

Aerobic respiration *refers to the complete oxidation of glucose in the presence of oxygen.*

champagne are caused by the same chemical process. Both champagne and bread dough also contain alcohol as a product of fermentation, although most of the alcohol that would be found in bread evaporates during the baking.

AEROBIC RESPIRATION

Aerobic respiration takes place in the presence of oxygen and involves the complete oxidation of glucose. Carbon dioxide, water, and 36 ATP are the products of aerobic respiration. The reaction below summarizes aerobic respiration:

$$C_6H_{12}O_6 + 6O_2 \longrightarrow 6CO_2 + 6H_2O + 36\ ATP$$

As in anaerobic respiration, glycolysis in aerobic respiration occurs within the cytoplasm, but, unlike anaerobic respiration, additional hydrogen atoms are removed from the glucose molecule. Each time an additional hydrogen is extracted, energy is released. The final products—carbon dioxide and water—are

low-energy compounds, unlike lactic acid and alcohol.

Cytochrome Enzyme System

The key to understanding why aerobic respiration provides greater amounts of energy than anaerobic respiration is found within the cytochrome enzyme system. Located in the mitochondria, a series of enzymes pass hydrogen and its electron along a series of progressively stronger electron acceptors. Each time hydrogen and its electron are passed from one electron acceptor to another, energy is released. The energy released by these oxidation reactions is used to make ATP.

Consider the fate of the hydrogen removed from the PGAL molecules during glycolysis. The hydrogen atoms are picked up by NAD, but unlike anaerobic respiration, in which NAD passes hydrogen and its electron back to pyruvic acid, the cytochrome enzyme system passes hydrogen and its electron to *flavin adenine dinucleotide*, or *FAD*, another coenzyme that acts as an electron acceptor. In turn, the FAD loses hydrogen and its electron to an even stronger electron acceptor, the cytochrome enzymes. The iron-containing cytochromes eventually give up hydrogen and its electron to the strongest electron acceptor in the body—oxygen. Oxygen acts as the terminal electron acceptor for aerobic respiration, combining with two hydrogen atoms and their electrons to form water.

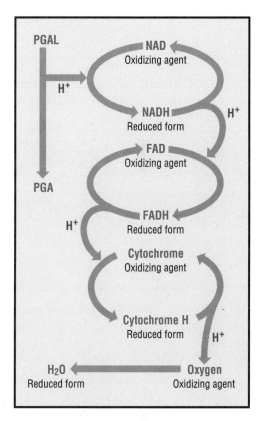

Figure 7.38

The cytochrome enzyme system. Each time hydrogen and its electron are passed from one electron acceptor to another, energy is released.

Without a constant supply of oxygen, the electrons could not be passed down the chain of oxidizing agents, and aerobic respiration would soon stop.

Each time hydrogen and its electron are transferred, energy is made available for the formation of ATP. Because two PGAL are converted to two PGA, two hydrogens are passed through the elec-

tron transport system. Each hydrogen and its electron moves through four transfers; however, the last transfer uses its oxidation energy to form water.

Krebs Cycle

In aerobic respiration, unlike anaerobic respiration, pyruvic acid does not serve as the final electron acceptor. Unstable pyruvic acid, in the presence of oxygen, breaks down to acetic acid. The acetic acid becomes attached to coenzyme A to form a complex called acetyl coenzyme A, or coA for short. Note that the conversion of pyruvic acid, a three-carbon compound, to acetyl coA, a two-carbon compound, requires the removal of carbon dioxide. During the reaction, hydrogen and its electron are removed from the pyruvic acid by NAD. If adequate amounts of oxygen are available, the hydrogen atom moves through the cytochrome system and additional ATP is formed.

The acetyl coenzyme A enters a series of chemical reactions referred to as the *Krebs cycle*, during which additional hydrogen and carbon dioxide molecules are extracted. The hydrogen atoms move through the electron transport chain within the mitochondria and eventually combine with oxygen atoms to form water. The final products of the Krebs cycle are carbon dioxide and water. Figure 7.40 summarizes the chemical changes that take place during the Krebs cycle.

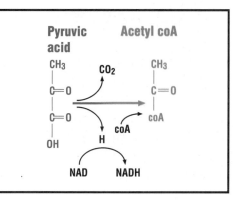

Figure 7.39

The conversion of pyruvic acid to acetyl coA.

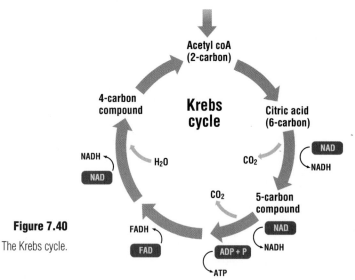

Figure 7.40

The Krebs cycle.

metabolism can also be used as raw materials to form fatty acids or amino acids. These biosynthesis pathways are indicated by blue arrows, while energy-releasing pathways are indicated by black arrows.

COMPARISON OF PHOTOSYNTHESIS AND RESPIRATION

Table 7.1 provides a comparison of photosynthesis and respiration.

Aerobic Respiration	Photosynthesis
Energy produced	Energy required
Oxidation	Reduction
High-energy reactants	Low-energy reactants
Low-energy products	High-energy products
Oxygen required	Oxygen released
Glucose required	CO_2 and H_2O required
CO_2 and H_2O produced	Glucose produced

INTERMEDIARY PATHWAYS

Glucose is the main source of energy, but many other organic compounds are consumed. How is energy derived from fats or proteins? Figure 7.41 indicates how excess fat can enter either glycolysis or the Krebs cycle. Similarly, proteins can be converted into amino acids and enter aerobic metabolic pathways. The products of

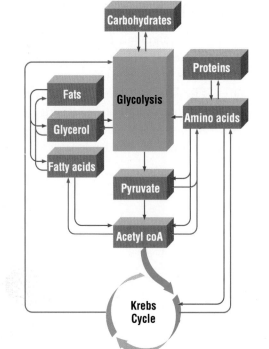

Figure 7.41

Intermediary metabolic pathways.

◼ REVIEW QUESTIONS ◼ ?

23 Contrast ATP production in anaerobic and aerobic respiration.

24 Why is the phosphorylation of glucose necessary?

25 How does the cytochrome enzyme system provide the energy for the synthesis of ATP?

26 How does the Krebs cycle provide additional ATP?

27 Under what conditions do plant cells undergo cellular respiration?

CAREER INVESTIGATION CAREER

Cells are essential to all forms of life on earth.
We must understand the mechanisms of cell function if we are to
properly utilize our life-form resources.

Cosmetologist

The science of skin and hair care is big business. Designing products that protect and preserve hair and skin requires a knowledge of how the hair and skin cells work, as well as how they are affected by environmental factors.

Veterinarian

A veterinarian needs to be familiar with many different species. Often diseases will have different symptoms and treatments for each species. To provide effective care to animals, a veterinarian must thoroughly understand how nutrients are exchanged in the animal.

Athlete

In order to obtain the greatest benefit from exercise and training, an athlete must know how the human body reacts to exercise and the nutrients the body requires to perform at peak levels. This requires a knowledge of how cells in the body exchange both information and material.

- Identify a career that requires a knowledge of cells and energy.
- Investigate and list the features that appeal to you about this career. Make another list of features that you find less attractive about this career.
- Which high-school subjects are required for this career? Is a postsecondary degree required?
- Survey the newspapers in your area for job opportunities in this career.

Pharmacist

A pharmacist's background in chemistry and biology comes into use daily. Many of the drugs we use were discovered accidentally. However, research into drugs designed specifically to combat certain diseases is now an important field.

SOCIAL ISSUE:
Interdependence of Cellular Respiration and Photosynthesis

The interaction of plants and animals has largely been ignored. The interdependence of respiration and photosynthesis illustrates the importance of plants. Any substantial changes in plant population must be considered in a wider sphere. Any reduction in forest areas will have an impact on human populations. Satellite sensing shows that tropical rain forests are disappearing at a rate of 171 000 km^2 each year.

Statement:

Legislation should be introduced to reduce fossil fuel emissions and the cutting of rain forests.

Point

- The constantly increasing human population has altered the balance between the carbon dioxide produced by combustion (oxidation) and the oxygen liberated by photosynthesis. The burning of fossil fuels is increasing levels of carbon dioxide. The climate is changing.
- Each year, the more than five billion people on our planet use about 40% of the organic matter fixed by plants. One estimate suggests that we consume two tonnes of coal and produce 150 kg of steel annually for every person on our planet. Carbon dioxide is produced by each of these reactions. Governments must institute a carbon tax to reduce carbon emissions.
- Wealthy countries should pay into a fund so that poor nations are not tempted to cut their trees.

Research the issue.
Reflect on your findings.
Discuss various viewpoints with others.
Prepare for the class debate.

Counterpoint

- The climate has changed in the past. Suggestions that increasing carbon dioxide levels have created an imbalance between the products of respiration and photosynthesis are not cause for alarm.

- The fact that some governments have considered taxing polluters does not mean that they understand the problem. The most important supply of oxygen is the ocean. Algae use carbon dioxide to synthesize sugars and oxygen. Should extreme measures be instituted, a country's economy would be seriously hindered.

- Education is the answer, not increasing taxes or bringing in legislation.

CHAPTER HIGHLIGHTS

- Chemical energy is stored in chemical bonds.
- Activation energy is required to begin many chemical reactions in cells.
- Chemical reactions within cells are regulated by enzymes. Enzymes are protein catalysts that lower activation energy.

- Enzymes permit chemical reactions within your body to proceed at low temperatures.
- Cofactors are inorganic molecules that help enzymes combine with substrate molecules.
- Coenzymes are organic molecules that help enzymes combine with substrate molecules.

- A competitive inhibitor has a shape complementary to a specific enzyme, thereby permitting it access to the active site of the enzyme. Inhibitors block chemical reactions.
- Feedback inhibition is the inhibition of the first enzyme in a metabolic pathway by the final product of that pathway.
- Precursor activity is the activation of the last enzyme in a metabolic pathway by the initial reactant.
- Photosynthesis is the process by which plants and some bacteria use chlorophyll, a green pigment, to trap sunlight energy. The energy is then used to synthesize carbohydrates.
- Photosynthetic reactions can be classified according to two main phases; light-dependent reactions and light-independent, or carbon-fixation, reactions.
- Cellular respiration is the process by which organisms obtain energy from the breakdown of high-energy compounds.
- The first law of thermodynamics states that energy can be neither created nor destroyed.
- The second law of thermodynamics states that all conversions of energy produce some heat, which is not useful energy.

- Electron transport systems are a series of progressively stronger electron acceptors. Each time an electron is transferred, energy is released.
- Oxidation occurs when an atom loses electrons. Reduction occurs when an element gains electrons.
- NAD is a strong electron acceptor important for electron transport systems in cellular respiration. NADH is the reduced form of NAD^+.
- $NADP^+$ is a strong electron acceptor important for electron transport systems in photosynthesis. NADPH is the reduced form of $NADP^+$.
- Chlorophyll is the green pigment found in plants. Chlorophyll traps sunlight energy for photosynthesis.
- Photolysis is the splitting of water by means of light energy.
- Cellular respiration is the oxidation of glucose to produce ATP.
- Phosphorylation is the process by which high-energy phosphate groups are added to molecules.
- Anaerobic respiration takes place in the absence of oxygen. Aerobic respiration refers to the oxidation of glucose in the presence of oxygen.

APPLYING THE CONCEPTS

1 Explain how enzymes work in the lock-and-key model. How has the "induced-fit" model changed the way in which biochemists describe enzyme activities?
2 Using the information that you have gained about enzymes, explain why high fevers can be dangerous.
3 Cyanide attaches to the active site of a cytochrome enzyme. How does cyanide cause death?
4 Use the metabolic pathway shown below to explain feedback inhibition.

5 What is the source of the oxygen released during photosynthesis? What is the source of carbon dioxide fixed during photosynthesis?
6 How does light intensity affect the rate of photosynthesis?
7 A photosynthesizing plant is exposed to radioactively labelled carbon dioxide. In which compound would the radioactive carbon first appear? (Consider NADPH, PGA, RuBP, and glucose.) Explain your answer.

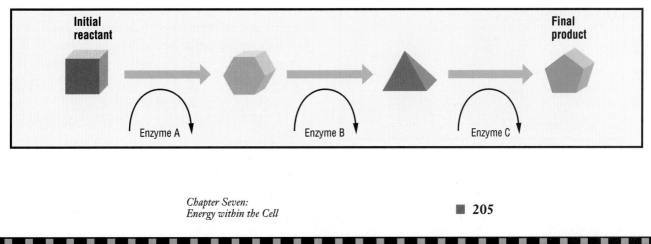

8 The removal of carbon dioxide from pyruvic acid distinguishes fermentation from lactic acid anaerobic respiration. What would occur if an enzyme in your body removed the carbon dioxide from pyruvic acid before lactic acid is formed?

9 Copy the chart below and place the correct answer in the appropriate column.

Characteristic	Anaerobic	Aerobic
1 Amount of ATP produced		
2 Terminal electron acceptor		
3 Site of activity in cell		
4 Final products		

10 Compare respiration and photosynthesis by completing a chart like the one below. Place a ✔ in the correct column.

Characteristic	Photosynthesis	Respiration
1 ATP used to initiate reaction		
2 ATP produced during reaction		
3 Oxidation reaction		
4 Reduction reaction		
5 Oxygen is a product		
6 Occurs in plant cells		
7 Occurs in animal cells		
8 Carbohydrate is reactant		

CRITICAL-THINKING QUESTIONS

1 In an attempt to study photolysis, a student uses a heavy isotope of oxygen in the form of water $(H_2{}^{18}O)$. If the following apparatus were used, indicate where you would expect to find the heavy isotope of oxygen. Explain your answer.

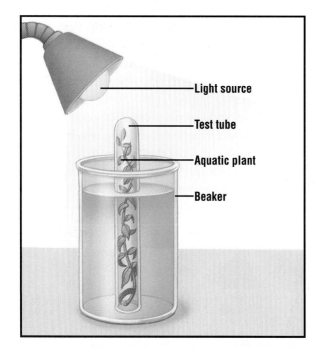

- Light source
- Test tube
- Aquatic plant
- Beaker

2 The following graph shows the rate at which products are formed for an enzyme-catalyzed reaction. By completing the graph, predict how a competitive inhibitor added at time X would affect the reaction.

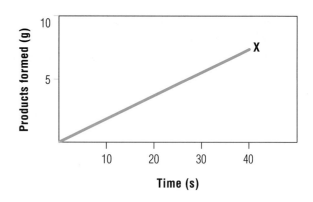

3 Antibiotic drugs act as competitive inhibitors for metabolic pathways in invading microbes but do not interfere with the metabolic pathways in human cells. Using this information, explain how cancer-preventing drugs are designed to work. Why is it difficult to develop a cancer-preventing drug?

4 The earth's early atmosphere was rich in carbon dioxide and low in oxygen. With the evolution of plankton, the composition of the atmosphere changed. Through photosynthesis, carbon dioxide in the atmosphere was fixed by the plankton, and oxygen was released. As plants became more abundant, carbon dioxide levels began to fall and oxygen levels began to rise. A number of scientists have suggested that increasing algae growth in the oceans may reverse increasing carbon dioxide levels and reduce atmospheric warming. One suggestion is to seed the oceans with iron and phosphates—nutrients that will increase plankton blooms. Comment on this strategy.

5 The following experiment was designed to demonstrate the relationship between plants and animals. Bromthymol blue indicator was placed in each of the test tubes. High levels of carbon dioxide will combine with water to form carbonic acid. Acids will cause the bromthymol blue indicator to turn yellow. The initial color of the test tubes is blue.

Predict the color change, if any, in each of the test tubes. Explain your predictions.

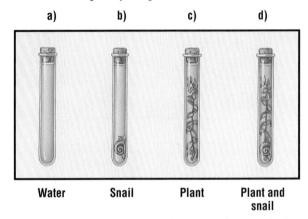

a)	b)	c)	d)
Water	Snail	Plant	Plant and snail

ENRICHMENT ACTIVITIES

Suggested reading:
- Doolittle, G. "Proteins." *Scientific American*, 253(4) (October 1988): pp. 88–99.
- Hinkle, P., and R. McCarthy. "How Cells Make ATP." *Scientific American* 238(23) (March 1978): p. 104.
- Kunzig, Robert. "Invisible Garden." *Discover* 11(4) (April 1990): p. 66.
- Lovelock, James. *The Ages of Gaia: A Biography of Our Living Earth*. New York: W.W. Norton Company, 1988.
- Miller, K. "Three-Dimensional Structure of Photosynthetic Membrane." *Nature* 300 (1982): p. 5887.
- Moore, P. "The Varied Ways Plants Trap the Sun." *New Scientist*, February 1981, p. 394.
- Steinberg, S. "Genes Shed Light on Photosynthesis." *Science News*, August 1983, p. 102.

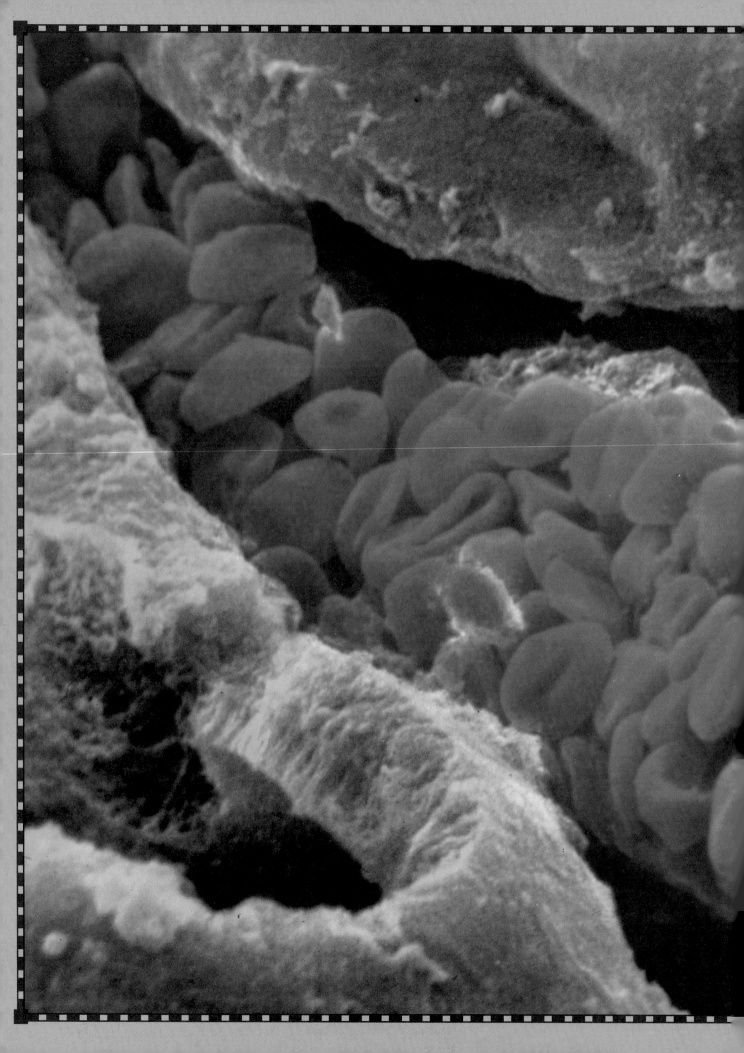

Exchange of Matter and Energy in Humans

■

UNIT

3

Organs and Organ Systems

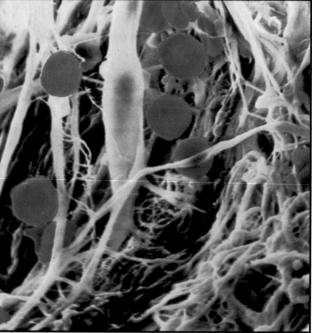

IMPORTANCE OF ORGANS AND ORGAN SYSTEMS

Your body is made up of trillions of cells. Unlike single-celled organisms, which appear somewhat alike, the cells of your body display remarkable variation. Your cells are organized into groups that have distinctive shapes and functions. Muscle cells, for example, are relatively long and contain special protein filaments that slide over each other when stimulated. These specialized filaments cause the muscle cell to shorten during muscle contractions. Fat cells contain unusually large vacuoles that store the high-energy fat compounds. Long, thin nerve cells, some of them nearly 1 m in length but less than 0.001 mm in diameter, are covered with a glistening sheath that helps provide insulation and speeds the movement of impulses. Specialized ends of the motor nerve cells either receive nerve messages or produce transmitter chemicals that permit the passing of the impulse to other nerves.

Like multicellular organisms, single-cell organisms must move, coordinate information from their surroundings, and store energy, but unlike multicellular

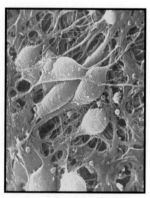

Nerve cells from the cerebral cortex. Nerve cells exist in varying sizes and shapes, but all have the same basic structure.

Figure 8.1

Human connective tissue, showing collagen fibers. Red blood cells are interspersed between the fibres. Connective tissue provides metabolic and structural support for other tissues and organs.

organisms they must accomplish these tasks without the benefit of specialization.

Imagine how difficult your life would be if specialists did not exist. Could you repair a car's transmission by yourself? Would you be able to diagnose your own illnesses? Would you be able to produce the food required for a well-balanced diet? The division of labor permits greater efficiency in function. By specializing, cells become very good at performing particular tasks. For example, muscle cells are well suited for contracting, but are poorly suited for other tasks.

LEVELS OF CELL ORGANIZATION

Tissues

Cells similar in shape and function work together to ensure the survival of the organism. These cells make up the body's **tissues.** Cells within a tissue are capable of recognizing similar cells and sticking to them. This point is well illustrated by a simple experiment that was performed on two types of sponges. Although sponges do not have organs or organ systems, their cells are arranged in two layers of similar tissue. The cells of red and yellow sponges were separated from neighboring cells by passing them through a fine-mesh sieve. Researchers noted that some cells adhered to each other, but not to other cells. With further investigation, the researchers were able to determine that cells from the red sponge only adhered to other cells from the red sponge. Similarly, cells from the yellow sponge formed clusters with other cells from the yellow sponge. As the clusters became larger, researchers noted that the cells began to arrange themselves into patterns characteristic of an intact sponge.

In the human body there are four primary kinds of tissue: epithelial, connec-tive, muscle, and nervous. **Epithelial tissue** is a covering tissue that protects not only the outer surface of your body, but your internal organs as well. It also lines the inside of your digestive tract and blood vessels. A cell's size, shape, and arrangement in the tissues is consistent with its designated function. For example, a single layer of cells is well suited for transporting nutrients or wastes within the body. Multiple layers form a barrier to protect organs from invading microbes and disease-causing agents. Layers of thin, flat cells, referred to as *squamous epithelium*, provide maximum coverage and protection for the outer layers of your skin. Similarly shaped cells also line blood vessels and comprise the air sacs of the lung. Cube-shaped epithelial tissue provides the framework for the kidney tubules, while pillar-shaped cells, referred to as *columnar epithelium*, line the stomach and the respiratory tract. The cube-shaped cells are well suited for providing strong structural support, while the much longer cells within columnar epithelium often produce secretions.

Connective tissue is made up of fibrous proteins and a material called the **matrix,** which is found between the cells. Cells called *fibroblasts* produce the solid, semi-solid, or fluid matrix and fibrous proteins. The components of the connective tissue are arranged according to function. For example, fibrous protein structures are commonly found in muscles, which move limbs; tendons, which anchor muscle to bone; and ligaments, which bind bones at the joints. Connective tissues are also found in blood, adipose (fat), cartilage, bone, and lymph tissue. Blood and lymph tissue contain a fluid matrix that is well-suited for transporting materials throughout the body. A more solid matrix is found in bone and cartilage, where support is necessary.

Tissues *are groups of similarly shaped cells that work together to carry out a similar function.*

Epithelial tissue *is a covering tissue that protects organs, lines body cavities, and covers the surface of the body.*

Connective tissue *provides support and holds various parts of the body together.*

The **matrix** *is the noncellular material secreted by the cells of connective tissue.*

Figure 8.2

Adipose connective tissue contains many cells with fat-storage vacuoles.

Muscle tissue is composed of cells containing special contractile proteins and accounts for approximately 40 to 50% of your total body mass. Muscle tissues shorten when these proteins contract. This contraction enables you to move.

Nerve tissue may be the most diverse and complex of all tissues. The brain, the spinal cord, and the sense organs are primarily composed of nerve tissue. Charged with the task of communication, nerve tissue is essential for the growth and development of all other tissues. Other tissues rely on nerve tissue for information about the environment in order to coordinate responses to environmental changes.

Organs

Organs are groups of different tissues specialized to carry out particular functions. The heart is an excellent example of a complex organ. The heart's outer structure is covered by epithelial tissue, which also lines the inside of the heart's chambers and blood vessels. Nerve tissue initiates and synchronizes cardiac muscle

contractions and relays information to the brain about the strength of these contractions and the relative condition of the muscle cells. The blood that pulses through the heart is connective tissue. Your hands, stomach, and kidneys are further examples of complex organs. Although each organ is composed of a variety of different tissues, the tissues act together to accomplish a common goal.

Organ Systems

Organ systems are groups of organs that have related functions. Although your heart can be considered the focal point of blood transportation, it is but one of the many important organs that make up your circulatory system. The circulatory system consists of the heart, the arteries (which carry blood rich in oxygen and other nutrients to the tissues), the capillaries (the site of nutrient and waste exchange), and the veins (which carry wastes away from the cell). Table 8.1 outlines the levels of cell organization for the body's important organ systems.

Organs are structures composed of different tissues specialized to carry out a specific function.

Organ systems are groups of organs that have related functions. Organ systems often interact.

Figure 8.3
Human organ systems.

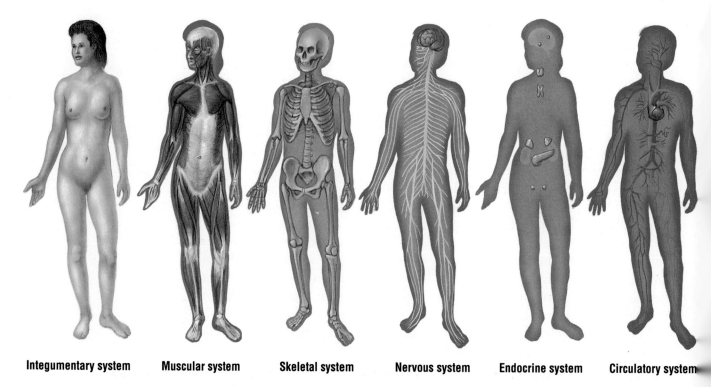

Integumentary system Muscular system Skeletal system Nervous system Endocrine system Circulatory system

Unit Three:
Exchange of Matter and Energy in Humans

Table 8.1 Levels of Cell Organization

Organ System	Organs	Tissues
Nervous	brain, spinal cord, eye, ear, peripheral nerves	nerve, connective, epithelial
Excretory	kidney, bladder, ureter, urethra	epithelial, nerve, connective, muscle
Circulatory	heart, blood vessels	epithelial, nerve, connective, muscle
Digestive	esophagus, stomach, intestines	epithelial, nerve, connective, muscle
Reproductive	testes, vas deferens, ovary, uterus, Fallopian tubes, glands	epithelial, nerve, connective, muscle
Respiratory	lungs, windpipe, blood vessels	epithelial, nerve, connective, muscle
Endocrine	pancreas, adrenal glands, pituitary	epithelial, nerve, connective

Organ systems can be classified in many different ways. Part of the difficulty in establishing a classification system stems from the way in which body systems interact with each other. For example, the body's circulatory system would not be able to function properly if the respiratory system did not provide adequate gas exchange. The functions of the circulatory and respiratory systems are so intertwined that some scientists choose to refer to a single cardiopulmonary system.

A second problem in classifying organ systems arises because some organs are classified according to anatomy rather than function. Consider the kidneys and large intestine, which both remove wastes from the body. The kidneys remove wastes from the blood, while the large intestine concentrates and stores undigested matter. Although the kidney is classified as part of the excretory system, the large intestine is often described as the last segment of the digestive system, even though it does not digest food. Figure 8.3 shows an overview of human organ systems.

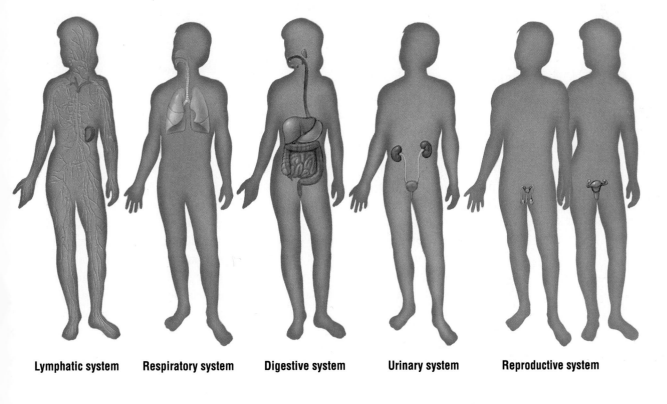

| Lymphatic system | Respiratory system | Digestive system | Urinary system | Reproductive system |

MONITORING ORGANS

The first X-ray photograph was taken in 1895 when the German physicist Wilhelm Roentgen used high-energy electromagnetic waves to take a picture of his wife's hand. X rays pass through soft tissue like muscle, but are absorbed by denser bone.

Walter Cannon was the first person to harness X rays to diagnose soft tissue. Cannon developed a stain that contained bismuth, a nontoxic mineral that is opaque to X rays. Because the X ray could not penetrate the stained organ, the organ became visible as a white image on a black background. For the first time, organ structures could be observed without surgery.

Today, scientists have combined X-ray technology with computer technology to view body organs in ways that Roentgen and Cannon never dreamed possible. Computer-assisted tomography, or CAT scans, allow doctors to view many different X rays from numerous angles. The computer interprets the X-ray images and reconstructs them to provide three-dimensional representations of any body organ.

In the CAT scan procedure, the X-ray machine rotates around the patient, taking hundreds of individual pictures. The images are stored in a computer along with their location and angle. The computer can reassemble the pictures to provide thin cross-sectional views, which the physician can rotate 180°. The computer organizes the pictures automatically, permitting three-dimensional imaging; the organ can then be viewed section by section.

The imaging of the CAT scan is so accurate that it can detect abrasions as small as one millimeter. The scanner can also distinguish between gases, fluids, and solid tissues, and is able to identify tumors imbedded in the brain or liver. CAT scans are particularly useful as a diagnostic tool

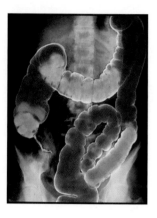

Figure 8.4

The penetrating properties of X rays, combined with stains, are used to monitor the state of internal organs. This X ray shows the large intestine.

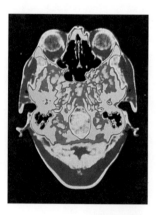

Figure 8.5

CAT scan of a cross section through the base of the skull, showing the eyeballs at the top.

to assess head injuries, which can often be life threatening because blood masses, called *hematomas*, impair the flow of blood through the blood vessels of the brain.

FRONTIERS OF TECHNOLOGY: NUCLEAR MEDICINE

Nuclear imaging is a valuable diagnostic tool that allows doctors to view a beating heart or detect bone cancer without resorting to surgery. Unlike the CAT scan, which uses external radiation to produce an image, nuclear imaging measures the radiation emitted from within the body. Nuclear imaging also provides information about the function of the organ as opposed to its structure.

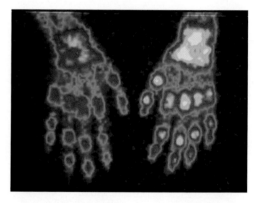

Figure 8.6

Hands of a person with extensive rheumatoid arthritis. The arthritic joints appear as brighter areas. The image records the distribution and intensity of gamma rays emitted by a tiny amount of radioisotope injected into the patient.

Nuclear imaging employs radioisotopes much like X rays use opaque dyes to identify organs. Radioisotopes, now called *radionuclides*, are unstable atoms that emit rays of energy. The radionuclides are injected into the body and collect in the target organs. A scanner, called a gamma camera, records the release of the energy from the radionuclides and produces a picture.

Various radionuclides can be used to identify specific organs. A thallium isotope is very valuable for heart imaging. Large amounts of the isotope collect in areas of damage, called *infarctions*, and produce a "hot spot." If the camera detects the hot spot, the physician knows that the damage has occurred within the past five days, the maximum amount of time in which the infarction will collect the radionuclides.

Nuclear magnetic resonance (NMR) technology is beginning to rival the CAT scan as a valuable diagnostic technique. Nuclear magnetic resonance works by subjecting the nucleus of a specific atom to a combination of magnetic forces and radio waves to determine whether or not the nucleus behaves normally. Because NMR does not use any external source of radiation, it is, at least theoretically, safer than the CAT scan. The newer NMR technique not only provides a map of various organs, but also reveals how the organs are functioning. It even shows promise for shedding light on some of the chemical processes that take place within a cell.

NMR technology is continually being refined and its applications extended. It is becoming a promising tool for determining if tumors are cancerous; and because it measures the spin of the hydrogen atom, it is very valuable for imaging soft tissues. Soft tissues contain great quantities of water, of which the hydrogen molecule is a component. However, despite the great promise of NMR, its use is not yet widespread as equipment costs can run up to $1.5 million.

ANATOMY TERMS

The human body can be divided into a number of body cavities, as shown in Figure 8.8. The brain is housed in the **cranial cavity.** The cranial cavity is composed of smaller cavities such as the mouth, nasal cavities, and auditory cavities, which contain the organs of balance and equilibrium. The spinal cavity runs from the cranial cavity down the back.

Nuclear imaging *techniques use radioisotopes to view organs and tissues of the body.*

Nuclear magnetic resonance (NMR) *techniques employ magnetic fields and radio waves to determine the behavior of molecules in soft tissue.*

The **cranial cavity** *is surrounded by the skull, which protects the brain, eyes, and inner ear.*

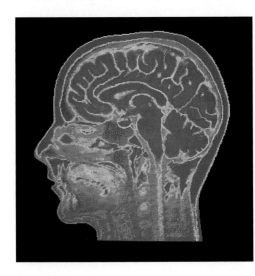

Figure 8.7

Magnetic resonance image of a section through the head of a normal 42-year-old female, showing structures of the brain, spine, and facial tissues.

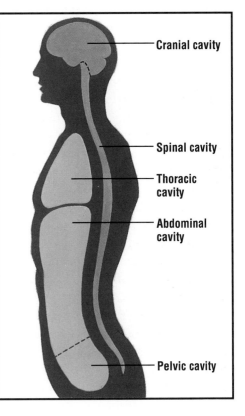

Cranial cavity

Spinal cavity

Thoracic cavity

Abdominal cavity

Pelvic cavity

Figure 8.8

The anterior body cavities are shown in orange; posterior cavities are shown in purple.

Protected by the ribs, the **thoracic cavity** houses the lungs and heart. A large band of muscle called the *diaphragm* separates the thoracic cavity from the **abdominal cavity.** Organs that occupy the abdominal cavity are often referred to as *viscera* and include the major digestive organs. The spleen, kidneys, and adrenal glands are also found in this area.

The body can also be divided into imaginary planes, which are often used as maps to help locate body parts. Like treasure maps, which use prominent features of the landscape to locate a hidden treasure, the imaginary planes of the body can be used to locate less obvious structures. Figure 8.9 provides directions for reading a map of the body. The *mid-sagittal* plane divides the body into left and right halves. The *transverse* plane divides the body into front (anterior) and back (posterior). The *frontal* plane divides the body into dorsal and ventral segments or posterior and anterior in humans.

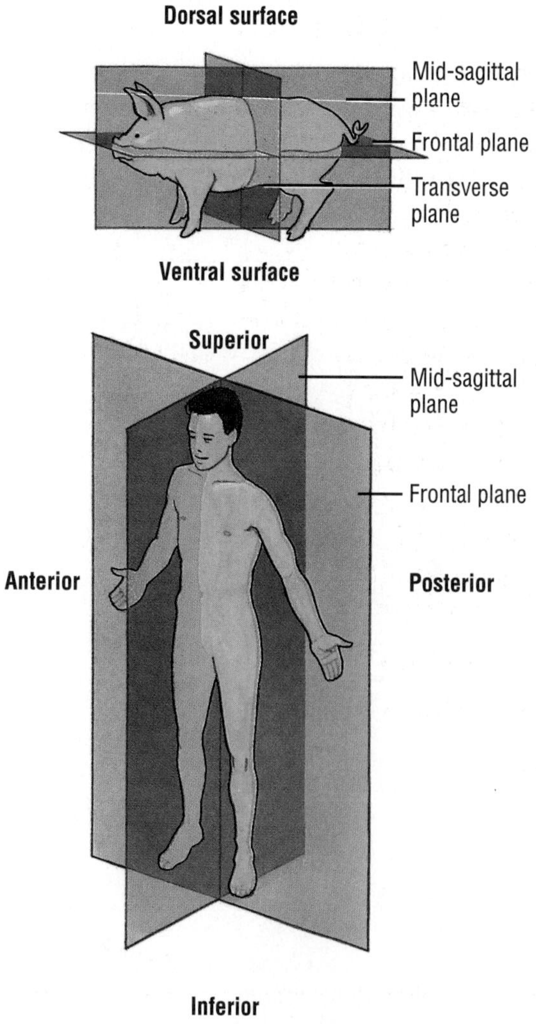

Table 8.2 provides a list of terms that will help you locate various anatomical structures.

*The **thoracic cavity** is the chest cavity. Heart and lungs are contained in the thoracic cavity.*

*The **abdominal cavity** extends from the diaphragm to the upper part of the pelvis. The stomach, liver, pancreas, small intestine, large intestine, and kidneys are contained in the abdominal cavity.*

Dorsal surface

Mid-sagittal plane

Frontal plane

Transverse plane

Ventral surface

Figure 8.9

Planes of symmetry in a pig and a human.

Superior

Mid-sagittal plane

Frontal plane

Anterior

Posterior

Inferior

Table 8.2   Some Anatomical Terms

Term	Definition
Anterior	The front section of the body; from the face to the feet
Posterior	The back section of the body; the area behind an organ or the area along the back of the body
Superior	The upper part of the body
Inferior	The lower part of the body
Medial	Close to the mid-line of the body
Lateral	Away from the mid-line of the body
Proximal	Close to the point of attachment; for example, the elbow is proximal to the forearm.
Distal	Away from the point of attachment; for example, the elbow is distal from the wrist.

◼ REVIEW QUESTIONS ◼ ?

1 Differentiate between a tissue, an organ, and an organ system. Provide examples of each in your explanation.

2 List the four main groupings of tissue and provide an example of each.

3 Why is blood considered to be a tissue?

4 Why do different classification schemes for organ systems exist?

5 Discuss the advantages and disadvantages associated with X rays, CAT scans, nuclear imaging, and NMR.

6 Draw a diagram of the human body and label the following cavities: cranial, thoracic, and abdominal.

LABORATORY
LAB #1: FETAL PIG DISSECTION

Background Information

Like humans, the pig is a placental mammal, meaning that the fetus receives nourishment from the mother through the umbilical cord. Because the anatomy of the fetal pig resembles that of other placentals, this laboratory serves two important functions. It provides a representative overview of vertebrate anatomy and provides the framework for understanding functioning body systems.

Read and follow the procedure carefully. Accompanying diagrams are designed to provide reference only. Use the appropriate dissecting instruments. This and the following lab have been designed to minimize the use of a scalpel.

Objective

To study the external anatomy and organ systems of the abdominal cavity of the fetal pig.

Materials

preserved pig	dissecting gloves
probe	lab apron
string	ruler
dissecting tray	forceps
scalpel	dissecting pins
scissors	safety goggles
dissecting microscope or hand lens	

Safety Notes

- **Wear safety goggles and an apron at all times.**
- **Wear plastic gloves when performing a dissection to prevent any chemicals from coming in contact with your skin.**
- **Fetal pigs are preserved in a formalin-based or other preservative solution. Wash all splashes off your skin and clothing immediately. If you get any chemical in your eyes, rinse for at least 15 min.**

- **Work in a well-ventilated area.**
- **When you are finished the activities, clean your work area, wash your hands thoroughly, and dispose of all specimens, chemicals, and materials as instructed by your teacher.**

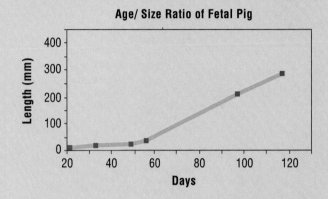

Procedure

Part I: External Anatomy

1 Place your pig in a dissecting tray. Using a ruler, measure the length of the pig from the snout to the tail. Use the graph below to estimate the age of the fetal pig.

Age/ Size Ratio of Fetal Pig

(Graph: Length (mm) on y-axis ranging from 0 to 400; Days on x-axis ranging from 20 to 120.)

a) Estimate and record the age of your fetal pig.

2 Identify the four regions of the pig's body: the head, the neck, the trunk, and the tail.

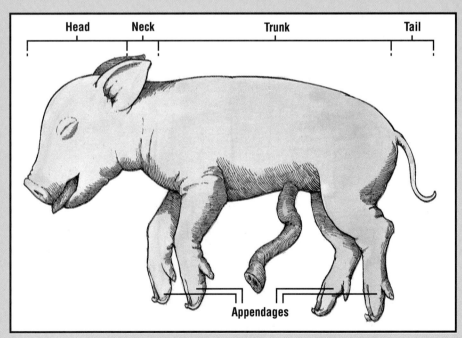

Head   Neck   Trunk   Tail

Appendages

Lateral view of fetal pig.

3 Place the pig on its back (dorsal surface) and observe the umbilical cord. Locate the paired rows of nipples along the ventral surface of the pig. Both males and females have these nipples.

b) What is the function of the umbilical cord?

4 Use the diagrams to determine the sex of your pig. In females, the urogenital opening is located immediately below the anus. A small, spiked tissue called the *genital papilla* projects from the urogenital opening. See the diagram of the female pig. In males, the scrotum containing the testes can be located just below the anus. The urogenital opening of the male is found immediately posterior to the umbilical cord. See the diagram of the male pig.

c) What is the sex of your pig?

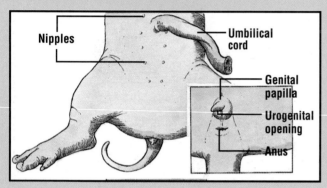

Ventral view of the female pig.

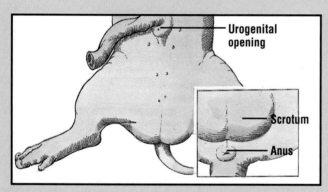

Ventral view of the male pig.

5 Examine the feet of the fetal pig.

d) Indicate the position and the number of toes.

Part II: The Abdominal Cavity

During the dissection you will be directed to examine specific organs as they become visible. Remove only those organs indicated by the dissection procedure. Proceed cautiously, to prevent damaging underlying structures.

6 With the pig still on its dorsal surface, attach one piece of string to the pig's ankle and another to its opposite wrist. Pull each piece under the dissecting pan and tie together. Repeat the procedure for the other wrist and pull once again so that the ventral surface is exposed.

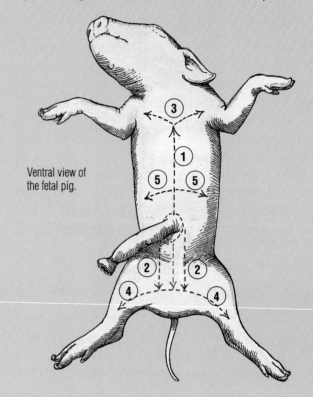

Ventral view of the fetal pig.

7 Using scissors, make an incision just in front of the umbilical cord and cut toward the anterior of the pig. Follow the incision shown by the diagram, which is indicated as #1.

8 Follow the incision markers, indicated in the diagram as #2, toward the posterior of the pig. Make an incision near the neck, indicated by #3, and then at the posterior portion of the abdominal cavity, indicated by #4. Make a lateral incision near the posterior portion of the ribs, indicated by #5. This incision runs parallel to the diaphragm, which separates the thoracic cavity at the anterior of the animal from the abdominal cavity, near the posterior of the animal.

9 Pull apart the flaps along incision #5, exposing the abdominal cavity. Use dissecting needles to tear the connective tissue (*peritoneum*) that holds the internal organs to the lining of the body cavity. Now pull apart the flaps of skin covering incision #4 to expose the posterior portion of the abdominal cavity. Use pins to hold back the flaps of skin.

10 Locate the liver near the anterior of the abdominal cavity. Using a probe, lift the lobes and locate the saclike gall bladder. Follow the thin duct from the gall bladder to the coiled small intestine. Bile salts, produced in the liver, are stored in the gall bladder. The bile duct conducts the fat-emulsifying bile salt to the small intestine.

e) How many lobes does the liver have?

f) Describe the location of the gall bladder.

11 Locate the J-shaped stomach beneath the liver. Using forceps and a probe, lift the stomach and locate the esophagus near the anterior junction. Locate the small intestine at the posterior junction of the stomach. The coiled small intestine is held in place by *mesentery*, a thin, somewhat transparent connective tissue. Note the blood vessels that transport digested nutrients from the intestine to the liver.

12 Using a dissecting needle and forceps, lift the junction between the stomach and small intestine, removing supporting tissue. Uncoil the junction and locate the creamy-white gland called the pancreas. The pancreas produces a number of digestive enzymes and a hormone called insulin, which helps regulate blood sugar.

g) Describe the appearance of the pancreas.

13 Locate the spleen, the elongated organ found around the outer curvature of the stomach. The spleen stores red and white blood cells. The spleen also removes damaged red blood cells from the circulatory system.

14 Using a scalpel, remove the stomach from the pig by making transverse cuts near the junction of the stomach and the esophagus, and near the junction of the stomach and small intestine. Make a cut along the mid-line of the stomach, open the cavity, and rinse. View the stomach under a dissecting microscope.

h) Describe the appearance of the inner lining of the stomach.

Laboratory Application Questions

1 State the function of the following organs.

Organ	Function
Stomach	
Liver	
Small intestine	
Gall bladder	
Pancreas	
Large intestine	
Spleen	

2 What is the function of the mesentery?

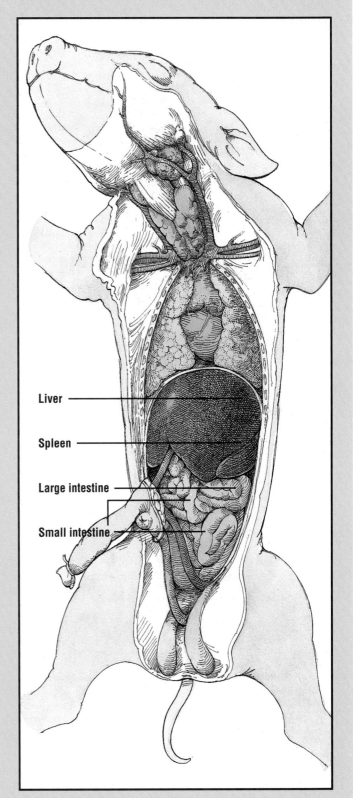

Abdominal cavity and thoracic cavity of the fetal pig. Organs of the digestive system and circulatory system are highlighted in the diagram.

LABORATORY

LAB #2: FETAL PIG DISSECTION

Objective

To study the thoracic cavity and urogenital system of the fetal pig.

Materials

preserved pig	dissecting tray
scissors	dissecting pins
scalpel	probe
forceps	ruler
dissecting gloves	dissecting microscope or hand lens
lab apron	safety goggles

Safety Notes

- **Wear safety goggles and an apron at all times.**
- **Wear plastic gloves when performing a dissection to prevent any chemicals from coming in contact with your skin.**
- **Fetal pigs are preserved in a formalin-based or other preservative solution. Wash all splashes off your skin and clothing immediately. If you get any chemical in your eyes, rinse for at least 15 min.**
- **Work in a well-ventilated area.**
- **When you are finished the activities, clean your work area, wash your hands thoroughly, and dispose of all specimens, chemicals, and materials as instructed by your teacher.**

Procedure

Part I: Thoracic Cavity

1 Carefully fold back the flaps of skin that cover the thoracic cavity. You may use dissecting pins to attach the ribs to the dissecting tray.

a) What organs are found in the thoracic cavity?

2 Locate the heart. Using forceps and a dissection probe, remove the *pericardium* from the outer surface of the heart. The large blood vessel that carries blood from the liver to the right side of the heart is called the *inferior vena cava*. (The right side refers to the pig's right side.)

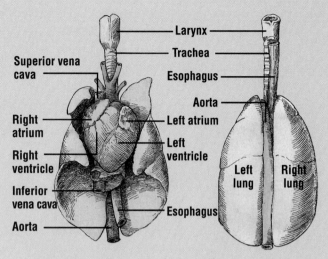

(a) Ventral view of heart and lungs. (b) Dorsal view of heart and lungs.

3 Blood from the head enters the right side of the heart through the *superior vena cava*. Both the superior and inferior venae cavae are considered to be veins because they bring blood to the heart.

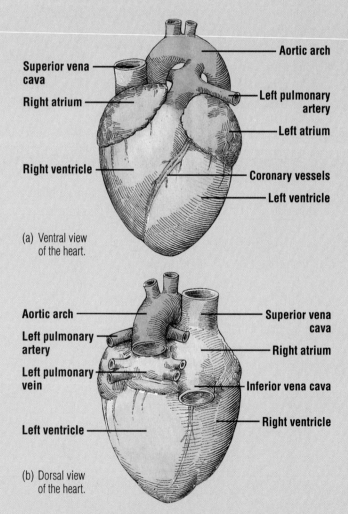

(a) Ventral view of the heart.

(b) Dorsal view of the heart.

4 Trace blood flow through the heart. Blood entering the right side of the heart collects in the *right atrium*. Blood from the right atrium is pumped into the *right ventricle*. Upon contraction of the right ventricle, blood flows to the lungs by way of the *pulmonary artery*. Arteries carry blood away from the heart. Blood, rich in oxygen, returns from the lungs by way of the pulmonary veins and enters the left atrium. Blood is pumped from the left atrium to the left ventricle and out the *aorta*.

5 Make a lateral incision across the heart and expose the heart chamber.

b) Compare the size of the wall of a ventricle with that of an atrium.

c) Why does the left ventricle contain more muscle than the right ventricle?

6 Locate the spongy lungs on either side of the heart and the *trachea* leading into the lungs.

d) Why do the lungs feel spongy?

7 Place your index finger on the trachea and push downward.

e) Describe what happens.

f) What function do the cartilaginous rings of the trachea serve?

Part II: Urogenital System

8 Using scissors, remove the intestines and what remains of the stomach.

9 Refer to the diagram of the thoracic cavity and urogenital system and locate the kidney. Using forceps and a scalpel, carefully remove the fat deposits that surround the kidney.

g) Describe the shape and color of the kidneys.

10 Locate the thin tube leading from the kidneys. The *ureter* carries urine from the kidney to the bladder. Cut into the kidney and note the large number of tubules.

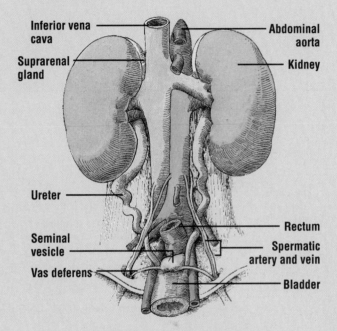

Urinary system of a fetal pig.

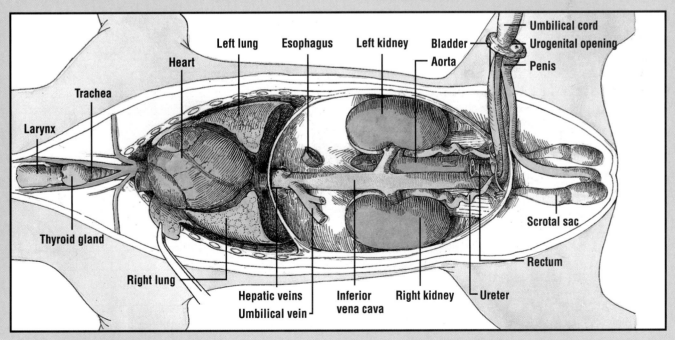

Thoracic cavity and urogenital system.

If your pig is a male, refer to steps 11 and 12. If your pig is a female, refer to steps 13 and 14. Make sure that you also view the organs of a pig of the opposite sex from your specimen.

11 Use the diagram of the male urogenital system to locate the testes, which produce the sperm cells. If your fetal pig is advanced in development, the testes may have descended into the scrotum; however, they will probably be found in the *inguinal canal*. Like the ovaries, the testes develop inside the body cavity. The lower temperatures of the scrotum promote the proper development of sperm cells.

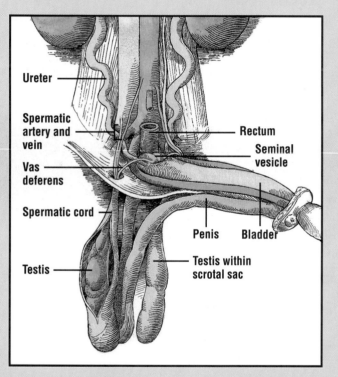

Urogenital system of a male pig.

12 Use the diagram to locate the vas deferens, which conducts sperm cells from the testes to the urethra.

13 Use the diagram of the female reproductive system to locate the ovaries, which produce the egg cells. The ovaries can be found immediately posterior to the kidneys.

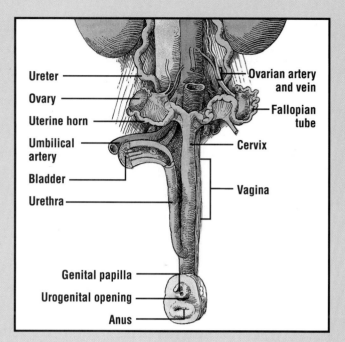

Ventral view of the female reproductive system.

14 Locate the Fallopian tubes leading from the ovaries. The Fallopian tubes are supported by the broad ligaments. Follow the Fallopian tubes, which meet to form the uterus, or womb. The Fallopian tube is the site of fertilization. Once fertilized, the egg travels to the uterus, the site of embryo and fetal development. Locate the vagina and follow the canal into the uterus. The constriction that marks the division between the vagina and uterus is known as the *cervix*.

Laboratory Application Questions

1 State the function of the following organs.

Organ	Function
Heart	
Kidney	
Ureter	
Urethra	
Testes	
Ovaries	
Uterus	

2 How is the abdominal cavity separated from the thoracic cavity?

3 Indicate why the male reproductive system is often referred to as the *urogenital system* while the female's is not. ■

HOMEOSTASIS AND CONTROL SYSTEMS

Your body works best at 37°C, with a 0.1% blood sugar level, and at a blood pH level of 7.35. However, the external environment does not always provide the ideal conditions for life. Air temperatures in Canada can fluctuate between −40°C and +40°C. Foods are rarely 0.1% glucose and rarely have a pH of 7.35. You also place different demands on your body when you take part in various activities, such as playing racketball, swimming, or digesting a large meal. Your body systems must adjust to these variations to maintain a reasonably constant internal environment. The system of active balance requires constant monitoring or feedback about body conditions. Information about blood sugar, body temperature, blood pressure, and oxygen levels, to name a few, are relayed to a coordinating center once they move outside the normal limits. From the coordinating center, regulators bring about the needed adjustments. An increase in the heart rate during exercise or the release of glucose from the liver to restore blood sugar levels are but a few of the adjustments made by regulators.

> **BIOLOGY CLIP**
> Sweating is an important homeostatic mechanism. Because evaporation requires heat, your body cools when you perspire. Following extreme exercise, sweat is produced all over the surface of your body. Fear or nervousness produces sweat mainly on the palms and soles.

The term **homeostasis** is most often used to describe the body's attempt to adjust to a fluctuating external environment. The word is derived from the Greek words *homoios*, meaning "similar" or "like," and *stasis*, which means "standing still." The term is appropriate because the body maintains a constant balance, or steady state, through a series of monitored adjustments.

Special receptors located in the organs of the body signal a coordinating center once an organ begins to operate outside its normal limits. The coordinating center relays the information to the appropriate regulator, which helps restore the normal balance. For example, pressure receptors located in the arteries of your neck become distended when blood pressure exceeds normal limits. A nerve is excited and a message is sent to the brain, which relays the information, by way of another nerve, to the heart. The pace and strength of heart contractions are reduced, thereby lowering blood pressure. A system of turning down or turning up the force and rate of heart contractions is used to maintain homeostasis.

Homeostasis *is a process by which a constant internal environment is maintained despite changes in the external environment.*

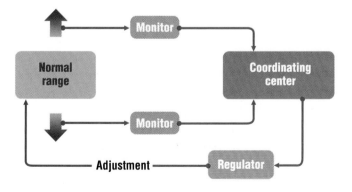

Figure 8.10

A control system.

FRONTIERS OF TECHNOLOGY: ARTIFICIAL BODY PARTS

In the 1987 movie *RoboCop*, a police officer is killed in the line of duty. Scientists use his remains to construct a cyborg—part man, part machine— that is virtually indestructible. Although the film may have been ahead of its time in showing how science can rebuild human body parts, it did reflect developments in a rapidly growing field.

Artificial hips became practical during the late 1960s, elbows during the mid-1970s, and wrists during the late 1970s. People with rheumatoid arthritis and degenerative bone disorders have been given new hope. Microelectronics has provided a new generation of artificial limbs for amputees. Special electrodes set in artificial limbs can be activated either subconsciously or consciously by the user.

Skin substitutes have provided a temporary solution for burn victims. Unlike skin transplants from pigs or cadavers, new artificial grafts do not present rejection problems. A polymer film, applied to burn victims as a gauze, provides a good seal, protecting the living cells from infections and offering a moist surface to promote new cell growth.

The University of Western Ontario has established itself as a world leader in the development of artificial eyes. Researchers placed a wafer containing 65 electrodes on the visual cortex of a blind man. By connecting wires from the electrodes to a computer, these researchers were able to produce visual images for the man, the first flashes of light he had seen since an accident blinded him. By altering the firing pattern, researchers were then able to produce letters and objects of various shapes. This experiment may well be the first step toward providing artificial sight.

Artificial organs have a long history. Dialysis machines have enabled individuals with severe kidney damage to filter wastes from their blood. Artificial hearts have also been used, but with much less success. Recently, drugs such as cyclosporin have increased success. Artificial voice boxes have been developed for people whose larynx has been removed because of cancer. In this operation, a 3 cm rubber tube is fitted into the person's windpipe. A small connection is made between the windpipe and the esophagus. A one-way valve permits air to enter the windpipe and prevents it from leaking back. To speak, the person closes a small hole in the neck and exhales. Air passes into the esophagus and makes a sound. Although the sound is somewhat different from that made by the larynx, it does allow the person to communicate.

RESEARCH IN CANADA

A revolutionary innovation called electrical stimulation (ES) therapy has given a number of people mobility without the use of wheelchairs or walking aids. Pioneered by Dr. Arthur Prochazka and his colleagues at the University of Alberta, the principle of ES therapy is similar to that used in pacemakers. Electrical impulses supplied by healthy nerves are replaced with electrical impulses supplied by wires. Muscles respond to electrical currents, whether supplied by nerves or wires.

In ES therapy, a small implanted wire is placed into the hip flexor muscle. The other end of the wire enters a small electrical control box. About the size of a Walkman, the electrical control box relays an electrical signal to the hip flexor muscle, causing it to contract. Sensors relay the information back to the control box indicating the state of muscle activation. Movement requires both contraction and relaxation of the hip flexor muscle. The true challenge is not just to make the muscle contract but to get it to contract and relax at the right time. The challenge becomes even greater when the coordination of artificial electrical signals must be synchronized with those of normal nerves. Many people who are paralyzed have retained some nerve control over muscles. Muscles that contract at the wrong time pull against other muscles. Not only does this prevent movement, but it can actually damage the muscle.

Dr. Prochazka and his research team have also used ES to control tremors in a patient with Parkinson's disease and a patient with multiple sclerosis. One of the most promising aspects of ES technology is its potential for helping people with spinal cord injuries walk.

Electric Therapy for the Paralyzed

DR. ARTHUR PROCHAZKA

SOCIAL ISSUE:
Artificial Organs

In 1966, a plastic artificial heart was implanted in a human at a hospital in Houston, Texas. An artificial pancreas was developed by researchers at the University of Toronto. Gordon Murray, a Toronto researcher, developed the first artificial kidney to be used successfully in North America. Artificial limbs, eyes, and skin are further examples of synthetic body organs that have been implanted, with varying degrees of success, in humans.

Although the prospects of successful implantation of organs may have sounded like science fiction two decades ago, many people with disabilities today are able to lead longer and more comfortable lives as a result of artificial organ technology. However, the union of science and technology raises many ethical issues that must be confronted by society. Are there boundaries to artificial organ research? What are the implications if artificial organs one day exceed the capabilities of nature's organs?

Statement:

Research into artificial organs should be carefully controlled by an impartial committee representing community standards.

Point

- If artificial organs that exceed the capacity of nature's organs ever become available, millions of people who just want to enhance their physical or mental abilities will demand to have them. The boundaries of research must be defined by members of society outside the medical and health-care establishment to ensure that developments are acceptable to the community.

- The cost of producing artificial organs and giving them to patients could bankrupt the health-care system. Only the rich will be able to afford them. Can our society as a whole bear the costs of research from which only a few people will benefit?

- The body was not designed to receive artificial organs. The side effects for patients could be serious. It can take years to discover that a procedure that was thought to be safe is actually dangerous.

Counterpoint

- Laws can always be passed to ensure that artificial organs are only available to those who genuinely need them. But if research is too strictly controlled, the good and useful work will never be done.

- The argument of cost ignores the long-term benefits of artificial organ research in improving the quality and length of life for people who might otherwise remain disabled or even die.

- Caution is always necessary, but the benefits outweigh the dangers. For example, not too many years ago artificial kidneys and heart pacemakers seemed impractical and even dangerous. Now they are an accepted part of medical treatment. We must explore the opportunities that technology offers us.

Research the issue.
Reflect on your findings.
Discuss the various viewpoints with others.
Prepare for the class debate.

CHAPTER HIGHLIGHTS

- Tissues are groups of similarly shaped cells that work together to carry out a similar function.
- Epithelial tissue is a covering tissue that protects organs and lines body cavities.
- Connective tissue is a group of cells that provides support and holds various parts of the body together.
- Organ systems contain organs that have related functions. Organ systems often interact.

- Nuclear imaging techniques use radioisotopes to view organs and tissues of the body.
- Nuclear magnetic resonance techniques employ magnetic fields and radio waves to determine the behavior of molecules in soft tissue.
- Homeostasis is the system by which a constant internal environment is maintained despite changes in the external environment.

APPLYING THE CONCEPTS

1 Describe the advantages associated with cell specialization.
2 Using epidermal tissue as an example, explain the relationship between cell shape and tissue function.
3 List seven organ systems and provide at least two examples of organs that belong to those systems.

4 Explain the advantages of the CAT scan over conventional X rays.
5 Explain the concept of homeostasis by describing how your body adjusts to cold environmental temperatures.

CRITICAL-THINKING QUESTIONS

1 The cost of nuclear medicine, CAT scans, and artificial body parts is extremely high and places a heavy financial burden on the health care system. Can we continue to support such expensive research projects?

2 Liposuction is a fat-reducing technique in which fat cells are mechanically sucked out of the lower layer of the skin. Fees for this kind of surgery can range from $750 to over $4000, depending on the type of procedure. Discuss the moral and economic aspects of medical procedures that are designed to improve appearance.

ENRICHMENT ACTIVITIES

1 Research career opportunities in the field of medical engineering.
2 Locate photographs of X rays, NMR, and nuclear imaging. Which organs can you identify?

3 Survey dissection guides for chordates other than a pig. What anatomical similarities and differences can you observe?
4 Research Leonardo da Vinci's pioneering work in the field of human anatomy.

*D*igestion

IMPORTANCE OF DIGESTION

Unlike plants, which make their own food, heterotrophs must consume organic compounds to survive. These organic compounds, called **nutrients,** are digested in the gastrointestinal tract, absorbed, and transported by the circulatory system to the cells of the body. Once inside the cells, the nutrients supply the body with energy or the raw materials for the synthesis of essential chemical compounds. The chemical compounds are used for growth and tissue repair.

The digestive system is responsible for the breakdown of large, complex organic materials into smaller components that are utilized by the tissues of the body. Every organ system depends on the digestive system for nutrients, but the digestive system also depends on other organ systems. Muscles and bones permit the ingestion of foods. The circulatory system transports oxygen and other needed materials to the digestive organs. The circulatory system complements the digestion process by transporting the absorbed foods to the tissues of the body. The nervous and endocrine systems coordinate and regulate the actions of the digestive

Nutrients *are chemicals that provide nourishment. Nutrients provide energy or are assimilated to form protoplasmic structures.*

Figure 9.1

The sight and aroma of food can activate the salivary glands, setting the digestive process into action.

organs. In many respects, the study of the digestive organs is a study of the interacting body systems.

ORGANS OF DIGESTION

In this chapter you will study four components of digestion. *Ingestion* involves the taking in of nutrients. *Digestion* involves the breakdown of organic molecules into smaller complexes. *Absorption* involves the transport of digested nutrients to the tissues of the body. *Egestion* involves the removal of materials from the food that the body cannot digest. The assimilation of nutrients within the cells of the body is dealt with in other chapters.

The digestive tract, or *alimentary canal*, is an open-ended muscular tube. Measuring between 6.5 and 9 m in adults, the digestive tract stores and breaks down organic molecules into simpler components. Physical digestion begins in the mouth, where food is chewed into a bolus (the Greek word for "ball"). Salivary gland secretions activate the taste buds and lubricate the passage of food. **Amylase enzymes** contained in the saliva break down starches to smaller-chain carbohydrates called *dextrins*. The watery fluids produced by the salivary glands also dissolve food particles. Flavor can only be detected if food particles are in solution. Food particles dissolved in solution penetrate the cells of the taste buds located along the tongue. Different types of

nerve cells respond to specific flavors. For example, sweet flavors are detected by taste buds near the tip of the tongue, while bitter flavors are detected by taste buds near the middle of the tongue. You can see the importance of dissolving foods by drying your tongue and then placing a few grains of sugar or salt on it. You will not detect any flavor until the crystals dissolve.

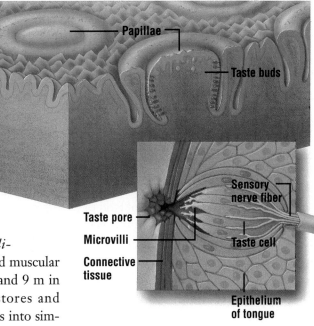

Papillae

Taste buds

Sensory nerve fiber

Taste pore

Microvilli

Connective tissue

Taste cell

Epithelium of tongue

Once swallowed, food travels from the mouth to the stomach by way of the **esophagus.** The bolus of food stretches the walls of the esophagus, activating smooth muscles, which set up waves of rhythmic contractions called **peristalsis.** Involuntary peristaltic contractions move food along the entire gastrointestinal tract. Voluntary control of food movement is only exercised during swallowing and during the last phase, egestion. Peristaltic action will move food or fluids from the esophagus to the stomach even if you stand on your head.

Amylase enzymes *hydrolyze complex carbohydrates.*

The esophagus *is a tube that carries food from the mouth to the stomach.*

Figure 9.2

Taste buds are located along the tongue.

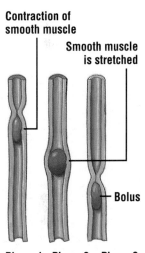

Contraction of smooth muscle

Smooth muscle is stretched

Bolus

Phase 1   Phase 2   Phase 3

Figure 9.3

Rhythmic contractions of the smooth muscle move food along the digestive tract.

Peristalsis *is the rhythmic, wavelike contraction of smooth muscle that moves food along the gastrointestinal tract.*

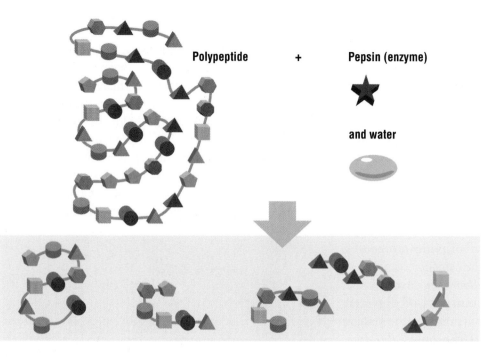

Polypeptide　　+　　Pepsin (enzyme)

and water

Figure 9.4

Proteins are composed of many amino acids. Digestion reactions are referred to as *hydrolysis reactions*. The water molecule is used to break bonds within large organic molecules.

STOMACH

The stomach is the site of food storage and initial protein digestion. **Sphincter** muscles regulate the movement of food to and from the stomach. These circular muscles act like the drawstrings of a purse. The contraction of the *cardiac sphincter* closes the opening to the stomach; its relaxation allows food to enter. A second sphincter, the *pyloric sphincter*, regulates the movement of foods and stomach acids to the small intestine.

The J-shaped stomach can store about 1.5 L of food. Millions of secretory cells line the inner wall of the stomach. Approximately 500 mL of gastric fluids are produced following a large meal. Mucous cells secrete a protective coating; parietal cells secrete hydrochloric acid. Peptic cells secrete a protein-digesting enzyme called *pepsinogen*. The active form of the enzyme, called **pepsin,** breaks proteins into peptones, shorter chains of amino acids.

The storage of a highly corrosive acid (hydrochloric acid) and a protein-digesting enzyme in a stomach that is composed of cells made of protein poses a major engineering problem. The pH inside the stomach normally ranges between 2.0 and 3.0, but may approach pH 1.0. Acids with a pH of 2.0 can dissolve fibers in a rug! Although the mechanism of HCl formation is not fully understood, a model has been proposed to explain its role in the digestive process. As shown in Figure 9.5, HCl forms in the *lumen*, or gut cavity. The protective mucous lining prevents the HCl from dissolving cells. In turn, the HCl destroys invading microbes.

Sphincters *are constrictor muscles that surround a tube-like structure.*

Pepsin *is a protein-digesting enzyme produced by the cells of the stomach.*

Figure 9.5

Proposed mechanism to explain how HCl is synthesized.

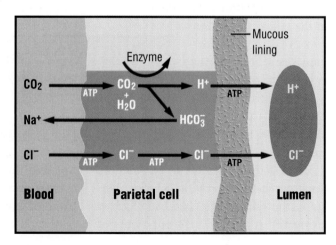

Unit Three:
Exchange of Matter and Energy in Humans

The storage of the protein-digesting enzyme is better understood. Pepsin is stored in an inactive form, called pepsinogen, which moves through the cell membrane and mucous lining. Upon entering the lumen, the enzyme is activated by HCl. Like most proteins, the geometry of pepsinogen is altered in an acidic environment. The new form, pepsin, is capable of breaking down proteins but cannot dissolve the cell, since it cannot penetrate the mucous lining.

cells are destroyed, an **ulcer** is formed. Beneath the thin layer of cells is a rich capillary network. As the acids irritate the cells of the stomach lining, histamine is released. The histamine increases blood flow to the damaged tissues, stimulates capillary permeability, and increases acid secretions. With this increased blood flow and acid secretion, more tissue is burned, and eventually the blood vessels begin to break down. In turn, even more histamine is released and the cycle continues.

Rennin *is an enzyme that coagulates milk proteins.*

Ulcers *are lesions along the surface of an organ.*

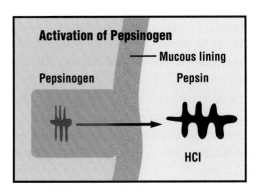

Activation of Pepsinogen

Pepsinogen — Mucous lining

Pepsin

HCl

A second enzyme found in the stomach, **rennin,** is important because it slows the movement of milk through the gastrointestinal tract, thereby permitting more time for the breakdown and absorption of nutrients. Rennin works in an alkaline or neutral environment but has little impact in the stomach of adults; it does, however, cause milk to curdle in the stomachs of infants. "Curds" and "whey," mentioned in the nursery rhyme "Little Miss Muffet," were made by adding a preparation from a calf's stomach to milk. The proteins are the milk curds; the watery portion is the whey.

When the protective mucous lining of the stomach breaks down, the cell membrane is exposed to the corrosive acid and the protein-digesting enzymes. When the

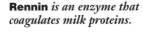

REVIEW QUESTIONS

1 Provide examples of how the digestive system interacts with other organ systems.
2 Define ingestion, digestion, absorption, and egestion.
3 Differentiate between physical and chemical digestion. Provide examples of each.
4 State the functions of the enzymes amylase, pepsin, and rennin.
5 What is the function of the mucous layer that lines the stomach?
6 What causes stomach ulcers?

Figure 9.6

Model showing how pepsinogen is activated to pepsin. (Note: The model does not represent the actual shape of either pepsinogen or pepsin.)

FRONTIERS OF TECHNOLOGY: ULCERS AND LASERS

In *War of the Worlds*, a book written at the turn of the century, H.G. Wells predicted the use of light rays by invading Martians. In 1916, Albert Einstein hypothesized that electrons, when excited, could emit light energy or photons of a particular wavelength. By 1960, Theodore Maiman

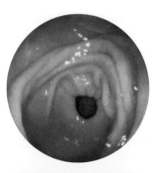

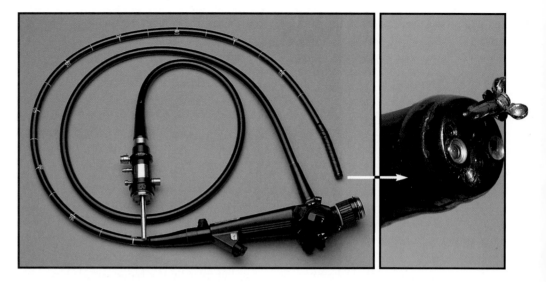

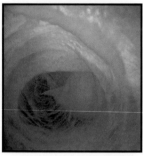

Figure 9.7

The endoscope can be used to provide a view of the stomach.

Endoscopes are instruments that view the interior of the body.

The duodenum is the first segment of the small intestine.

had built the first laser. Today lasers are used in a wide variety of applications—from reproducing music on compact discs to scanning food containers at supermarket checkouts.

Lasers emit an intense light beam, but unlike normal white light, which vibrates along many planes, laser light vibrates in a single plane and is composed of a single wavelength. This ordered packing of laser light waves delivers more energy, size for size, than other sources.

Lasers are made up of a series of fine rods whose ends are coated with a mirror-like metal. The rods are fitted in a tube that is filled with a gas such as argon or carbon dioxide. A flash of light directed at the rods is used to start the laser. Electrons within the surrounding gas become excited and move to a higher energy level. Because the high-energy electrons are unstable, they tend to fall back to lower energy levels; as they do so, light is released. The light, reflected by mirrors at either end, is amplified as it moves along the rods. Moving in tightly bunched waves, the single-colored laser light leaves the machine in narrow, concentrated beams.

Laser beams have many medical applications. They can be used to remove damaged tissues such as those created by

stomach ulcers. The laser beam is thinner than most scalpels, and provides the added advantage of sealing small blood vessels. In addition, the laser may reduce the need for surgery. A device called an **endoscope** can be fitted with a light-emitting glass fiber and then positioned inside a patient's body. The endoscope can then be used to view such things as stomach ulcers. Tiny forceps, fitted in the endoscope, can even extract small pieces of tissue.

SMALL INTESTINE AND PANCREAS

In many ways the name small intestine is a misnomer. The small intestine is actually longer than the large intestine. Measuring up to 7 m in length, the small intestine is only 2.5 cm in diameter. The large intestine, by comparison, is only 1.5 m in length, but 7.6 cm in diameter. The name small intestine, therefore, is derived from its comparatively narrow diameter.

The majority of digestion occurs in the first 25 to 30 cm of the small intestine, an area known as the **duodenum.** The second and third components of the small intestine are called the *jejunum* and *ileum.*

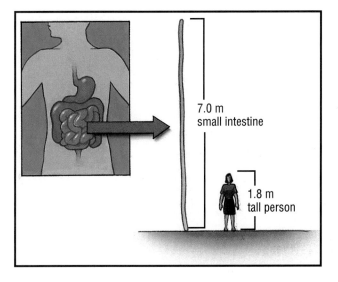

7.0 m
small intestine

1.8 m
tall person

Figure 9.8
A comparison of the length of the small intestine with the height of a tall person.

The three segments are differentiated on the basis of cell shape and will not be investigated further in this chapter.

As you already know, food moves from the stomach to the small intestine. A second engineering problem arises once the food enters the small intestine. Partially digested foods reach the small intestine already soaked in HCl and pepsin. How are the cells of the small intestine protected? To answer this question, you must look beyond the small intestine to the pancreas. Once acids enter the small intestine, a chemical called prosecretin is converted into **secretin.** Secretin is absorbed into the bloodstream and carried to the pancreas, where it signals the release of bicarbonate ions. Bicarbonate ions are carried by way of the pancreatic duct to the small intestine, where they buffer the HCl from the stomach. This rush of bicarbonate ions raises the pH of fluids from about 2.5 to 9.0. Pepsin, the protein-digesting enzyme from the stomach, becomes inactive in the alkaline fluids of the small intestine. If the secretion of these bicarbonate ions is inhibited, a duodenal ulcer can form.

The pancreatic secretions also contain enzymes that promote the breakdown of the three major components of foods: proteins, carbohydrates, and lipids. A pro-tein-digesting enzyme, called *trypsinogen*, is released from the pancreas. Once the trypsinogen reaches the small intestine, an enzyme called **enterokinase** converts the inactive trypsinogen into **trypsin,** which acts on the partially digested proteins. Under the influence of trypsin, the long-chain peptones are broken down into shorter-chain peptones. A second group of enzymes, the **erepsins,** complete the final stages of protein digestion. Erepsins, discharged from the pancreas and small intestine, cleave the bonds between short-chain peptones, thus releasing individual amino acids. Protein digestion has been completed.

The pancreas also releases amylase enzymes, which continue the digestion of carbohydrates initiated in the mouth. The intermediate-size chains are fractured into disaccharides. A series of disaccharide enzymes from the small intestine completes the digestion of carbohydrates.

Many people are unable to digest lactose (milk sugar) because their bodies do not produce sufficient quantities of the enzyme lactase. Normally, the disaccharide lactose is broken down into two monosaccharides, which are then absorbed into the blood. People who are lactose intolerant are not able to absorb lactose in the small intestine. As the disaccharide moves to the large intestine, water is drawn in by osmosis, causing diarrhea.

Lipases, which are lipid-digesting enzymes, are also released from the pancreas. Three different types of lipid-digesting enzymes are known. Pancreatic lipase, the most common, breaks down triglycerides into fatty acids and glycerol. Cholesterol lipase removes a single

Secretin *is a hormone that stimulates pancreatic and bile secretions.*

Enterokinase *is an enzyme of the small intestine that converts trypsinogen to trypsin.*

Trypsin *is a protein-digesting enzyme.*

Erepsins *are enzymes that complete protein digestion by converting small-chain peptones to amino acids.*

Lipases *are lipid-digesting enzymes.*

fatty acid from a steroid cholesterol molecule. The third type of lipase acts on phospholipids, breaking them down into glycerol and fatty-acid components. The phosphate remains attached to one of the fatty acids.

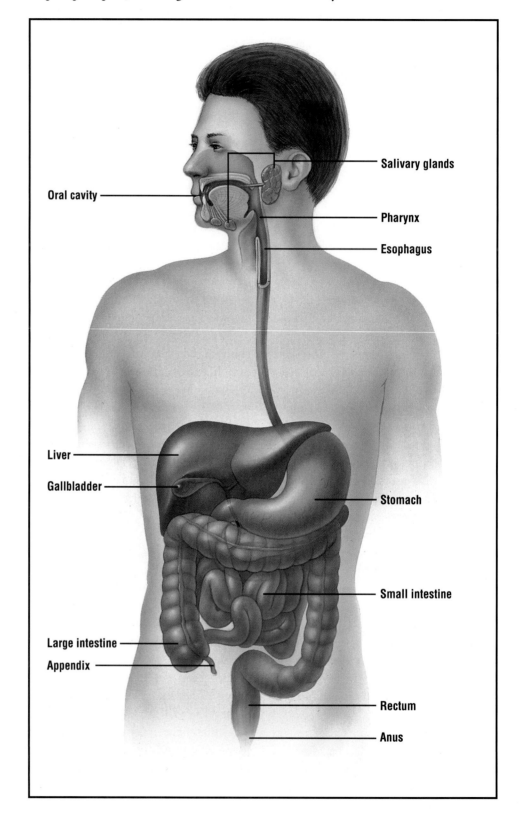

Figure 9.9

The human digestive system and accessory organs.

Oral cavity

Salivary glands

Pharynx

Esophagus

Liver

Gallbladder

Stomach

Small intestine

Large intestine

Appendix

Rectum

Anus

LIVER AND GALLBLADDER

The liver has a number of different functions, two of which relate to digestion. **Bile salts** are produced in the liver, stored in the gall bladder, and carried by way of the bile duct to the small intestine. Once in the gall bladder, water is reabsorbed and the salts are concentrated about tenfold.

The presence of fats in the small intestine causes the release of a hormone that is carried in the blood to the gall bladder. The hormone triggers the release of bile salts from the gall bladder. Once inside the small intestine, the bile salts act as detergents, emulsifying large fat globules. It is important to note that the breakdown of large fat globules into smaller droplets is not chemical, but physical, digestion—chemical bonds are not broken. During the emulsion, the surface area of the fat is increased, speeding chemical digestion. Pancreatic lipase works along the surface of the fat droplets. If the fat droplets are large, all the fat molecules found in the interior of the fat droplets are unaffected. As fat is broken into smaller droplets, the enzymes are exposed to a greater number of substrate molecules.

The production and concentration of bile can result in certain problems. Cholesterol, an insoluble component of bile, acts as a binding agent for the salt crystals found in bile. The crystals precipitate and aggregate into larger crystals, called **gallstones.** When the gallstone lodges in the bile duct, not only is fat digestion impaired, but the blockage gives rise to considerable pain.

Bile pigments are composed of the byproducts of aged red blood cells. The respiratory pigment, hemoglobin, is divided into protein and heme components. Iron is extracted from the heme portion of the molecule, leaving four ring-shaped molecules. These ring-shaped molecules are processed by the liver, and their products are stored in the gall bladder for removal. The characteristic brown color of the feces is caused by the bile pigments formed from hemoglobin breakdown. Any obstruction of the bile duct or accelerated destruction of red blood cells can give rise to **jaundice,** which comes from the French word meaning "yellow."

The liver also acts as a storehouse for glycogen and vitamins such as vitamin A, B_{12}, and D. In addition, the liver **detoxifies** many substances in the body. Harmful chemicals are converted into more soluble chemicals that can be dissolved in the blood and eliminated in the urine. One of the more common poisons is alcohol, which the liver breaks down to usable materials and wastes. Unfortunately, alcohol, like many other harmful agents, can destroy liver tissue. Damaged liver cells are replaced by connective tissue and fat, which are not able to carry out the multitude of normal liver duties. This condition is referred to as **cirrhosis** of the liver.

COLON

Chemical digestion is complete by the time food reaches the large intestine. The **colon,** the largest part of the large intestine, must store wastes long enough to reabsorb water. Some inorganic salts are also absorbed with the water. A small pouch, called the *appendix,* is found near

Bile salts *are the components of bile that emulsify fats.*

Gallstones *are crystals of bile salts that form in the gallbladder.*

Jaundice *is the yellowish discoloration of the skin and other tissues brought about by the collection of bile salts in the blood.*

Detoxify *means to remove the effects of a poison.*

Cirrhosis *of the liver is a chronic inflammation of liver tissue characterized by an increase of nonfunctioning fibrous tissue and fat.*

The **colon** *is the largest segment of the large intestine. Water reabsorption occurs in the colon.*

the junction of the small and large intestine. The appendix, which serves no apparent function, is only noticed when it becomes infected. An inflamed appendix may rupture, sending microbes into the body cavity or peritoneum. Infections within the body cavity can be extremely dangerous.

The large intestine also houses bacteria, which use waste materials to synthesize vitamins B and K. Cellulose, the long-chain carbohydrate characteristic of plant cell walls, is found in the large intestine. Although cellulose cannot be broken down, it may serve one of the most important functions of all organic chemicals consumed by humans. Cellulose provides bulk. As wastes build up in the large intestine, receptors in the wall of the intestine provide information to the central nervous system, which, in turn, prompts a bowel movement. The bowel movement ensures the removal of potentially toxic wastes from the body. However, individuals who do not eat sufficient amounts of roughage or fiber have fewer bowel movements. This means that wastes and toxins remain in their bodies for longer periods of time. Scientists have determined that cancer of the colon can be related to diet. Individuals who eat mostly processed, highly refined foods are more likely to develop cancer of the colon.

Medical reports that roughage may also reduce cholesterol levels (specifically LDLs) have been embraced by many food manufacturers. Although the actual benefits of bran, a form of roughage, in reducing cholesterol can be debated, its value in providing a balanced diet is undeniable.

Table 9.1 Summary Table: Digestive System

Organ	Secretion	Function
Mouth, salivary glands	salivary amylase	Initiates the breakdown of polysaccharides to dextrins and disaccharides.
Stomach	hydrochloric acid	Converts pepsinogen to pepsin, kills microbes.
	pepsinogen	Converted to pepsin, which initiates the digestion of proteins.
	mucus	Protects the stomach.
	rennin	Coagulates proteins in milk.
Pancreas, small intestine	bicarbonate ions	Neutralizes HCl from the stomach.
	trypsinogen	Activated to trypsin; converts long-chain peptones into short-chain peptones.
	lipase	Breaks down fats to glycerol and fatty acids.
	erepsin	Completes the breakdown of proteins.
	disaccharases	Breaks down disaccharides into monosaccharides. Most absorption of food takes place in the small intestine.
Liver	bile	Emulsifies fat.
Gallbladder		Stores concentrated bile from the liver.
Large intestine	mucus	Helps movement of food. Stores wastes. Absorbs water.
		Produces vitamin K and B_{12} by bacteria.

LABORATORY
EFFECT OF pH ON PROTEIN DIGESTION

Objective

To investigate how pH affects the rate of protein digestion.

Materials

hydronium pH paper
6 large test tubes
  (22 × 175 mm)
10 mL graduated cylinder
10% pepsin solution
knife
tweezers or forceps

wax pen
1.0 mol NaOH solution
1.0 mol HCl solution
distilled water
hard-boiled egg
metric ruler
plastic gloves

Safety Notes

- **Wear safety goggles and an apron at all times.**
- **Wear plastic gloves to prevent any chemical from coming in contact with your skin.**
- **Sodium hydroxide and hydrochloric acid are very corrosive. Wash all splashes off your skin and clothing immediately. If you get any chemical in your eyes, rinse for at least 15 min.**
- **Work in a well-ventilated area.**
- **When you are finished the activities, clean your work area, wash your hands thoroughly, and dispose of all specimens, chemicals, and materials as instructed by your teacher.**

Procedure

1. Using a wax pen, label the test tubes 1 to 6.
2. Measure and cut twelve 2 cm³ cubes of boiled egg white. Place two cubes in each of the test tubes.
3. Using the 10 mL graduated cylinder, measure 5 mL of distilled water and pour it into test tube #1. Repeat for test tube #2. Repeat the procedure, but measure and pour 5 mL of HCl into test tubes #3 and #4. Rinse the graduated cylinder. Repeat again, but this time measure and pour 5 mL of NaOH into test tubes #5 and #6.
   a) Why was the graduated cylinder rinsed between adding the HCl and NaOH solutions?
4. Place a small piece of hydronium pH paper in each of the test tubes.
   b) Record the pH of each solution.
5. Add 1 mL of pepsin solution to test tubes #2, #4, and #6. Place stoppers on the test tubes and view after 24 hours.
   c) Using tweezers or forceps, measure each of the egg white cubes.
   d) Compare the amount of digestion in the test tubes.

Laboratory Application Questions

1. Which test tubes served as controls for the experiment? Note: This experiment uses more than one control.
2. At what pH does pepsin work best?
3. After interpreting the data from the experiment, a student concludes that HCl causes the digestion of proteins. Do you agree with this conclusion? Give your reasons.
4. Using the data collected in the experiment, predict how an alkaline environment in the stomach would affect the digestion of carbohydrates, fats, and proteins.

ABSORPTION

The stomach absorbs some water, specific vitamins, and alcohol. Although additional water and vitamins are absorbed in the large intestine, most absorption takes place within the small intestine. Long fingerlike tubes called **villi** (singular: "villus") greatly increase the surface area of the small intestine (see figures 9.10 and 9.11). One estimate suggests that villi account for a tenfold increase in surface area for absorption. Without villi your small intestine would have to be 70 m long to maintain normal absorption. Imagine your appearance if your waist size were to increase to accommodate such a long small intestine.

Villi *are small fingerlike projections that extend into the small intestine. Villi increase surface area for absorption.*

The outer cell membranes of the cells that line the small intestine are folded to increase surface area. This folded outer-membrane structure is referred to as the **microvilli.** Each villus is supplied with a capillary network that intertwines with lymph vessels called *lacteals.* Some nutrients are absorbed by diffusion, but cells of the small intestine expend energy to actively transport materials from the gut. Carbohydrates and amino acids are absorbed into the capillary networks; fats are absorbed into the lacteals.

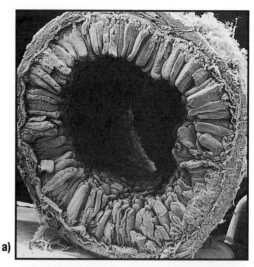

a)

Figure 9.10

Villi greatly increase the surface area for absorption of nutrients. Each villus has blood capillaries and lymph vessels.

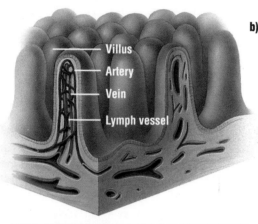

Villus
Artery
Vein
Lymph vessel

b)

Figure 9.11

(a) Villi in the mammalian intestine. (b) Microvilli on the surface of an epithelial cell.

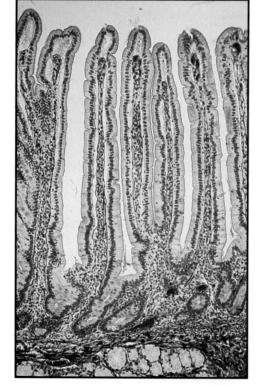

BEAUMONT AND ST. MARTIN

On June 6, 1822, a blast from a shotgun pierced the stomach of Alexis St. Martin, a Canadian trapper. An American army surgeon, William Beaumont, attended the trapper. On seeing the damage done by the gunshot, Beaumont predicted that St. Martin would not survive the night.

The edges of St. Martin's stomach had become attached to the outer edge of the wound, sealing the body cavity from invading microbes. This probably saved St. Martin's life. Since St. Martin was still

*Unit Three:
Exchange of Matter and Energy in Humans*

left with a hole in his stomach, Beaumont seized the occasion to conduct experiments on digestion. He tied small pieces of food with silk thread and dropped them into St. Martin's stomach. Beaumont later removed the pieces and recorded the amount of digestion that had occurred.

Beaumont talked the grateful St. Martin into accompanying him to the United States to help demonstrate his medical findings. Eventually, tired of being a living laboratory, St. Martin ran away and returned to Canada. Beaumont, stripped of his source of experiments, saw his popularity decline.

After four years of searching Canada, Beaumont located St. Martin. This time, Beaumont placed the trapper in the U.S. army and under his control. St. Martin did escape a second time and lived to the age of 82.

■ REVIEW QUESTIONS ▊▊ ?

7 Outline the mechanism by which stomach acids are buffered by pancreatic secretions.

8 What is the function of bile salts?

9 What is cirrhosis of the liver?

10 Why is cellulose considered to be an important part of your diet?

11 What are villi and what function do they serve?

CONTROL OF DIGESTION

The Russian physiologist, Ivan Pavlov (1849–1936), performed a series of experiments on dogs to show how digestive secretions are regulated. Digestive secretions are influenced by three stimuli. Pavlov noted that dogs began to salivate if they were able to see, smell, or taste food.

Pavlov suspected that digestion must be partly controlled by sensory stimuli.

Pavlov first cut a hole in the dog's esophagus to prevent food from entering the stomach. A device designed to collect and measure gastric juice was placed in the dog's stomach. Food placed in the mouth initiated gastric secretions, despite the fact that no food entered the stomach. These observations appeared to confirm his initial suspicions. When the vagus nerve (a branch of the parasympathetic system) was cut, gastric secretions were reduced. The mechanical stimulus of swallowing created peristaltic motions, which, in turn, also stimulated the production of gastric juices.

In a second experiment, Pavlov connected the circulatory systems between two dogs. He noted that when he fed the first dog, the second dog began to produce gastric secretion, despite the fact that no food was in the second dog's stomach. In this case neither the sensory stimuli of sight, smell, or taste nor the mechanical stimulus of swallowing could account for the production of gastric secretions.

A hormone called gastrin is produced when partially digested proteins are present in the stomach. The hormone was absorbed in the blood and carried by way of the circulatory system to Pavlov's second dog, where it initiated the release of gastric juices. The response was confirmed when a graft from the stomach was attached to the back of a dog.

A second hormone, enterogastrone is released when fats enter the small intestine. Fats are more difficult to digest. Enterogastrone slows peristalsis, thereby allowing greater time for fat digestion. As you read earlier, the pancreas also responds to hormones. Secretin protects the duodenum from stomach acids. The release of bile salts from the gallbladder is also under hormonal control.

RESEARCH IN CANADA

Digestion and Nutrition

Dr. Patrick du Souich and Dr. Gilles Caillé, researchers at the Université de Montréal, are attempting to determine the role of the gastrointestinal tract and the lungs in the body's metabolism of drugs. It had previously been assumed that drugs administered directly into the blood (intravenously) were only affected by the liver. However, Drs. du Souich and Caillé believe that the high metabolic activity of the lungs and the gastrointestinal tract may affect the amount of drug reaching the circulatory system. Their research will provide clearer guidelines for determining dosages of certain drugs.

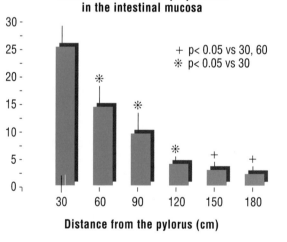

Rate of elimination of propranolol in the intestinal mucosa

Rate constant of elimination (min $^{-1}$/mg protein x 10^{-4})

+ $p < 0.05$ vs 30, 60
＊ $p < 0.05$ vs 30

Distance from the pylorus (cm)

**DR. PATRICK DU SOUICH
DR. GILLES CAILLÉ**

A number of Canadian researchers are investigating other aspects of digestion and nutrition. At McMaster University in Hamilton, Ontario, Jacqueline Lewis Halton is researching nutrition in childhood cancer, and Rhona Hanning is investigating amino acid requirements and metabolism in infants. Mark Tarnopolsky, also of McMaster, is doing research on the protein requirements of athletes. Laurence Fraher of the University of Western Ontario in London is investigating the potential of vitamins A and D to improve the ability of the elderly to combat infections.

SOCIAL ISSUE:
Fad Diets

Dieting is big business. An array of low-calorie food products and specialized diet plans are competing in an ever-expanding market. Diet plans like the Scarsdale Diet are high on proteins and low on carbohydrates, while the Beverly Hills Diet recommends low protein and high carbohydrate consumption. Some weight-loss plans include appetite suppressants such as amphetamines, as well as laxatives.

It is now well documented that people who are overweight are more prone to certain diseases such as atherosclerosis and diabetes. Less well known is the fact that being underweight can also cause problems, such as fatigue and increased proneness to illness and injury.

Statement:

Specialized diet plans may actually contribute to malnutrition in people who use them consistently.

Point

- Some diets emphasize high-calorie fatty foods such as steaks, cheese, and milk, which can increase cholesterol levels. Liquid protein diets provide only about 400 calories per day, whereas most people need about 1200 calories per day.

Counterpoint

- Low-calorie, nonfattening foods are carefully monitored by nutritionists. Prepared products are not the only answer to good eating. People must take responsibility for maintaining a healthy balance in their food intake. Dieting alone can't be expected to perform miracles.

Research the issue.
Reflect on your findings.
Discuss the various viewpoints with others.
Prepare for the class debate.

CHAPTER HIGHLIGHTS

- Organic compounds, or nutrients, are digested in the gastrointestinal tract.
- Both physical and chemical digestion occur in the mouth.
- The stomach is a J-shaped tube that initiates the digestion of proteins.
- Peristaltic waves move food along the gastrointestinal tract.

- The pancreas and liver produce secretions that aid in digestion. The pancreas produces enzymes that digest carbohydrates, lipids, and proteins. The liver produces bile salts, which emulsify fats.
- Nutrients are absorbed through the villi found in the small intestine.
- Digestion is regulated by nerves, mechanical stimuli, and hormones.

APPLYING THE CONCEPTS

1 Why are you able to drink water while standing on your head?

2 Why does the stomach not digest itself? (Give two reasons.)

3 How is the duodenum protected against stomach acids? Why does pepsin not remain active in the duodenum?

4 Why are pepsin and trypsin stored in inactive forms? Why can erepsins be stored in active forms?

5 Explain why the backwash of bile salts into the stomach can lead to stomach ulcers.

6 Why do individuals with gallstones experience problems digesting certain foods?

7 Why might individuals with an obstructed bile duct develop jaundice?

8 In cases of extreme obesity, a section of the small intestine may be removed. What effect do you think this procedure has on the patient?

9 Trace the digestion of a spaghetti dinner from the mouth to the colon. Describe the enzymes and hormones involved in digestion.

10 The following experiment was designed to investigate factors that affect lipid digestion. A pH indicator is added to three test tubes as shown below. (This indicator turns pink when the pH is 7 or above and clear when the pH is below 7.) The initial color of each test tube is pink. Predict the color changes for each of the test tubes shown in the diagram and provide your reasons.

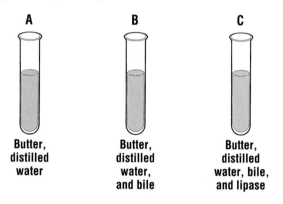

A — Butter, distilled water

B — Butter, distilled water, and bile

C — Butter, distilled water, bile, and lipase

CRITICAL-THINKING QUESTIONS

1 Coal-tar dyes are used to enhance the color of various foods. Some dyes, however, have been found to be carcinogenic (cancer causing). Many of the artificial food colors used in Europe are banned in North America, and many used in North America are banned in Europe. How is it possible that two groups of scientists could test a food dye and come to different conclusions?

2 The incidence of colon cancer is highest in countries where people eat the greatest quantities of animal proteins and fats. Individuals who live in countries where cereal grains form the basic diet have a much lower incidence of colon cancer. What conclusion might you draw from these data? Can colon cancer be eliminated by a change in diet?

3 Comment on the research performed by William Beaumont on Alexis St. Martin. Would similar experiments be tolerated today? Give your reasons.

4 Salivary amylase breaks down starch to small-chain polysaccharides and disaccharides. The presence of these products can be detected by Benedict's solution. In the procedures shown on the next page, 5 mL of starch has been added to each of the test tubes. Benedict's solution changes color in the presence of reducing sugars.

Which of the procedures would best determine the optimal pH for the digestion of starch by salivary amylase? Critique the procedures not chosen.

5 Modify the experimental design provided above to investigate how pepsin concentration affects the rate of protein digestion.

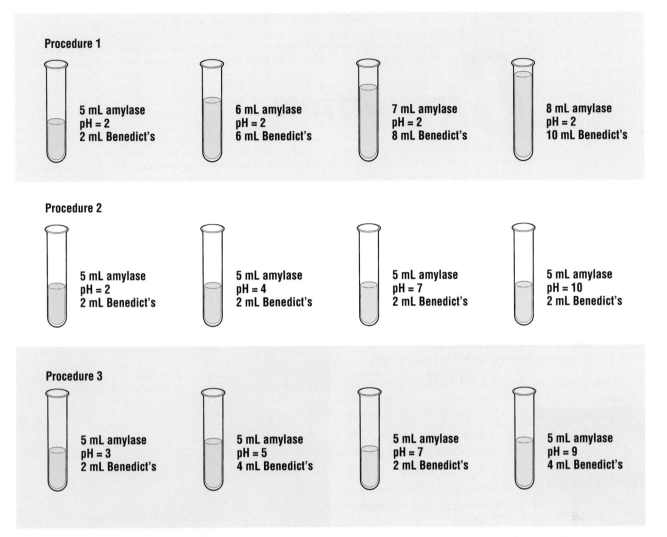

Procedure 1

5 mL amylase
pH = 2
2 mL Benedict's

6 mL amylase
pH = 2
6 mL Benedict's

7 mL amylase
pH = 2
8 mL Benedict's

8 mL amylase
pH = 2
10 mL Benedict's

Procedure 2

5 mL amylase
pH = 2
2 mL Benedict's

5 mL amylase
pH = 4
2 mL Benedict's

5 mL amylase
pH = 7
2 mL Benedict's

5 mL amylase
pH = 10
2 mL Benedict's

Procedure 3

5 mL amylase
pH = 3
2 mL Benedict's

5 mL amylase
pH = 5
4 mL Benedict's

5 mL amylase
pH = 7
2 mL Benedict's

5 mL amylase
pH = 9
4 mL Benedict's

ENRICHMENT ACTIVITIES

1 Prepare a report on anorexia nervosa. Why do you think females suffer from this disorder more often than males?

2 Consult Canada's Food Guide for information on providing a well-balanced diet. Keep a record of everything that you eat and drink for three consecutive days. Analyze your diet using Canada's Food Guide. How could you improve your eating habits? What factors might be included in a more detailed diet analysis?

*C*irculation

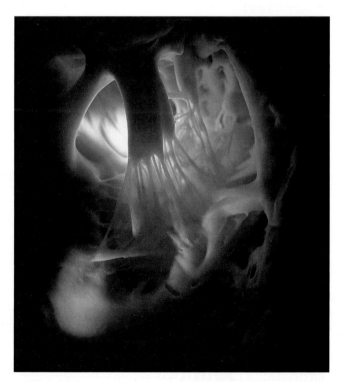

IMPORTANCE OF THE CIRCULATORY SYSTEM

All living things require nutrients to grow and reproduce. The earliest organisms lived in the sea. Composed of single cells, these primitive organisms were continually bathed by seawater. Oxygen and other nutrients in the seawater diffused through the cell membranes of these primitive cells. Once inside, nutrients were carried to all areas of the cell by way of the endoplasmic reticulum. Wastes produced from the breakdown of nutrients diffused through the same cell membrane. Once again, the surrounding seawater acted as the transport system, carrying potentially toxic materials away from cells.

With the evolution of multicellular organisms, a transport system became more important. Organisms composed of two cell layers could still be bathed by seawater, but relationships between cells had changed—cells no longer acted independently. Although some cells specialize in feeding, all cells require nutrients. Nutrients must be transported from feeding cells to support cells, which are charged with

Figure 10.1

From embryonic development until the moment of death, the heart beats continuously during every second of life.

other functions. Similarly, wastes produced in all cells must be transported to cells specialized for waste removal. Intercellular transport systems allow cells to specialize and work together.

The importance of a circulatory system can be emphasized by examining higher organisms, which have three distinct cell layers. Because the middle layer of cells cannot receive nutrients directly from seawater, an efficient circulatory system is required. Your circulatory system is composed of 96 000 km of blood vessels that supply the estimated 60 trillion cells of your body with nutrients. No cell is further than two cells away from a blood vessel. Every minute, five liters of blood cycle from the heart to the lungs, where oxygen is received, and then back to the heart. The heart pumps oxygen-rich blood to the tissues of the body. Here, oxygen and nutrients are given up, and wastes return to the heart. Glucose, the primary fuel of cells, is broken down by oxygen into carbon dioxide and water. The conversion of high-energy glucose into low-energy compounds releases energy, which is used by cells to build new materials, repair pre-existing structures, or for a variety of other energy-consuming reactions. It is apparent that the conversion of cell energy depends on an adequate supply of glucose and oxygen. It is here that the value of the circulatory system becomes evident. The more effective the delivery system, the greater the energy available, as shown in Table 10.1.

Your circulatory system carries nutrients to cells, wastes away from cells, and chemical messengers from cells in one part of the body to distant target tissues. It distributes heat throughout the body and, along with the kidneys, maintains acceptable levels of body fluid. Defense against invading organisms is also associated with the circulatory system.

Table 10.1 Various Types of Transport Systems

Organism	Function	Oxygen used (μg/g body mass/hr)
Jellyfish	Two cell layers, bathed by seawater	5
Earthworm	Three cell layers, 5 primitive hearts	60
Cockroach	Open circulatory system	450
Goldfish	Closed circulatory system, two-chambered heart	420
Mouse (resting)	Closed circulatory system, four-chambered heart	2000

Whether simple or complex, a circulatory system is vital to an organism's survival. In this chapter, you will investigate human circulation. Then, in the next chapter, you will study the circulating fluid, the blood, in greater detail.

EARLY THEORIES OF CIRCULATION

The ancient Greeks believed that the heart was the center of human intelligence, an "innate heat" that generated four humors: black and yellow bile, phlegm, and blood. Galen, the personal physician of Roman emperor Marcus Aurelius in the second century A.D., influenced early physiology. Although he provided many enlightening theories, Galen is best known for steering scientists in the wrong direction. Galen believed that blood did not circulate. Although he believed that blood might ebb like the tides, he never thought of the heart as a pump. Galen's theory was generally accepted until the 17th century.

Some science historians have suggested that his failure to consider the pumping action of the heart could be attributed to a lack of a technical model: the water pump had not been invented when Galen applied his theory.

William Harvey (1578–1657), the great English physiologist, questioned Galen's hypothesis. Harvey, like many Europeans during that period, was influenced by the astronomer Galileo. Galileo's new principles of dynamics became the foundation of Harvey's work. By applying Galileo's theories of fluid movement to that of blood, Harvey reasoned that blood must circulate.

Figure 10.2
William Harvey (1578–1657).

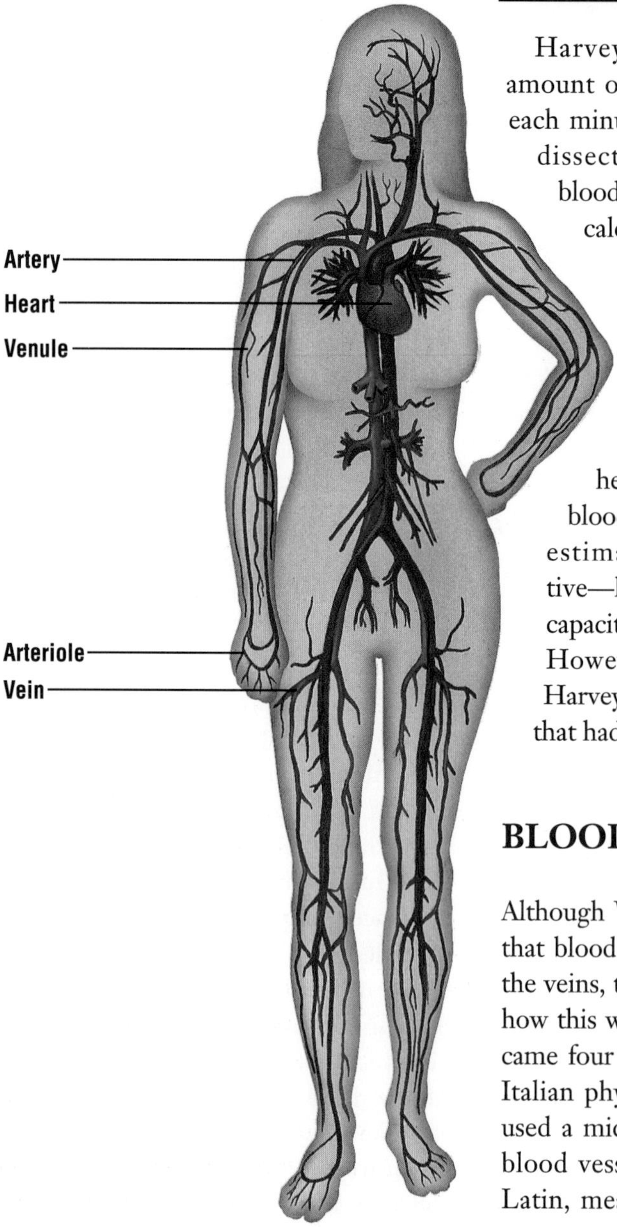

Figure 10.3
Circulatory system.

Artery
Heart
Venule

Arteriole
Vein

> **BIOLOGY CLIP**
> Evidence from an ancient Egyptian papyrus discovered in the 19th century suggests that the Egyptians correctly mapped the flow of blood from the heart 3300 years before William Harvey.

Harvey attempted to quantify the amount of blood pumped by the heart each minute. He began his research by dissecting cadavers and observing blood vessels. Using mathematics, he calculated that the heart contains approximately 57 mL of blood. Harvey then concluded that 14.8 L must be pumped from the heart each hour. However, much less blood could be found in the body; the heart must be pumping the same blood over and over again. Harvey's estimates were at best conservative—he greatly underestimated the capacity of the heart to pump blood. However, by using empirical data, Harvey tested and challenged a theory that had been accepted for 1400 years.

BLOOD VESSELS

Although William Harvey was convinced that blood must pass from the arteries to the veins, there was no visible evidence of how this was accomplished. The solution came four years after his death, when an Italian physiologist, Marcello Malpighi, used a microscope to observe the tiniest blood vessels, the capillaries (from the Latin, meaning "hairlike"). Harvey had speculated that blood vessels too small to be seen by the human eye might explain how blood circulates. Now, with Malpighi's observations, Harvey's theory of circulation was confirmed.

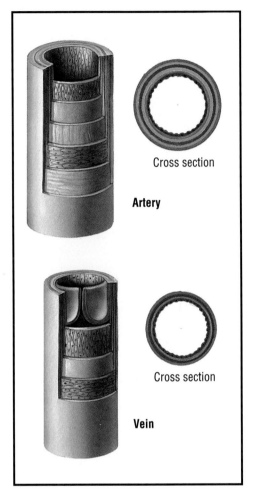

Figure 10.4

Simplified diagrams of an artery and a vein. Arteries have strong walls capable of withstanding great pressure. The middle layer of arteries contains both muscle tissue and elastic connective tissue. The low-pressure veins have a thinner middle layer.

Arteries

Arteries are the blood vessels that carry blood away from the heart. They have thick walls composed of three distinct layers. The outer and inner layers are primarily rigid connective tissue. The middle layer is made up of muscle fibers and elastic connective tissue. Every time the heart contracts, blood surges from the heart and enters the arteries. The arteries stretch to accommodate the inrush of blood. The **pulse** you can feel near your wrist and on either side of your neck is created by the changes in the diameter of the artery near the surface of your body following heart contractions. Heart contraction is followed by a relaxation phase. During this phase, pressure drops and elastic fibers in the walls of the artery recoil. It is interesting to note that the many cells of the artery are themselves supplied with blood vessels that provide nourishment.

A birth defect or injury can cause the inner wall of the artery to bulge. Known as an **aneurysm,** this condition is infrequent in young people, but can lead to serious problems. In much the same way as the weakened wall of an inner tube begins to bulge, the weakened segment of the artery protrudes as blood pulses through. The problem escalates as the thinner wall offers less support and eventually ruptures. A weakened artery in the brain is one of the conditions that can lead to a stroke. Cells die because less oxygen and nutrients are delivered to the tissues.

Blood from the arteries passes into smaller arteries, called **arterioles.** The middle layer of arterioles is composed of elastic fibers and smooth muscle. The diameter of the arterioles is regulated by nerves from the autonomic nervous system. A sympathetic nerve impulse causes smooth muscle in the arterioles to contract, reducing the diameter of the blood vessel. This process is called **vasoconstriction.** Vasoconstriction decreases blood flow to tissues. Relaxation of the smooth muscle causes dilation of the arterioles, and blood flow increases. This process is called **vasodilation.** Vasodilation increases the delivery of nutrients to tis-

Arteries *are high-pressure blood vessels that carry blood away from the heart.*

A **pulse** *is caused by blood being pumped through an artery.*

An **aneurysm** *is a fluid-filled bulge found in the weakened wall of an artery.*

Arterioles *are fine branches from arteries.*

Vasoconstriction *is the narrowing of a blood vessel. Less blood goes to the tissues when the arterioles constrict.*

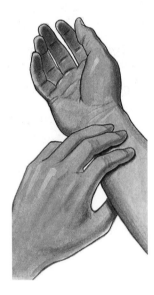

Figure 10.5

Arteries near the surface of the body provide a pulse. Galen believed that 27 different types of pulses could be produced; however, he never related pulse to the heart or to the movement of blood.

sues. This, in turn, increases the capacity of the cells in that localized area to perform energy-consuming tasks.

Precapillary sphincter muscles regulate the movement of blood from the arterioles into capillaries. Blushing is caused by vasodilation of the arterioles leading to skin capillaries. Red blood cells close to the surface of the skin produce the pink color. Have you ever noticed someone's face turn a paler shade when they are frightened? The constriction of the arteriolar muscle diverts blood away from the outer capillaries of the skin toward the muscles. The increased blood flow to the muscles provides more oxygen and glucose for energy to meet the demands of the "flight-or-fight" response.

Arterioles leading to capillaries open only when cells in that area require blood. It has been estimated that 200 L of blood would be required if the arterioles could not selectively open and close gateways to the capillaries. Although the majority of brain capillaries remain open, as few as 1/50 of the capillaries in resting muscle remain open.

Fat in the Arteries: Atherosclerosis

Anyone who has ever washed dishes is aware of how fat floats on water. You may have noticed that when one fat droplet meets another they stick together and form a larger droplet. Unfortunately, the same thing can happen in your arteries. As fat droplets grow into larger and larger blockages, they slowly close off the opening of the blood vessel. Calcium and other minerals deposit on top of the lipid, forming a fibrous net of plaque. This condition, known as **atherosclerosis,** can narrow the artery to one-quarter of its original diameter and lead to high blood pressure. To make matters worse, blood clots, a natural *life-saving* property of blood, form around the fat deposits. As fat droplets accumulate, adequate

amounts of blood and oxygen cannot be delivered to the heart muscle, resulting in chest pains.

Every year heart disease kills more Canadians than any other disease. Lifestyle changes must accompany any medical treatment. A low-fat diet, plus regular, controlled exercise are keys to prevention.

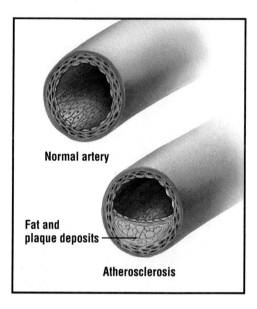

Normal artery

Fat and plaque deposits

Atherosclerosis

Figure 10.6

Fat deposits have narrowed the passageway.

Capillaries

The **capillary,** composed of a single layer of cells, is the site of fluid and gas exchange between blood and body cells. No cell is further than two cells away from a capillary, and many active cells, such as muscle cells, may be supplied by more than one capillary. Most capillaries extend between 0.4 and 1.0 mm, but have a diameter of less than 0.005 mm. The diameter is so small that red blood cells must travel through capillaries in single file.

The single cell layer, although ideal for diffusion, creates problems. Capillary beds are easily destroyed. High blood pressure or any impact, such as that caused by a punch, can rupture

Vasodilation *is the widening of the diameter of the blood vessel. More blood moves to tissues when arterioles dilate.*

Atherosclerosis *is a degeneration of the blood vessel caused by the accumulation of fat deposits along the inner wall.*

Capillaries, *tiny blood vessels that connect arteries and veins, are the site of fluid and gas exchange.*

the thin-layered capillary. Bruising occurs when blood rushes into the **interstitial spaces.**

Oxygen diffuses from the blood into the surrounding tissues through the thin walls of the capillaries. Oxygenated blood, which appears red in color, becomes a purple-blue color as it leaves the capillary. The deoxygenated blood collects in small veins called venules and is carried back to the heart. Some protein is also exchanged, but the process is believed to involve endocytosis and exocytosis rather than diffusion. Water-soluble ions and vitamins are believed to pass through the spaces between capillary cells. The fact that some spaces are wider than others may explain why some capillaries seem to be more permeable than others.

Veins

Capillaries merge and become progressively larger vessels, called **venules.** Unlike capillaries, the walls of venules are lined with smooth muscle. Venules merge into **veins,** which have greater diameter. Gradually the diameter of the veins increases as blood is returned to the heart. However, the very return of blood to the heart poses a problem. Blood flow through the arterioles and capillaries is greatly reduced. The passage of blood through incrementally narrower vessels reduces fluid pressure. By the time blood enters the venules, the pressure is between 15 mm Hg and 20 mm Hg. These pressures, however, are not enough to drive the blood back to the heart, especially from the lower limbs.

How, then, does blood get back to the heart? Let us return to William Harvey's experiments to answer that question. In one of his experiments, Harvey tied a band around the arm of one of his subjects, restricting venous blood flow. The veins soon became engorged with blood, and swelled. Harvey then placed his fin-

ger on the vein and pushed blood toward the heart. The vein collapsed. Harvey repeated the procedure, but this time he pushed the blood back toward the hand. Bulges appeared in the vein at regular intervals. What caused the bulges? Dissection of the veins confirmed the existence of valves.

The valves open in one direction, steering blood toward the heart. By attempting to push blood toward the hand, Harvey closed the valves, causing blood to pool in front of the valve. The pooling of blood caused the vein to become distended. However, by directing blood toward the heart, Harvey opened the valves, and blood flowed from one compartment into the next.

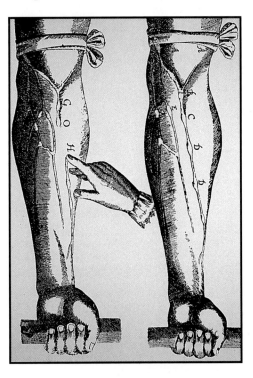

Figure 10.7
The one-way valves direct blood flow back to the heart. William Harvey's teacher, Hieronymus Fabricius, had already discovered the valves, but he did not fully investigate their function. He, like many others, still believed in Galen's ebb-and-flow theory.

Interstitial spaces *are the spaces between the cells.*

Venules *are small veins.*

Veins *carry blood back to the heart.*

Skeletal muscles also aid venous blood flow. Venous pressure increases when skeletal muscles contract and push against the vein. The muscles bulge when they contract, thereby reducing the vein's diameter. Pressure inside the vein increases and the valves open, allowing blood to flow toward the heart. Sequential contractions of skeletal muscle create a massaging action that moves blood back to the heart. This may explain why you feel like stretching first thing in the morning. It also provides clues as to why some soldiers faint after standing at attention for long periods of time. Blood begins to pool in the lower limbs. The movement of the leg muscles is required to move blood back to the heart.

The veins serve as more than just low-pressure transport canals—they are also important blood reservoirs. As much as 50% of your total blood volume can be found in the veins. During times of stress, venous blood flow can be increased to help you meet increased energy demands. Nerve impulses cause smooth muscle in the walls of the veins to contract, increasing fluid pressure. Increased pressure drives more blood to the heart, increasing heart filling.

Unfortunately, veins, like other blood vessels, are subject to problems. Large volumes of blood can distend the veins. In most cases, veins return to normal diameter, but if the pooling of blood occurs over a long period of time, the one-way valves are damaged. Without proper functioning of the valves, gravity carries blood toward the feet and greater pooling occurs. Surface veins gradually become larger and begin to bulge. The disorder is known as **varicose veins.** Although a genetic link to a weakness in the vein walls exists, lifestyle can accelerate the damage. Prolonged standing, especially with restricted movement, increases pooling of blood. Prolonged compression of the superficial veins in the leg can contribute to varicose veins.

Figure 10.8

Venous valves and skeletal muscle work together in a low-pressure system to move blood back to the heart.

Varicose veins *are distended veins.*

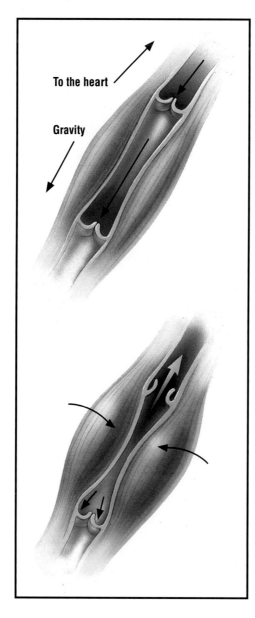

■ REVIEW QUESTIONS ■ ?

1 Why do multicellular animals need a circulatory system?
2 Explain the importance of William Harvey's theory that blood circulated?
3 How do arteries differ from veins?
4 What causes a pulse?
5 Why are aneurysms dangerous?
6 Define vasodilation and vasoconstriction.
7 Why are fat deposits in arteries dangerous?
8 What is the function of capillaries?
9 Fluid pressure is very low in the veins. Explain how blood gets back to the heart.

LABORATORY
EFFECTS OF TEMPERATURE ON PERIPHERAL BLOOD FLOW

Objective

To monitor blood flow in different parts of the body.

Background Information

A recently developed technique for monitoring skin temperature involves the use of liquid crystals. Referred to as liquid crystal thermography, this approach is based on the color changes that accompany structural changes of crystals at different temperatures. Cool temperatures can be identified by a brown color, while progressively warmer temperatures produce yellow, green, and blue colors.

You will be given different liquid crystal disks that are sensitive to different temperatures. The 28°C disk is the most sensitive and the 30°C, 32°C, and 34°C disks are progressively less sensitive.

Materials

liquid crystal disks	two large beakers
masking tape	ice

Procedure

1 Place the 34°C disk on the back of your hand. (You may wish to secure the disk with a few strips of masking tape.) Leave it for at least 30 s and observe any color change.

2 If the disk does not change color, continue by replacing it with progressively more sensitive disks.

a) Record the color displayed by the 34°C , 32°C , and 30°C disks.

3 Repeat the procedure to record the temperature of your neck, forehead, and index finger.

b) Record your results in a data table like the one below.

Area of the body	Temperature of disk used	Colors observed
Neck		
Forehead		
Index finger		

4 Fill a large beaker with ice water. Attach a 30°C disk to the back of your hand and then immerse your fingers in the cold water. Do not submerge the disk.

c) Observe and record changes in the color of the disk for about 3 min, or **for as long as the fingers can be comfortably immersed in the water.**

5 Fill a second large beaker with warm water (approximately 35°C). Place your fingers into the warm beaker.

d) Observe and record changes in the color of the disk for about 3 min.

Laboratory Application Questions

1 According to your results, which areas of the body have the greatest blood flow? Provide an explanation for your results.

2 How might your data have been affected if you had exercised before the experiment? Give your reasons.

3 A subject places a sensitive thermography disk on her forehead. A bright blue appears. The subject then begins smoking and the color changes to green, and then to yellow. What conclusions can you draw from this simple experiment?

4 How does temperature affect peripheral circulation? Explain the homeostatic adjustment mechanism for immersion in both warm and cold water.

5 Scientists have long known that cancer cells are more active than normal cells. Cancer cells divide many times faster. Explain how liquid crystals have been used to diagnose cancerous tumors. ■

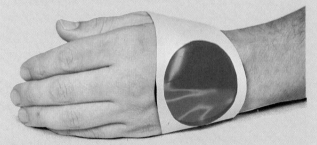

B I O L O G Y   C L I P
Many people believe that a drink of alcohol will warm them up on a cold day. Alcohol causes dilation of the arterioles leading to the skin capillaries, causing the sensation of warmth. However, the sensation is misleading. The dilation of these arterioles actually speeds cooling.

THE HEART

The heart is surrounded by a fluid-filled membrane called the **pericardium.** The fluid bathes the heart, preventing friction between its outer wall and the covering membrane. No larger than the size of your fist, and with a mass of about 300 g, the heart beats about 70 times each minute from the beginning of life until death. During an average lifetime, the heart pumps enough blood to fill two large ocean tankers.

Walking or mild exercise will increase your heart rate by 20 to 30%. If you are in good health, your heart rate during exercise can increase to an incredible 200 beats per minute to meet the increased energy demands associated with extreme exercise. Although few individuals can sustain such a rapid heart rate, it indicates the capacity of the heart to meet changing situations.

The heart is not a single pump, but two parallel pumps separated by a wall of muscle, called the *septum.* Although blood from the right heart pump is continually separated from the blood in the left pump, the pumping action is synchronized. Muscle contractions on the right side mirror contractions on the left side. The pump on the right receives deoxygenated blood from the body tissues and pumps it to the lungs. Vessels that carry blood to and from the lungs comprise the **pulmonary circulatory system.** The pump on the left receives oxygenated blood from the lungs and pumps it to the cells of the body. Vessels that carry blood to and from the body cells comprise the **systemic circulatory system.**

The four-chambered human heart is composed of two thin-walled atria and two thick-walled ventricles. The **atria** act as holding chambers for blood entering the heart from either the systemic or pulmonary circulatory systems. The

*The **pericardium** is a saclike membrane that protects the heart.*

*The **pulmonary circulatory system** carries deoxygenated blood to the lungs and oxygenated blood back to the heart.*

*The **systemic circulatory system** carries oxygenated blood to the tissues of the body and deoxygenated blood back to the heart.*

***Atria** are thin-walled heart chambers that receive blood from veins.*

***Ventricles** are muscular, thick-walled heart chambers that pump blood through the arteries.*

*The **venae cavae** carry deoxygenated blood back to the heart.*

*The **pulmonary veins** carry oxygenated blood from the lung to the heart.*

***Atrioventricular (AV) valves** prevent the backflow of blood from the ventricles into the atria.*

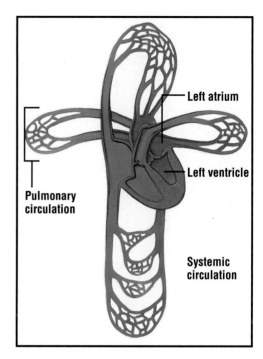

Figure 10.9

The systemic circulatory circuit carries oxygenated blood to the tissues of the body and transports deoxygenated blood back to the heart. The pulmonary circulatory circuit carries deoxygenated blood to the lungs, and the oxygenated blood back to the heart.

stronger, more muscular **ventricles** pump the blood to distant tissues.

ONE-WAY BLOOD FLOW

Blood is carried to the heart by veins. The **superior vena cava** carries deoxygenated blood from the head to the right atrium. The **inferior vena cava** carries blood from the tissues of the body to the same atrium. Oxygenated blood flowing from the lung enters the left atrium by way of the **pulmonary veins.** Blood on both sides of the heart fills the atria and is eventually pumped into the larger ventricles.

Valves, referred to as **atrioventricular,** or **AV valves,** separate the atria from the ventricles. In much the same way as the valves within veins ensure one-direction flow, the AV valves prevent

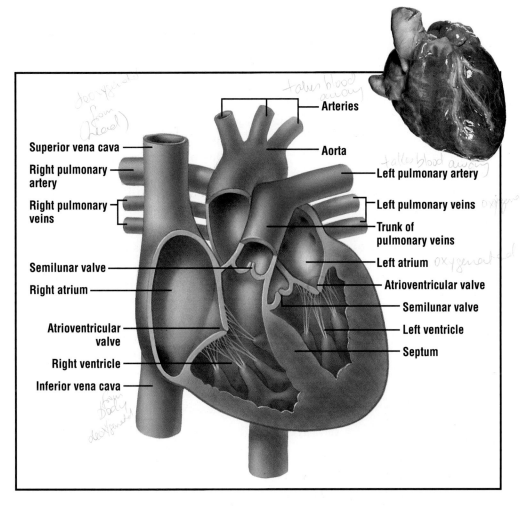

Arteries

Superior vena cava

Right pulmonary artery

Right pulmonary veins

Semilunar valve

Right atrium

Atrioventricular valve

Right ventricle

Inferior vena cava

Aorta

Left pulmonary artery

Left pulmonary veins

Trunk of pulmonary veins

Left atrium

Atrioventricular valve

Semilunar valve

Left ventricle

Septum

Figure 10.10

Mammalian heart showing atria and ventricles.

Figure 10.11

Deoxygenated blood enters the right atrium. Blood from the right atrium is pumped to the right ventricle, which, in turn, pumps it to the lungs. Oxygenated blood moves from the lungs to the left atrium. Blood from the left atrium is pumped to the left ventricle, which, in turn, pumps it to the cells of the body.

Chordae tendinae *support the AV valves.*

Semilunar valves *prevent the backflow of blood from arteries into the ventricles.*

The **pulmonary artery** *carries deoxygenated blood from the heart to the lungs.*

The **aorta** *carries oxygenated blood to the tissues of the body.*

Coronary arteries *supply the cardiac muscle with oxygen and other nutrients.*

Angina *literally means to suffocate. It is often used to describe the chest pain produced by heart attack.*

blood from flowing from the ventricles back to the atria. The AV valves are supported by bands of connective tissue called **chordae tendinae.** A second set of valves, called **semilunar valves,** are found at the areas in which blood vessels attach to the ventricles. These half-moon-shaped (hence, the name "semilunar") valves prevent blood that has entered one of the exit blood vessels from flowing back into the ventricles.

Blood is carried away from the heart by arteries. The **pulmonary artery** carries deoxygenated blood from the heart to the lungs. Once in the lungs, the blood receives oxygen by diffusion and returns

BIOLOGY CLIP
Nitroglycerine, a chemical explosive, is also used to treat angina. First used in this way in 1879, nitroglycerine continues to be one of the most effective drugs for angina.

to the left side of the heart. Oxygenated blood is carried away from the heart by the **aorta,** the largest artery in your body.

One of the most important branches of the aorta, the **coronary arteries** supply the heart with oxygenated blood. The importance of proper coronary circulation is illustrated by a blocked artery. Chest pains, or **angina,** occur when too little oxygen reaches the heart. The heart, unlike other organs that slow down if they cannot receive enough nutrients, must continue beating no matter what demands are placed on it. It has been estimated that the heart may use 20% of the blood oxygen during times of stress.

As in other arteries, fat deposits and plaque can collect inside coronary arteries. Drugs are often used to increase blood flow, but in severe situations blood flow must be rerouted. A coronary bypass operation involves removing the patient's leg vein and grafting the vein into position in the heart. However, in order to graft the vein, the heart must be temporarily stopped. During the operation, the patient's heart is cooled and a heart-lung machine is used to supply oxygen and push blood to the tissues of the body.

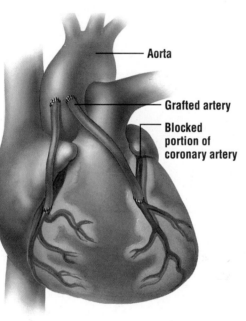

Aorta

Grafted artery

Blocked portion of coronary artery

Figure 10.12

Coronary bypass operation. Blood flow is rerouted around the blockage.

correct disease. An easy way to determine whether or not a patient is suffering from coronary artery problems is to perform surgery. Unfortunately, surgery is not without risks. Clearly, other means for diagnosing the problem would be helpful.

One of the most useful techniques is that of cardiac catheterization. In this procedure, a small, thin plastic tube, called a catheter, is passed into a leg vein as the patient lies on an examination table. Dye, which is visible on X-ray film, is then injected into the catheter. The dye travels through the blood vessels and into the heart while its image is traced by means of a fluoroscope. The image can also be projected on a television monitor.

A longer catheter can be drawn up from the vein into the heart so that the dye can be injected into a coronary artery. As the blood and dye move through the artery, an X-ray movie can be made. An area of restricted blood flow pinpoints the region of blockage. The catheter helps direct the surgeon to the problem prior to the surgery.

FRONTIERS OF TECHNOLOGY: CARDIAC CATHETERIZATION

Figure 10.13

The image collected from an obstructed coronary artery.

At one time doctors had to rely on external symptoms to detect coronary artery blockage. An inability to sustain physical activity, rapid breathing, and a general lack of energy are three of the symptoms of coronary distress. However, these same symptoms can also indicate a wide variety of other circulatory and respiratory diseases. One of the greatest challenges in medicine is matching symptoms with the

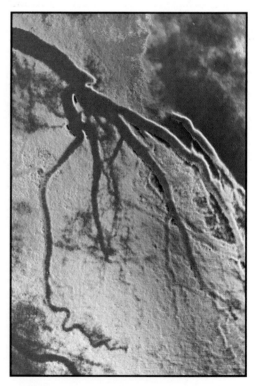

Blood samples can also be taken with the catheter to determine how much oxygen is in the blood in the different chambers. This tells the physician how well the blood is being oxygenated in the lungs. Low levels of oxygen in the left side of the heart can provide information about the teamwork of the circulatory and respiratory systems. The catheter can even be used to monitor pressures in each of the heart chambers.

SETTING THE HEART'S TEMPO

Heart or cardiac muscle differs from other types of muscle. Like skeletal muscle, cardiac muscle appears striated when viewed under a microscope. But, unlike skeletal muscle, cardiac muscle displays a branching pattern. The greatest difference stems from the ability of this muscle to contract without external nerve stimulation. Muscle with this ability is called **myogenic muscle.** This latter ability explains why the heart will continue to beat, at least for a short time, when removed from the body.

The remarkable capacity of the heart to beat can be illustrated by a simple experiment. A frog's heart is removed and placed in a salt solution that simulates the minerals found within the body. The heart is then sliced into small pieces. Incredible as it may seem, each of the pieces continues to beat, although not at the same speed. Muscle tissue from the ventricles follows a slower rhythm than muscle tissue from the atria. Muscle tissue closest to the entry port of the venae cavae has the faster tempo. The unique nature of the heart becomes evident when two separated pieces are brought together. The united fragments assume a single beat. The slower muscle tissue assumes the tempo set by the muscle tissue that beats more rapidly.

The heart's tempo or beat rate is set by the **sinoatrial,** or **SA node.** This bundle of specialized nerves and muscle is located where the venae cavae enter the right atrium. The sinoatrial node acts as a pacemaker, setting a rhythm of about 70 beats per minute for the heart. Nerve impulses are carried from the pacemaker to other muscle cells by modified muscle tissue. Originating in the atria, the contractions travel to a second node, the *atrioventicular,* or *AV node.* The AV node serves as a conductor, passing nerve impulses along special tracts through the dividing septum toward the ventricles. Both right and left atria contract prior to the contraction of the right and left ventricles.

Figure 10.14

The heart is composed of cardiac muscle. The branching pattern is unique to cardiac muscle.

Myogenic muscle *tissue contracts without external nerve stimulation. Cardiac muscle sets a beat.*

The **sinoatrial node** *is the heart's pacemaker.*

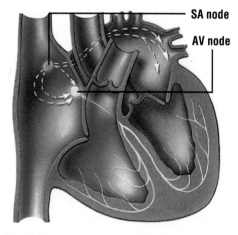

Figure 10.15

The pacemaker initiates heart contractions. Modified muscle tissue passes a nerve impulse from the pacemaker down the dividing septum toward the ventricles.

One of the greatest challenges for surgeons performing open-heart surgery is to make incisions at the appropriate location. A scalpel placed in the wrong spot could cut conducting fibers.

Electrical fields within the heart can be mapped by a device called the **electrocardiograph.** Electrodes placed on the body surface are connected to a recording device. The electrical impulses are displayed on a graph called an electrocardiogram. Changes in electrical current reveal normal or abnormal events of the cardiac cycle. The first wave, referred to as the P wave, monitors atrial contraction. The larger spike, referred to as the QRS wave, records ventricular contraction. A final T wave signals that the ventricles have recovered.

Doctors use electrocardiograph tracings to diagnose certain heart problems. A patch of dead heart tissue, for example, will not conduct impulses, and produces abnormal line tracings. By comparing the tracings, doctors are able to locate the area of the heart that is damaged.

The electrocardiograph is especially useful for monitoring the body's response to exercise. Stress tests are performed by monitoring a subject who is riding a sta-tionary bike or running on a treadmill. Some heart malfunctions remain hidden during rest, but can be detected during vigorous exercise.

Heart rate is influenced by autonomic or automatic nerves. Two regulatory nerves—the sympathetic and parasympathetic nerves—conduct impulses from the brain to the pacemaker. Stimulated during times of stress, the sympathetic nerve increases heart rate. This increases blood flow to tissues, enabling the body to meet increased energy demands. Conditions in which the heart rate exceeds 100 beats per minute are referred to as *tachycardia*. Tachycardia can result during exercise or from the consumption of such drugs as caffeine or nicotine. During times of relaxation, the parasympathetic nerve is stimulated. The parasympathetic nerve slows heart rate. The condition in which the heart beats very slowly is referred to as *bradycardia*.

An **electrocardiograph** *is an instrument that monitors the electrical activity of the heart.*

Figure 10.16

Electrocardiograph tracing of a single normal heartbeat.

Figure 10.17

An abnormal electrocardiograph tracing. Can you determine what has gone wrong?

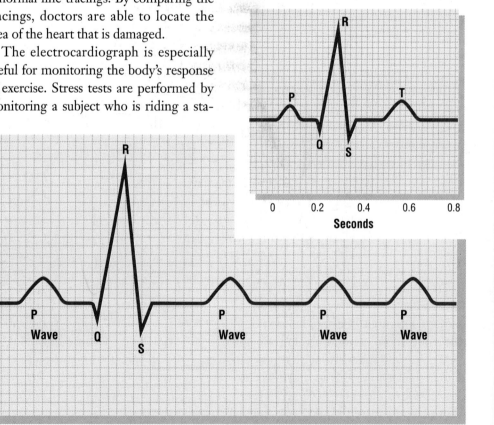

RESEARCH IN CANADA

Cardiac Research

Dr. Alexandra Lucas, a researcher at the University of Alberta, is one of Canada's new breed of cardiac researchers. Equipped with an arsenal of new devices, Dr. Lucas is studying atherosclerotic plaque, one of the main causes of heart attack. Dr. Lucas uses fluorescent light to evaluate the content of plaque, identifying substances such as unsaturated cholesterol and carotenoid. The fluorescence emission is excited by low-power (ultraviolet) laser and is measured by fluorescence emission spectroscopy. The laser-induced fluorescence was originally developed as a guiding system for other lasers to clear clogged arteries, a procedure called angioplasty. However, a cutting laser presents many problems. Without guiding lasers the catheter might perforate the wall of the artery. The perforation, although less than life-threatening in the leg, becomes very serious in an artery near the heart. Most angioplasty is done with a balloon, which is placed inside a catheter. Once the catheter is inserted inside the blocked artery, the balloon is inflated. The balloon pushes the plaque back and the artery widens.

Dr. Lucas is concentrating on using the fluorescence emission laser for diagnostic purposes. Her goal is to develop diagnostic markers that identify changes in the plaque. Any changes in the growth of plaque or changes in fat or fiber content could signal future problems. The prospects of warning patients about the potential of heart attacks may go a long way toward directing preventative action such as changes in diet and lifestyle.

DR. ALEXANDRA LUCAS

HEART SOUNDS

The familiar "lubb-dubb" heart sounds are caused by the closing of the heart valves. Contraction of the muscular walls of the atria increases pressure and forces blood through AV valves into the ventricles. As stated before, the cardiac muscle contraction proceeds from the atria to the ventricles. Atria begin to fill with blood as they relax. The term **diastole** is used to describe this relaxation. As the ventricles begin to contract, blood is forced up the sides of the ventricles and the AV valves close, producing a lubb sound. Ventricular contraction increases pressure in the chambers, forcing blood through the semilunar valves and out of the arteries. The term **systole** is used to describe this contraction. As the ventricles begin to relax, the volume of the chambers increases. With increased volume, pressure in the ventri-

cles begins to decrease and blood is drawn from the arteries toward the area of lower pressure. However, the blood is prevented from re-entering the ventricles by the semilunar valves. The blood causes the semilunar valves to close, creating the dubb sound.

Occasionally, the valves do not close completely. This condition, referred to as a heart **murmur,** occurs when blood leaks past the closed heart valve because of an improper seal. The AV valves, especially the left AV valve, or *mitral valve*, must withstand great pressure and are especially susceptible to defects. The rush of blood from the ventricle back into the atrium produces a gurgling sound that can be detected by a stethoscope. Blood flowing back toward the atrium is not directed to the systemic or pulmonary systems, but the hearts of individuals who experience murmurs do compensate for

Diastole *refers to heart relaxation.*

Systole *refers to heart contraction.*

Murmurs *are caused by faulty heart valves, which permit the backflow of blood into one of the heart chambers.*

> **BIOLOGY CLIP**
> In 1816 Rene Laennec, a young physician, was examining a patient for heart distress. The common practice at the time was for the doctor to place his ear on the patient's chest and listen for the lubb-dubb sounds. However, Laennec found that because of the patient's heavy bulk, the heart sounds were muffled. As the examination became more and more awkward, Laennec decided to try another avenue. He rolled up a paper and placed it to the patient's chest. Much to his relief, the heart sounds became clearer. Later, wooden cylinders were used, eventually to be replaced by the modern Y-shaped stethoscope, which literally means "chest-viewer."

Figure 10.18

Right and left atria contract in unison, pushing blood into the right and left ventricles. Ventricular contractions close the AV valves and open the semilunar valves. The relaxation of the ventricles lowers pressure and draws blood back to the chamber. The closing of the semilunar valves prevents blood from re-entering the ventricles.

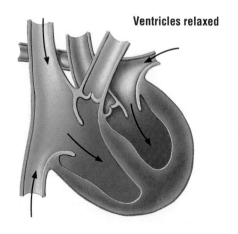

Ventricles relaxed

Ventricles contracted

decreased oxygen delivery by beating faster. You will learn more about this compensating mechanism in the next section.

A second compensatory mechanism helps increase blood flow. Like an elastic band, the more cardiac muscle is stretched, the stronger is the force of contraction. When blood flows from the ventricle back into the atrium, blood volume in the atrium increases. The atrium accepts the normal filling volumes but, in addition, accepts blood from the ventricle. This causes the atrium to stretch and the blood to be driven to the ventricle with greater force. Subsequently, increased blood volume in the ventricle causes the ventricle to contract with greater force, driving more blood to the tissues.

CARDIAC OUTPUT

Cardiac output is defined as the amount of blood that flows from each side of the heart per minute. Unless some dysfunction occurs, the amount of blood pumped from the right side of the heart is equal to the amount of blood pumped from the left side of the heart. Two factors affect cardiac output: stroke volume and heart rate.

Stroke volume is the quantity of blood pumped with each beat of the heart. The stronger the heart contraction, the greater the stroke volume. Approximately 70 mL of blood per beat leave each ventricle while you are resting. *Heart rate* is the number of times the heart beats per minute. The equation below shows how cardiac output is determined by stroke volume and heart rate.

$$\text{Cardiac output} = \text{heart rate} \times \text{stroke volume}$$
$$= 70 \text{ beats/min} \times 70 \text{ mL/beat}$$
$$= 4900 \text{ mL/min}$$

Individuals who have a mass of 70 kg must pump approximately 5 L of blood per minute. Smaller individuals require less blood and therefore have lower cardiac outputs. Naturally, cardiac output must be adjusted to meet energy needs. During exercise, heart rate may increase to 150 or 180 beats per minute to meet increased energy demands.

The cardiac output equation provides a basis for comparing individual fitness. Why do two people with the same body mass have different heart rates? If you assume that both people are at rest, both should require the same quantity of oxygen each minute. For example, Tom, who has a heart rate of 100 beats per minute, has a lower stroke volume. Lee, who has a heart rate of 50 beats per minute, has a higher stroke volume.

Cardiac output = stroke volume × heart rate

Tom	Lee
5 L = 50 mL/beat × 100 beats/min	5 L = 100 mL/beat × 50 beats/min

Lee's lower heart rate indicates a higher stroke volume. People who have well-developed hearts can pump greater volumes of blood with each beat. This is why athletes often have low heart rates. Those with weaker hearts are unable to pump as much blood per beat, but compensate by increasing heart rate to meet the body's energy demands. It is impor-

> **BIOLOGY CLIP**
> Due to greater stroke volume, some athletes have much slower heart rates. The tennis player Bjorn Borg once demonstrated a resting heart rate of 35 beats/min.

Cardiac output *is the amount of blood pumped from the heart each minute.*

Stroke volume *measures the quantity of blood pumped with each beat of the heart.*

tant to recognize that heart rate is only one factor that determines physical fitness. You may also find that your pulse rate will fluctuate throughout the day. Various kinds of food, stress, or a host of other factors can affect your heart rate.

BLOOD PRESSURE

Blood surges through the arteries with every beat of the heart. Elastic connective tissue and smooth muscle in the walls of the arteries stretch to accommodate the increase in fluid pressure. The arterial walls recoil much like an elastic band as the heart begins the relaxation phase characterized by lower pressure. Even the recoil forces help push blood through arterioles toward the tissues.

Blood pressure can be measured indirectly with an instrument called a **sphygmomanometer.** A cuff with an air bladder is wrapped around the arm. A small pump is used to inflate the air bladder, thereby closing off blood flow through the brachial artery, one of the major arteries of the arm. A stethoscope is placed below the cuff and air is slowly released from the bladder until a low-pitched sound can be detected. The sound is caused by blood entering the previously closed artery.

Each time the heart contracts, the sound is heard. A gauge on the sphygmomanometer measures the pressure that blood exerts during ventricular contraction. This pressure is called *systolic blood pressure.* Normal systolic blood pressure for young adults is about 120 mm Hg. The cuff is then deflated even more, until the sound disappears. At this point, blood flows into the artery during ventricular relaxation or filling. This pressure is called *diastolic blood pressure.* Normal diastolic blood pressure for young adults is about 80 mm Hg. Reduced filling, such as

Figure 10.19

This sphygmomanometer is calibrated in the nonmetric units of millimeters of mercury.

Figure 10.20

Fluid pressure decreases the further blood moves from the heart.

A **sphygmomanometer** *is a device used to measure blood pressure.*

that caused by an internal hemorrhage, will cause diastolic blood pressure to fall.

Figure 10.20 shows that fluid pressure decreases with distance from the ventricles. The aorta records the greatest pressure readings, despite the fact that they have the largest diameter. To help you understand why, imagine two hoses connected to a tap; the first hose is 1 m in length, the second is 100 m in length. The same amount of water is released from the tap into each hose. As water passes through the hoses, friction is created, slowing down its movement. The longer the hose, the greater the amount of friction and the slower the stream of water. This explains why the pulse in the carotid artery is stronger than the pulse detected near your wrist and why blood-pressure readings are not the same in all arteries.

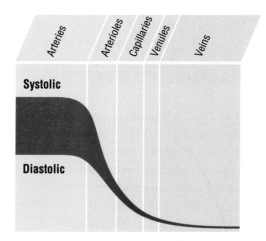

The artery acts as a reservoir for blood. Two factors regulate the amount of blood in the reservoir. The first is cardiac output. Any factor that increases cardiac output will increase blood pressure. The second factor is arteriolar resistance. You may recall that the diameter of the arterioles is regulated by coiling, smooth muscles. Constriction of the smooth muscles surrounding the arterioles closes the opening and reduces blood flow through the arteriole. With this reduced blood

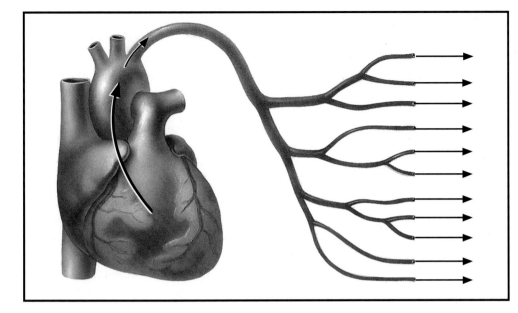

Figure 10.21

Blood pressure is measured in arteries. Two factors, cardiac output and arteriolar resistance, affect blood pressure.

flow, more blood is left in the artery. The increased blood volume in the artery produces higher blood pressure. Conversely, factors that cause arteriolar dilation increase blood flow from the arterioles, thereby reducing blood pressure.

The smooth muscles in the walls of the arterioles respond to nerve and endocrine controls that regulate blood pressure. However, the diameter of the arterioles also adjusts to metabolic products. For example, the cellular breakdown of sugar with oxygen produces carbon dioxide and water. The cellular breakdown of sugar in the absence of oxygen yields lactic acid. Carbon dioxide and lactic acid are two of the metabolic products that cause relaxation of smooth muscles in the walls of the arterioles, causing the arterioles to open. The dilation of the arterioles increases blood flow to local tissues. Arteriolar dilation, in response to increased metabolic products, provides a good example of homeostasis (see Figure 10.22). Because these products accumulate in the most active tissues, the increased blood flow helps provide greater nutrient supply, while carrying the potentially toxic materials away. Tissues that are less active produce fewer meta-

bolic products. These arterioles remain closed until the products accumulate.

Table 10.2 Factors that Affect Arteriolar Resistance

Factor	Effect
Epinephrine	arteriolar constriction, except to the heart, muscles, and skin
Sympathetic nerve stimulation	arteriolar constriction, except in skeletal and cardiac muscle
Acid accumulation	arteriolar dilation
CO_2 accumulation	arteriolar dilation
Lactic acid accumulation	arteriolar dilation

Homeostatic Adjustment

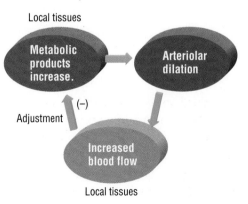

Figure 10.22

Dilation of the arterioles increases blood flow to local tissues.

10 Draw and label the major blood vessels and chambers of the heart. Trace the flow of deoxygenated and oxygenated blood through the heart.

11 Differentiate between the systemic circulatory system and the pulmonary circulatory system.

12 What causes the characteristic heart sounds?

13 What are coronary bypass operations and why are they performed?

14 Explain the function of the sinoatrial node.

15 What is an electrocardiogram?

16 Differentiate between systolic and diastolic blood pressure.

17 Define cardiac output, stroke volume, and heart rate.

18 How do metabolic products affect blood flow through arterioles?

REGULATION OF BLOOD PRESSURE

Regulation of blood pressure is essential. Low blood pressure reduces your capacity to transport blood. The problem is particularly acute for tissue in the head where blood pressure works against the force of gravity. High blood pressure creates equally serious problems. High fluid pressure can weaken an artery and eventually lead to the rupturing of the vessel.

Special **blood pressure receptors** are located in the walls of the aorta and the carotid arteries, which are major arteries found on either side of the neck. These receptors, known as *baroreceptors*, are sensitive to high pressures. When blood pressure exceeds acceptable levels, the baroreceptors respond to the increased pressure on the wall of the artery. A nerve message travels to the medulla oblongata, the blood pressure regulator located at the stem of the brain. The sympathetic nerve—the "stress nerve"—is turned down and the parasympathetic nerve—the "slow down nerve"—is stimulated. By decreasing sympathetic nerve stimulation, arterioles dilate, increasing outflow of blood from the artery. Stimulation of the parasympathetic nerve causes heart rate and stroke volume to decrease. The decreased cardiac output slows the movement of blood into the arteries and consequently lowers blood pressure.

Low blood pressure is adjusted by the sympathetic nerve. Without nerve information from the pressure receptors of the carotid artery and aorta, the sympathetic nerve will not be "turned off." Under the influence of the sympathetic nerve, cardiac output increases and arterioles constrict. The increased flow of blood into the artery accompanied by decreased outflow raises blood pressure to acceptable levels.

> **Blood pressure receptors** *are specialized nerve cells that are activated by high blood pressure.*

Figure 10.23

Decreased cardiac output and arteriolar dilation decrease blood volume in arteries. The lower volume of blood in the artery decreases blood pressure.

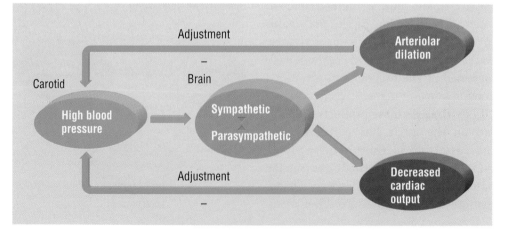

Unit Three:
Exchange of Matter and Energy in Humans

LABORATORY

EFFECTS OF POSTURE ON BLOOD PRESSURE

Objective

To determine how posture affects blood pressure.

Materials

sphygmomanometer   stethoscope
watch with second hand   alcohol

Procedure

1 Ask your partner to sit quietly for one minute. Clean the earpieces of the stethoscope with alcohol. Expose the arm of your partner and place the sphygmomanometer just above the elbow.

2 Close the valve on the rubber bulb on the end of the sphygmomanometer, and inflate it by squeezing the rubber ball until a pressure of 180 mm Hg registers.

CAUTION: Do not leave the pressure on for longer than 1 min. If you are unsuccessful, release the pressure and try again.

3 Place the stethoscope bell, or diaphragm, on the inside of the arm immediately below the cuff.

4 Slowly release the pressure by opening the valve on the rubber ball of the sphygmomanometer. Listen for a low-pitched sound.

   a) Record the reading on the sphygmomanometer. This is the systolic blood pressure.

5 Continue releasing the pressure until the sound can no longer be heard.

   b) Record the reading on the dial of the sphygmomanometer when the sound disappears. This is the diastolic pressure.

6 Completely deflate the sphygmomanometer and take your partner's pulse. Place your index and middle fingers on the arm near the wrist. Count the number of pulses in one minute.

   c) Record the pulse rate while seated.

7 Repeat the procedure while your partner is in a standing position and then in a lying position.

   d) Record your results in a table like the one below.

Position	Systolic B P (mm Hg)	Diastolic B P (mm Hg)	Pulse rate (beats/min)
Standing			
Sitting			
Lying			

Laboratory Application Questions

1 Would you expect blood pressure readings in all the major arteries to be the same? Explain your answer.

2 Why should the lowest systolic pressure be recorded while you are lying down?

3 Atherosclerosis, or hardening of the arteries, is a disorder that causes high blood pressure. Provide an explanation for this condition.

4 Predict how exercise would affect systolic blood pressure. Provide your reasons.

5 Why might diastolic blood pressure decrease as heart rate increases?

6 Design a procedure to investigate the role of exercise in influencing blood pressure. ■

ADJUSTMENT OF THE CIRCULATORY SYSTEM TO EXERCISE

Your body's response to exercise is an excellent example of a homeostatic mechanism. The demands placed on the circulatory system by tissues during exercise are considerable. The circulatory system does not act alone in monitoring the needs of tissues or in ensuring that adequate levels of oxygen and other nutrients are delivered to the active cells. The nervous and endocrine systems also play important roles in adjustment mechanisms.

The sympathetic nerve stimulates the adrenal glands. During times of stress, the hormone epinephrine is released from the adrenal medulla and travels in the blood to other organs of the body. Epinephrine stimulates the release of red blood cells from the spleen, a storage site. Although the significance of the response is not yet understood, it is clear that increased numbers of red blood cells aid oxygen delivery. Epinephrine and direct stimulation from the sympathetic nerve increase heart rate and breathing rate. The increased heart rate provides for faster oxygen transport, while the increased breathing rate ensures that the blood contains higher levels of oxygen. Both systems work together to improve oxygen delivery to the active tissues. A secondary, but equally important, function is associated with more effective waste removal from the active tissues.

Blood cannot flow to all capillaries of the body simultaneously. The effect of dilating all arterioles would be disastrous—blood pressure would plunge. Epinephrine causes vasodilation or widening of the arterioles leading to the heart, brain, and muscles. At the same time, epinephrine causes the constriction of blood vessels leading to the kidney, stomach, and intestines. The most active tissues receive priority in times of stress. Blood flow is diverted to the muscles and heart, enabling the organism to perform responses associated with flight-or-fight reactions. Organs from the digestive system and kidney are deprived of much of their required nutrients until the stress situation has been overcome.

CAPILLARY FLUID EXCHANGE

It has been estimated that nearly every tissue of the body is within 0.1 mm of a capillary. Earlier in the chapter you learned how capillaries provide cells with oxygen, glucose, and amino acids. Capillaries are also associated with fluid exchange between the blood and surrounding **extracellular fluid (ECF).** Most fluids simply diffuse through capillaries. The capillary cell membranes are permeable to oxygen and carbon dioxide. Water and certain ions are thought to pass through the clefts between the cells of the capillary. Larger molecules and a very small number of proteins are believed to be exchanged by endocytosis or exocytosis. This section will focus on the movement of water molecules.

Two forces regulate the movement of water between the blood and ECF: fluid pressure and osmotic pressure. The force that blood exerts on the wall of a capillary is about 35 mm Hg pressure at the arteriole end of the capillary and approximately 15 mm Hg pressure at the venous end. The reservoir of blood in the arteries creates pressure on the inner wall of the capillary. Much lower pressure is found in the ECF. Although fluids bathe cells, no force drives the extracellular fluids. Water moves from an area of high pressure—the capillary—into an area of low pressure—the ECF. The outward flow of water and small minerals ions is known as **filtration.** Because capillaries are selectively permeable, large materials such as proteins, red blood cells, and white blood cells remain in the capillary.

The movement of fluids from the capillary must be balanced with a force that moves fluid into the capillary. The fact that large proteins are found in the blood but not in the ECF may provide a hint as to the nature of the second force. Osmotic pressure draws water back into the capillary. The large protein molecules of the blood and dissolved minerals are primarily responsible for the movement of fluid into capillaries. The movement of

Extracellular fluids (ECF) *occupy the spaces between cells and tissues.*

Filtration *is the selective movement of materials through capillary walls by a pressure gradient.*

fluid into capillaries is called **absorption. Osmotic pressure** in the capillaries is usually about 25 mm Hg, but it is important to note that the concentration of solutes can change with fluid intake or excess fluid loss caused by perspiration, vomiting, or diarrhea.

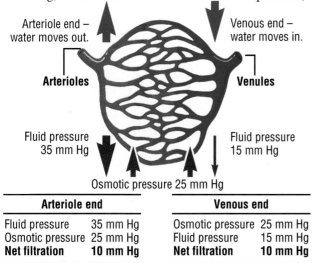

Arteriole end – water moves out.

Arterioles

Venous end – water moves in.

Venules

Fluid pressure 35 mm Hg

Fluid pressure 15 mm Hg

Osmotic pressure 25 mm Hg

Arteriole end		Venous end	
Fluid pressure	35 mm Hg	Osmotic pressure	25 mm Hg
Osmotic pressure	25 mm Hg	Fluid pressure	15 mm Hg
Net filtration	**10 mm Hg**	**Net filtration**	**10 mm Hg**

Application of the capillary exchange model provides a foundation for understanding homeostatic adjustments to a variety of problems. The balance between osmotic pressure and fluid pressure is upset during hemorrhage. The decrease in blood volume resulting from the hemorrhage affects blood pressure. The force that drives fluid from the capillaries diminishes, but the osmotic pressure, which draws water into the capillaries, is not altered. Although proteins are lost with the hemorrhage, so are fluids. Fewer proteins are present, but the concentration has not been changed. The force drawing water from the tissues and ECF is greater than the force pushing water from the capillary. The net movement of water into the capillaries provides a homeostatic adjustment. As water moves into the capillaries, fluid volumes are restored.

Individuals who are suffering from starvation often display tissue swelling, or **edema.** Plasma proteins are often mobilized as one of the last sources of energy. The decrease in concentration of plasma proteins has a dramatic effect on osmotic pressure, which draws fluids from the tissues and ECF into the capillaries. The decreased number of proteins lowers osmotic pressure, thereby decreasing absorption. More water enters the tissue spaces than is pulled back into the capillaries, causing swelling.

Why do tissues swell during inflammation or allergic reactions? When you eat a food to which you are allergic, endangered cells—or cells that "believe" they are endangered—release a chemical messenger, called *bradykinin*, which stimulates the release of another chemical stimulator, *histamine*. Histamine changes the cells of the capillaries, thereby increasing permeability. The enlarged capillary causes the area to redden. Proteins and white blood cells leave the capillary in search of the foreign invader, but, in doing so, they alter the osmotic pressure. The proteins in the ECF create another osmotic force that opposes the osmotic force in the capillaries. Less water is absorbed into the capillary, and tissues swell.

Figure 10.24

Fluid movement into and out of the capillaries.

Absorption *is the movement of fluids in the direction of a diffusion, or osmotic, gradient.*

Osmotic pressure *is the pressure exerted on the wall of a semipermeable membrane resulting from differences in solute concentration. (In this case, the more concentrated the plasma proteins, the greater is the osmotic pressure.)*

Edema *is tissue swelling caused by decreased osmotic pressure in the capillaries.*

Figure 10.25

The balance between osmotic pressure and fluid pressure is upset during hemorrhage, starvation, or inflammation.

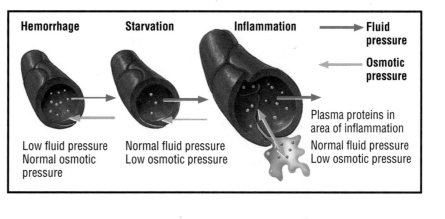

Hemorrhage

Starvation

Inflammation

Fluid pressure

Osmotic pressure

Low fluid pressure
Normal osmotic pressure

Normal fluid pressure
Low osmotic pressure

Plasma proteins in area of inflammation

Normal fluid pressure
Low osmotic pressure

THE LYMPHATIC SYSTEM

Normally, a small amount of protein leaks from capillaries to the tissue spaces. Despite the fact that the leak is very slow, the accumulation of proteins in the ECF would create a major problem: osmotic pressure would decrease and tissues would swell.

Figure 10.26

Lymph vessels (shown in green) are open-ended vessels.

The proteins are drained from the ECF and returned to the circulatory system by way of another network of vessels: the lymphatic system. **Lymph,** a fluid similar to blood plasma, is transported in open-ended lymph vessels in much the same way as veins. The low-pressure return system operates by slow muscle contractions against the vessels, which are supplied with flaplike valves that prevent the backflow of fluids. Eventually, lymph is returned to the venous system. In the previous chapter you read about specialized lymph vessels that carry fats from the small intestine. Called *lacteals*, these vessels provide the products of fat digestion access to your circulatory system.

Enlargements called **lymph nodes** are located at intervals along the lymph vessel. These house phagocytotic white blood cells that filter bacteria from lymph. Should bacteria be present, the phagocytotic white blood cells engulf and destroy them. The lymph nodes also fil-

ter damaged cells and debris from the lymph. Have you ever experienced swelling of the lymph nodes? The lymph nodes sometimes swell when you have a sore throat.

Figure 10.26

Lymph vessels (shown in green) are open-ended vessels.

Figure 10.27

Lymph, or ECF, vessels carry lymph back to the circulatory system. Lymph nodes filter debris from the lymph. Insert shows lymph flow through a lymph node.

Lymph nodes *contain white blood cells that filter lymph.*

Lymph *is the fluid found outside capillaries. Most often, the lymph contains some small proteins that have leaked through capillary walls.*

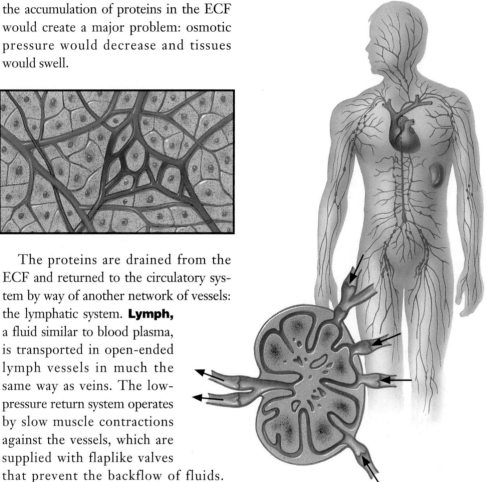

■ REVIEW QUESTIONS ■ ?

19 How do blood pressure regulators detect high blood pressure?

20 Outline homeostatic adjustment to high blood pressure.

21 What two factors regulate the exchange of fluids between capillaries and ECF?

22 Why does a low concentration of plasma protein cause edema?

23 What are lymph vessels and how are they related to the circulatory system?

SOCIAL ISSUE:
Heart Care

Heart disease is the number-one killer of North Americans. Although congenital heart defects account for some of the problems, the incidence of heart disease can often be traced to lifestyle. The relationship between heart problems and smoking, excessive alcohol consumption, stress, high-cholesterol diets, and high blood pressure has long been established.

Statement:

People who refuse to alter lifestyles that are dangerous to their health should not be permitted equal access to health care.

Point

- People who deliberately ignore a doctor's advice and continue to place themselves at risk should not be permitted equal access to the health system. High-cost health care should be reserved for those who deserve it most.
- People who have a high-risk lifestyle should pay higher medical insurance and be placed lower on waiting lists for heart transplants. How can we be assured that these people will not abuse their transplanted heart?

Research the issue.
Reflect on your findings.
Discuss the various viewpoints with others.
Prepare for the class debate.

Counterpoint

- People cannot always be held responsible for their behavior. Some people have a genetic link to obesity and a predisposition to alcoholism, indicating that behavior could be determined by factors other than environment.
- Health care should always be given on the basis of need. Medical authorities should decide who needs a transplant, regardless of the person's previous lifestyle or reliability.

CHAPTER HIGHLIGHTS

- Multicellular organisms need a circulatory system.
- The heart pumps blood into the arteries of the pulmonary and systemic circulatory systems. Blood travels from arteries to arterioles, capillaries, venules, and finally veins, which carry blood back to the heart.
- Arteries are high-pressure blood vessels that carry blood away from the heart. Arteries stretch to accommodate high pressure, producing a pulse.
- The capillaries are vessels composed of a single layer of cells. The diffusion of materials between the blood and extracellular fluid occurs in the capillaries.
- Veins are low-pressure blood vessels that carry blood back to the heart. Valves located in the veins prevent the backflow of blood.
- Cardiac output is a function of stroke volume and heart rate.
- Sympathetic nerve stimulation and the hormone epinephrine increase heart rate and cardiac output. Parasympathetic nerve stimulation decreases heart rate and cardiac output.

- Blood pressure is measured in arteries. Systolic blood pressure occurs when blood surges into arteries following ventricular contraction. Diastolic blood pressure occurs while the heart is relaxing.
- Blood pressure is regulated by cardiac output and arteriolar resistance.

- Arterioles dilate in response to metabolic products.
- Lymph vessels complement the circulatory system by restoring osmotic pressure and transporting protein and other solutes back into the blood.

APPLYING THE CONCEPTS

1 Agree or disagree with the following statement and give reasons for your views: Oxygenated blood is found in all arteries of the body.
2 Why does the left ventricle contain more muscle than the right ventricle?
3 Why does blood pressure fluctuate in an artery?
4 Which area of the graph represents blood in a capillary? Explain your answer.

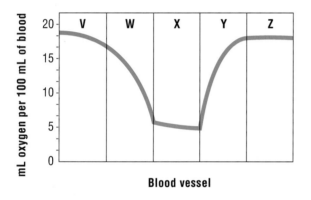

Blood vessel

5 Using a capillary exchange model, explain why the intake of salt is regulated for patients who suffer from high blood pressure. (Hint: The salt is absorbed from the digestive system into the blood.)
6 Why do some soldiers faint after standing at attention for a long time?
7 Explain why someone who suffers a severe cut might develop a rapid and weak pulse. Why might body temperature begin to fall?

8 Why does the blockage of a lymph vessel in the left leg cause swelling in that area?
9 A fetus has no need for pulmonary circulation. Oxygen diffuses from the mother's circulatory system into that of the fetus through the placenta. Therefore, the movement of blood through the heart is highly modified: blood flows from the right atrium through an opening in the septum to the left atrium and then to the left ventricle. The opening between the right and left atria becomes sealed at birth. Explain why any failure to seal the opening results in what has been termed a "blue baby."
10 A person's blood pressure is taken in a sitting position before and after exercise. Compare blood pressure readings before and after exercise as shown in the following chart.

Condition	Systolic B P (mm Hg)	Diastolic B P (mm Hg)	Pulse rate (beats/min)
Resting	120	80	70
After exercise	180	45	160

a) Why does systolic blood pressure increase after exercise?
b) Why does diastolic blood pressure decrease after exercise?

CRITICAL-THINKING QUESTIONS

1 a) Nicotine causes the constriction of arterioles. Using the information that you gained about fetal circulation from question 9 in Applying the Concepts, explain why pregnant women are advised not to smoke.

b) Mothers who smoke give birth to babies who are, on average, 1 kg smaller than normal. Speculate on the relationship between the effects of nicotine on the mother's circulatory system and the lower body mass of babies.

2 Coronary heart disease is often related to lifestyle. Stress, smoking, alcohol consumption, and poor diet are considered to be contributing factors to heart problems. Should all people have equal access to heart transplants? Should people born with genetic heart defects be treated differently from people who have abused their hearts?

3 Heart disease is currently the number-one killer of middle-aged males, accounting for billions of dollars every year in medical expenses and productivity loss. Should males be required, by law, to undergo heart examinations?

4 Caffeine causes heart rate to accelerate; however, a scientist who works for a coffee company has suggested that blood pressure will not increase due to coffee consumption. This scientist states that homeostatic adjustment mechanisms ensure that blood pressure readings will remain within an acceptable range. Design an experiment that will test the scientist's hypothesis. For what other reasons do you think the scientist might have suggested that caffeine does not increase blood pressure?

5 It has been estimated that for every extra kilogram of fat a person carries, an additional kilometer of circulatory vessels is required to supply the tissues with nutrients. Indicate why obesity has often been associated with high blood pressure. Do only overweight people suffer from high blood pressure? Explain your answer.

ENRICHMENT ACTIVITIES

1 Suggested reading:

- Brand, David. "Searching for Life's Elixir." *Time*, December 12, 1988, p. 60.
- Eisenberg, M.S., et al. "Sudden Cardiac Death." *Scientific American* 25(5) (1986): p. 33.
- Franklin, D. "Steroids Heft Heart Risk In Iron Pumpers." *Science News* 126(July 21, 1984): p. 38.
- Fritz, Sand. "Drugs and Olympic Athletes." *Scholastic Science World*, 40(16) (April 13, 1984): pp. 7–14.
- Harper, Alfred. "Killer French Fries." *Sciences*, Jan/Feb 1988, pp. 21–27.
- Hastings, Paul. *Medicine: An International History.* London: Ernest Benn, 1974.
- Kusinitz, Beryl. "The Artificial Heart." *Science World* 39(16) (1983): p. 8.
- Miller, Jonathan. *The Body in Question.* London: Macmillan, 1978.
- Monmaney, Terrence, and Karen Springen. "The Cholesterol Connection." *Newsweek*, February 8, 1988, pp. 56–58.
- Robinson, T.E. et al. "The Heart as a Suction Pump." *Scientific American* 254(6) (1986): p. 84.
- Sperryn, Perry. "Drugged and Victorious: Doping and Sport." *New Scientist*, August 1984, pp. 16–18.
- ——. "New Medicine." *National Geographic*, January 1987.

2 View the following films:
- TVOntario. *Medical.* BPN 179110.
- National Geographic Society. *Circulatory and Respiratory System.* 51305. This excellent video examines the human transport system.

Blood and Immunity

IMPORTANCE OF BLOOD AND IMMUNITY

In the last chapter you discovered that blood can be associated with many different functions. In addition to its transport functions, blood helps maintain the water balance of organ systems, body temperature, and pH balance. Blood is also an important part of the human immune system, protecting the body against a host of invaders.

To appreciate the importance of the immune system, consider the story of David, "the boy in the plastic bubble." David was born without an immune system, which meant that his body was unable to produce the cells necessary to protect him from disease. As a result, David had to live in a virtually germ-free environment. People who came in contact with him had to wear plastic gloves. Eventually, David received a bone marrow transplant from his sister. Unfortunately, a virus was hidden in the bone marrow. The sister, who had a functioning immune system, was able to protect herself from the virus, but David was not.

Figure 11.1

David, the "boy in the plastic bubble," had severe combined immunodeficiency syndrome.

Because blood cells are suspended in a watery fluid, blood has been described as a fluid tissue. Like other tissues, the individual cells in the blood work together for a common purpose. The watery nature of blood provides an interesting reminder of the origins of human life. Blood has the same ions in approximately the same relative concentration as the ancient seas from which humans evolved.

COMPONENTS OF BLOOD

The average 70 kg individual is nourished and protected by about 5 L of blood. Approximately 55% of the blood is fluid; the remaining 45% is composed of blood cells. The fluid portion of the blood is referred to as the **plasma.** Although it is approximately 90% water, the plasma also contains blood proteins, glucose, vitamins, minerals, dissolved gases, and waste products of cell metabolism.

The large plasma proteins play special roles in maintaining homeostasis. One group of proteins, the *albumins*, along with inorganic minerals, establishes an osmotic pressure that draws water back into capillaries and helps maintain body fluid levels. A second group of proteins, the *globulins*, produces antibodies that provide protection against invading microbes. You will learn more about antibodies later in the chapter. *Fibrinogens*, the third group of plasma proteins, are important in blood clotting. Table 11.1 summarizes the types of plasma proteins and their functions.

Erythrocytes

The primary function of red blood cells is the transport of oxygen. Referred to as erythrocytes (from the Greek *erythros*, meaning "red"), the red blood cells are packed with a respiratory pigment that absorbs oxygen. Oxygen diffuses from the air into the plasma, but the amount of oxygen that can be carried by the plasma is limited. At body temperature, 1 L of blood would carry about 3 mL of oxygen. An iron-containing respiratory pigment called **hemoglobin** greatly increases the capacity of the blood to carry oxygen. When hemoglobin is present, 1 L of blood is capable of carrying 200 mL of oxygen, a 70-fold increase. Without hemoglobin, your red blood cells would supply only enough oxygen to maintain life for approximately 4.5 s. With hemoglobin, life can continue for five minutes. Five minutes is not very long, but remember that the blood returns to the heart and is pumped to the lungs, where oxygen supplies are replenished. Anyone deprived of oxygen for longer than five minutes starts to experience cell death. This might indicate why people survive even when the heart stops for short periods of time. Children who have been immersed in cold water for longer than five minutes have survived with comparatively minor cell damage. Colder temperatures slow body metabolism, thereby decreasing oxygen demand.

Table 11.1 Plasma Proteins

Type	Function
Albumins	osmotic balance
Globulins	antibodies, immunity
Fibrinogen	blood clotting

Hematocrit

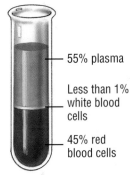

- 55% plasma
- Less than 1% white blood cells
- 45% red blood cells

Figure 11.2
The percentage of red blood cells in the blood is called the hematocrit.

Plasma *is the fluid portion of the blood.*

Hemoglobin *is the pigment found in red blood cells.*

An estimated 280 million hemoglobin molecules are found in a single red blood cell. The hemoglobin is composed of *heme*, the iron-containing pigment, and *globin*, the protein structure. Four iron molecules attach to the folded protein structure and bind with oxygen molecules. The oxyhemoglobin complex gives blood its red color. (Actually, a single red blood cell appears pale orange—the composite of many red blood cells produces the red color.) Once oxygen is given up to cells of the body, the shape of the hemoglobin molecule changes, causing the reflection of blue light. This explains why blood appears blue in the veins of your arms and hands.

Red blood cells appear as biconcave (meaning concave on both sides) disks approximately 7.0 μm in diameter. The folded disk shape provides a greater surface area for gas exchange—between 20 and 30% more surface area than a similar sphere. Red blood cells do not contain a nucleus when mature; they are said to be **enucleated.** The absence of a nucleus provides more room for the cell to carry hemoglobin. This enucleated condition raises two important questions. Since cells by definition contain a nucleus, are red blood cells actually cells? The second question addresses cell reproduction. How do cells without a nucleus and chromosomes reproduce? Since red blood cells live only about 120 days, cell reproduction is essential. One estimate suggests that at least five million red blood cells are produced every minute of the day.

The answer to both of the above questions can be found in the bone marrow, the site of red blood cell reproduction, or **erythropoiesis** (the suffix *poiesis* means "to make"). Red blood cells begin as stem cells, which do contain a nucleus. The cells divide and shrink as they take up hemoglobin. Eventually, the nucleus disappears and the cells are discharged into the blood. The mature red blood cell cannot undergo mitosis, but the immature form may divide many times.

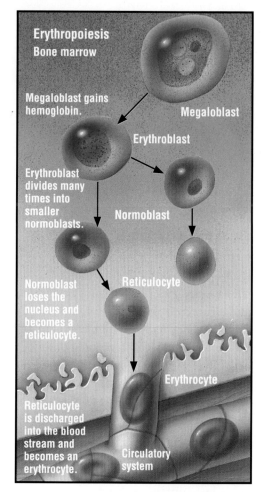

The average male contains about 5.5 million red blood cells per milliliter of blood, while the average female has about 4.5 million red blood cells per milliliter of blood. Individuals living at high altitudes can have as many as 8 million red blood cells per milliliter of blood. How does the body count red blood cells to ensure that adequate numbers are maintained? The problem is an immense one, just from a bookkeeping point of view. The outer membranes of red blood cells become brittle with age, causing them to rupture as they file through the narrow capillaries. Specialized white blood cells,

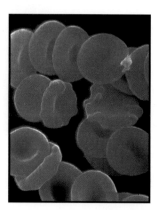

Figure 11.3
Red blood cells move through the capillaries in single file.

Figure 11.4
Immature red blood cells are found in the bone marrow. Stem cells in the bone marrow are continually dividing and forming megaloblasts, cells destined to develop into red blood cells.

Enucleated *cells do not contain a nucleus.*

Erythropoiesis *is the process by which red blood cells are made.*

located primarily in the spleen and liver, also monitor the age of red blood cells and remove debris from the circulatory system. Following the breakdown of red blood cells, the hemoglobin is released. Iron is recovered and stored in the bone marrow for later use. The heme portion of the hemoglobin is transformed into bile pigments.

The short lifespan of red blood cells and the body's changing needs mean that oxygen delivery must continually be monitored. However, red blood cell numbers are not monitored directly. Red blood cell reproduction is directed by oxygen levels. Any condition that lowers blood oxygen levels causes an increase in the rate of erythropoiesis. The kidneys respond to low levels of oxygen by releasing REF, or renal erythropoietic factor. The REF combines with liver globulins and forms erythropoietin, a chemical messenger that stimulates red blood cell production in the bone marrow. Individuals who live at high altitudes compensate for lower oxygen levels in the air by increasing red blood cell production. In a similar manner, red blood cell production is stimulated following blood transfusions or hemorrhaging.

A deficiency in hemoglobin or red blood cells decreases oxygen delivery to the tissues. This condition, known as **anemia,** is characterized by low energy levels. The most common cause of a low red blood cell count is hemorrhage. Physical injury or internal bleeding caused by ulcers or hemorrhage in the lungs associated with tuberculosis can cause anemia. If more than 40% of the blood is lost, the body is incapable of coping. Anemia may also be associated with a dietary deficiency of iron, which, you will recall, is an important component of hemoglobin. The red blood cells must be packed with sufficient numbers of hemoglobin molecules to ensure adequate oxygen delivery. Raisins and liver are two foods rich in iron.

Leukocytes

White blood cells, or leukocytes, are much less numerous than red blood cells. It has been estimated that red blood cells outnumber white blood cells by a ratio of 700 to 1. White blood cells have a nucleus, making them easily distinguishable from red blood cells. In fact, the shape and size of the nucleus, along with the granules in the cytoplasm, have been used to identify different types of leukocytes. Figure 11.6 shows the different types of leukocytes. The *granulocytes* are classified according to small cytoplasmic granules that become visible when stained. The *agranulocytes* are white blood cells that do not have a granular cytoplasm. Granulocytes are produced in the bone marrow. Agranulocytes are also produced in the bone marrow, but are modified in the lymph nodes. Some leukocytes destroy invading microbes by phagocytosis,

Anemia *refers to the reduction in blood oxygen due to low levels of hemoglobin or poor red blood cell production.*

Figure 11.5

Low levels of blood oxygen stimulate the production of erythropoietin, which, in turn, triggers the production of red blood cells. With an increase in red blood cell production, oxygen delivery is improved.

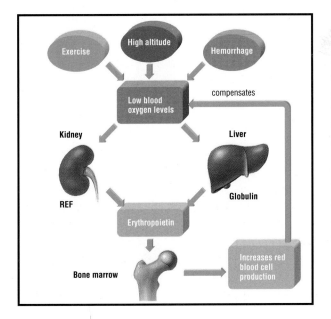

squeezing out of capillaries and moving toward the microbe like an amoeba. This process is known as **diapedesis.** Once the microbe has been engulfed, the leukocyte releases enzymes that digest the microbe and the leukocyte itself. Fragments of remaining protein from the white blood cell and invader are called **pus.** Other white blood cells form special proteins, called *antibodies*, which interfere with foreign invading microbes and toxins. You will learn more about antibodies later in this chapter.

Platelets

Platelets, like red blood cells, do not contain a nucleus, and are produced from large nucleated cells in the bone marrow. Small fragments of cytoplasm break from the large megakaryocyte to form platelets. The irregularly shaped platelets move through the smooth blood vessels of the body but rupture if they strike a sharp edge, such as that produced by a torn

blood vessel. The fragile platelets initiate blood-clotting reactions. You will learn more about blood-clotting reactions later in the chapter.

■ REVIEW QUESTIONS ?

1 Why is blood considered to be a tissue?
2 Name the two major components of blood.
3 List three plasma proteins and indicate the function of each.
4 What is the function of hemoglobin?
5 What is erythropoiesis?
6 List factors that initiate red blood cell production.
7 What is anemia?
8 How do white blood cells differ from red blood cells?
9 State two major functions associated with leukocytes.
10 What is the function of platelets?

Diapedesis *is the process by which white blood cells squeeze through clefts between capillary cells.*

Pus *is formed when white blood cells engulf and destroy invading microbes. The white blood cell is also destroyed in the process. The remaining protein fragments are known as pus.*

Figure 11.6

Stem cells of the bone marrow give rise to blood cells. Two classes of white blood cells are shown. The agranulocytes include the monocytes and lymphocytes. The granulocytes include the eosinophils, basophils, and neutrophils.

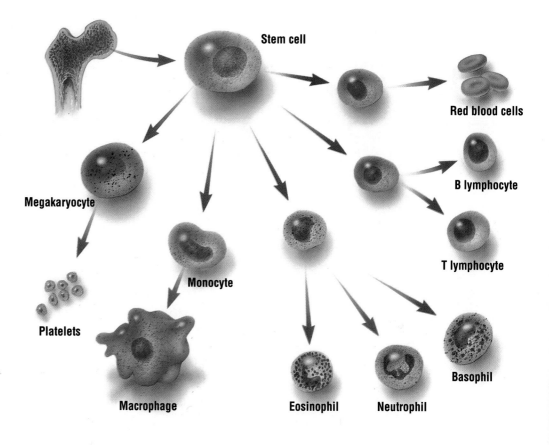

Stem cell

Red blood cells

B lymphocyte

T lymphocyte

Basophil

Neutrophil

Eosinophil

Macrophage

Platelets

Monocyte

Megakaryocyte

Unit Three:
Exchange of Matter and Energy in Humans

LABORATORY

MICROSCOPIC EXAMINATION OF BLOOD

Objective

To examine red and white blood cells.

Materials

prepared slide of fish blood   light microscope
prepared slide of human blood

Procedure

1 Before beginning the investigation, clean all microscope lenses with lens paper and rotate the nosepiece to the low-power objective. Place the slide of fish blood on the stage and focus under low power. Locate an area in which individual blood cells can be seen.

2 Rotate the revolving nosepiece to the medium-power objective, and focus. Red blood cells greatly outnumber white blood cells. Locate a single red blood cell in the center of the field of view and rotate the nosepiece to the high-power objective. Note the nucleus found in the red blood cells of the fish.

   a) Diagram the red blood cell of the fish.

   b) Estimate the size of the red blood cell.

3 Repeat the same procedure with the human blood slide.

   c) Diagram a single human red blood cell.

   d) Estimate the size of the human red blood cell.

4 Scan the field of view for different white blood cells. Using the classification of leukocytes provided in the chart below, classify the leukocytes and record your results in a table similar to the one below.

5 Repeat the procedure by scanning 10 different visual fields. Record your data in your chart.

Laboratory Application Questions

1 The red blood cells of fish contain a nucleus, while the human red blood cells do not. Indicate the advantage of the mammalian type of red blood cell over that of the fish.

Blood tests are used to help diagnose different diseases. The chart below shows a few representative diseases. Use the chart to answer questions 2 to 4.

Leukocyte change	Associated conditions
Increased eosinophils	Allergic condition, chorea, scarlet fever, granulocyte leukemia
Increased neutrophils	Toxic chemical, newborn acidosis, hemorrhage, rheumatic fever, severe burns, acidosis
Decreased neutrophils	Pernicious anemia, protozoan infection, malnutrition, aplastic anemia
Increased monocytes	Tuberculosis (active), monocyte leukemia, protozoan infection, mononucleosis
Increased lymphocytes	Tuberculosis (healing), lymphocyte leukemia, mumps

2 Why would a physician not diagnose leukemia on the basis of a single blood test?

3 What information might a blood test provide a physician about a patient being treated for the lung disease tuberculosis? Why would blood tests be taken even after the disease has been diagnosed?

4 Leukemia can be caused by the uncontrolled division of cells from two different sites: the bone marrow or lymph nodes. Indicate how blood tests could be used to determine which of the sites harbors the cancerous tumor. ■

Classification of Leukocytes

Type	Description	Number	%
Granulocyte	Granular cytoplasm		
Neutrophil	Three-lobed nucleus, 10 nm (Wright's stain: purple nucleus, pink granules)		
Eosinophil	Two-lobed nucleus, 13 nm (Wright's stain: blue nucleus, red granules)		
Basophil	Two-lobed nucleus, 14 nm (Wright's stain: blue-black nucleus, blue-black granules)		
Agranulocyte	Nongranular cytoplasm		
Monocyte	U-shaped nucleus, 15 nm (Wright's stain: light bluish-purple nucleus, no granules)		
Lymphocyte (small)	Large nucleus, 7 nm (Wright's stain: dark bluish-purple nucleus, no granules)		
Lymphocyte (large)	Large nucleus, 10 nm (Wright's stain: dark bluish-purple nucleus, no granules)		

CASE STUDY

DIAGNOSIS USING HEMATOCRITS

Objective

To use hematocrits to diagnose various disorders.

Materials

metric ruler

Hematocrits

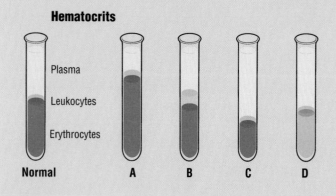

Plasma

Leukocytes

Erythrocytes

Normal   A   B   C   D

Procedure

1 Determine the normal hematocrit by using the following formula:

$$\text{Hematocrit} = \frac{\text{red blood cell volume}}{\text{total blood volume}} \times 100$$

a) Calculate the hematocrit of the normal subject.

2 A device called a hemacytometer is used to measure the amount of hemoglobin present. Red blood cells have the ability to concentrate hemoglobin to about 34 g/100 mL of blood. Readings below 15 g/100 mL of blood indicate anemia. Blood appears pale if hemoglobin levels are low.

b) Which subject do you believe has a low level of hemoglobin: A, B, C, or D?

Case-Study Application Questions

1 Cancer of the white blood cells is called leukemia. Like other cancers, leukemia is associated with rapid and uncontrolled cell production. Using the data in the case study, predict which subject might be suffering from leukemia. Give your reasons.

2 Although hematocrits provide some information about blood disorders, most physicians would not diagnose leukemia on the basis of one test. What other conditions might explain the hematocrit reading you chose in question 1? Give your reasons.

3 Lead poisoning can cause bone marrow destruction. Which of the subjects in the case study might have lead poisoning? Give your reasons.

4 Which subject lives at a high altitude? Give your reasons.

5 Recently, athletes have begun to take advantage of the benefits of extra red blood cells. Two weeks prior to a competition, a blood sample is taken and centrifuged, and the red blood cell component is stored. A few days before the event, the red blood cells are injected into the athlete. Why would athletes remove red blood cells only to return them to their body later? What problems could be created should the blood contain too many red blood cells? Give your reasons. ■

BLOOD CLOTTING

Blood clotting maintains homeostasis by preventing the loss of blood from torn or ruptured blood vessels. Blood clots also forestall the rupture of weakened blood vessels by providing additional support.

Trillions of fragile platelets move through the blood vessels. Should the platelet strike a rough surface, such as that created by a torn blood vessel in a cut or abrasion, the platelet breaks apart and releases a protein called *thromboplastin*. The thromboplastin, along with calcium ions present in the blood, activates a plasma protein called *prothrombin*. Prothrombin, along with another plasma protein, called fibrinogen, is produced by the liver. Under the influence of thromboplastin, prothrombin is transformed into thrombin. In turn, thrombin acts as an enzyme by splicing two amino acids from the fibrinogen molecule. Fibrinogen is converted into fibrin threads, which wrap around the damaged area, sealing the cut in the skin with a clot. Invading microbes cannot gain access. Although the threads prevent red blood cells from passing into the damaged area, they provide a framework that white blood cells crawl over.

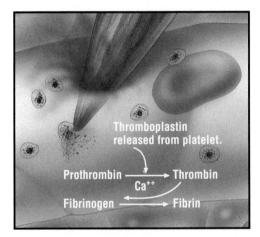

Although blood clotting preserves life, it can also result in life-threatening situations. A **thrombus** is a blood clot that seals a blood vessel. Because blood will not pass through the area, local tissues are not supplied with oxygen and nutrients. If a clot forms in the brain, cerebral thrombosis can cause a stroke. Coronary thrombosis—a clot in the coronary artery of the heart—can be equally dangerous.

Should a blood clot dislodge, it becomes an **embolus.** The embolus may travel through the body to lodge in a vital organ. Cerebral embolisms, coronary embolisms, and pulmonary embolisms can be life-threatening. What causes an embolus or thrombus is not completely understood, but scientists believe that genetic factors may be involved. It is known, however, that the incidence of thrombosis and embolisms becomes greater as people get older.

BLOOD GROUPS

In the 17th century, Jean-Baptiste Denis performed the first blood transfusion by injecting lamb's blood into a young boy. The youth survived, but a repeat of the experiment, on an older man, proved disastrous—the man died almost immediately. Denis attempted to explain what went wrong, but he lacked crucial information. (It was eventually revealed that the older man was being poisoned by his wife; Denis was exonerated.)

The idea that young people are healthier than older people became linked with the notion that older people have older blood. Occasionally, the transfusion of blood from a younger person to an older person worked: the greater oxygen-carrying capacity provided more energy. However, the recipient often died. Why do some transfusions help, while others kill?

At the turn of the 20th century, Karl Landsteiner discovered that different blood types exist. Therefore, the secret to successful transfusions was the correct matching of blood types. Special markers, called **glycoproteins,** are located on the membrane of some of the red

Figure 11.7

Platelets burst when they strike a sharp surface. The thromboplastin released from the platelet initiates a series of reactions that produce a blood clot.

A **thrombus** *is a blood clot that forms within a blood vessel.*

An **embolus** *is a blood clot that dislodges and is carried by the circulatory system to vital organs.*

Glycoproteins *are large chemical complexes composed of carbohydrates and protein. Glycoproteins can be found on cell membranes.*

blood cells. Individuals with blood type A have a special glycoprotein, the A marker, attached to their cell membrane. Individuals with blood type B have a special glycoprotein, the B marker, attached to their cell membrane. Individuals with blood type AB have both A and B markers attached to their cell membrane. Blood type O has no special marker.

Should an individual with blood type O receive blood from an individual with blood type A, the individual with blood type O would recognize the A marker as a foreign invader. The A marker acts as an **antigen** in the body of the individual with blood type O. Special proteins, called **antibodies,** are produced in response to a foreign invader. The antibodies attach to the antigen markers and cause the blood to clump. It is important to note that antigen A would not cause the same immune response if transfused into the body of an individual with blood type A. The marker associated with blood type A is not a foreign invader because it is part of the genetic makeup of that individual. A-type antigens are found on that individual's red blood cells. Table 11.2 summarizes the antigens and antibodies for the various blood groups.

The antibodies produced by the recipient act on the invading antigens. As shown in Figure 11.9, the antibodies cause the blood to **agglutinate,** or clump. The importance of the correct transfusion is emphasized by the fact that agglutinated blood can no longer pass through the tiny capillaries. The agglutinated blood therefore clogs local tissues and prevents the delivery of oxygen and nutrients. Individuals with type AB blood, in fact, possess both antigens and, therefore, are able to receive blood from any donor. Blood type AB is the universal acceptor. Blood type O is referred to as the universal donor because it can be received by individuals of all blood types. Since blood type O contains no antigen, it contains no special features not found in any of the other blood types. Although antibodies will not be produced against type O, individuals with blood type O can recognize antigens on other blood cells. In some ways, blood types O and AB provide a paradox. Blood type O, despite being the universal donor, may only accept blood from individuals with blood type O. Blood type AB, despite being the universal acceptor, may only donate blood to individuals with blood type AB.

*An **antigen** is a substance, usually protein in nature, that stimulates the formation of antibodies.*

***Antibodies** are proteins formed within the blood that react with antigens.*

***Agglutination** refers to the clumping of blood cells caused by antigens and antibodies.*

Figure 11.8

Individuals with blood type A can receive blood type O during a transfusion. However, individuals with blood type O cannot receive blood type A during a transfusion.

Table 11.2 Antigens and Antibodies Found in Blood Groups

Blood group	Antigen on red blood cell	Antibody in serum
O	none	A and B
A	A	B
B	B	A
AB	A and B	none

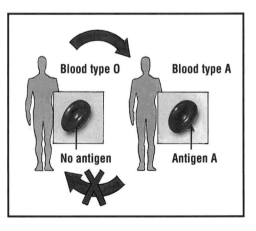

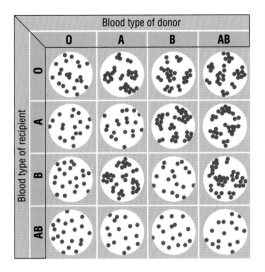

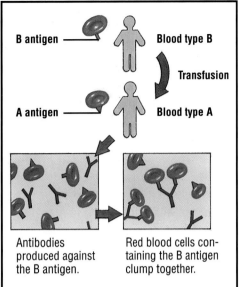

Figure 11.9

Agglutination response of blood type A (recipient) to blood type B (donor).

Antibodies produced against the B antigen.

Red blood cells containing the B antigen clump together.

RHESUS FACTOR

During the 1940s scientists discovered another antigen on the red blood cell—the rhesus factor. Like the ABO blood groups, the rhesus factor is inherited. Individuals who have this special antigen are said to be Rh+. Approximately 85% of Canadians have the antigen. The remaining 15% of individuals who do not have the antigen are said to be Rh–. Individuals who are Rh– may donate blood to Rh+ individuals, but should not receive their blood. The human body has no natural antibodies against Rh factors, but antibodies can be produced following a transfusion. Although antibodies are produced in response to antigens, it should be pointed out that the immune reaction is subdued compared with that of the ABO group.

Rhesus-factor incompatibilities become important for Rh+ babies of Rh– mothers. If the baby inherits the Rh+ factor from the father, a condition called **erythroblastosis fetalis** can occur with the second and subsequent pregnancies. The first child is spared because the blood of the mother and blood of the baby are separated by the **placenta,** a thin membrane found between the developing embryo and mother. Located within the uterus, the placenta permits the movement of materials between mother and baby. Nutrients and oxygen move from the mother's blood into the baby, while wastes diffuse from the baby's blood into the mother for disposal. Although capillary beds from the mother and baby intertwine in the placenta, blood flows do not mix until birth. During birth, the placenta is removed from the uterus. Capillary beds rupture, and, for the first time, the blood of the baby comes in contact with the blood of the mother. The mother's immune system recognizes the Rh+ antigens and triggers the production of antibodies. By the time the antibodies are produced, the first baby, no longer connected to the placenta, has escaped the potentially dangerous environment.

A second pregnancy presents problems if the child is Rh+. The mother retains many of the antibodies from her first encounter with Rh+ blood. Should some of the antibodies move across the placenta, they attach to the antigen on the baby's red blood cells, causing them to clump. In this condition, red blood cells

Erythroblastosis fetalis, *or "blue baby," occurs when the mother's antibodies against Rh+ blood enter the Rh+ blood of her fetus.*

The **placenta** *is an organ made from the cells of the baby and the cells of the mother. It is the site of nutrient and waste exchange between mother and baby.*

are unable to pass through the narrow capillaries. Red blood cells jam the capillary entrances, and oxygen delivery is severely reduced. The body attempts to compensate by increasing red blood cell production, but even these new red blood cells contain antigens. Once again, the antibodies that have seeped into the baby's circulatory system agglutinate the new red blood cells. Low oxygen levels often cause the baby to turn blue, hence the name "blue baby."

Treatment for erythroblastosis fetalis involves a transfusion of Rh– blood. The antibodies from the mother will not attack the Rh– blood because it carries no antigens. Eventually, the baby begins producing Rh+ blood, but by then some of the mother's antibodies will have broken down—no protein lives forever. The effects of erythroblastosis fetalis may be decreased, if not eliminated altogether, by injecting the mother with a drug that inhibits the formation of antibodies against Rh+ antigens. The injection is given immediately after the woman's first birth. The mother's blood can then be monitored during the second pregnancy to determine whether or not antibodies are being developed.

FRONTIERS OF TECHNOLOGY: ARTIFICIAL BLOOD

On March 1, 1982, a precedent-setting legal case brought attention to an emerging medical technology. A man and woman, trying to push their car, were critically injured when they were hit by another car. Because of their religious beliefs, the couple refused a blood transfusion. During the legal dispute that ensued, the wife died, and the courts ruled that action must be taken to save the husband's life. Five liters of

Fluosol—artificial blood—were transfused into the man over a period of five days. Doctors believed that the artificial blood could maintain adequate oxygen levels until the man's bone marrow began replenishing red blood cells.

Fluosol, a nontoxic liquid that contains fluorine, was developed in Japan. Fluosol carries both oxygen and carbon dioxide. It requires no blood matching and, when frozen, can be stored for long periods of time. Artificial blood, unlike human blood, does not have to undergo expensive screening procedures before being used in transfusions. Artificial blood will not carry the HIV or hepatitis viruses. However, despite its advantages, artificial blood is not as good as the real thing. Although it carries oxygen, it is ill-suited for many of the other functions associated with blood, such as blood clotting and immunity. The real value of artificial blood is that it provides time until natural human blood can be administered. It could also serve as a supplement for patients with diseases like thalassemia (Cooley's anemia) or aplastic anemia, which require multiple transfusions. Artificial blood might also help prevent hepatitis or an overload of iron.

IMMUNE RESPONSE

The body's first line of defense against foreign invaders is largely physical. The skin provides a protective barrier; only a few bacteria and parasites are specialized enough to break through the skin's barrier. In the respiratory passage, invading microbes and foreign debris become trapped in a layer of mucus and are swept away from the lungs by tiny hairlike structures called *cilia*.

Chemical protection is provided by the stomach, which secretes a strong acid that destroys many of the marauding cells.

Complementary Proteins

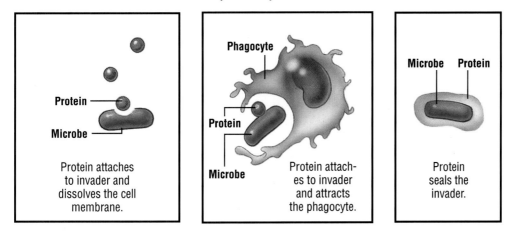

Protein attaches to invader and dissolves the cell membrane.

Protein attaches to invader and attracts the phagocyte.

Protein seals the invader.

Figure 11.10

Proteins aid the immune response.

Lysozyme, a special enzyme secreted in human tears, destroys the cell walls of bacteria. Without support from the cell wall, the bacteria burst.

A second line of defense can be mobilized if the invader takes up residence within the body. Leukocytes seek out and destroy any potentially destructive agents. Neutrophils, eosinophils, and monocytes engulf the foreign invader. As you read earlier, this process is called phagocytosis.

The appearance of foreign proteins in the body activates plasma proteins, often referred to as **complementary proteins.** About 20 different types of complementary proteins are known. Under normal conditions they are present in the circulatory system in an inactive form. Marker proteins from invading microbes activate the complementary proteins, which, in turn, serve as messengers. Some of the activated proteins trigger the formation of a protective coating around the invader. A second group dissolves the cell membrane, and a third group attracts phagocytes. Moving much like an amoeba, the phagocytotic white blood cell is able to leave the blood vessels in search of the invader. Once the invader is engulfed, the white blood cell releases a packet of enzymes that not only destroy the invader but the white blood cell as well.

Another specialized group of white blood cells, called **lymphocytes,** produces antibodies. You will recall that antibodies are protein molecules that protect the body from invaders. All cells have special markers located on their cell membrane. Normally, the immune system does not react to the body's own markers. However, intruding cells or foreign proteins activate the immune response. The cell wall of a bacterium or the outer coat of a virus contain many different antigens. The antigen may even be a toxic poison produced by molds, bacteria, or algae. The poison presents a danger to the cells of the body because it interferes with normal cell metabolism.

Complementary proteins *help phagocytotic cells engulf foreign cells.*

Lymphocytes *are antibody-producing white blood cells.*

Antigen Markers

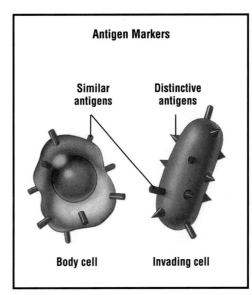

Similar antigens

Distinctive antigens

Body cell

Invading cell

Figure 11.11

Sugar-protein complexes, located on the cell membrane, act as markers. T lymphocytes distinguish the markers on the body's cells from those of invading cells. Markers that are found on foreign cells are recognized by the T cells, and antibody formation is triggered.

Two different types of lymphocytes are found in the immune system. The first is the **T cell,** which is produced in the bone marrow and stored in a tiny organ called the *thymus gland*, from which the T cell receives its name. The T cell's mission is to seek out the intruder and signal the attack. Acting much like a sentry, some T cells identify the invader by its antigen markers, which are located on the cell membrane. Once the antigen is identified, another T cell passes this information on to the antibody-producing **B cells.**

B cells multiply and produce the chemical weapon—the antibodies. Each B cell produces a single type of antibody, which is displayed along the cell membrane. Eventually, the B cells are released from the bone marrow and enter the circulatory system. It should be noted that some B cells differentiate into super-antibody-producing cells called *plasma cells*. These plasma cells can produce as many as 2000 antibody molecules every second.

ANTIGEN-ANTIBODY REACTIONS

Antibodies are Y-shaped proteins engineered to target foreign invaders. Antibodies are specific; this means that an antibody produced against the influenza virus is not effective against HIV, the virus that causes AIDS. The tails of the Y-shaped proteins are very similar, regardless of the type of antibody. Variations only exist at the outer edge of each of its arms, the area in which the antibody combines with the antigen. The antigen mark-

ers found on the influenza virus are different from those found on HIV. Each antibody has a shape that is complementary to its specific antigen. Thus the combining site of an antibody produced in response to the influenza virus will not combine with HIV.

Many different antigen markers are located on the membrane of a virus or bacterium. Although different antibodies can attach to the invader, each antibody attaches only to its complementary marker. The attachment of antibodies to the antigens increases the size of the complex, making the antigen much more conspicuous. Thus, the larger antigen-antibody complex is more easily engulfed and destroyed by the wandering **macrophages.** In Figure 11.12, antibody type 2 attaches to two different antigens, causing the foreign cells to clump together. The invaders can no longer gain access to other cells, and are prevented from infecting them. The immobilized invader is often clumped into a soluble mass, far too large to pass through the cell membrane of its prospective host. Tagged for destruction, the invader merely awaits its fate.

T-cell lymphocytes are produced in the lymph nodes. Different types of T cells regulate an immune response.

B-cell lymphocytes make antibodies.

Macrophages are phagocytotic white blood cells found in lymph nodes or in the blood (in bone marrow, spleen, and liver).

Figure 11.12

Antibody type 1 will only combine with the appropriate antigen.

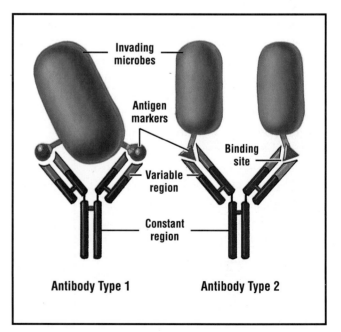

How do antibodies prevent poisons from destroying cells? Toxins or poisons have a specialized geometry that allows them to become attached to the **recep-tor sites** on cell membranes. Specialized receptor sites are found on different cells, which may explain why some poisons affect the nervous system, while others affect the digestive or circulatory system. The receptor site is designed to accommodate either a hormone or a specific nutrient. Unfortunately, the poison has a shape similar to the hormone or nutrient. Once attached, the poison is engulfed by the cell, which assumes that the poison is actually a needed substance. Figure 11.13 shows how antibodies prevent the poison from entering the cell.

Viruses also use the receptor sites as entry ports. The virus injects its hereditary material into the cell, but most often leaves the outer protein coat in the entry port. Different viruses come to rest in distinct locations. For example, the outer coat of the cold virus has a geometry that enables it to attach to lung cells. HIV has a shape that provides access to the T-cell lymphocytes. However, unlike most other viruses, HIV is engulfed by the T cell. Does this pro-

vide a clue as to why the body has difficulty defeating HIV?

Antibodies that attach themselves to the invading viruses alter their shape, thereby preventing access to the entry ports. Misshapen viruses float around the body, unable to find an appropriate entry port. Occasionally, the outer coat of the invader will change shape due to mutations. The mutated microbes may still gain access to the receptor site, but are not tied up by the antibody.

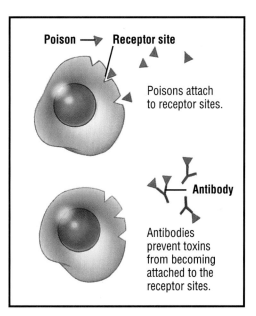

Poison ⟶ Receptor site

Poisons attach to receptor sites.

Antibody

Antibodies prevent toxins from becoming attached to the receptor sites.

Figure 11.13

Different receptor sites are found along cell membranes. In the top diagram, the poison attaches to the appropriate receptor site and masquerades as a nutrient or hormone. Once attached, the toxin is engulfed by the cell. In the bottom diagram, the poison is tied up by antibodies. The shape of the toxin has been altered so that it can no longer gain access to the receptor site.

Receptor sites *act as ports along cell membranes. Nutrients and other needed materials fit into specialized areas along cell membranes.*

BIOLOGY CLIP
Your body contains antibodies for about 100 000 different antigens.

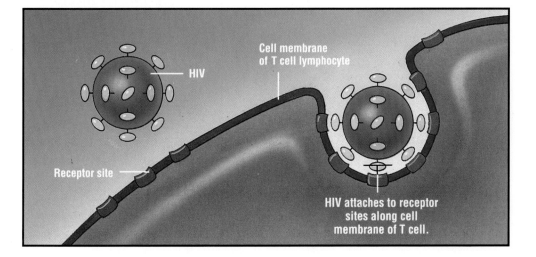

HIV

Cell membrane of T cell lymphocyte

Receptor site

HIV attaches to receptor sites along cell membrane of T cell.

Figure 11.14

HIV (human immunodeficiency virus) attaches to the receptor sites of the T-cell lymphocytes. Once attached, the T cell engulfs the virus, creating another problem for the immune system. Antibody production requires a blueprint of the invader. The protein coat of the virus hides inside the very cells assigned as sentries for invading antigens.

HOW THE BODY RECOGNIZES HARMFUL ANTIGENS

The T cells scout the body in search of foreign invaders that pose a threat to your survival. The macrophages attack the invaders and engulf them. As mentioned earlier, the antigen markers are not destroyed with the invader but are pushed toward the cell membrane of the macrophage. Pressing the antigens into its cell membrane, the macrophage cou-

Figure 11.15

The body recognizes harmful antigens.

Helper T cells *identify antigens.*

Lymphokine *is a protein produced by the T cells that acts as a chemical messenger between cells.*

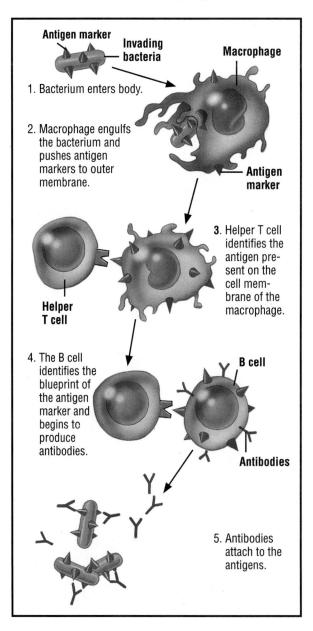

Antigen marker
Invading bacteria
Macrophage

1. Bacterium enters body.

2. Macrophage engulfs the bacterium and pushes antigen markers to outer membrane.

Antigen marker

3. Helper T cell identifies the antigen present on the cell membrane of the macrophage.

Helper T cell

4. The B cell identifies the blueprint of the antigen marker and begins to produce antibodies.

B cell

Antibodies

5. Antibodies attach to the antigens.

ples with the T cells, also referred to as **helper T cells.** The T cells read the antigen's shape and release a chemical messenger called **lymphokine.** The lymphokine causes the B cells to divide into identical cells called *clones*. Later, a second message is sent from the helper T cells to the B cells, signalling the production of antibodies. Each B cell produces a specific type of antibody. By the time the B cells enter the circulatory system, many antibodies are attached to the cell membrane.

The helper T cells activate an additional defender, the **killer T cell.** As the name suggests, these lymphocytes carry out search-and-destroy missions. Once activated, the killer T cells puncture the cell membrane of the intruder, should it be a fungus, protozoan parasite, or bacterium. Viruses, however, are much more insidious, because they hide within the familiar confines of the host cell. Here, the true value of the killer T cells is demonstrated. Once the viral coat is found attached to the cell membrane of the cell, the T cell attacks the infected cell. By destroying the infected body cell, the killer T cell prevents the virus from reproducing.

Killer T cells also destroy mutated cells. This is an extremely important process because some of the altered cells may be cancerous. Many experts believe that everyone develops cancerous cells, but in most cases the T cells eliminate the problem before a tumor forms. Whether or not you develop cancer depends on the success of your killer T cells.

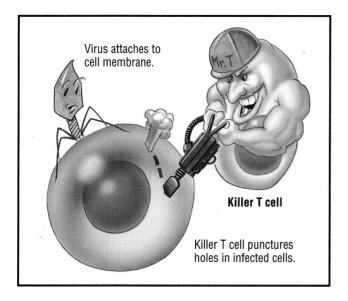

Virus attaches to cell membrane.

Killer T cell

Killer T cell punctures holes in infected cells.

and the suppressor T cells ensures that the body maintains adequate numbers of antibodies to contain the invading antigen. Most of the B cells and T cells will die off within a few days after the battle, but a small contingent will remain long after, to guard the site. Phagocytes survey the area, cleaning up the debris left from dead and injured cells. Tissues begin the work of repair and replacement.

Figure 11.16

Viruses use a cell's machinery for reproducing virus particles. Once the cell is destroyed, the viral cycle is stopped.

Killer T cells may also account for the body's rejection of organ transplants. Different individuals have different antigen markers on their cell membrane. Once the foreign markers of the transplanted tissue are recognized, their killer T cells initiate an assault. Immunosuppressant drugs, like cyclosporin, can slow the killer T cells. Unfortunately, slowing the killer T cells can result in a new set of complications. Individuals who receive these drugs become susceptible to bacterial infections. One of the leading causes of death for an organ transplant patient is pneumonia.

Once the battle against foreign invaders has been won, another T cell, the **suppressor T cell,** signals the immune system to shut down. Communication between the helper T cells

THE IMMUNE SYSTEM'S MEMORY

The native population of Hawaii was nearly annihilated by measles. In North America, the native population has been devastated by epidemics of small pox. Because neither group had been exposed to these viruses before, they had no antibodies to fight against infection. Europeans, unlike the native populations of Hawaii and North America, had already been exposed to many different types of viruses and were better able to produce antibodies to fight them. As you read earlier, your helper T cells must read a blueprint of the invader before B cells produce antibodies. This blueprint is stored even after the invader is destroyed

Killer T cells *puncture the cell membranes of cells infected with foreign invaders, thereby killing the cell and the invader.*

Suppressor T cells *turn off the immune system.*

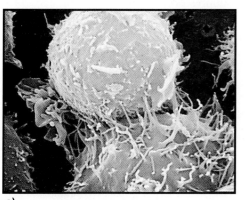

a)   b)

Figure 11.17

(a) Killer T cell binds with a tumor cell. (b) The target cell is perforated and cytoplasm leaks out.

so that subsequent infections can be destroyed before the microbe gains a foothold. Immunity is based on maintaining an adequate number of antibodies.

It is believed that a **memory T cell** is generated during the infection. Like helper T cells, the memory T cells hold an imprint of the antigen or antigens that characterize the invader. Most of the T cells and B cells produced to fight the infection die off within a few days; however, the memory T cells remain. The memory T cells identify the enemy and quickly mobilize antibody-producing B cells. Invading pathogens are defeated before they become established. As long as the memory T cell survives, you are immune. That is why you do not usually catch chicken pox more than once.

Table 11.3 The Body's Defense System

Cell type	Function
Lymphocytes	Produce antibodies.
Helper T cells	Act as sentries to identify foreign invading substances.
B cells	Produce antibodies.
Killer T cells	Puncture cell membranes of infected cells, thereby killing the cell.
Suppressor T cells	Turn off the imune system.
Memory T cells	Retain information about the geometry of the antigen.

Allergies

Allergies occur when your immune system mistakes harmless cells for harmful invaders. If you are allergic to peanuts, your immune system recognizes one of the proteins in the peanut as dangerous. Although the protein is quite safe, you mobilize the antibody strike-force against the peanut. Increased tissue swelling and mucous secretion, and sometimes constricted air passages, are all part of the immune response. Dust, ragweed, strawberries, and leaf molds do not pose any direct threat to life, but the immune response itself can sometimes be so violent that it becomes life-threatening.

Autoimmune Disease

The immune system can make mistakes. Problems arise when the immune system goes awry, and antibodies are directed toward the body's own cells. Renegade lymphocytes treat the body's own cells as aliens and fashion antibodies to attach to their own cell membranes. Many researchers believe that most people have mutated T cells and B cells that are capable of attacking the body; however, the renegade cells are usually held in check. Suppressor T cells play an important role in recognizing and intercepting the renegade T and B cells. One theory suggests that the suppressors secrete a substance that tells the macrophages to engulf the renegade cells.

Failure of the suppressor T cells to control the renegade cells can be seen in rheumatoid arthritis, where an immune response is mounted against the bones and connective tissues surrounding the joints. The scarring of the heart muscle provides a reminder of rheumatic fever, another autoimmune disorder.

Drugs or serious infections can weaken the suppressor T cells, leaving the body vulnerable to autoimmune diseases. Scientists have learned that the number of suppressor T cells decline with age. This increases the incidence of rheumatoid arthritis and other autoimmune diseases. However, autoimmune disease is not restricted to the elderly. Some individuals are born with defective suppressor T cells, and battle these diseases throughout their lives. Although no single cure exists, artificial immunosuppressant drugs can reduce the intensity of the attack by the renegade cells.

11 Outline the biochemical pathway involved in blood clotting.

12 Define antigen and antibody.

13 Explain why type O blood is considered to be the universal donor. Why is type AB considered to be the universal acceptor?

14 How does Rh+ blood differ from Rh– blood?

15 Explain why erythroblastosis fetalis may affect a woman's second and third child, but not her first?

16 List the advantages and disadvantages associated with using artificial blood.

17 How do T-cell lymphocytes differ from B-cell lymphocytes?

18 Differentiate between macrophages and lymphocytes.

19 Explain how B-cell, helper T-cell, and killer T-cell lymphocytes provide immunity.

20 How do antibodies defeat antigens?

21 How do memory T cells provide continuing immunity?

22 Indicate why the suppressor T cells are important.

VACCINES: THE NEEDLE OF HOPE

It is difficult to understand what it must have been like before there were vaccines. Large families were the only defense against epidemics of the bubonic plague and smallpox. Microbes killed both the rich and poor without discrimination. Mary II, Queen of England in the 17th century, was a victim of smallpox. Over 60 million Europeans died from the smallpox virus during the 18th century.

Vaccines are one of the most effective methods of preventing disease. Ironically, the principle of vaccination involves introducing the deadly microbe into the system, rather than avoiding it. The key to developing a successful vaccine is to make sure that the microbe has been weakened. A microbe that is full strength might reproduce in the body faster than the immune system could produce antibodies against it. Microbes that are capable of spreading very rapidly are referred to as *virulent*. Such a microbe would cause the disease before providing immunity.

Contrary to popular belief, vaccines are not a western invention. The ancient Chinese scraped dried bits of skin from smallpox victims. The powder was blown into the noses of healthy people. Records show that Greeks and Turks experimented with inoculations by pricking the sores of smallpox patients with a needle, and then poking it into the skin of healthy people. The people who received the needle usually demonstrated some resistance to the disease, but not everyone was so fortunate—the introduction of live, active smallpox was occasionally fatal. Scratching the skin with unsterilized needles often created as many problems as it solved. A partial immunity from one microbe often meant the introduction of another.

Smallpox and the First Vaccine

An English country doctor named Edward Jenner (1749–1823) developed the first vaccine in 1796. Noticing that rural dwellers seemed less susceptible to smallpox than urban residents, Jenner began to formulate a hypothesis. He believed that the immunity of the country people must be related to their environment. Jenner noted that dairy maids had a particularly low incidence of smallpox, but a high incidence of a much less harmful disease called cowpox. Cowpox produces many of the same symptoms as smallpox.

Jenner reasoned that exposure to cowpox must have provided some immunity to the more virulent smallpox virus. To test his theory, Jenner injected the pus from a festering wound on the arm of a dairy maid, Sarah Nelms, into a young boy, James Phipps. Not surprisingly, James developed cowpox. However, he recovered from the mild infection very quickly. Two months later, Jenner inoculated James once again, but this time with the more virulent smallpox. When James failed to develop smallpox, Jenner declared that a successful vaccine had been developed.

The cowpox virus is similar in shape to that of smallpox. Therefore, when the helper T cells identify cowpox, they signal the B cells to produce antibodies against them. Fortunately, the antibodies for cowpox also prove successful against smallpox. If Jenner had injected smallpox first, James might have died.

Pasteur and the Rabies Vaccine

Rabies, although never a mass killer like smallpox, has been greatly feared for its devastating effects. The tiny rabies virus migrates from the blood into the nervous system, where it destroys cells, causing convulsions and great suffering.

The technique used to develop a vaccine for smallpox could not be applied to rabies. Most virulent microorganisms do not have a harmless twin like cowpox. Injecting a full-strength virus would mean that the subject would develop the disease. The French chemist, Louis Pasteur (1822–1895), was able to grow the rabies virus in tissue cultures, and, by using trial-and-error testing, found ways to weaken the virus.

Figure 11.18
The cowpox virus is very similar to the smallpox virus.

Because the rabies virus remains dormant in the body for 14 days, a weakened virus would stimulate the production of antibodies that would lie in wait for the virulent rabies virus to develop. Once the dormant virus emerged, the antibodies would destroy it before it could gain a foothold in the body.

In 1885, Pasteur administered a rabies vaccine to a nine-year-old boy named Joseph Meisner, who had been bitten by a rabid dog. Joseph lived to become the gatekeeper at the Pasteur Institute. His devotion to Pasteur was so great that he committed suicide rather than open Pasteur's crypt to the invading Nazi army in 1940.

Jonas Salk and the Polio Vaccine

The polio virus was responsible for many deaths before an American physician, Jonas Salk, introduced a polio vaccine in 1955. Jackie Parker, formerly a quarterback of the Canadian Football League, is an athlete who had childhood polio and went on to live a normal life.

Salk, the son of an immigrant garment worker, became an overnight celebrity with the development of the polio vaccine. By killing the virus in a bath of formaldehyde, Salk was able to inject the polio into test animals without causing the disease. The viral coats stimulate the production of antibodies. The polio virus would now have to deal with the antibodies before it could gain a foothold. The vaccine is 60% successful against type 1 polio and 90% successful against types 2 and 3.

CHEMICAL CONTROL OF MICROBES

The first disease to be controlled by a chemical agent was syphilis, a crippling venereal disease. Mercury was used to treat it as early as 1495; however, mercury proved to be toxic to both the microbe and the patient.

In 1910 Paul Erlich used a drug called Salvarsan to control syphilis. This arsenic compound selectively blocked chemical reactions of the microbe without interfering with those of the patient to any great extent. Using a chemical to treat a patient was a revolutionary and controversial practice that was not initially accepted by the scientific community. While vaccines stimulated the immune system, which then destroyed the microbe, chemical therapy attacked the organism directly.

Chemical therapy was largely ignored until 1935, when a drug called sulfanilamide was found to selectively block the action of a chemical reaction essential to bacteria. Figure 11.19 shows the drug attaching itself to the enzyme that speeds up the reaction.

ANTIBIOTICS

Antibiotics are a special kind of chemical agent usually obtained from living organisms. It has long been known that living things compete with each other for food and space. Some types of microbes gain an advantage over others by producing substances toxic to their competitors. Some of these substances are useful in treating disease in humans.

Long before European and North American physicians hailed penicillin as a wonder drug, the Chinese used moldy soybean curds to treat boils. For many years people buried diseased organisms. In 1924 soil organisms called actinomycetes were identified as the producers of a bacteria-killing substance called actinomycetin.

In 1929, the British bacteriologist Alexander Fleming observed that some agar plates inoculated with bacteria had become contaminated with mold. The mold gradually overgrew the bacteria on the plate. The mold produced a bacteria-destroying secretion, dubbed penicillin, which was destined to become the miracle drug of the 1940s.

Penicillin interferes with the production of bacterial cell walls: as the bacteria cells divide, the new cells are unable to make new cell walls. The weakened walls

Figure 11.20

Stained glass window in Paddington, London, depicting Sir Alexander Fleming in his laboratory.

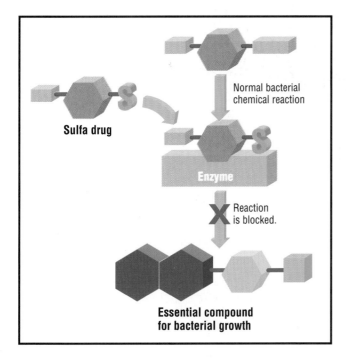

Normal bacterial chemical reaction

Sulfa drug

Enzyme

Reaction is blocked.

Essential compound for bacterial growth

Figure 11.19

With the enzyme tied up by the drug, normal chemical reactions are halted, and the cell dies.

are incapable of withstanding the pressure created by their own cytoplasms. Like a flawed dam, the cell walls of the bacteria eventually burst and the cell dies.

Since the 1940s many new antibiotics have been introduced. Ideally, an antibiotic should kill the invading bacteria without interfering with a person's normal body functions. However, individuals with very sensitive helper T cells may identify the antibiotic as a harmful antigen, activating the immune response. Allergic reactions to an antibiotic can often be more life-threatening than the invading bacteria. Everything from stomach upset to a puffiness around the eyes may indicate an allergic reaction to an antibiotic. If an antibiotic causes an allergic reaction, a physician should be contacted immediately.

Antibiotic Resistance

Some bacteria remain unaffected by certain antibiotics. Some strains of *streptococcus* bacteria, for example, have a gene that destroys penicillin. Microbes that reproduce sexually pass the genetic information on to the other microbes. However, these strains are very sensitive to other antibiotics, such as streptomycin and chloramphenicol.

■ REVIEW QUESTIONS ■   ?

23 Comment on the statement that vaccines originated in the far east.
24 Why did Edward Jenner inject cowpox viruses into a patient?
25 What would have happened had Jenner first injected his patient with smallpox instead of cowpox?
26 Why did Pasteur inject dead rabies into a patient who had contracted the living virus?
27 Why do people receive booster shots for polio?

28 Why do physicians not prescribe penicillin for the common cold?
29 Why is sulfanilamide not classified as an antibiotic?
30 What chemical characteristics make sulfanilamide a desirable chemical therapy agent? What are its limitations?

THE GUIDED MISSILES: MONOCLONAL ANTIBODIES

When a harmful invader enters the body, your immune system releases a strike force of antibodies. However, of the great number of antibodies released into the system only a few may have the correct shape to tie up the invader. Your immune system releases many antibodies on the assumption that other invaders may be lurking in the body.

A new technology is able to make a large number of antibodies in pure form by taking advantage of cloning, a technique that permits the exact duplication of identical cells. Developed in 1975 by George Kohler and Cesar Milstein of Cambridge University, England, monoclonal antibodies have become a promising weapon in the fight against cancer and other diseases. In this procedure, an antibody-producing cell from a mouse is fused with a cancer cell. The new cell, called a *hybridoma*, inherits the characteristics of the two individual cells—producing antibodies like a lymphocyte, while reproducing at the rate of a cancer cell. Each of the clones produces a specific antibody.

The possibilities of monoclonal antibodies are intriguing. Antibodies could be grown in tissue cultures, collected, and given to patients who have a particular disease. For example, one culture could produce antibodies against hepatitis,

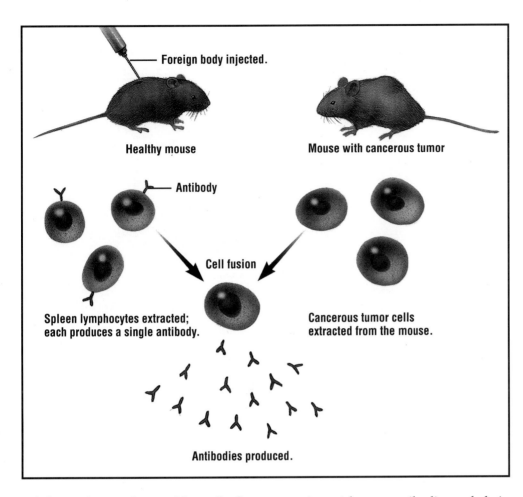

Foreign body injected.

Healthy mouse

Mouse with cancerous tumor

Antibody

Cell fusion

Spleen lymphocytes extracted;
each produces a single antibody.

Cancerous tumor cells
extracted from the mouse.

Antibodies produced.

Figure 11.21

The production of hybridomas.

while another might provide antibodies for specific cancer cells. Cancer cells have marker antigens on their cell membranes that distinguish them from normal cells. If the monoclonal antibody could identify the cancer antigen it could carry a poison to the cancer cell. The potential of using monoclonal antibodies to carry radioisotopes, drugs, or killer toxins to cancerous tumors has sparked a great deal of research. Since cancer cells and normal cells are very similar, many of the undesirable side effects of chemotherapy involve the interference of normal cell action. The promise of having a cancer drug that can be delivered to cancer cells without destroying normal cells now looms on the horizon.

While some scientists are seeking to identify the unique markers on the cancer cell membrane, others are experimenting with new antibodies and their production, and others are investigating new poisons to couple with monoclonal antibodies. A highly toxic poison called *ricin* has been successfully attached to monoclonal antibodies to kill leukemia in mice.

Monoclonal antibodies have also been used for diagnostic purposes such as identifying the hepatitis B virus in blood. This technique is especially important for blood banks because conventional methods can sometimes fail to identify chronic hepatitis in donors. Pregnancy tests using monoclonal antibodies are reported to be much more accurate than standard tests. Monoclonal antibodies could also attach themselves to specific kinds of white blood cells so that they can be counted. This technique could be used to identify AIDS and leukemia.

RESEARCH IN CANADA

Dr. Wanda Wenman, a pediatrician and molecular biologist at the University of Alberta Hospital, is working on a vaccine for the most common, yet least known, venereal disease. *Chlamydia trachomatis* is more common than herpes, gonorrhea, or AIDS. It has been estimated that one million Canadians and 500 million people worldwide have been infected with the microbe.

Vaccines and Venereal Disease

Although chlamydia can be cured by antibiotics, the disease is particularly difficult to identify because the symptoms can be subtle. If the microbe goes undetected, it can cause pelvic infections, which can lead to infertility in women. Babies born to women with chlamydia can develop eye infections and/or pneumonia.

The challenge of providing a cure is really twofold. First, comparatively little is known about chlamydia, because the organism is difficult to grow in the laboratory. Second, the disease is difficult to diagnose. Dr. Wenman may have an answer to both problems. She has identified two proteins from chlamydia that attach to the host cell. The two proteins provide indicators for diagnosis, and will serve as the basis for the development of a vaccine.

DR. WANDA WENMAN

SOCIAL ISSUE:
Compulsory AIDS Testing

In 1991, David Acer, a Florida dentist with AIDS, was thought to have infected five of his former patients with HIV. It is not known whether all were infected by the dentist himself, or whether he failed to use new gloves and sterile instruments after treating other patients who were HIV positive.

This widely publicized case has prompted demands for compulsory HIV testing for people in the health-care professions. Some people are concerned that an HIV-positive health-care worker might be accidentally cut during a medical procedure and bleed into a patient, thereby transmitting the virus.

While there is no requirement at the moment that medical workers be tested, many health-care practitioners are concerned about possible mandatory testing, since a positive test might mean the end of a career.

Mandatory testing would mean taking blood samples from thousands of health-care workers regularly, and testing them for the presence of HIV antibodies. (The body reacts to HIV infection by producing antibodies in the blood, which appear between 3 and 14 weeks after infection, depending on the individual.) Testing would cost millions of dollars annually.

Statement:

AIDS testing should be mandatory for health-care workers.

Point

- Patients have a right to know if health-care workers are HIV positive, so that they can choose whether or not to accept the risk of infection. They must be given the chance to refuse treatment by an HIV-positive worker.

- If health-care workers knew their HIV status, they could take extra precautions to protect patients, such as double-gloving.

- The cost of testing health-care workers has no bearing on the issue. If even one life can be saved through mandatory testing, any cost is justified.

Counterpoint

- A doctor's other medical conditions, from hepatitis-B to alcoholism, can pose much greater threats to patients than HIV infection. No one is making demands for disclosure of these problems; there is no reason to single out HIV.

- Health-care workers should already be following strict infection control procedures that successfully protect patients against many types of infection, some far easier to transmit than HIV.

- The risk of a patient being infected by a surgeon is about the same as the risk of dying in a car accident on the way to hospital. The cost of regular testing for health-care workers is no more justifiable than the cost of police escorts to protect all ambulances.

Research the issue.
Reflect on your findings.
Discuss the various viewpoints with others.
Prepare for the class debate.

CHAPTER HIGHLIGHTS

- The blood is composed of two components: a cellular portion and plasma.
- Red blood cells are designed to transport oxygen and carbon dioxide. Hemoglobin is the pigment found in red blood cells.
- Erythropoiesis is the process by which red blood cells are made.
- Anemia refers to the reduction in blood oxygen due to low levels of hemoglobin or poor red blood cell production.
- Leukocytes defend the body against invading microbes either by phagocytosis or by the formation of antibodies.
- Blood clots seal ruptured blood vessels or forestall the rupture of weakened vessels.

- An antigen is a substance, usually protein in nature, that stimulates the formation of antibodies within the body.
- Antibodies are proteins formed within the blood that react with antigens.
- Blood types are inherited. ABO blood types are classified on the basis of antigens on the red blood cell.
- Some blood types are not compatible. Recipients will form antibodies against invading antigens, causing them to agglutinate.
- Agglutination refers to the clumping of cells with antigens.
- Receptor sites act as ports along cell membranes. Nutrients and other needed materials fit into specialized areas along cell membranes.

APPLYING THE CONCEPTS

1 Sodium citrate and sodium oxalate are used by blood banks when storing blood. Both chemicals tie up calcium, which is present in the blood. Why do blood banks use these chemicals?

2 A physician notes fewer red blood cells and prolonged blood clotting times in a patient. White blood cell numbers appear to have increased, but further examination reveals that only the granulocyte numbers have increased, while the agranulocytes have decreased. In an attempt to identify the cause of the anomaly, the physician begins testing the bone marrow. Why did the physician suspect the bone marrow? Predict what might have caused the problem. Provide the reasons behind your prediction.

3 What would happen if blood type A was transfused into people with blood types B, A, O, and AB? Provide an explanation for each case.

4 The following illustrates how scientists determine blood type by cross matching. Heparin, an anticoagulant, is added to both a known and unknown blood sample to prevent blood clotting. The samples are then placed in a centrifuge and separated into components. The red blood cells from the known sample, type A, are mixed with the serum of the unknown sample. (Plasma without clotting factors is called *serum*.) In the next step, the red blood cells from the unknown sample are mixed with the serum from type A blood. The data are provided in the following diagram.

Predict the blood type of the unknown sample. Provide your reasons.

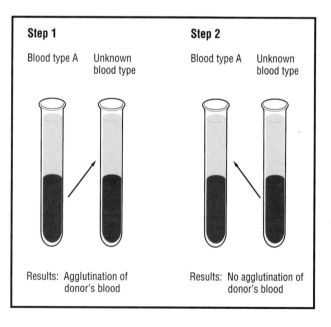

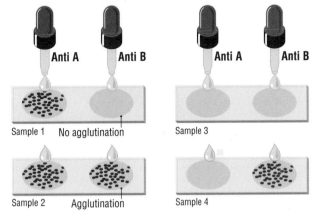

5 The serum containing antibody A will cause agglutination of donor blood that has antigen A. The serum containing antibody B will agglutinate donor blood that has antigen B. Use the data to the right to predict each of the blood types.

6 Why might people who have serious liver disease as a result of excessive alcohol consumption display prolonged blood clotting times?

7 Why does exposure to the influenza virus not provide immunity to other viruses?

Sample 1   No agglutination   Sample 3

Sample 2   Agglutination   Sample 4

Prediction for samples 1, 2, 3, and 4.

CRITICAL-THINKING QUESTIONS

1 Individuals who work in a chemical plant are found to have unusually high numbers of leukocytes. A physician calls for further testing. Hypothesize about the physician's reasons for concern. Why might the physician check both bone marrow and lymph node areas of the body?

2 AZT (azidothymidine) is an experimental drug used against HIV (human immunodeficiency virus), the virus that causes AIDS. To date, the drug has been somewhat successful in arresting the infection, but is not considered to be a cure. A great deal of controversy has erupted over whether or not drugs that prolong life but do not cure the disease should be used, even experimentally. What is your opinion? Provide your reasons.

3 The incidence of hepatitis and AIDS has increased the need for blood screening. Although all blood samples in Canada are screened for HIV and many other diseases, the same cannot be said about other countries. Do you think Canada should export screened blood to these countries? Give your reasons.

4 While HIV is able to gain access to the T cells, it is not able to infect other cells of the body. High concentrations of HIV are found in the blood, semen, and vaginal fluids of infected people. HIV has been found in lower concentrations in saliva, urine, and tears, but these are not believed to cause infection. AIDS can be spread through sexual intercourse, intravenous drug use, or from an infected mother to her baby during pregnancy or at birth.

A person cannot be infected with AIDS from casual contact such as shaking hands, kissing, coughing or sneezing, or from swimming pools, toilet seats, eating utensils, food, mosquitoes, or animals. Strict screening procedures for blood donations have been in place in Canada since November 1985.

The risk of HIV infection can be lowered by practicing safer sex (avoiding contact with semen, vaginal fluids, or blood), using latex rubber condoms, and avoiding intravenous drug use.

Despite the availability of this information, many individuals still fear AIDS needlessly. Why do you think people fear AIDS? How can the fear of AIDS be reduced?

ENRICHMENT ACTIVITIES

1 Read the case study on AIDS in the chapter, "Protein Synthesis."

2 Suggested reading:
- Ferry, Georgina. "New Cells for Old Brains." *New Scientist*, March 24, 1988, p. 54.

- Gamlin, Linda. "The Human Immune System." *New Scientist*, March 10, 1988, p. 1.
- Kolata, Gina. "Immune Boosters." *Discover*, September 1987, p. 68.

CHAPTER 12

Breathing

IMPORTANCE OF BREATHING

Composition of the Earth's Atmosphere

Carbon dioxide 0.03%

Oxygen 21.01%

Nitrogen 78.02%

Argon and other gases 0.94%

Nitrogen and oxygen are the two most common components of the atmosphere.

You live in a sea of air. Nitrogen, oxygen, carbon dioxide, and trace gases are taken into and expelled from your body with every breath. Seventy-eight percent of the earth's atmosphere is nitrogen, 21% is oxygen, a little less than 1% is argon, with carbon dioxide and other gases making up the remainder. Although nitrogen from the air appears to have little impact on the body, the second most common component, oxygen, is vital to life. Cells obtain energy by oxidizing organic compounds, and the most abundant oxidizing agent on the planet is oxygen. Although energy can be obtained in the absence of oxygen, the anaerobic conditions cannot maintain life processes in humans. Humans need oxygen to survive.

Oxygen is so essential to survival that any prolonged shortage will cause almost immediate death. By comparison, individuals can live for many days without water or several weeks without food. It has been estimated that an average adult utilizes 250 mL of oxygen every minute while resting. Oxygen consumption may increase up to 20 times with strenuous exercise.

Figure 12.1

Astronaut Edwin Aldrin, Jr., is photographed by Neil Armstrong walking near the lunar module.

In this chapter you will investigate the mechanism of breathing. You will also look at the way in which gases are exchanged between the atmosphere and the blood, and between the blood and the cells. Cellular respiration and energy transformation are presented in the chapter Energy within the Cell.

THE HUMAN RESPIRATORY SYSTEM

Air enters the respiratory system either through the two nasal cavities or the mouth. Large foreign particles are prevented from entering the nasal cavities by tiny hairs that line the passageways. The hairs act as a filtering system. The nasal cavities also contain mucus, which traps particles and keeps the cells lining the cavities moist.

The nasal cavities open into an air-filled channel in the mouth, called the **pharynx.** Two openings branch from the pharynx—the **trachea,** or windpipe, and the *esophagus*, which carries food to the stomach. Ciliated cells, which produce mucus, line the trachea. The mucus traps debris that may have escaped the filters in the nasal passage. The **cilia** (singular: "cilium") sweep the debris from the windpipe. The wall of the trachea is supported by cartilage rings, which keep the trachea open. An enlarged segment of cartilage supports the **epiglottis,** a flaplike structure that covers the opening of the trachea when food is being swallowed. When food is

> **BIOLOGY CLIP**
> Did you ever wonder how cough medicines work? Coughing is a protective reflex. Small foreign particles are swept from the respiratory tract by cilia. Larger particles activate many cilia, triggering a cough. Cough medicines act as sedatives and decrease this reflex; however, the reflex should never be completely inhibited. This is why manufacturers of cough medicines recommend specific dosages.

chewed, it is forced to the roof of the mouth and pushed backward. This motion initiates a reflex arc, which closes the epiglottis, allowing food to enter the esophagus rather than the trachea. If you have ever taken in food or drink too quickly, you will know how it feels to bypass this reflex. As food or drink enters the trachea, it stimulates the cilia. If the food particles are too large to be swept out of the respiratory tract, a second more powerful reflex is summoned—a violent cough usually expels the debris. One of the more dangerous foods in this respect is the hot dog. Every year people choke to death because of hot dogs, whose shape is almost ideally suited for lodging in the trachea.

Air is carried down the windpipe or trachea to the **larynx,** or voice box. The larynx is composed of two thin sheets of elastic ligaments, called the *vocal cords*. The vocal cords vibrate as air is forced from the lungs toward the pharynx. Different sounds are produced by a change in tension on the vocal cords. Your larynx is protected by a thick band of cartilage commonly known as the Adam's apple. Following puberty, the cartilage and larynx of males increase in size and thickness. In the same way as a larger drum creates a lower-pitched sound, the larger voice box in males produces a deeper sound. Rapid growth of the larynx creates problems for adolescent boys who have difficulty controlling the pitch of their voices. Have you ever noticed how your voice lowers when you have a cold? Inflammation of the vocal

*The **pharynx** is the passage for both food and air.*

*The **trachea** is the windpipe.*

Cilia *are tiny hairlike protein structures found in eukaryotic cells. Cilia sweep foreign debris from the respiratory tract.*

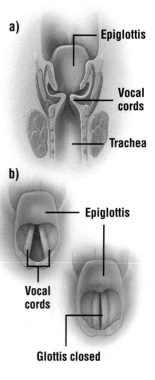

Figure 12.2

(a) Larynx, showing the vocal cords. (b) Position of the vocal cords when the glottis is open and closed.

*The **epiglottis** is a structure that covers the opening of the trachea (glottis) during swallowing.*

*The **larynx** is the voice box.*

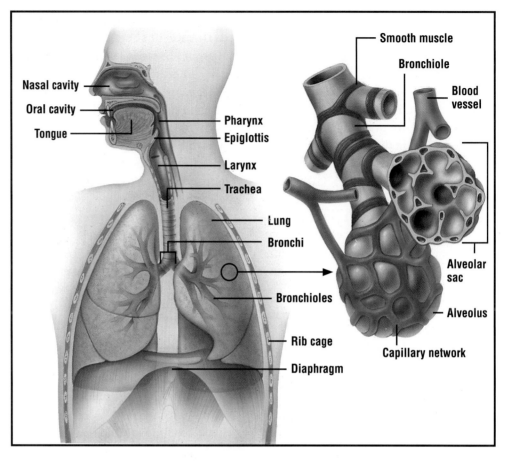

Figure 12.3

The human respiratory system.

Bronchioles *are the smallest passageways of the respiratory tract.*

Alveoli *are blind-ended sacs of the lung. The exchange of gases between the atmosphere and the blood occurs in the alveoli.*

cords causes swelling and produces lower-frequency vibrations. Should the infection become severe, and result in a condition referred to as *laryngitis*, you may temporarily lose your voice.

Inhaled air moves from the trachea into two bronchi (singular: "bronchus"), which, like the trachea, contain cartilage rings. The bronchi carry air into the right and left lungs, where they branch into many smaller airways called **bronchioles.** Unlike the trachea and bronchi, the bronchioles do not contain cartilaginous rings. Smooth muscle in the walls of the bronchioles can decrease their diameter. Any closing of the bronchioles increases the resistance of air movement and can produce a wheezing

> **BIOLOGY CLIP**
> Have you ever noticed that your nose fills after crying? This phenomenon is caused by the tear glands draining into the nasal cavities.

sound. Air moves from the bronchioles into tiny blind-ended sacs called **alveoli** (singular: "alveolus"). Measuring between 0.1 and 0.2 mm in diameter, each alveolus is surrounded by capillaries. In the alveoli, gases diffuse between the air and blood according to concentration gradients. Not surprisingly, the alveoli are composed of a single layer of cells, permitting more rapid gas exchange. Each lung contains about 150 million alveoli. That provides enough surface area to cover half a tennis court, or about 40 times the surface area of the human body.

Have you ever tried to pull the cover slip from a microscope slide, only to discover that the cover slip and slide seemed fused together? This phenomenon is caused by water molecules adhering to the glass. A similar problem faces the alveoli. During inhalation, the alveoli appear bulb-shaped, but during exhalation the tiny sacs collapse. The two membranes touch but are prevented from sticking together by a film of lipoprotein. This film lines the alveoli, allowing them to pop open during inhalation. Some newborn babies, especially premature babies, do not produce enough of the lipoprotein. Extreme force is required to overcome the surface tension created, and the baby experiences tremendous difficulty inhaling. This condition, referred to as *respiratory distress syndrome*, often results in death.

The outer surface of the lungs is surrounded by a thin membrane called the *pleural membrane*, which also lines the inner wall of the chest cavity. The pleural membrane is filled with fluids that reduce the friction between the lungs and the chest cavity during inhalation. *Pleurisy*, the inflammation of the pleural membranes and the build-up of fluids in the chest cavity, is most often caused when the two membranes rub together. This build-up of fluids puts great pressure on the lungs, making expiration easier, but inspiration much more difficult.

BREATHING MOVEMENTS

Pressure differences between the atmosphere and the chest cavity determine the movement of gases into and out of the lung. Atmospheric pressure remains relatively constant, but the pressure in the chest cavity may vary. An understanding

of gas exchange hinges on an understanding of gas pressures.

Gases move from an area of high pressure to an area of low pressure. Inspiration occurs when pleural pressure is less than that of the atmosphere. Air moves from the external environment into the lungs. Expiration occurs when pleural pressure is greater than that of the atmosphere. Air moves from the lungs to the external environment.

The **diaphragm,** a band of muscle that separates the chest cavity from the abdominal cavity, regulates the pressure in the chest cavity. When the muscle contracts, the diaphragm flattens and pulls downward. As the chest volume increases, pressure decreases. The atmospheric pressure is now greater than the pleural pressure, and air moves into the lungs. When the muscle relaxes, the diaphragm returns to a dome shape, due to the force exerted by the organs from the abdomen. As the chest volume decreases, pressure increases. The pleural pressure is now greater than the atmospheric pressure, and air moves out of the lungs. Have you ever found yourself gasping for air after receiving a blow to the solar plexus (the bottom of the rib cage)? The blow drives abdominal organs upward, causing the dome shape of the diaphragm to be exaggerated. Volume in the chest cavity is further reduced, and a large quantity of air is expelled.

Think of the diaphragm as a piston. As the piston moves down, the volume in the chest cavity increases and pressure begins to fall—air is drawn into the lungs. As the piston moves up, the volume in the chest cavity decreases and pressure begins to increase—air is forced out of the lungs.

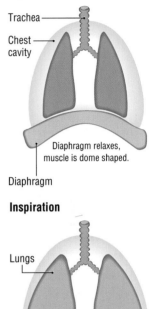

Expiration

Trachea

Chest cavity

Diaphragm relaxes, muscle is dome shaped.

Diaphragm

Inspiration

Lungs

Diaphragm contracts, muscle flattens.

Diaphragm

Figure 12.4

The diaphragm can be compared to a piston.

*The **diaphragm** is a sheet of muscle that separates the organs of the chest cavity from those of the abdominal cavity.*

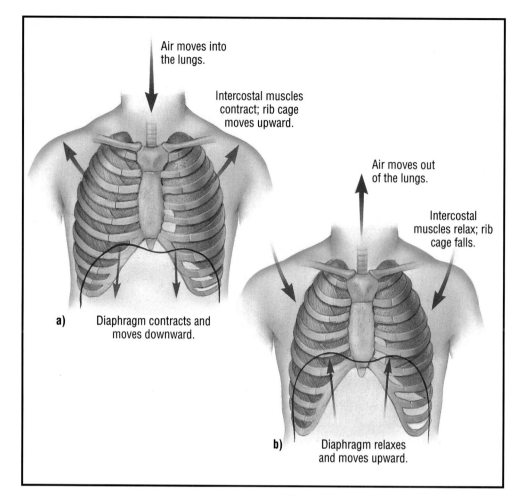

Air moves into
the lungs.

Intercostal muscles
contract; rib cage
moves upward.

Air moves out
of the lungs.

Intercostal
muscles relax; rib
cage falls.

a) Diaphragm contracts and
moves downward.

b) Diaphragm relaxes
and moves upward.

Figure 12.5

Changes in chest volume during inspiration and expiration. (a) The intercostal muscles contract and the rib cage pulls upward. Because pressure in the chest cavity is lower than atmospheric pressure, air moves into the lung. (b) The intercostal muscles relax and the rib cage falls. Because pressure in the chest cavity is higher than atmospheric pressure, air moves out of the lung.

External intercostal muscles *raise the rib cage, decreasing pleural pressure.*

Internal intercostal muscles *pull the rib cage downward, increasing pleural pressure.*

The diaphragm is assisted by the movement of the ribs. Have you ever noticed how you raise your ribs when you inhale? The ribs are hinged to the vertebral column, allowing them to move up and down. Bands of muscle, the **external intercostal muscles,** are found between the ribs. A nerve stimulus causes the external intercostal muscles to contract, pulling the ribs upward and outward. This increases the volume of the chest and lowers the pleural pressure. Air moves into the lungs. If the intercostals are not stimulated, the muscle relaxes and the rib cage falls. The fluids inside the pleural membrane push against the lungs with greater pressure, and air is forced out of the lungs. A second set of intercostals, the **internal intercostal muscles,** pulls the rib cage downward during times of extreme exercise. The internal intercostal muscles are not employed during normal breathing.

The importance of establishing pressure differences between the lungs and the atmosphere is best illustrated by a *pneumothorax*, or collapsed lung. A bullet wound or a stab wound to the ribs creates a hole in the pleural cavity, making it impossible for the chest cavity to establish pressure differences. When the diaphragm contracts and the rib cage is raised, the pressure inside the chest cavity is reduced. However, much less air is drawn into the lungs. Much of the air flows directly through the hole left by the wound. In most cases, the lungs remain collapsed. The chest cavity must be sealed and any excess air or fluid withdrawn before the lungs are capable of functioning normally.

FRONTIERS OF TECHNOLOGY: USING COMPUTER DATA BASES

Is the incidence of lung disorders greater among certain social or economic groups? A McGill University research team made up of Margaret Becklake, Pierre Ernst, and Richard Menzies believes that despite universally accessible health care, lung disorders are more common among certain economically disadvantaged groups. Their research program proposes to study the characteristics of young people and their environment that place them at risk for lung infections. Specifically, the researchers will investigate the group's propensity to develop allergies and their history of childhood infections. Data about environmental factors such as crowding, fumes from cooking, and second-hand smoke will be stored by computer. The computer will be an important ally, because the research team has no way of knowing which factor or combination of factors is most prevalent. The computer is able to search and compare different sets of data in incredibly short periods of time.

The research team hopes that information about lifestyle and environmental factors and their links to lung problems will serve as a basis for future prevention of lung disorders.

■ REVIEW QUESTIONS ■ ?

1 Which gases are exchanged during breathing?
2 Differentiate between breathing and cellular respiration.
3 Trace the pathway of an oxygen molecule from the atmosphere to its attachment to a binding site on a hemoglobin molecule.
4 Why does a throat infection cause your voice to produce lower-pitched sounds?
5 Describe the function of cilia in the respiratory tract.
6 Describe the movements of the ribs and the diaphragm during inhalation and exhalation.

REGULATION OF BREATHING MOVEMENTS

Breathing movements are controlled by nerves from the medulla oblongata in the brain. Information about the accumulation of carbon dioxide and acids and the need for oxygen is detected by **chemoreceptors.** Two different types of receptors are oxygen chemoreceptors and carbon dioxide or acid chemoreceptors. The carbon dioxide receptors are the most sensitive and are the main regulators of breathing movements.

Carbon dioxide (CO_2) dissolves in the blood and forms an acid. Should the carbon dioxide accumulate, special chemoreceptors in the medulla oblongata become activated. Once activated, the medulla oblongata relays messages to the muscles of the diaphragm and ribs to begin breathing movements. The acceleration of breathing rate decreases the levels of CO_2 in the blood. Once CO_2 levels fall, the chemoreceptors become inactive, and breathing rate returns to normal.

It should be noted that the thinking area of the brain can also influence the medulla oblongata. That may explain why you are able to deliberately hold your breath for extended periods. However, should CO_2 build up to dangerous levels, the medulla oblongata takes over and breathing movement resumes. Little children sometimes express their anger by holding their breath. However, they can-

Chemoreceptors are specialized nerve receptors that are sensitive to specific chemicals.

Body cell uses oxygen to break down organic molecules. Carbon dioxide levels rise in the cell and CO_2 diffuses into the blood.

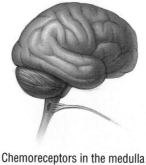

Carbon dioxide is expelled faster when breathing movements increase.

Figure 12.6

Carbon dioxide control.

Chemoreceptors in the medulla oblongata detect high levels of carbon dioxide. A nerve message is sent to the ribs and diaphragm to increase breathing movements.

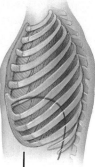

Negative feedback

not do so indefinitely. Eventually the medulla oblongata will take over breathing movements. Some drugs, such as morphine and barbiturates, make the CO_2 receptors of the medulla oblongata less sensitive, causing breathing rates to fall. Excessive intake of such drugs can actually slow breathing movements to a level that makes life impossible.

A second monitoring system, which relies on chemoreceptors sensitive to oxygen, is found in the carotid and aortic arteries. Referred to as the *carotid* and *aortic bodies*, these specialized receptors are primarily responsible for detecting low levels of oxygen. When stimulated, the oxygen receptors send a nerve message to the medulla oblongata. Once activated, the medulla sends nerve messages to the ribs and diaphragm to begin breathing movements. Increased ventilation increases blood oxygen, thereby compensating for low levels of oxygen. A secondary function of these bodies is to detect high blood CO_2 or high levels of acidity, although the medulla oblongata is the more sensitive receptor of CO_2 and acids.

Because the CO_2 receptors are more sensitive to changes in blood chemistry,

the oxygen receptors act as a backup system. The oxygen receptors are only called into action when O_2 levels fall and CO_2 levels remain within the normal range. For example, when you hold your breath, CO_2 levels rise and oxygen levels drop—the high CO_2 levels would initiate breathing movements. However, the situation differs at high altitudes, where the air is thinner and fewer oxygen molecules are found. Since low oxygen levels are not accompanied by higher CO_2 levels, the chemoreceptors in the carotid and aorta stimulate breathing movements. Increased ventilation helps establish normal blood oxygen levels.

Carbon monoxide poisoning is another example of how falling blood oxygen levels stimulate increased breathing rate. Carbon monoxide (CO) competes with oxygen for the active site on the hemoglobin molecule. Unfortunately, CO gains faster access. As more and more hemoglobin molecules bind with CO, less and less oxygen is carried to the tissues. The carbon dioxide level tends not to increase. Eventually, the low oxygen level is detected by the chemoreceptors and breathing movements increase.

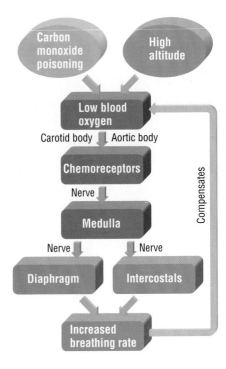

the inflammation of the bronchioles. In both conditions, greater effort is required to exhale than to inhale. This occurs because lower pleural pressure is exerted on the lungs and bronchioles during inspiration. The expansion of the lungs not only opens the alveoli but also the small bronchioles, increasing the diameter of the bronchioles and decreasing resistance to airflow. Conversely, during expiration, increased pleural pressure compresses the lungs and the bronchioles, decreasing the diameter of the bronchioles and increasing resistance to airflow. Consequently, less air leaves the lungs. The imbalance between the amount of air entering the lung and the amount of air leaving the lung must be met by increasing the exertion of expiration.

Although all the causes of asthma are not yet known, the condition is often associated with allergies. As you read in the previous chapter, allergic reactions initiate the swelling of the tissues. In the case of asthma, the tissues that line the walls of the bronchioles swell. It is also suspected that the allergic reaction sends the muscles that line the walls of the bronchioles into spasms. Both responses greatly increase the resistance of airflow out of the lung. The discomfort experienced during an asthmatic attack is associated not with getting air into the lungs, but with getting air out of the lungs.

Emphysema

Emphysema is associated with chronic bronchitis. Like bronchitis, emphysema is characterized by increased resistance to airflow through the bronchioles. Although air flows into the alveoli fairly easily, the decreased diameter of the bronchioles creates resistance to the movement of air out of the lungs. Air pressure builds up in the lungs (the word *emphysema* means "over-inflated"). Unable to support the building pressure, the thin walls of the alveoli

Figure 12.7

Low blood oxygen levels are detected by special oxygen chemoreceptors in the aorta and carotid arteries. Once stimulated, the receptors send a message to the medulla oblongata, the respiratory center of the brain. In turn, nerves from the medulla trigger breathing movements. Increased ventilation increases blood oxygen levels.

Bronchitis *is an inflammation of the bronchioles.*

Bronchial asthma *is characterized by a reversible narrowing of the bronchial passage.*

Emphysema *is an overinflation of the alveoli. Continued overinflation can lead to the rupture of the alveoli.*

DISORDERS OF THE RESPIRATORY SYSTEM

All respiratory disorders share one common characteristic: they all decrease oxygen delivery to the tissues.

Bronchitis

Bacterial or viral infections, as well as reactions to environmental chemicals, cause the mucous cells that line the respiratory pathways to secrete more mucus. The irritation may initiate an inflammation response and bring about tissue swelling. This condition, called **bronchitis,** refers to a wide variety of ailments characterized by a narrowing of the air passages. As mucous secretions increase, air movement decreases. The condition becomes even more serious in the bronchioles. Unlike the trachea and the bronchi, the bronchioles are not supported by rings of cartilage that help keep them open.

Two conditions, **bronchial asthma** and **emphysema,** are associated with stretch and eventually rupture. Fewer alveoli means less surface area for gas exchange, which, in turn, leads to decreased oxygen levels. In the body's attempt to maintain homeostasis, the breathing rate increases and exhalation becomes more labored. The circulatory system adjusts by increasing the heart rate.

The destruction of alveoli also means the destruction of the adjoining pulmonary capillaries. Fortunately, blood clotting prevents any major internal hemorrhaging, but the healing process creates a secondary problem. Because the elastic tissue of the lung is replaced by scar tissue, the lung is less able to expand and, therefore, it holds less air. The broken-down cells of the alveoli do not regenerate, leaving less surface area for diffusion.

Lung Cancer

Lung cancer is the most common cause of death from cancers for Canadian men. For Canadian women lung cancer is the third most common cause of death from cancer, although it is increasing at an alarming rate. Like other cancers, lung cancer is characterized by the uncontrolled growth of cells. The solid mass of cancer cells in the lungs greatly decreases the surface area for diffusion. Tumors may actually block bronchioles, thereby reducing airflow to the lungs, potentially causing the lung to collapse.

a)

b)

Figure 12.8

(a) A lung scan reveals a cancer of the left lung. The colors representing the right, healthy lung indicate normal ventilation. The absence of color on the left side indicates a non-functioning lung. (b) Smoke descends toward the lungs.

Figure 12.9

Lung cancer. The inset shows a post-mortem specimen of a human lung, showing a cancerous tumor of the upper lobe as a black and white area. The entire lung is permeated with black, tarry deposits suggesting a history of heavy cigarette smoking.

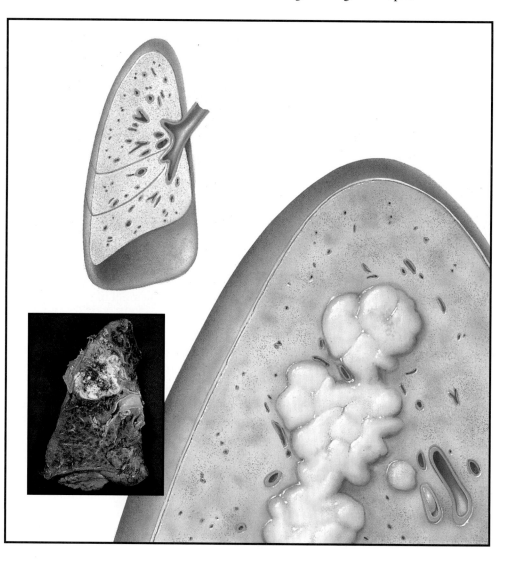

LABORATORY
MONITORING LUNG VOLUME

Objective
To measure lung capacity.

Materials
respirometer with disposable mouthpieces

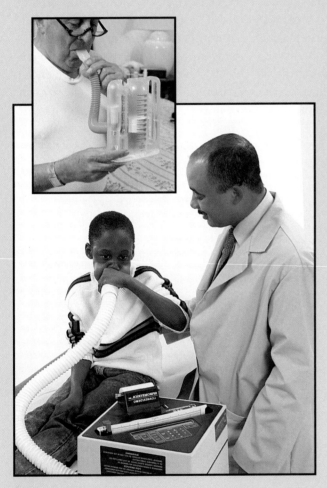

Procedure

1 Place a new mouthpiece in the respirometer and set the gauge to zero.

2 Inhale normally, then place your mouth around the respirometer and exhale normally. Read the gauge on the respirometer.
   a) Record the volume exhaled as *tidal volume*.

3 Reset the respirometer to zero. Inhale normally, then place your mouth around the respirometer and exhale as forcibly as you can. Read the gauge on the respirometer.
   b) Record the value as *expiratory reserve volume*.

4 Reset the respirometer to zero. Inhale as much air as possible and then exhale forcibly into the respirometer. Read the gauge on the respirometer.
   c) Record the value as *vital capacity*.

5 Determine your *inspiratory reserve volume* by using the following formula:

Vital capacity = Inspiratory + Expiratory + Tidal
                 reserve   reserve   volume
                 volume   volume

Inspiratory
reserve volume = ?

6 Repeat the procedure for an additional two trials.

Laboratory Application Questions

1 Predict how the tidal volume and vital capacity of a marathon runner might differ from that of the average Canadian.

2 How might bronchitis affect your expiratory reserve volume? Provide your reasons.

3 Predict how the respiratory volumes for a person with emphysema would differ from those collected during the laboratory.

4 Why might respiratory volumes be measured during exercise? (Provide a list of things that could be investigated or diagnosed.) ■

B I O L O G Y   C L I P
Approximately 20% of the air taken into the lungs is not exchanged. Some air remains in the lung and respiratory tract even after you exhale.

GAS EXCHANGE AND TRANSPORT

An understanding of gas exchange in the human body is tied to an understanding of the physical nature of gases. The Italian scientist Amedeo Avogadro (1776–1856) stated that equal volumes of gases at the same temperature and pressure contain the same number of molecules. Avogadro's hypothesis provides the basis for calibrating the volume occupied by different gases in a mixture. If the equal volumes of gas are maintained at the same temperature, the pressure exerted by a gas determines the speed of diffusion. In other words, the diffusion of a gas occurs from an area of high pressure to an area of low pressure.

Dalton's law of partial pressure states that each gas in a mixture exerts its own pressure, which is proportional to the total volume. The pressure of each gas in the atmosphere can be calculated using this law. Consider that atmospheric pressure at sea level is 101 kPa. The partial pressure exerted by oxygen can be calculated since we know that 21% of the air is oxygen. Therefore, 21% ×101 kPa = 21.21 kPa. By similar calculations, carbon dioxide exerts a partial pressure of 0.04 kPa.

The graph in Figure 12.11 shows partial pressures of oxygen and carbon dioxide in the body. Gases diffuse from an area of high partial pressure to an area of lower partial pressure. The highest partial pressure of oxygen is found in atmospheric air. Oxygen diffuses from the air (21 kPa) into the lungs (13.3 kPa for alveoli). The largest change in partial pressure of oxygen is observed as oxygen travels between arteries (12.6 kPa) and veins (5.3 kPa). This change should not be surprising, since oxygen diffuses from the capillaries into the tissues. (It is important to remember that oxygen is continuously used by the cells of the body to provide energy. Oxygen will never accumulate in the cells.) Carbon dioxide, the product of cell respiration, follows an opposite pattern. Partial pressure of carbon dioxide is highest in the tissues and venous blood. The partial pressure of nitrogen, although not shown in the graph, remains relatively constant. Atmospheric nitrogen is not involved in cellular respiration.

Figure 12.10

Diagram showing a barometer.

Dalton's law of partial pressure *states that each gas in a mixture exerts its own pressure, which is proportional to the total volume.*

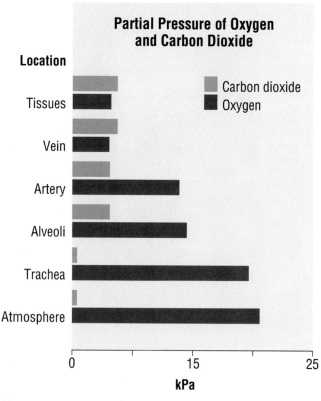

Figure 12.11

The partial pressure of oxygen is lowest in the veins and tissues, but highest in atmospheric air. Conversely, the partial pressure of carbon dioxide is highest in the tissues and veins, but lowest in atmospheric air.

Oxygen Transport

Oxygen moves from the atmosphere, the area of highest partial pressure, to the alveoli. Boosted by its partial pressure, oxygen diffuses from the alveoli into the blood and dissolves in the plasma. Only a limited amount of oxygen can dissolve in the blood—about 0.3 mL of oxygen per 100 mL of blood. However, even at rest, the body requires approximately 10 times as much oxygen as could be supplied in this manner. Hemoglobin, the respiratory pigment of red blood cells, greatly increases oxygen-carrying capacity. When oxygen dissolves into the plasma, hemoglobin forms a weak bond with the oxygen molecule to form *oxyhemoglobin*. Once oxyhemoglobin is formed, other oxygen molecules can dissolve in the plasma. With hemoglobin, the blood can carry 20 mL of oxygen per 100 mL of blood, a 70-fold increase.

The amount of oxygen that combines with hemoglobin is dependent on partial pressure. The partial pressure in the lungs is approximately 13.3 kPa. Thus, blood leaving the lungs is nearly saturated with oxygen. As blood enters the capillaries, which support the tissues of the body, the partial pressure drops to about 5.3 kPa. This drop in partial pressure causes the dissociation of oxygen from the hemoglobin, and oxygen diffuses into the tissues. The graph in Figure 12.12 shows an oxygen-hemoglobin dissociation curve. Notice that very little oxygen is released from the hemoglobin until the partial pressure of oxygen reaches 5.3 kPa. This ensures that most of the oxygen remains bound to the hemoglobin until it gets to the tissue capillaries. It is important to note that venous blood still carries a rich supply of oxygen. Approximately 70% of the hemoglobin is saturated when blood returns to the heart.

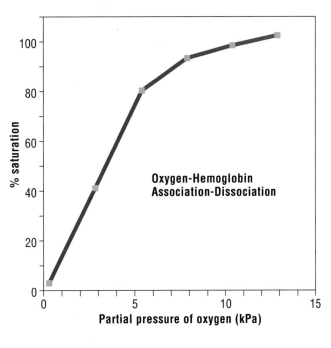

Oxygen-Hemoglobin Association-Dissociation

Figure 12.12

Hemoglobin gives up very little oxygen until the partial pressure of oxygen reaches 5.3 kPa in the surrounding tissues. As the partial pressure of oxygen rises, hemoglobin picks up more oxygen.

Carbonic anhydrase *is an enzyme found in red blood cells. The enzyme speeds the conversion of CO_2 and H_2O to carbonic acid.*

Carbon Dioxide Transport

Carbon dioxide is about 20 times more soluble than oxygen. About 9% of the carbon dioxide produced by the tissues of the body is carried in the plasma. Approximately 27% of the body's carbon dioxide combines with hemoglobin to form *carbaminohemoglobin*. The remaining 64% of the body's carbon dioxide combines with water from the plasma to form carbonic acid (H_2CO_3). An enzyme called **carbonic anhydrase** increases the rate of this chemical reaction by about 250 times. The rapid conversion of free carbon dioxide into carbonic acid decreases the concentration of carbon dioxide in the plasma. This maintains a low partial pressure of CO_2 in the blood, ensuring that CO_2 continues to diffuse into the blood.

However, the formation of acids can create problems. Because acids can change the pH of the blood and eventually bring about death, they must be buffered. This is where the second function of hemoglobin comes into effect. Being unstable, the carbonic acid dissociates to bicarbonate ions (HCO_3^-) and hydrogen ions (H^+). The hydrogen ions help dislodge oxygen from the hemoglobin, and then combine with the hemoglobin to form reduced hemoglobin. By removing the hydrogen ions from solution, hemoglobin serves as a buffer. Meanwhile the bicarbonate ions are transported into the plasma. Oxygen is released from its binding site and is now free to move into the body cells.

Once the venous blood reaches the lungs, oxygen dislodges the hydrogen ions from the hemoglobin binding sites. Free hydrogen and bicarbonate ions combine to form carbon dioxide and water.

The highly concentrated CO_2 diffuses from the blood into the alveoli and is eventually eliminated during exhalation.

■ REVIEW QUESTIONS   ?

7   How do CO_2 levels regulate breathing movements?

8   Why does exposure to carbon monoxide (CO) increase breathing rates?

9   How does emphysema affect the lungs?

10 How does partial pressure affect the movement of oxygen from the alveoli to the blood?

11 How is CO_2 transported in the blood?

12 Describe the importance of hemoglobin as a buffer.

Figure 12.13

Under the influence of carbonic anhydrase, an enzyme found in red blood cells, CO_2 combines with H_2O to form H^+ and HCO_3^- ions.

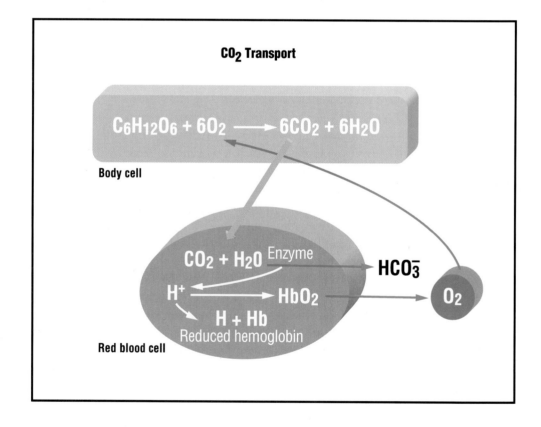

CO$_2$ Transport

$$C_6H_{12}O_6 + 6O_2 \longrightarrow 6CO_2 + 6H_2O$$

Body cell

$$CO_2 + H_2O \quad \text{Enzyme}$$

$$H^+ \longrightarrow HbO_2$$

$$H + Hb$$
Reduced hemoglobin

HCO_3^-

O_2

Red blood cell

RESEARCH IN CANADA

Although the number of Canadians dying from most types of cancer has been declining, the number who die from lung and respiratory cancers is increasing every year. Cancer of the lung and respiratory tract is by far the leading cause of cancer in men and the third most common cause of cancer in women, slightly behind breast and colo-rectal cancer. Dr. Barbara Campling is conducting research on lung cancer at Queen's University in Kingston, Ontario. According to Dr. Campling, about 25% of lung cancer cases are a type called small-cell lung cancer, which can be treated by chemotherapy. However, in spite of an initial response rate to chemotherapy of close to 90%, patients are seldom cured. It would appear that the major failure of chemotherapy can be traced to the development of drug resistance by the cancer cells.

Lung Cancer Research

There are a number of ways in which cancer cells can develop resistance to multiple drug dosages. The most common has been linked with the production of a sugar-protein complex, P-gp (glycoprotein), which is pumped out of tumor cells. However, P-gp is not detected with small-cell lung cancer. Does this mean that this type of cancer responds to chemotherapy differently, or does it simply mean that current detection methods are inadequate? Dr. Campling believes that small-cell lung cancer behaves much like other types of lung cancer but that monitoring techniques are not sensitive enough. To support her hypothesis, Dr. Campling will use two highly sensitive methods on a number of cancer-cell lines in which the relative sensitivity to a number of chemicals has already been predetermined.

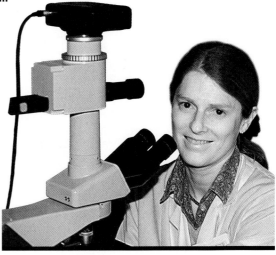

DR. BARBARA CAMPLING

CASE STUDY

SMOKING AND LUNG CANCER

Objective

To investigate how smoking can bring about lung cancer.

Procedure

1 Cancer usually begins in the bronchus or bronchioles. Components of cigarette smoke contribute to the development of cancerous tumors. Study the four diagrams below, which show the development and progression of lung cancer.

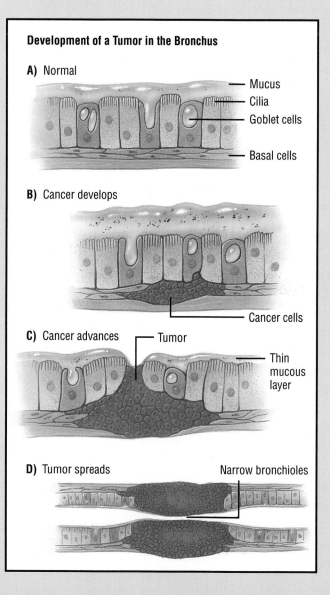

Development of a Tumor in the Bronchus

A) Normal
— Mucus
— Cilia
— Goblet cells
— Basal cells

B) Cancer develops
— Cancer cells

C) Cancer advances — Tumor
— Thin mucous layer

D) Tumor spreads   Narrow bronchioles

2 Cigarette smoke travels through the bronchi and irritates the cells. Goblet cells produce mucus, which is designed to trap foreign particles. Compare the mucous layers in diagrams A and B.

   a) How has cigarette smoke affected the mucous layer?

3 Ciliated cells line the bronchi. Cilia sweep away the debris trapped by the mucus. Unfortunately, the tar found in cigarette smoke slows the action of the cilia. The sludge-like tar becomes trapped in the mucus.

   b) Why does the build-up of tar in the bronchi limit airflow?

4 Diagram B shows the beginning of a cancerous tumor.

   c) In what area does the tumor begin to develop?

5 Diagram C shows how the tumor begins to grow.

   d) Why has the mucous layer in diagram C decreased in size?

6 While the cancerous tumor is still walled in by the basal membrane in diagram C, the tumor breaks through the membrane in diagram D. At this stage the tumor may spread to other parts of the body. Cancer cells often use lymph vessels as pathways to others parts of the body, where they continue to divide.

   e) Why does this characteristic make cancer especially dangerous?

Case-Study Application Questions

1 Based on the information that you gained in the case study, explain why smokers are often plagued by a cough.

2 Why is the slowing of the cilia lining the bronchi especially dangerous?

3 Survey different smokers and calculate the amount of tar taken in each day. Most cigarettes contain about 15 mg of tar, but it should be noted that only 75% of the tar is absorbed. Show your calculations.

4 Nicotine, one of the components of cigarettes, slows cilia lining the respiratory tract, causes blood vessels to constrict, and increases heart rate. Another component of cigarette smoke is carbon monoxide. As you have read, carbon monoxide competes with oxygen for binding sites on the hemoglobin molecule found in red blood cells. Analyze the data presented, and describe the potential dangers associated with smoking. ■

SOCIAL ISSUE:
Smoking

The correlation between cigarette smoking and lung cancer has been established. Smoking has also been linked to high blood pressure, heart disease, and a host of other diseases. Individuals who smoke two or more packs of cigarettes a day have about a 25 times greater chance of contracting lung cancer than a nonsmoker. With increasing evidence of the dangers of second-hand smoke, nonsmokers' rights groups and health officials have successfully lobbied for a ban on smoking in the workplace and other public areas.

In 1986, the Canadian tobacco industry employed 6000 full-time and 51 000 seasonal workers in addition to the thousands of wholesale and retail workers who derive profits from the sale of tobacco products. Both federal and provincial taxes on tobacco products account for 67% of the retail price of a pack of cigarettes. In 1986 the federal government received over $1.8 billion in taxes from the sale of tobacco products, with the provinces earning a similar amount.

In 1992, workers in the tobacco industry protested that with its policy of increased taxes on tobacco products, the government was costing workers their jobs. The government responded that the cost of health care for people with tobacco-related illnesses far surpasses the revenues it generates from tobacco taxes. The government claims that by increasing the price of cigarettes, more people will stop smoking and the general health of the populace will improve.

Statement:

The government should ban the sales of tobacco products.

Point

- Smoking leads some people to addiction. Such people have great difficulty stopping smoking. Making tobacco unavailable would help them to quit. It would also save them a lot of money.
- According to one estimate, the costs of treating patients with respiratory and circulatory diseases brought on by smoking exceed one billion dollars every year. Smokers also miss more days of work and tend to be less productive on the job.
- The tobacco industry provides industry high profit returns for investment. This wealthy industry has a formidable lobby group.

Counterpoint

- Individuals in our society value freedom. If governments force people to give up smoking, what other actions might they take to restrict our civil liberties?
- The major portion of the cost of cigarettes is in the form of government taxes. These taxes help finance health and other government programs. If governments are concerned about our health, perhaps they should outlaw alcohol or overeating as well.
- Tobacco companies have been disclosing the dangers of smoking for many years. People who still choose to smoke should be permitted to do so. The legitimate profits made by the industry keep farmers on the farm and workers on the job.

> **Research the issue.**
> **Reflect on your findings.**
> **Discuss the various viewpoints with others.**
> **Prepare for the class debate.**

CHAPTER HIGHLIGHTS

- Respiration involves the intake, transportation, and utilization of oxygen by cells to produce energy.
- Air passes from the atmosphere to the pharynx, trachea, bronchi, bronchioles, and finally to the alveoli.
- Cilia are tiny hairlike protein structures found in eukaryotic cells. Cilia sweep foreign debris from the respiratory tract.
- The larynx, also called the voice box, is responsible for producing sound.
- Rib and diaphragm movements change pleural pressure. Inhalation and exhalation are regulated by changes in pleural pressure. The diaphragm is a sheet of muscle that separates the organs of the chest cavity from those of the abdominal cavity.
- Breathing movements are regulated by CO_2 and O_2 chemoreceptors. CO_2 receptors are more sensitive.
- Bronchial asthma is characterized by a narrowing of the bronchial passage.
- Emphysema involves an overinflation of the alveoli. Continued overinflation can lead to the rupture of the alveoli.

APPLYING THE CONCEPTS

1 Why do breathing rates increase in crowded rooms?

2 A patient is given a sedative that inhibits nerves leading to the pharynx, including those that control the epiglottis. What precautions would you take with this patient? Give your reasons.

3 A man has a chest wound. The attending physician notices that the man is breathing rapidly and gasping for air.
  a) Why does the man's breathing rate increase? Why does he gasp?
  b) What could be done to restore normal breathing?

4 During mouth-to-mouth resuscitation, exhaled air is forced into the victim's trachea. As you know, exhaled air contains higher levels of CO_2 than atmospheric air. Would the higher levels of CO_2 create problems or would they be beneficial? Provide your reasons.

5 Prior to swimming underwater, a diver breathes deeply and rapidly for a few seconds. How has hyperventilating helped the diver hold her breath longer?

Use the following data table to answer questions 6, 7, and 8.

Individual	Breathing rate (breaths/ min)	Hemoglobin (g/100 mL blood)	O₂ Content (mL/100 mL)
A (normal)	15	15.1	19.5
B	21	8	13.7
C	12	17.9	22.1
D	22	16.0	14.1

6 Which individual has recently moved from Calgary to Halifax? (Note: Halifax is at or near sea level, while Calgary is at a much higher altitude.) Give your reasons.

7 Which individual is suffering from dietary iron deficiency? Give your reasons.

8 Which individual has been exposed to low levels of CO? Give your reasons.

Use the graph to answer question 9. The graph compares fetal and adult hemoglobin.

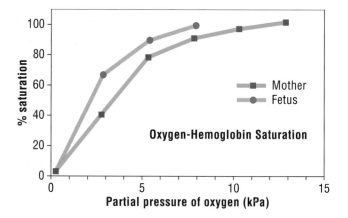

9 Which hemoglobin is more effective at absorbing oxygen? What adaptive advantage is provided by a hemoglobin that readily combines with oxygen?

CRITICAL-THINKING QUESTIONS

1 A scientist sets up the following experimental design.
Sodium hydroxide absorbs carbon dioxide. Limewater turns cloudy when it absorbs carbon dioxide.

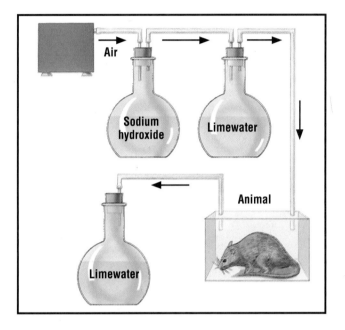

a) Indicate the purpose of the experiment.

b) Which flask acts as a control?
c) Why is sodium hydroxide used?

2 The following chart shows a comparison of inhaled and exhaled gases. On the basis of the experimental data, a scientist concludes that nitrogen from the air is used by the cells of the body. Critique the conclusion.

| | Percentage by volume | |
Component	Inhaled air	Exhaled air
Nitrogen	78.62%	74.90%
Oxygen	20.85%	15.30%
Carbon dioxide	0.03%	3.60%
Water (vapor)	0.50%	6.20%

3 Nicotine causes blood vessels, including those in the placenta, to constrict. Babies born to women who smoke are, on average, about 1 kg smaller than normal. This may be related to decreased oxygen delivery. Speculate about other problems that may face developing embryos due to the constriction of blood vessels in the placenta.

ENRICHMENT ACTIVITIES

1 Allan Becker, of the University of Manitoba, is studying dogs to learn more about how asthma works in people. Asthma is frequently associated with allergies, especially in children. Dr. Becker is especially interested in studying how allergic reactions affect airway function. Research how allergies have been linked with asthma. Indicate the benefits of using modelling experiments on dogs. Are there any disadvantages?

2 Suggested reading:
- Arehart-Treichel, Joan. "Lubricating Distressed Lungs." *Science News*, March 5, 1983, p. 150.
- Kanigel, Robert. "Nicotine Becomes Addictive." *Science Illustrated*, Oct/Nov, 1988, p. 32.
- Fletcher, William. "Asbestos." *Medical Self Care*, Nov/Dec, 1988, p. 32.

*K*idneys and Excretion

IMPORTANCE OF THE KIDNEYS

The cells of your body obtain energy by converting complex organic compounds into simpler compounds. Unfortunately, many of these simpler compounds can be harmful. To maintain life processes, the body must eliminate waste products. The lungs eliminate carbon dioxide, one of the products of cellular respiration. The large intestine removes toxic wastes from the digestive system. The liver transforms ingested toxins such as alcohol and heavy metals into soluble compounds that can be eliminated by the kidneys. The liver also transforms the hazardous products of protein metabolism into metabolites, which are then eliminated by the kidneys. In this chapter you will concentrate on the role that the kidneys play in removing waste, balancing blood pH, and maintaining water balance.

The average Canadian consumes more protein than is required to maintain tissues and promote cell growth. Excess protein is often converted into carbohydrates.

Figure 13.1

Kidney transplants are performed on patients with chronic renal failure. Transplants extend the life expectancy for these patients. However, survival is limited by the rejection of the transplanted organ by the immune system.

Proteins, unlike carbohydrates, contain a nitrogen molecule. The nitrogen and two attached hydrogen molecules, characteristic of amino acids, must be removed. (Recall that amino acids are the building blocks of proteins.) This process, referred to as **deamination,** occurs in the liver. The by-product of deamination is ammonia. Ammonia, like carbon dioxide, is a water-soluble gas. However, ammonia is extremely toxic—as little as 0.005 mg can kill you. Fish are able to avoid ammonia build-up by continually releasing it through their gills. Land animals, however, do not have the ability to release small quantities of ammonia throughout the day—wastes must be stored. Once again, the liver is called into action. In the liver, two molecules of ammonia combine with another waste product, carbon dioxide, to form **urea.** Urea is 1000 times less toxic than ammonia. The blood can dissolve 33 mg of urea per 100 mL of blood. A second waste product, **uric acid,** is formed by the breakdown of nucleic acids.

The kidneys also help maintain water balance. Although it is possible to survive for weeks without food, humans cannot survive more than a few days without water. Humans deplete their water reserves faster than their food reserves. The average adult loses about 2 L of water every day through urine, perspiration, and exhaled air. Greater volumes are lost when physical activity increases. For the body to maintain water balance, humans must consume 2 L of fluids daily. A drop in fluid intake by as little as 1% of your body mass will cause thirst, a decrease of 5% will bring about extreme pain and collapse, while a decrease of 10% will cause death.

> **BIOLOGY CLIP**
> Uric acid is only found in the urine of humans, higher apes, and Dalmatian dogs. The uric acid molecule has a structure similar to that of caffeine.

URINARY SYSTEM

Renal arteries branch from the aorta and carry blood to the paired kidneys. With a mass of about 0.5 kg, the fist-shaped kidneys may hold as much as one-quarter of the body's blood at any given time. Wastes are filtered from the blood by the kidneys and conducted to the urinary bladder by **ureters.** A sphincter muscle located at the base of the urinary bladder acts as a valve, permitting the storage of urine. When approximately 200 mL of urine has been collected, the bladder stretches slightly, and nerves signal the brain about the condition of the bladder. Should the bladder fill to 400 mL, more stretch receptors are activated and the message becomes more urgent. If you continue to ignore the messages, the bladder continues to fill, and after about 600 mL of urine accumulates, voluntary control is lost. The sphincter relaxes, urine enters the **urethra** and is voided.

Differences between males and females become evident in this last structure of the urinary system. In males, the urethra exits by way of the penis, providing a common pathway for sperm and urine from the body. In females, the reproductive and excretory functions remain distinct. The urethra of females lies within the vulva, the external genital organ, but has no connection to reproductive organs. The urethra of males is longer than that found in women. This may account for the fact that females are more prone to bladder infections than are males.

A cross section of a kidney reveals three different structures. An outer layer of connective tissue, the **cortex,** encircles the kidney. An inner layer, the

Deamination *is the removal of an amino group from an organic compound.*

Urea *is a nitrogen waste formed from two molecules of ammonia and one molecule of carbon dioxide.*

Uric acid *is a waste product formed from the breakdown of nucleic acids.*

Ureters *are tubes that conduct urine from the kidneys to the bladder.*

The **urethra** *is a tube that carries urine from the bladder to the exterior of the body.*

The **cortex** *is the outer layer of the kidney.*

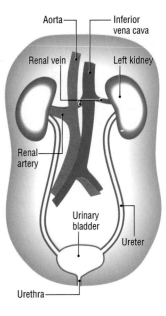

Figure 13.2

Simplified diagram of the human urinary system.

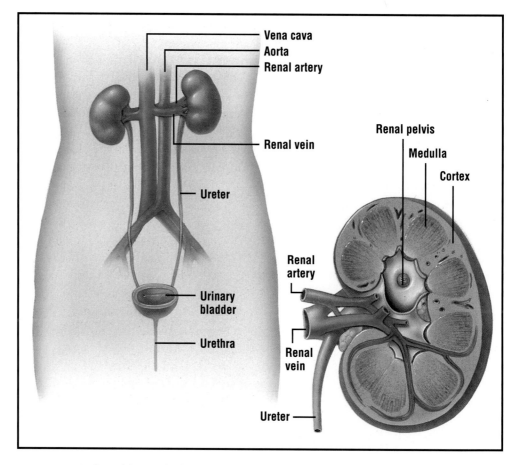

Figure 13.3

Structure of the human excretory system (left). Microscopic view of the kidney.

*The **medulla** is the area inside of the cortex.*

*The **renal pelvis** is the area in which the kidney joins the ureters.*

***Nephrons** are the functional units of the kidneys.*

*The **afferent arteriole** carries blood to the glomerulus.*

*The **glomerulus** is a high-pressure capillary bed that is surrounded by Bowman's capsule. The glomerulus is the site of filtration.*

*The **efferent arteriole** carries blood away from the glomerulus to a capillary net.*

***Bowman's capsule** is a cuplike structure that surrounds the glomerulus. The capsule receives filtered fluids from the glomerulus.*

medulla, is found beneath the cortex. A hollow chamber, the **renal pelvis,** joins the kidney with the ureter.

NEPHRONS

Approximately one million slender tubules, called the **nephrons,** are the functional units of the kidneys. Small branches from the renal artery, the **afferent arterioles,** supply the nephrons with blood. The afferent arterioles branch into a capillary bed, called the **glomerulus.** Unlike other capillaries, the glomerulus does not transfer blood to a venule. Blood leaves the glomerulus by way of another arteriole,

> **BIOLOGY CLIP**
> Pregnant women often experience anxious moments when the fetus pushes downward on top of the bladder. The reflex arc, designed to protect the bladder against overfilling, can cause the sphincter at the base of the bladder to relax if the fetus compresses the bladder.

the **efferent arteriole.** Blood is carried from the efferent arteriole to a capillary net that wraps around the kidney tubule.

The glomerulus is surrounded by a funnel-like part of the nephron, called **Bowman's capsule.** Bowman's capsule, the afferent arteriole, and the efferent arteriole are located in the cortex of the kidney. Fluids to be processed into urine enter Bowman's capsule from the blood. The capsule tapers to a thin tubule, called the **proximal tubule.** Urine is carried from the proximal tubule to the *loop of Henle,* which descends into the medulla of the kidney. Urine moves through the **distal tubule,** the last seg-

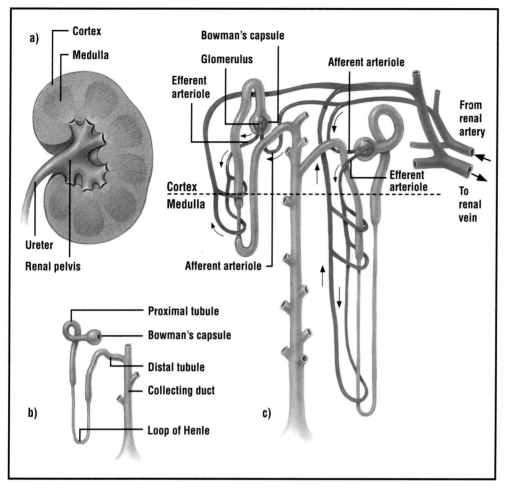

a)
- Cortex
- Medulla
- Ureter
- Renal pelvis

Bowman's capsule
Glomerulus
Efferent arteriole
Afferent arteriole
From renal artery
Cortex
Medulla
Efferent arteriole
To renal vein
Afferent arteriole

b)
- Proximal tubule
- Bowman's capsule
- Distal tubule
- Collecting duct
- Loop of Henle

c)

Figure 13.4

(a) Anatomy of the kidney showing cortex and medulla. (b) Diagram of the nephron. (c) Blood vessels associated with the kidney.

*The **proximal tubule** is a section of the nephron joining Bowman's capsule with the loop of Henle. The proximal tubule is found within the cortex of the kidney.*

*The **distal tubule** conducts urine from the loop of Henle to the collecting duct.*

*The **collecting duct** receives urine from a number of nephrons and carries urine to the pelvis.*

***Filtration** occurs during the movement of fluids from the glomerulus into Bowman's capsule.*

*A **micropipette** is a thin glass rod that can be used to extract urine from the nephron.*

ment of the nephron, and into the **collecting ducts.** As the name suggests, the collecting ducts collect urine from many nephrons, which, in turn, merge in the pelvis of the kidney.

FORMATION OF URINE

Urine formation depends on three functions: filtration, reabsorption, and secretion. Filtration is accomplished by the movement of fluids from the blood into Bowman's capsule. Reabsorption involves the transfer of essential solutes and water from the nephron back into the blood. Secretion involves the movement of materials from the blood back into the nephron.

Filtration

Each nephron has an independent blood supply. Blood moves through the afferent arteriole into the glomerulus, a high-pressure filter. Normally, pressure in a capillary bed is about 2 kPa. The pressure in the glomerulus is 8 kPa. Dissolved solutes pass through the walls of the glomerulus into Bowman's capsule. While materials move from areas of high pressure to areas of low pressure, not all materials enter the capsule. Scientists have extracted fluids from the glomerulus and Bowman's capsule using a thin glass tube called a **micropipette.** Table 13.1 compares sample solutes extracted from the glomerulus and Bowman's capsule. Which solutes are filtered? Provide a hypothesis that explains selective permeability.

Table 13.1 Comparison of Solutes

Solute	Glomerulus	Bowman's capsule
Water	yes	yes
NaCl	yes	yes
Glucose	yes	yes
Amino acids	yes	yes
Hydrogen ions	yes	yes
Plasma proteins	yes	no
Erythrocytes	yes	no
Platelets	yes	no

Plasma protein, blood cells, and platelets are too large to move through the walls of the glomerulus. Smaller molecules pass through the cell membranes and enter the nephron.

Reabsorption

The importance of reabsorption can be emphasized by examining changes in the concentrations of fluids as they move through the kidneys. On average, about 600 mL of fluid flows through the kidneys every minute. Approximately 20% of the fluid, or about 120 mL, is filtered into the nephron. Imagine what would happen if none of the filtrate was reabsorbed. You would form 120 mL of urine each minute. You would also have to consume at least 1 L of fluids every 10 min to maintain homeostasis. Much of your day would be concerned with regulating water balance. Fortunately, only 1 mL of urine is formed for every 120 mL of fluids filtered into the nephron. The remaining 119 mL of fluids and solutes are reabsorbed.

Selective reabsorption occurs by both active and passive transport. Carrier molecules move Na^+ ions across the cell membranes of the cells that line the

Figure 13.5

Carrier molecules transport Na^+ ions from the nephron to the blood.

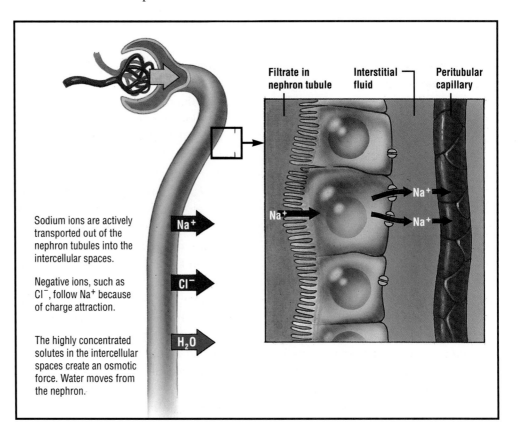

Sodium ions are actively transported out of the nephron tubules into the intercellular spaces.

Negative ions, such as Cl^-, follow Na^+ because of charge attraction.

The highly concentrated solutes in the intercellular spaces create an osmotic force. Water moves from the nephron.

Filtrate in nephron tubule

Interstitial fluid

Peritubular capillary

Na^+

Cl^-

H_2O

nephron. Negative ions, such as Cl^- and HCO_3^-, follow the positive Na^+ ions by charge attraction. Numerous mitochondria supply the energy necessary for active transport. However, the energy supply is limited. Reabsorption occurs until the **threshold level** of a substance is reached. Excess NaCl remains in the nephron and is excreted with the urine.

Other molecules are actively transported from the proximal tubule. Glucose and amino acids attach to specific carrier molecules, which shuttle them out of the nephron and into the blood. However, the amount of solute that can be reabsorbed is limited. For example, excess glucose will not be shuttled out of the nephron by the carrier molecules. This means that individuals who have diabetes mellitus, a disease characterized by high blood glucose, will lose excess glucose in their urine. Individuals who consume large amounts of simple sugars also excrete some of the excess glucose.

The solutes that are actively transported out of the nephron create an osmotic gradient that draws water from the nephron. A second osmotic force, created by the proteins not filtered into the nephron, also helps reabsorption. The proteins remain in the bloodstream and draw water from the intercellular spaces into the blood. As water is reabsorbed from the nephron, the remaining solutes become more concentrated. Molecules such as urea and uric acid will diffuse from the nephron back into the blood, although less is reabsorbed than was originally filtered.

Secretion

Secretion is the movement of wastes from the blood into the nephron. Nitrogen-containing wastes, histamine, excess H^+ ions, and other minerals are balanced by secretion. Even drugs such as penicillin can be secreted. Cells loaded with mito-

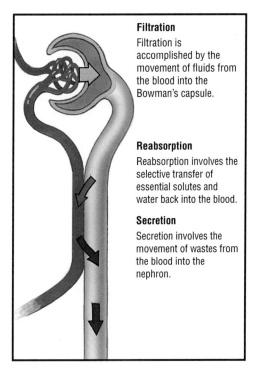

Filtration

Filtration is accomplished by the movement of fluids from the blood into the Bowman's capsule.

Reabsorption

Reabsorption involves the selective transfer of essential solutes and water back into the blood.

Secretion

Secretion involves the movement of wastes from the blood into the nephron.

Threshold level, *in terms of reabsorption, is the maximum amount of material that can be moved across the nephron.*

Figure 13.6

Overview of the steps in urine formation: filtration, reabsorption, and secretion.

chondria line the distal tubule. Like reabsorption, tubular secretion occurs by active transport, but, unlike reabsorption, molecules are shuttled from the blood into the nephron.

■ REVIEW QUESTIONS ?

1 List the three main functions of the kidneys.

2 What is deamination and why is it an important process?

3 How does the formation of urea prevent poisoning?

4 Diagram and label the following parts of the excretory system: kidney, renal artery, renal vein, ureter, bladder, and urethra. State the function of each organ.

5 Diagram and label the following parts of the nephron: Bowman's capsule, proximal tubule, loop of Henle, distal tubule, and collecting duct. State the function of each of the parts.

6 List and describe the three processes involved in urine formation.

Objective

To compare solutes along the nephron.

Procedure

Micropipettes were used to draw fluids from Bowman's capsule, the glomerulus, the loop of Henle, and the collecting duct. The data are displayed in the table below. Unfortunately, some of the data were not recorded. The absence of data is indicated on the table.

Solute	Glomerulus	Bowman's capsule	Loop of Henle	Collecting duct
Protein	8.0	0	0	0
Urea	0.05	0.05	1.50	2.00
Glucose	0.10	no data	0	0
Chloride	0.37	no data	no data	0.6
Ammonia	0.0001	0.0001	0.0001	0.04
Substance X	9.15	0	0	0

Note: Quantities are recorded in grams per 100 mL.

Case-Study Application Questions

1 Which of the solutes was not filtered into the nephron? Explain your answer.

2 Unfortunately, the test for glucose was not completed for the sample taken from the glomerulus. Predict whether or not glucose would be found in the glomerulus. Provide reasons for your prediction.

3 Why do urea and ammonia levels increase after filtration occurs?

4 Chloride ions (Cl^-) follow actively transported Na^+ ions from the nephron into the blood. Would you not expect the Cl^- concentration to decrease as fluids are extracted along the nephron? What causes the discrepancy noted? ■

WATER BALANCE

You adjust for increased water intake by increasing urine output. Conversely, you adjust for increased exercise or decreased water intake by decreasing urine output. The kidneys are involved in regulating body fluid levels. The adjustments involve the interaction of the body's two communication systems: the nervous system and the endocrine system.

Regulating ADH

A hormone called the **antidiuretic hormone (ADH)** helps regulate the osmotic pressure of body fluids by acting on the kidneys to increase water reabsorption. When ADH is released, a more concentrated urine is produced, thereby conserving body water. ADH is produced by specialized nerve cells in the brain in an area called the hypothalamus. ADH moves along specialized fibers from the hypothalamus to the pituitary gland. The pituitary stores and releases ADH into the blood.

Specialized nerve receptors, called *osmoreceptors*, located in the hypothalamus detect changes in osmotic pressure. When you decrease water intake or

*The **antidiuretic hormone (ADH)** acts on the kidneys to increase water reabsorption.*

increase water loss, by sweating for example, the solutes of the blood become more concentrated. This increases the blood's osmotic pressure. Consequently, water moves into the bloodstream, causing the cells of the hypothalamus to shrink. When this happens, a nerve message is sent to the pituitary, signalling the release of ADH, which is carried by the bloodstream to the kidneys. By reabsorbing more water, the kidneys produce a more concentrated urine, preventing the osmotic pressure of the body fluids from increasing any further.

The shrinking of the hypothalamus also initiates a behavioral response: the sensation of thirst. If more water is taken in, it is absorbed by the blood and the concentration of solutes in the blood decreases. The greater the volume of water consumed, the lower the osmotic pressure of the blood. As the blood becomes more dilute, fluids move from the blood into the hypothalamus. The hypothalamus swells, and nerve messages to the pituitary stop. Less ADH is released, and thus less water is reabsorbed from the nephrons.

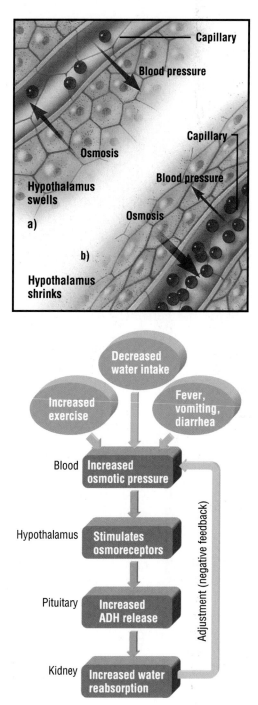

a)
Hypothalamus swells

b)
Hypothalamus shrinks

Figure 13.7

(a) Solutes are dilute. Osmotic pressure is balanced by blood pressure, and the hypothalamus will not lose water. (b) If the solute concentration increases, osmotic pressure of the blood increases, more water is drawn into the blood, and the cells of the hypothalamus shrink. ADH is released when the hypothalamus shrinks.

Figure 13.8

Physiological response to increased osmotic pressure of body fluids. High concentrations of solutes will cause the release of ADH. By increasing water reabsorption in the kidney, ADH helps conserve body water.

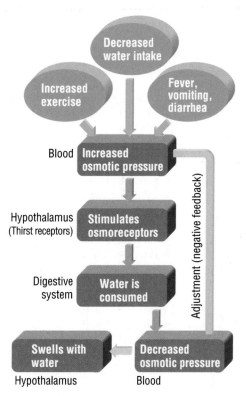

Figure 13.9

The special receptors within the hypothalamus cause the thirst response. If water is consumed, the concentration of blood solutes decreases.

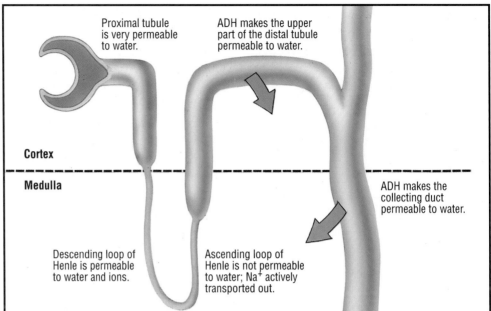

Proximal tubule is very permeable to water.

ADH makes the upper part of the distal tubule permeable to water.

Cortex

Medulla

ADH makes the collecting duct permeable to water.

Descending loop of Henle is permeable to water and ions.

Ascending loop of Henle is not permeable to water; Na^+ actively transported out.

Figure 13.10

ADH acts on the upper section of the distal tubule and collecting duct by increasing permeability to water. Regulation of water reabsorption in these two areas allows the kidney to balance the osmotic pressure of body fluids.

Aldosterone *is a hormone that increases Na^+ reabsorption from the distal tubule and collecting duct.*

ADH and the Nephron

Approximately 85% of the water filtered into the nephron is reabsorbed in the proximal tubule. Although the proximal tubule is very permeable to water, this permeability does not extend to other segments of the nephron. The descending loop of Henle is permeable to water and ions, but the ascending tubule is only permeable to NaCl. Active transport of Na^+ ions from the ascending section of the loop concentrates solutes within the medulla of the kidney. Without ADH, the rest of the tubule remains impermeable to water, but continues to actively transport Na^+ ions from the tubules. The remaining 15% of the water filtered into the nephron will be lost if no ADH is present.

ADH makes the upper part of the distal tubule and collecting duct permeable to water. When ADH makes the cell membranes permeable, the high concentration of NaCl in the intercellular spaces

> **BIOLOGY CLIP**
>
> Alcohol consumption decreases the release of ADH. Some of the symptoms experienced the day following excessive alchohol consumption can be attributed to increased water loss by urine and decreased body fluid levels.

creates an osmotic pressure that draws water from the upper section of the distal tubule and collecting duct. As water passes from the nephron to the intercellular spaces and the blood, the urine remaining in the nephron becomes more concentrated. It is important to note that the kidneys only control the last 15% of the water found in the nephron. By varying water reabsorption, the kidneys regulate the fluid volumes of the body.

KIDNEYS AND BLOOD PRESSURE

The kidneys also play a role in the regulation of blood pressure by adjusting for blood volumes. A hormone called **aldosterone** acts on the nephrons to increase Na^+ reabsorption. The hormone is produced in the *adrenal cortex*, which lies in the outer core of the adrenal gland, above the kidneys. Not surprisingly, as NaCl reabsorption increases, the osmotic gradient increases and more water moves out of the nephron by osmosis.

Conditions that lead to increased fluid loss can decrease blood pressure. Blood pressure receptors in the *juxtaglomerular apparatus* detect low blood pressure. The juxtaglomerular apparatus, as the name suggests, is found near the glomerulus. Specialized cells within the structure release *renin*, an enzyme that converts angiotensinogen into angiotensin. *Angiotensinogen* is a plasma protein produced by the liver. *Angiotensin*, the activated form, has two important functions. First, the activated enzyme causes constriction of blood vessels. Blood pressure increases when the diameter of blood vessels is reduced. Second, angiotensin initiates the release of aldosterone from the adrenal gland. The aldosterone is then carried in the blood to the kidneys, where it acts on the cells of the distal tubule and collecting duct to increase Na^+ transport. This causes the fluid level of the body to increase.

KIDNEY DISEASE

Proper functioning of the kidneys is essential if the body is to maintain homeostasis. The multifunctional kidneys are affected when other systems break down. Similarly, kidney dysfunctions have an impact on other systems. A variety of kidney disorders can be detected by urinalysis.

Diabetes Mellitus

Diabetes mellitus is caused by inadequate secretion of insulin from islet cells in the pancreas. Without insulin, blood sugar levels tend to rise. The cells of the proximal tubule are supplied with enough ATP to reabsorb 0.1% blood sugar, but in diabetes mellitus much higher blood sugar concentrations are found. The excess sugar remains in the nephron. This excess sugar provides an osmotic pressure that opposes the osmotic pressure created by other solutes that have been actively transported out of the nephron. Water remains in the nephron and is lost with the urine. Individuals with diabetes mellitus void large volumes of sweet urine. This explains why individuals with diabetes mellitus are often thirsty. The water lost with the excreted sugar must be replenished.

Diabetes Insipidus

The destruction of the ADH-producing cells of the hypothalamus or the destruction of the nerve tracts leading from the hypothalamus to the pituitary gland can cause diabetes insipidus. Without ADH to regulate water reabsorption, urine output increases dramatically. In

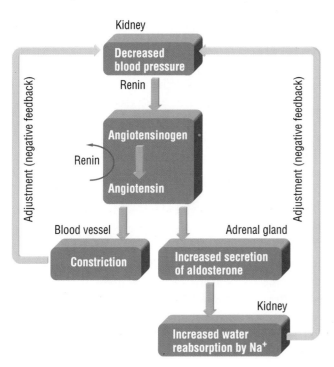

Figure 13.11

The hormone aldosterone maintains homeostasis by increasing water reabsorption.

extreme cases, as much as 20 L of urine can be produced each day, creating a strong thirst response. A person with diabetes insipidus must drink large quantities of water to replace what he or she has not been able to reabsorb.

Bright's Disease

Named after Richard Bright, a 19th-century English physician, this disease is also called *nephritis*. Nephritis is not a single disease, but a broad description of many diseases characterized by inflammation of the nephrons. One type of nephritis affects the tiny blood vessels of the glomerulus. It is believed that toxins produced by invading microbes destroy the tiny blood vessels, altering the permeability of the nephron. Proteins and other large molecules are able to pass into the nephron. Because no mechanism is designed to reabsorb protein, the proteins remain in the nephron and create an osmotic pressure that draws water into the nephron. The movement of water into the nephron increases the output of urine.

Kidney Stones

Kidney stones are caused by the precipitation of mineral solutes from the blood. Kidney stones can be categorized into two groups: alkaline and acid stones. The sharp-sided stones may become lodged in the renal pelvis or move into the narrow ureter. Delicate tissues are torn as the stone moves toward the bladder. The stone may move farther down the excretory passage and lodge in the urethra, causing excruciating pain as it moves.

> **BIOLOGY CLIP**
> Operations to remove kidney stones were performed in the time of Hippocrates, the Greek physician considered to be the father of medicine (ca 460–ca 377 B.C.).

FRONTIERS OF TECHNOLOGY: BLASTING KIDNEY STONES

The traditional treatment for kidney stones has been surgical removal, followed by a period of convalescence. A technique developed by Dr. Christian Chaussy in Germany has greatly improved prospects for kidney stone patients. The technique utilizes high-energy sound waves to smash the kidney stones. No surgery is required, and recovery time is greatly reduced. Sound waves bounce off soft tissue but penetrate the stone. The collision of waves that bounce off soft tissue with those that penetrate the stone cause the stone to explode. After a few days, tiny granules from the stone can be voided through the excretory system.

REVIEW QUESTIONS ?

7 What is ADH and where is it produced? Where is ADH stored?

8 Describe the mechanism that regulates the release of ADH.

9 Where is the thirst center located?

10 Describe the physiological adjustment to increased osmotic pressure in body fluids.

11 Describe the behavioral adjustment to increased osmotic pressure in body fluids.

12 Discuss the mechanism by which aldosterone helps maintain blood pressure.

13 What are kidney stones?

RESEARCH IN CANADA

Dr. Linda Peterson, a research scientist at the University of Ottawa, is studying the way in which the kidneys regulate water balance. The aim of Dr. Peterson's research is to investigate the interactions of hormones, such as ADH and aldosterone, and electrolytes, such as sodium, calcium, and chloride ions, in the conservation of salt by the ascending limb of the nephron.

Dr. Peterson's earlier studies have shown that ADH stimulates salt conservation by the ascending limb. One of the major questions yet to be answered is how calcium affects the ascending limb. Although a number of other studies have indicated that high blood calcium levels impair the conservation of water, the exact mechanism is still unknown. Dr. Peterson has demonstrated that high blood calcium levels increase the production of a hormone-like substance called prostaglandin E2. Since prostaglandin inhibits the effects of ADH in other segments of the nephron, high levels of the hormone may be responsible for impairing the activity of cells in the ascending limb.

A better understanding of water conservation will offer scientists a more complete picture of how humans maintain a constant internal environment.

Kidneys and Water Balance

DR. LINDA PETERSON

LABORATORY
DIAGNOSIS OF KIDNEY DISORDERS

Objective

To identify kidney disorders by urinalysis.

Materials

Combustrip
10 mL graduated cylinder
distilled water in wash bottle
4 small test tubes
4 urine samples (simulated) W, X, Y, and Z

Procedure

1 Using a graduated cylinder, measure 2 mL of urine sample W and place it in a test tube. Label the test tube W. Rinse the graduated cylinder with distilled water and repeat the procedure for samples X, Y, and Z.
   a) Why was the graduated cylinder washed?
2 Place separate Combustrips in each of the test tubes and leave for one min.
3 Compare the color bars for glucose, protein, and pH with the chart provided in the package. (Note: Different companies use slightly different charts.)

b) Record the values for each of a samples in a table similar to the one below.

Sample	Glucose	Protein	pH
W			
X			
Y			
Z			

Laboratory Application Questions

1 Which subject do you suspect has diabetes mellitus? Provide your reasons.
2 Which subject do you suspect has diabetes insipidus? Provide your reasons.
3 Which subject do you suspect has Bright's disease? Provide your reasons.
4 Which subject do you suspect has lost a tremendous amount of body water while exercising? Provide your reasons. ■

CAREER INVESTIGATION CAREER

Our aging population will require a larger and more complete health care system than is in place today. The emphasis in the field is expected to shift from the treatment of diseases to the prevention of disease and the improvement in the quality of life for the elderly. In the long run, it will be more efficient and less costly to prevent problems than to cure them with drugs and expensive surgery.

Geriatric specialist
Geriatrics is the branch of medicine that deals with health problems related to old age. The population of Canada is expected to age dramatically over the next few decades. A greater percentage of the population will be retired, and will require specialized health care.

Dietician/Nutritionist
Our society has never been more aware that a healthy diet is a major part of a healthy lifestyle. Dieticians and nutritionists are required by hospitals, retirement homes, and food manufacturers.

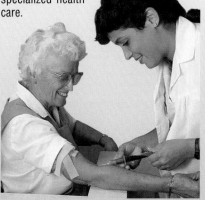

Home health aide
More people are choosing to recover from injuries and illnesses in their own home rather than in a hospital. Not only does this save money, but the familiar and comfortable surroundings of a person's home also speed recovery. This practice should lead to a high demand for home health-care aides. A home health aide requires a general understanding of the human body and a knowledge of emergency procedures such as cardiopulmonary resuscitation (CPR).

- Identify a career that requires a knowledge of the exchange of energy and matter in humans.
- Investigate and list the features that appeal to you about this career. Make another list of features that you find less attractive about this career.
- Which high-school subjects are required for this career? Is a postsecondary degree required?
- Survey the newspapers in your area for job opportunities in this career.

Physiotherapist
Physiotherapists help people who have been in accidents or who have degenerative diseases regain their health through treatments such as ultrasound, laser therapy, and massage. Physiotherapy allows people to lead more active lives during their recovery period and decreases the amount of time needed for rehabilitation. A physiotherapist must have a thorough understanding of how the body works.

SOCIAL ISSUE:
Fillings and Kidney Disease

*Experiments by Drs. Murray Vimy and Fritz Lorscheider at the University of Calgary Medical School point out the potential risks of using mercury fillings for teeth. Studies carried out on sheep indicated that half of their kidneys became dysfunctional within 30 days after 12 mercury amalgam fillings were placed in their mouths. The results of this study were published in the prestigious journal, **The Physiologist,** in August 1990, and presented to the American Physiology Society in October of that year. The two researchers believe that mercury in the amalgam filling can leak into the bloodstream, where it can do harm to the kidneys.*

Statement:

Alloy fillings that contain mercury should be banned, pending further research.

Point

- Mercury has long been perceived as dangerous when used in dental fillings. Dentists should discontinue its use until conclusive evidence proves otherwise.

- Dentists can use alternative substances for fillings, such as gold, ceramic, and different kinds of resins.

Counterpoint

- Not all researchers agree with the interpretations of Drs. Vimy and Lorscheider. Their studies were not carried out on humans. Amalgam fillings have been used for many years on humans, the vast majority of whom have well-functioning kidneys.
- How do we know that the replacements are less harmful? Dr. Bill Long, past president of the Calgary and District Dental Society, has stated that the amalgam fillings are safe and that replacement materials are not as effective.

Research the issue.
Reflect on your findings.
Discuss the various viewpoints with others.
Prepare for the class debate.

CHAPTER HIGHLIGHTS

- The liver processes wastes, making them soluble.
- The excretory system rids the body of wastes carried in the blood.
- The kidneys filter wastes from the blood, help regulate blood pH, and regulate water balance.
- The nephron is the functional unit of a kidney.
- Urine formation involves filtration, reabsorption, and secretion.

- ADH regulates water balance by controlling absorption of the remaining 15% of the water filtered into the nephron. ADH acts on the cells of the distal tubule and collecting duct.
- Aldosterone regulates the reabsorption of sodium in the distal tubule and collecting duct.
- A number of diseases affect proper kidney function including diabetes mellitus and Bright's disease.

APPLYING THE CONCEPTS

1 Why is the formation of urea by the liver especially important for land animals?

2 Predict how a drop in blood pressure would affect urine output. Give reasons for your prediction.

3 A drug causes dilation of the afferent arteriole and constriction of the efferent arteriole. Indicate how the drug will affect urine production.

4 Why do the walls of the proximal tubule contain so many mitochondria?

5 In an experiment, the pituitary gland of a dog is removed. Predict how the removal of the pituitary gland will affect the dog's regulation of water balance.

6 How does excessive salt intake affect the release of ADH from the pituitary gland?

7 A drug that inhibits the formation of ATP by the cells of the proximal tubule is introduced into the nephron. How will the drug affect urine formation? Provide a complete physiological explanation.

8 A blood clot lodges in the renal artery and restricts blood flow to the kidney. Explain why this condition leads to high blood pressure.

9 For every 100 mL of salt water consumed, 150 mL of body water is lost. The solute concentration found in seawater is greater than that found in the blood. Provide a physiological explanation to account for the loss of body water. (Hint: Consider the threshold level for salt reabsorption by the cells of the nephron.)

10 Explain why the presence of proteins in the urine can lead to tissue swelling, or *edema*.

CRITICAL-THINKING QUESTIONS

1 Alcohol is a diuretic, a substance that increases the production of urine. It also suppresses the production and release of ADH. Should individuals who are prone to developing kidney stones consume alcohol? Explain.

2 In an experiment, four subjects each consumed four cups of black coffee per hour for a two-hour duration. A control group each consumed four cups of water per hour over the same two hours. It was noted that the group consuming coffee had a greater urine output than the control group. Provide a hypothesis that accounts for the data provided. Note: Formulate your hypothesis with an "If ... then ..." statement.

3 Design an experiment that would test the hypothesis developed in question 2.

4 Diseases such as syphilis, a venereal disease, can cause glomerular nephritis. (Other factors can also give rise to this disorder.) Glomerular nephritis is associated with the destruction of the high-pressure capillaries that regulate filtration. This occurs when proteins and other large molecules enter the filtrate. Once the filter is destroyed, it cannot be repaired. When a significant number of nephrons are destroyed, kidney function fails. Patients who suffer kidney failure can be treated by artificial dialysis or kidney transplants, two extremely expensive techniques. Should everyone who needs a kidney transplant be given the same priority?

ENRICHMENT ACTIVITIES

1 Research how kidney machines remove toxins from the blood.

2 The following article will provide you with more information on fillings:
- Wilkerfield, Irene, and Hawley Truax. "The Fear of Fillings." *Environmental Action*, Nov/Dec 1987, p. 10.

3 Using dialysis tubing, construct a working model of an artificial kidney machine. Does your kidney machine use force filtration?

Coordination and Regulation in Humans

∎

FOURTEEN
The Endocrine System and Homeostasis

FIFTEEN
The Nervous System and Homeostasis

SIXTEEN
Special Senses

U N I T

4

*T*he Endocrine System and Homeostasis

IMPORTANCE OF THE ENDOCRINE SYSTEM

The trillions of cells of the body all interact with each other—no cell operates in isolation. The integration of body functions depends on chemical controls. **Hormones** are chemical regulators produced by cells in one part of the body that affect cells in another part of the body. Chemicals produced in glands and secreted directly into the blood are referred to as **endocrine hormones.** The circulatory system carries these hormones to the various organs of the body.

Hormones can be classified according to their activation site. Hormones such as growth hormone, which regulates the development of the long bones; insulin, which regulates blood sugar; and epinephrine (adrenaline), which is produced in times of stress, affect many cells throughout the body. These hormones are referred to as *nontarget hormones.* Hormones such as parathyroid hormone, which regulates calcium levels in the body, and gastrin, which stimulates cells of the stomach to produce digestive enzymes, affect specific cells, or **target tissues.**

Hormones *are chemicals released by cells that affect cells in other parts of the body. Only a small amount of a hormone is required to alter cell metabolism.*

Endocrine hormones *are chemicals secreted by glands directly into the blood.*

Target tissues *have specific receptor sites that bind with the hormones.*

Figure 14.1

A nurse teaching a child with diabetes mellitus to inject herself with insulin.

CHEMICAL CONTROL SYSTEMS

Along with the nervous system, the endocrine system provides integration and control of the various organs and tissues. The malfunction of one organ affects other organs. However, an animal can continue to function because of compensations made by the two control systems. The nervous system enables the body to adjust quickly to changes in the environment. The endocrine system is designed to maintain control over a longer duration. Growth hormone and the various sex hormones, for example, regulate and sustain development for many years.

The division between the nervous system and endocrine system is most obscure in an area of the brain called the *hypothalamus*. The hypothalamus regulates the pituitary gland, referred to as the master gland of the body, through nerve stimulation. However, the endocrine glands, stimulated by the pituitary, secrete chemicals that affect the nerve activity of the hypothalamus.

The name hormone comes from the Greek *hormon*, meaning to excite or set into motion. Hormones do not serve the body as chemical products, but rather as regulators, speeding up or slowing down certain bodily processes. Long before scientists began to study the clinical effects of hormones, farmers knew that castrated bulls (steers) produced better meat. Today we know that chemicals within the testes of the bull are associated with the animal's aggressive nature and with tougher meat. Steers have greater value because of their docility and increased body mass.

This relationship between chemical messengers and the activity of organ systems within the body was established by experiment. In 1899, Joseph von Merring and Oscar Minkowski showed that a chemical messenger produced in the pancreas was responsible for the regulation of blood sugar. After removing the pancreases from a number of dogs, the two scientists noticed that the animals began to lose weight very quickly. Within a few hours the dogs became fatigued and displayed some of the symptoms that are now associated with human sugar diabetes. By chance, the two scientists also noted that ants began gathering in the kennels where the sick dogs were kept. No ants, however, were found in the kennels of healthy dogs. What caused the ants to gather? The scientists analyzed the urine of the sick dogs and found that it contained glucose, a sugar, while the urine of the healthy dogs did not. The ants had been attracted to the sugar. The experiment provided evidence that a chemical messenger, produced by the pancreas, was responsible for the regulation of blood sugar. This chemical is the hormone called **insulin.**

The von Merring and Minkowski experiment typified classical approaches to uncovering the effect of specific hormones. In many cases a gland or organ was removed and the effects on the organism were monitored. Once the changes in behavior were noted, chemical extracts from the organ were often injected into the animal, and the animal's activities monitored. By varying dosages of the identified chemical messenger, scientists hoped to determine how it worked. Although effective to a certain degree, classical techniques were limited. No hormonal response works independently. The concentrations of other hormones often increase to help compensate for a disorder. Some glands produce many different hormones. Therefore, the effect cannot be attributed to a single hormone. For example, early experimenters who attempted to uncover the function of the thyroid gland unwittingly

Insulin *is a hormone produced by the islets of Langerhans in the pancreas. Insulin is secreted when blood sugar levels are high.*

removed the parathyroid gland along with the thyroid. It might be expected that these tiny glands are part of the thyroid and are related to its function. However, although the parathyroid is embedded in the tissue of the thyroid, it is a separate organ. What the scientists failed to realize is that many illnesses they associated with low thyroid secretions were actually created by the parathyroid.

The main problem for early researchers was obtaining and isolating the actual messenger among the other chemicals found within the removed organ. Most hormones are found in very small amounts. Furthermore, the concentration of hormone varies throughout the day. The prediction of site and time was often a matter of mere luck.

In the past few years, technological improvements in chemical analysis techniques and microscopy have vastly increased our knowledge of the endocrine system. Radioactive tracers enable scientists to follow messenger chemicals from the organ in which they are produced to the target cells. The radioactive tracers also allow researchers to discern how the chemical messenger is broken down into other compounds and removed as waste. With new chemical analysis equipment, scientists can determine and measure the concentration of even the smallest amounts of a hormone as the body responds to changes in the external environment. In addition, high-power microscopes provide a clearer picture of the structure of cell membranes and along with it a better understanding of how chemical messengers attach themselves to target sites.

Steroid hormones are made from cholesterol. This group includes male and female sex hormones and cortisol.

Protein hormones are composed of chains of amino acids. This group includes insulin, growth hormone, and epinephrine.

Cyclic AMP is a secondary chemical messenger that directs the synthesis of protein hormones by ribosomes.

BIOLOGY CLIP

Protein hormones that do not attach to receptor molecules on target cells are removed from the body by the liver or kidney. The presence of these hormones can be monitored by urinalysis.

CHEMICAL SIGNALS

How do hormones signal cells? First, it is important to note that hormones do not affect all cells. Cells may have receptors for one hormone but not another. The number of receptors found on individual cells may also vary. For example, liver cells and muscle cells have many receptor sites for the hormone insulin. Fewer receptor sites are found in less active cells such as bone cells and cartilage cells.

Second, there are two different types of hormones that differ in chemical structure and action. **Steroid hormones,** which include both male and female sex hormones and cortisol, are made from cholesterol, a lipid compound. Steroid molecules are composed of complex rings of carbon, hydrogen, and oxygen molecules. Steroid molecules are not soluble in water, but are soluble in fat. The second group, **protein,** or **protein-related hormones,** includes insulin, growth hormone, and epinephrine. These hormones contain chains of amino acids of varying length and are soluble in water.

Steroid hormones diffuse from the capillaries into the interstitial fluid, and then into the target cells, where they combine with receptor molecules located in the cytoplasm. The hormone-receptor complex then moves into the nucleus and attaches to a segment of chromatin that has a complementary shape. The hormone activates a gene that sends a message to the ribosomes in the cytoplasm to begin producing a specific protein.

Protein hormones exhibit a different action. Unlike steroid hormones, which diffuse into the cell, protein hormones combine with receptors on the cell mem-

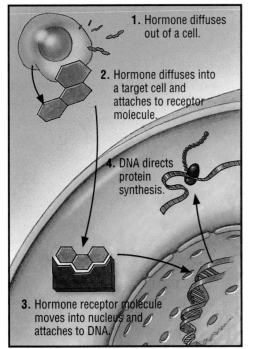

Figure 14.2

The steroid hormone molecule passes into the cell, combines with a receptor molecule, and then activates a gene in the nucleus. The gene directs the production of a specific protein.

1. Hormone diffuses out of a cell.

2. Hormone diffuses into a target cell and attaches to receptor molecule.

4. DNA directs protein synthesis.

3. Hormone receptor molecule moves into nucleus and attaches to DNA.

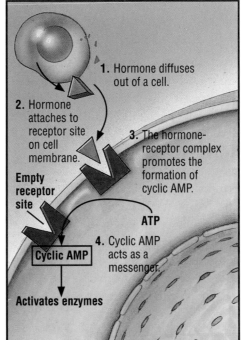

1. Hormone diffuses out of a cell.

2. Hormone attaches to receptor site on cell membrane.

3. The hormone-receptor complex promotes the formation of cyclic AMP.

Empty receptor site

ATP

Cyclic AMP

4. Cyclic AMP acts as a messenger.

Activates enzymes

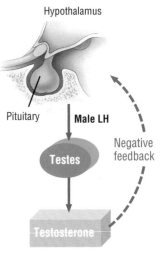

Figure 14.3

The protein hormone combines with specific receptor sites and triggers the formation of cyclic AMP from ATP. Cyclic AMP acts as a secondary messenger, activating enzymes within the cell.

Hypothalamus

Pituitary   **Male LH**

Testes   Negative feedback

Testosterone

Figure 14.4

High levels of testosterone inhibit the release of LH from the pituitary.

A **negative-feedback** *system is a control system designed to prevent chemical imbalances in the body. The body responds to changes in the external or internal environment. Once the effect is detected, receptors are activated and the response is inhibited, thereby maintaining homeostasis.*

brane. Specific hormones combine at specific receptor sites. The hormone-receptor complex activates the production of an enzyme called *adenyl catalase*. The adenyl catalase causes the cell to convert ATP (adenosine triphosphate), the primary source of cell energy, into **cyclic AMP** (adenosine monophosphate). The cyclic AMP functions as a messenger, activating enzymes in the cytoplasm to carry out their normal functions. For example, when thyroid-stimulating hormone (TSH) attaches to the receptor sites in the thyroid gland, cyclic AMP is produced in thyroid cells. Cells of the kidneys and muscles are not affected because they have no receptors for thyroid-stimulating hormone. The cyclic AMP in the thyroid cell activates enzymes, which begin producing thyroxine, a hormone that regulates metabolism. You will learn more about thyroxine later in the chapter.

NEGATIVE FEEDBACK

Hormone production must be regulated. Once the hormone produces the desired effect, hormone production must be decreased. Consider, for example, testosterone, the male sex hormone, which is responsible for the development of secondary male characteristics. The development of facial hair, sex drive, and lowering of the voice are all associated with the production of the hormone. The hormone itself is regulated by a hormone from the pituitary gland, called the male luteinizing hormone (LH), which activates the testosterone-producing cells of the male testes. Once LH (the male hormone) is produced, testosterone secretions begin. However, once testosterone reaches acceptable levels, secretions must be turned off. Testosterone exerts a negative effect on male LH: high levels of testosterone inhibit the release of male LH. This feedback control system is referred to as **negative feedback.**

The regulation of other hormones such as growth hormone and epinephrine (adrenaline) are equally important. Gigantism results when the production of growth hormone fails to turn off after adequate levels have been reached. Unregulated epinephrine production is equally dangerous. Epinephrine enables the body to respond to stress situations. The hormone causes heart and breathing rates to accelerate and blood sugar levels to rise, among other responses. All actions are designed to allow the body to respond to stress in what has come to be known as the "flight-or-fight" response. However, once the stressful situation is gone, the body returns to normal resting levels. Once again, negative-feedback action is required to restore homeostasis. Throughout this chapter you will study other specific examples of negative feedback.

produced by the hypothalamus. Antidiuretic hormone acts on the kidneys and helps regulate body water. Oxytocin initiates strong uterine contractions during labor. The hormones travel by way of specialized nerve cells from the hypothalamus to the pituitary. The pituitary gland stores the hormones, releasing them into the blood when necessary.

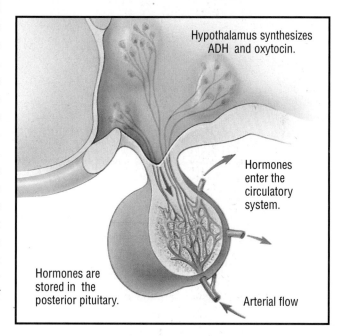

Hypothalamus synthesizes ADH and oxytocin.

Hormones enter the circulatory system.

Hormones are stored in the posterior pituitary.

Arterial flow

Figure 14.5

Hormones secreted by nerve cells of the hypothalamus are stored in the posterior pituitary.

THE PITUITARY: THE MASTER GLAND

The pituitary gland is often referred to as the "master gland" because it exercises control over other endocrine glands. This small saclike structure is connected by a stalk to the hypothalamus, the area of the brain associated with homeostasis. The interaction between the nervous system and endocrine system is evident in this hypothalamus-pituitary complex.

The pituitary gland is actually composed of two separate lobes: the posterior lobe and the anterior lobe. The posterior lobe of the pituitary stores and releases hormones such as antidiuretic hormone (ADH) and oxytocin, which have been

The anterior lobe of the pituitary, unlike the posterior lobe, produces its own hormones. However, like the posterior lobe, the anterior lobe of the pituitary is richly supplied with nerves from the hypothalamus. The hypothalamus regulates the release of hormones from the anterior pituitary. Hormones are secreted from the nerve ends of the cells of the hypothalamus and transported in the blood to the pituitary gland. Most of the hormones target specific cells in the pituitary, causing the release of pituitary hormones, which are then carried by the blood to target tissues. Two hypothalamus-releasing factors inhibit the release of hormones from the anterior lobe of the pituitary. The releasing factor dopamine inhibits the secretion of prolactin, a pituitary hormone that stimulates milk production in pregnant women. The hormone somatostatin inhibits the secretion of somatotropin, a pituitary hormone associated with growth of the long bones.

A number of different regulator hormones are stored in the anterior lobe of the pituitary gland. Thyroid-stimulating hormones (TSH), as the name implies, stimulate the thyroid gland to produce its hormone thyroxine. The anterior pituitary also releases reproductive-stimulating hormones, growth-stimulating hormones, prolactin, as well as the hormone that stimulates the adrenal cortex. Table 14.1 summarizes the hormones released from the pituitary gland.

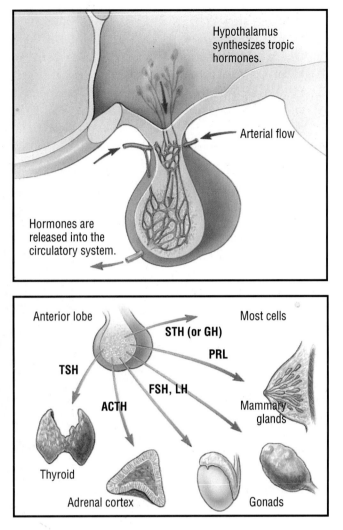

Figure 14.6

Releasing hormones secreted by nerve cells of the hypothalamus regulate hormones secreted by the anterior pituitary.

Figure 14.7

Regulator hormones of the anterior lobe of the pituitary and their target organs.

Table 14.1 Pituitary Hormones

Hormone	Target	Primary function
Anterior lobe		
Thyroid-stimulating hormone (TSH)	Thyroid gland	Stimulates release of thyroxine from thyroid. Thyroxine regulates cell metabolism.
Corticotropin adrenal steroid (ACTH)	Adrenal cortex	Stimulates the release of hormones involved in stress responses.
Somatotropin (STH), or growth hormone (GH)	Most cells	Promotes growth.
Follicle-stimulating hormone (FSH)	Ovaries, testes	In females, stimulates follicle development in ovaries. In males, promotes the development of sperm cells in testes.
Luteinizing hormone (LH)	Ovaries, testes	In females, stimulates ovulation and formation of the corpus luteum. In males, stimulates the production of the male sex hormone, testosterone.
Prolactin (PRL)	Mammary glands	Stimulates and maintains milk production in females.
Posterior lobe		
Oxytocin	Uterus	Initiates strong contractions.
	Mammary glands	Triggers milk production.
Antidiuretic hormone (ADH)	Kidney	Increases water reabsorption by kidney.

Growth hormone *is produced by the cells of the anterior pituitary. Prior to puberty, the hormone promotes growth of the long bones.*

The adrenal medulla *is found at the core of the adrenal gland. The adrenal medulla produces epinephrine and nor-epinephrine.*

Figure 14.8
The effects of low and high secretions of growth hormone are apparent in people who have the conditions called gigantism and dwarfism.

REVIEW QUESTIONS ?

1 Define the term hormone.
2 What are target tissues or organs?
3 How are the nervous system and endocrine system specialized to maintain homeostasis?
4 Why is the pituitary called the "master gland"?
5 Describe the signalling action of steroid hormones and protein hormones.
6 What is negative feedback?

GROWTH HORMONE

The effects of **growth hormone,** or somatotropin, are most evident when the body produces too much or too little of it. Low secretion of growth hormone during childhood can result in dwarfism; high secretions during childhood can result in gigantism. Although growth hormone affects most of the cells of the body, the effect is most pronounced on cartilage cells and bone cells. If the production of growth hormone continues after the cartilaginous growth plates have been fused,

other bones respond. Once the growth plates have fused, the long bones can no longer increase in length, but bones of the jaw, forehead, fingers, and toes increase in width. The disorder, referred to as *acromegaly*, causes a broadening of facial features.

Figure 14.9
Acromegaly is caused by high secretion of growth hormone during adulthood. Note the progressive widening of bones in the face.

ADRENAL GLANDS

The adrenal glands are located above each kidney. (The word adrenal comes from the Latin *ad*, meaning "to" or "at," and *renes*, meaning "kidneys.") Each adrenal gland is made up of two glands encased in one shell. The inner gland, the adrenal medulla, is surrounded by an outer casing called the adrenal cortex. The medulla is regulated by the nervous system, while the adrenal cortex is regulated by hormones.

The **adrenal medulla** produces two hormones: epinephrine (adrenaline) and norepinephrine (noradrenaline). The link between the nervous system and the adrenal

Unit Four:
Coordination and Regulation in Humans

medulla lies in the fact that both produce epinephrine. The hormone-producing cells within the adrenal gland are stimulated by sympathetic nerves in times of stress.

In a stress situation, **epinephrine** and **norepinephrine** are released from the adrenal medulla into the blood. Under their influence, the blood sugar level rises. Glycogen, a carbohydrate storage compound in the liver and muscles, is converted into glucose, a readily usable form of energy. The increased blood sugar level ensures that a greater energy reserve will be available for the tissues of the body. The hormones also increase heart rate, breathing rate, and cell metabolism. Blood vessels dilate, allowing more oxygen and nutrients to reach the tissues. Even the iris of the eye dilates, allowing more light to enter the retina—in a stress situation, the body attempts to get as much visual information as possible.

The **adrenal cortex** produces three different types of hormones: the glucocorticoids, the mineralocorticoids, and small amounts of sex hormones. The glucocorticoids are associated with blood glucose levels. One of the most important of the glucocorticoids, **cortisol,** increases the level of amino acids in the blood in an attempt to help the body recover from stress. The amino acids are converted into glucose by the liver, thereby raising the level of blood sugar. Increased glucose levels provide a greater energy source, which helps cell recovery. Any of the amino acids not converted into glucose are available for pro-

tein synthesis. The proteins can be used to repair damaged cells.

Negative-feedback control of cortisol is shown in Figure 14.10. Stressful situations are identified by the brain. The hypothalamus sends a releasing hormone to the anterior lobe of the pituitary, stimulating the pituitary to secrete corticotropin, or, as it is also called, adrenocorticotropin hormone (ACTH). The ACTH is carried by the blood to the target cells in the adrenal cortex. Under the influence of ACTH, the cells of the adrenal cortex secrete cortisol, which is carried to target cells in the liver and muscles. As cortisol levels rise, cells within the hypothalamus and pituitary decrease the production of regulatory hormones, and eventually the levels of cortisol begin to fall.

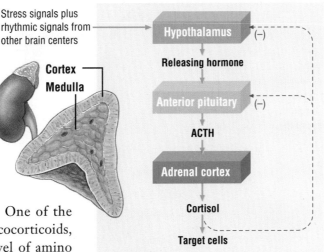

Stress signals plus rhythmic signals from other brain centers

Cortex
Medulla

Hypothalamus (−)

Releasing hormone

Anterior pituitary (−)

ACTH

Adrenal cortex

Cortisol

Target cells

Aldosterone is the most important of the mineralocorticoids, the second major group of hormones produced by the adrenal cortex. Secretions of aldosterone increase sodium retention and water reabsorption by the kidney and thereby help maintain body fluid levels.

Epinephrine *is a hormone produced by the adrenal medulla that initiates the flight-or-fight response.*

Norepinephrine *is a hormone produced by the adrenal medulla that initiates the flight-or-fight response.*

The **adrenal cortex** *is the outer region of the adrenal gland. It produces glucocorticoids and mineralocorticoids.*

Cortisol *is a hormone that stimulates the conversion of amino acids to glucose by the liver.*

Aldosterone *is a hormone produced by the adrenal cortex. It helps regulate water balance in the body by increasing sodium and water reabsorption by the kidneys.*

Figure 14.10

Negative feedback of cortisol.

INSULIN AND THE REGULATION OF BLOOD SUGAR

The pancreas contains two types of cells: one type produces digestive enzymes, the other type produces hormones. The hormone-producing cells are located in structures called the *islets of Langerhans*, named after their discoverer, Paul Langerhans. Over 2000 tiny islets scattered throughout the pancreas are responsible for the production of two hormones: insulin and glucagon.

Insulin is produced in the beta cells of the islets of Langerhans and is released when the blood sugar level is high. Insulin increases glucose utilization by making many cells of the body permeable to glucose. After a meal, when the blood sugar level rises, insulin is released. This insulin causes cells of the liver and muscles, among other organs, to become permeable to the glucose. In the liver, the glucose is converted into glycogen, the primary storage form for glucose. By storing excess glucose in the form of glycogen, the insulin enables the blood sugar level to return to normal. Insulin helps maintain homeostasis.

Glucagon and insulin work in complementary fashion. Insulin causes a decrease in blood sugar level, and glucagon causes an increase in blood sugar level. Produced by the alpha cells of the islets of Langerhans, glucagon is released when blood sugar levels are low. After periods of fasting, when blood sugar dips below normal levels, glucagon is released. Glucagon promotes the conversion of glycogen to glucose, which is absorbed by the blood. As glycogen is converted to glucose in the liver, the blood sugar level returns to normal.

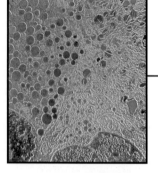

Figure 14.11

Microscopic view of the islets of Langerhans.

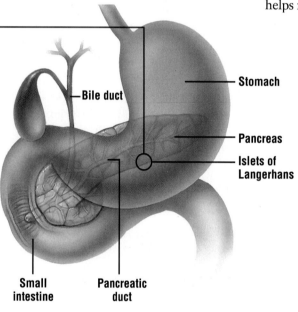

Bile duct

Stomach

Pancreas

Islets of Langerhans

Small intestine

Pancreatic duct

Figure 14.12

Insulin is released when blood sugar levels are high. Insulin increases the permeability of cells to glucose. Glucose is converted into glycogen within the liver, thereby restoring blood sugar levels. Glucagon is released when blood sugar levels are low. Glucagon promotes the conversion of liver glycogen into glucose, thereby restoring blood sugar levels.

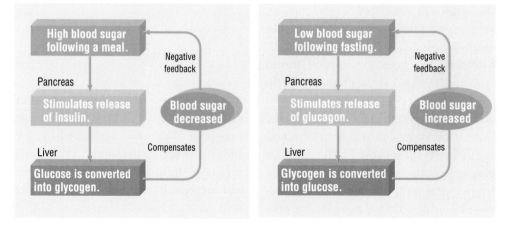

High blood sugar following a meal.

Negative feedback

Pancreas

Stimulates release of insulin.

Blood sugar decreased

Liver

Compensates

Glucose is converted into glycogen.

Low blood sugar following fasting.

Negative feedback

Pancreas

Stimulates release of glucagon.

Blood sugar increased

Liver

Compensates

Glycogen is converted into glucose.

SUGAR DIABETES

Diabetes mellitus is a genetic disorder associated with inadequate production of insulin. Although the exact cause of diabetes is not known, it is generally believed to occur when insulin-producing cells within the islets of Langerhans deteriorate. Without adequate levels of insulin, blood sugar levels tend to rise very sharply following meals. This condition is known as *hyperglycemia*, or high blood sugar (from *hyper*, meaning "too much," *glyco*, meaning "sugar," and *emia* referring to a condition of the blood).

A variety of symptoms are associated with high blood sugar. Because the kidney is unable to reabsorb all of the blood glucose that is filtered through it, glucose appears in the urine. The appearance of sugar in the urine accounts for the name *diabetes mellitus*, which literally means "going through honey-sweet." The loss of glucose in the urine creates yet another symptom. Since the excretion of glucose draws water from the body by osmosis, diabetics excrete unusually large volumes of urine and are often thirsty.

Diabetics also experience low energy levels. Remember, insulin is required for the cells of the body to become permeable to glucose. Despite the abundance of glucose in the blood, little is able to move into the cells of the body. Cells of diabetics soon become starved for glucose and must turn to other sources of energy. Fats and proteins can be metabolized for energy, but, unlike carbohydrates, they are not an easily accessible energy source. The switch to these other sources of energy creates a host of problems for diabetics. Acetone, an interme-

diary product of fat metabolism can change blood pH. In severe cases, doctors are able to smell acetone on the breath of people with diabetes.

There are at least two different types of diabetes mellitus: juvenile, or early-onset diabetes, and adult, or maturity-onset, diabetes. Juvenile diabetes is caused by the early degeneration of the beta cells in the islets of Langerhans and can only be treated by insulin injections. Maturity-onset diabetes is associated with decreased insulin production. Insulin-producing cells do not disappear as they do in juvenile diabetes, but become less effective. People with maturity-onset diabetes can be helped by oral drugs known as sulfonamides. It is believed that these drugs, which are not effective for juvenile diabetes, stimulate the residual function of the islets of Langerhans in older people.

> **BIOLOGY CLIP**
> Not all of the body's cells depend on insulin. Nerve cells and blood cells are able to absorb glucose without insulin. Muscle cells, which make up nearly 50% of your body mass, depend on insulin.

Glucagon *is a hormone produced by the pancreas. When blood sugar levels are low, glucagon promotes the conversion of glycogen to glucose.*

Diabetes mellitus *is a genetic disorder characterized by high blood sugar levels.*

■ REVIEW QUESTIONS ▨ ?

7 How would decreased secretions of growth hormone affect an individual?

8 How would an increased secretion of growth hormone affect an individual after puberty?

9 Name two regions of the adrenal gland, and list two hormones produced in each area.

10 How would high levels of ACTH affect secretions of cortisol from the adrenal glands? How would high levels of cortisol affect ACTH?

11 Where is insulin produced?

12 How does insulin regulate blood sugar levels?

13 How does glucagon regulate blood sugar levels?

RESEARCH IN CANADA

Frederick Banting (1891–1941) served in World War I as a doctor. On returning from the war, he became interested in diabetes research. At the time, diabetes was thought to be caused by a deficiency of a hormone located in specialized cells of the pancreas. However, extracting the hormone from the pancreas presented a problem since the pancreas also stores digestion enzymes capable of breaking down the protein hormone.

The Discovery of Insulin

In 1921, Banting approached John J.R. MacLeod, a professor at the University of Toronto, with his idea for isolating the hormone. MacLeod assigned him a makeshift laboratory as well as an assistant, Charles Best, who was a graduate student in biochemistry. Banting and Best tied the pancreatic duct of experimental dogs, and waited seven weeks for the pancreas to shrivel. Although the cells producing digestive enzymes had deteriorated, cells from the islets of Langerhans remained. The hormone was then extracted from the pancreas. When the hormone was injected into dogs who had had their pancreases removed, symptoms of diabetes ceased. Banting and Best wanted to call the hormone "isletin," but MacLeod insisted that it be called "insulin." In 1923 Banting and MacLeod were awarded the Nobel Prize for physiology and medicine. Banting was furious. Charles Best, his co-worker had not been included and MacLeod, the professor who had contributed laboratory space, had been included.

DR. FREDERICK BANTING
DR. CHARLES BEST

LABORATORY

IDENTIFICATION OF HYPERGLYCEMIA

Objective

To use urinalysis techniques to identify diabetes mellitus.

Materials

4 samples of simulated urine (labelled A, B, C, D)

hot plate	wax pen
400 mL beaker	beaker tongs
Clinitest tablets	forceps
Benedict's solution	medicine dropper
pipettes	distilled water
10 mL graduated cylinder	test-tube clamp
test-tube rack	goggles
lab apron	

> CAUTION: Benedict's solution is toxic and an irritant. Avoid skin and eye contact. Wash all splashes off your skin and clothing thoroughly. If you get any chemical in your eyes, rinse for at least 15 min and inform your teacher.

Procedure

Part I: Benedict's Test

Benedict's solution identifies reducing sugars. Cupric ions in the solution combine with sugars to form cuprous oxides, which produce color changes. See the table below for color changes.

Benedict Test

Color of solution	Glucose concentration
Blue	0.0%
Light green	0.15% – 0.5%
Olive green	0.5% – 01.0%
Yellow-green to yellow	1.0% – 1.5%
Orange	1.5% – 2.0%
Red to red-brown	2.0%

1 Label the four test tubes A, B, C, and D. Using a 10 mL graduated cylinder, measure 5 mL of Benedict's solution into each test tube.

2 Using a medicine dropper, add 10 drops of urine from sample A to test tube A. Rinse the medicine dropper and repeat for samples B, C, and D.
   a) Why must the medicine dropper be rinsed?

3 Fill a 400 mL beaker with approximately 300 mL of tap water. Using beaker tongs, position the beaker on a hot plate. The beaker will be used as a hot-water bath. Using the test-tube clamp, place the test tubes in the hot-water bath for 5 min.

4 Using the test tube clamp, remove the samples and record the final colors of the solutions.
   b) Provide your data.

Part II: Clinitest Tablet Method

The reducing sugar in the urine will react with copper sulfate to reduce cupric ions to cupric oxide. The chemical reaction is indicated by a color change. The table below provides quantitative results.

Color of solution	Glucose concentration
Blue	0.0%
Green	0.25% – 0.5%
Green to green-brown	0.5% – 1.0%
Orange	2.0% – greater

5 Clean the four test tubes and place 10 drops of distilled water into each of the test tubes.

6 Add 5 drops of urine to each of the appropriately labelled test tubes. Place each of the test tubes in a test-tube rack.

7 Using forceps, place one Clinitest tablet in each of the test tubes. Observe the color.
   c) Provide your data.

Laboratory Application Questions

1 Why is insulin not taken orally?

2 Explain why diabetics experience the following symptoms: low energy levels, large volumes of urine, the presence of acetone on the breath, and acidosis (blood pH becomes acidic).

3 Why might the injection of too much insulin be harmful?

4 How would you help someone who had taken too much insulin? Explain your answer. ■

FRONTIERS OF TECHNOLOGY: ISLET TRANSPLANTS

Juvenile diabetes is the second leading cause of blindness in Canada. Other side effects of the disease, such as kidney and heart failure, stroke, and peripheral nerve damage, affect over 50 000 Canadians.

Why do doctors not just replace defective cells from the islets of Langerhans with ones that are working properly? In theory, islet transplants sound rather straightforward, but nothing could be further from the truth. A number of important questions must be addressed. First, how can cells be removed from the donor? Organs can be fixed into place, but what about cells? Can the islets develop a new blood supply? As in all living cells, nutrients are required and wastes must be eliminated. Will the transplanted islet cells be rejected by the recipient? Will the transplanted islet cells actually produce insulin?

Researchers around the world have been busily searching for answers to these questions. Unlike insulin therapy, islet transplants have the potential to reverse the effects of diabetes. Although insulin injection provides some regulation of blood sugar, it will not necessarily prevent many of the serious complications, such as blindness and stroke. Insulin therapy requires monitoring of blood sugar level and balancing injections of insulin with carbohydrate intake and exercise. Transplanted islet cells, however, could replace the body's natural mechanism for monitoring and producing insulin.

On February 24, 1989, Jim Connor became Canada's first patient to receive transplanted islets of Langerhans cells. Dr. Garth Warnock, a surgeon at the University of Alberta Hospital, transplanted millions of isolated cells from the pancreas of a donor. A month later, the procedure was once again successfully completed. Although other transplants had taken place before, none had reported the same degree of success. Why had the team at the University of Alberta Hospital triumphed where others failed? First, Dr. Ray Rajotte, the team leader, had developed frozen preservation methods used for tissue banks. Second, Dr. Garth Warnock devised more successful ways to harvest and purify islets. Third, Dr. Norman Kneteman improved immunosuppression therapy, thereby preventing the rejection of the transplanted cells.

It is important to note that cell transplants, although a significant step above insulin therapy, do not provide a cure for juvenile diabetes. A cure would prevent diabetes from occurring. Transplant therapy, like insulin therapy, controls the effects. Currently, the transplants are only being done on people who must undergo kidney transplants and therefore must already take immunosuppressive drugs. The researchers are guardedly optimistic about the future of cell transplants, but also warn that transplant therapy is not suitable for all patients.

Thyroxine *is a hormone secreted by the thyroid gland that regulates the rate of body metabolism.*

THYROID GLAND

The thyroid gland is located at the base of the neck, immediately in front of the trachea or windpipe. Two important thyroid hormones, thyroxine and triiodothyronine, regulate body metabolism and the growth and differentiation of tissues. Approximately 65% of thyroid secretions are thyroxine; however, both hormones appear to have the same function. Most of this discussion will focus on the principal hormone, **thyroxine.**

Have you ever wondered why some people seem to be able to consume fantastic amounts of food without any weight change, while others appear to gain weight at the mere sight of food? Part of this anomaly can be explained by thyroxine and the regulation of metabolic rate. Individuals who secrete higher levels of thyroxine oxidize sugars and other nutrients at a faster rate. Approximately 50% of the glucose oxidized in the body is released as heat (which explains why these individuals usually feel warm). The remaining 50% is converted to ATP, the storage form for cell energy. This added energy reserve is often consumed during activity. Therefore, these individuals tend not to gain weight.

Individuals who have lower levels of thyroxine do not oxidize nutrients as quickly, and therefore tend not to break down sugars as quickly. Excess blood sugar is eventually converted into liver and muscle glycogen. However, once the glycogen stores are filled, excess sugar is converted into fat. It follows that the slower the blood sugar is used, the faster the fat stores are built up. People who secrete low amounts of thyroxine tend to be less active, intolerant of cold, and have skin that tends to dry out quickly. It is important to note that not all types of weight gain are due to *hypothyroidism* (low thyroid secretions). In many cases, weight gain reflects a poor diet.

Control of thyroid hormones, like many other hormones, is accomplished by negative feedback. Should the metabolic rate decrease, receptors in the hypothalamus are activated. Nerve cells release thyroid-releasing factor (TRF), which stimulates the pituitary to release thyroid-stimulating hormone (TSH). Thyroid-stimulating hormone is carried by way of the blood to the thyroid gland, which, in turn, releases thyroxine. Thyroxine raises metabolism by stimulating increased sugar utilization by the cells of the body. Higher levels of thyroxine cause the pathway to be "turned off." Thyroxine inhibits the release of the thyroid-releasing factor from the hypothalamus, thereby turning off the production of TSH from the pituitary.

THYROID DISORDERS

Iodine is an important component of both thyroid hormones. A normal component of your diet, iodine is actively transported from the blood into the follicle cells of the thyroid. The concentration of iodine in the cells can be 25 times greater than that of the blood. Problems are created when iodine levels begin to fall. When adequate amounts of iodine are not obtained from the diet, the thyroid enlarges, producing a **goiter.**

The presence of a goiter emphasizes the importance of a negative-feedback control system. Without iodine, thyroid production and secretion of thyroxine drops. This causes more and more TSH to be produced, and, consequently, the thyroid is stimulated more and more. Under the relentless influence of TSH, cells of the thyroid continue to develop, and the thyroid enlarges. Goiters were once prevalent in areas where the soil lacked iodine salts. This is why table salt has been iodized.

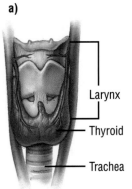

a)

Larynx

Thyroid

Trachea

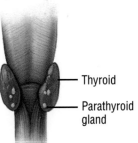

b)

Thyroid

Parathyroid gland

Figure 14.13

(a) Anterior view of thyroid gland. (b) Posterior view of thyroid gland.

Figure 14.14

Control of thyroid hormones.

*A **goiter** is an enlargement of the thyroid gland.*

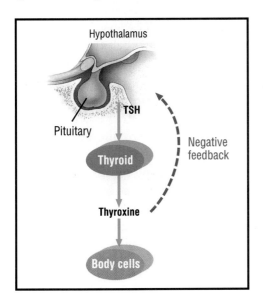

Hypothalamus

TSH

Pituitary

Thyroid

Negative feedback

Thyroxine

Body cells

CASE STUDY
THE EFFECTS OF HORMONES ON BLOOD SUGAR

Objective

To use experimental data to investigate the effects of various hormones on blood sugar levels.

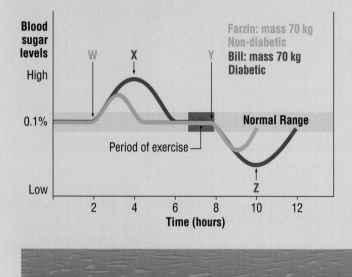

Procedure

Blood sugar levels of a diabetic and a nondiabetic patient were monitored over a period of 12 h. Both ate an identical meal and performed 1h of similar exercise. Use the data in the graph to answer the Case-Study Application Questions.

Case-Study Application Questions

1 Which hormone injection did Bill receive at the time labelled X? Provide your reasons.
2 What might have happened to Bill's blood sugar level if hormone X had not been injected? Provide your reasons.
3 Explain what happened at time W for Bill and Farzin?
4 Explain why blood sugar levels begin to fall after time Y?
5 What hormone might Bill have received at time Z? Explain your answer.
6 Why is it important to note that both Farzin and Bill had the same body mass?
7 What differences in blood sugar levels are illustrated by the data collected from Bill and Farzin.
8 Why do Bill and Farzin respond differently to varying levels of blood sugar? ■

PROSTAGLANDINS

Prostaglandins *are hormones that have a pronounced effect in a small localized area.*

Local responses to changes in the immediate environment of cells are detected by mediator cells, which produce **prostaglandins.** More than 16 different types of prostaglandins alter cell activity in a manner that counteracts or adjusts for the change. Generally, prostaglandins are secreted in low concentrations by mediator cells, but secretions increase when changes take place. Most of the molecules released during secretions, even in time of change, tend to be absorbed rapidly by surrounding tissues. Few of the prostaglandin molecules are absorbed by capillaries and carried in the blood.

Two different prostaglandins can adjust blood flow in times of stress. Stimulated by the release of epinephrine, the hormones increase blood flow to local tissues. Other prostaglandins respond to stress by triggering the relaxation of smooth muscle in the passages leading to the lung. Prostaglandins are also released during allergic reactions.

■ REVIEW QUESTIONS ?

14 How does thyroxine affect blood sugar?
15 List the symptoms associated with hypothyroidism and hyperthyroidism.
16 How do the pituitary and hypothalamus interact to regulate thyroxine levels?
17 What is a goiter?
18 What are prostaglandins and what is their function?

SOCIAL ISSUE:
Growth Hormone–The Anti-aging Drug

*In July 1990, Dr. Daniel Rudman published a study in the prestigious **New England Journal of Medicine** proposing that injections of growth hormone could slow the aging process. According to Dr. Rudman's findings, injections of growth hormone in older people actually increase the development of muscle tissue and the disappearance of fat. In many ways, growth hormone appears to be able to reverse the effects of decades of aging.*

Although researchers warn that the drug's long-term effects have not been documented and that the drug may not be suitable for everyone, speculation about the potentials of an anti-aging drug abounds in both scientific and nonscientific communities.

Statement:

Widespread use of anti-aging drugs could cause serious problems for society.

Point

- At 1990 prices, injections of growth hormone for a 70 kg man would be about $14 000 per year. Only the richest people in society would be able to pay for such treatments.

- It is very possible that growth hormone, like other steroids, might have negative long-term effects on individuals.

- People are already living longer and thus putting a strain on our health services and pension plans. If more people live even longer, the social system could collapse.

Counterpoint

- Although growth hormone is expensive today, the price will likely come down once the procedure becomes established. Plastic surgery was once available only to the rich, and it has become commonplace.

- With proper studies and control of the use of growth hormone, potential users would be protected and well informed. Ultimately, it is up to the individual to choose whether or not to risk any possible negative effects.

- Far from causing the social system to collapse, anti-aging therapy would produce economic benefits. People would work longer and generally experience a better quality of life.

Research the issue.
Reflect on your findings.
Discuss the various viewpoints with others.
Prepare for the class debate.

CHAPTER HIGHLIGHTS

- Hormones are chemicals released by cells that affect cells in other parts of the body. Only small amounts of hormones are required to alter cell metabolism.
- Endocrine hormones are secreted from glands directly into the blood.
- Integration of organs is accomplished by the nervous system and endocrine system. Chemical transmitters act on specific tissues. Target tissues have specific receptor sites that bind with the hormones.
- A negative-feedback system is a control system designed to prevent chemical imbalances in the body. The body responds to changes in the external or internal environment. Once the effect is detected, receptors are activated and the response is inhibited.
- Growth hormone is produced by the cells of the anterior lobe of the pituitary. Prior to puberty, the hormone promotes growth of the long bones.

- The adrenal gland secretes hormones. The adrenal medulla, found at the core of the adrenal gland, produces epinephrine and norepinephrine. The adrenal cortex, the outer region of the adrenal gland, produces glucocorticoids and mineralocorticoids.
- Insulin is a hormone produced by the pancreas. Secreted when blood sugar levels are high, insulin promotes carbohydrate storage.
- Glucagon is a hormone produced by the pancreas. Secreted when blood sugar levels are low, glucagon promotes the conversion of glycogen to glucose.
- Diabetes mellitus is a genetic disorder characterized by high blood sugar levels.
- Thyroxine is a hormone secreted by the thyroid gland that regulates the rate of body metabolism.
- Prostaglandins are hormones that have a pronounced effect in a small localized area.

APPLYING THE CONCEPTS

1 Referring to the interaction between the hypothalamus and pituitary, indicate how the nervous system and endocrine system complement each other.

2 With reference to the adrenal glands, explain how the nervous system and endocrine system interact in times of stress.

3 A disorder called testicular feminization syndrome occurs when the receptor molecules to which testosterone binds are defective. Predict the effect of testicular feminization syndrome and explain how normal steroid hormone action is altered.

4 A rare virus destroys cells of the anterior lobe of the pituitary. Predict how the destruction of the pituitary cells would affect blood sugar. Explain.

5 Why do insulin levels increase during times of stress?

6 Provide an explanation for the following symptoms associated with diabetes mellitus: lack of energy, increased urine output, and thirst.

7 A tumor on a gland can increase the gland's secretions. Explain how increases in the following hormones affect blood sugar levels: insulin, epinephrine, and thyroxine.

8 A physician notes that her patient is very active and remains warm on a cold day even when wearing a light coat. Further discussion reveals that even though the patient's daily food intake exceeds that of most people, the patient remains thin. Why might the doctor suspect a hormone imbalance? Which hormone might the doctor suspect?

9 With reference to negative feedback, provide an example of why low levels of iodine in your diet can cause goiters.

10 Three classical methods have been used to study hormone function:
- The organ that secretes the hormone is removed. The effects are studied.
- Grafts from the removed organ are placed within a gland. The effects are studied.
- Chemicals from the extracted gland are isolated and injected back into the body. The effects are studied.

Explain how each of these procedures could be used to investigate how insulin affects blood sugar.

CRITICAL-THINKING QUESTIONS

1 A physician notes that individuals with a tumor on the pancreas secrete unusually high levels of insulin. Unfortunately, insulin in high concentrations causes blood sugar levels to fall below the normal acceptable range. In an attempt to correct the problem, the physician decides to inject the patient with cortisol. Why would the physician give the patient cortisol? What problems could arise from this treatment?

2 Some scientists have speculated that certain young female Olympic gymnasts may have been given growth hormone inhibitors. Why might the gymnasts have been given growth inhibitors? Do you think hormone levels should be altered to regulate growth patterns?

3 Cattle are given various steroid hormones to increase meat production. Recently some scientists have expressed concern that animal growth stimulators might have an effect on humans. Comment on the practice of using hormones in cattle. What potential problems might be associated with such procedures?

4 Negative-feedback control systems influence hormonal levels. The fact that some individuals have higher metabolic rates than others can be explained in terms of the response of the hypothalamus and pituitary to thyroxine. Some feedback systems turn off quickly. Sensitive feedback systems tend to have comparatively lower levels of thyroxine; less sensitive feedback systems tend to have higher levels of thyroxine. One hypothesis attempts to link different metabolic rates with differences in the number of binding sites in the hypothalamus and pituitary. How might the number of binding sites for molecules along cell membranes affect hormonal levels? How would you go about testing the theory?

ENRICHMENT ACTIVITIES

1 Draw a diagram of the human body showing the location of the glands that produce the various hormones studied in this chapter.

2 Suggested reading:
- Berridge, M. "The Molecular Basis of Communication within the Cell." *Scientific American*, October 1985, pp. 142–52.
- Fellman, B. "A Clockwork Gland." *Science 85* (June 1985): pp. 76–81.
- Franklin, D. "Growing Up Short." *Science News*, 125 (1984): pp. 92–94.
- Synder, S. "The Molecular Basis of Communication between Cells." *Scientific American*, October 1985, pp. 132–41.

The Nervous System and Homeostasis

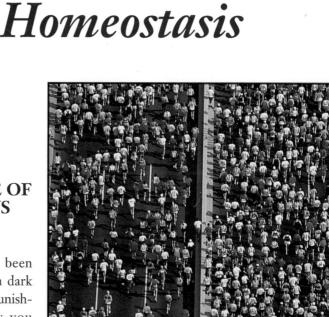

IMPORTANCE OF THE NERVOUS SYSTEM

Prisoners have often been isolated and placed in dark rooms as a means of punishment. Imagine how you would be affected if you didn't know whether it was day or night, or if you couldn't hear a sound for days.

Even in these extreme conditions, however, your nervous system remains active. Information about your depth of breathing, the physical condition of the breathing muscles, and the amount of water contained in the respiratory tract is continually relayed to the brain for processing and storage. Other nerve cells detect air temperature, light intensity, and odors. Pressure receptors in your skin inform you of the fit of your clothes and can detect an insect scurrying across your legs. Blinking your eyes or scratching your nose requires coordinated nerve impulses. Dreams of past happy days and hopes for your future reside in the nervous system.

The nervous system, along with the endocrine system, coordinates the actions of your body. Through a series of adjustments, all systems of the body are regu-

Figure 15.1

Whether you are alone or surrounded by many people, your nervous system is constantly active, responding to both internal and external stimuli.

lated to ensure effective behavior and to maintain the internal environment within safe limits. Responses to internal and external environments are made possible by either electrochemical messages relayed from the brain, or by a series of chemical messengers, many of which are carried by the blood. The chemical messengers, called hormones, are produced by glands and require more time for response than do nerves.

The nervous system is an elaborate communication system that contains more than 100 billion nerve cells in the brain alone. That number exceeds the number of visible stars in the Milky Way galaxy. Although all animals display some type of response to the environment, the development of the nervous system seems to reach its pinnacle in humans. Indeed, what separates us from other animals is often described in terms of the nervous system. Memory, learning, and language are all components of the human nervous system.

VERTEBRATE NERVOUS SYSTEMS

The nervous system can be divided into two main divisions: the central nervous system and the peripheral nervous system. The central nervous system, or CNS, contains the nerves of the brain and spinal cord. The central nervous system acts as a coordinating center for incoming and outgoing information. The peripheral nervous system, or PNS, carries information between the organs of the body and the central nervous system.

The peripheral nervous system can be further subdivided into somatic and autonomic nerves. *Somatic nerves* control the skeletal muscle, bones, and skin. Sensory somatic nerves relay information about the environment to the central nervous system, while motor somatic nerves initi-

ate an appropriate response. The *autonomic nerves* are special motor nerves that are designed to control the internal organs of the body. The two divisions of the autonomic system—the sympathetic and parasympathetic systems—often operate as "on-off" switches.

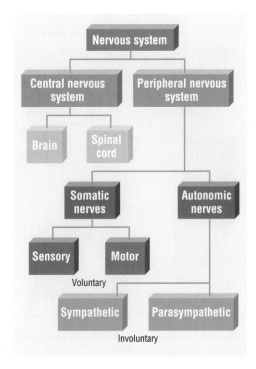

Figure 15.2

The main divisions of the nervous system.

Neurons *are cells that conduct nerve impulses.*

Sensory neurons *carry impulses to the central nervous system.*

NEURONS

Two different types of cells—glial cells and neurons—are found in the nervous system. Glial cells, often called neuroglial cells, are nonconducting cells and are important for the structural support and metabolism of the nerve cells. **Neurons** are the functional unit of the nervous system. These specialized nerve cells can be categorized into three groups: the sensory neurons, interneurons, and motor neurons. **Sensory neurons** relay information about the environment to the central nervous system for processing. For example, special sensory receptors in your eyes respond to light, while some in your skin respond to pressure; still others respond to either warm or cold temperatures. **Interneurons,** as the name suggests, link neurons within the body. Found predominantly throughout the brain and spinal cord, the interneurons integrate and interpret the sensory information and connect neurons to outgoing motor neurons. **Motor neurons** relay information to the effectors. Muscles and glands are classified as effectors because they cause things to happen.

All neurons contain cell bodies, axons, and dendrites. The **dendrites** receive the information, either from specialized receptors, as in the case of sensory neurons, or from other nerve cells, as in the case of motor neurons. Dendrites conduct nerve impulses toward the cell body. Like all living cells, nerve cells contain a cell body with a nucleus. An extension of cytoplasm, called the **axon,** projects from the cell body. In some cases it extends to nearly one meter in length. In humans the axon is extremely thin. More than 100 axons could be placed inside the shaft of a single human hair. The axon carries the nerve impulse toward other neurons or to effectors. A close examination of most nerves shows that they are composed of many neurons held together. Large composite nerves appear much like a telephone cable that contains many branches of incoming and outgoing lines.

Interneurons carry impulses within the central nervous system.

Motor neurons carry impulses from the central nervous system to effectors.

Dendrites are projections of cytoplasm that carry impulses toward the cell body.

An axon is an extension of cytoplasm that carries nerve impulses away from the dendrites.

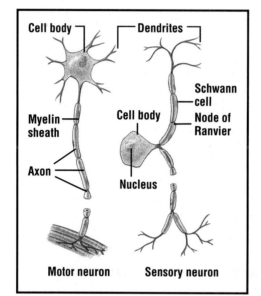

Figure 15.3

Structure of a motor and a sensory neuron.

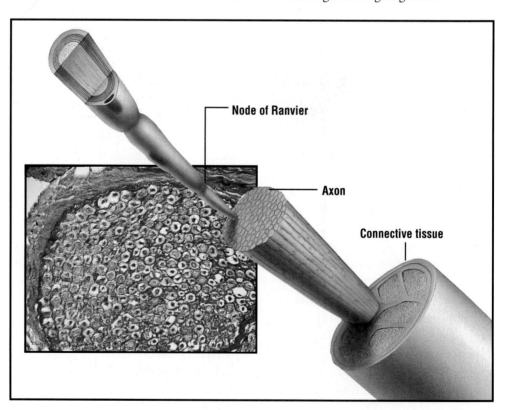

Figure 15.4

Most nerves are composed of many different neurons.

352 ■

Unit Four:
Coordination and Regulation in Humans

Many axons are covered with a glistening white coat of a fatty protein called the **myelin sheath,** which acts as insulation for the neurons. Formed by the Schwann cells, the myelin sheath acts as an insulator by preventing the loss of charged ions from the nerve cell. The areas between the sections of myelin sheath are known as the **nodes of Ranvier.** Axons that have a myelin covering are said to be *myelinated.* Nerve impulses jump from one node to another, thereby speeding the movement of nerve impulses. Not surprisingly, nerve impulses move much faster along myelinated nerve fibers than unmyelinated ones. The speed of the impulse along the nerve fiber is also affected by the diameter of the axon. Generally, the smaller the diameter of the axon, the faster the speed of the nerve impulse.

All nerve fibers found within the peripheral nervous system contain a thin membrane, called the **neurilemma,** that surrounds the axon. The neurilemma promotes the regeneration of damaged axons. This explains why feeling gradually returns to your finger following a paper cut—severed neurons can be rejoined. However, not all nerve cells contain a neurilemma and a myelin sheath. Nerves within the brain that contain myelinated fibers and a neurilemma are called *white matter.* Other nerve cells within the brain and spinal cord, referred to as the *gray matter,* lack a myelin sheath, and will not be regenerated after injury. Damage to the gray matter is permanent.

BIOLOGY CLIP

Multiple sclerosis is caused by the destruction of the myelin sheath that surrounds the nerve axons. The myelinated nerves in the brain and spinal cord are gradually destroyed as the myelin sheath hardens and forms scars, or plaques. The destruction of the sheath results in short-circuits. Often referred to as MS, multiple sclerosis can produce symptoms of double vision, speech difficulty, jerky limb movements, and partial paralysis of the voluntary muscles.

NEURAL CIRCUITS

If you have ever touched a hot stove, you probably didn't think about how your nervous system told you it was hot! The sensation of heat is detected by specialized temperature receptors in your skin, and a nerve impulse is carried to the spinal cord. The sensory nerve passes the impulse on to an interneuron, which, in turn, relays the impulse to a motor neuron. The motor nerve causes the muscles in the hand to contract and the hand to pull away. All this happens in a split second, long before the information travels to the brain. A few seconds later, the sensation of pain becomes noticeable and you may let out a scream. Reflexes are involuntary and often unconscious. Imagine how badly you would have burned yourself if you had to wait for the sensation of pain before removing your hand. The damage would have been much worse if you had attempted to gauge the intensity of pain and to contemplate the appropriate action. Time is required for nerve impulses to move through the many circuits of the brain.

The simplest nerve pathway is the *reflex arc.* Reflexes occur without brain coordination. They contain five essential components: the receptor, the sensory neuron, the interneuron in the spinal cord, the motor neuron, and the effector. Figure 15.5 on the following page shows a reflex arc. The touch receptor in the finger detects the tack. Sensory information is transmitted to the CNS.

*The **myelin sheath** is a fatty covering over the axon of a nerve cell.*

*The **nodes of Ranvier** are the regularly occurring gaps between sections of myelin sheath along the axon.*

*The **neurilemma** is the delicate membrane that surrounds the axon of nerve cells.*

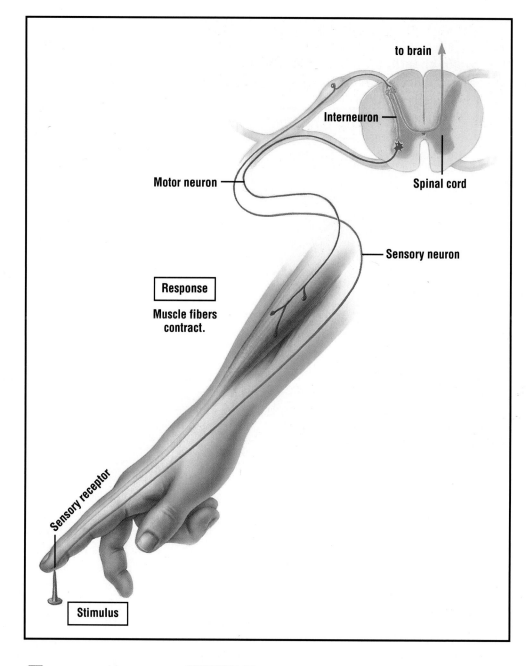

Figure 15.5

Sensory information is relayed from the sensory neuron (in purple) to the spinal cord. Interneurons in the spinal cord receive the information from the sensory neuron and relay it to the motor neuron shown in red. The motor neuron activates the muscle cell, causing it to contract.

to brain

Interneuron

Motor neuron

Spinal cord

Sensory neuron

Response

Muscle fibers contract.

Sensory receptor

Stimulus

■ REVIEW QUESTIONS ■ ?

1 Differentiate between the PNS and CNS.

2 What are neurons?

3 Differentiate between sensory and motor nerves.

4 Briefly describe the function of the following parts of a neuron: dendrites, myelin sheath, Schwann cells, cell body, and axon.

5 What is the neurilemma?

6 Name the essential components of a reflex arc, and state the function of each of the components.

REFLEX ARCS AND REACTION RATE

Objective

To investigate different reflex arcs and reaction rate.

Materials

rubber reflex hammer   penlight or microscope light
30 cm ruler

Procedure

Knee-jerk

1 Have your subject sit with his or her legs crossed on a chair. The subject's upper leg should remain relaxed.

2 Locate the position of the kneecap on the upper leg and feel the large tendon below the midline of the knee cap.

3 Using a reflex hammer, gently strike the tendon below the kneecap.

   a) Describe the movement of the leg.

4 Ask the subject to clench a book with both hands, then strike the tendon of the upper leg once again.

   b) Compare the movement of the leg while the subject is clenching the book with the movement in the previous procedure.

Achilles Reflex

5 Remove the subject's shoe. Ask your subject to kneel on a chair so that the feet hang over the edge of the chair. Push the toes toward the legs of the chair and then lightly tap the Achilles tendon with the reflex hammer.

   c) Describe the movement of the foot.

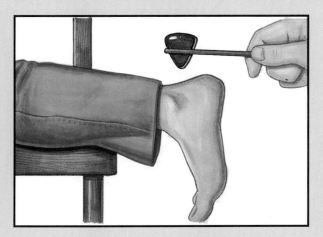

Babinski Reflex

6 Ask the subject to remove a shoe and sock. Have the subject sit in a chair, placing the shoeless foot on another chair for support. Quickly slide the reflex hammer across the sole of the subject's foot, beginning at the heel and moving toward the toes.

   d) Describe the movement of the toes.

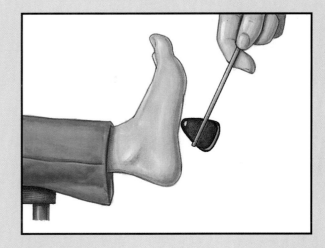

Pupillary Reflex

7 Have the subject close one eye for approximately one minute. Ask him or her to open the closed eye; now compare the size of the pupils.

   e) Which pupil is larger?

8 Ask the subject to close both eyes for one minute. Have the subject open both eyes; now shine a penlight in one of the eyes.

   f) Describe the changes you observe in the pupil.

Reaction Rate

9 Ask your subject to place his or her forearm flat on the surface of a desk. The subject's entire hand should be extended over the edge of the desk.

10 Ask the subject to place the index finger and thumb approximately 2 cm apart. Place a 30 cm ruler between the thumb and forefinger of the subject. The lower end of the ruler should be even with the top of the thumb and forefinger.

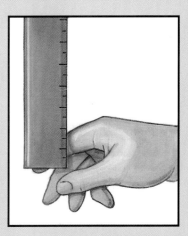

11 Indicate when ready, and release the ruler within the next 30 s. Measure the distance the ruler falls before being caught with the subject's thumb and forefinger. Repeat the procedure for the left hand. Record your data in a table similar to the one below.

g)

Trial	Distance–right hand (cm)	Distance–left hand (cm)
1		
2		
3		
Average		

Laboratory Application Questions

1 A student touches a stove, withdraws his or her hand, and then yells. Why does the yelling occur after the hand is withdrawn? Does the student become aware of the pain before the hand is withdrawn?

2 The neuron is severed at point X. Explain how the reflex arc would be affected.

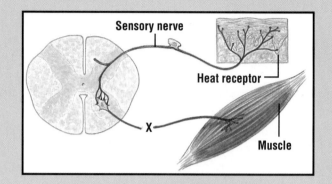

3 Explain how the knee-jerk and Achilles reflexes are important in walking.

4 While examining the victim of a serious car accident, a physician lightly pokes the patient's leg with a needle. The light pokes begin near the ankle and gradually progress toward the knee. Why is the physician poking the patient? Why begin near the foot?

5 Why is the knee-jerk reflex exaggerated when the subject is clenching the book? ∎

ELECTROCHEMICAL IMPULSE

In 1791, the Italian scientist Luigi Galvani discovered that the calf muscle of a dead frog could be made to twitch under electrical stimulation. Galvani concluded that the "animal electricity" was produced by the muscle. Although Galvani's conclusion was incorrect, it spawned a flood of research that led to the development of theories about how electrical current is generated in the body. In 1840, Emil DuBois-Reymond set about refining instruments that would enable him to detect the passage of currents in nerves and muscles. By 1906, the Dutch physiologist Willem Einthoven began recording the movement of electrical impulses in heart muscle. The electrocardiogram, or ECG, has been refined many times since 1906 and is still used today to diagnose heart problems. In 1929, Hans Berger placed electrodes on the skull and measured electrical changes that accompany brain activity. The electroencephalograph, or EEG, is used to measure brain-wave activity.

Figure 15.6

Mapping electrical current has diagnostic value.

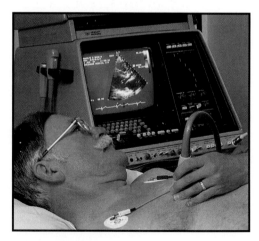

As research continued, the difference between electricity and nerves soon became evident. Current travels along a wire much faster than the impulse travels across a nerve. In addition, the cytoplasmic core of a nerve cell offers great resistance to the movement of electrical current. Unlike electrical currents, which diminish as they move through a wire, nerve impulses remain as strong at the end of a nerve as they were at the beginning. One of the greatest differences between nerve impulses and electricity is that nerves use cellular energy to generate current. By comparison, the electrical wire relies on some external energy source to push electrons along its length. As early as 1900, Julius Berstein suggested that nerve impulses were an electrochemical message created by the movement of ions through the nerve cell membrane. Evidence supporting Berstein's theory was provided in 1939 when K.S. Cole and J.J. Curtis placed a tiny electrode inside the large nerve cell of a squid. A rapid change in charge across the membrane was detected every time the nerve became excited. The **resting membrane** normally had a negative charge somewhere near −70 mV; however, when the nerve became excited, the charge on the inside of the membrane registered +40 mV. This

reversal of charges is described as an **action potential.** Cole and Curtis noticed that the +40 mV did not last more than a few milliseconds before the charge on the inside of the nerve cell returned to −70 mV.

Nerve cells are different from other cells in that no other cell is charged. How do nerve cell membranes become charged? To find the answer, we must examine the nerve cell on a molecular level. Unlike most cells, neurons have a rich supply of positive and negative ions both inside and outside the neuron. Although it might seem surprising, the negative ions do little to create a charged membrane. The electrochemical event is caused by an unequal concentration of positive ions across the nerve membrane. A potassium pump, located in the cell membrane, pulls potassium ions into the nerve

Resting membranes *maintain a steady charge difference across the cell membrane. These membranes are not being stimulated.*

Action potentials *are nerve impulses. The reversal of charge across a nerve cell membrane initiates an action potential.*

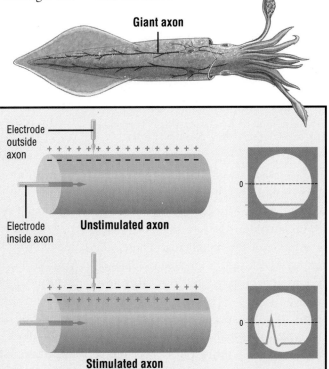

Giant axon

Electrode outside axon

Electrode inside axon

Unstimulated axon

Stimulated axon

Figure 15.7

The squid has a large axon. A miniature electrode is placed inside the giant axon of a squid. The inside of the resting membrane is negative with respect to the outside of the membrane. When stimulated, the charges across the nerve membrane temporarily reverse.

cell, while a sodium pump relays sodium ions outside the nerve cell. It has been estimated that for every 150 potassium ions on the inside of the cell membrane there are 5 potassium ions on the outside of the cell membrane. The sodium pump is nearly as effective: for every 150 sodium ions on the outside of the nerve cell membrane, 15 can be found on the inside of the nerve cell membrane. The highly concentrated potassium ions on the inside of the nerve cell have a tendency to diffuse outside the nerve cell. Similarly, the highly concentrated sodium ions on the outside of the nerve cell have a tendency to diffuse into the nerve cell. As potassium diffuses outside of the neuron, sodium diffuses into the neuron. Therefore, positively charged ions move both into and out of the cell. However, the diffusion of sodium ions and potassium ions is not equal. In a resting membrane, more channels are open to the potassium ions. Consequently, more potassium ions diffuse out of the nerve cell than sodium ions diffuse into the nerve cell.

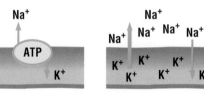

Figure 15.8

The potassium and sodium pumps of the nerve cell are highly effective.

Sodium is pumped outside of the nerve cell, while potassium is pumped into the nerve cell.

Potassium diffuses out of the nerve cell faster than sodium diffuses into the nerve cell.

The more rapid diffusion of potassium ions outside of the nerve membrane means that the area outside of the membrane becomes positive relative to the area inside of the membrane. Therefore, the nerve cell loses a greater number of positive ions than it gains. Biologists now believe that ion gates control the movement of ions across the membrane. According to popular theory, the resting membrane has more potassium gates open for diffusion than sodium gates.

Excess positive ions accumulate along the outside of the nerve membrane, while excess negative ions accumulate along the inside of the membrane. The resting membrane is said to be charged, or **polarized**. The separation of electrical charges by a membrane has the potential to do work, which is expressed in millivolts (mV). A charge of -70 mV indicates the difference between the number of positive charges found on the inside of the nerve membrane relative to the outside. (A charge of -90 mV on the inside of the nerve membrane would indicate even fewer positive ions inside the membrane relative to the outside.)

Upon excitation, the nerve cell membrane becomes more permeable to sodium than potassium. Scientists believe that sodium gates are opened in the nerve membrane, while potassium gates close. The highly concentrated sodium ions rush into the nerve cell by diffusion and charge attraction. The rapid inflow of sodium causes a charge reversal, or **depolarization**. Once the voltage inside of the nerve cell becomes positive, the sodium gates slam closed and the inflow of sodium is halted. The potassium gates now open and potassium ions once again begin to diffuse out of the nerve cell. Eventually, the flow of potassium out of the nerve cell restores the original polarity of the membrane. However, the Na^+ and K^+ are now on the side of the resting membrane opposite to their position before depolarization occurred. Once again, the sodium-potassium pump will restore the condition of the resting membrane by transporting Na^+ out of the neuron while moving K^+ ions inside the neuron. The energy supply from ATP maintains the polarized membrane. The process of restoring the original polarity of the nerve membrane is referred to as **repolarization**. The process usually takes about 0.001 s.

Nerves conducting an impulse cannot be activated until the condition of the resting membrane is restored. The period of depolarization must be completed and the nerve must repolarize before a second action potential can be conducted. The period of time required for the nerve cell to become repolarized is called the **refractory period.** The refractory period usually lasts 1 to 10 ms.

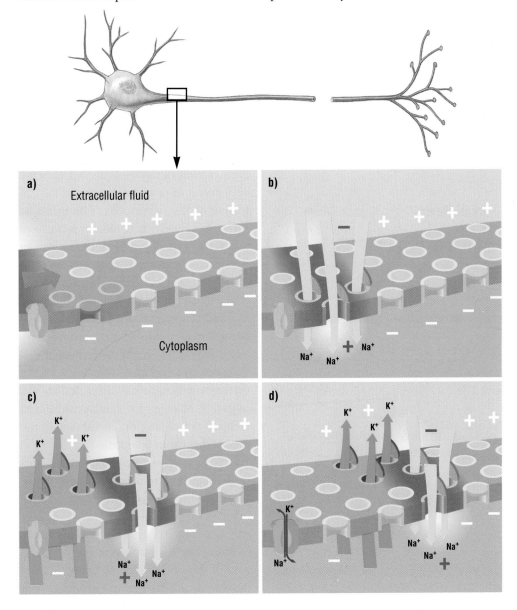

*The **refractory period** is the recovery time required before a neuron can produce another action potential.*

Figure 15.9

(a) The resting membrane is more permeable to potassium than sodium. Potassium ions diffuse out of the nerve faster than sodium ions diffuse into the nerve. The outside of the nerve cell becomes positive relative to the inside. (b) A strong electrical disturbance, shown by the darker coloration of the cell membrane, moves across the cell membrane. The disturbance opens sodium ion gates, and sodium ions rush into the nerve cell. The membrane becomes depolarized. (c) Depolarization causes the sodium gates to close, while the potassium gates are opened once again. Potassium follows the concentration gradient and moves out of the nerve cell by diffusion. Adjoining areas of the nerve membrane become permeable to sodium ions, and the action potential moves away from the site of origin. (d) The electrical disturbance moves along the nerve membrane in a wave of depolarization. The membrane is restored as successive areas once again become more permeable to potassium. The sodium and potassium pumps restore and maintain the polarization of the membrane.

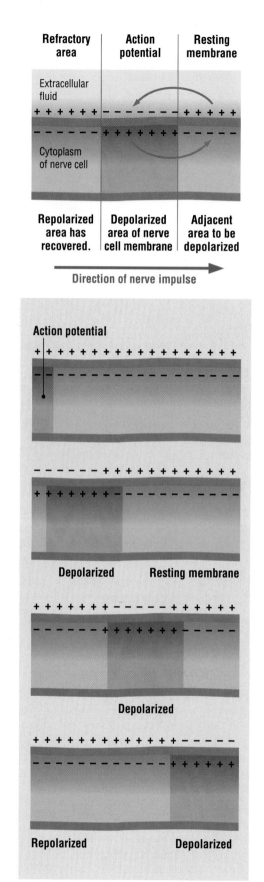

Refractory area | **Action potential** | **Resting membrane**

Extracellular fluid

+ + + + +   – – – – –   + + + +

– – – – –   + + + + +   – – – –

Cytoplasm of nerve cell

Repolarized area has recovered. | **Depolarized area of nerve cell membrane** | **Adjacent area to be depolarized**

→ **Direction of nerve impulse**

Figure 15.10

The movement of a nerve impulse.

Action potential

+|+ + + + + + + + + + + + + + + + +

– – – – – – – – – – – – – – – – – –

– – – – – + + + + + + + + + + +

+ + + + + + – – – – – – – – – –

Depolarized   **Resting membrane**

+ + + + + + – – – – – + + + + + +

– – – – – + + + + + + – – – – –

Depolarized

+ + + + + + + + + + + + – – – – –

– – – – – – – – – – – + + + + + +

Repolarized   **Depolarized**

Figure 15.11

The action potential moves along the nerve cell membrane, creating a wave of depolarization and repolarization.

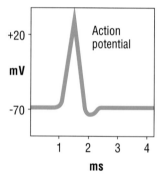

+20

mV

–70

Action potential

1   2   3   4

ms

Movement of the Action Potential

The movement of sodium ions into the nerve cell causes a depolarization of the membrane and signals an action potential in that area. However, for the impulse to be conducted along the axon, the impulse must move from the zone of depolarization to adjacent regions.

It is important to understand the action potential. The action potential is characterized by the opening of the sodium channels in the nerve membrane. Sodium ions rush into the cytoplasm of the nerve cell, diffusing from an area of high concentration (outside of the nerve cell) to an area of lower concentration (inside of the nerve cell). The influx of the positively charged sodium ion causes a charge reversal, or depolarization, in that area. The positively charged ions that rush into the nerve cell are then attracted to the adjacent negative ions, which are aligned along the inside of the nerve membrane. A similar attraction occurs along the outside of the nerve membrane. The positively charged Na^+ ions of the resting membrane are attracted to the negative charge that has accumulated along the outside of the membrane in the area of the action potential.

The flow of positively charged Na^+ ions from the area of the action potential toward the adjacent regions of the resting membrane causes a depolarization in the adjoining area. The electrical disturbance causes sodium channels to open in the adjoining area of the nerve cell membrane and the movement of the action potential. As a wave of depolarization moves along the nerve membrane, the initiation point of the action potential enters a refractory period as the membrane once again becomes more permeable to K^+ ions. Depolarization of the membrane causes the sodium channels to close and the potassium channels to reopen. The wave of depolarization is followed by a wave of repolarization.

Threshold Levels and the All-or-None Response

A great deal of information about nerve cells has been acquired through laboratory experiments. Nerve cells respond to changes in pH, changes in pressure, and to specific chemicals. However, mild electrical shock is most often used in experimentation because it is easily controlled and its intensity can be regulated.

In a classic experiment, a single neuron leading to a muscle is isolated and a mild electrical shock is applied to the neuron. A special recorder measures the strength of muscle contraction. For this example, stimuli less than 2 mV will not produce any muscle contraction. A potential stimulus must be above a critical value in order to produce a response. The critical intensity of the stimulus is known as the **threshold level**. Stimuli below threshold levels do not initiate a response. A threshold level of 2 mV is required to produce a response in the data shown in Figure 15.12; however, threshold levels are different for each neuron.

A second, but equally important, conclusion can be drawn from the experimental data. Increasing the intensity of the stimuli above the critical threshold value will not produce an increased response—the intensity of the nerve impulse and speed of transmission remain the same. In what is referred to as the **all-or-none response**, neurons either fire maximally or not at all.

How do animals detect the intensity of stimuli if nerve fibers either fire completely or not at all? Experience tells you that you are capable of differentiating between a warm object and one that is hot. To explain the apparent anomaly, we must examine the manner in which the brain interprets nerve impulses. Although stimuli above threshold levels produce nerve impulses of identical speed and intensity, variation with respect to frequency does occur. The more intense the stimulus, the greater the frequency of impulses. Therefore, when a warm glass rod is placed on your hand, sensory impulses may be sent to the brain at a slow rate. A hot glass rod placed on the same tissue will also cause the nerve to fire, but the frequency of impulses is greatly increased. The brain interprets the frequency of impulses.

The different threshold levels of neurons provide a second way in which the intensity of stimuli can be detected. Each nerve is composed of many individual nerve cells or neurons. A glass rod at 40°C may cause a single neuron to reach threshold level, but the same glass rod at 50°C will cause two or more neurons to fire. The second neuron has a higher threshold level. The greater the number of impulses reaching the brain, the greater the intensity of the response.

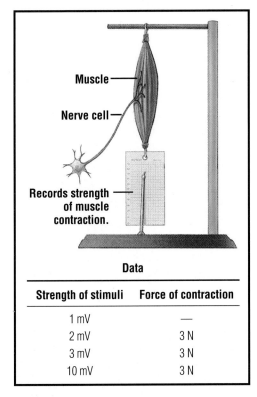

Figure 15.12

The threshold level for this neuron is 2 mV. Different neurons have different threshold levels.

Data	
Strength of stimuli	**Force of contraction**
1 mV	—
2 mV	3 N
3 mV	3 N
10 mV	3 N

Threshold level *is the minimum level of a stimulus required to produce a response.*

*The **all-or-none response** of a nerve or muscle fiber means that the nerve or muscle responds completely or not at all to a stimulus.*

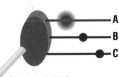

Glass rod 40°C

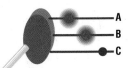

Glass rod 50°C

Figure 15.13

A glass rod at 40°C will elicit a response from neuron A. A glass rod heated to 50°C will elicit a response from neuron A and neuron B. Neuron B has a higher threshold level than neuron A, and will not fire until the glass rod is heated above 40°C. The brain interprets both the number of neurons excited and the frequency of impulses.

THE SYNAPSE

Small spaces between neurons or between neurons and effectors are known as **synapses.** A single neuron may branch many times at its end plate and join with many different neurons. Synapses rarely involve just two nerves. Small vesicles containing transmitter chemicals are located in the end plates of axons. The impulse moves along the axon and releases transmitter chemicals from the end plates. The transmitter chemicals are released from the **presynaptic neuron** and diffuse across the synapse, creating a depolarization of the dendrites of the **postsynaptic neuron.** Although the space between neurons is very small—approximately 20 nm (nanometers)—the nerve transmission slows across the synapse. Diffusion is a slow process. Not surprisingly, the greater the number of synapses, the slower the speed of transmission over a specified distance. This may explain why you react so quickly to a stimulus in a reflex arc, which has few synapses, while solving biology problems requires greater time.

Acetylcholine is a typical transmitter chemical found in the end plates of many nerve cells. Acetylcholine can act as an excitatory transmitter chemical on many postsynaptic neurons by opening the sodium ion channels. Once the channels are opened, the sodium ions rush into the postsynaptic neuron, causing depolarization. The reversal of charge causes the action potential. However, acetylcholine also presents a problem. By opening the sodium channels, the postsynaptic neuron would remain in a constant state of depolarization. How would the nerve ever respond to a second impulse if it did not recover? The release of the enzyme **cholinesterase** follows the acetylcholine and destroys it. Once acetylcholine is destroyed, the sodium channels are closed, and the neuron begins its recovery phase. Many insecticides take advantage of the synapse by blocking cholinesterase. The heart of an insect, unlike the human heart, is totally under nerve control. The next time you use an insecticide, consider that the insect's heart responds to the nerve message by contracting, but it never relaxes.

Figure 15.14

(a) Branching end plates synapse with the dendrites from many different neurons. (b) Synaptic vesicles in the end plates of the presynaptic neuron release transmitter chemicals by exocytosis. (c) The transmitter chemical attaches itself to the postsynaptic membrane, causing it to depolarize. The action potential continues along the postsynaptic neuron.

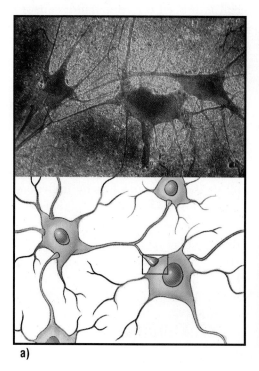

a)

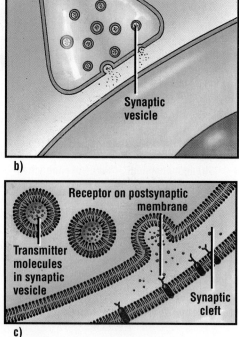

b)

Synaptic vesicle

Receptor on postsynaptic membrane

Transmitter molecules in synaptic vesicle

Synaptic cleft

c)

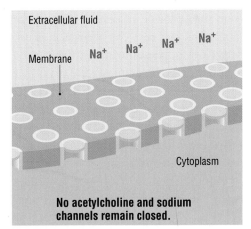

No acetylcholine and sodium channels remain closed.

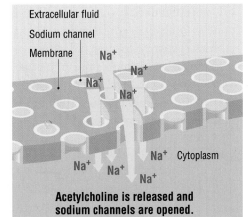

Acetylcholine is released and sodium channels are opened.

Figure 15.15

Model of an excitatory synapse. Acetylcholine opens channels for Na⁺ ions in the postsynaptic membrane.

Not all transmissions across a synapse are excitatory. While acetylcholine may act as an excitatory transmitter chemical on one postsynaptic membrane, it may act as an inhibitory transmitter chemical on another. It is believed that many inhibitory transmitter chemicals make the postsynaptic membrane more permeable to potassium. By opening even more potassium gates, the potassium ions on the inside of the neuron follow the concentration gradient and diffuse out of the neuron. The rush of potassium out of the cell increases the number of positive ions on the outside of the cell relative to the number found on the inside of the cell. Such neurons are said to be **hyperpolarized** because the resting membrane is even more negative. More sodium channels must now be opened to achieve depolarization and an action potential. As their name suggests, inhibitory transmitter chemicals prevent postsynaptic neurons from becoming active.

Figure 15.16 shows a model of a typical neural pathway. Transmitter chemicals released from neurons A and B are both excitatory; however, neither neuron is capable of causing sufficient depolarization to initiate an action potential in neuron D. However, when both neurons A and B fire at the same time, a sufficient amount of transmitter chemical is released to cause depolarization of the postsynaptic

membrane. The production of an action potential in neuron D requires the sum of two excitatory neurons. This principle is referred to as **summation**.

The transmitter chemical released from neuron C produces a dramatically different response. Neuron D becomes more negatively charged when neuron C is activated. You may have already concluded that neuron C is inhibitory. But data reveal even more striking information: transmitter chemicals other than acetylcholine must be present. A number of transmitter chemicals such as serotonin, dopamine, gamma-aminobutyric acid (GABA), and glutamic acid have been identified in the central nervous system. Another common transmitter chem-

Hyperpolarized *membranes are much more permeable to potassium than usual. The inside of the nerve cell membrane becomes even more negative.*

Summation *is the effect produced by the accumulation of transmitter chemicals from two or more neurons.*

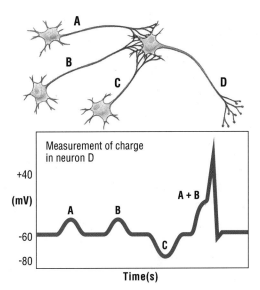

Measurement of charge in neuron D

Figure 15.16

Action potentials must occur simultaneously in A and B to reach threshold level in D.

ical, norepinephrine (noradrenaline) is found in both the central and peripheral nervous systems. To date, all effects of norepinephrine in the peripheral nervous system appear to be excitatory, while those of the central nervous system can be excitatory or inhibitory.

The interaction of excitatory and inhibitory transmitter chemicals is what allows you to throw a ball. As the triceps muscle on the back of your upper arm receives excitatory impulses and contracts, the biceps muscle on the front of your arm receives inhibitory impulses and relaxes. The triceps muscle straightens the arm, while the biceps muscle bends the arm. Inhibitory impulses in your central nervous system are even more important. Sensory information is received by the brain and is prioritized. Much of the less important information is inhibited so that you can devote your attention to the most important sensory information. For example, during a biology lecture, your sensory information should be directed at the sounds coming from your teacher, the visual images that appear on the chalkboard, and the sensations produced as you move your pen across the page. Although your temperature receptors may signal a slight chill in the air, and the pressure receptors in your skin may provide reassuring information about the fact that you are indeed wearing clothes, the information from these sensory nerves is suppressed.

Various disorders have been associated with transmitter chemicals. Parkinson's disease, characterized by involuntary muscle contractions and tremors, is caused by inadequate production of dopamine. Alzheimer's disease, associated with the deterioration of memory and mental capacity, has been related to decreased production of acetylcholine.

■ REVIEW QUESTIONS   ?

7 What evidence suggests that nerve impulses are not electricity but electrochemical events?

8 Why was the giant squid axon particularly appropriate for nerve research?

9 What is a polarized membrane?

10 What causes the inside of a neuron to become negatively charged?

11 Why does the polarity of a cell membrane reverse during an action potential?

12 What changes take place along a nerve cell membrane as it changes from a resting membrane to an action potential and then into a refractory period?

13 Why do nerve impulses move faster along myelinated nerve fibers?

14 What is the all-or-none response?

15 Use the model of the synapse to explain why nerve impulses move from neuron A to neuron B, but not from neuron B back toward neuron A.

16 Explain the functions of acetylcholine and cholinesterase in the transmission of nerve impulses.

17 Use a synapse model to explain summation.

RESEARCH IN CANADA

Memorial University of Newfoundland has one of Canada's busiest brain research centers. One of the most promising projects there is being conducted by Dr. John McLean. According to Dr. McLean, three cell groups in the brain produce transmitter chemicals that play important roles in development and normal brain function. Imbalances of these chemicals (i.e., serotonin, acetylcholine, and noradrenaline) can cause brain disorders. For example, serotonin imbalances have been linked to constrictions of blood vessels. During fetal development, imbalances of serotonin have been related to problems such as Down syndrome and autism. In adults, serotonin imbalances have been associated with hallucinations, suicidal depression, and schizophrenia. Noradrenaline imbalances have been linked to sleep disorders and learning problems. Acetylcholine imbalances have been associated with learning difficulties and the deterioration of memory processes in patients with Alzheimer's disease.

Dr. McLean is using cell markers to determine if and how these systems affect brain development when they are depleted. Using animal studies, Dr. McLean also intends to investigate factors that will enhance growth of these systems during development.

Investigations into how brain tissue can be repaired after injury are being carried out by Dr. Sergey Fedoroff at the University of Saskatchewan. Dr. Fedoroff is studying star-shaped cells of the brain and spinal cord. Employing tissue culture techniques, he has been able to grow the star-shaped cells, called *astrocytes*, and to collect some of the substances they produce. Dr. Fedoroff believes that many of these substances play an important role in the response of brain cells to disease or injury.

Brain Research

**DR. JOHN MCLEAN
DR. SERGEY FEDOROFF**

HOMEOSTASIS AND THE AUTONOMIC NERVOUS SYSTEM

The autonomic nervous system is part of the peripheral nervous system. All **autonomic nerves** are motor nerves that regulate the organs of the body without conscious control. **Somatic nerves**, by contrast, lead to muscles and are regulated by conscious control. Rarely do you consciously direct your breathing movements. Blood carbon dioxide and oxygen levels are monitored throughout the body. Once carbon dioxide or oxygen levels exceed or drop below the normal range, autonomic nerves act to restore homeostasis. The autonomic system maintains the internal environment of the body by adapting to the changes and demands of an external environment. During emergencies, your autonomic nervous system diverts blood flow from your stomach to the skeletal muscles, increases your heart and breathing rates, and increases your visual field by causing the pupils of your eyes to dilate.

The autonomic system is made up of two distinct and often opposing units, the **sympathetic** and **parasympathetic** systems. The sympathetic system prepares the body for stress, while the parasympathetic system restores normal balance. Sympathetic and parasympathetic nerves also differ in anatomy. Sympathetic nerves have a short preganglionic nerve and a longer postganglionic nerve; the parasympathetic nerves have a long preganglionic nerve and a shorter postganglionic nerve. The preganglionic nerves of both systems release acetylcholine, but the postganglionic nerve from the sympathetic system releases norepinephrine. The postganglionic nerves from the parasympathetic system release acetylcholine. The sympathetic nerves come from the thoracic vertebrae (ribs) and lumbar vertebrae (small of the back). The parasympathetic nerves exit directly from the brain or from either the cervical (the neck area) or caudal (tailbone) sections of the spinal cord. Nerves that exit directly from the brain are referred to as *cranial nerves*. An important cranial nerve is the **vagus nerve** (vagus meaning "wandering"). Branches of the vagus nerve regulate the heart, bronchi of the lungs, liver, pancreas, and the digestive tract.

Table 15.1 Effects of the Autonomic Nervous System

Organ	Sympathetic	Parasympathetic
Heart	Increases heart rate.	Decreases heart rate.
Digestive	Decreases peristalsis.	Increases peristalsis.
Liver	Increases the release of glucose.	Stores glucose.
Eye	Dilates pupil.	Constricts pupil.
Bladder	Relaxes sphincter.	Contracts sphincter.
Skin	Increases blood flow.	Decreases blood flow.
Adrenal gland	Causes release of epinepherine.	No effect

BIOLOGY CLIP

Lie detectors (polygraphs) monitor changes in the activity of the sympathetic nervous system. One of the components of a lie detector, the galvanic skin response, checks for small changes in perspiration. Because the perspiration contains salt, any amount of sweating will increase the flow of current. In theory, a stressful situation, like lying, would cause the stimulation of sympathetic nerves, which, in turn, would activate the sweat glands. Increased breathing and pulse rates are also monitored by lie detectors. Even the pupils of the eye dilate, at least a little, when a person is subjected to a stressful situation such as lying. Because lie detectors cannot always differentiate between lying and other stressful situations, they are not considered 100% accurate.

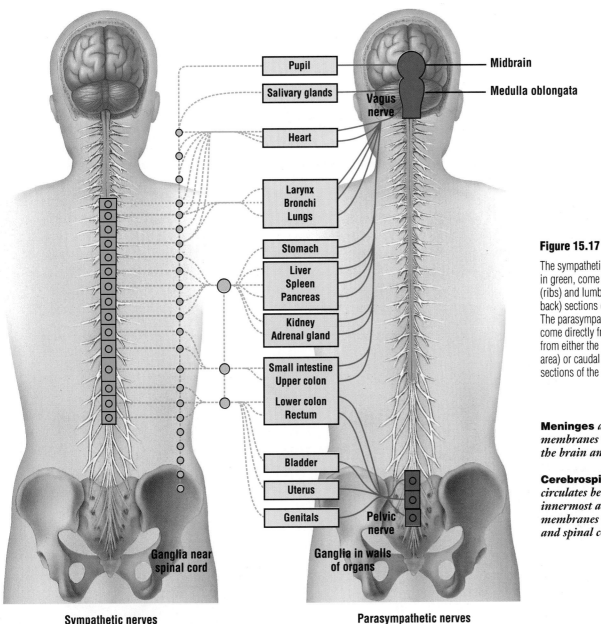

Pupil	Midbrain
Salivary glands	Medulla oblongata
	Vagus nerve
Heart	
Larynx Bronchi Lungs	
Stomach	
Liver Spleen Pancreas	
Kidney Adrenal gland	
Small intestine Upper colon	
Lower colon Rectum	
Bladder	
Uterus	
Genitals	Pelvic nerve

Ganglia near spinal cord

Ganglia in walls of organs

Sympathetic nerves

Parasympathetic nerves

Figure 15.17

The sympathetic nerves, shown in green, come from the thoracic (ribs) and lumbar (small of the back) sections of the spinal cord. The parasympathetic nerves come directly from the brain or from either the cervical (the neck area) or caudal (tailbone) sections of the spinal cord.

Meninges *are protective membranes that surround the brain and spinal cord.*

Cerebrospinal fluid *circulates between the innermost and middle membranes of the brain and spinal cord.*

CENTRAL NERVOUS SYSTEM

The central nervous system comprises the brain and spinal cord. The brain is formed from a concentration of nerve tissue in the anterior portion of animals, and acts as the coordinating center of the nervous system. Enclosed within the skull, the brain is surrounded by three protective membranes known as **meninges.** (Meningitis is caused by a bacterial or viral infection of the outer membranes of the brain.) The outer membrane is called the *dura mater*, the middle layer is the *arachnoid*, and the inner layer is the *pia mater*. **Cerebrospinal fluid** circulates between the innermost and middle meninges of the brain and through the central canal of the spinal cord. The cerebrospinal fluid acts both as a shock absorber and a transport medium, carrying nutrients to cells of the brain while relaying wastes from the cells to the blood. Physicians can extract cerebrospinal fluid from the spinal cord to diagnose bacterial or viral infection. The technique, referred to as a *spinal tap*, is used to identify poliomyelitis and meningitis.

Spinal Cord

The spinal cord carries sensory nerve messages from receptors to the brain and relays motor nerve messages from the brain to muscles and glands. Emerging from the skull through an opening called the *foramen magnum*, the spinal cord extends downward through a canal within the backbone. A cross section of the spinal cord reveals two types of nerve tissue: white matter and gray matter. The central gray matter consists of unmyelinated axons and cell bodies of motor neurons. The surrounding white matter is composed of unmyelinated nerve fibers of *interneurons*. The interneurons are organized into nerve tracts that connect the spinal cord with the brain. A dorsal nerve tract brings sensory information into the spinal cord, while a ventral nerve tract carries motor information from the spinal cord to the peripheral muscles and organs.

Figure 15.18

The spinal cord is protected by the vertebral column. Sensory nerves enter the spinal cord through the dorsal root ganglion, and motor nerves leave through the ventral root ganglion.

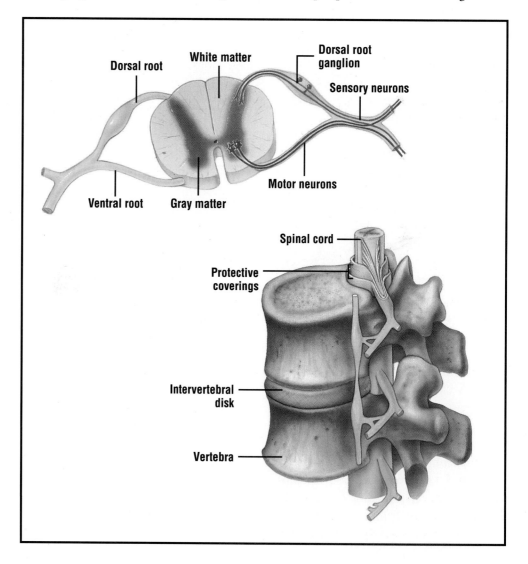

Brain

Brain complexity is what distinguishes humans from other animals. Clearly, what makes *Homo sapiens* unique can be related to superior brain development. Humans lack the strength and agility of other mammals of comparable size. Human hearing, vision, and sense of smell are unimpressive when compared with that of many other species. Unlike insects, the success of the species does not rest with reproductive capacity—humans reproduce relatively slowly. Human distinctiveness is related to intellect and reason—functions of the brain. However, despite its apparent uniqueness, the human brain has developmental links with other chordates. Like primitive vertebrates, the human brain is composed of three distinct regions: the forebrain, the midbrain, and the hindbrain.

The forebrain contains paired **olfactory lobes,** which are centers that receive information about smell. The **cerebrum** is also contained within the forebrain. These two giant hemispheres act as the major coordinating center from which sensory information and accompanying motor actions originate. Speech, reasoning, memory, and even personality reside within these paired cerebral hemispheres. The surface of the cerebrum is known as the **cerebral cortex.** Composed of gray matter, the cortex has many folds that increase surface area. Deep folds are known as *fissures.*

Recent research has demonstrated that information stored in one side of the brain is not necessarily present in the other. The right side of the brain has been associated with visual patterns or spatial awareness; the left side of the brain is linked to verbal skills. Your ability to learn may be related to the dominance of one of the hemispheres. A bundle of nerves referred to as the

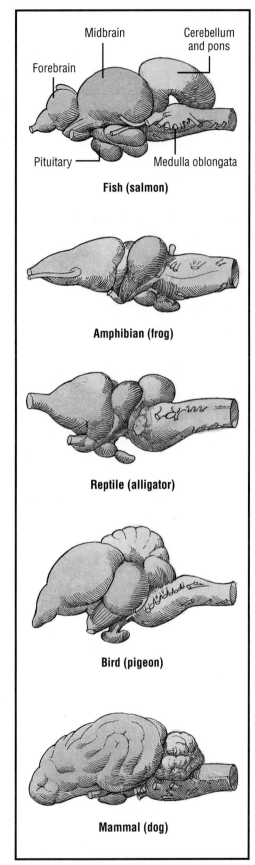

Fish (salmon)

Forebrain · Midbrain · Cerebellum and pons · Pituitary · Medulla oblongata

Amphibian (frog)

Reptile (alligator)

Bird (pigeon)

Mammal (dog)

Figure 15.19

The greatest changes in the evolution of the human brain occurred within the forebrain. Colored in blue, the forebrain is the site of reason, intellect, memory, language, and personality.

Olfactory lobes *are areas of the brain that detect smell.*

The **cerebrum** *is the largest and most highly developed part of the human brain. The cerebrum stores sensory information and initiates voluntary motor activities.*

The **cerebral cortex** *is the outer lining of the cerebral hemispheres.*

corpus callosum allows communication between the two hemispheres. Each hemisphere can be further subdivided into four lobes: the frontal lobe, the temporal lobe, the parietal lobe, and the occipital lobe. Table 15.2 lists the functions for each of the lobes.

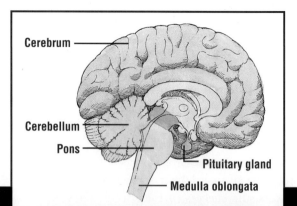

The **corpus callosum** *is a nerve tract that joins the two cerebral hemispheres.*

Figure 15.20

Sagittal section through the human brain. Note the many folds of the cerebral cortex. The corpus callosum connects the two cerebral hemispheres.

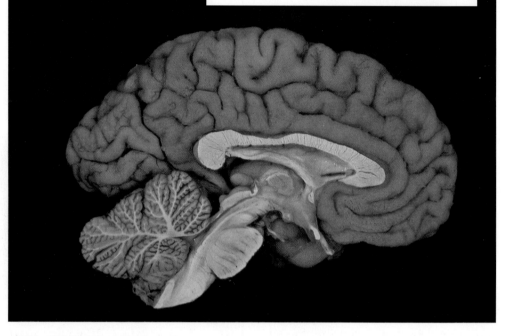

Table 15.2 The Lobes of the Cerebrum

Lobe	Function
1. Frontal lobe	Motor areas control movement of voluntary muscles (e.g., walking and speech). Association areas are linked to intellectual activities and personality.
2. Temporal lobe	Sensory areas are associated with vision and hearing. Association areas are linked to memory and interpretation of sensory information.
3. Parietal lobe	Sensory areas are associated with touch and temperature awareness. Association areas have been linked to emotions and interpreting speech.
4. Occipital lobe	Sensory areas are associated with vision. Association areas interpret visual information.

Below the cerebrum is the thalamus, which coordinates and interprets sensory information. Immediately below the thalamus is the hypothalamus. A direct connection between the hypothalamus and the pituitary gland unites the nervous system with the endocrine system.

The midbrain is less developed than the forebrain. Consisting of four spheres of gray matter, the midbrain acts as a relay center for some eye and ear reflexes. The hindbrain, as the name suggests, is found posterior to the midbrain and joins with the spinal cord. The cerebellum, pons, and medulla oblongata are the major regions of the hindbrain. The **cerebellum,** located immediately beneath the two cerebral hemispheres, is the largest section of the hindbrain. The cerebellum controls limb movements, balance, and muscle tone. Have you ever considered the number of coordinated muscle actions required to pick up a pencil? Your hand must be opened before you touch the pencil. The synchronous movement of thumb and fingers requires coordination of both excitatory and inhibitory nerve impulses.

The **pons,** meaning "bridge," is largely a relay station that passes information between the two regions of the cerebellum and between the cerebellum and the medulla. The posterior region of the hindbrain is the **medulla oblongata.** Nerve tracts from the spinal cord and higher brain centers run through the medulla, which acts as the connection between the peripheral and central nervous systems. As you read earlier, the medulla oblongata regulates involuntary muscle action. Breathing movements, the diameter of the blood vessels, and heart rate are but a few things regulated by this area of the hindbrain. The medulla oblongata acts as the coordinating center for the autonomic nervous system.

Brain Mapping

Once considered the seat of intellect and reason, the human brain is one of the most valued organs of the body. However, the brain was not always thought of in such lofty terms. Aristotle believed that the primary function of the brain was to cool the blood. His vision of the brain as a glorified air conditioner persisted for centuries. Galen, the great physician of ancient Rome, speculated that the brain functioned as a hydraulic system, in which fluid spirits were shunted through a series of ducts. The French philosopher, René Descartes, showed Galen's influence by describing the brain in terms of the flow of "animal spirits." However, unlike Galen, Descartes did associate the brain with coordinated sensory and motor activity. He believed that sensory information was stored in the brain and that motor function completed the loop. He even speculated that the pineal gland of the brain was the site of the soul.

By 1811, Franz Gall, an Austrian physician, suggested that sensory nerves terminate in specific sections of the gray matter of the brain. The idea that specific parts of the brain carry out specific functions led some of Gall's followers to suggest that mental ability or personality could be studied by investigating brain development. This train of thought led to the study of a false science called phrenology. Assuming that the brain case could reveal changes in brain development, phrenologists attempted to determine mental capacity by feeling for bumps on the head. Unfortunately, the relationship between skull size and mental capacity has been used to reinforce racist notions of the superiority of one race over another. Brain size is more often commensurate with body size than with intelligence. Ironically, the fact that Gall himself had a small head seemed to escape the phrenologists. The chart on the next page shows the relationship between brain size and body mass.

*The **cerebellum** is the region of the brain that coordinates muscle movement.*

*The **pons** is the region of the brain that acts as a relay station by sending nerve messages between the cerebellum and the medulla.*

*The **medulla oblongata** is the region of the hindbrain that joins the spinal cord to the cerebellum. The medulla is the site of autonomic nerve control.*

Table 15.3 Comparison of Brain Size to Body Mass

Animal	Brain size	Ratio of brain size to body mass
Chimpanzee	400 g	1/150
Gorilla	540 g	1/500
Human	1500 g	1/50
Elephant	6000 g	1/1000
Whale	9000 g	1/10 000

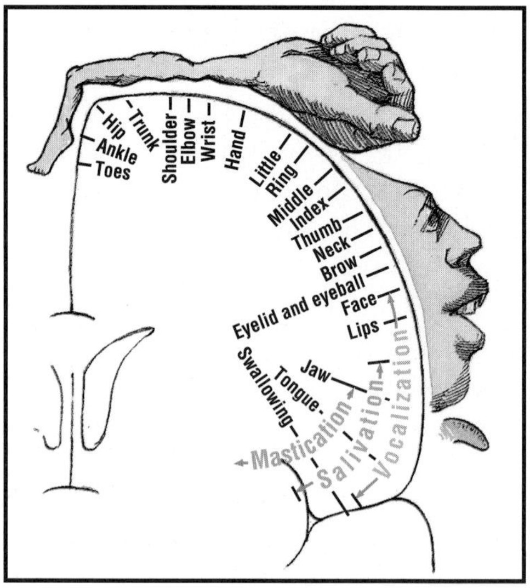

Figure 15.22

Regions of the body are drawn in proportion to the area of the motor cortex required to control the region.

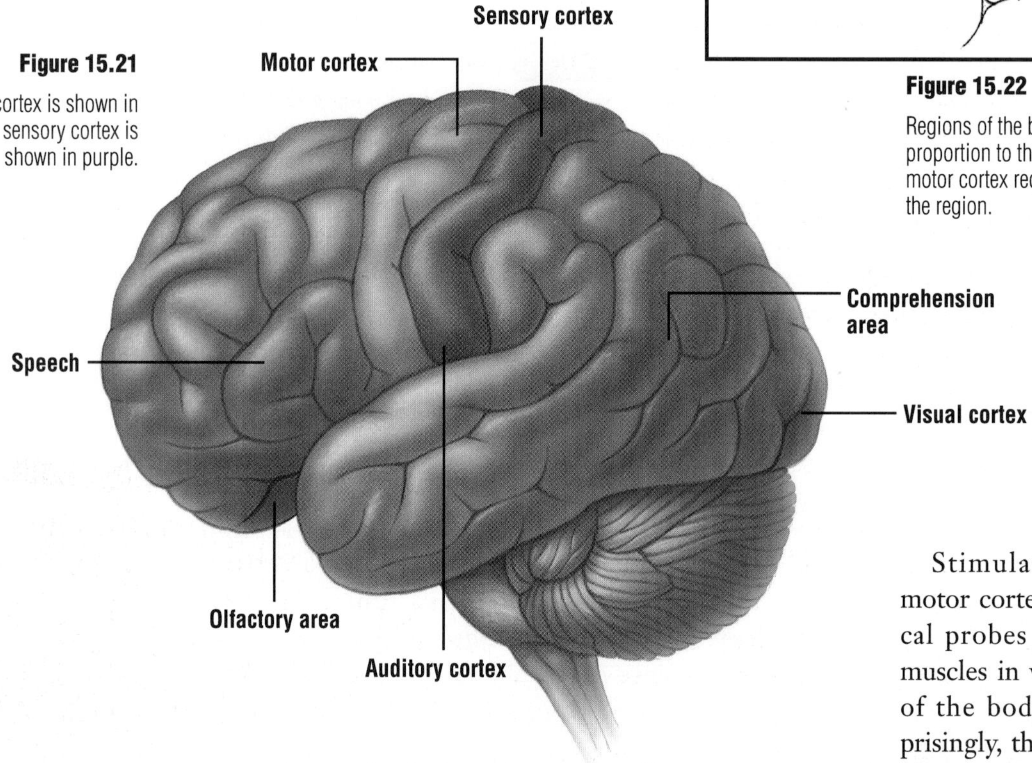

Figure 15.21

The motor cortex is shown in yellow and the sensory cortex is shown in purple.

Patients who had strokes eventually provided scientists with evidence that certain areas of the brain have specific functions. Any factor that reduces blood flow to cells of the brain can cause a stroke. Strokes that occur in the right side of the motor cortex cause paralysis of the left side of the body. A patient's recovery from a stroke provides evidence that other cells can be trained to assume the function of previously damaged cells. With time, speech, hearing, and some limb movements may partially be restored.

Stimulation of the motor cortex by electrical probes can trigger muscles in various parts of the body. Not surprisingly, the number of nerve tracts leading to the thumb and fingers is greater than those leading to the arms or legs, since the thumb and fingers are capable of many fine motor movements. Wrist and arm movements, by contrast, are limited and, therefore, regulated by fewer nerves. Figure 15.22 shows parts of the human body drawn in proportion to the number of motor nerves that control them. Note the size of the tongue and mouth. Human speech depends on subtle changes in the position of the tongue and mouth.

RESEARCH IN CANADA

Wilder G. Penfield (1891–1976), founder of the Montreal Neurological Institute, was the foremost pioneer in brain mapping. Using electrical probes, Penfield located three speech areas within the cerebral cortex. Interestingly, the predominant speech areas reside on the left side of the brain. Penfield's finding dismissed the once-held notion that the two hemispheres were mirror images of each other.

Brain Mapping

In a classic experiment, Penfield applied an electrode to a particular speech area and then showed the subject a picture of a foot. Although the subject was able to recognize the foot, the word would not come. Once electrical stimulation ceased, the subject said "foot." Penfield concluded that the thought processes associated with recognizing the foot reside within a particular location.

Penfield spent a great deal of his time mapping the cerebral cortex of epileptics. Epilepsy is often associated with injuries to the cerebral cortex; electrical "storms" spread across the damaged tissue, creating anything from a tingly sensation to violent convulsions. Penfield developed a surgical technique by which he removed a section of the skull and probed the brain with electrodes to locate the diseased area. Since the brain does not contain any sensory receptors, the surgeon can be guided by the conscious patient—only a local anesthetic is required. The damaged tissue can be removed, but adjacent functional tissues must not be extracted. During the mapping process, the surgeon stimulates various areas of the brain, and the patient talks to an observer to ensure preservation of functional tissue. Penfield noted that some patients would begin to laugh as they recalled past events. One woman heard songs as clearly as if a record player were in the operating room. Incredibly, the song stopped once the electrode was removed from that area of the brain.

DR. WILDER G. PENFIELD

CASE STUDY

PHINEAS GAGE

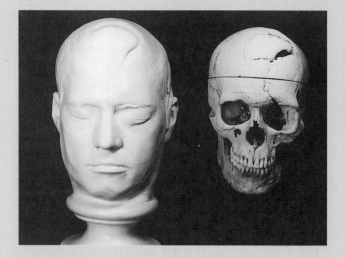

The tamping iron entered the skull immediately above the left eye and exited through the back of the skull of Phineas Gage.

Objective

To investigate brain mapping through a case study.

Background Information

In 1948, a thunderous explosion vibrated throughout a Vermont mine. A quarry worker, Phineas Gage, lay on the ground impaled by a tamping iron. Apparently Gage had set off blasting caps by tamping them with a large iron bar. A closer examination revealed that the bar had entered his skull immediately above the left eye and exited through the back of the skull. Incredible as it may seem, Phineas Gage recovered from the explosion and lived for another 12 years. He showed no signs of physical impairment. His vision, hearing, balance, and speech remained intact. However, he did experience one change: the once quiet and thoughtful Phineas became irresponsible and short-tempered. Spontaneous temper tantrums would send him into a fit of profanity. What could have triggered such changes?

Case-Study Application Questions

1 Which lobe of Gage's brain had been damaged?
2 Provide a hypothesis to explain why Phineas Gage's personality changed. How would you test your hypothesis?
3 In one operation, first performed by Antonio Muniz, some of the nerve tract between the thalamus and the frontal lobes was severed. Why might a physician attempt such an operation? ■

FRONTIERS OF TECHNOLOGY: BEAMs

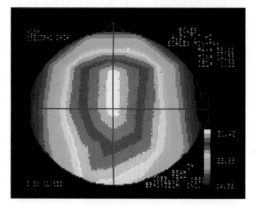

Brain electrical activity maps (BEAMs) use color pictures of the brain to identify malfunctions. The BEAM tracings look much like topographical land maps. Unlike Wilder G. Penfield's early mapping techniques, which required surgery, BEAM technology maps brain-wave activity on the surface of the skull. The technique combines two existing technologies: the EEG (electroencephalograph) and the computer. After 20 electrodes are applied to the scalp, the normal EEG translates the electrical activity near each electrode into a series of spike potentials. Unfortunately, when thousands of nerves fire, small abnormalities are often difficult to identify. The BEAM technique feeds the data gathered by an EEG into a computer that produces and displays a series of colored contours of brain activity. Different colors and shades represent different degrees of electrical activity. The assembled sequences of brain activity provide, in effect, a movie of brain action.

NATURAL AND ARTIFICIAL PAINKILLERS

Have you ever heard a runner talk about the euphoria he or she feels while exercising? This feeling is produced by a group of natural painkillers, called **endorphins** and **enkaphalins,** which are manufactured by the brain. Apparently, pain is interpreted by specialized cells in the *substantia gelatinosa*, or *SG*. When stimulated, the SG cells produce a transmitter chemical that informs the injured organ or tissue of the damage. The greater the amount of pain transmitter attached to the injured organ, the greater is the perception of pain. However, when endorphins and enkaphalins occupy the receptor sites on the SG cell, the pain transmitter is not produced and pain is reduced.

Opiates such as heroin, codeine, and morphine work in much the same way as endorphins. Opiates attach to the neurons in the central nervous system, preventing the production of the pain transmitters. Heroin and opium not only reduce pain but also create a feeling of tranquility. The intake of opiates causes the production of the body's natural painkillers to decrease. Therefore, to achieve a consistent effect, the user must continue to take the drug. When use of the drug stops, the SG receptors are soon vacant and the pain transmitter is produced in abundance.

Drugs that act as depressants, such as Valium and Librium, appear to enhance the action of inhibitory synapses. The synthesis of inhibitory transmitter chemicals like GABA often increases under the influence of these drugs. Alcohol differs from the drugs mentioned above in that it does not act directly on the synapse. Correctly categorized as a depressant, alcohol acts directly on the plasma membrane to increase threshold levels.

Dr. James Reynolds from Memorial University of Newfoundland is investigating the actions of commonly used sedative-hypnotic drugs on the central nervous system. Drugs such as ethanol, barbiturates, and benzodiazepines are among the most widely used and abused of this class of drugs. However, little information exists on how long-term use might cause serious effects such as tolerance, dependence, drug withdrawal seizures, or brain damage. A more complete understanding of the molecular activities of such drugs is essential before the effects of long-term exposure can be explained.

■ REVIEW QUESTIONS ■ ?

18 Compare the structure and function of autonomic and somatic nerves.

19 State the two divisions of the autonomic nervous system and compare them by both structure and function.

20 What is the vagus nerve?

21 State the function of the meninges and the cerebrospinal fluid of the brain.

22 Describe the functions of the following: cerebrum, olfactory lobes, corpus callosum, cerebellum, thalamus, pons, and medulla oblongata.

23 List the four regions of the cerebral cortex and state the function of each.

24 Explain the medical importance of brain-mapping experiments.

25 What are endorphins? How do they work?

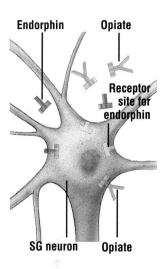

Figure 15.23

Opiates have a similar geometry to endorphins. This allows opiates to take up neuron receptor sites designed for endorphins.

Endorphins *belong to a group of chemicals classified as neuropeptides. Containing between 16 and 31 amino acids, endorphins are believed to reduce pain.*

Enkephalins *belong to a group of chemicals classified as neuropeptides. Containing five amino acids, the enkephalins are produced by the splitting of larger endorphin chains.*

SOCIAL ISSUE:
A New View of Drug Addiction

Have you ever wondered why some people appear to be more prone to alcohol or drug addiction? Alcoholics are not necessarily those individuals who drink the most, but rather individuals who are affected most by drinking. Although environmental factors weigh greatly on the tendency toward alcohol and drug abuse, they may not be the only cause. Some researchers have hypothesized that chemical imbalances may make certain individuals more prone to drug and alcohol abuse. Genetic factors may also come into play. Scientists have already been able to breed animals that crave opiates or stimulants. Predictably, the offspring of these animals show the same dependencies.

The gene-related approach to understanding drug dependency took a giant step forward when a gene related to alcoholism was located in April 1990 at the University of Texas and the University of California at Los Angeles. The gene is linked to the receptors for a nerve transmitter called dopamine. High levels of dopamine in particular areas of the brain produce the sensation of pleasure. Could alcoholics suffer from low levels of dopamine? An incredible 77% of the alcoholics studied had the defective gene. The researchers suggest that alcohol corrects the chemical imbalance brought about by the defective gene.

Their hypothesis is supported by the fact that some kinds of depression are related to low levels of dopamine. Some antidepressants act by increasing dopamine levels. It has been suggested that people who suffer depression may be more susceptible to drugs that increase dopamine production.

Statement:

There is a genetic link between chemical imbalances and drug dependency.

Point

- The fact that alcoholics and drug addicts are often found in the same family supports the hypothesis that gene-related chemical imbalances are at the root of the problem.

- The prospect of finding a cure for chemical imbalances is on the horizon. By curing the chemical imbalance with a nonaddictive drug, substance abuse could be eliminated.

Counterpoint

- Environmental factors such as poverty and stress are the real cause of substance abuse. The fact that alcoholism and drug abuse can be associated with some families does not support the gene hypothesis. Members of the same family are exposed to similar environmental factors.

- The idea that certain people carry genes that make them susceptible to drug dependency or alcoholism is dangerous. Companies might refuse to employ individuals who carry those genes even if they do not drink or take drugs. Insurance companies might refuse coverage to those individuals they deem genetically inferior and, therefore, at high risk.

Research the issue.
Reflect on your findings.
Discuss the various viewpoints with others.
Prepare for the class debate.

CHAPTER HIGHLIGHTS

- The nervous system coordinates the activities of other body systems. Nerves enable the body to respond to stimuli and maintain homeostasis.
- Neurons conduct nerve impulses. Sensory neurons carry impulses to the central nervous system. Interneurons carry impulses within the central nervous system. Motor neurons carry impulses from the central nervous system to effectors.
- The neuron is composed of functionally distinct components. Dendrites are projections of cytoplasm that carry impulses toward the cell body. An axon is an extension of cytoplasm that carries nerve impulses away from the cell body. The myelin sheath is a fatty covering over the axon of a nerve cell. The nodes of Ranvier are the regularly occurring gaps between sections of myelin sheath along the axon.
- The nerve impulse results from the movement of ions across the nerve membrane.
- Threshold is the minimum level of a stimulus required to produce a response.
- The all-or-none response of a nerve or muscle fiber indicates that the nerve or muscle responds completely to a stimulus or not at all.
- Synapses are the regions between neurons or between neurons and effectors. Synapses may be excitatory or inhibitory.
- Summation is the effect produced by the accumulation of transmitter chemicals from two or more neurons.
- Somatic nerves lead to skeletal muscle and are under conscious control.
- Autonomic nerves are motor nerves designed to maintain homeostasis. Autonomic nerves are not under conscious control.
- Sympathetic nerves are a division of the autonomic nervous system and prepare the body for stress.
- Parasympathetic nerves are a division of the autonomic nervous system and are designed to return the body to normal resting levels following adjustments to stress.
- The central nervous system consists of the brain and spinal cord. The brain is responsible for the coordination of sensory and motor nerve activity.
- Meninges are protective membranes that surround the brain and spinal cord. Cerebrospinal fluid circulates between the innermost and middle membranes of the brain and spinal cord.
- The cerebrum is the largest and most developed part of the human brain. The cerebrum stores sensory information and initiates voluntary motor activities.
- The corpus callosum is a nerve tract that joins the two cerebral hemispheres.
- The cerebellum is the region of the brain that coordinates muscle movement.
- The pons is the region of the brain that acts as a relay station by sending nerve messages between the cerebellum and the medulla.
- The medulla oblongata is the region of the hindbrain that joins the spinal cord to the cerebellum. The medulla is the site of autonomic nerve control.
- Endorphins belong to a group of chemicals classified as neuropeptides. Containing between 16 and 31 amino acids, endorphins are believed to reduce pain.

APPLYING THE CONCEPTS

1 How did Luigi Galvani's discovery that muscles twitch when stimulated by electrical current lead to the discovery of the ECG and EEG?
2 During World War I, physicians noted a phenomenon called "phantom pains." Soldiers with amputated limbs complained of pain or itching in the missing limb. Use your knowledge of sensory nerves and the central nervous system to explain this phenomenon.
3 Explain the importance of a properly functioning myelin sheath by outlining the pathology of multiple sclerosis.
4 Explain why damage to the gray matter of the brain is permanent, while minor damage to the white matter may only be temporary.
5 Use information that you have gained about threshold levels to explain why some individuals can tolerate more pain than others.

6 Use the diagram below to answer the following question.

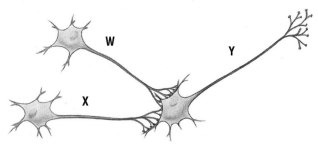

The neurotransmitter released from nerve X causes the postsynaptic membrane of nerve Y to become more permeable to sodium. However, the neurotransmitter released from nerve W causes the postsynaptic membrane of nerve Y to become less permeable to sodium but more permeable to potassium. Why does the stimulation of neuron X produce an action potential in neuron Y, but fail to produce an action potential when neuron X and W are stimulated together?

7 Botulism (a deadly form of food poisoning) and curare (a natural poison) inhibit the action of acetylcholine. What symptoms would you expect to find in someone exposed to botulism or curare? Provide an explanation for the symptoms.

8 Nerve gas inhibits the action of cholinesterase. What symptoms would you expect to find in someone exposed to nerve gas? Provide an explanation for the symptoms.

9 A patient complains of losing his sense of balance. A marked decrease in muscle coordination is also mentioned. On the basis of the symptoms provided, which area of the brain might a physician look to for the cause of the symptoms?

10 In a classic experiment performed by Robert Sperry, two patients viewed the word "cowboy." The first subject acted as a control. The second subject had his corpus callosum severed. Note the right and left hemispheres have been supplied with different stimuli. Predict the results of the experiment and provide an explanation for your predictions.

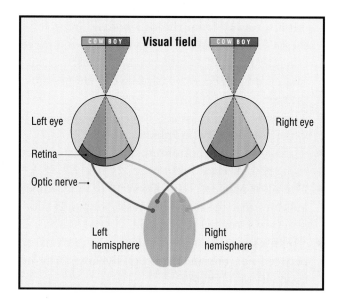

CRITICAL-THINKING QUESTIONS

1 People with Parkinson's disease have low levels of the nerve transmitter dopamine. In 1982, a group of Swedish scientists grafted cells from a patient's adrenal glands into the patient's brain. The adrenal glands produce dopamine. In July 1987, Swedish scientists announced the first transplant of human fetal brain cells into other animals. In September 1987, officials at several hospitals in Mexico City announced the transplant of brain tissue from dead fetuses into patients with Parkinson's disease. Although this research is very new, it would appear that the fetal cells begin producing dopamine. Patients with Parkinson's disease who have received the transplanted tissue have demonstrated remarkable improvement. Comment on the ethics of this research.

2 Research by Dr. Bruce Pomerantz of the University of Toronto has revealed a link between endorphins and acupuncture. Dr. Pomerantz believes that acupuncture needles stimulate the production of pain-blocking endorphins. Although acupuncture is a time-honored technique in the east, it is still considered to be on the fringe of modern western medical practice. Why have western scientists been so reluctant to accept acupuncture?

3 Painkillers are big business. Television commercials tell us that the answer to our headaches, sore backs, or sore muscles is a pill. Should people be encouraged to take pills? What are the implications for individuals who refuse to give their children pills for pain?

4 Aspirin and Tylenol were both marketed long before their action was defined. Although both drugs are generally considered safe, they are not without side effects. Aspirin in high dosages can destroy the protective lining of the stomach. Tylenol in high dosages can cause liver damage. Should the biochemical mechanism of drug action be defined before a drug is marketed? Should drugs that produce side effects be sold over the counter? What tests would you like to see before drugs are approved for sale?

5 The EEG has been used to legally determine death. Although the heart may continue to beat, the cessation of brain activity signals legal death. Ethical problems arise when some brain activity remains despite massive damage. Artificial resuscitators can assume the responsibilities of the medulla oblongata and regulate breathing movements. Feeding tubes can supply food, and catheters can remove wastes when voluntary muscles can no longer be controlled. The question of whether life should be sustained by artificial means has often been raised. Should a machine like the EEG be used to define the end of life?

6 According to the study of craniometry, males were once considered more intelligent than females because they have larger skulls. Outline some of the ethical dangers associated with using skull size to define intelligence.

ENRICHMENT ACTIVITIES

1 Inexpensive galvanic skin-response detectors can be purchased from scientific companies. Although the unit is not a complete lie detector (polygraph), you may wish to test its accuracy.

2 Suggested reading:
- Allport, Susan. *Explorers of the Black Box: The Search for the Cellular Basis of Memory*. New York: W.W. Norton, 1986.
- Gould, Stephen Jay. *Mismeasure of Man*. New York: W.W. Norton, 1981. This book takes a critical view of the pseudoscience of craniometry. Craniometry attempted to link intelligence with skull size and was often inspired by racism. Gould shows how the selective manipulation of data can be used either intentionally or unintentionally to promote untruths.

- Maranto, Gina. "The Mind within the Brain." *Discover*, May 1984, p. 36.
- Montgomery, Geoffrey. "The Mind in Motion." *Discover*, March 1989, p. 58.
- ——. "Molecules of Memory." *Discover*, December 1989, p. 46.
- Penfield, Wilder. *The Mysteries of the Mind: A Critical Study of Consciousness and the Human Mind*, Princeton University Press, 1975.
- Restak, Richard. *The Brain: The Last Frontier*, Warner Books, 1979.

*S*pecial *Senses*

IMPORTANCE OF SENSORY INFORMATION

Sensory neurons supply the central nervous system with information about the external environment and the quality of our internal environment. Specialized chemoreceptors in the carotid artery provide the central nervous system with information about blood carbon dioxide and oxygen levels. Special osmoregulators in the hypothalamus monitor water concentration in the blood, and highly modified stretch receptors monitor blood pressure in arteries.

Environmental stimuli are conveyed to the central nervous system through sensory nerves. The sounds of thunder, the chill of a cold day, and the smells of food are relayed to the brain by sensory neurons. David Hume, the great Scottish philosopher, once concluded that humans are nothing more than the sum of their experiences. Our experiences, or what some philosophers call reality, exist because of a sensory nervous system. Each of us experiences sensations as highly personal events. Your perception of color is not the same as your friend's. You hear different things while listening to the same song. Indeed, while some individuals enjoy a particular song, others would rather not hear it.

Figure 16.1

The central nervous system processes environmental stimuli. Sound waves can be interpreted as music or noise.

As you learned in the previous chapter, information about the environment is transmitted to the brain along neurons. A neuron that carries visual information from the eye functions in essentially the same manner as one that carries auditory information from the ear. Both are electrochemical impulses. Your ability to differentiate between visual and auditory information depends on the area of the brain that receives the impulse. The brain interprets your reality. Visual information is stored in the posterior portion of the occipital lobe of the cerebrum, while auditory information is stored in the temporal lobe of the cerebrum. The similarities between neurons would be evident if a neuron carrying visual information could be rerouted from the occipital lobe to the temporal lobe or if an auditory neuron could be rerouted from the temporal lobe to the occipital lobe. A bolt of lightning across the sky would be heard, while the crashing thunder that follows would be seen. No doubt, the prospects of hearing lightning and seeing thunder would be confusing.

Clearly, what enables you to distinguish visual and auditory information resides in the sensory receptors. Light-sensitive receptors within the retina of the eye are stimulated by light, not sound. A group of specialized temperature receptors in the skin identify cold, while other ones identify heat. How do different receptors respond to different stimuli? How are different stimuli converted into electrochemical events? How do you identify the intensity of different stimuli? How does the brain interpret stimuli?

> **BIOLOGY CLIP**
> Occasionally a sensory receptor can be activated by stimuli that it was not designed to detect. Boxers who receive a blow to the eye often see stars. The pressure of the blow stimulates the visual receptors at the back of the eye, and the blow is interpreted as light. Similarly, a blow near the temporal lobe can often be interpreted as a bell ringing.

WHAT ARE SENSORY RECEPTORS?

A stimulus is a form of energy. Sensory receptors convert one source of energy into another. For example, taste receptors in your tongue convert chemical energy into a nerve action potential, a form of electrical energy. Light receptors of the eye convert light energy into electrical energy, and balance receptors of the inner ear convert gravitational energy and mechanical energy into electrical energy.

Sensory receptors are highly modified ends of sensory neurons. Often, different sensory receptors and connective tissues are grouped within specialized sensory organs, such as the eye or ear. This grouping of different receptors often amplifies the energy of the stimulus to ensure that the stimulus reaches threshold levels. Table 16.1 lists different types of sensory receptors found within the body.

Sensory receptors *are modified ends of sensory neurons that are activated by specific stimuli.*

Table 16.1 The Body's Sensory Receptors

Receptor	Stimulus	Information provided
Taste	Chemical	Taste buds identify specific chemicals.
Smell	Chemical	Olfactory cells detect presence of chemicals.
Pressure	Mechanical	Movement of the skin or changes in the body surface.
Proprioceptors	Mechanical	Movement of the limbs.
Balance (ear)	Mechanical	Body movement
Outer ear	Sound	Signals sound waves.
Eye	Light	Signals changes in light intensity, movement, and color.
Thermo-regulators	Heat	Detect the flow of heat.

A network of touch, high-temperature, and low-temperature receptors are found all over the skin. A simple experiment indicates that sensations occur in the brain and not the receptor itself. When the transmitter chemical released by the sensory neuron is blocked, the sensation stops. The brain registers and interprets the sensation. This phenomenon is supported by brain-mapping experiments. When the sensory region of the cerebral cortex is excited by mild electrical shock at the appropriate spot, the sensation returns even in the absence of the stimulus.

Despite an incredible collection of specialized sensory receptors, much of your environment remains undetected. What you detect are stimuli relevant to your survival. For example, consider the stimuli from the electromagnetic spectrum. You experience no sensation from radio waves or from infrared or ultraviolet wavelengths. Humans can only detect light of wavelengths between 350 and 800 nm. Your range of hearing, compared with that of many other species, is also limited.

Temperature receptors do not work as thermometers, which detect specific temperatures. Heat and cold receptors are adapted to signal changes in environmental temperatures. The following simple demonstration emphasizes how the nervous system responds to change. Fill three bowls with hot, room-temperature, and cold water respectively. Place your right hand in the cold water and your left hand in the hot water. Allow the hands to adjust to the temperature and then transfer both hands to the bowl that contains room-temperature water. The right hand, which had adjusted to the cold water, now

feels warm. The left hand, which had adjusted to the hot water, now feels cold.

Most animals can tolerate a wide range of temperatures, but are often harmed by rapid temperature changes. For example, a rapid change in temperature of 4 °C will kill some fish. Even people have succumbed to an unexpected plunge in very cold or hot water. This principle can be dramatized by what has come to be known as the "hot frog" experiment. If a frog is placed in a beaker of water above 40 °C, the frog will leap out immediately. However, if the frog is placed in room-temperature water, and the temperature is slowly elevated, it will remain in the beaker. The frog's skin receptors have time to adjust.

Sensory adaptation occurs once the receptor becomes accustomed to the stimulus. The neuron ceases to fire even though the stimulus is still present. This would seem to indicate that the new environmental condition is not dangerous. The same principle of adaptation can be applied to touch receptors in the skin. Generally, the receptors are only stimulated when clothes are put on or taken off. Sensory information assuring you that your clothes are still on your body is usually not required.

Sensory adaptation *occurs once you have adjusted to a change in the environment.*

Figure 16.2

The right hand will adjust to the cold water and the left hand will adjust to the warm water.

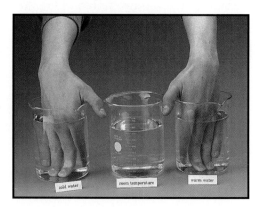

TASTE AND SMELL

Taste receptors enable you to differentiate between foods and non-edible matter. One theory that attempts to explain the extinction of the dinosaur suggests that it was poorly developed taste buds that eventually accounted for the disappearance of the giant reptiles. According to this theory, dinosaurs were not able to identify the bitter taste of the newly evolving poisonous plants.

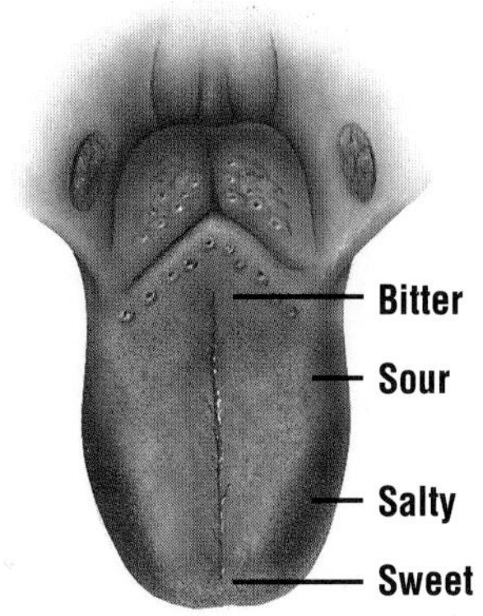

Taste receptors are found in different locations in different species throughout the animal kingdom. For example, octopuses have taste receptors located on their tentacles, crayfish have taste receptors situated on their antennae, and insects have taste receptors on their legs. Human taste receptors are centralized within the taste buds of the tongue. Once dissolved, specific chemicals stimulate the receptors within the taste buds. Scientists believe that four types of taste (salty, sweet, sour, and bitter) can be detected by specialized taste receptors, each of which is located in a specific area of the tongue. Figure 16.3 shows the four regions. Many individuals place an Aspirin near the back of the tongue believing that the bitter-tasting pill will spend less time in the mouth. Why might you suggest placing the pill near the tip of the tongue?

Olfactory cells are located in the nasal cavity (*olfactory* refers to the sense of smell). Airborne chemicals combine with the finely branched receptor ends on olfactory cells to create an action potential. Scientists believe that chemicals with specific geometry gain access to a specific receptor site so that the chemicals combine with complementary receptors. The impulse is carried to the frontal lobe of the brain for interpretation.

Experience tells you that your sense of taste and smell work together. Have you ever noticed that a cold reduces the taste of your food? Clogged nasal passages reduce the effectiveness of the olfactory cells. Since you use both smell and taste receptors to experience food, the diminished taste is actually the result of your reduced capacity to smell the food. The cooperation of the senses of smell and taste is well known to wine tasters. Have you ever noticed how wine tasters smell the wine before sipping it? You may have used the same test on milk that you believe has soured.

Bitter
Sour
Salty
Sweet

Figure 16.3

The human tongue.

Figure 16.4

(a) Specialized hairlike receptor ends in the tongue detect specific chemicals. Axons from the sensory neurons are shown in yellow. (b) Taste buds of a rabbit tongue. The taste bud appears as a bulblike structure.

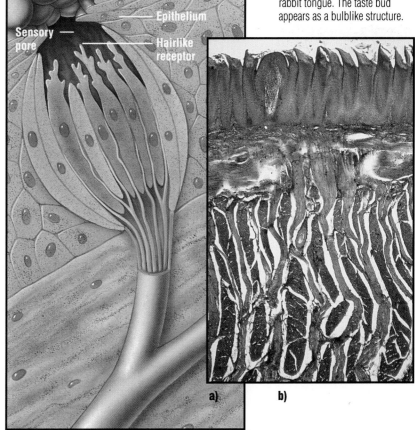

Epithelium
Sensory pore
Hairlike receptor

a)

b)

Olfactory cells demonstrate the phenomenon of sensory adaptation. You may have noticed how a sharp smell tends to disappear after you have been exposed to it for a long period of time. People who live in cities with pulp mills often comment on how they adjust to the odor after a while. Similarly, people who visit the ocean are at first impressed by its characteristic odor. However, the odor seems to fade after a short period of time.

The ... three separate layers: t... ... layer, and the retina. ... ...ermost layer of the eye. ... ...ve layer, the white fibro... the eye's shape. The front ... covered by a clear, bulging ... which acts as the window to the ey... bending light toward the *pupil*. Like all tis-

■ REVIEW QUESTIONS ■ ?

1 Identify a sensory receptor for each of the following stimuli: chemical energy, mechanical energy, heat energy, light energy, and sound energy.

2 Do sensory receptors identify all environmental stimuli? Give examples to back up your answer.

3 Explain the concept of "sensory adaptation" by using examples of olfactory stimuli and auditory stimuli.

4 Draw a diagram of the various chemoreceptors on the tongue.

5 Why are you less able to taste food when you have a cold?

*The **sclera** is the outer covering of the eye that supports and protects the eye's inner layers.*

*The **cornea** is a transparent tissue that refracts light toward the pupil.*

*The **aqueous humor** supplies the cornea with nutrients and refracts light.*

*The **choroid layer** is the middle layer of the eye. Pigments prevent scattering of light in the eye by absorbing stray light. Many blood vessels are found in this layer.*

*The **iris** regulates the amount of light entering the eye.*

> **B I O L O G Y   C L I P**
>
> Despite the wide range of eye colors, the iris contains a single pigment called *melanin*. Eyes appear blue because of a lack of activated melanin. The blue color is produced as light penetrates the clear aqueous humor. Only the shortest wavelength of visible light, blue light, is scattered. With increases in melanin, combinations of other wavelengths of light are reflected and other colors are produced.
>
> You may have noticed that all babies have blue eyes at birth. Within a few months, the melanin migrates toward the surface of the iris and the eyes change color.

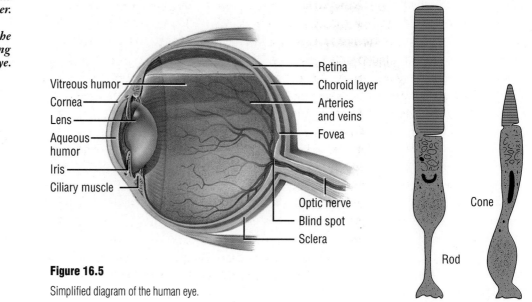

Figure 16.5

Simplified diagram of the human eye.

Vitreous humor
Cornea
Lens
Aqueous humor
Iris
Ciliary muscle
Retina
Choroid layer
Arteries and veins
Fovea
Optic nerve
Blind spot
Sclera
Cone
Rod

sues, the cornea requires oxygen and nutrients. However, the cornea is not supplied with blood vessels—blood vessels would cloud the transparent cornea. Most of the oxygen is absorbed from gases dissolved in tears. Nutrients are supplied by the **aqueous humor,** a chamber of transparent fluid behind the cornea.

The middle layer of the eye is called the **choroid layer.** Pigmented granules within the layer prevent light that has entered the eye from scattering. Toward the front of the choroid layer is the **iris.** The iris is composed of a thin circular muscle that acts as a diaphragm, controlling the size of the pupil opening.

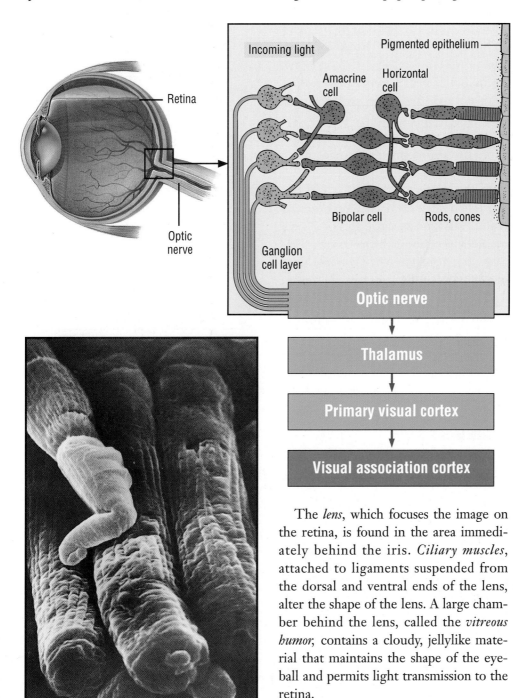

Figure 16.6

The pathway leading from the retina to the brain.

The *lens,* which focuses the image on the retina, is found in the area immediately behind the iris. *Ciliary muscles,* attached to ligaments suspended from the dorsal and ventral ends of the lens, alter the shape of the lens. A large chamber behind the lens, called the *vitreous humor,* contains a cloudy, jellylike material that maintains the shape of the eyeball and permits light transmission to the retina.

The innermost layer of the eye is the **retina,** which is composed of three different layers of cells: light-sensitive cells, bipolar cells, and cells from the optic nerve. The light-sensitive cell layer is positioned next to the choroid layer. Two different types of light-sensitive cells are the rods and cones. The **rods** respond to low-intensity light; the **cones,** which require high-intensity light, identify color. Both rods and cones act as the sensory receptors. Once excited, the nerve message is passed from the rods and cones to the bipolar cells, which, in turn, relay the message to the cells of the optic nerve. The optic nerve carries the impulse to the central nervous system.

In the center of the retina is a tiny depression referred to as the **fovea centralis.** The most sensitive area of the eye, the fovea centralis contains cones packed very close together. When you look at an object, most of the light rays fall on the fovea centralis. The rods surround the cones. This may explain why you may see an object from the periphery of your visual field without identifying its color. There are no rods and cones in the area in which the optic nerve comes in contact with the retina. Because of this absence of photosensitive cells, this area is appropriately called the **blind spot.**

*The **retina** is the innermost layer of the eye.*

***Rods** are photoreceptors used for viewing in dim light.*

***Cones** are photoreceptors that identify color.*

*The **fovea centralis** is the most sensitive area of the retina and contains only cones.*

*The **blind spot** is the area in which the optic nerve attaches to the retina.*

Table 16.2 Parts of the Eye

Structure	Function
Outer layer:	
Sclera	Supports and protects delicate photocells.
Cornea	Refracts light toward the pupil.
Middle layer:	
Aqueous humor	Supplies cornea with nutrients and refracts light.
Choroid	Contains pigments that prevent scattering of light in the eye by absorbing stray light. Also contains blood vessels.
Iris	Regulates the amount of light entering the eye.
Vitreous humor	Maintains the shape of the eyeball and permits light transmission to the retina.
Lens	Focuses the image on the retina.
Pupil	Functions as the hole in the iris.
Inner layer:	
Retina	Contains photoreceptors.
Rods	Used for viewing in dim light.
Cones	Identify color.
Fovea centralis	Most sensitive area of the retina; contains only cones.
Blind spot	The area in which the optic nerve attaches to the retina.

RESEARCH
IN CANADA

Dr. Roseline Godbout from the Cross Cancer Institute at the University of Alberta is investigating gene expression in the developing retina. Dr. Godbout is one of Canada's most promising scientists in a relatively new field: molecular neurobiology.

The retina is a highly specialized neurological network. In much the same way as information is processed in the body's most complicated nerve network—the brain—the retina continuously receives sensory information and adjusts to environmental stimuli during your waking hours.

Genes and the Retina

The conversion of light energy to electrochemical nerve impulses is due to the partnership of two types of cells that make up the retina: the neurons and glial cells. Although both types of cells differ in both structure and function, the neuron and glial cells originate from a common ancestor. Dr. Godbout is interested in determining how cells in the immature retina differentiate to become either glial cells or neurons in the retina. She hopes to be able to identify and characterize genes that regulate cell development in the retina. By identifying the genes in the normal retina, researchers may eventually be able to identify genes that are abnormally expressed in various eye disorders.

DR. ROSELINE GODBOUT

LIGHT AND VISION

The Greek philosopher Democritus first speculated about how the eye worked during the fifth century B.C. In his hypothesis, matter was composed of indivisible particles, which he called atoms. He reasoned that once the atoms touched the eye, they were carried to the soul and therefore could be viewed.

Empedocles, a contemporary of Democritus, proposed a different theory of vision. He believed that matter was composed of four essential elements—earth, air, water, and fire. Vision must be linked to fire because it provided light. According to Empedocles, light radiated from the eye and struck objects, making them visible.

Galen, a Roman physician in the second century A.D., combined Democritus' theory of the eye and the soul with Empedocles' notion that light was emitted from the eye. Galen believed that the optic nerve conducted visual spirits from the brain. The spiritual link between the soul and vision remains today, for example, in the expression "evil eye."

Today scientists accept two complementary theories of light first proposed by Sir Isaac Newton and Christiaan Huygens. Particles of light (photons) travel in waves of various lengths. Light enters the eye as it is reflected or transmitted from objects. In many ways the eye operates by the same principle as a camera. Both camera and eye are equipped with lenses that focus images. The diaphragm of a camera opens and closes to regulate the amount of light entering the camera. The iris of the eye provides an equivalent function.

The image of the camera is focused on a chemical emulsion—the film. Similarly, the image of the eye is focused on the retina.

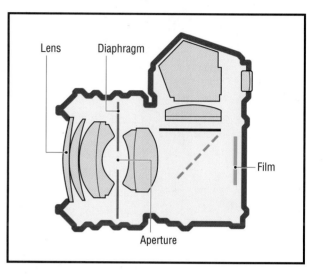

Figure 16.7

Light rays are bent by the lens and an inverted image is projected on the film.

In the 1880s Wilhelm Kuhne, a German physiologist, performed a series of bizarre experiments that underscored the similarities between the actions of the retina and photographic emulsion. Kuhne placed a rabbit in a dark room and fixed its vision on a barred window, the only source of light in the room. Using a dark cloth, Kuhne covered the rabbit's eyes for 10 min and then once again fixed them on the window for 2 min. Immediately, Kuhne decapitated the rabbit and placed its eyes in an alum solution, which fixed the image. The following day, Kuhne examined the eye and found an imprint of the window on the animal's retina. Kuhne repeated the experiments, using progressively more complicated images. Although he was certain that the human retina would behave in much the same manner, he had no experimental proof. In November 1880, Kuhne was presented with the opportunity to test his theory: he was able to secure the head of a criminal who had been beheaded. Following the same procedure as he had with the rabbit, Kuhne was able to fix the criminal's final image.

Afterimages

Have you ever noticed a trailing blue or green line that stays in your vision after you look into a camera flash? What you see is an afterimage. There are two different types of afterimages: positive and negative. The positive afterimage occurs after you look into a bright light and then close your eyes. The image of the light can still be seen even though your eyes are closed. As you read earlier, Wilhelm Kuhne described the positive afterimage as an image printed on the retina. The more dramatic negative afterimage occurs when the eye is exposed to bright colored light for an extended period of time.

Stare at the cross in Figure 16.8 for 30 s, and then stare at a bright white surface. The colors will reverse. The afterimage is believed to be caused by fatigue of that particular type of cone in that area of the retina. The horizontal red cones become fatigued, but the complementary green cones continue to fire. The opposite effect occurs for the vertical bar.

FOCUSING THE IMAGE

As light enters the eye, it is first bent toward the pupil by the cornea. Light waves normally travel in straight lines and slow down when they strike more dense materials like the cornea. The slowing of light by a denser medium causes bending, or refraction. The cornea directs light inward toward the lens, resulting in further bending. Because the biconvex lens is symmetrically thicker in the center than at its outer edges, light is bent to a focal point. An inverted image is projected on the light-sensitive retina.

Ciliary muscles control the shape of the lens, and suspensory ligaments maintain a constant tension. When close objects are viewed, the ciliary muscle contracts, the tension on the ligaments decreases, and the lens becomes thicker. The thicker lens provides additional bending of light for near vision. For objects that are farther away, relaxation of the ciliary muscles increases the tension of the ligaments on the lens, and the lens becomes thinner. The adjustment of the lens to objects near and far is referred to as **accommodation.** Objects 6 m from the viewer need no accommodation.

Figure 16.8

The red bar produces a green afterimage; the green bar produces a red afterimage.

Accommodation *reflexes are adjustments made by the lens and pupil for near and distant objects.*

Figure 16.9

Distant objects are viewed through a flatter lens. The focal point is farther from the lens and the object appears smaller. Near objects are viewed through a thicker lens. The focal point is closer to the lens and the object appears larger.

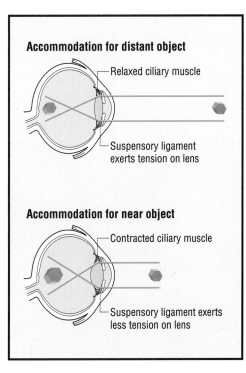

Accommodation for distant object

— Relaxed ciliary muscle

— Suspensory ligament exerts tension on lens

Accommodation for near object

— Contracted ciliary muscle

— Suspensory ligament exerts less tension on lens

The importance of the accommodation reflex becomes more pronounced with age. Layers of transparent protein covering the lens increase throughout your life, making the lens harder. As the lens hardens, it loses its flexibility. By the time you reach 40, near-point accommodation has diminished and may begin to hinder reading.

You may have noticed a secondary adjustment during the accommodation reflex. When objects are viewed from a distance, the pupil dilates in an attempt to capture as much light as possible. When objects are viewed close up, the pupil constricts in an attempt to bring the image into sharp focus. Test this for yourself by looking at the print on this page with one eye. Move your head toward the book until the print gets very blurry. Now crook your finger and look through the small opening. Gradually make the opening smaller. The image becomes sharper. Light passes through a small opening and falls on the most sensitive part of the retina, the fovea centralis. The Inuit were aware of this principle when they made eyeglasses for their elders by drilling holes in whalebone. Light passing through the narrow openings resulted in a sharper focus.

> **BIOLOGY CLIP**
> An estimated 10 million molecules of rhodopsin are found in each of the 160 million rods in your eyes.

CHEMISTRY OF VISION

An estimated 160 million rods surround the color-sensitive cones in the center of the retina. The rods contain a light-sensitive pigment called **rhodopsin,** or "visual purple." The cones contain similar pigments, but are less sensitive to light. Rhodopsin is composed of a form of vitamin A and a large protein molecule called opsin. When a single photon, the smallest unit of light, strikes a rhodopsin molecule, it divides into two components: retinene, the pigment portion, and opsin, the protein portion. This division alters the cell membrane of the rods and produces an action potential. Transmitter substances are released from the end plates of the rods, and the nerve message is conducted across synapses to the bipolar cells and the neuron of the optic nerve. For the rods to continue to work, rhodopsin levels must be maintained. Long-term vitamin A deficiency can cause permanent damage of the rods.

The extreme sensitivity of rhodopsin to light creates a problem. In bright light, rhodopsin breaks down faster than it can be restored. The opsins used for color vision are much less sensitive to light and, therefore, operate best with greater light intensity. The fact that images appear as shades of gray during periods of limited light intensity reinforces the fact that only the rods are active. Not surprisingly, the rods are most effective at dusk and dawn.

Color Vision
Vitamin A combines with three different protein opsins in the cones, each of which is sensitive to one of the three primary colors of source light: red, blue, and green. (Do not confuse the primary col-

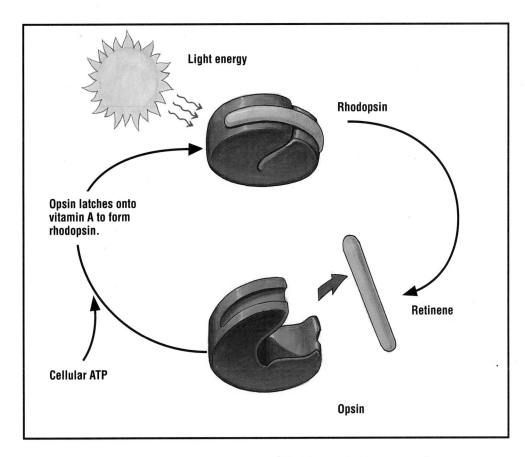

Light energy

Rhodopsin

Opsin latches onto vitamin A to form rhodopsin.

Cellular ATP

Retinene

Opsin

Figure 16.10

Light causes rhodopsin to break down. In the absence of light retinene changes into vitamin A. Cell energy is required to recombine vitamin A with the large opsin protein to restore rhodopsin levels.

ors of source light with the primary colors of reflected light: red, blue, and yellow.) Each pigment is located in a different cone. Other colors are seen by the stimulation of different combinations of the three types of cones. For example, yellow is produced by the stimulation of cones sensitive to both green and red wavelengths. Purple is produced by the stimulation of cones sensitive to both red and blue wavelengths. Cyan is produced by the combination of blue and green wavelengths. Stimulation of all of the cones in equal proportions produces white light.

Color blindness occurs when one or more types of cones are defective. The most common type of color blindness, red-green color blindness, occurs when the cones containing the red-sensitive pigment fail to work properly. The defect is genetic and more common in males than females.

B I O L O G Y C L I P

Long before biochemists knew the chemical components of rhodopsin, low levels of vitamin A were linked with poor night vision. Historically, carrots, a rich source of vitamin A, have been eaten to improve vision. Actually, vitamin A is found in most fruits and vegetables.

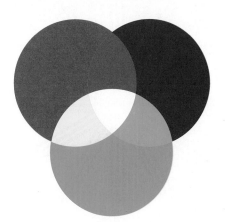

Figure 16.11

Three different types of cones are found in the retina, each sensitive to one of the three primary colors of source light: blue, red, and green. Photocells that are sensitive to the same three colors are found in a color television set.

VISION DEFECTS

Glaucoma is caused by an increased build-up of aqueous humor in the anterior chamber of the eye. Although a small amount of the fluid is produced each day, under normal conditions tiny ducts drain any excess. Blockage of these drainage ducts causes the fluid pressure to collapse blood vessels in the retina. Without a constant supply of nutrients and oxygen, neurons soon die and blindness results.

The lens can cause many problems. Occasionally the lens becomes opaque and prevents some of the light from passing through. The condition is known as a **cataract.** A traditional solution to the problem has been to remove the lens and to fit the patient with strong eyeglasses. A more recent solution is to replace the natural lens with a plastic one.

In most people, the lens and cornea are symmetrical. Incoming light is refracted along identical angles for both the dorsal and ventral surfaces, forming a sharp focal point. In some individuals, however, the lens or cornea is irregularly shaped. This condition is called **astigmatism.** The chart below will help you determine whether or not you have astigmatism. If you have cornea astigmatism, the lines along one plane will appear sharp, but those at right angles will appear fuzzy.

Two of the more common vision defects are nearsightedness (myopia) and farsightedness (hyperopia). **Nearsightedness** occurs when the eyeball is too long. The lens cannot flatten enough to project the image on the retina, and so distant images are brought into focus in front of the retina. Someone who is nearsighted is able to focus on close objects, but has difficulty seeing objects that are distant. Nearsightedness can be corrected by glasses that contain a concave lens. **Farsightedness** is caused by an eyeball that is too short, causing distant images to be brought into focus behind the retina. A farsighted person is able to focus on distant objects, but has trouble focusing on objects that are close up. Farsightedness can be corrected by glasses that have a convex lens.

Glaucoma *is a disorder of the eye caused by the build-up of fluid in the anterior chamber to the lens.*

Cataracts *occur when the lens or cornea become clouded.*

Astigmatism *is a vision defect caused by the abnormal curvature of the surface of the lens or cornea.*

Nearsightedness *occurs when the image is focused in front of the retina.*

Farsightedness *occurs when the image is focused behind the retina.*

Figure 16.12

Visual defects can be improved with corrective lenses.

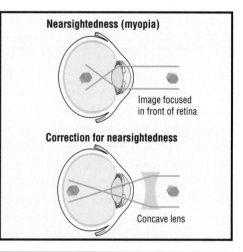

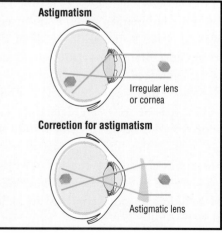

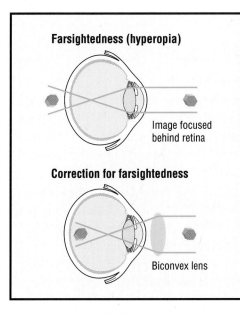

Farsightedness (hyperopia)

Image focused
behind retina

Correction for farsightedness

Biconvex lens

from the outer edge of the cornea toward the center in a spokelike arrangement. The incisions flatten the outer cornea, altering its curvature. After about 15 min, the eye is patched and the patient returns home. Once recovery is complete, the other eye is corrected.

Although the prospects of correcting myopia through surgery might appear promising, the Canadian Medical Association wants to ensure that the new procedure is completely understood before it allows Canadian doctors to perform it. Although the short-term benefits appear to be well documented, long-term risks must first be evaluated.

FRONTIERS OF TECHNOLOGY: RADIAL KERATOTOMY

A simple surgical procedure for curing shortsightedness might seem futuristic to most North Americans, but such an operation is already well established in Russia. The procedure was developed by Dr. Svyatoslav Fyodorov, director of the Moscow Scientific Laboratory of Experimental and Clinical Problems of Eye Surgery. Dr. Fyodorov was inspired by a young Russian teenager who had suffered a misfortune while fighting. The teenager's glasses shattered during the fight, badly cutting his cornea. Remarkably, the eye healed and the boy's myopia seemed cured. An alteration of the cornea corrected the myopia.

Fyodorov's clinic looks more like an automobile assembly line than a hospital. Patients in beds move toward the physicians along conveyer belts. Video cameras check for quality control and help direct the physicians in this high-tech facility.

The procedure for correcting myopia involves 8 to 16 incisions along the cornea. A diamond blade scalpel cuts

▓ REVIEW QUESTIONS ▓   ?

6 List the three layers of the eye and outline the function of each layer.

7 Indicate the function of each of the following parts of the eye: vitreous humor, aqueous humor, cornea, pupil, iris, rods, cones, fovea centralis, and blind spot.

8 How did the experiment performed by Wilhelm Kuhne in the 1880s provide an answer for the phenomenon of "positive afterimages"?

9 What are accommodation reflexes?

10 Outline the mechanism by which rhodopsin converts light energy into a nerve impulse.

11 Why is vitamin A essential for proper vision?

12 Why do the rods not function effectively in bright light?

13 What three color-sensitive pigments are found in the cones?

14 Identify the causes for each of the following eye disorders: glaucoma, cataract, astigmatism, nearsightedness, and farsightedness.

LABORATORY
VISION

Objective

To investigate aspects of human vision.

Materials

Snellen eye chart   scissors
paper cup   paper file card
ruler   coin
pencil

Procedure

Part I: Visual Acuity

Your ability to distinguish detail or visual acuity can be determined with a Snellen eye chart. Have you ever heard someone say that he or she had 20/20 vision? This means that a person is able to read the "20" line on an eye chart from a distance of 20 feet. (Charts that have been converted to metric measurements show normal vision to be "6/6." This indicates that the line marked "6" can be read from a distance of 6 m.) A person with 20/100 can only read line "100" from 20 feet, which is very poor vision. A person with 20/15 vision has better than 20/20 vision.

1 Stand 20 feet from the eye chart. (Stand 6 m away if a metric chart is available.) Cover your right eye and begin reading letters from each line.

  a) Indicate the visual acuity of your left eye.

2 Repeat the procedure by covering your left eye.

  b) Indicate the visual acuity of your right eye.

Part II: Blind Spot

3 Hold this text approximately 50 cm from your left eye and view the black cross. Close your right eye and begin moving the text toward your left eye while staring at the cross. Do not allow your eye to leave the cross! At some point you will not see the circle. The blind spot occurs when the image falls on the place where the optic nerve attaches to the retina.

  c) Measure the distance of the book from your left eye.

4 Repeat the procedure and determine the blind spot for the right eye.

  d) Measure the distance of the book from your right eye.

Part III: Dominant Eye

5 Face the corner of the room and view the upper corner. Holding a pencil in either your right or left hand, extend that arm, bringing the top of the pencil into the middle of your field of view. Align the pencil with the edge of the walls.

6 Close your right eye and note the location of the pencil.

  e) Describe what happened to the position of the pencil.

7 Close your left eye and note the location of the pencil.

  f) Describe what happened to the position of the pencil.

  g) Which is the dominant eye? (The object appears to move the least in the dominant eye.)

  h) Survey the people in your class to determine if right-handed people have dominant right eyes or if left-handed people have dominant left eyes.

Part IV: Stereoscopic Vision

Most of what you see with your right eye is also seen with the left eye. Visual fields overlap. Each eye sees the image from a slightly different angle. The image is created in each retina and stored on the opposing cerebral hemispheres. Information is relayed between the two hemispheres and the two images are superimposed on each other, creating a three-dimensional image. By working together, the eyes provide slightly different angles of view that permit the brain to estimate distance.

8 Stare directly at a distant object. Touch the fingertips of your right and left hands, but spread your fingers enough to allow light to pass between them. Extend your arms and slowly bring the fingertips in front of your face, but continue looking at the distant object. Look through your fingers.

  i) Describe what you see.

9 Place a plastic or paper cup on a table, and stand about 3 m away. Cover one eye.

10 Ask a friend to hold a coin at arm's length not quite above the cup.

11 Observing only the cup, direct the volunteer's hand until it is above the cup. Check your accuracy by asking the volunteer to drop the coin into the cup.

12 Repeat the procedure for 5 trials. Then close the opposite eye and record your results for another 5 separate trials.

j)

Trial	Left eye open	Right eye open
1		
2		
3		
4		
5		

13 Hold a card, with a hole punched in it, at arm's length.

14 In your other hand, hold a pencil at arm's length. Close one of your eyes and attempt to bring the point of the pencil into the hole in the card. Move the pencil and the paper. Repeat the procedure for 5 trials with each eye.

Trial	Left eye open	Right eye open
1		
2		
3		
4		
5		

k) Record your results.

Laboratory Application Questions

1 In an attempt to map the position of the blind spot, a student rotates a book 180° and follows the same steps described in part II. How would rotating the text help the student map the position of the blind spot?

2 In part III of the laboratory you discovered that the pencil moved when you viewed it with a different eye. Offer an explanation for your observation.

3 Horses and cows have eyes on the sides of their head; their visual fields overlap very little. What advantage would this kind of vision have over human vision?

4 Like humans, squirrels have eyes on the front of their face. The visual field of the squirrel's right eye overlaps with the visual field of its left eye. Why does a squirrel need this type of vision?

5 Students place a large piece of cardboard between their eyes while attempting to read two different books at the same time. Will they be able to read two books at once? Provide an explanation. ■

> **B I O L O G Y   C L I P**
> Leonardo da Vinci (1452–1519) speculated about contact lenses long before A.E. Flick fitted the first pair on a patient in 1887. The original contacts, made of glass, were designed to protect the eyes of a man who had cancerous eyelids. The early contact lenses were thick and cumbersome.

STRUCTURE OF THE EAR

The ear is associated with two separate functions: hearing and equilibrium. Sensory cells for both functions are located in the inner ear. These tiny hair cells contain from 30 to 150 cilia, which respond to mechanical stimuli. Movement of the cilia causes the nerve cell to generate an impulse.

> **B I O L O G Y   C L I P**
> The ear ossicles are the smallest bones in the body. They are fully developed at birth.

The ear can be divided into three sections for study: the outer ear, the middle ear, and the inner ear. The outer ear is composed of the **pinna,** the external ear flap, which collects the sound, and the

*The **pinna** is the outer part of the ear. The pinna acts like a funnel, taking sound from a large area and channeling it into a small canal.*

auditory canal, which carries sound to the eardrum. The auditory canal is lined with specialized sweat glands that produce earwax, a substance that traps foreign invading particles and prevents them from entering the ear.

The middle ear begins at the eardrum, or **tympanic membrane,** and extends toward the oval and round windows. The air-filled chamber of the middle ear contains three small bones, called the ear **ossicles,** which include the malleus (the hammer), the incus (the anvil), and the stapes (the stirrup). Sound vibrations that strike the eardrum are first concentrated within the solid malleus and then transmitted to the incus, and finally to the stapes. The stapes strikes the membrane covering the oval window in the inner wall of the middle ear. The amplification of sound is in part met by concentrating the sound energy from the large tympanic membrane to a smaller **oval window.** The surface area of the tympanic membrane is 64 mm^2, while that of the oval window is about 3.2 mm^2.

The **eustachian tube** extends from the middle ear to the air in the mouth and chambers of the nose. Approximately 40 mm in length and 3 mm in diameter, the eustachian tube permits the equaliza-tion of air pressure. Have you ever noticed how your ears seem to pop when you go up in a plane? The lower pressure on the tympanic membrane can be equalized by reducing air pressure in the eustachian tube. Yawning, swallowing, and chewing gum allow air to leave the middle ear. An ear infection can cause the build-up of fluids in the eustachian tube and create inequalities in air pressure. Discomfort, temporary deafness, and poor balance can result.

The inner ear is made up of three distinct areas: the vestibule and the semicircular canals, which are involved with balance, and the cochlea, which is connected with hearing. The **vestibule,** connected to the middle ear by the oval window, houses two small sacs, the *utricle* and *saccule*, which establish head position (static equilibrium). Three **semicircular canals** are arranged at different angles. The movement of fluid in these canals helps you identify body movement (dynamic equilibrium). The **cochlea** is shaped like a spiralling snail's shell and contains two rows of specialized hair cells that run the length of the inner canal. The hair cells identify and respond to sound waves of different frequencies and intensities.

Table 16.3 Parts of the Ear

Structure	Function
External Ear	
Pinna	Outer part of the external ear. Amplifies sound by funneling it from a large area into the narrower auditory canal.
Auditory canal	Carries sound waves to the tympanic membrane.
Middle Ear	
Ossicles	Tiny bones that amplify and carry sound in the middle ear.
Tympanic membrane	The eardrum.
Oval Window	Receives sound waves from the ossicles.
Eustachian tube	Air-filled tube of the middle ear that equalizes air pressure in the external ear.
Inner Ear	
Vestibule	A chamber at the base of the semicircular canals; concerned with static equilibrium.
Semicircular canals	Fluid-filled structures that provide information concerning dynamic equilibrium.
Cochlea	A coiled tube within the inner ear that identifies sound waves and converts them into nerve impulses.

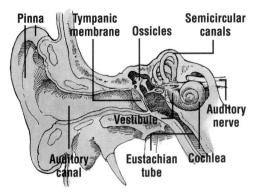

Pinna Tympanic Ossicles Semicircular
membrane canals

Auditory
nerve

Vestibule

Auditory Eustachian Cochlea
canal tube

HEARING AND SOUND

Sound is a form of energy. Like light, thermal energy, and various forms of chemical energy, sound energy must be converted into an electrical impulse before you can interpret it. The sensitivity of the ear can be illustrated by the fact that you can hear a mosquito outside your window, even though the energy reaching your ear is less than one quadrillionth of a watt. The average light in your house uses 60 W of energy.

The nature of sound became clearer after an experiment by the British physicist Robert Boyle in 1660. Boyle wondered whether sound could be produced in a vacuum. In his experiment, Boyle suspended a watch with an alarm inside a bell jar. The alarm was set and the air was pumped out of the jar. Boyle waited for the time for which he had set the alarm to approach, but to his amazement the alarm could not be heard.

What Boyle learned from this experiment was that sound must travel through a medium. Although air is the most common medium, sound also travels through water and solids. Sound travels through water (1480 m/s at 20°C) four times faster than it does through air (370 m/s at 20°C). In fact, dolphins and porpoises rely on sound more than sight, especially in murky water. Solids transmit sound even more rapidly than liquids do; for example, sound travels in steel at a speed

of 6096 m/s at 20°C. This explains how movie cowboys were able to hear the approach of distant trains by putting their ears to a railway track.

Hearing begins when sound waves push against the eardrum, or tympanic membrane. The vibrations of the eardrum are passed on to the three bones of the middle ear: the malleus, the incus, and the stapes. The three bones are arranged in a lever system held together by muscles and ligaments. The bones concentrate and amplify the vibrations received from the tympanic membrane. The ossicles move a shorter distance, but exert greater force by concentrating the energy in a very small area.

Muscles that join the bones of the middle ear act as a safety net protecting the inner ear against excessive noise. Intense sound causes the tiny muscles—the smallest in your body—to contract, restricting the movement of the malleus and reducing the intensity of movement. At the same time, a second muscle contracts, pulling the stapes away from the oval window, thereby protecting the inner ear from powerful vibrations. Occasionally, the safety mechanism doesn't work quickly enough. The sudden blast from a firecracker can send the ossicles into wild vibrations before the protective reflex can be activated. The ossicles can triple the force of vibration from the eardrum.

Figure 16.13

Anatomy of the human ear. The *outer ear* is composed of the pinna and auditory canal. The *middle ear* is composed of the tympanic membrane, the ear ossicles (middle ear bones), and the eustachian tube. The *inner ear* includes the vestibule, semicircular canals, oval window, round window, and cochlea.

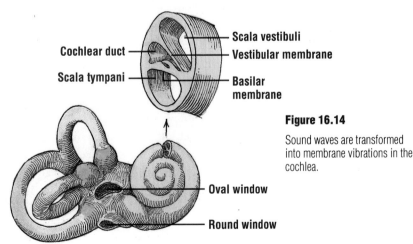

Cochlear duct —— —— Scala vestibuli
—— Vestibular membrane
Scala tympani —— —— Basilar
membrane

Oval window

Round window

Figure 16.14

Sound waves are transformed into membrane vibrations in the cochlea.

Figure 16.15

Hair cells within the cochlea are anchored to a basilar membrane. The free end of the hair cells is embedded in a jellylike substance. Vibrations of the basilar membrane cause the hair cells to bend. Movement of the hair cells stimulates nerves in the basilar membrane.

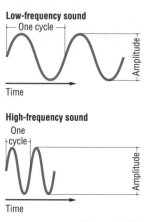

Figure 16.16

High-frequency sound produces higher pitches. Frequency is measured in cycles per second, or Hertz (Hz). Amplitude is a measure of the loudness that can be calibrated in decibels (dB). The louder the sound, the greater is the amplitude.

The oval window, a flaplike bone embedded in the wall of the inner ear, receives vibrations from the ossicles. As the oval window is pushed inward, the round window, located immediately below the oval window, moves outward. This triggers waves of fluid within the inner ear. A coiled tube—the cochlea—receives the fluid waves and converts them into electrical impulses, which you interpret as sound. The hearing apparatus within the cochlea is known as the **organ of Corti,** and is composed of a single inner row and three outer rows of specialized hair cells, anchored to the **basilar membrane.** Covered with a gelatinous coating, the hair cells respond to vibrations of the basilar membrane. Fluid vibrations move the basilar membrane, and the hair cells bend. The movement of the hair cells, in turn, stimulates sensory nerves in the basilar membrane. Auditory information is then sent to the brain.

The inner ear is able to identify both pitch and loudness. The cochlea is narrowest at the middle ear, and the hairlike fibers are rigid. This area is activated by high-frequency sound waves, which contain enough energy to move the rigid hair receptors. The high-frequency waves are transformed into basilar membrane vibrations, which, in turn, cause the hair cells to move. The receptor hair cells trigger an action potential, which is carried to the area of the brain that registers high-pitched sounds. The high-pitched noise of a police siren dies out quickly in the narrow, rigid part of the cochlea. However, low-frequency waves move further along the cochlea, causing the hair cells in the wider, more elastic area to vibrate. This may explain why you can actually feel the vibrations from a bass guitar. The stimulation of nerve cells in different parts of the cochlea enables you to differentiate sounds of different pitch. Each frequency or pitch terminates in a specific part of the auditory section of the brain.

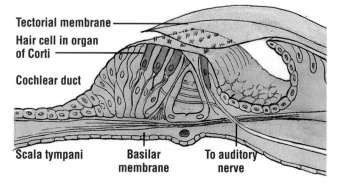

In addition to responding directly to sound energy, the basilar membrane can respond directly to mechanical stimulation. A jarring blow to the skull sets up vibrations that are passed on toward the cochlea. Aside from the sound created by the blow, the resulting mechanical vibrations of the skull can also be interpreted as sound.

EQUILIBRIUM

Balance consists of two components: static equilibrium and dynamic equilibrium. Static equilibrium involves movement along one plane such as horizontal or vertical. Head position is maintained by two fluid-filled sacs called the **saccule** and the **utricle.** Tiny hairlike receptors are found within the saccule and utricle. Cilia from the hair cells are suspended in a jellylike material that contains small calcium carbonate granules called **otoliths.** When the head is in the normal position, the otoliths do not move; however, when the head is bent forward, gravitational force acts on the otoliths, pulling them downward. The otoliths cause the gelatinous material to shift, and the hair receptors to bend. The movement of the hair receptors stimulates the sensory nerve, and information about head position is relayed to the brain for interpretation.

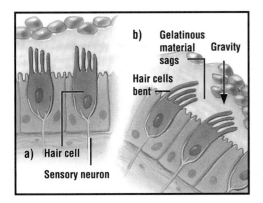

b) Gelatinous material sags Gravity

Hair cells bent

a) Hair cell

Sensory neuron

The second aspect of balance, referred to as dynamic equilibrium, provides information during movement. While you are moving, balance is maintained by the three fluid-filled semicircular canals. Each of the canals is equipped with a pocket called an ampulla (plural: "ampullae"). Rotational stimuli cause the fluid in the semicircular canals to move, bending the cilia attached to hair cells in the ampullae. Once the hair cells bend, they initiate nerve impulses, which are carried to the brain. It is believed that rapid continuous movement

of the fluids within the semicircular canals is the cause of motion sickness.

■ REVIEW QUESTIONS ▨ ?

15 Briefly outline how the external ear, middle ear, and inner ear contribute to hearing.

16 What function do the tympanic membrane, ossicles, and oval window serve in sound transmission?

17 Categorize the following structures of the inner ear according to whether their functions relate to balance or hearing: organ of Corti, cochlea, vestibule, saccule, ampulla, semicircular canals, oval window, round window.

18 In 1660, Robert Boyle placed an alarm clock in a bell jar from which air was then removed. Why was Boyle not able to hear the alarm clock?

19 Explain how the ligaments, tendons, and muscles that join the ossicles of the middle ear provide protection against high-intensity sound.

20 Differentiate between static and dynamic equilibrium.

21 How do the saccule and utricle provide information about head position?

22 Describe how the semicircular canals provide information about body movement?

Figure 16.17

(a) When the head is in the erect position, the cilia from the hair cells remain erect. (b) Movement of the head causes movement of the hair cells. Any movement of the cilia from the hair cells initiates nerve impulses.

Otoliths *are tiny stones of calcium carbonate found within the saccule and utricle. The tiny stones are embedded in a gelatinous coating. Gravity causes the otoliths to slide downward as the head is lowered or raised.*

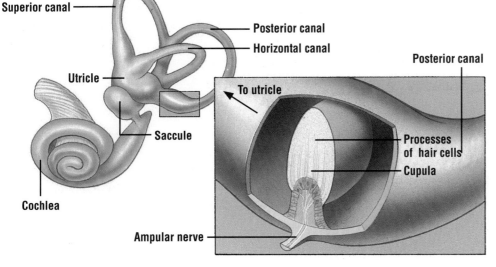

Superior canal

Posterior canal

Horizontal canal

Posterior canal

Utricle

To utricle

Saccule

Processes of hair cells

Cupula

Cochlea

Ampular nerve

Figure 16.18

Three semicircular canals provide information about motion. Cilia attached to hair receptor cells in the ampulla respond to the movement of fluid in the semicircular canals.

LABORATORY
HEARING AND EQUILIBRIUM

Objective

To investigate environmental factors that affect both hearing and equilibrium.

Materials

tuning fork
meter stick

rubber hammer
swivel chair

Procedure

Part I: Factors That Affect Hearing

1 Strike a tuning fork with a rubber hammer and listen to the sound. Place the tuning fork on your forehead. Place the palm of your free hand over your right ear.
   a) From which direction does the sound appear to be coming?
   b) Describe any changes in the intensity of the sound.
2 Repeat the procedure, but this time place your free hand over your left ear.
   c) From which direction does the sound appear to be coming?
3 Repeat the procedure a third time, but ask your lab partner to cover both of your ears.
   d) Describe any changes in the intensity of the sound.
4 Strike the tuning fork with a rubber hammer and hold the tuning fork approximately 1 m from your ear.
5 Ask your lab partner to place a meter stick on the bony process immediately behind your ear. Then ask him or her to place the stem of the tuning fork on the meter stick.
   e) Describe any changes in the intensity of the sound.

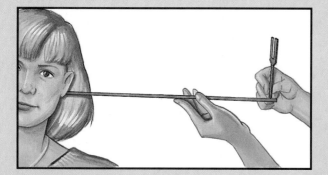

Part II: Equilibrium

6 Ask your lab partner to sit in a swivel chair. Have him or her elevate the legs and begin slowly rotating the chair in a clockwise direction. After 20 rotations, have the subject stand. (Be prepared to support your partner!)
   f) In which direction did the subject lean?
7 After a 3-min recovery period, repeat the process, but this time rotate the swivel chair in a counterclockwise direction.
   g) In which direction did the subject lean?
8 Ask your lab partner to tilt his or her head to the right, and begin a clockwise rotation of the swivel chair. After 20 rotations, ask the subject to hold his or her head erect and to stand up. (Be prepared to catch your lab partner—most people attempt to sit very quickly after they stand.)
   h) Ask the subject to describe the sensation.

Laboratory Applications Questions

1 Provide explanations for the data collected in observation questions a, b, c, and d.
2 Using the data collected, provide evidence to suggest that sound intensity is greater in fluids than in air.
3 Using the data collected, provide evidence to suggest that the fluid in the semicircular canals continues to move even after rotational stimuli cease.
4 What causes the falling sensation produced in step 8?
5 Describe the manner in which the semicircular canals detect changes in motion during a roller-coaster ride. ■

> **B I O L O G Y   C L I P**
> Sound frequency, or pitch, is measured in Hertz (Hz). The normal hearing range in humans is between 2000 and 4000 Hz. However, the human ear can occasionally detect frequencies as low as 16 Hz and as high as 20 000 Hz.

CAREER INVESTIGATION

The human body is an amazingly versatile tool. Many careers depend on a knowledge of how the body performs and how it reacts under stress.

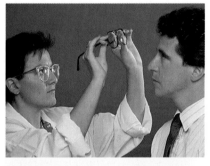

Choreographer

A choreographer should have an intimate knowledge of the workings of the human body, its capabilities and limitations. Routines must be designed to be both aesthetically pleasing and physically impressive, yet within the capabilities of the dancers who will perform them.

Ophthalmologist/Optometrist/Optician

An ophthalmologist is a physician who specializes in the function, structure, and diseases of the eye. An optometrist examines the eye for defects and prescribes corrective glasses or contact lenses. An optician fits customers with glasses or contact lenses as prescribed by an optometrist or ophthalmologist.

Advertising Executive

Good advertising, particularly television advertising, must communicate with the audience on several levels. An advertising executive must know how people perceive and interpret sounds and images in order to create an ad that will sell products.

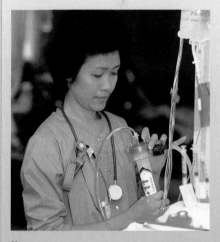

- Identify a career that requires a knowledge of coordination and regulation in humans.
- Investigate and list the features that appeal to you about this career. Make another list of features that you find less attractive about this career.
- Which high-school subjects are required for this career? Is a postsecondary degree required?
- Survey the newspapers in your area for job opportunities in this career.

Nurse

Nurses require not only a thorough knowledge of human anatomy, but also a knowledge of drugs and how to administer them. They must also be sensitive to the psychological needs of their patients. Nurses work in hospitals, in doctors' offices, or in senior citizens homes.

SOCIAL ISSUE:
Rock Concerts and Hearing Damage

In a recent interview, Pete Townshend stated that after years of playing guitar with the rock group The Who, he had developed partial hearing loss. For concertgoers, the results of listening to loud rock music for almost two hours can include ringing in the ears or impaired hearing for days after.

The most common type of hearing loss is caused by the destruction of the cilia on the hair cells of the cochlea. Although the cilia gradually wear away with aging, high-intensity sounds, such as loud rock music, can literally tear the cilia apart.

The sound level of a normal conversation ranges between 60 and 75 dB, heavy street traffic can reach 80dB, and the crash of thunder usually registers about 100 dB. Pain begins to be felt at 120 dB. At rock concerts, if the audience is close enough, the sound level can register well beyond 130 dB.

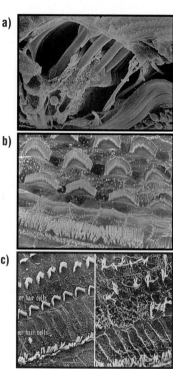

(a) Hair cells embedded in the basilar membrane. (b) Normal rows of hair cells in the cochlea. (c) Destruction of hair cells following 24-h exposure to high-intensity rock music, with noise levels approaching 120 dB at 2000 cycles per second.

Statement:

The sound levels at rock concerts should be regulated to protect the audience and performers.

> **Research the issue.**
> **Reflect on your findings.**
> **Discuss the various viewpoints with others.**
> **Prepare for the class debate.**

Point

- The sound levels at rock concerts can register up to 130 dB, well above the level of 85 dB that can cause permanent hearing loss. Governments have a responsibility to protect the health of their citizens by regulating the level of noise to which we are exposed.
- Individuals who knowingly risk hearing damage by attending high-decibel rock concerts are irresponsible, and raise the costs of treatment under our medical insurance plans.

Counterpoint

- Listening to tapes at high volume on a Walkman for extended periods of time is likely to be much more dangerous than occasional exposure to high-decibel levels at rock concerts.

- Earplugs are available to anyone who wants to take extra precautions at a rock concert.

CHAPTER HIGHLIGHTS

- A stimulus is a form of energy detected by the body.
- Sensory receptors are modified ends of sensory neurons that are activated by specific stimuli. Receptors convert one form of energy into another (electrochemical impulses).

- The eye converts light energy into electrical impulses, which are interpreted by the brain.
- Cones, which are specialized nerve cells found in the retina of the eye, contain pigments that are sensitive to light of three different wavelengths.

- The ear detects sound and regulates balance.
- Sound energy is detected by the eardrum and concentrated by the ossicles of the middle ear. The vibrating ossicles push against the oval window of the inner ear, setting up fluid waves that move toward the cochlea. Specialized hair cells in the cochlea detect fluid vibrations.

APPLYING THE CONCEPTS

1 A person steps from a warm shower and feels a chill. Upon stepping into the same room, another person says that the room is warm. What causes the chill?

2 If a frog is moved from a beaker of water at 20 °C and placed in a beaker at 40 °C, it will leap from the beaker. However, if the temperature is slowly heated from 20 to 40 °C, the frog will remain in the beaker. Provide an explanation for the observation.

3 The retina of a chicken is composed of many cones very close together. What advantages and disadvantages would be associated with this type of eye?

4 Why do people often require reading glasses after they reach the age of 40?

5 Indicate how the build-up of fluids in the eustachian tube may lead to temporary hearing loss.

6 A scientist replaces ear ossicles with larger, lightweight bones. Would this procedure improve hearing? Support your answer.

7 When the hearing of a rock musician was tested, the results revealed a general deterioration of hearing as well as total deafness for particular frequencies. Why is the loss of hearing not equal for all frequencies?

CRITICAL-THINKING QUESTIONS

1 One theory suggests that painters use less purple and blue as they age because layers of protein are added to the lens in their eyes and it gradually becomes thicker and more yellow. The yellow tint causes the shorter wavelength to be filtered. How would you test the theory?

2 Should individuals who refuse to wear ear protectors while working around noisy machinery be eligible for medical coverage for the cost of hearing aids? What about rock musicians or other individuals who knowingly play a part in the loss of their hearing?

ENRICHMENT ACTIVITIES

1 Use the diagram below to make a functional model of the ear.

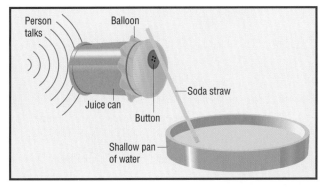

Person talks
Balloon
Juice can
Button
Soda straw
Shallow pan of water

2 Suggested reading:
- Barlow, Robert. "What the Brain Tells the Eye." *Scientific American*, April 1990, p. 90.
- Franklin, Deborah. "Crafting Sound from Silence." *Science News* (October 20, 1984): p. 252.
- Hudspeth, A. "The Hair Cells of the Inner Ear." *Scientific American*, January 1983, p. 58.
- Stryer, L. "The Molecules of Visual Excitation." *Scientific American*, July 1987, p. 42.

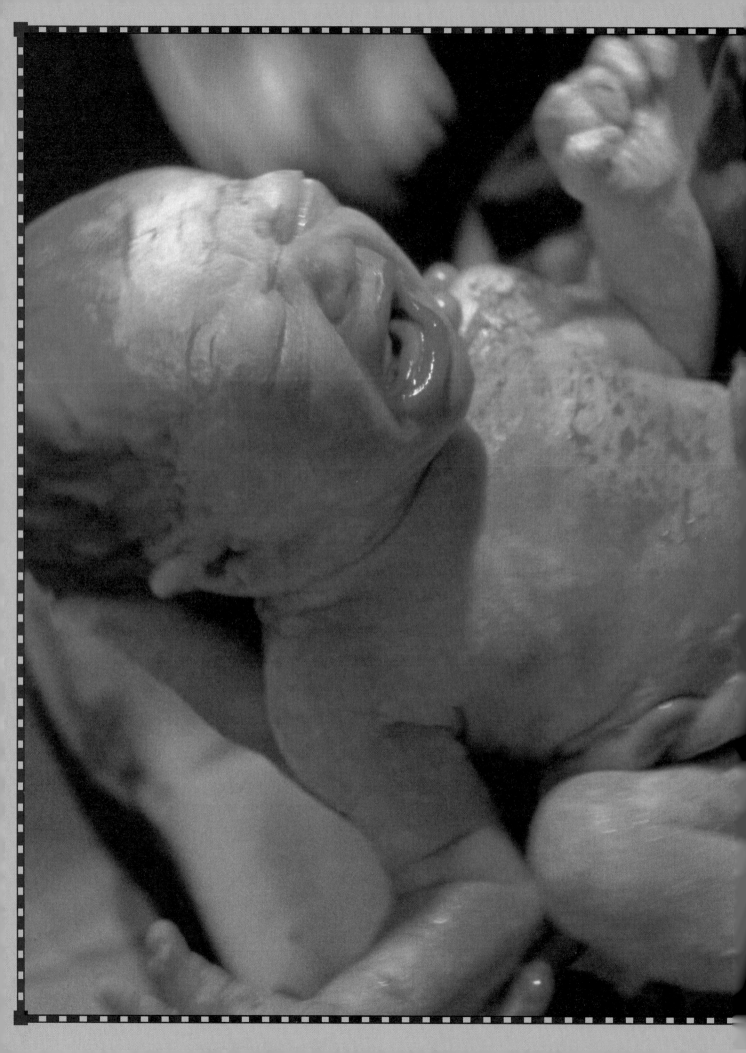

*C*ontinuity of Life

■

U N I T 5

*T*he Reproductive System

IMPORTANCE OF REPRODUCTION

Reproduction ensures the survival of a species. Sexual reproduction involving the fusion of male and female sex cells creates new gene combinations. The diversity produced by new gene combinations provides a basis for natural selection: only the best-adapted survive. Species survival is based on providing numerous and varied offspring.

Female oysters produce an estimated 115 million eggs for each spawning. Each year, female frogs produce hundreds of thousands of eggs for fertilization. In contrast, human females have 400 000 egg cells, of which only 400 mature throughout the reproductive years—from about age 12 to age 50. According to one source, the greatest number of offspring ever born to one woman was 57. The limited capacity of females to produce sex cells is contrasted with that of males. The average male, beginning at about age 13 to well into his eighties and nineties, can produce as many as one billion sex cells every day.

The human reproductive system involves separate male and female reproductive systems. The male gonads, the **testes** (singular: "testis"), produce male sex

Figure 17.1

From the smallest one-celled organisms to the largest mammals, all living things reproduce, ensuring the survival of the species.

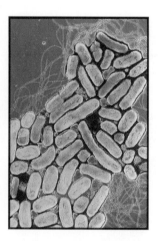

Testes *are the male gonads, or primary reproductive organs. Male sex hormones and sperm are produced in the testes.*

cells called *sperm*. The female gonads, the **ovaries,** produce "eggs." The fusion of a male and a female sex cell, in a process called **fertilization,** produces a **zygote.** The zygote divides many times to form an embryo, which in turn continues to grow into a fetus.

THE MALE REPRODUCTIVE SYSTEM

Male and female sex organs originate in the same area of the body—the abdominal cavity—and are almost indistinguishable until about the third month of embryonic development. At that time, the genes of the sex chromosomes cause differentiation. During the last two months of fetal development, the testes descend through a canal into the **scrotum,** a pouch of skin located below the pelvic region. A thin membrane forms over the canal, thereby preventing the testes from re-entering the abdominal cavity. Occasionally, an injury may cause the rupture of the membrane, producing an inguinal hernia. The hernia can be dangerous because a segment of the small intestine can be forced into the scrotum. The small intestine creates pressure on the testes, and blood flow to either the testes or small intestine may become restricted.

The temperature in the scrotum is a few degrees cooler than that of the abdominal cavity. The cooler temperatures are important, since sperm will not develop at body temperature. Should the testes fail to descend into the scrotum, the male will not be able to produce viable sperm. This makes the male sterile.

Ovaries *are the female gonads, or reproductive organs. Female sex hormones and egg cells are produced in the ovaries.*

Fertilization *occurs when a male and a female sex cell fuse.*

A **zygote** *is the cell resulting from the union of a male and female sex cell.*

The **scrotum** *is the sac that contains the testes.*

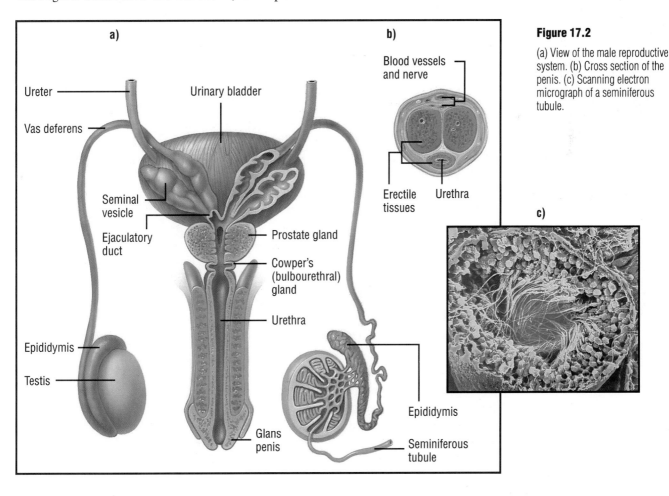

a)

Ureter

Urinary bladder

Vas deferens

Seminal vesicle

Ejaculatory duct

Epididymis

Testis

Prostate gland

Cowper's (bulbourethral) gland

Urethra

Glans penis

Epididymis

Seminiferous tubule

b)

Blood vessels and nerve

Erectile tissues

Urethra

c)

Figure 17.2

(a) View of the male reproductive system. (b) Cross section of the penis. (c) Scanning electron micrograph of a seminiferous tubule.

A tube called the **vas deferens** carries sperm from the testes to the *urethra*. Any blockage of the vas deferens will prevent the movement of sperm from the testes to the external environment. A surgical procedure, called a vasectomy, can be performed on males as a means of birth control. The *ejaculatory duct* regulates the movement of sperm and fluids, called **semen,** into the urethra, which also serves as a channel for urine. A **sphincter** regulates the voiding of urine from the bladder. Both regulatory functions work independently, and are never open at the same time. At any given time, the urethra conducts either urine or semen, but never both.

During sexual excitement, the erectile tissue within the penis fills with blood. Stimulation of the *parasympathetic nerve* causes the arteries leading to the penis to dilate, thereby increasing blood flow. As blood moves into the penis, the sinuses swell, compressing the veins that carry blood away from the penis. Any damage to the parasympathetic nerve can cause impotency, in which the penis fails to become erect. (Other causes, such as hormone imbalance and stress, have also been associated with impotency.)

TESTES AND SPERMATOGENESIS

In many ways the sperm cell is an example of mastery in engineering design. Built for motion, the sperm cell is streamlined with only a small amount of cytoplasm surrounding the nucleus. Although reduced cytoplasm is beneficial for a cell that must move, it also presents a problem. Limited cytoplasm means a limited energy reserve. A support cell, the **Sertoli cell,** nourishes the developing sperm cell. Energy-transforming organelles, the *mitochondria*, are located next to the *flagellum*, the organelle that propels the sperm cell. An entry capsule, called the **acrosome,** caps the head of the sperm cell. Filled with special enzymes that dissolve the outer coating surrounding the egg, the acrosome allows the sperm cells to penetrate the egg.

The **vas deferens** *are tubes that conduct sperm toward the urethra.*

A **sphincter** *is a ring of smooth muscle in the wall of a tubular organ. Contraction of the sphincter closes the opening.*

Semen *(seminal fluid) is a secretion of the male reproductive organs that is composed of sperm and fluids.*

Sertoli cells *nourish sperm cells.*

The **acrosome** *is the cap found on sperm cells. It contains packets of enzymes that permit the sperm cell to penetrate the gelatinous layers surrounding the egg.*

Figure 17.3

A human sperm cell.

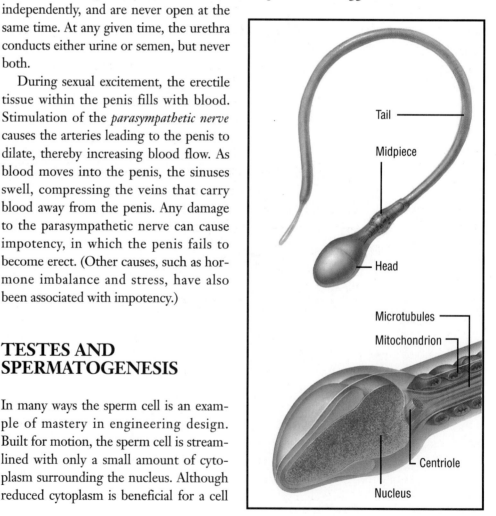

Tail

Midpiece

Head

Microtubules

Mitochondrion

Centriole

Nucleus

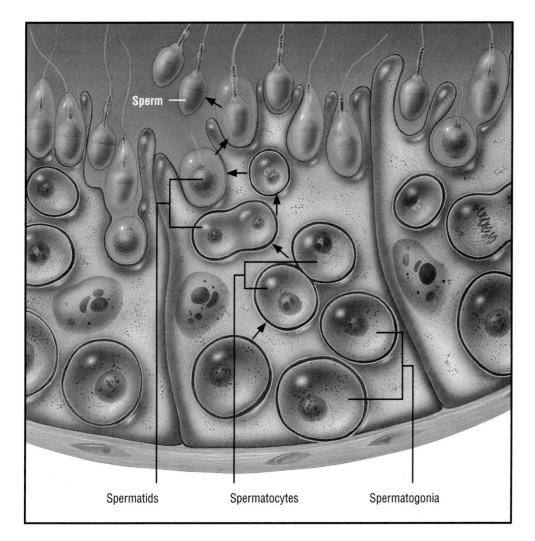

| Spermatids | Spermatocytes | Spermatogonia |

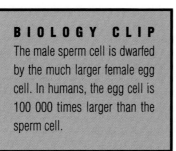

Figure 17.4

Development of sperm cells inside the seminiferous tubule.

The inside of each testis is filled with twisting tubes, called **seminiferous tubules,** that measure approximately 250 m in length. The seminiferous tubules are lined with sperm-producing cells called *spermatogonia*. These immature sperm cells contain a full complement of 46 chromosomes. During a process called *meiosis*, the spermatogonia divide into *spermatocytes*, which contain 23 chromosomes. (Meiosis is described in detail in the chapter Sexual Cell Reproduction.) Human sex cells of both males and females carry 23 chromosomes, which unite at fertilization to restore the original number of 46. Within 9 to 10 weeks, the spermatocytes differentiate into sperm cells. Although sperm cells are produced in the testes, they mature in the **epididymis,** a compact, coiled tube attached to the outer edge of the testis. Sperm cells develop their flagella and begin swimming motions within four days. It is believed that some defective sperm cells are destroyed by the immune system during their time in the epididymis.

BIOLOGY CLIP
The male sperm cell is dwarfed by the much larger female egg cell. In humans, the egg cell is 100 000 times larger than the sperm cell.

Seminiferous tubules *are coiled ducts found within the testes, where immature sperm cells divide and differentiate.*

The **epididymis** *is located along the posterior border of the testis and consists of coiled tubules that store sperm cells.*

SEMINAL FLUID

Fluid is secreted by three glands along the vas deferens and the urethra: the seminal vesicle, the prostate gland, and the Cowper's (bulbourethral) gland. Every time a man ejaculates, between 3 to 4 mL of fluid, containing approximately 500 million sperm cells, are released. The fluid, referred to as semen, provides a swimming medium for the flagellated sperm cells. Fluids from the **seminal vesicles** contain fructose and prostaglandins. The fructose provides a source of energy for the sperm cell. Recall that the sperm cell carries little energy reserves in its drastically reduced cytoplasm. Prostaglandins act as a chemical signal in the female system, triggering the rhythmic contraction of smooth muscle along the reproductive tract. It is believed that the contraction of muscles along the female reproductive pathways assists the movement of sperm cells toward the egg. The **prostate gland** secretes an alkaline buffer that protects sperm cells against the acidic environment of the vagina.

Cowper's (bulbourethral) gland secretes mucus-rich fluids prior to ejaculation. It is believed that the fluids protect the sperm cells from the acids found in the urethra associated with the passage of urine. The fluid may also assist sperm movement.

Although sperm cells can exist for many weeks in the epididymis, they have a much reduced life span when they come in contact with the various fluids in the semen. At body temperature, sperm cells will live only 24 to 72 hours. When stored at $-100°C$, sperm cells have been known to remain viable for many years.

HORMONAL CONTROL OF THE MALE REPRODUCTIVE SYSTEM

Ancient herdsmen discovered that the removal of the testes, known as castration, increased the body mass of their animals, making their meat more tender and savory. The disposition of the castrated males also changed. Steers, which are castrated bulls, tend not to be very aggressive. The castrated animals also lack a sex drive and are sterile.

The male sex hormones—androsterone and **testosterone**—are produced in the *interstitial cells* of the testes. As the name suggests, the interstitial cells are found between the seminiferous cells. Although both hormones carry out many functions, testosterone is the more potent and abundant. Testosterone stimulates spermatogenesis, the process by which spermatogonia divide and differentiate into mature sperm cells. Testosterone also influences the development of secondary male sexual characteristics at puberty, stimulating the maturation of the testes and penis. Testosterone levels have also been associated with sex drive. Evidence comes from ancient times, when eunuchs—males who had had their testes

---

Seminal vesicles *contribute to the seminal fluid (semen), a secretion that contains fructose and prostaglandins.*

The **prostate gland** *contributes to the seminal fluid (semen), a secretion containing buffers that protect sperm cells from the acidic environment of the vagina.*

Cowper's (bulbourethral) gland *contributes a mucus-rich fluid to the seminal fluid (semen).*

Testosterone *is the male sex hormone produced by the interstitial cells of the testes.*

Anabolic steroids *are strength-enhancing drugs.*

Gonadotropic hormones *are produced by the pituitary gland and regulate the functions of the testes and ovaries.*

BIOLOGY CLIP

Boy sopranos are renowned for the beauty of their voices. However, as boys reach puberty, their larynxes begin to change and their voices become lower. During the 17th and 18th centuries, adult male sopranos were very popular. These singers, called *castrati*, had had their testes removed before puberty so that their voices would remain high. In 1878, Pope Leo XVII ended the inhumane practice of castrating boys for the papal choir.

---

removed—were used to guard the harems and households of these rulers. Because they no longer were able to produce testosterone, the eunuchs had a decreased sex drive.

The male sex hormone also promotes the development of facial and body hair, the growth of the larynx, which causes the lowering of the voice, and the strengthening of muscles. In addition, testosterone increases the secretion of body oils and has been linked to the development of acne in males as they reach puberty. Once males adjust to higher levels of testosterone, skin problems decline. The increased oil production can also create body odor. Testosterone, or testosterone-related compounds, are used in the production of **anabolic steroids,** the strength-building drugs often associated with athletes.

The production of sperm and male sex hormones in the testes is controlled by the hypothalamus and the pituitary gland in the brain. Negative-feedback systems ensure that adequate numbers of sperm cells and constant levels of testosterone are maintained. The pituitary gland produces and stores the **gonadotropic hormones,** which regulate the functions of the testes; the male **follicle-stimulating hormone (FSH),** which stimulates the production of sperm cells in the seminiferous tubules; and the male **luteinizing hormone (LH),** which promotes the production of testosterone by the interstitial cells.

At puberty, the hypothalamus secretes the **gonadotropin-releasing hormone (GnRH).** GnRH activates the pituitary gland to secrete and release FSH and LH. The FSH acts directly on the sperm-producing cells of the seminiferous tubules, while LH stimulates the interstitial cells to produce testosterone. In turn, the testosterone itself increases sperm production. Once high levels of

testosterone are detected by the hypothalamus, a negative-feedback system is activated. Decreased GnRH production slows the production and release of LH, leading to less testosterone production. Testosterone levels thus remain in check. The feedback loop for sperm production is not well understood. It is believed that FSH acts on Sertoli cells, which produce a peptide hormone that sends a feedback message to the pituitary, inhibiting production of FSH.

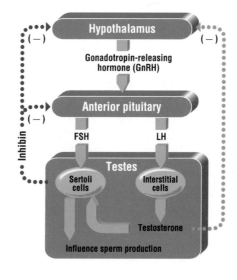

Figure 17.5

Negative feedback regulatory system for FSH and LH hormones. Testosterone inhibits LH production by the pituitary by deactivating the hypothalamus. The hypothalamus will release less GnRH, leading to decreased production of LH. The feedback mechanism for FSH is less understood. It has been suggested that a signalling chemical produced by the Sertoli cells inhibits both GnRH and FSH production.

*The **follicle-stimulating hormone (FSH)** increases sperm production in males.*

*The **luteinizing hormone (LH)** regulates the production of testosterone in males.*

*The **gonadotropin-releasing hormone (GnRH)** is a chemical messenger from the hypothalamus that stimulates secretions of FSH and LH from the pituitary.*

■ REVIEW QUESTIONS ■   ?

1 Name the primary male and female reproductive organs.
2 What would happen if the testes failed to descend into the scrotum?
3 Describe the function of the following structures: Sertoli cells, seminiferous tubules, and epididymis.
4 What is semen? Where is it found? What function does it serve?
5 What is spermatogenesis?
6 Outline the functions of testosterone.
7 How do gonadotropic hormones regulate spermatogenesis and testosterone production?
8 Using examples of LH and testosterone, explain the mechanism of negative feedback.

DEMONSTRATION

MICROSCOPIC VIEW OF THE TESTES

Objective

To view structures within the testes.

Materials

lens paper   prepared slide of testes (cross section)
light microscope   pencil

Procedure

1 Using lens paper, clean the ocular and all the objective lenses of the microscope. Rotate the revolving nosepiece so that the low-power objective is in place. Position the prepared slide on the stage of the microscope and view the cross section of the testes under low power.

a) Estimate the number of seminiferous tubules seen under low-power magnification.

2 Center the slide on a single seminiferous tubule and rotate the revolving nosepiece to the medium-power objective. Use only the fine adjustment to focus the cells. Locate an interstitial cell and a seminiferous tubule.

b) Estimate the size of the seminiferous tubule and the interstitial cell.

3 Rotate the nosepiece to the high-power objective lens and view the immature sperm cells within the seminiferous tubules.

c) Diagram five different cells viewed in the seminiferous tubule.

d) Would you expect to find mature sperm cells in the seminiferous tubule? Give your reasons. ■

THE FEMALE REPRODUCTIVE SYSTEM

Menopause *marks the termination of the female reproductive years.*

The **oviduct**, *or* **Fallopian tube**, *is the passageway through which an ovum moves from the ovary to the uterus, or womb.*

In many ways the female reproductive system is more complicated than that of the male. Once sexual maturity is reached, males continue to produce sperm cells at a somewhat constant rate. By contrast, females follow a complicated sexual cycle, in which one egg matures approximately every month. Hormonal levels fluctuate through the reproductive years that end at **menopause.**

During fetal development, paired ovaries (flattened, olive-shaped organs) form near the kidneys. Like the similarly shaped testes, the ovaries migrate along a canal, but unlike the testes, which come to rest outside of the abdominal cavity, the ovaries remain in the pelvic region.

Egg cells, or *ova* (singular: "ovum"), are found within the ovary. The ovary is also responsible for the production of female sex hormones.

An **oviduct,** or **Fallopian tube** (named after Gabriello Fallopio, a 16th-century Italian anatomist), is found next to each of the ovaries. Once mature, an ovum will enter the oviduct through wide, open ends called *fimbria*. As the ovum is moved along the oviduct by cilia, it goes through its final stages of development. Fertilization of the ovum occurs in the oviduct. However, unless the ovum is fertilized, the cell will deteriorate within 48 hours and die. The paired oviducts join a hollow, inverted, pear-shaped organ called the **uterus,** or **womb.** The length of time required for the fertilized ovum to travel the

10–12 cm oviduct to the uterus is between three and five days.

The uterus, the site where the **embryo** and **fetus** develop, is composed of two major tissues. The muscular outer lining of the uterus, known as the *myometrium*, provides support for the developing embryo. During the last phase of pregnancy, strong muscular contractions help move the baby into the birth canal. The glandular inner lining of the uterus, known as the **endometrium,** provides nourishment for the developing embryo. If pregnancy does not occur, the endometrium is shed. The process is called **menstruation.** Normally, the embryo embeds itself in the rich glandular tissue of the endometrium located in the uterus; however, the embryo will occasionally embed in a less-developed layer of the endometrium that extends into the oviduct. This type of pregnancy, called an *ectopic pregnancy*, can be dangerous. Not only is the amount of glandular tissue and nutrients limited, but the delicate oviduct is unable to stretch to accommodate a growing embryo.

The vagina connects the uterus with the outer environment. Sexual intercourse occurs within the vagina, which also serves as the birth canal. The vagina is strongly acidic, which creates a hostile environment for microbes that might attempt to enter the female reproductive system. A muscular band, called the **cervix,** separates the vagina from the uterus and is designed to hold the fetus in place. Dilation of the cervix during birth permits the fetus to enter the birth canal.

Cancer of the cervix is one of the major forms of cancer in females. Fortunately, early detection makes the prospects of arresting the tumor favorable. A *Pap test* provides physicians with a sample of cells from the cervix. Like skin cells and the cells that line your mouth, cells will slough off the cervix; therefore, the procedure requires no surgery. Physicians simply use a swab to collect epithelial cells from the cervix.

*The **uterus (womb)** is the female organ in which the fertilized ovum normally becomes embedded and in which the embryo and fetus develop.*

***Embryo** refers to the early stages of an animal's development. In humans, the embryo stage lasts until the ninth week of pregnancy.*

***Fetus** refers to the later stages of an unborn offspring's development. In humans, the embryo is called a fetus after the ninth week of development.*

*The **endometrium** is the glandular lining of the uterus that prepares the uterus for the embryo.*

***Menstruation** is the shedding of the endometrium.*

*The **cervix** is a muscular band that prevents the fetus from prematurely entering the birth canal.*

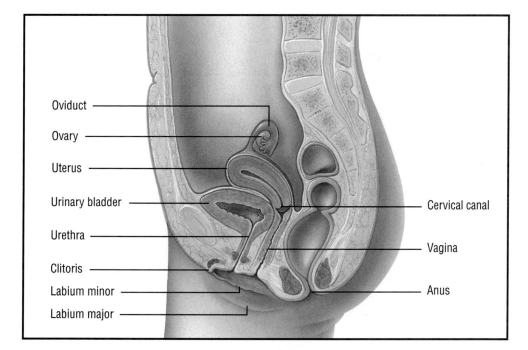

Figure 17.6

Female reproductive anatomy, side view. Note that two ovaries and two oviducts are present.

OOGENESIS AND OVULATION

The ovary contains fibrous connective tissue and small groups of cells called **follicles.** The follicles are composed of two types of cells: the *primary oocyte* and the *granulosa cells*. The oocyte, containing 46 chromosomes, undergoes meiosis and is transformed into a mature oocyte, or *ovum*. The granulosa cells provide nutrients for the oocyte.

Unlike the testes, which replenish sex cells, the female ovary undergoes continual decline after the onset of puberty. As mentioned earlier, the ovary contains about 400 000 follicles at puberty. Approximately 1000 follicles develop during each female reproductive cycle, but usually only a single follicle becomes dominant and reaches maturity. The remaining follicles deteriorate and are reabsorbed within the ovary. Between the ages of 12 and 50 in a woman's life about 400 eggs will mature. By the time a

woman reaches menopause, few follicles remain. It has been suggested that the higher incidence of genetic defects in children produced by older women can be linked to the age of the follicles. The longer the follicle lives, the greater is the chance of genetic damage. Because female sex hormones are produced within the ovary, menopause not only marks the end of a female's reproductive life, but also signals a drop in the production of female hormones.

Follicle development is controlled by a hormone produced by the pituitary gland. Nutrient follicle cells surrounding the primary oocyte begin to divide. As the primary oocyte undergoes meiosis I, the majority of cytoplasm and nutrients move to one of the poles and form a secondary oocyte. The secondary oocyte contains 23 chromosomes. The remaining cell, referred to as the *polar body*, receives little cytoplasm and dies. As the secondary cells surrounding the secondary oocyte develop, a fluid-filled cavity forms.

Follicles *are structures in the ovary that contain the egg and secrete estrogen.*

Figure 17.7

The process of ovulation. Pituitary hormones regulate the events of follicle development, ovulation, and the formation of the corpus luteum.

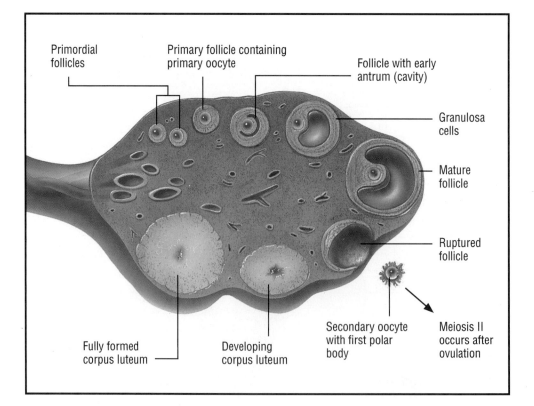

Primordial follicles

Primary follicle containing primary oocyte

Follicle with early antrum (cavity)

Granulosa cells

Mature follicle

Ruptured follicle

Fully formed corpus luteum

Developing corpus luteum

Secondary oocyte with first polar body

Meiosis II occurs after ovulation

Eventually the dominant follicle pushes outward, ballooning the outer wall of the ovary. Blood vessels along the distended outer wall of the ovary collapse, and the wall weakens. The outer surface of the ovary wall bursts and the secondary oocyte is released. This process is referred to as **ovulation.** Surrounding follicle cells remain within the ovary and are transformed into the **corpus luteum,** which secretes hormones essential for pregnancy; however, if pregnancy does not occur, the corpus luteum degenerates after about 10 days. All that remains is a scar, referred to as the *corpus albicans.* The secondary oocyte enters the oviduct and undergoes meiosis II. Once again, the division of cytoplasm and nutrients is unequal; the cell that retains most of the cytoplasm and nutrients is referred to as the mature oocyte, or ovum. As in meiosis I, the polar body deteriorates.

MENSTRUAL CYCLE

The human female menstrual cycle, which is repeated throughout a woman's reproductive lifetime, takes an average of 28 days, although great variation in this cycle is not uncommon. The menstrual cycle can be divided into four distinct phases: flow phase, follicular phase, ovulatory phase, and luteal phase. The **flow phase** is marked by the shedding of the endometrium, or menstruation. This is the only phase of the female reproductive cycle that can be determined externally. For this reason, the flow phase is used to mark the beginning of the menstrual cycle. Approximately five days are

> **BIOLOGY CLIP**
> The English word "hysteria," meaning emotional excitability, derives from *hystera,* the Greek word for uterus. It was originally thought that women were more prone to hysteria than men. The surgical operation, hysterectomy, literally means removal of the uterus.

required for the uterus to shed the endometrium.

The **follicular phase** is characterized by the development of follicles within the ovary. As follicle cells develop, the hormone **estrogen** is secreted. Estrogen promotes the development of secondary female sex characteristics, which include development of the breasts and body hair, and increased thickening of the endometrium. As follicles continue to develop, estrogen concentration in the blood increases. The follicular phase normally takes place between days 6 and 13 of the female menstrual cycle.

During ovulation, the third phase of the female menstrual cycle, the egg bursts from the ovary, and follicular cells differentiate into the corpus luteum.

The development of the corpus luteum marks the beginning of the **luteal phase.** Estrogen levels begin to decline when the oocyte leaves the ovary, but are restored somewhat when the corpus luteum forms. The corpus luteum secretes both estrogen and progesterone. **Progesterone** continues to stimulate the endometrium and prepares the uterus for an embryo. It also inhibits further ovulation. Does this provide any clues about why birth control pills contain high concentrations of progesterone? In addition, progesterone prevents uterine contractions. Thus, progesterone, as used in birth control pills, prevents ovulation. Should progesterone levels fall, uterine contractions would begin. The luteal phase, which occurs between days 15 and 28, prepares the uterus to receive a fertilized egg. Should fertilization of an ovum not

Ovulation *involves the release of the egg from the follicle held within the ovary.*

The **corpus luteum** *is made up of the follicle cells of the ovary following ovulation. The corpus luteum secretes estrogen and progesterone.*

The **flow phase** *of the menstrual cycle is marked by the shedding of the endometrium.*

The **follicular phase** *is marked by the development of the ovarian follicles prior to ovulation.*

Estrogen *is a female sex hormone.*

The **luteal phase** *is characterized by the formation of the corpus luteum following ovulation.*

Progesterone *is a female sex hormone.*

occur, the concentrations of estrogen and progesterone will decrease, thereby causing weak uterine contractions. These weak uterine contractions cause the endometrium to pull away from the uterine wall. The shedding of the endometrium marks the beginning of the next flow phase.

Table 17.1 Female Menstrual Cycle

Phase	Description of events	Hormone produced	Days
Flow	Menstruation		1–5
Follicular	Follicles develop in ovaries. Endometrium is restored.	Estrogen produced by follicle cells.	6–13
Ovulation	Oocyte bursts from ovary.		14
Luteal	Corpus luteum forms: endometrium thickens.	Estrogen and progesterone	15–28

Figure 17.8

The thickness of the endometrium increases from the beginning of the follicular phase to the end of the luteal phase. The development of blood vessels and glandular tissues helps prepare the uterus for a developing embryo. Should no embryo enter the uterus, menstruation occurs, and the menstrual cycle begins again.

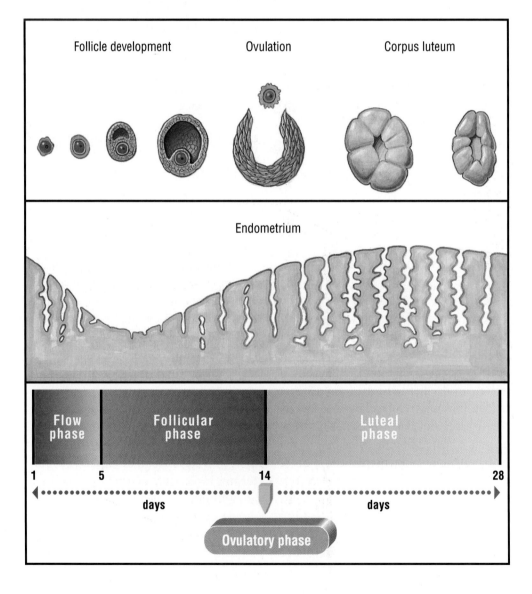

HORMONAL CONTROL OF THE FEMALE REPRODUCTIVE SYSTEM

The production of estrogen and progesterone—hormones of the ovary—is regulated by the hypothalamus-pituitary complex. Gonadotropins—female **FSH** (follicle-stimulating hormone) and **LH** (luteinizing hormone)—regulate the control of hormones produced by the ovaries: estrogen and progesterone. In turn, the gonadotropins are regulated by ovarian hormones as part of a complex negative-feedback mechanism.

The onset of female puberty is signalled by the release of **GnRH (gonadotropin-releasing hormone)** from the hypothalamus. GnRH activates the pituitary gland, which is the storage site of FSH and LH. During the follicular phase of the menstrual cycle, FSH secretions are carried by the blood to the ovary, where follicle development is stimulated. The follicles within the ovary secrete estrogen, which is carried in the blood, stimulating the development of secondary female characteristics and initiating the development of the endometrium. As estrogen levels rise, a negative-feedback message is sent to the pituitary gland to turn off secretions of FSH. The follicular phase of the menstrual cycle has come to an end. Simultaneously, the rise in estrogen exerts a positive message on the LH-producing cells of the pituitary gland. LH secretion rises and ovulation occurs.

After ovulation, the remaining follicular cells, under the influence of LH, are transformed into a functioning corpus luteum. The luteal phase of the menstrual cycle has begun. Cells of the corpus luteum secrete both estrogen and progesterone. The build-up of estrogen and progesterone will further increase the development of the endometrium. As

progesterone and estrogen build up within the body, a second negative-feedback mechanism is activated. Progesterone and estrogen work together to inhibit the release of both FSH and LH. Without gonadotropic hormones, the corpus luteum begins to deteriorate, slowing estrogen and progesterone production. The drop in ovarian hormones signals the beginning of menstruation.

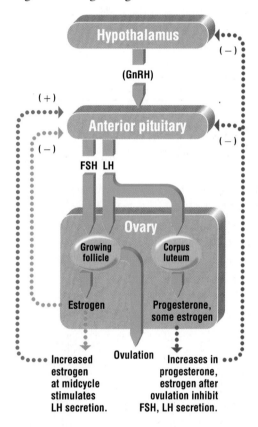

Figure 17.9

Feedback loop showing the regulation of ovarian hormones.

FSH *(follicle-stimulating hormone) is a gonadotropin that promotes the development of the follicles in the ovary.*

LH *(luteinizing hormone) is a gonadotropin that promotes ovulation and the formation of the corpus luteum.*

The similarities between male and female systems extend beyond the secretion of FSH and LH. Androgens (male sex hormones) and estrogen (the female sex hormone) can be produced by either gender. Male characteristics result not because androgens are the only hormones produced, but because the levels of androgens exceed the levels of estrogen. Males are ensured of maintaining low levels of female hormones by excreting them at an accelerated rate. This may explain why the urine of a stallion contains high levels of estrogen. The importance of balancing androgen and estrogen levels has been demonstrated with roosters. The removal of the testes from a rooster will caponize it. Injections of estrogen will bring about the same effect. In humans, the secretions of androgens will stimulate the development of the male's prostate gland, but injections of estrogen will slow the process. This may explain why cancerous tumors of the male prostate can be slowed down by injections of estrogen-like compounds.

Table 17.2 Female Reproductive Hormones

Hormone	Location	Description of function
Estrogen	Follicle cells (ovary)	Inhibits growth of facial hair, initiates secondary female characteristics, and thickening of the endometrium.
Progesterone	Corpus luteum (ovary)	Inhibits ovulation, inhibits uterine contractions, stimulates the endometrium.
FSH	Pituitary	Stimulates the development of the follicle cells in the ovary.
LH	Pituitary	Stimulates ovulation, and the formation and maintenance of the corpus luteum.

BIOLOGY CLIP

Acne is a common inflammatory disease of skin areas where oil glands are largest and most numerous. Acne is attributable to the effect of androgen hormones on hair follicles and their oil glands. It is very common at puberty and affects more than 80% of teenagers to some degree. Certain foods seem to aggravate it, particularly chocolate, nuts, and cola drinks. Before, during, or following the menstrual period flare ups are common, but interestingly, estrogen treatment is rarely effective.

REVIEW QUESTIONS ?

9 Diagram the female reproductive system and label the following parts: vagina, ovaries, cervix, oviducts, uterus, and endometrium.

10 Can a woman who has reached menopause ever become pregnant? Explain your answer.

11 What is menstruation? Why is it important?

12 Explain why ectopic pregnancies are dangerous.

13 What is a Pap test?

14 Describe the process of ovulation. Differentiate between primary oocytes, secondary oocytes, and mature ova.

15 Describe how the corpus luteum forms in the ovary.

16 Describe the events associated with the flow phase, follicular phase, and luteal phase of menstruation.

17 Outline the functions of estrogen and progesterone.

18 How do gonadotropic hormones regulate the function of ovarian hormones?

19 Predict how low secretions of GnRH from the hypothalamus would affect the female menstrual cycle.

20 With reference to the female reproductive system, provide an example of a negative-feedback control system.

CASE STUDY

HORMONE LEVELS DURING THE MENSTRUAL CYCLE

Temperature °C		
Days	Ovulation occurs	No ovulation occurs
5	36.4	36.3
10	36.2	35.7
12	36.0	35.8
14	38.4	36.2
16	37.1	36.1
18	36.6	36.0
20	36.8	36.3
22	37.0	36.3
24	37.1	36.4
28	36.6	36.5

Objective

To investigate how hormone levels regulate the female menstrual cycle.

Procedure

1 Ovarian hormones are regulated by gonadotropic hormones. Study the feedback loop shown in the diagram below.

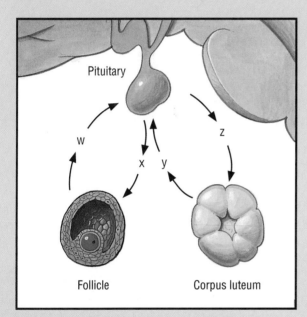

Pituitary

w

z

x   y

Follicle   Corpus luteum

a) Identify as w, x, y, or z, the two gonadotropic hormones represented in the diagram.

b) Identify the ovarian hormones shown in the diagram.

c) Which two hormones exert negative-feedback effects?

2 Body temperatures of two women were monitored during their menstrual cycles. One woman ovulated; the other did not.

d) Graph the data provided. Plot changes in temperature along the y-axis (vertical axis) and the days of the menstrual cycle along the x-axis (horizontal axis).

e) Assuming this menstrual cycle represents the average 28-day cycle, label the ovulation day on the graph.

f) Describe changes in temperature prior to and during ovulation.

g) Compare body temperatures with and without a functioning corpus luteum.

3 The graph below shows changes in the thickness of the endometrium throughout the female menstrual cycle.

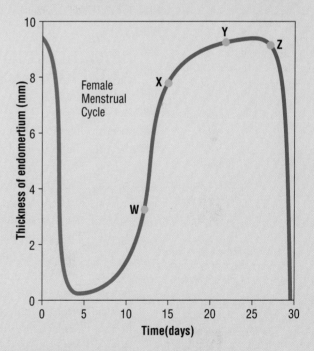

h) Identify the events that occur at times *X* and *Z*.

i) Identify by letter the time when follicle cells produce estrogen.

j) Identify by letter the time when the corpus luteum produces estrogen and progesterone.

4 Levels of gonadotropic hormones are monitored throughout the female reproductive cycle. Levels are recorded in relative units.

k) How does LH affect estrogen and progesterone?

Case-Study Application Questions

1 Explain why birth control pills often contain high concentrations of progesterone and estrogen.

2 Why would a woman not take birth control pills for the entire 28 days of the menstrual cycle? On which days of the menstrual cycle would the pill not be taken? ■

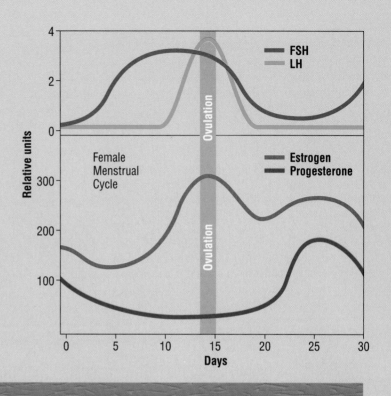

FERTILIZATION AND PREGNANCY

Figure 17.10

Human sperm cell attached to ovum.

Blastocyst *is an early stage of embryo development.*

Implantation *is the attachment of the embryo to the endometrium.*

Between 150 million and 300 million sperm cells travel through the cervix into the uterus upon ejaculation during intercourse. However, only a few hundred actually reach the oviducts. Although several sperm become attached to the outer edge of the mature ovum, only a single sperm cell fuses with the ovum.

The fertilized egg, now referred to as a zygote, undergoes many divisions as it travels by way of the oviduct toward the uterus. By the time it reaches the uterus, in about six days, the single fertilized egg cell has been transformed into a cell mass, called a **blastocyst.** Once in the uterus, the blastocyst becomes attached to the wall of the endometrium, a process referred to as **implantation.**

For pregnancy to continue, menstruation cannot occur. Any shedding of the

endometrium would mean the dislodging of the embryo from the uterus. However, maintaining the endometrium presents a problem for the hormonal system. High levels of progesterone and

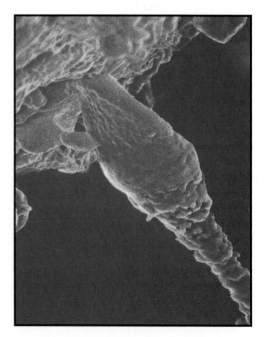

estrogen have a negative-feedback effect on the secretion of gonadotropic hormones. LH levels must remain high to sustain the corpus luteum. Should the corpus luteum deteriorate, the levels of estrogen and progesterone would drop, stimulating uterine contractions and the shedding of the endometrium. For pregnancy to continue, progesterone and estrogen levels must be maintained.

The outer layer of the developing cell mass forms a hormone called **HCG** (human chorionic gonadotropic hormone), which maintains the corpus luteum for the first three months of pregnancy. The functioning corpus luteum continues producing progesterone and estrogen, which in turn maintain the endometrium. The endometrium and embryo thus remain in the uterus. Pregnancy tests identify HCG levels in the urine of women.

Cells from the embryo and endometrium combine to form the placenta, through which materials are exchanged between the mother and developing embryo. At approximately the fourth month of pregnancy, the placenta begins to produce estrogen and progesterone. High levels of progesterone prevent further ovulation. This means that once a woman is pregnant, she cannot become pregnant again during that pregnancy.

PRENATAL DEVELOPMENT

The outer layer of the blastocyst gives rise to two cell membranes. The outer membrane is the **chorion,** which produces HCG. The inner membrane, the **amnion,** develops above the embryo.

The amnion evolves into a fluid-filled sac that insulates the embryo, and later the fetus, protecting it from infection, dehydration, impact, and changes in temperature. By the fourth week of pregnancy the yolk sac, which is a vestigial formation in humans, forms below the embryo.

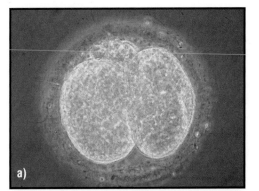

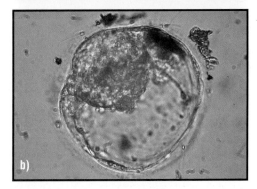

Cells of the fetus and cells of the endometrium comprise the **placenta.** The placenta is richly supplied with blood vessels. Projections called *chorionic villi* ensure that many blood capillaries of the mother are exposed to a large number of blood capillaries of the fetus. A third membrane, the **allantois,** provides blood vessels in the placenta. However, unlike the chorion and amnion, the allantois does not envelop the fetus. The placenta provides a lifeline between mother and fetus. Nutrients and oxygen diffuse from the mother's blood into the

Figure 17.11

(a) Two-cell stage.
(b) Blastocyst after 4 to 6 days.

HCG *is a placental hormone that maintains the corpus luteum.*

The **chorion** *is the outer membrane of a developing embryo.*

The **amnion** *is a fluid-filled embryonic membrane.*

The **placenta** *is the site for the exchange of nutrients and wastes between mother and fetus.*

The **allantois** *is an embryonic membrane.*

blood of the developing fetus. Wastes diffuse in the opposite direction, moving from the fetus to the mother. The **umbilical cord** connects the embryo with the placenta.

The nine months of pregnancy are divided into three trimesters. The **first trimester** extends from fertilization to the end of the third month. By the second week of development, three germ layers begin to form: the ectoderm, the mesoderm, and the endoderm. Each of the organs shown in Table 17.3 develops from one of the germ layers.

The **umbilical cord** *connects the fetus to the placenta.*

The **first trimester** *extends from conception until the third month of pregnancy.*

Figure 17.12

Formation of the membranes that protect the embryo.

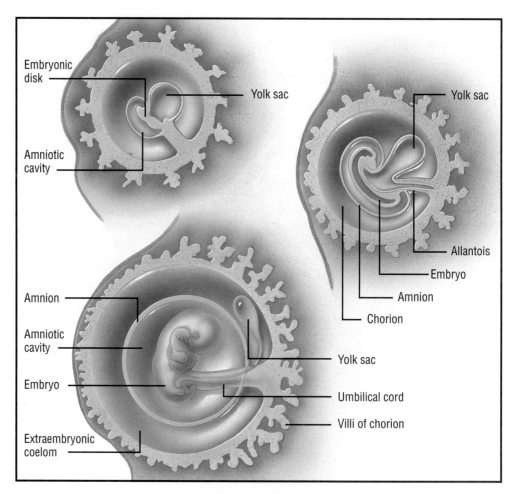

Table 17.3 Organs from Germ Layers

Germ Layer	Organ and Accessory Structures
Ectoderm	Skin, hair, finger nails, sweat glands
	Nervous system, brain, peripheral nerves
	Lens, retina, and cornea
	Inner ear, cochlea, semicircular canals
	Teeth, and inside lining of mouth
Mesoderm	Muscles (skeletal, cardiac, and smooth)
	Blood vessels and blood
	Kidneys and reproductive structures
	Connective tissue, cartilage, and bone
Endoderm	Liver, pancreas, thyroid, parathyroid
	Urinary bladder
	Lining of digestive system
	Lining of the respiratory tract

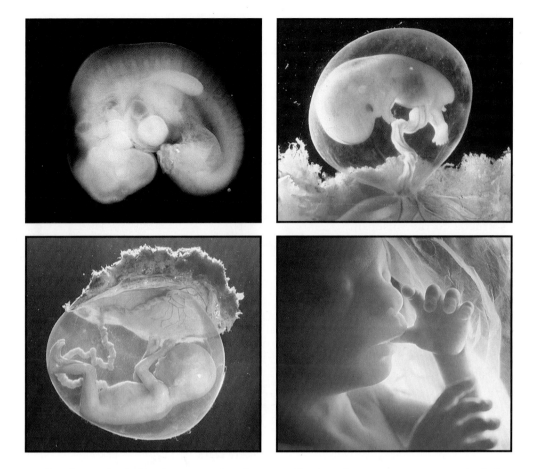

Figure 17.13

Human embryo at four weeks, and the fetus at nine weeks, sixteen weeks, and at eighteen weeks.

By the end of the first month, the 7 mm embryo is 500 times larger than the fertilized egg. The four-chambered heart has formed, a large anterior brain is visible, and limb buds with tiny fingers and toes have developed. By the ninth week, the embryo is referred to as a fetus. Arms and legs begin to move and a sucking reflex is evident.

By the **second trimester,** the 57 mm fetus moves enough to make itself known to the mother. All of its organs have formed and the fetus begins to look more like a human infant. As in other mammals, soft hair begins to cover the entire body. By the sixth month eyelids and eyelashes form. Most of the cartilage that formed the skeleton has been replaced by bone cells. Should the mother go into labor at the end of the second trimester, there is a chance that the 350 mm, 680 g fetus will survive.

During the **third trimester,** the baby grows rapidly. Organ systems have been established during the first two trimesters; all that remains is for the body mass to increase and the organs to become more developed. At birth, the average human infant is approximately 530 mm long and weighs about 3400 g.

BIRTH

Approximately 266 days after implantation, uterine contractions signal the beginning of labor. The cervix thins and begins to dilate. The amniotic membrane is forced into the birth canal. The amniotic membrane often bursts and amniotic fluid lubricates the canal (a process referred to as the breaking of the water). As the cervix dilates, uterine contractions move the baby through the birth canal.

*The **second trimester** extends from the third month to the sixth month of pregnancy.*

*The **third trimester** extends from the seventh month of pregnancy until birth.*

Hormones play a vital role in the birthing process. **Relaxin,** a hormone produced by the placenta prior to labor, causes the ligaments within the pelvis to loosen, providing a more flexible passageway for the baby during delivery. Although the actual mechanism is not completely understood, it is believed that a decreased production of progesterone is crucial to the onset of labor. **Oxytocin,** a hormone from the posterior pituitary gland, causes strong uterine contractions. Prostaglandins, which are also believed to trigger strong uterine contractions, appear in the mother's blood prior to labor.

LACTATION

Breast development is stimulated from the onset of puberty by estrogen and progesterone. During pregnancy, elevated levels of estrogen and progesterone prepare the breasts for milk production. Each breast contains about 20 lobes of glandular tissue, each supplied with a tiny duct that carries fluids toward the nipple. A hormone called **prolactin,** produced by the pituitary gland, is believed to be responsible for stimulating glands within the breast to begin producing fluids. Although small concentrations of prolactin are secreted throughout pregnancy, the levels rise dramatically after birth has occurred. The fact that the rise of prolactin levels coincides with rapid decreases in both estrogen and progesterone levels has led scientists to speculate that the female sex hormones suppress prolactin. Prolactin causes the production of a fluid called *colostrum,* a fluid that closely resembles breast milk. Colostrum contains milk sugar and milk proteins, but lacks the milk fats found in breast milk. A few days after birth, the prolactin stimulates the production of milk.

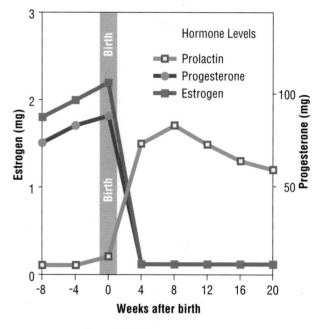

Figure 17.15

A dramatic increase in prolactin is attributed to lowering levels of estrogen and progesterone. Estrogen and progesterone levels drop following childbirth. Measurements for prolactin are in relative amounts.

Although prolactin increases milk production, the milk does not flow easily. Milk produced in the lobes of glandular tissue must be forced into the ducts that lead to the nipple. The suckling action of the newborn stimulates nerve endings in the areola of the breast. Sensory nerves carry information to the pituitary gland, causing the release of oxytocin. The hormonal reflex is completed as oxytocin is carried by the blood to the breasts and uterus. Within the breast, oxytocin causes weak contractions of smooth muscle, forcing milk into the ducts. Within the uterus, oxytocin causes weak contractions of smooth muscle, allowing the uterus to slowly return to its pre-pregnancy size and shape.

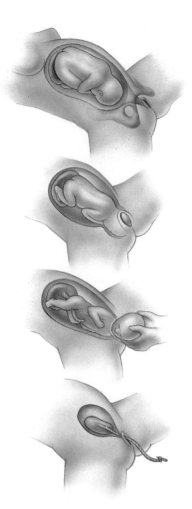

Figure 17.14

Movement of the baby through the birth canal.

Relaxin, *a hormone produced by the placenta prior to labor, causes the ligaments within the pelvis to loosen.*

Oxytocin, *a hormone from the posterior pituitary gland, causes strong uterine contractions.*

Prolactin *is a hormone produced by the pituitary and is associated with milk production.*

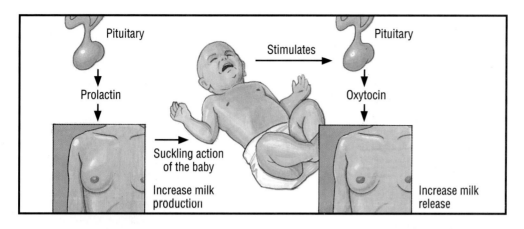

Figure 17.16

The hormone prolactin stimulates the breast to produce milk. The suckling action of the baby initiates a hormonal reflex involving the hormone oxytocin.

Although most North American mothers prefer to end breast-feeding once their youngster begins to develop teeth, women in some countries, especially where sources of protein are scarce, often continue to breast-feed for four or five years. Milk production causes a metabolic drain on the mother. At the height of lactation a woman can produce as much as 1.5 L of milk each day. A mother producing that much milk would lose approximately 50 g of fat and up to 100 g of lactose sugar. In addition, a breast-feeding mother would have to replace some 2 to 3g of calcium phosphate each day. To maintain adequate levels of calcium and phosphate, the parathyroid glands enlarge and bones decalcify. Failure to replace the needed calcium results in a progressive deterioration of the skeleton and teeth. Mother's milk also supplies an important source of antibodies, a topic that is dealt with in the chapter Blood and Immunity.

FRONTIERS OF TECHNOLOGY: *In Vitro* FERTILIZATION

Approximately 13% of Canadian couples are unable to conceive a child. Sterility, hormonal imbalances, and the destruction of reproductive organs by infections are the leading causes of infertility. **In vitro** fertilization, in which the egg is fertilized outside of the body, has provided renewed hope for many people who are unable to have children. On July 25, 1978, Drs. Patrick Steptoe and Robert Edwards of Cambridge, England, announced the first successful *in vitro* fertilization. The baby, Louise Brown, became the world's first so-called "test-tube baby." The name "test-tube baby" is actually a misnomer, since neither the egg nor sperm spend any time in a test tube.

In this procedure, a device, called a *laparoscope* is inserted into the woman's abdomen. An optical device within the instrument enables the physician to locate the ovary. A suction apparatus in the laparoscope allows the extraction of eggs from the ovary. The eggs are placed in a glass petri dish and fertilized by the sperm. Following a brief incubation period, one or more of the embryos is transferred into the uterus by a small catheter. If one of the embryos implants, a baby will be born nine months later.

A second technique, involving embryo freezing, has been combined with *in vitro* fertilization. Because *in vitro* fertilization uses multiple eggs, of which only a limited number can be placed in the uterus, freezing permits storage of any unused embryos and implantation without undergoing the laparoscope procedure. In 1990, Dr. Peter Leung of Toronto's East General Hospital was the first doctor in Canada to implant a frozen embryo.

In vitro fertilization occurs outside of the female's body. In vitro is Latin for "in glass."

RESEARCH IN CANADA

Pregnancy loss, birth defects, and mental retardation have been linked with chromosome abnormalities in sperm and eggs. Much of the scientific research has focused on abnormalities in the egg. Until recently, studies of developing sperm cells have been difficult; however, a new technique has opened the door to many research possibilities. One research group has shown that sperm cells can carry heavy metals. Another research team has provided data that may indicate a link between cocaine abuse and chromosome abnormalities in sperm cells.

Sperm Abnormalities

At the forefront of this research is a group headed by Dr. Renée Martin from the Alberta Children's Hospital in Calgary. Dr. Martin's research indicates that the sperm for normal men has a frequency of 10% chromosomal abnormality. Men who have undergone radiotherapy have much higher frequencies of abnormal sperm. According to Dr. Martin, some of these abnormalities could cause miscarriages. One of the most important questions to be answered is whether or not any of these abnormal sperm cells actually fertilize an egg. Is there some selection process? The research will provide valuable information on birth defects and miscarriages.

DR. RENÉE MARTIN

SOCIAL ISSUE:
Fetal Alcohol Syndrome

Fetal alcohol syndrome (FAS) is the third most common cause of mental retardation in babies born in Canada and the United States. FAS can also cause heart defects and disorders of the nervous system.

The placenta is a selective barrier that prevents the mother's blood cells from entering the circulatory system of the fetus. Smaller molecules, however, including some harmful substances such as alcohol, are able to move across the membrane. When a pregnant woman drinks, the alcohol crosses the placenta and enters the blood of the fetus.

Statement:

Women should refrain from drinking alcohol throughout their pregnancy.

Point

- Pregnant women have no right to endanger the health of the fetus, so they should abstain completely from drinking alcohol.
- Blood tests to detect the presence of alcohol in the bloodstream should be mandatory for all pregnant women.

Research the issue.
Reflect on your findings.
Discuss the various viewpoints with others.
Prepare for the class debate.

Counterpoint

- Both mothers and fathers should voluntarily recognize their responsibility to the fetus.

- The idea of suspending the rights of all pregnant women is insupportable. Legislation that forces women to have blood tests is not the answer. Changes in attitudes are accomplished best through education.

CHAPTER HIGHLIGHTS

- Sexual reproduction provides the basis for genetic diversity.
- The male gonads, the testes, produce sperm cells; the female gonads, the ovaries, produce egg cells.
- The scrotum is a sac that contains the testes.
- The vas deferens is a tube that conducts sperm toward the urethra.
- The Sertoli cells nourish sperm cells.
- The seminiferous tubules are coiled ducts found within the testes. Immature sperm cells divide and differentiate in the seminiferous tubules.
- The epididymis, which stores sperm cells, consists of coiled tubules along the posterior border of the testes.

- Semen is a secretion of the male reproductive organs and is composed of sperm and fluids.
- Gonadotropins are produced by the pituitary gland and regulate the functions of the testes and ovaries.
- The testes produce sperm cells and testosterone, the male sex hormone.
- Estrogen and progesterone are the female sex hormones.
- The ovaries produce ova as well as estrogen and progesterone. Estrogen promotes the development of secondary female sex characteristics and initiates the development of the endometrium. Progesterone also promotes the development of the endometrium and inhibits uterine contractions and ovulation.

- Fusion of male and female gametes creates a zygote.
- Menopause marks the termination of the female reproductive cycle.
- The oviduct, or Fallopian tube, is the passageway through which the ovum moves from the ovary to the uterus, or womb.
- The uterus is the female organ in which the fertilized egg develops.

- The endometrium is the glandular lining of the uterus. It prepares the uterus for the embryo.
- Menstruation is the shedding of the endometrium.
- Ovulation involves the release of the egg from the follicle held within the ovary.
- The corpus luteum is made from follicle cells of the ovary following ovulation. The corpus luteum secretes estrogen and progesterone.

APPLYING THE CONCEPTS

1 Predict the consequences if the testes failed to descend from the abdominal cavity during embryo development.

2 Anabolic steroids act in a similar fashion to testosterone by turning off secretions of gonadotropic hormones. What effects do anabolic steroids have on secretions of testosterone? Explain your answer.

3 The seminal vesicle, prostate gland, and Cowper's gland contribute secretions to the semen. How would reproduction be affected should secretions from these glands be inhibited?

4 An experiment was performed in which the circulatory systems of two mice with compatible blood types were joined. The data collected from the experiment are expressed in the table to the right. (Note: + indicates found; − indicates not found.) Explain the data collected.

Animal	Testosterone	LH	FSH	Sperm in urethra
A	+	+	+	−
B	+	+	+	+

5 Draw a diagram of the female reproductive system. Label the sites of ovulation, fertilization, and implantation.

6 The graph below shows estrogen and progesterone levels during a menstrual cycle.

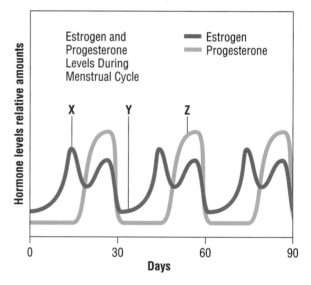

a) On which day (X, Y, or Z) would ovulation occur? Explain your answer.

b) On which day (X, Y, or Z) would you expect to find a functioning corpus luteum? Explain your answer.

7 The following graph shows hormone levels during the 40 weeks of pregnancy. Explain the changes in HCG, estrogen, and progesterone levels during pregnancy.

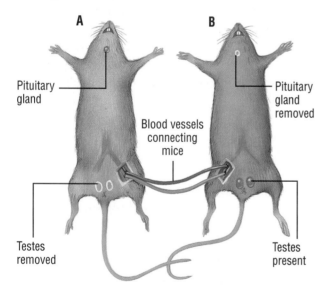

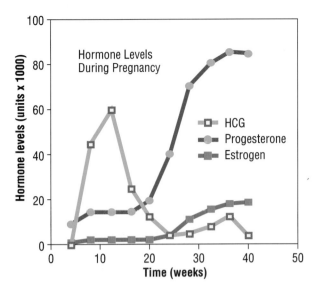

Hormone Levels During Pregnancy

Hormone levels (units x 1000) vs Time (weeks)

- HCG
- Progesterone
- Estrogen

8 During the eighth month of pregnancy, the ovaries of a woman are surgically removed. How will the removal of the ovaries affect the fetus? Explain your answer.

9 Explain why only one corpus luteum may be found in the ovaries of a woman who has given birth to triplets.

CRITICAL-THINKING QUESTIONS

1 Design an experiment to demonstrate the independent functions of the interstitial cells and seminiferous tubules of the testes.

2 Design an experiment to show how female gonadotropic hormones are regulated by ovarian hormones.

3 A 1988 article reported that prenatal testing had been banned in South Korea following a rise in male births. Seoul hospitals noted the change in the ratio of male to female births from 102.5 males for every 100 females, to 117 males for every 100 females. One statistician claimed that the change in ratio could be attributed to the selection of male fetuses following prenatal testing. In India, doctors have been told not to disclose the sex of a child following amniocentesis testing. Consider the implications for a society that attempts to select the sex of unborn children.

4 The embryo is sensitive to drugs, especially during the first trimester. In the late 1950s a drug called thalidomide was introduced in Europe to help prevent morning sickness. Unfortunately, the drug altered the genes regulating limb bud formation in developing embryos. Before the drug could be withdrawn from pharmacies, children were born with lifelong disabilities. Although drugs such as thalidomide are no longer available, a great debate centers around drugs that may have less pronounced effects on the fetus. Some evidence suggests that tranquilizers may cause improper limb bud formation. Even some acne drugs have been linked to facial deformities in newborns, and the antibiotic streptomycin has been associated with hearing problems. Explain why the embryo is so sensitive to drugs.

ENRICHMENT ACTIVITIES

Suggested reading:
- Kolata, Gina. "Operating on the Unborn." *New York Times Magazine*, May 14, 1989, p. 34.
- Krantorvitz, Barbara. "Preemies." *Newsweek*, May 16, 1988.
- Nilson, Leonard. "The First Days of Creation." *Life*, August 1990, p. 26.

- Poole, William. "The First 9 Months of School." *Hippocrates*, July/August 1987, p. 68.
- Vines, Gail. "Test Tube Embryos." *New Scientist*, November 19, 1987, pp. 1–4.
- Weiss, Richard. "Genetic Gender Gap." *Science News*, May 20, 1989, p. 312.
- Wolinsky, Howard. "Transplants from the Unborn." *American Health*, April 1988, p. 47.

Asexual Cell Reproduction

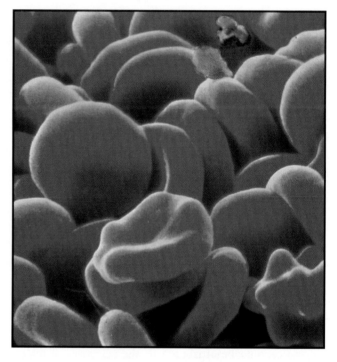

Figure 18.2

A normal person contains about 25 trillion red blood cells, yet the average life span of a red blood cell is only 20 to 120 days.

Figure 18.1

Early stages of cell division in a frog.

IMPORTANCE OF CELL DIVISION

For evidence of the importance of cell division you need look no further than your own body. The estimated 100 trillion cells that make up your body began from a single fertilized egg. In addition, your body is continually producing new cells. Red blood cells die and are replaced at a rate of one million every second. By the time you complete this biology course you will have totally replaced all of the red blood cells that are currently in your body.

The importance of cell division cannot be underestimated. Have you ever peeled the skin from your back after a sunburn? Imagine what you would look like if new cells did not replace the dead skin cells that had been sloughed off! Without cell division, scratches would never heal and blemishes would never disappear. If your red blood cells did not divide and reproduce, you would die.

Cell division is one of the most studied, yet least understood, areas of biology. Through painstaking hours of observation, scientists have collected a great deal of information about cell division. Yet despite all that has been observed, many questions remain unanswered. How do cells know when to divide? Why does an egg cell divide so rapidly after fertilization? How does the rate of cell division change (in the formation of calluses, for example)? Why do red blood cells divide at enormous rates, while adult brain cells do not? Why do cancer cells divide at uncontrolled rates?

PRINCIPLES OF MITOSIS

The first part of the cell theory states that all living things are made up of one or more cells. It was nearly 40 years after the discovery of the cell that the second part of the modern cell theory was formulated. Scientists concluded that cells are formed from pre-existing cells by cell division. It was earlier believed that some single-cell organisms developed from nonliving objects. The linking of life to reproduction was an important step in the development of modern biology.

Despite great differences in the forms and structures of living things, most cells show remarkable similarities in the manner in which they divide. Cell division occurs in very primitive forms of life, such as bacteria, which split into two identical daughter cells by a process known as **binary fission.** Multicellular organisms follow a similar mechanism. The initial cell, referred to as the mother cell, divides into two identical daughter cells. The division involves two phases: the division of nuclear materials and the division of cytoplasm. During the first phase, the genetic material, located along double-stranded chromosomes, divides and moves to opposite ends of the cell. The second phase of cell division is marked by the separation of cytoplasm, along with its constituent organelles, into equal parts.

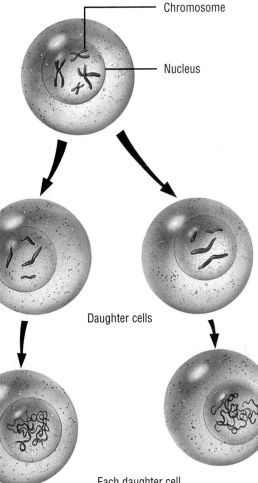

Chromosome

Nucleus

Daughter cells

Each daughter cell synthesizes a duplicate strand and chromosomes once again become double-stranded.

Figure 18.3

In binary fission the nucleus divides by mitosis and the cytoplasm divides into two equivalent parts.

The process, known as **mitosis,** repeats itself millions of times each day, replacing worn-out cells with new ones.

During mitosis the double-stranded chromosomes separate into single strands. Each of the daughter cells gets one of the single-stranded pairs of genetic material. The daughter cells use the single strands of genetic material to synthesize a complementary strand, and the chromosome again becomes double-stranded. The cell of a cow with 60 chromosomes will divide by mitosis to daughter cells of 60 chromosomes. Each

Binary fission *is a form of asexual reproduction in which one cell divides into two equal cells.*

Mitosis *is a type of cell division in which daughter cells receive the same number of chromosomes as the parent cell.*

cell in your body contains 46 chromosomes; all succeeding generations of cells will also contain 46 chromosomes. The duplication of complementary strands of genetic information ensures that the daughter cells are identical to each other and to the mother cell. The duplication of genetic information also ensures future cell divisions. Each daughter cell is a potential mother cell for the next generation.

Because all cells in the human body are duplicates of the same fertilized egg, all cells have the same genetic information. A muscle cell, for example, has all of the chromosomes of a cell found in your brain, or of a cell found in your heart. However, not all cells in the human body have the same shape or carry out the same functions. One of the most puzzling questions that confronts scientists who study cells is why different cells do different jobs. What makes a brain cell conduct nerve impulses, and what makes a muscle cell contract? How do specialized cells know which genes to use? These questions remain to be answered.

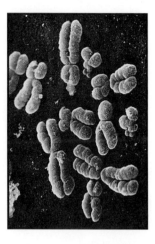

Figure 18.4
Double-stranded chromosomes just before mitosis.

Centrioles *are small protein bodies that are found in the cytoplasm of animal cells.*

Spindle fibers *are protein structures that guide chromosomes during cell division.*

STAGES OF CELL DIVISION

Cell division is often described as taking place in phases. The phases help scientists describe the events of mitosis. Cell division, however, does not pause after each phase—it is a continuous process.

Interphase
Interphase describes the processes of cell activity between cell divisions. Most cells, even the rapidly dividing, immature red blood cells, spend the majority of their time in this phase. During interphase, cells grow, carrying out the chemical activities that sustain life during this period. Cells make structural proteins that repair damaged parts, move nutrients from one organelle to another, and eliminate wastes. During interphase, cells prepare themselves for mitosis by building proteins. These proteins are used for new cell membranes as well as for constructing enzymes that aid chemical reactions including those that control the duplication of genetic information. The single-stranded chromosomes once again become double strands during interphase. If the cell did not duplicate its chromosomes during interphase, it would never again be able to divide.

Prophase
Prophase is the first true phase of cell division. Under the microscope, the contents of the nucleus become visible as the chromosomes shorten and thicken. In animal cells, a small body in the cytoplasm separates and moves to opposite poles of the cell as the chromosomes become visible. These tiny structures, called **centrioles,** provide attachment for **spindle fibers.** The spindle fibers serve as guide wires for the attachment and movement of the chromosomes during cell division. Small fragments of the spindle fibers, called *asters,* radiate out from the centrioles. The name aster, from the Greek word for "star," is appropriate because the centrioles appear as stars within the cytoplasm. Most plant cells do not have centrioles, but spindle fibers still form and serve a similar purpose. Not surprisingly, plant cells without centrioles do not have asters.

The chromosomes consist of two **chromatids,** held together by a tiny structure called a **centromere.** The centromere helps anchor the chromosomes to the spindle fibers. During prophase, the nuclear membrane appears to fade when viewed under the microscope; in effect, it is dissolving.

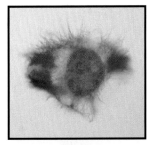

a) Interphase (before mitosis)

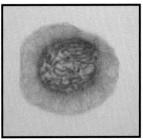

b) Early prophase

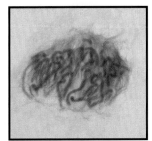

c) Prophase

Figure 18.5

Mitosis in a plant cell.

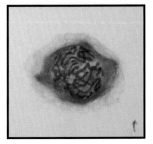

d) Late prophase

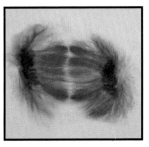

e) Transition to metaphase

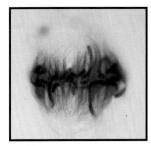

f) Metaphase

g) Anaphase

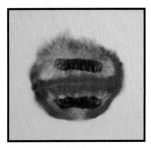

h) Telophase

i) Late telophase

Metaphase

The second phase of cell division is referred to as metaphase. During this phase, chromosomes composed of sister chromatids move toward the center of the cell. This center area is called the *equatorial plate*, because, like the equator of the world, it is an imaginary line that divides the cell and its chromosomes. The chromosomes appear as dark, thick masses that are attached to the spindle fibers. Even though they are most visible at this stage, it is still very difficult to count the number of chromosomes in most cells. Chromatids can become intertwined during metaphase. Occasionally, segments of the chromatids will break apart once anaphase begins.

Anaphase

Anaphase is the third phase of mitosis. During this phase the centromeres divide and the chromatids move to opposite poles of the cell. If mitosis proceeds correctly, the same number of single-stranded chromosomes will be found at each pole.

Telophase

The last phase of cell division is telophase. During this phase the chromosomes reach the opposite poles of the cell, and once again begin to lengthen and intertwine. The spindle fibers dissolve and a nuclear membrane begins to form around each mass of **chromatin.** Telophase is followed by the division of the cytoplasm.

Chromatids *are single strands of a chromosome which remain joined by a centromere.*

Centromeres *are structures that hold chromatids together.*

Chromatin *is the material found in the nucleus and is composed of protein and DNA.*

Late prophase

Figure 18.6

Mitosis is the nuclear division mechanism that maintains the parental chromosome number in each daughter nucleus. Shown here is a diploid animal cell with pairs of homologous chromosomes derived from two parents.

For the sake of clarity, only two pairs of homologues are shown and the spindle apparatus is simplified. With rare exceptions, the picture is more involved than this, as indicated by the photomicrographs on the previous page.

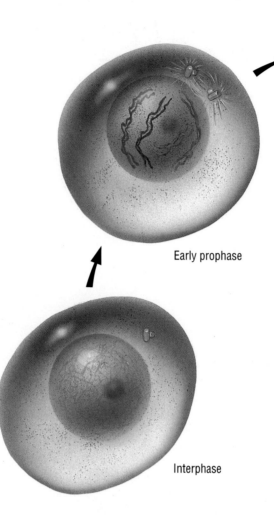

Early prophase

Interphase

Cytoplasmic Division

Once the chromosomes have moved to opposite poles, the cytoplasm begins to divide. **Cytokinesis,** or the division of the cytoplasm, appears to be quite separate from the nuclear division. The division of the cytoplasm must provide organelles for each of the newly established cells. In plant cells, the separation is accomplished by a cell plate that forms between the two chromatin masses. The cell plate will develop into a new cell wall, eventually sealing the contents of the new

Cytokinesis *refers to the division of cytoplasm.*

cells from each other. Animal cells, however, do not have cell walls to use as barriers. The more elastic animal cells pinch off in the center as the cytoplasm moves to opposite poles.

◼ REVIEW QUESTIONS ◼   ?

1 Compare the daughter cells with the original mother cell.

2 Describe the structure and function of the spindle fibers.

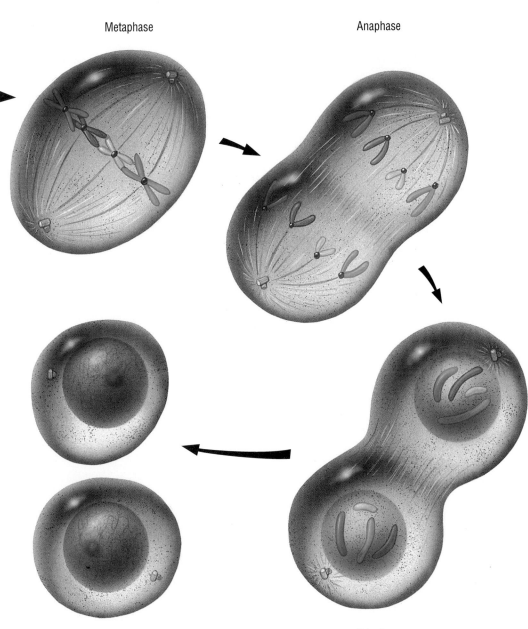

Metaphase

Anaphase

Telophase

Interphase

3 During interphase, what event must occur for the cell to be capable of undergoing future divisions?

4 Describe the events of prophase, metaphase, anaphase, and telophase.

5 Compare and contrast cell division in plant and animal cells.

6 What would happen if you ingested a drug that prevented mitosis?

B I O L O G Y   C L I P
A giant cell in the bone marrow undergoes cytoplasmic division without nuclear division. Small pieces of cytoplasm pinch off the cell to form blood-clotting platelets.

LABORATORY
MITOSIS

Objectives

To identify cells in the four phases of mitosis and to calculate the rate of cell growth.

Materials

microscope
prepared slides of onion root tip
lens paper
prepared slides of whitefish blastula

Procedure

Observing Dividing Cells

1 Obtain an onion root tip slide and place it on the stage of your microscope. View the slide under low power magnification. Focus using the coarse adjustment.

2 Center the root tip and then rotate the nosepiece to the medium-power objective lens. Focus the image using the fine-adjustment focus. Observe the cells near the root cap. This area is referred to as the *meristematic* region of the root.

3 Move the slide away from the root tip and observe the cells. These are the mature cells of the root.
   a) How do the cells of the meristematic area differ from the mature cells of the root?

4 Return the slide to the meristematic area and center the root tip. Rotate the nosepiece to the high-power objective lens. Use the fine adjustment to focus the image.

5 Locate and observe cells in each of the phases of mitosis. It will be necessary to move the slide to find each of the four phases. Use the pictures on the previous pages as guides.
   b) Draw, label, and title each of the phases. It is important to draw and label only the structures that you can actually see under the microscope.

6 Return your microscope to the low-power objective lens and remove the slide of the onion. Place the slide of the whitefish embryo on the stage and focus, using the coarse adjustment. Repeat the procedure that you followed for the onion and locate dividing animal cells under high-power magnification.
   c) Compare the appearance of the animal cells with that of the plant cells.

Determining the Rate of Cell Division

7 Count 20 adjacent whitefish embryo cells and record whether the cells are in interphase or are dividing.
   d) Cells in interphase   = ___
   Cells actively dividing = ___
   e) Calculate the percentage of cells that are undergoing mitosis:

$$\frac{\text{Number of cells dividing}}{20} \times 100 = \_\_\_ \text{ \% dividing}$$

8 Repeat the same procedure for the meristematic region of the plant cell.
   f) Cells in interphase   = ___
   Cells actively dividing = ___
   g) Calculate the percentage of cells that are undergoing mitosis:

$$\frac{\text{Number of cells dividing}}{20} \times 100 = \_\_\_ \text{ \% dividing}$$

   h) Compare the percentage of animal cells that are undergoing mitosis with the percentage of plant cells that are undergoing mitosis.

Laboratory Application Questions

1 Predict what will happen if both sister chromatids move to the same pole during mitosis.

2 A cell with 10 chromosomes undergoes mitosis. Indicate how many chromosomes would be expected in each of the daughter cells.

3 Predict what will happen if a small mass of cells breaks off from a human blastula.

4 Herbicides like 2,4-D and 2,4,5-T stimulate cell division. Why does the stimulation of cell division make these chemicals effective herbicides? ■

CLONING

Cloning is the process in which identical offspring are formed from a single cell or tissue. Because the clone originates from a single parent cell, it is identical to, or nearly identical to, the parent. Although some clones show accidental changes in genetic information, the majority do not show the variation of traits expected from the combination of male and female sex cells. For that reason, cloning is referred to as asexual or nonsexual reproduction.

Figure 18.7

Hydra reproduce asexually by budding. Note the small bud beginning to form on the side of the body.

The word "clone" comes from the Greek word referring to plant cutting. Some plants and animals reproduce naturally by this asexual method. The tiny *Hydra* shown in Figure 18.7 is a good example. The small outgrowth from the parent's body is called a bud. Eventually the bud will break off from the body of the parent and form a separate, but identical, organism. Plants, like strawberries, that reproduce by runners, produce clones. Unlike plants that grow from seeds, these plants have the added advantage of relying on the parent for nutrients. This gives the clone a head start.

Figure 18.8

The strawberry plant reproduces by sending out runners from its main stem.

Fredrick Stewart excited the scientific world in 1958 when he revealed a plant created from a single carrot cell. Stewart's research has had profound economic importance. Most orchids, for example, are produced from clones. Unlike plants that reproduce sexually, cloned plants are identical to their parents. This allows the production of strains of plants with predictable characteristics.

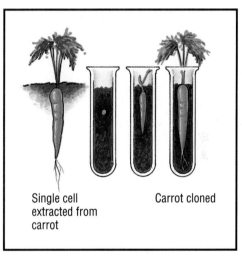

Single cell extracted from carrot

Carrot cloned

Figure 18.9

It is possible for a single cell isolated from a carrot to give rise to an entire carrot plant.

Plant tissue culture and cloning techniques have laid the groundwork for genetic engineering. While carrots, ferns, tobacco, petunias, and lettuce respond well to cloning, grass and legume families do not. So far, scientists do not know why. The secret seems to be hidden somewhere in the genetic makeup of the plant. Each regenerated cell contains the full complement of chromosomes and genes from the parent. Yet some cells specialize and become roots, stems, or leaves.

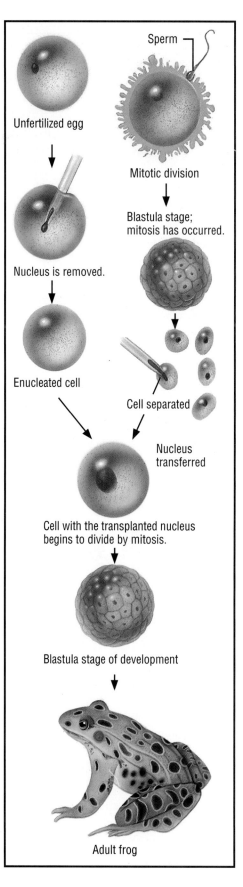

Figure 18.10

An experiment conducted by Briggs and King placed a nucleus from a frog cell in the blastula stage into an unfertilized, enucleated egg cell.

Enucleated *cells do not contain a nucleus.*

Blastula *refers to the early stage of development in which the cells of the dividing embryo form a hollow ball of cells.*

A **totipotent** *cell has the ability to support the development of an organism from egg to adult.*

Unfertilized egg

Sperm

Mitotic division

Nucleus is removed.

Blastula stage; mitosis has occurred.

Enucleated cell

Cell separated

Nucleus transferred

Cell with the transplanted nucleus begins to divide by mitosis.

Blastula stage of development

Adult frog

The cells within the leaf use only certain parts of their DNA, while the root cells use other segments of the DNA. Huge sections of the genetic information remain dormant in specialized plant cells. The trick in cloning plant cells appears to be in delaying specialization or differentiation.

While plant cloning experiments were being conducted, Robert Briggs and Thomas King were busy investigating nuclear transplants in frogs. Working with the common grass frog, the scientists extracted the nucleus from an unfertilized egg cell by inserting a fine glass tube, or *micropipette*, into the cytoplasm. The resulting cell without a nucleus is said to be **enucleated.**

Next, a nucleus from a cell of another frog in the **blastula** stage of development was extracted and inserted into the enucleated cell. The egg cell with the transplanted nucleus began to divide much like any normal fertilized egg cell and eventually grew into an adult frog. Careful analysis proved that the adult was a clone of the frog that donated the nucleus.

However, Briggs and King obtained different results when they transplanted the nucleus of a cell in the later gastrula stage of development. This time the nucleus did not bring the enucleated cell from the single-cell stage to the adult. The cell did not divide. Biologists use the term **totipotent** to describe a nucleus that is able to bring a cell from egg to adult. Something must be missing from the nuclear material of these cells. Since the nucleus of cells in the gastrula stage, unlike the cells of the earlier blastula stage, have begun to specialize, perhaps some regulatory mechanism is turning the genes off.

In another experiment, a scientist took the nuclei from the gut cells of tadpoles of the African clawed toad and inserted them into egg cells whose nuclei had been destroyed by ultraviolet radiation. Although many of these cells failed to develop, some grew into adults. Analysis of the adults confirmed that they were clones. The nuclei of the cells from the gut of the toad remain totipotent much longer than do those of the grass frog.

Scientists have also been able to clone mammal cells. However, successful cloning of mice does not mean that the cloning of humans is just around the corner. Adult mammal cells are not totipotent. However, clones can be obtained by splitting cells of a developing embryo. It would appear that cells must be taken before the eight-cell stage of development to ensure that their nuclei are totipotent. After the eight-cell stage, the cells specialize, and the genetic switch is turned off.

At present, the prospect of cloning humans belongs more to the realm of science fiction than true science, but few scientists would deny that it could ever happen. Genetics has progressed faster than even the most imaginative scientists could have predicted 25 years ago. Who knows what lies in store 25 years from now? Will we ever clone humans? Will we be able to use our clones for organ transplants? Will we be able to clone skin for burn victims? Today's answer is no, but tomorrow's answer may be yes. However, the social and moral consequences of cloning will have to be debated before such technology is embraced wholeheartedly.

FRONTIERS OF TECHNOLOGY: LIVER TISSUE TRANSPLANTS

A revolutionary medical procedure performed on a 21-month-old girl may provide a new direction in transplant surgery. Alyssa Smith has biliary atresia, a common and fatal childhood liver disease.

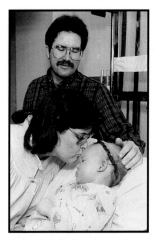

Figure 18.11

Alyssa Smith, shown with her mother and father, received transplanted cells from her mother's liver in a living liver donor transplant operation.

Figure 18.12

Cloning mammals.

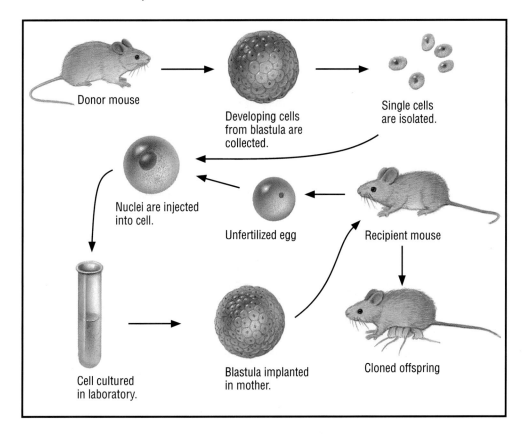

Donor mouse

Developing cells from blastula are collected.

Single cells are isolated.

Nuclei are injected into cell.

Unfertilized egg

Recipient mouse

Cell cultured in laboratory.

Blastula implanted in mother.

Cloned offspring

The liver performs many important functions. It helps maintain constant levels of blood sugar, detoxifies poisons, extracts wastes from the blood, produces bile salts for fat digestion, and stores minerals and vitamins. In this new procedure, doctors removed the left lobe of her mother's liver, and implanted fragments into Alyssa. Because the liver tissue is regenerative, cells from the transplanted liver fragments will continue to divide and grow into a normal liver.

Doctors hope that the transplanted liver will prevent illness as the diseased liver begins to fail. The idea of using a living donor has many benefits, the most obvious being a much wider accessibility to donors.

Figure 18.13

The top photograph shows identical twins; the bottom photograph shows fraternal twins.

IDENTICAL TWINS: NATURE'S CLONES

Identical twins originate from a single egg cell. During mitosis, one of the cells breaks free and a second embryo begins to develop. Should the cell masses remain separated, two people with identical gene structures will develop. Identical twins are nature's clones. Barring the possibility of genetic accidents, identical twins are the same sex, have the same blood type, and have similar facial structures.

Fraternal twins originate from two different eggs and are fertilized by different sperm cells. Although both eggs of the fraternal twins carry the same number of chromosomes, they do not have the same genes. Fraternal twins are no more similar than any other sisters or brothers. They merely share the uterus or womb at the same time.

Identical twins provide scientists with an opportunity to study which human characteristics are influenced by genes and which are influenced by environmental factors. In one case, two identical twin

brothers, who were separated at birth, both gravitated to careers in firefighting. They were both married in the same year and had families of four. Both men became fathers in the same year and lived in similar houses. A coincidence, you say? Perhaps—but other studies show similar findings.

■ REVIEW QUESTIONS ■ ?

7 What is cloning?

8 Discuss the economic importance of plant tissue cloning.

9 What is a nuclear transplant?

10 Define totipotency.

11 Why is it difficult to clone human cells?

12 Why are identical twins often called "nature's clones"?

CELL DEATH AND THE AGING PROCESS

People today are living longer. The number of people over the age of 65 accounts for approximately 10% of our total population. By the year 2030, about 20% of Canada's population will be over the age of 65. The question "Why do cells age?" is becoming central to a growing field of research into the aging process.

Why do some cells in the body divide faster than others? Red blood cells divide at an incredible rate. They live a mere 120 days, while the nerve cells in the adult brain do not divide at all. The giant redwood trees are some of the oldest organisms on earth, yet their oldest living cells are only about 30 years old—the vast majority of the cells in the tree are dead. An understanding of old age must be explored in terms of cell lineages, rather than cell age, since few living things go through life with the same cells.

Research on cells grown in tissue culture seems to indicate that a biological clock regulates the number of cell divisions available to cells. When immature heart cells maintained in tissue culture are frozen, they reveal an internal memory. If a cell has undergone 20 divisions before freezing, it will complete another 30 divisions once it is thawed. However, after the thirtieth division, the cell dies. When a cell is frozen after 10 divisions, it completes another 40 divisions when thawed. The magic number of 50 divisions is maintained, no matter how long the freezing or at what stage the division is suspended.

Not all cells of the body have the same ability to reproduce. Skin cells, blood cells, and the cells that line the digestive tract reproduce more often than do the more specialized muscle cells, nerve cells, and secretory cells. Only two cell types in the human body seem to escape aging: the sperm-producing cells, called spermatocytes, and the cells of a cancerous tumor. Males are capable of producing as many as one billion sperm cells a day from the onset of puberty well into old age. However, once the spermatocyte specializes and becomes a sperm cell, it loses its ability to undergo further divisions. Cancer cells divide at such an accelerated rate that they do not have adequate time for specialization. For example, white blood cells that divide too rapidly are not able to carry out their normal functions. Therefore, people who have leukemia—white blood cell cancer—find themselves with a reduced ability to fight infections. Cells need time to specialize. It would appear that the more specialized a cell is, the less able it is to undergo mitosis. The fertilized egg cell is not a specialized cell; it only begins to differentiate after many divisions. Interestingly, it is at that point that the alarm clock within the cell is turned on. Specialized cells have a limited life span.

Once again, new information raises even more questions. Do all cells age in a similar fashion? Does specialization cause aging? Will an understanding of cancer be linked to the study of aging?

The basic question "What causes aging?" has not yet been answered, but the theory that a clock exists in different cells has provided the groundwork for testing many different hypotheses. One hypothesis suggests that aging is caused by mutations to the genetic messages. The longer the exposure to environmental chemicals, the greater is the probability of damage to the genetic code. However, work with cells grown in tissue cultures does not appear to support this theory. Cells grown in controlled environments still die. There are a finite number of divisions. It has been estimated that a human who did not encounter disease would only live to be about 115. After 115 years, new cells cannot be generated to replace worn-out cells—mitosis has run its course. This estimate seems to be supported by historical records. Even though the average person lives longer today, the maximum longevity of humans has changed little since the days of ancient Rome.

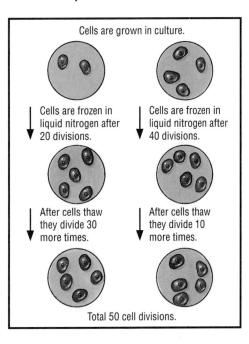

Cells are grown in culture.

Cells are frozen in liquid nitrogen after 20 divisions.

Cells are frozen in liquid nitrogen after 40 divisions.

After cells thaw they divide 30 more times.

After cells thaw they divide 10 more times.

Total 50 cell divisions.

Figure 18.14

Cell division appears to be regulated by a biological clock. For immature heart cells in tissue culture, the total number of divisions does not exceed 50.

A second hypothesis proposes that there are aging genes that shut down chemical reactions within the body. This theory would explain the graying of the hair by a gene message that turns off melanin production. However, the graying of the hair, although associated with aging, does not provide an accurate indicator of age. In some individuals the hair begins to gray in their early twenties or thirties, while some people show only traces of gray hair well into their seventies.

A third theory suggests that aging occurs when cell lineages die out. According to this theory, hair turns gray because the melanin-producing cells no longer divide, and worn-out cells are not replaced. Melanin cells might only live for short periods of time.

If aging is due to a cellular time limit, would decreased activity reduce the wear and tear on tissues? Would inactive people live longer? The answer is no. Evidence from astronauts in weightless environments indicates that more rapid cell deterioration took place. Hospital studies of bedridden patients also support the finding that activity not only helps recovery, but may even reduce aging.

> **BIOLOGY CLIP**
> Cancer is not exclusive to humans. It is found in plants and animals alike. Sunflowers and tomatoes often display a form of cancer called a *gall*. These plant tumors are caused by invading bacteria, fungi, or insects. Evidence of cancerous tumors has been found in dinosaur bones and even in the cells found in the wrapped linen of ancient mummies.

The graph in Figure 18.15 shows some of the organ deterioration that occurs with aging.

ABNORMAL CELL DIVISION: CANCER

Cancer is a broad group of diseases characterized by the rapid, uncontrolled growth of cells. Unlike most diseases, which are associated with cell or tissue death, cancer can be described in terms of too much life. Cancer cells are extremely active, dividing at rates that far exceed those of normal cells.

You began life as a single cell that divided and then specialized to become stomach cells, nerve tissue, and blood. Your cells now only divide to replace damaged cells. A balance between cell destruction and cell replacement maintains a healthy organism. You already know that cells communicate information about the body's needs from one cell to another. The growth of a callus on your hand after hours of gardening is an example of how cellular communication meets the changing needs of the body. In this example, accelerated cell divisions not only replace damaged cells, but by increasing the cell numbers of the protective outer layers of the skin, they also shield the delicate nerve and blood vessels that occupy the skin's inner layer.

Normal cells cannot divide when isolated from one another. Cell-to-cell communication is essential for cellular reproduction. Cancer cells, on the other hand, are capable of reproducing in isolation. A cancer cell growing in an artificial

Figure 18.15

The graph shows the percentage function remaining in two organ systems after the age of 30.

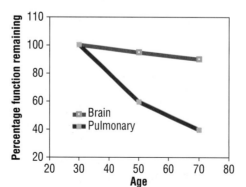

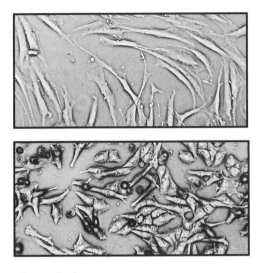

Figure 18.16

Cancerous cells can often be identified by an enlarged nucleus and reduced cytoplasm.

culture is capable of dividing once every 24 h. At this rate of division, a single cancer cell would generate over one billion descendants in a month. Fortunately, cancer cells do not reproduce that quickly in the body of an organism. Such rapid growth would cause a tumor containing trillions of cells, with an estimated mass of 10 kg, within 6 weeks. Even the rapidly dividing cells of the embryo are not capable of such proliferation.

The cancer cell does not totally ignore messages from adjacent cells that regulate its rate of reproduction. Scientists have discovered that some cancers progress at a very slow rate, often stopping for many years, only to become active at a later date. However, all cancer cells demonstrate the ability to reproduce without directions from adjacent cells.

Most normal cells adhere to one another. No doubt this attraction can be associated with the communication that occurs between cells in a given tissue. Kidney, liver, and heart cells, even when isolated from each other in the laboratory, show a remarkable quality of attraction when placed close together. Cancer cells, however, do not adhere to other cancer

cells; nor do they stick to normal cells. This ability to separate from other cell masses is what makes cancer growth so dangerous. Cancer cells can dislodge from a tumor and move to another area. This movement, called **metastasis,** makes the source of cancer difficult to locate and control. For example, when Canadian athlete Terry Fox was diagnosed with a rare form of bone cancer, amputation of his leg did not succeed in confining the cancer. During Fox's "Marathon of Hope" run across Canada, cancer was discovered in his lungs.

Another important difference between normal cells and cancer cells is that cancer cells do not change shape as they mature; they are said to lack the ability to differentiate. It has been estimated that there are about 100 different types of cells in the human body. Each cell type has a unique shape that enables it to carry out a specialized function. Since cancer cells do not mature and specialize like normal cells they are inefficient. Therefore, another threat arises because the cancer cells cannot carry out some of the functions of the normal cells.

■ REVIEW QUESTIONS   ?

13 Do all the cells of your body divide at the same rate? Explain.

14 What evidence suggests that cells contain a "biological counter"?

15 What evidence supports the theory that the human body is only capable of living about 115 years, even if disease were eliminated?

16 In what ways does cellular communication regulate the division of normal cells?

17 How rapidly can a cancer cell grown in tissue culture divide? Is it likely that the cancer cell could divide that quickly in the human body? Explain.

Figure 18.17

Terry Fox, during his "Marathon of Hope" run.

Metastasis *occurs when a cancer cell breaks free from the tumor and moves into another tissue.*

LABORATORY

IDENTIFICATION OF A CANCER CELL

Objectives

To identify cancerous cells and to recognize the differences between cancerous and noncancerous cells.

Materials

light microscope
lens paper
prepared slide of squamous cell carcinoma

Procedure

1 Clean the microscope lenses with lens paper before beginning. Rotate the revolving nosepiece to the low-power objective.

2 Place the slide of the carcinoma on the stage of the microscope and bring the image into focus using the coarse-adjustment knob.

   a) Locate the dermal and epidermal layers. Draw a line diagram showing the position of the epidermal and dermal cell layers.

   b) Are the cells of the epidermis invading the dermis?

3 Rotate the revolving nosepiece to medium-power magnification and locate a cancerous cell. Use the fine-adjustment focus to bring the image into view. Note how cells of the carcinomas have a much larger nucleus. They appear pink in color and often have an irregular shape.

4 Rotate the nosepiece to high-power magnification, and bring the image into focus. Remember to use only the fine-adjustment focus, and always focus away from the slide.

   c) Estimate the size of a cancerous cell in micrometers (μm).

   d) Estimate the size of the nucleus of a cancerous cell in micrometers (μm).

   e) Determine the nucleus-to-cytoplasm ratio:

   $$\text{Ratio} = \frac{\text{nucleus (μm)}}{\text{cytoplasm (μm)}}$$

5 Locate a normal cell. Using the same procedure as in step 4, determine the nucleus-to-cytoplasm ratio.

   f) Record the nucleus-to-cytoplasm ratio for the normal cell.

6 Compare the cancerous and normal cells in a table similar to the one below.

   g)

	Cell size	Nuclear shape	Nuclear size
Normal cell			
Cancerous cell			

Laboratory Application Questions

1 Cancerous cells are often characterized by a large nucleus. Based on what you know about cancer and cell division, provide an explanation for the enlarged nucleus.

2 Why are malignant (cancerous) tumors a greater threat to life than benign tumors?

3 Provide a hypothesis that explains why the skin is so susceptible to cancer.

4 A scientist finds a group of irregularly shaped cells in an organism. The cells demonstrate little differentiation, and the nuclei in some of the cells stain darker than others.

   a) Based on these findings, would it be logical to conclude that the organism has cancer? Justify your answer.

   b) What additional tests might be required to confirm or disprove the hypothesis that the cells are cancerous? ∎

RESEARCH IN CANADA

One of Canada's leading cancer researchers is Dr. Margarida Krause from the University of New Brunswick. Dr. Krause is conducting research on cervical tumors, one of the leading causes of cancer in women. The papilloma viruses have been linked with small tumors on the surface lining of tissues such as the cervix. Dr. Krause's research will provide more information about how viruses interfere with the genetic information to cause cancer.

Dr. Grant McFadden from the Cross Cancer Institute of the University of Alberta is investigating how a normal healthy cell changes into a cancerous cell. Dr. McFadden is studying a virus called Shope fibroma virus which has been linked with tumors in wild animals. It would appear that once the virus infects a cell, the genetic machinery of the host cell shuts down, and the viral genetic information takes over. Dr. McFadden's research will provide important clues to how one type of cancer begins.

Dr. Nancy Simpson from Queen's University in Kingston is currently working on a rare hereditary cancer known as multiple endocrine neoplasia (MEN IIA). Dr. Simpson has traced the cancer gene to the tenth chromosome by using a technique called chromosome markers. The hereditary cancer, which causes tumors of the thyroid, is passed on to half of the children of an affected parent. Dr. Simpson's next stage of research will be carried on by a former student, Dr. Paul Goodfellow, at the University of British Columbia. He is attempting to clone the cancer-causing gene.

Cancer Research

DR. MARGARIDA KRAUSE

DR. NANCY SIMPSON

DR. GRANT MCFADDEN

SOCIAL ISSUE:
Biotechnology and Agriculture

Biotechnology in agriculture includes a variety of techniques designed to increase food production and protect existing strains of plants and trees. Cell fusion techniques can produce hybrid plants that grow faster and have greater resistance to disease. Tissue cloning provides offspring that are identical to the parents—a single parent can be selected to produce offspring with desired traits. Sexual reproduction, by contrast, permits variation, as different genes combine.

Agriculture Canada preserves genetic material for future use in the breeding of new varieties of crops, principally through seed storage. Currently some 82 000 seed samples, mainly barley and wheat, are stored at the Plant Research Centre in Ottawa. Smithfield Experimental Farm in Ontario stores over 1000 varieties of fruit trees, shrubs, and plants to ensure that their genes are preserved after the plants are no longer grown commercially.

Statement:
Tissue cloning and seed banks will provide cheaper and higher-quality food for future generations.

Point

- Tissue cloning provides offspring that are identical to their parents. A single parent can be selected to produce offspring with desired traits. Sexual reproduction, by contrast, permits variation, as different genes combine.

- Biotechnology can provide more food at lower cost. With half the world's population going hungry, we need new sources of cheaper, plentiful food.

Counterpoint

- Biotechnology such as tissue cloning may provide offspring with desirable characteristics, but it also reduces the number of genes that will be expressed. The lack of variation could be disastrous should the climate change or a disease attack a specific gene combination.

- Although biotechnology can provide more food at lower costs, can we trust scientists to control gene combinations? Some random gene combinations may prove less beneficial, other gene combinations could produce new strains of plants with better resistance to disease and better yield.

Research the issue.
Reflect on your findings.
Discuss the various viewpoints with others.
Prepare for the class debate.

CHAPTER HIGHLIGHTS

- Mitosis produces new cells for cell growth and for the replacement of worn-out cells in the body.
- Mitosis involves a series of steps that produce two identical daughter cells. Two divisions occur during mitosis: nuclear division and cytoplasmic division.
- During interphase, genetic material is duplicated.
- Cells can only divide a finite number of times.

- Cloning permits the production of offspring with characteristics identical to those of their parents.
- Totipotency is the ability of a cell to support the development of an egg to the adult form of a multicellular organism.
- Cell division helps us understand aging and cancer.
- Cancer is abnormal cell division.

APPLYING THE CONCEPTS

1 Consider what would happen if you remained a single cell from the onset of life. How would your life as a single-cell organism differ from that of a multicellular organism?

2 How does cell division in plants and animals differ? What is responsible for the differences?

3 How does the mechanism of mitosis support the theory that many living things come from common ancestors?

4 What evidence suggests that one of your nerve cells carries the same number of chromosomes and the same genetic information as one of your muscle cells?

5 Explain why orchids cloned from tissue cultures are so similar.

6 Discuss some of the economic benefits of plant tissue cloning.

7 Explain why a better understanding of the mechanism of cell division may enable scientists to regenerate limbs.

8 How might the puzzle of aging be unlocked by an understanding of the mechanisms of cell division?

9 How might the puzzle of cancer be unlocked by an understanding of the mechanisms of cell division?

10 Hypothesize about why some types of cancer are more difficult to detect than others.

CRITICAL-THINKING QUESTIONS

1 X rays and other forms of radiation break chromosomes apart. Using the information gained in this chapter, assess some of the medical problems caused by the atomic bombs dropped on the Japanese cities of Hiroshima and Nagasaki in World War II.

2 Predict some of the potential problems that might arise if all plant reproduction were controlled by tissue cloning. Would the same problems exist if all domestic animals were restricted to cloning? Explain your answer.

3 A girl suffering from a malignant melanoma—a cancer of the pigment cells of the skin—had one of her cancer cells grafted into the skin of her mother. In theory, scientists believed that the invading cancer cell from the dying child would stimulate the production of antibodies within the mother.

Antibodies would destroy the foreign invading cancer cells. Scientists hoped to harvest the antibodies from the mother and transfuse some of them into the blood of the dying girl. The experiment did not work, and the young girl died a few days after the procedure. The true failure of the experiment was not known until some months later. The mother was diagnosed as having skin cancer. The melanoma also killed the mother. This incident raises moral and ethical questions. Should untried procedures like this be attempted on humans? Explain.

4 Chemotherapy treatments given to cancer patients are designed to destroy the rapidly dividing cancer cells. From what you know about normal cell division, predict what other body cells could be affected by the chemotherapy. What side effects might you expect such a treatment to produce?

ENRICHMENT ACTIVITIES

1 Observe prepared slides of cells that have been treated with radiation.

2 Research and prepare a paper on amphibian limb regeneration.

*S*exual Cell Reproduction

IMPORTANCE OF MEIOSIS

Meiosis is the process by which sex cells are formed. Sex cells, unlike the daughter cells formed by mitosis, do not contain the same number of chromosomes as the original mother cell. Sex cells are often called **gametes.** During meiosis the chromosome number is reduced by one-half. A human cell containing 46 chromosomes will undergo meiosis and produce sex cells, or gametes, that have 23 chromosomes. The 46 chromosome number is referred to as the **diploid** chromosome number and is written as $2n$. The 23 chromosome number is referred to as the **haploid** chromosome number and is given the symbol n. The union of the 23-chromosome sperm cell and the 23-chromosome egg cell produces a 46-chromosome zygote or fertilized egg. The fertilized egg contains a full set of chromosomes, or a diploid chromosome number. If development proceeds normally, the zygote will begin dividing by mitosis and will produce a multicellular human baby. Thus, meiosis allows sexual reproduction to take place.

Meiosis is a two-stage cell division in which the chromosome number of the parental cell is reduced by half. Meiosis is the process by which sex cells are formed.

Figure 19.1

While gathering pollen and nectar, the bee transfers pollen from the anther of one flower to the stigma of another. If compatible, fertilization and seed production follow.

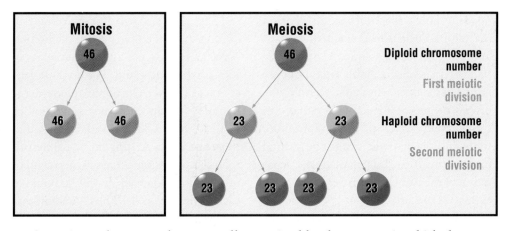

Figure 19.2

Comparison of mitosis and meiosis in humans. Mitosis produces two diploid daughter cells from one diploid cell. Meiosis produces four haploid cells from one diploid cell.

Organisms that reproduce sexually show a greater range of characteristics within the species. Because the male and female sex cells, in most organisms, come from different individuals, sexual reproduction ensures a recombination of genes. Each offspring carries genetic information from each of the parents. This explains why you might have your father's eyes but your mother's hair. Although you may look more like one parent than another, you receive the same amount of genetic information from each parent. Your father gives you a chromosome with genes that code for eye color, but so does your mother. Each of the 23 chromosomes that you receive from your father is matched by 23 chromosomes from your mother. The paired chromosomes are called **homologous chromosomes** because they are similar in shape, size, and gene arrangement. Your appearance is deter-

mined by the manner in which the genes from homologous chromosomes interact.

In animals, meiosis takes place in the testes and ovaries. The testes produce sperm cells and the ovaries produce egg cells. Plants also form sex cells, or gametes, by meiosis. The pollen cells of flowering plants are male sex cells. The egg cells of plants are stored in a variety of structures. Once fertilized, the eggs will become either spores or seeds and produce new generations of plants. Like animal sex cells, plant gametes contain a haploid chromosome number. The fusion of male and female gametes restores the diploid chromosome number.

BIOLOGY CLIP

The common housefly is capable of reproducing both sexually and asexually. When food is abundant, female houseflies produce diploid eggs, which do not require fertilization. The larvae develop inside the mother fly's body and devour her tissues as they grow. Completely skipping the usual metamorphosis, the larvae emerge with eggs of their own. The cycle continues as the eggs within the new mothers begin to consume the mothers' tissues. Eventually, the asexual cycle of reproduction is broken by some environmental factor, and for the first time normal males are produced. The males and females produce haploid gametes and the flies begin sexual reproduction.

Gametes *are sex cells. They have a haploid chromosome number.*

Diploid *chromosome number refers to the full complement of chromosomes. Every cell of the body, with the exception of sex cells, contains a diploid chromosome number.*

Haploid *chromosome number refers to one-half of the full complement of chromosomes. Sex cells have haploid chromosome numbers.*

Homologous chromosomes *are similar in shape, size, and gene arrangement.*

STAGES OF MEIOSIS

Meiosis involves two cell divisions that produce four haploid cells. During the first division, called meiosis I, the homologous chromosomes move to opposite poles. It is during this division that diploid cells separate into two haploid cells. Meiosis I is often called *reduction division* because the chromosome number is reduced by half. The diploid, or *2n*, chromosome number is reduced to the haploid, or *n*, chromosome number following the first division. The second phase, or meiosis II, is marked by a separation of the two chromatids.

Meiosis I

The phases used to describe the events of mitosis can also be used to describe meiosis. During prophase I, the nuclear membrane begins to dissolve, the centriole splits and moves to opposite poles within the cell, and spindle fibers are established. The chromosomes come together in homologous pairs. Because each pair is composed of four chromatids, this structure is referred to as a **tetrad.**

As the tetrads come closer together, the chromatids often intertwine. The pairing of homologous chromosomes is referred to as **synapsis.** Sometimes the intertwined chromatids break and exchange segments. This process is called **crossing-over.** Crossing-over permits the exchange of genetic material between homologous pairs of chromosomes.

Metaphase I follows prophase I. During this stage of meiosis I, the homologous chromosomes attach themselves to the spindle fibers and line up along the equatorial plate. During anaphase I, the homologous chromosomes move toward opposite poles. The process is known as **segregation.** At this point of meiosis reduction division occurs. One member of each homologous pair will be found in each of the new cells. The diploid mother cell becomes two haploid daughter cells. Each chromosome remains double stranded.

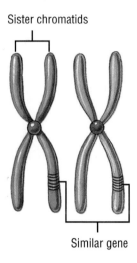

Homologous chromosomes

Sister chromatids

Similar gene

Figure 19.3

Sister chromatids are attached to form a single chromosome.

Tetrads *contain four chromatids.*

Synapsis *is the pairing of homologous chromosomes.*

Crossing-over *is the exchange of genetic material between two homologous chromosomes.*

Segregation *is the separation of paired genes during meiosis.*

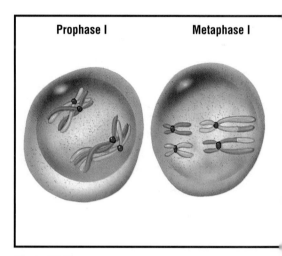

Prophase I	Metaphase I

Figure 19.5

During meiosis I, homologous chromosomes are segregated.

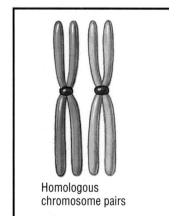
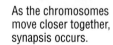

Homologous chromosome pairs

As the chromosomes move closer together, synapsis occurs.

Chromatids break, and genetic information is exchanged.

Figure 19.4

Crossing-over occurs between homologous pairs of chromosomes during prophase I of meiosis.

During telophase I, the cytoplasm divides, forming two cells. A nuclear membrane begins to form around the chromosomes within each of the two cells. However, as is not the case in mitosis, the two cells are not identical. Each of the daughter cells contains one member of the chromosome pair. Although homologous chromosomes are similar, they are not identical. They do not carry exactly the same information. Complete interkinesis follows telophase I and the two distinct cells separate and prepare themselves for meiosis II.

Meiosis II

Meiosis II occurs at approximately the same time in each of the haploid daughter cells. However, for simplicity, consider the events in only one of the cells. During meiosis II, pairs of chromatids will separate and move to opposite poles.

Prophase II signals the beginning of the second division. During this stage of meiosis II, the nuclear membrane dissolves and the spindle fibers begin to form. Metaphase II follows prophase II. It is signalled by the arrangement of the chromosomes, each with two chromatids, along the equatorial plate. The chromatids remain pinned together by the centromere.

Anaphase II can be identified by the movement of the sister chromatids to the opposite poles. Anaphase II ends when the nuclear membrane begins to form around what are now called the chromosomes. At this point, the cell enters its final stage of meiosis, telophase II. During this stage, the second division of cytoplasm is completed. Four daughter cells are produced from each meiotic division.

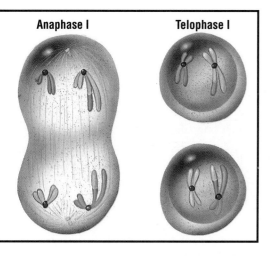

Anaphase I   Telophase I

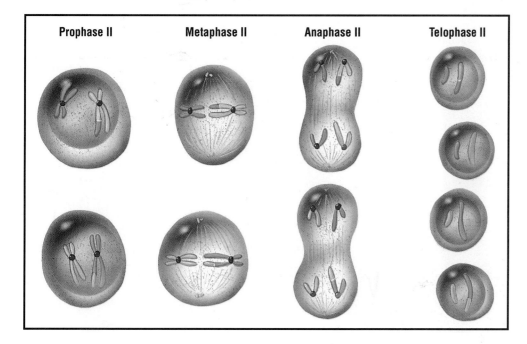

Prophase II   Metaphase II   Anaphase II   Telophase II

Figure 19.6

During meiosis II, sister chromatids separate.

DEVELOPMENT OF MALE AND FEMALE GAMETES

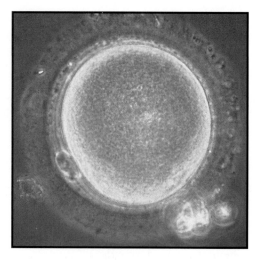

The formation of sex cells during meiosis is referred to as **gametogenesis.** Although human male and female gametes both follow the general plan of meiosis, some differences do exist. The cytoplasm of the female gametes does not divide equally after each division. As shown in Figure 19.7, one of the daughter cells, called the **ootid,** receives most of the cytoplasm. The other cells, called the **polar bodies,** die, and the nutrients are absorbed by the body. Only one egg cell is produced from meiosis. In contrast, the sperm cells show equal division of cytoplasm. Because of their function sperm cells have much less cytoplasm than egg cells. Sperm cells are specially designed for movement. They are streamlined and, like a rocket or a sports car, cannot carry excess weight. The egg, by comparison, does not move. Egg cells require the nutrients and organelles carried within the cytoplasm to fuel future cell divisions in the event that the egg cell becomes fertilized.

Males make many more sex cells than females. The diploid spermatocytes—the cells that give rise to sperm cells—are capable of many mitotic divisions before meiosis ever begins. Males can produce one billion sperm cells every day. Females are born with about 400 000 egg cells, of which only about 400 ever mature. About 1000 egg cells mature within the ovary approximately every 28 days, but only a single egg cell leaves the ovary. The others break down and are absorbed by the body. The oocyte—the cell that produces egg cells—does not continue to divide after a woman reaches puberty. As a woman ages, the number of egg cells in the ovary declines until about age 50 or 60, at which time no eggs remain in the ovary. The time at which no eggs remain in the ovary is called *menopause.*

In 1906, the American geneticist, Thomas Hunt Morgan, while observing meiosis in the testes cell of a fruit fly, noted that one of the chromosomes appeared hook-shaped. The last chromosome pair did not appear to have a homologous member. Did

Gametogenesis *refers to the formation of sex cells in animals.*

Ootids *are unfertilized egg cells.*

Polar bodies *are formed during meiosis. These cells contain all the genetic information of a haploid egg cell but lack sufficient cytoplasm to survive.*

Figure 19.7

Generalized picture of spermatogenesis and oogenesis in humans.

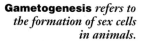

Human Gametogenesis

Spermatogenesis — Oogenesis

Spermatocyte — Oocyte

46 — Chromosome number — 46

First meiotic division

23 — 23 — 23 — 23 First polar body

Second meiotic division

23 — 23 — 23 — 23 — 23 — 23 — 23 — 23

Four sperm cells — Ootid — Polar bodies

this mean that the fruit fly carried a genetic mutation? Morgan examined another male and found that it also had a small hook-shaped chromosome. Within the female cell, all the chromosomes were rod-shaped. After observing different male and female cells during meiosis, Morgan concluded that the last pair of chromosomes differed in males and females. Females always contained two rod-shaped chromosomes of identical length, while the male had one rod-shaped chromosome and a smaller hook-shaped chromosome. Morgan described the last pair of chromosomes as **sex chromosomes.** Chromosomes that are not sex chromosomes are referred to as **autosomes.**

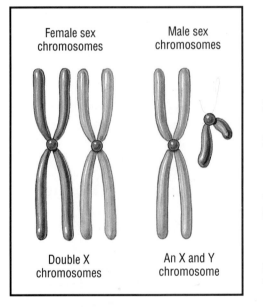

Figure 19.8

Sex chromosomes.

The rod-shaped chromosomes were called X chromosomes, while the hook-shaped chromosome was called a Y chromosome. Females have two homologous X chromosomes and males have one X and one Y chromosome.

Many male animals, including humans, have a hook-shaped Y chromosome that identifies the individual as a male.

■ REVIEW QUESTIONS   ?

1   A muscle cell of a mouse contains 22 chromosomes. Indicate the number of chromosomes you would expect to find in the following cells of the same mouse:
   a)   daughter cell formed from mitosis
   b)   skin cell
   c)   egg cell
   d)   fertilized egg cell
2   Differentiate between haploid and diploid cells in humans.
3   In what ways does the first meiotic division differ from the second meiotic division?
4   What is a tetrad?
5   Explain why synapsis may lead to the exchange of genetic information between chromosomes.
6   What are homologous chromosomes?
7   Do homologous chromosomes have the same number of genes? Explain.
8   Do homologous chromosomes have identical genes? Explain.
9   Compare the mechanisms of gametogenesis in males and females.

Sex chromosomes *are pairs of chromosomes that determine the sex of an individual.*

Autosomes *are chromosomes not involved with sex determination.*

DEMONSTRATION
COMPARING MEIOSIS AND MITOSIS

Objective

To use models to compare the events of mitosis and meiosis.

Materials

2 sheets of blank paper
pencil
ruler
scissors
4 paper clips

10 cm strands of blue yarn (4)
10 cm strands of red yarn (4)
5 cm strands of blue yarn (4)
5 cm strands of red yarn (4)

Procedure

Mitosis

1 Draw a solid line down the center of one of the sheets of blank paper.

2 Place two 10 cm strands of red yarn side by side, and hold in place with a paper clip. Repeat the procedure with two 5 cm strands of red yarn.
   a) What cell structures do the strands of yarn and the paper clips represent? What area of the cell does the center line represent?

3 Repeat the same procedure with the 5 cm and 10 cm lengths of blue yarn.

4 Line the four chromosomes up along the center line that you have drawn on the sheet of paper.
   b) What is the diploid chromosome number of this cell?

5 Remove the paper clip and move each of the single strands of yarn to the opposite ends of the paper.
   c) What structure do the single strands of yarn represent in a true cell?

6 Each single strand synthesizes a duplicate strand. Add a duplicate strand of yarn to each of the new daughter cells.
   d) How many chromosomes are in each of the daughter cells?
   e) Compare the daughter cells with the mother cell.

Meiosis

7 Place two 10 cm strands of red yarn side by side and join them by placing a paper clip in the center. Repeat the procedure with two 5 cm strands of red yarn.

8 Make homologous chromosomes by using blue yarn.

9 To simulate metaphase I, place the 10 cm red and blue double-stranded chromosomes on either side of the equatorial plate. Repeat the same procedure with the 5 cm red and blue strands of yarn.
   f) On what basis are the simulated chromosomes considered to be homologous?
   g) What is the diploid chromosome number?
   h) How does metaphase I of meiosis differ from metaphase I of mitosis?
   i) What important event occurs during synapsis?

10 Move the homologous chromosomes to opposite ends of the paper.
   j) What is the haploid chromosome number?

11 Choose one of the haploid daughter cells and line the chromosomes up along the equatorial plate. Remove the paper clip, and move chromatids to opposite poles.
   k) Compare the resulting daughter cells of mitosis and meiosis. ■

ABNORMAL MEIOSIS: NONDISJUNCTION

Meiosis, like most processes of the body, is not immune to mistakes. *Nondisjunction* occurs when two homologous chromosomes move to the same pole during meiosis. The result is that one of the daughter cells will be missing one of the chromosomes while the other will retain an extra chromosome. Cells that lack genetic information or have too much information will not function properly. Although nondisjunction can occur in any cell during cell division, the effects are most devastating during the formation of sex cells.

Nondisjunction is associated with many different human genetic disorders. In humans, nondisjunction produces gametes with 22 and 24 chromosomes. The gamete with 24 chromosomes has both chromosomes from one of the homologous pairs. If that gamete joins with a normal gamete of 23 chromosomes from the opposite sex, a zygote containing 47, rather than 46, chromosomes will be produced; the zygote will then have three chromosomes in place of the normal pair. This condition is referred to as **trisomy.** However, if the sex cell containing 22 chromosomes joins with a normal gamete, the resulting zygote will have 45 chromosomes; the zygote will have only one of the chromosomes rather than the homologous pair. This condition is called **monosomy.** Once the cells of the trisomic or monosomic zygotes begin to divide, each cell of the body will contain the abnormal number of chromosomes.

NONDIS-JUNCTION DISORDERS

Compare the chromosomes of a normal male shown in Figure 19.9(a) with the chromosomes of a

female who has **Down syndrome** shown in Figure 19.9(b). Notice how the chromosomes are arranged in pairs. Such a picture of the chromosomes is known as a **karyotype.** In about 95% of cases, a child with Down syndrome has an extra chromosome in the chromosome pair number 21. The trisomic Down syndrome disorder is produced by nondisjunction—the person has too much genetic information. Down syndrome is generally associated with mental retardation, although people with this condition retain a wide range of mental abilities.

Trisomy *is the presence of three homologous chromosomes in every cell of an organism.*

Monosomy *is the presence of a single chromosome in place of a homologous pair.*

Down syndrome *is a trisomic disorder in which a zygote receives three homologous chromosomes for chromosome pair number 21.*

Karyotypes *are pictures of chromosomes arranged in homologous pairs.*

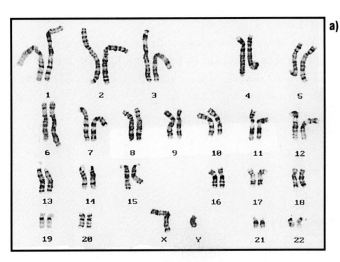

a)

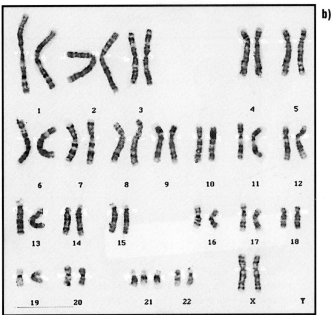

b)

Figure 19.9

(a) Karyotype of a normal male. Notice that the chromosome pair number 23 is not homologous. Males contain an X and a Y chromosome.
(b) Down syndrome female. Note trisomy of pair number 21. Down syndrome affects both males and females.

The trisomic condition is referred to as a *syndrome* because it involves a group of disorders that occur together. People with Down syndrome can be identified by several common traits, regardless of race: a round, full face; enlarged and creased tongue; short height; and a large forehead. It has been estimated that one in 600 babies is born with Down syndrome. Statistics indicate that the risk of having a baby with Down syndrome increases with the age of the mother. A woman in her forties has a one in 40 chance of having a child with Down syndrome. This rate is 25 times greater than for a woman in her twenties (see Figure 19.10).

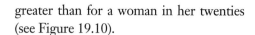

BIOLOGY CLIP
It was not until 1956 that scientists established that the diploid human chromosome number was 46. Normal humans have 44 autosomal chromosomes and 2 sex chromosomes. Three years after this discovery, the trisomy of chromosome pair 21 was linked with Down syndrome.

Turner's syndrome
occurs when sex chromosomes undergo nondisjunction. This monosomic disorder produces a female with a single X chromosome. In the egg cell, both homologous X chromosomes move to the same pole during meiosis I. When the egg with no X chromosome is fertilized by a normal sperm cell, a zygote with 45 chromosomes is produced. Females with Turner's syndrome do not usually develop sexually, and tend to be short and have thick, widened necks. About one in every 10 000 births produces a Turner's syndrome baby. Most Turner's syndrome fetuses are miscarried before the twentieth week of pregnancy; however it should be noted that some children can go on to lead highly productive lives.

Klinefelter syndrome
is caused by nondisjunction in either the sperm or egg. The child inherits two XX chromosomes—characteristic of females—and a single Y chromosome—characteristic of males. The child appears to be a male at birth; however, as he enters sexual maturity, he begins producing high levels of female sex hormones. Males with Klinefelter syndrome are sterile. It has been estimated that Klinefelter syndrome occurs in one in every 1000 births.

Figure 19.10
Incidence of Down syndrome.

Figure 19.11
Nondisjunction disorders in humans.

Turner's syndrome *is a monosomic disorder in which a female has a single X chromosome.*

Klinefelter syndrome *is a trisomic disorder in which a male carries an XXY condition.*

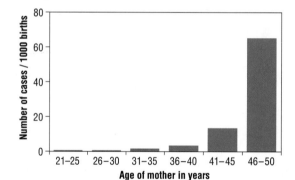

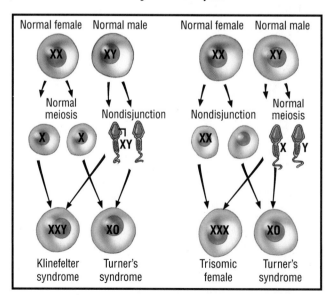

RESEARCH IN CANADA

Down syndrome was first described by the English doctor John Down in 1866. By the early 1930s, another geneticist, L.S. Penrose, had shown a link between the incidence of Down syndrome and the mother's age. The Canadian researcher Dr. Irene Uchida was intrigued by the higher incidence of nondisjunction disorders in older women, specifically why older mothers had proportionally more children with Down syndrome.

Dr. Uchida found a 1930s study on fruit flies, which showed that exposures to high dosages of radiation increased the frequency of nondisjunction disorders. Although meiosis in fruit flies is not identical to that in humans, Dr. Uchida decided to pursue the radiation link for clues in her research on Down syndrome. During her first set of experiments, Dr. Uchida verified the original fruit fly experiments. She found that the greater the exposure to radiation, the higher was the incidence of chromosome abnormalities. Assuming that the results for fruit flies would hold true for other chromosomes, Dr. Uchida hypothesized that older women have been exposed to radiation over a longer period of time and therefore are more likely to suffer chromosome damage.

A 1960 survey, conducted by Dr. Uchida at the Children's Hospital in Winnipeg, indicated that women who had been exposed to radiation prior to conception were more likely to have children with Down syndrome. Similar studies carried out in other hospitals supported Dr. Uchida's findings. In 1968, Dr. Uchida carried out a larger study with women who had been exposed to radiation. Using a list of patients who had had abdominal X-ray examinations or fluoroscopies from the Winnipeg General Hospital between 1956 to 1959, Dr. Uchida compiled a second list of women who had given birth to a child with Down syndrome. The larger study supported the link between X rays and Down syndrome.

Down Syndrome

DR. IRENE UCHIDA

FRONTIERS OF TECHNOLOGY: AMNIOCENTESIS

Recent developments in medical technology now make it possible for doctors to detect genetic disorders like Down syndrome even before the baby is born. By using ultrasound or echo location techniques, physicians are able to locate the position of the developing fetus within the mother's womb. Another technique, called **amniocentesis,** involves the use of a syringe to draw fluid from the sac surrounding the developing fetus. An analysis of the fluid can identify certain types of disorders. The fluid, called amniotic fluid, also contains cells from the developing fetus. When these cells are treated with special stains, the chromosomes can be made visible for microscopic examination.

A camera mounted to the microscope is often used to take a picture of the chromosomes. A complementary technique, called karyotyping, discussed earlier in the chapter, compares the number, size, and shape of homologous chromosomes. A chromosome count of 47, for example, would forewarn of a nondisjunction disorder. By comparing homologous chromosomes, physicians can identify the specific disorder or syndrome.

A more recent advance is **chorionic villus sampling (CVS).** This technique draws cells from the outer membrane surrounding the embryo. Unlike amniocentesis, this procedure can be performed as early as eight weeks into pregnancy.

Amniocentesis, like many other advances in reproductive technology, raises many moral and ethical questions. Will the ability to diagnose genetic disorders prior to birth increase the incidence of abortion? Because amniocentesis allows physicians to differentiate between males and females, some people fear that, in the wrong hands, the technique could be used to select the sex of offspring. Do you think the technique should be used? Would you place any limits or restrictions on its use?

Amniocentesis *is a technique used to identify certain genetic defects in a fetus or embryo.*

Chorionic villus sampling (CVS) *is a prenatal diagnosis technique that secures cells from the outer membrane of the embryo for analysis.*

Figure 19.12

Ultrasound can be used to locate the position of the fetus within the uterus.

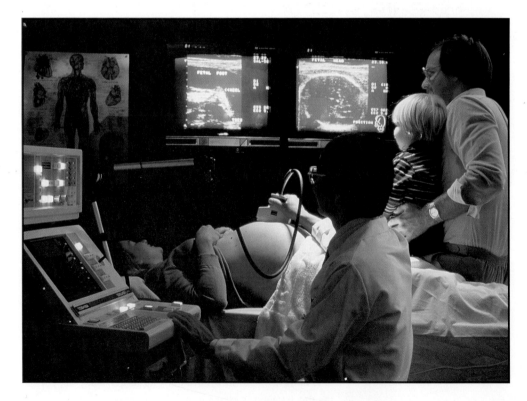

LABORATORY
HUMAN KARYOTYPES

Objective

To use a karyotype chart to diagnose genetic disorders.

Materials

human karyotype plate
scissors
blank paper
transparent tape

Procedure

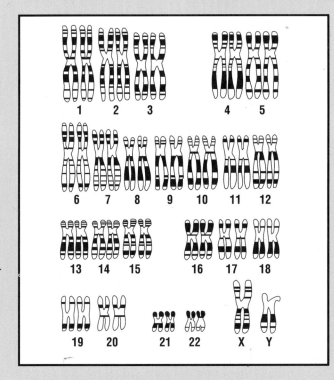

1 Examine the karyotype chart above.
   a) On what basis are the chromosomes arranged in pairs?
   b) In what ways does chromosome pair 2 differ from pair 20?
   c) What is the sex of the individual shown in the chart above?

2 Obtain a duplicate of the karyotype chart from your teacher and carefully cut out each of the chromosomes.

3 Match the paired chromosomes, and prepare a karyotype chart by taping the chromosomes to the unlined paper. Use the karyotype chart provided above for reference.
   d) How many chromosomes are found in the individual?
   e) What is the sex of the individual?

4 Use the chart below to identify the genetic disorder of the individual represented in the karyotype plate.

Chromosome abnormality	Syndrome	Effect
XO	Turner	Sterile female, 45 chromosomes, short stature, sometimes below normal intelligence.
XXY	Klinefelter	Sterile male, 47 chromosomes, often long arms and legs, often below-normal intelligence.
Trisomy 21	Down	Lower mental ability, 47 chromosomes, distinct facial features.
Trisomy 18	Edward	Severe defects, 47 chromosomes, life expectancy about 10 weeks.
Trisomy 13	Patau	Severe defects, 47 chromosomes, non-functioning eyes, very short life expectancy.

   f) On the basis of the information shown in the karyotype chart, provide a diagnosis.

Laboratory Application Questions

1 Would the diagnosis of Turner's syndrome in a single cell necessarily mean that every cell of the body would contain 45 chromosomes? Explain your answer.

2 Is it possible for two people who have Down syndrome to give birth to a normal child? Explain your answer.

3 Would it ever be possible to produce a baby who has 48 chromosomes? Explain your answer.

4 More males than females suffer from color blindness. Speculate as to why females who have Turner's syndrome have a similar incidence of color blindness as males. ■

THE MYTH OF CRIMINAL CHROMOSOMES

The killer's face stares out at you from the front page of your newspaper. On the TV news, his astonished neighbors express shock to learn that someone they thought they knew well turned out to be a criminal. You say to yourself, "He doesn't even look like a criminal."

What does a criminal look like? Are shifty eyes and heavy brows really signs of a felon? Can criminal genes be inherited? Some of the early work with karyotypes seemed to support this idea. Scientists discovered that some males carried a double Y chromosome. Unlike normal males, the XYY males had 47 chromosomes. Since the Y chromosome codes for maleness, these XYY males were dubbed "supermales." However, the characteristics associated with the supermale were not at all flattering.

In 1965, Patricia Jacobs, a well-respected geneticist, published an article on XYY males that rocked the scientific community and fueled discussions for years to come. Jacobs suggested that XYY males were prone to aggressive behavior. Drawing from a survey of 196 men from a high-security hospital in Scotland, Jacobs linked the XYY condition with subnormal intelligence and tendencies for violence and crime. Testing indicated that 3.5% of the violent men had the XYY condition—a frequency 20 times greater than that found in the normal population. Jacobs noted that the XYY men also tended to be somewhat taller than the general population. Jacobs' results were supported by other researchers who confirmed the disproportionate numbers of XYY men in prison. By the late 1960s, the belief that criminals were being driven by some sinister genes had gained wide acceptance.

The linking of a genetic condition to violent behavior was quickly embraced by those who believed that criminals are born, not made. Some lawyers even suggested that their clients were not guilty of their wrongdoings because of their genetic predisposition to crime. Sensationalized trials and misreported scientific data led to a flood of paranoia. Some scientists and political leaders even called for public screening to protect against genetic criminals, and Canada did allow mass testing of male infants for XYY.

By the mid-1970s, however, scientists began to question the scientific basis for linking the double Y condition with crime. It is now accepted that most XYY people show no greater tendency toward violence than the normal population. The higher incidence of people with XYY in prison and mental institutions has been explained by learning difficulties that may have contributed to frustration and antisocial behavior. In addition, some XYY children may have been victims of a tragic self-fulfilling prophecy. Today, a tendency toward tallness is the only characteristic of the XYY condition accepted by scientists.

The idea of determining behavior from an analysis of genetic information has long been abandoned. The importance of environment is now emphasized.

Figure 19.13

Richard Speck, the killer of eight Chicago nurses in 1968, was thought to have been an XYY criminal. It was later proved that Speck did not have the XYY condition.

FRONTIERS OF TECHNOLOGY: REPRODUCTIVE TECHNOLOGY

A new technique, known as **artificial involution,** is revolutionizing the cattle industry. This technique permits a low-grade cow to give birth to a high-grade calf. Under normal circumstances, the cow only releases one egg from the ovary during each reproductive cycle. This means that the cow only produces one calf. However, when a high-grade cow is injected with a fertility drug, it produces more egg cells than normal.

> **BIOLOGY CLIP**
> The world record for offspring is an estimated 220 000 progeny sired by artificial insemination by Soender Jylland Jens, a Danish black and white bull.

The cow treated with fertility drugs is anesthetized and a special catheter removes the eggs from the cow's Fallopian tube. The eggs are then fertilized in a petri dish or test tube with the sperm of a high-grade bull. This method is called *in vitro* fertilization, which literally means fertilization "in glass." A nutrient medium provides nourishment for each fertilized egg, which begins to divide in the glass vessel in the same manner it would in the Fallopian tube of the cow. In the final step, the fertilized egg is transferred into the uterus of a low-grade **surrogate** cow. The newborn calf contains none of the surrogate cow's genes.

Artificial involution *is a process by which egg cells are extracted from a donor and placed in a nongenetic mother.*

Surrogate *means "substitute." The nongenetic mother is a surrogate mother.*

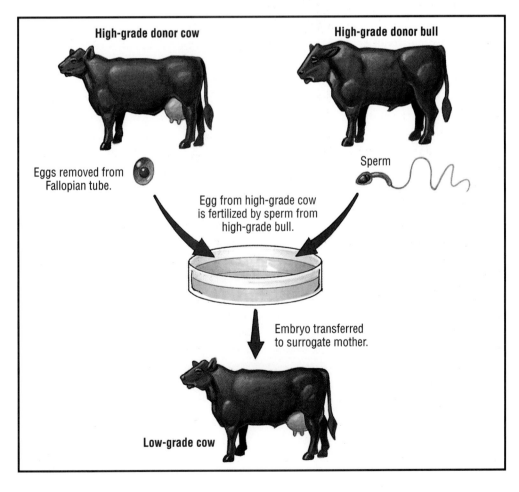

High-grade donor cow

High-grade donor bull

Eggs removed from Fallopian tube.

Sperm

Egg from high-grade cow is fertilized by sperm from high-grade bull.

Embryo transferred to surrogate mother.

Low-grade cow

Figure 19.14

Artificial involution.

REDEFINING MOTHERHOOD

Can a woman who lacks ovaries give birth to a baby? A few short years ago the answer to this question would have been an emphatic no. Today, however, a woman without ovaries can give birth to a baby if eggs from another woman are fertilized and transferred to the first woman's womb. In most cases, the sperm of the first woman's husband is used. In effect, the child has three parents: the father, the genetic mother, and the mother who carries the child.

Another reproductive technology has also redefined the concept of motherhood. A woman with a defective cervix or womb may ask another woman to give birth to her genetic child. In this procedure, the woman's egg is fertilized *in vitro* and then transferred to a surrogate, who carries the baby to term and then returns it to the genetic parents.

These two procedures raise some fundamental ethical questions: Who is the real mother—the one who gives birth or the one who contributes the genetic information? Will the woman who contributes the egg have to adopt her genetic child?

◼ REVIEW QUESTIONS　?

10 What is nondisjunction?
11 Use a diagram to illustrate how nondisjunction in meiosis I differs from nondisjunction in meiosis II.
12 Differentiate between monosomy and trisomy.
13 What is Down syndrome?
14 What is a karyotype?
15 What is Turner syndrome?
16 Indicate some of the benefits of amniocentesis.
17 What is *in vitro* fertilization?

Figure 19.15

In vitro fertilization permits a woman who lacks ovaries to give birth to a baby.

Figure 19.16

A surrogate mother can give birth to a child for a woman who has a defective cervix or womb.

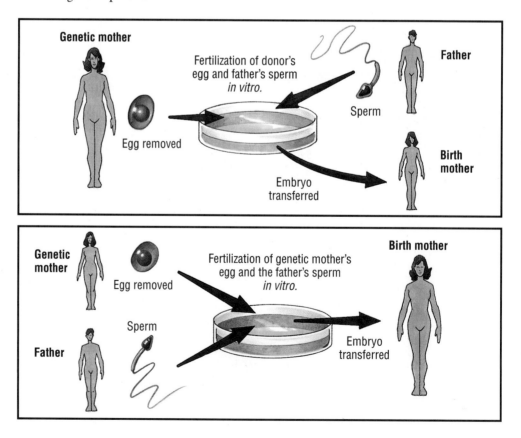

462 ◼

*Unit Five:
Continuity of Life*

INVESTIGATION

Our economic system depends on the production and utilization of living organisms such as plants and animals. To arrange an equitable distribution of these resources among a changing human population, we must be aware of the mechanisms and problems inherent in the continuity of life.

Animal breeder
Animal breeders use techniques such as artificial insemination and fertility drugs to obtain high-quality livestock. Such techniques require the identification of the best breeding period as well as desirable traits.

Family planning counselor
Family planning counselors must be aware of the mechanisms of human reproduction as well as the methods available to slow its rate.

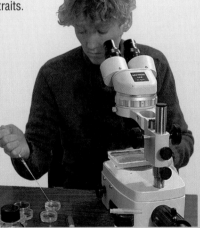

Child care worker
Working with children often involves attending to the special needs of those with genetic disorders such as Down syndrome.

Florist
A florist has to have a constant supply of flowers and plants ready for sale. This involves a detailed knowledge of the plants' different reproductive cycles, growing habits, and resistance to disease.

- Identify a career that requires a knowledge of the continuity of life.
- Investigate and list the features that appeal to you about this career. Make another list of features that you find less attractive about this career.
- Which high-school subjects are required for this career? Is a postsecondary degree required?
- Survey the newspapers in your area for job opportunities in this career.

SOCIAL ISSUE:
Limits on Reproductive Technology

Reproductive technologies include a variety of procedures that have allowed previously childless couples to have children. Debates surrounding two reproductive technologies in particular—surrogate motherhood and "test-tube" fertilization—have received wide media attention in the last several years.

It has been estimated that between 12 and 15% of couples are unable to have children. The demands on scientists to use reproductive technology have been increasing. However, does everyone have the right to produce children? The moral and ethical questions surrounding reproductive technology are far-reaching.

Statement:

Research in reproductive technology should be encouraged.

Point

- A technology like artificial involution was not developed out of any harmful motives. The technology was designed to produce better-quality beef cattle.
- Louise Brown and the other test-tube babies are the very human products of this technology. Clearly, the good outweighs any potential harm: couples who were once childless are now able to have children.

Counterpoint

- Although the technology may not have been developed for harmful purposes, it provides enormous potential for harm. The technology could be exploited to produce humans with selected traits.
- Reproductive technology is advancing faster than moral and ethical questions are being resolved. The fact that the concept of "motherhood" must be redefined is just one example. Scientific research should not determine our moral pathway.

> *Research the issue.*
> *Reflect on your findings.*
> *Discuss the various viewpoints with others.*
> *Prepare for the class debate.*

CHAPTER HIGHLIGHTS

- Meiosis involves the formation of sex cells.
- Cells pass through two divisions in meiosis.
- All gametes produced by meiosis have haploid chromosome numbers.
- Homologous chromosomes are similar in shape, size, and gene arrangement.

- Crossing-over is the exchange of genetic material between chromosomes that occurs during meiosis.
- Sex chromosomes are pairs of chromosomes that determine the sex of an organism.
- Abnormal meiosis, or nondisjunction, produces gametes with irregular chromosome numbers.

APPLYING THE CONCEPTS

1 Explain why sexual reproduction promotes variation.
2 Predict what might happen if the polar body were fertilized by a sperm cell.
3 A microscopic water animal called *Daphnia* reproduces from an unfertilized egg. This form of reproduction is asexual because male gametes are not required. Indicate the sex of the offspring produced.
4 Indicate which of the following body cells would be capable of meiosis. Provide a brief explanation for your answers.
   a) brain cells
   b) fat cells
   c) cells of a zygote
   d) sperm-producing cells of the testes
5 Compare the second meiotic division with mitotic division.

6 King Henry VIII of England beheaded his wives when they did not produce sons. Indicate why a little knowledge of meiosis might have been important for Henry's wives.
7 Explain how it is possible to produce a trisomic female XXX.
8 Abnormal cell division can produce an XYY condition for males. Diagram the nondisjunction that would cause a normal male and female to produce an XYY offspring.
9 A number of important genes are found on the X chromosome. Explain why many biologists suggest that genetic differences account for the fact that more male than female babies die shortly after birth.
10 Why might a physician decide to perform amniocentesis on a pregnant woman who is 45 years of age?

CRITICAL-THINKING QUESTIONS

1 According to one report, the incidence of infertility appears to be increasing in countries like Canada. The sperm count in males has fallen more than 30% in the last half-century and is continuing to fall. Although there is no explanation for this phenomenon at present, environmental pollution is suspected. Suggest other reasons for decreased fertility in males and females.
2 A couple who are unable to have children decide to hire another woman to carry the fetus. The procedure will cost them $10 000. They want absolute legal rights to the child, both while it is in the womb and when it is born. Discuss the ethical implications of hiring surrogate mothers.

3 Advances in modern medicine have increased the number of people who carry genetic disorders. People who might have died at an early age because of their disorder are living longer, marrying and producing children, and thus passing on their defective genes. Nondisjunction disorders could be eliminated by screening prospective sperm and egg cells. Sperm and egg banks could all but eliminate many genetic disorders. Comment on the implications to society of the systematic elimination of genetic disorders in humans.
4 A technique called egg fusion involves the union of one haploid egg cell with another. The zygote contains the full *2n* chromosome number and is always a female. Discuss the implications for society if this technique were to be employed for humans.

ENRICHMENT ACTIVITIES

1 Contact the department of agriculture in your area and inquire about innovative genetic technology projects. How is farming changing?

2 Research a genetic disorder caused by nondisjunction. How is it detected? What are the physiological effects of the disorder? and psychological?

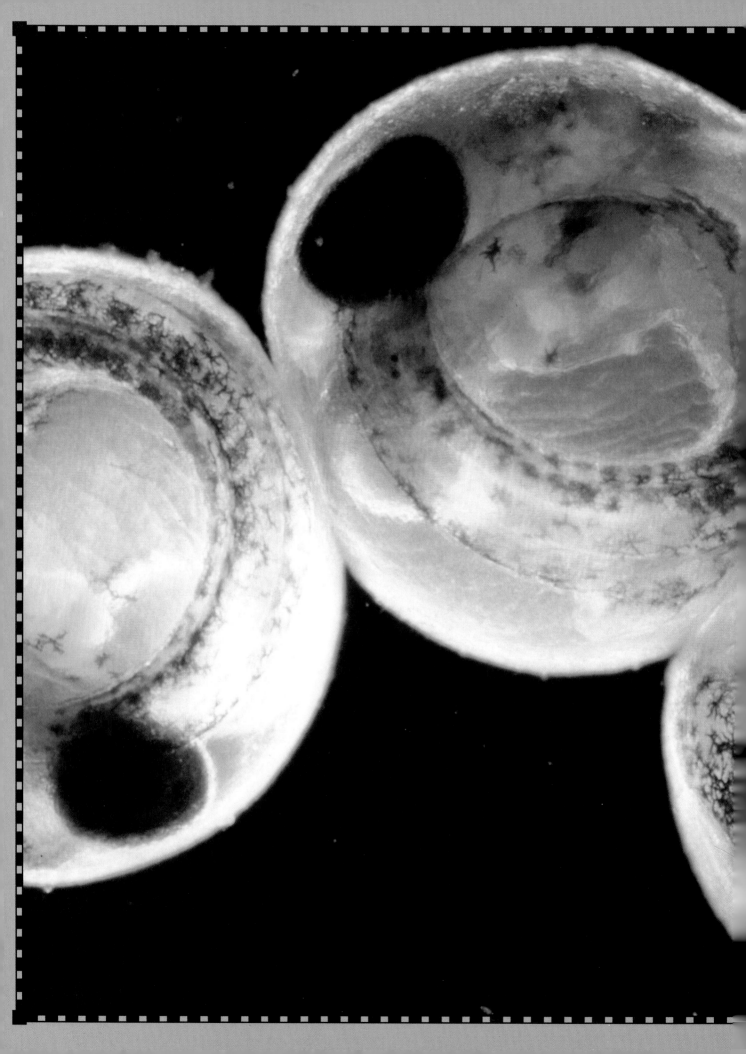

*H*eredity

■

U N *6* I T

*G*enes and Heredity

Heredity *is the passing of traits from parents to offspring.*

Genes *are units of instruction, located on chromosomes, that produce or influence a specific trait in the offspring.*

IMPORTANCE OF GENETICS

The study of genetics examines the inheritance of biological traits. Have you ever been able to identify a stranger as a member of a particular family? Red hair, high cheekbones, or a prominent nose can often be traced through family lineages. The observation that a young child resembles her grandmother suggests that physical characteristics are inherited. Similar observations can be made in the world of plants and animals. Flowers with white petals most often produce seeds that germinate and produce white-colored flowers. Palomino horses produce other palomino horses. Characteristics appear to be repeated from generation to generation. The passing of traits from parents to offspring is called **heredity.**

In the pattern of continuity of biological traits, however, there are examples of diversity. How is it possible for two parents with dark black hair to have a child with red hair? Why might hemophilia—a blood-clotting disorder—skip a generation? How is it possible for parents who both have type A blood to have an offspring who has type O blood?

Your biological traits are controlled by **genes** located on the chromosomes that are found in every cell

Figure 20.1

The observation that offspring resemble their parents gave rise to the speculation that characteristics are inherited.

of your body. Because you inherited half of your chromosomes from your mother and the other half from your father, your traits are a result of the interactions of the genes of both parents. Although you contain half of the genetic information from each parent, your genes and traits are uniquely your own. It has been estimated that more than eight million combinations are possible from the 23 chromosomes that you inherit from your mother and the 23 chromosomes that you inherit from your father.

The ancient Egyptians encouraged the intermarriage of royalty to preserve blood lines. For example, Cleopatra married her younger brother. Plato, a Greek philosopher of the early 4th century B.C., called for the segregated mating of the elite. To maintain a line of strong warriors, the ancient Spartans practised infanticide by killing babies who had undesirable characteristics.

EARLY BELIEFS

The idea that biological traits are inherited existed long before the mechanisms of gene interaction were understood. Stone tablets crafted by the Babylonians 6000 years ago show the pedigrees of successive generations of champion horses. Other carvings from the same period show the artificial cross-pollination of date palms. Early records kept by Chinese farmers provide evidence of the methods used for improving different varieties of rice. The selection of desired traits was based on keen observation and, to a large extent, on trial-and-error investigation.

Early naturalists hypothesized about some incredible cross-species, or **hybrids.** The giraffe, for example, was thought to have been a cross between a leopard and a camel. (The fact that camels and leopards are not compatible did not seem to deter the theory.) The banana was thought to have been a hybrid of the acacia and the palm.

PIONEER OF GENETICS: GREGOR MENDEL

One of the classic scientific experiments was performed by an Austrian monk named Gregor Mendel (1822–1884) during the mid-19th century. Mendel's work with garden peas not only explained the mechanism of gene inheritance for plants, but provided a basis for understanding heredity. When Mendel's work was discovered many years later, it provided the missing piece in the theory of evolution by natural selection.

Why did Mendel choose to do his work on the garden pea? First, he observed that garden peas have a number of different characteristics, each of which is expressed in one of two ways. For example, some garden peas produce green seeds, while others produce yellow seeds. Some plants are tall, while others are short. Mendel also noticed different flower positions on the stem and different flower colors.

A second reason for using garden peas is related to the way the plant reproduces. Garden peas are both *self-fertilizing* and *cross-fertilizing*. During self-fertilization the pollen produced from the male sta-

Figure 20.2

Gregor Mendel was an Austrian monk whose experiments with garden peas laid the foundation for the science of genetics.

Hybrids *are offspring that differ from their parents in one or more traits. Interspecies hybrids result from the union of two different species.*

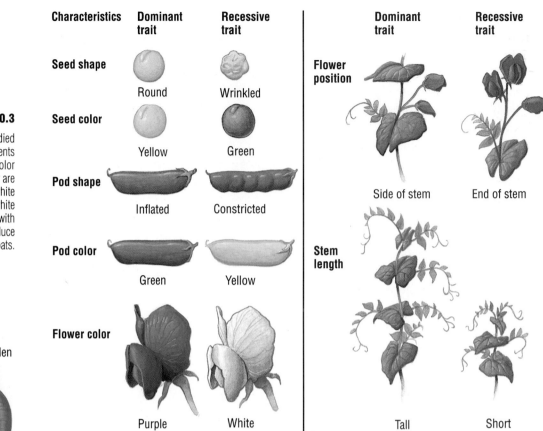

Characteristics	Dominant trait	Recessive trait		Dominant trait	Recessive trait
Seed shape	Round	Wrinkled	Flower position	Side of stem	End of stem
Seed color	Yellow	Green			
Pod shape	Inflated	Constricted	Stem length		
Pod color	Green	Yellow			
Flower color	Purple	White		Tall	Short

Figure 20.3

The seven characteristics studied by Mendel in his experiments with garden peas. Flower color and seed-coat color are correlated. Plants with white seeds produce seeds with white seed coats, and plants with violet-purple flowers produce seeds with gray seed coats.

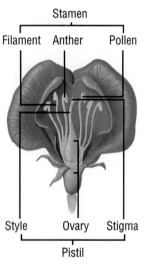

Stamen

Filament   Anther   Pollen

Style   Ovary   Stigma

Pistil

Figure 20.4

The structure of a flower.

Figure 20.5

The process of cross-pollination. To prevent the possibility of self-fertilization, the anthers are removed from the plant chosen as the seed-parent before they are fully mature. Pollen from the plant chosen as the pollen-parent is transferred to the stigma of the seed-parent flower at the appropriate time.

mens attaches to the pistil inside the flower. This process is called pollination. The pistil consists of the stigma, style, and ovary, and is said to be the female part of the plant. The pollen grains fertilize the egg cells in the ovary. Because male and female sex cells come from the same plant, the parentage of the offspring rarely changes.

Mendel fertilized plants by cross-pollination rather than self-fertilization. He removed the stamen from one plant and transferred the pollen to the pistil of another plant, thus combining the male and female sex cells of different plants. To ensure that the seeds were produced from cross-pollinated plants, Mendel first removed the stamen from the recipient plant. The pollen present must then originate from the donor plant.

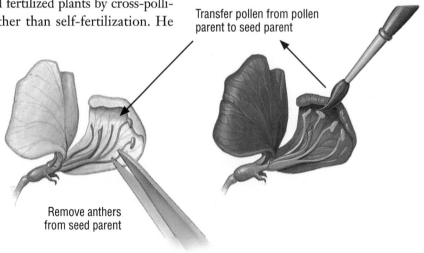

Transfer pollen from pollen parent to seed parent

Remove anthers from seed parent

Unit Six: Heredity

MENDEL'S EXPERIMENTS

When Mendel crossed the pollen from a plant that produced round seeds with the eggs of one that produced wrinkled seeds, he discovered that the offspring were always round. Mendel's predecessors had speculated that the crossing of different traits would create a blend. Rather, the offspring in Mendel's experiment did not have slightly wrinkled seed coats, but round ones. Did this mean that the pollen determines the seed coat? Mendel crossed the pollen from a wrinkled-seed-coat plant with the eggs from a round-seed-coat plant. Once again, all the offspring were round. In fact, the round trait dominated, regardless of parentage.

Mendel repeated the procedure for other characteristics. He discovered that one trait dominated another, whether the sex cell came from the male or female part of the plant. Tall plants produced tall offspring when cross-pollinated with short plants; likewise, plants that had yellow seed coats produced yellow offspring when cross-pollinated with plants that had green seed coats. Mendel reasoned that things called *factors* controlled the traits of a plant. The factors were later known as genes. He concluded that some genes, such as the ones for round and yellow seed coats and tall stem height, were **dominant.** The other genes were **recessive.**

MENDEL'S LAWS OF HEREDITY

Mendel cross-pollinated many plants and kept track of all the results. He recorded the number of dominant and recessive offspring for each cross. Then, applying his knowledge of mathematics, he came up with the following explanations:

1 Inherited characteristics are controlled by factors, known as genes, that occur in pairs. A self-fertilizing pea plant with round seeds receives a gene for round seed coats from the pollen and a complementary gene for round seed coats from the egg. Similarly, a self-fertilizing pea plant with wrinkled seed coats has two genes for wrinkled seed coats. During cross-fertilization each parent contributes one of its genes.

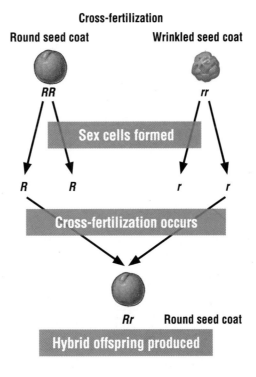

Cross-fertilization

Round seed coat   Wrinkled seed coat

RR   *rr*

Sex cells formed

R   *R*   *r*   *r*

Cross-fertilization occurs

Rr   Round seed coat

Hybrid offspring produced

2 One factor, or gene, masks the effect of another. This process is known as the principle of dominance. The gene for round seed coats masks the effect of the gene for wrinkled seed coats. The dominant gene is indicated by an upper-case letter, in this case, *R.* The recessive trait is designated by the lower-case letter, *r.*

3 A pair of factors, or genes, separate or segregate during the formation of sex cells. This process is often referred to as the law of **segregation.** Mendel

Figure 20.6

Cross-fertilization. The result of crossing a pea plant with round seeds and a pea plant with wrinkled seeds.

Dominant *genes determine the expression of the genetic trait in offspring.*

Recessive *genes are overruled by dominant genes, which determine the genetic trait.*

Segregation *refers to the separation of paired genes during meiosis.*

concluded that sex cells must contain only one member of the pair of factors, or genes. Mendel came to the correct conclusion, despite the fact that he did not know the mechanism of meiosis. The principles of sex cell formation were not discovered until 25 years after Mendel had completed his experiments.

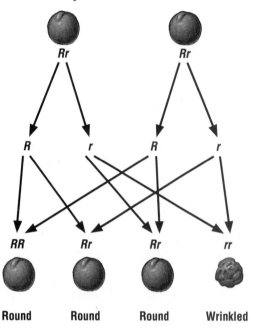

Meiosis occurs. Each gamete has one of the homologous chromosomes.

F₂ generation inherits genes from the gametes of the F₁ generation.

Round   Round   Round   Wrinkled

Figure 20.7

The result of crossing two hybrid pea plants with round seeds from the first generation.

Hybrids *result from crosses between parents that are genetically dissimilar.*

Genotype *refers to the genes an organism contains.*

Mendel continued his experimentation by crossing two **hybrid** plants with round seed coats from the first generation. He referred to the first generation as filial generation one, or F₁ generation. The word *filial* comes from the Latin for "son." Because both round plants are hybrid, they contain *R* and *r* genes. Remember, the *R* is used to identify the round gene, while the *r* is used to identify the wrinkled gene.

You might predict an equal number of round and wrinkled offspring in the second, or F₂, generation. However, this is not the ratio Mendel discovered when he did the cross. He was astonished to find that 75% of the offspring expressed the dominant trait, while only 25% expressed the wrinkled trait. How can these results

be explained? Figure 20.7 shows what happens when the sex cells, or gametes, from the F₁ generation recombine to form an F₂ generation. Both members of the F₁ generation are round, but wrinkled offspring reappear during the F₂ generation.

Any members of the F₂ generation with an *R* gene will appear round. The round gene is dominant over the wrinkled gene. To be wrinkled, the offspring must have two recessive *r* genes. Figure 20.7 illustrates why the expected ratio in the F₂ generation is three round to one wrinkled.

SIGNIFICANCE OF MENDEL'S WORK

You may be surprised to learn that you know more about genetics than Gregor Mendel did. Mendel did not know that what he called factors were located on chromosomes. He had not observed cells during cell division. However, Mendel was one of the first biologists to perform careful experiments, and to record and interpret quantitative data. He often repeated procedures many times to support his conclusions. Prior to Mendel, the application of mathematical concepts to biology was not common. Gregor Mendel is considered to be the father of genetics. Even today, biologists refer to the study of a particular type of inheritance as Mendelian genetics.

GENETIC TERMS

The following terms will help you read about and describe heredity.

Genotype refers to the genes that an organism contains for a particular trait. One trait of the genes is inherited from each of the parents. Since the genes occur in pairs, various combinations are possible. A tall-stem pea plant could have two different genotypes, *TT* and *Tt*. A short-stem pea plant can have only one genotype, *tt*, since the gene for tall stems is dominant over the gene for short stems.

Phenotype refers to the observable traits of an individual. Since a pea plant can be tall or short, there are only two possible phenotypes for this characteristic. The tall phenotype may have two different genotypes, *TT* or *Tt*.

Homozygous is a term used to describe a trait in which an organism contains two genes that are the same. Homozygous means that the organism is pure for that trait. The homozygous condition for a tall-stem plant would be *TT*. The homozygous condition for a short-stem plant would be *tt*.

Heterozygous is a term used to describe a trait in which an organism contains two genes that are different. The only heterozygous condition possible for stem length in a garden pea is *Tt*. The phenotype for this heterozygous genotype is a tall-stem plant.

Alleles are two or more alternate forms of a gene. The pea plant may have a gene for tall stem length *(T)* and a gene for short stem length *(t)*. *T* is one allele and *t* is a second allele. Each of the alleles is located at the same position on one of the pairs of homologous chromosomes.

SINGLE-TRAIT INHERITANCE

The combining of single contrasting traits is referred to as a **monohybrid cross.** For example, the cross between a *TT* tall pea plant and a *tt* short pea plant is a monohybrid cross. Crossing members of the F_1 generation, *Tt* tall pea plants, to find an F_2 generation is another example of a monohybrid cross.

A special chart, referred to as a **Punnett square,** helps geneticists organize the results of a cross between the sex cells of two individuals. The Punnett square resembles an "X" and "O" chart. The chart can be used to predict the genotypes and the phenotypes of the offspring.

Consider a cross between a pea plant that is heterozygous for a round seed and a pea plant that has a wrinkled seed. The allele for a round seed is found to be dominant over that for a wrinkled seed. The uppercase letter *R* can be used to indicate the round dominant allele and *r* to indicate the wrinkled recessive allele. Because the plant with the round seeds is heterozygous, the genotype for the plant is *Rr*. Since the gene for wrinkled seeds is recessive, the plant with wrinkled seeds must contain two wrinkled alleles. The genotype of this plant is *rr*.

Phenotype *refers to the observable traits of an organism that arise because of the interaction between genes and the environment.*

Homozygous *refers to a genotype in which both genes of a pair are identical.*

Heterozygous *refers to a genotype in which the gene pairs are different.*

Alleles *are two or more alternate forms of a gene.*

A **monohybrid cross** *involves one gene pair of contrasting traits.*

A **Punnett square** *is a chart used by geneticists to show the possible combinations of alleles in offspring.*

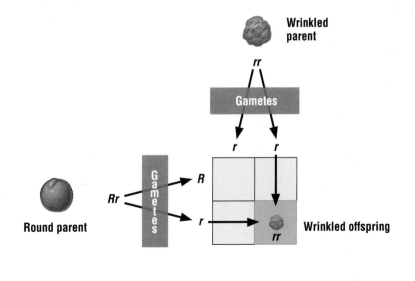

Figure 20.8

A partially completed Punnett square for a cross between a pea plant that is heterozygous for a round seed and a pea plant that has a wrinkled seed. The *rr* genotype shown in the chart is only one of the four possible combinations. When the chart is completed, all the possible genotypes are shown.

The Punnett square can be used to determine the genotypes of the offspring resulting from the cross between these two plants. The symbols for the gametes are written across the top and along the left side of the square. By drawing a line from each gamete at the top and another line from each gamete along the side, the possible gene combinations of the offspring can be determined.

Figure 20.9

A completed Punnett square for a cross between a heterozygous round-seed pea plant and a wrinkled-seed pea plant.

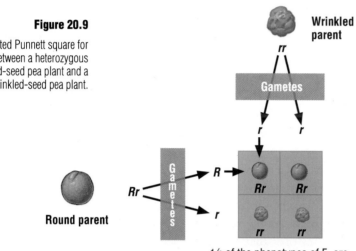

½ of the phenotypes of F₁ are round and ½ are wrinkled.

From the Punnett square, it is also possible to determine the probability of each phenotype occurring in the offspring. For Figure 20.9 chances are that one-half, or 50%, of the offspring would produce wrinkled seeds.

Figure 20.10

A Punnett square showing the results of a cross between two heterozygous plants with round seeds.

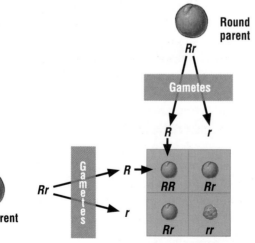

¾ of the phenotypes of F₁ are round and ¼ are wrinkled.

1 For Labrador retrievers, black fur color is dominant to yellow. Explain how a homozygous black dog can have a different genotype than a heterozygous black dog. Could the heterozygous black dog have the same genotype as a yellow-haired dog?

2 A pea plant with round seeds is cross-pollinated with a pea plant that has wrinkled seeds. For the cross, indicate each of the following:

   a) the genotypes of the parents if the round-seed plant were heterozygous

   b) the gametes produced by the round and wrinkled-seed parents

   c) the genotypes and the phenotypes of the F₁ generation

   d) the F₂ generation if two round plants from the F₁ generation were allowed to cross-pollinate

3 For Dalmatian dogs, the spotted condition is dominant to non-spotted.

   a) Using a Punnett square, show the cross between two heterozygous parents.

   b) A spotted female Dalmatian dog mates with an unknown father. From the appearance of the pups, the owner concludes that the male was a Dalmatian. The owner notes that the female had six pups, three spotted and three non-spotted. What are the genotype and phenotype of the unknown male?

4 For Mexican hairless dogs, the hairless condition is dominant to hairy. A litter of eight pups is found; six are hairless and two are hairy. What is the genotype of their parents?

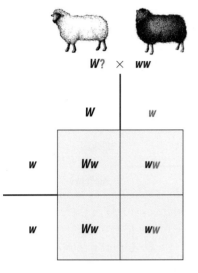

	W	w
w	Ww	ww
w	Ww	ww

½ of the offspring are
black and ½ are white.

	W	W
w	Ww	Ww
w	Ww	Ww

All of the offspring
are white.

Figure 20.11

A test cross involves crossing an individual of unknown genotype, showing a dominant trait, with a recessive. If any offspring show the recessive trait, then the individual must be hybrid. If all offspring show the dominant trait, then the individual must be homozygous.

TEST CROSS

You have probably heard someone referred to as the "black sheep of the family." Do you know why black sheep are not preferred? Black wool tends to be brittle and is very difficult to dye. How can a sheep rancher avoid getting black sheep? By using a homozygous white ram—the male of the species—a rancher can ensure that all of the flock will have white hair. But how would you know that a ram is homozygous for the white phenotype? Heterozygous white sheep will produce some black wool progeny.

A test cross is often performed to determine the genotype of a dominant phenotype. The test cross is always performed between the unknown genotype and a homozygous recessive genotype. In this case, the homozygous recessive individual would be a black sheep. If one-half of the offspring of the cross are black and the other half are white, then the unknown genotype must be heterozygous white. However, if all the offspring are white, then the unknown genotype must be homozygous dominant.

MULTIPLE ALLELES

For each of the traits studied by Mendel, there were only two possible alleles. The dominant allele controlled the trait. It is possible, however, to have more than two different alleles for one trait. In fact, many traits with multiple alleles exist in nature.

Geneticists who study the tiny fruit fly called *Drosophila* have noted that many different eye colors are possible. The red, or wild-type, is the most common, but apricot, honey, and white colors also exist. Although a fruit fly can only have two different genes at any one time, more than two alleles are possible. A fruit fly may have an allele for wild-type eyes and another for white. Its prospective mate may have an allele for apricot-colored eyes and another for honey-colored eyes. The dominance hierarchy is as follows: wild type is dominant to apricot, is dominant to honey, is dominant to white. In the case of multiple alleles, it is no longer appropriate to use upper- and lower-case letters. Capital letters and superscript numbers are used to express the different genes and their combinations.

Table 20.1 Dominance Hierarchy and Symbols for Eye Color in *Drosophila*

Phenotype	Genotypes	Dominant to
Wild type	E^1E^1, E^1E^2, E^1E^3, E^1E^4	Apricot, honey, white
Apricot	E^2E^2, E^2E^3, E^2E^4	Honey, white
Honey	E^3E^3, E^3E^4	White
White	E^4E^4	

The dominance and symbols for eye color in *Drosophila* are shown in Table 20.1

Consider the mating of the following *Drosophila*:

E^1E^4 (wild-type eye color) $\times$ E^2E^3 (apricot eye color)

The phenotypic ratio of the F$_1$ offspring is two wild-type eye color to one apricot eye color to one honey eye color. The Punnett square for this cross is shown in Figure 20.13.

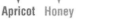

Figure 20.13

A cross between a fruit fly with wild-type eye color and one with apricot eyes.

Figure 20.12

(a) *Drosophila melanogaster*, the fruit fly, is widely used for genetics studies. (b) Wild-type, or red, is the normal, or most common, eye color. It is the dominant allele.

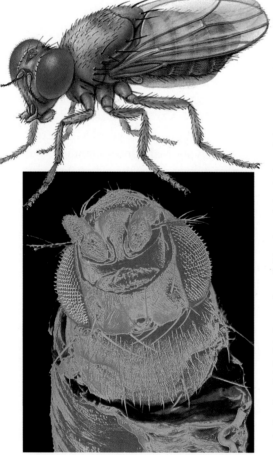

a)

b)

INCOMPLETE DOMINANCE

Prior to Mendel's studies, many scientists believed that hybrids would have a blending of traits. Although Mendel never found any examples of new traits being produced by the combinations of different genes, many do exist in nature. When two genes are equally dominant, they interact to produce a new phenotype. This lack of a dominant gene is known as *incomplete dominance.*

For example, if red snapdragons are crossed with white snapdragons, all of the F$_1$ offspring are pink. The pink color is produced by the interaction of red and white alleles. This type of incomplete dominance is often called *intermediate inheritance.* If the F$_1$ generation is allowed to self-fertilize, the F$_2$ generation produces a surprising ratio of one red to two pink to one white. The Punnett square in Figure 20.14 helps to explain this result.

Another type of incomplete domi-nance is referred to as *codominance*. In this type of gene interaction, both genes are expressed at the same time. Shorthorn cattle provide an excellent example of codominance. A red bull crossed with a white cow produces a roan calf. The roan calf has intermingled white and red hair. The roan calf would also be produced if a white bull were crossed with a red cow.

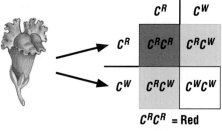

	C^R	C^W
C^R	$C^R C^R$	$C^R C^W$
C^W	$C^R C^W$	$C^W C^W$

$C^R C^R$ = Red
$C^R C^W$ = Pink
$C^W C^W$ = White

Figure 20.14

Color in snapdragons is an example of incomplete dominance. Red-flowering and white-flowering snapdragons combine to produce pink-flowering plants in F_1. The F_2 generation produces one red to two pink to one white.

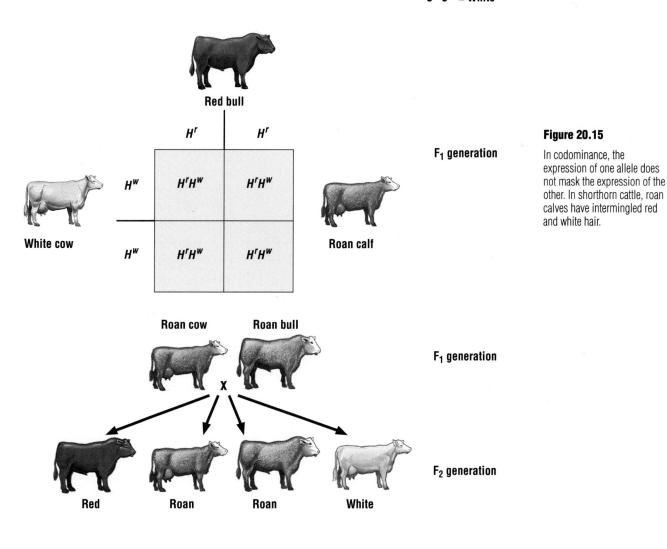

Red bull

	H^r	H^r
H^W	$H^r H^W$	$H^r H^W$
H^W	$H^r H^W$	$H^r H^W$

White cow

F_1 generation

Roan calf

Figure 20.15

In codominance, the expression of one allele does not mask the expression of the other. In shorthorn cattle, roan calves have intermingled red and white hair.

Roan cow   **Roan bull**

F_1 generation

X

Red   **Roan**   **Roan**   **White**

F_2 generation

5 Multiple alleles control the intensity of pigment in mice. The gene D^1 designates full-color, D^2 designates dilute color, and D^3 is deadly when homozygous. The order of dominance is $D^1 > D^2 > D^3$. When a full-color male is mated to a dilute-color female, the offspring are produced in the following ratio: two full color to one dilute to one dead. Indicate the genotypes of the parents.

6 Multiple alleles control the coat color of rabbits. A gray color is produced by the dominant allele C. The C^{cb} allele produces a silver-gray color when present in the homozygous condition, $C^{cb}C^{cb}$, called chinchilla. When C^{cb} is present with a recessive gene, a light silver-gray color is produced. The allele C^b is recessive to both the full-color allele and the chinchilla allele. The C^b allele produces a white color with black extremities. This coloration pattern is called Himalayan. An allele C^a is recessive to all genes. The C^a allele results in a lack of pigment, called albino. The dominance hierarchy is $C > C^{cb} > C^b > C^a$. The table below provides the possible genotypes and phenotypes for coat color in rabbits. Notice that four genotypes are possible for full-color but only one for albino.

Phenotypes	Genotypes
Full color	CC, CC^{ch}, CC^h, CC^a
Chinchilla	$C^{ch}C^{ch}$
Light gray	$C^{ch}C^h, C^{ch}C^a$
Himalaya	C^hC^h, C^hC^a
Albino	C^aC^a

a) Indicate the genotypes and phenotypes of the F$_1$ generation from the mating of a heterozygous Himalayan-coat rabbit with an albino-coat rabbit.

b) The mating of a full-color rabbit with a light-gray rabbit produces two full-color offspring, one light-gray offspring, and one albino offspring. Indicate the genotypes of the parents.

c) A chinchilla-color rabbit is mated with a light-gray rabbit. The breeder knows that the light-gray rabbit had an albino mother. Indicate the genotypes and phenotypes of the F$_1$ generation from this mating.

d) A test cross is performed with a light-gray rabbit, and the following offspring are noted: five Himalayan-color rabbits and five light-gray rabbits. Indicate the genotype of the light-gray rabbit.

7 A geneticist notes that crossing a round-shaped radish with a long-shaped radish produces oval-shaped radishes. If oval radishes are crossed with oval radishes, the following phenotypes are noted in the F$_2$ generation: 100 long, 200 oval, and 100 round radishes. Use symbols to explain the results obtained for the F$_1$ and F$_2$ generations.

8 Palomino horses are known to be caused by the interaction of two different genes. The allele C^r in the homozygous condition produces a chestnut, or reddish-color, horse. The allele C^m produces a very pale cream coat color, called cremello, in the homozygous condition. The palomino color is caused by the interaction of both the chestnut and cremello alleles. Indicate the expected ratios in the F$_1$ generation from mating a palomino with a cremello.

CASE STUDY

A MURDER MYSTERY

Objective

To solve a murder mystery using genetics.

Background Information

There are four different blood types. The alleles for blood type A and B are codominant but are dominant to O.

Phenotypes	Genotypes
Type A	$I^A I^A$, $I^A I^O$
Type B	$I^B I^B$, $I^B I^O$
Type AB	$I^A I^B$
Type O	$I^O I^O$

The rhesus factor is another blood factor that is regulated by genes. The Rh+ gene is dominant to the Rh− gene.

The Case Study

As a bolt of lightning flashed above Black Mourning Castle, a scream echoed from the den of Lord Hooke. When the upstairs maid peered through the door, a freckled arm reached for her neck. Quickly, the maid bolted from the doorway, locked herself in the library, and telephoned the police.

Inspector Holmes arrived to find a frightened maid and the dead body of Lord Hooke. Apparently, the lord had been strangled. The inspector quickly gathered evidence. He noted blood on a letter opener, even though Lord Hooke did not have any cuts or abrasions. The blood sample proved to be type O, Rh negative. The quick-thinking inspector gathered all the members of the family and began taking blood samples. The chart below shows the relatives who were in the castle at the time of Lord Hooke's murder.

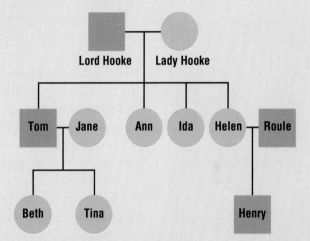

The inspector gathered the following information. Some of the family members were wearing long-sleeved shirts, so the inspector found it difficult to determine whether or not freckles were present on the arms. Note that the gene for freckles is dominant to the gene for no freckles.

The crafty inspector drew the family close together and, while puffing on his pipe, indicated that he had found the murderer. He explained that Lady Hooke had been unfaithful to her husband. One of the heirs to the fortune was not really a *sibling*. The murder was committed to preserve a share of the fortune.

Who was the murderer? State the reasons for your answer. How did the inspector eliminate the other family members? ■

Family member	Blood type	Rh factor	Freckles
Lord Hooke	AB	+	no
Lady Hooke	A	+	no
Helen	A	+	no
Roule	O	+	no
Henry	Refused blood test		?
Ida	A	−	?
Ann	B	+	?
Tom	O	−	no
Jane	A	+	?
Beth	O	−	?
Tina	A	+	yes

SCIENCE AND POLITICS: LYSENKO AND VAVILOV

During the 1940s, Trofim Denisovich Lysenko was considered the father of genetics in the Soviet Union. His theory for explaining genetic inheritance challenged the views of most western geneticists. Lysenko believed that acquired characteristics could be inherited. He maintained that, by exposing wheat to very cold or very arid conditions, he could produce a resistance that would affect genetic structure and be passed on to the next generation. A wheat plant resistant to cold or specially adapted to dry conditions would then arise.

Lysenko's theories were popular in the Soviet Union because he made his arguments fit Marxist philosophy. Joseph Stalin seized the opportunity to add science to Marx's economic and philosophical theories. Great political pressure was applied to silence opponents of Lysenko. Geneticists from the west were accused of fabricating research.

Despite an overwhelming flood of evidence that opposed Lysenko's theories, political pressure ensured that his views dominated Soviet genetics. An academic argument between Lysenko and the noted Soviet geneticist, Nikolai Vavilov, became political when Stalin sided with Lysenko. On August 6, 1940, Vavilov was arrested and subsequently sentenced to death for subversive acts. By his opposition to the theory of the inheritance of acquired characteristics, Vavilov became a threat to communist economic and social philosophy. His sentence was later reduced to ten years' imprisonment.

Lysenko retained a dominant position in Soviet science long after his theories had been discredited. It was not until 1965 that he was removed from his post as director of the Institute of Genetics, a position he had held since 1940.

Figure 20.16

Trofim Denisovich Lysenko (1898–1976) was a Soviet geneticist who believed that acquired characteristics could be inherited.

DIHYBRID CROSSES

Mendel also studied two separate traits with a single cross by following the same procedure he had used for studying single traits. He cross-pollinated pure-breeding plants that produce yellow, round seed coats with pure-breeding plants that produce green, wrinkled seed coats, in order to study the inheritance of two traits. Note that pure-breeding plants always produce identical offspring. The laws of genetics that apply for single-trait inheritance (monohybrid crosses) also apply for two-trait inheritance (*dihybrid* crosses).

Figure 20.17 shows a dihybrid cross. The pure-breeding round coat is indicated by the symbol *RR*, and the pure-breeding wrinkled coat by the recessive alleles *rr*. The pure yellow is indicated by the alleles *YY* and the green by *yy*. The genotype for the pure yellow, round parent is *RRYY*, while the genotype for the green, wrinkled parent is *rryy*. The F₁ offspring produced from such a cross are heterozygous for both the yellow and round genotypes.

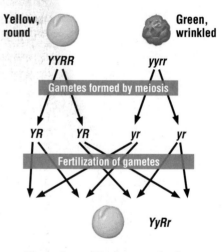

All members of the F₁ generation have the same genotype and phenotype.

Figure 20.17

A dihybrid cross between a pure-breeding pea plant with yellow, round seeds and a plant with green, wrinkled seeds.

Consider a cross between a pure-breeding green, round pea plant and a pure-breeding yellow, wrinkled pea plant. Figure 20.18 shows the resulting offspring. Inheritance of the gene for color is not affected by either the wrinkled or round alleles. By doing other crosses, Mendel soon discovered that the genes assort independently. Today, this phenomenon is referred to as *the law of independent assortment*. The genes that govern pea shape are inherited independently of the ones that control pea color.

with round and wrinkled genes in equal frequency. The number of gametes containing Yr will be equal to the number of gametes containing the yr alleles.

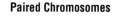

Paired Chromosomes

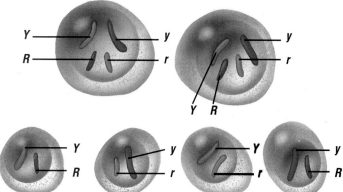

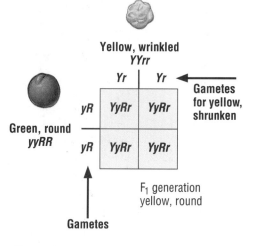

Figure 20.18

The inheritance of the gene for pea color is not affected by the gene for pea shape.

Mendel allowed plants of the F_1 generation to self-fertilize in order to produce an F_2 generation. Each heterozygous, yellow, round plant can produce four different phenotypes. Remember the law of independent assortment, which indicates that if they are located on separate chromosomes, they are inherited independently of each other. As the homologous chromosomes move to opposite poles during meiosis, the yellow gene will segregate with the round and wrinkled genes in equal frequency. This means that the sex cells containing YR will equal the number of sex cells containing yR. Similarly, the green genes will segregate

A modified Punnett square (see below) can be used to predict the genotypes and phenotypes of the F_2 generation.

Gametes	YR	yR	Yr	yr
YR	YYRR	YyRR	YYRr	YyRr
yR	YyRR	yyRR	YyRr	yyRr
Yr	YYRr	YyRr	YYrr	Yyrr
yr	YyRr	yyRr	Yyrr	yyrr

The phenotypes of the F_2 generation shown in the Punnett square in Figure 20.20 are as follows:

9/16 yellow, round   3/16 green, round
3/16 yellow, wrinkled   1/16 green, wrinkled

Figure 20.20

A Punnett square that has been modified to find the F_2 generation of a dihybrid cross.

PROBABILITY

Probability is the study of the outcomes of events or occurrences. For example, you can calculate the probability of getting heads when you toss a coin. Because there are only two possibilities—heads and tails—the chances of getting heads is 1/2. Probability can be expressed by the following formula:

$$\text{Probability} = \frac{\text{number of chances for an event}}{\text{number of possible combinations}}$$

Two important rules will help you understand probability:

- *The rule of independent events.* This rule states that chance has no memory. This means that previous events will not affect future events. For example, if you tossed two heads in a row, the probability of tossing heads once again would still be 1/2. Each event or occurrence must be regarded as an individual event.
- *The product rule.* The product rule states that the probability of independent events occurring simultaneously is equal to the product of these events occurring separately. For example, the chances of tossing heads after two tosses is 1/2, but the chances of tossing heads three times in a row is:

$$1/2 \times 1/2 \times 1/2 = 1/8$$

The genotypic and phenotypic ratios are determined by the chance, or probability, of inheriting a certain trait. In humans, free ear lobes are controlled by the dominant allele E, and attached ear lobes by the recessive allele e. The widow's peak hairline is regulated by the dominant allele W, while the straight hairline is controlled by the allele w. Consider the mating of the following genotypes:

$$EeWw \times EeWw$$

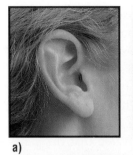
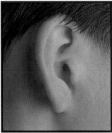

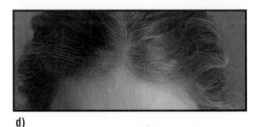

a)   b)

c)

d)

Figure 20.21

The shape of both ear lobes and hairline are inherited characteristics in humans. (a) The free ear lobe is dominant to the attached ear lobe (b), and the widow's peak (c) is dominant to a straight hairline (d).

Dihybrids can be treated as two monohybrids, as shown in Figure 20.22. Isolate the gene for ear lobes and work with it as a monohybrid. The cross between the heterozygous parents can be determined by the F_1 generation. The phenotype probabilities are: free ear lobes 3/4, attached ear lobes 1/4.

Now consider the F_1 generation from the cross between the heterozygous parents for the hairline trait. The phenotype probabilities are: widow's peak 3/4, straight hairline 1/4.

	½ E	½ e
½ E	¼ EE	¼ Ee
½ e	¼ Ee	¼ ee

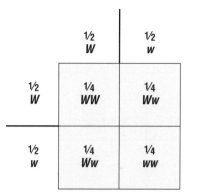

Figure 20.22

Punnett squares showing two monohybrid crosses between heterozygous parents for free earlobes and for a widow's peak.

The monohybrids can now be combined to calculate the probabilities of a dihybrid cross. For example, the chances of producing an F_1 offspring from the mating of $EeWw \times EeWw$ who has

- a widow's peak and free ear lobes is: $3/4 \times 3/4$, or 9/16;
- a straight hairline and free ear lobes is: $1/4 \times 3/4$, or 3/16;
- a widow's peak and attached ear lobes is: $3/4 \times 1/4$, or 3/16;
- a straight hairline and attached ear lobes is: $1/4 \times 1/4$, or 1/16.

The calculation indicates that the dihybrid cross is equivalent to two separate monohybrid crosses. To determine the probability that the first child from the mating of the $EeWw \times EeWw$ parents would be a male with a widow's peak and attached ear lobes, consider each of the separate probabilities:

- The probability of producing a male is 1/2.
- The probability of widow's peak is 3/4.
- The probability of attached ear lobes is 1/4.
- Therefore, the probability of producing a male with a widow's peak and attached ear lobes is $1/2 \times 3/4 \times 1/4$, or 3/32.

9 In guinea pigs, black coat color (*B*) is dominant to white (*b*), and short hair length (*S*) is dominant to long (*s*). Indicate the genotypes and phenotypes from the following crosses:

   a) Homozygous for black, heterozygous for short-hair guinea pig crossed with a white, long-hair guinea pig.

   b) Heterozygous for black and short-hair guinea pig crossed with a white, long-hair guinea pig.

   c) Homozygous for black and long-hair crossed with a heterozygous black and short-hair guinea pig.

10 Black coat color (*B*) in cocker spaniels is dominant to white coat color (*b*). Solid coat pattern (*S*) is dominant to spotted pattern (*s*). The pattern arrangement is located on a different chromosome than the one for color, and its gene segregates independently of the color gene. A male that is black with a solid pattern mates with three females. The mating with female A, which is white, solid, produces four pups: two black, solid, and two white, solid. The mating with female B, which is black, solid, produces a single pup, which is white, spotted. The mating with female C, which is white, spotted, produces four pups: one white, solid; one white, spotted; one black, solid; one black, spotted. Indicate the genotypes of the parents.

11 For human blood type, the alleles for types A and B are codominant, but both are dominant over the type O allele. The Rh factor is separate from the ABO blood group and is located on a separate chromosome. The Rh+ allele is dominant to Rh−. Indicate the possible phenotypes from the mating of a woman, type O, Rh−, with a man, type A, Rh+.

LABORATORY

GENETICS OF CORN

Objective

To investigate the inheritance of traits using corn.

Materials

dihybrid corn ears (sample A, sample B)

Procedure

1 Obtain a sample A corn ear from your instructor. The kernels display two different traits that are located on different chromosomes.

   a) Indicate the two different traits.

   b) Predict the dominant phenotypes.

   c) Predict the recessive phenotypes.

2 Assume that the ear of corn was from the F_2 generation. The parents were pure-breeding homozygous for each of the characteristics. Assign the letters P and p to the alleles for color, and S and s to the alleles for shape. Use the symbols $PPss \times ppSS$ for the parent generation.

   d) Indicate the phenotype of the $PPss$ parent.

   e) Indicate the phenotype of the $ppSS$ parent.

   f) Indicate the expected genotypes and phenotypes of the F_1 generation.

   g) Use a Punnett square to show the expected genotypes and the phenotypic ratio of the F_2 generation.

3 Count 100 of the kernels in sequence, and record the actual phenotypes in a table similar to the one below.

Phenotype	Number	Ratio
Dominant genes for color and shape		
Dominant gene for color, but recessive for shape		
Recessive gene for color, but dominant gene for shape		
Recessive genes for color and shape		

4 Obtain sample B. Assume that this ear was produced from a test cross. Count 100 kernels in sequence and record your results.

   h) Indicate the phenotypic ratio of the F_1 generation (the kernel).

   i) Give the phenotype of the unknown parent.

Laboratory Application Questions

1 Why are test crosses important to plant breeders?

2 A dihybrid cross can produce 16 different combinations of alleles. Explain why 100 seeds were counted rather than 16.

3 A dominant allele Su, called starchy, produces kernels of corn that appear smooth. The recessive allele su, called sweet, produces kernels of corn that appear wrinkled. The dominant allele P produces purple kernels, while the recessive p allele produces yellow kernels. A corn plant with starchy, yellow kernels is cross-pollinated with a corn plant with sweet, purple kernels. One hundred kernels from the hybrid are counted, and the following results are obtained: 52 starchy, yellow kernels and 48 starchy, purple kernels. What are the genotypes of the parents and the F_1 generation? ■

SELECTIVE BREEDING

Farmers and ranchers have used **selective breeding** processes to improve domestic varieties of plants and animals. Early farmers identified plants with desirable characteristics. Rust-resistant wheat; sweet, full-kernel corn; and canola, which germinates and grows rapidly in colder climates, have improved food harvests. Selection of specific traits from wild cabbage has produced green and red varieties of cabbage, broccoli, and cauliflower.

You are probably familiar with the term "purebreds." Many dogs and horses are considered to be purebreds, or thoroughbreds. Genotypes of these animals are closely regulated by a process called **inbreeding,** in which similar phenotypes are selected for breeding. The desirable traits vary from breed to breed. For example, Irish setters are bred for their long, narrow facial structure and long, wispy hair. The bull terrier was originally bred for fighting. Quick reflexes and strong jaws were chosen as desirable phenotypes. Some geneticists have complained that inbreeding has caused problems for the general public as well as for the breed itself.

New varieties of plants and animals can be developed by hybridization. This process is the opposite to that of inbreeding. Rather than breed plants or animals with similar traits, the hybridization technique attempts to blend desirable but different traits. Corn has been hybridized extensively. The hybrids tend to be more vigorous than either parent. Figure 20.23 shows the most common method used.

Selective breeding *is the crossing of desired traits from plants or animals to produce offspring with both characteristics.*

Inbreeding *is the process whereby breeding stock is drawn from a limited number of individuals possessing desirable phenotypes.*

Figure 20.23

Hybridization can be used to produce a more vigorous strain of corn.

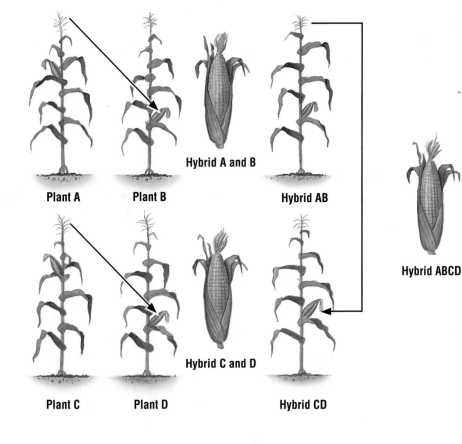

Plant A   Plant B   Hybrid A and B   Hybrid AB

Plant C   Plant D   Hybrid C and D   Hybrid CD   Hybrid ABCD

RESEARCH IN CANADA

The Plant Breeders

**SIR CHARLES
SAUNDERS
DR. KEITH DOWNEY**

Canada boasts a long history of contributions in plant genetics. One of our first great plant breeders was Sir Charles Saunders. Born in Toronto in 1837, Saunders was noted for developing and introducing Marquis wheat. Marquis is a high-quality, bread-producing wheat that was derived from a cross between the Red Calcutta and Red Fife varieties of wheat. Prior to the Marquis strain, Red Fife was the most common variety of wheat grown in Canada. However, the slower-maturing Red Fife was often damaged by early frosts, especially in Saskatchewan and Alberta. The Marquis variety matures at least a week earlier than Red Fife and also provides better yields.

Keith Downey, formerly of the University of Saskatchewan, and Baldur Stefansson, from the University of Manitoba, are recognized as world leaders in plant genetics. Downey was born in Saskatoon, Saskatchewan, in 1927, and Stefansson was born in Manitoba in 1917.

Their research has helped bring wealth to their respective provinces. Downey and Stefansson developed a rape seed with low levels of erucic acid and glucosinolate. The rape seed was transformed into a high-quality oil-seed crop, renamed and now known worldwide as canola.

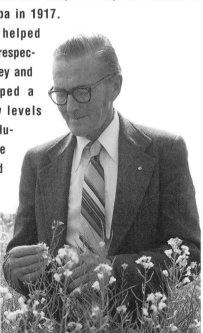

DR. BALDUR STEFANSSON

GENE INTERACTION

The traits studied by Mendel are controlled by one gene. However, some traits are regulated by more than one gene. Many of your characteristics are determined by several pairs of independent genes. Skin color, eye color, and height are but a few of your characteristics that are **polygenic.**

Coat color in dogs provides an example of genes that interact. The allele *B* produces black coat color, while the recessive allele *b* produces brown coat color. However, a second gene, located on a separate chromosome, also affects coat color. The allele *W* prevents the formation of pigment, thereby preventing color. The recessive allele *w* does not prevent color. The genotype *wwBb* would be black, but the genotype *WwBb* would appear white. The *W* allele masks the effect of the *B* color gene. Genes that interfere with the expression of other genes are said to be **epistatic.**

Another type of gene interaction, called *complementary interaction*, occurs when two different genotypes interact to produce a phenotype that neither is capable of producing by itself. One of the best examples of complementary interaction can be seen in the combs of chickens. The *R* allele produces a rose comb. Another allele, *P*, located on a different chromosome, produces a pea comb. When the *R* and *P* alleles are both present, they combine to produce a walnut comb. The absence of the rose and pea alleles results in an individual with a single comb.

Polygenic *traits are inherited characteristics that are affected by more than one gene.*

Epistatic *genes mask the expression of other genes.*

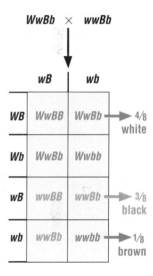

Figure 20.24

A cross between a white dog (*WwBb*) and a black dog (*wwBb*).

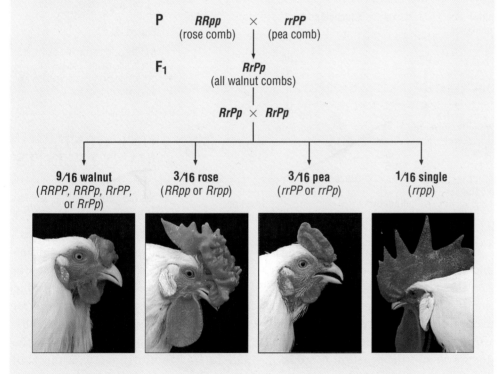

Figure 20.25

Sometimes two gene pairs cooperate to produce a phenotype that neither can produce alone. A cross between a chicken with a rose comb on the crest of its head and one with a pea comb results in a chicken with a walnut comb.

Genes regulate chemical reactions, providing the instructions for cells to build enzymes. Enzymes are protein molecules that speed up chemical reactions that occur in all cells. In some cases, enzymes help molecules combine to form larger molecules, and in other cases, they help larger molecules break down into less complicated molecules. The white clover plant is an excellent example of how genes interact in chemical reactions. The plant's production of cyanide is regulated by two genes. Figure 20.26 summarizes the biochemical pathway.

appears in the homozygous condition, the correct enzyme cannot be produced, and the reaction ends at glucoside. The genotype *GgHh* will produce cyanide, but the genotypes *ggHH* and *GGhh* are not capable of producing cyanide. However, a cross between a *ggHH* and *GGhh* plant will yield offspring capable of producing cyanide.

Figure 20.26

The production of cyanide in the white clover plant is regulated by two genes.

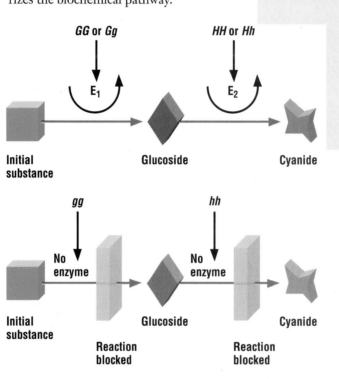

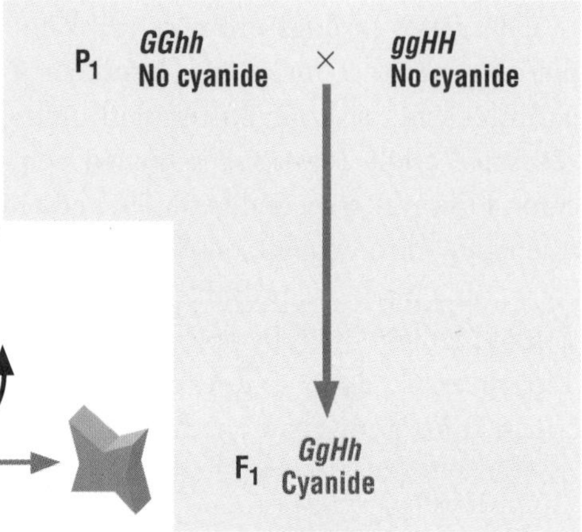

Figure 20.27

A cross between two white clover plants that are not capable of producing cyanide yields offspring that produce cyanide.

Pleiotropic genes are genes that affect many characteristics.

ONE GENE, MANY EFFECTS

The first allele, *G*, provides the genetic information for the production of enzyme #1 (E_1). If the recessive mutation, *g*, appears in the homozygous condition, the proper enzyme (E_1) is not produced. Unfortunately, the mutated enzyme will not work, and the reaction is stopped. The *gg* genotype will prevent the formation of the glucoside. Another allele, *H*, is also required. The allele *H* produces enzyme #2 (E_2), which converts glucoside to its final product, cyanide. If the allele *h*

Some genes, called **pleiotropic genes,** affect many different characteristics. Sickle-cell anemia, a blood disorder, is an example of how one gene can cause many devastating symptoms. Hemoglobin, the pigment that carries oxygen, is found in all red blood cells. Normal hemoglobin is produced by the allele Hb^A. Unfortunately, a mutated gene, Hb^S, causes difficulties when it appears in the homozygous condition. Although the hemoglobin pigment produced by the Hb^S gene can still carry oxygen, it creates problems once it gives oxygen to the cells of the body. The oddly shaped hemoglobin molecules begin to interlock with one another. The new arrangement of molecules changes the shape of the red blood cells, which become bent into a sickle shape.

The mutated red blood cells do not pass through the capillaries. Oxygen delivery is halted in the area of the blockage, and normal organ function is impaired. People with sickle-cell anemia can suffer from fatigue and weakness, an enlarged spleen, rheumatism, and pneumonia. Patients often show signs of heart, kidney, lung, and muscle damage.

Marfan's syndrome, an inability to produce normal connective tissue, is also associated with a single gene. Because connective tissue is found in many organs, the symptoms of Marfan's syndrome show up as eye, skeleton, and cardiovascular defects. Many historians and geneticists have speculated that Abraham Lincoln suffered from Marfan's syndrome. A picture of the late U.S. president shows a severe curling of the hands and an infolding of the limbs, characteristic of advanced stages of the disease.

■ REVIEW QUESTIONS   ?

12 In mice, the gene C causes pigment to be produced, while the recessive gene c makes it impossible to produce pigment. Individuals without pigment are albino. Another gene, B, located on a different chromosome, causes a chemical reaction with the pigment and produces a black coat color. The recessive gene, b, causes an incomplete breakdown of the pigment, and a tan, or light-brown, color is produced. The genes that produce black or tan coat color rely on the gene C, which produces pigment, but are independent of it. Indicate the phenotypes of the parents and provide the genotypic and phenotypic ratios of the F_1 generation from the following crosses.

a) $CCBB \times Ccbb$ **b)** $ccBB \times CcBb$
c) $CcBb \times ccbb$ **d)** $CcBb \times CcBb$

13 The mating of a tan mouse and a black mouse produces many different offspring. The geneticist notices that one of the offspring is albino. Indicate the genotype of the tan parent. How would you determine the genotype of the black parent?

14 The gene R produces a rose comb in chickens. An independent gene, P, which is located on a different chromosome, produces a pea comb. The absence of the dominant rose comb gene and pea comb gene ($rrpp$) produces birds with single combs. However, when the rose and pea comb genes are present together, they interact to produce a walnut comb ($R\_P\_$).

Indicate the phenotypes of the parents and give the genotypic and phenotypic ratios of the F_1 generation from the following crosses.

a) $rrPP \times RRpp$ **b)** $RrPp \times RRPP$
c) $RrPP \times rrPP$ **d)** $RrPp \times RrPp$

Figure 20.28

Abraham Lincoln is believed to have suffered from Marfan's syndrome.

EFFECTS OF ENVIRONMENT ON PHENOTYPE

All genes interact with the environment. At times, it is difficult to identify how much of the phenotype is determined by the genes (nature) and how much is determined by the environment (nurture). Fish of the same species show variable numbers of vertebrae if they develop in water of different temperatures. Primrose plants are red if they are raised at room temperature, but become white when raised at temperatures above 30°C.

Himalayan rabbits are black when raised at low temperatures, but white when raised at high temperatures.

The water buttercup, *Ranunculus aquatilis*, provides another example of how genes can be modified by the environment. The buttercup grows in shallow ponds, with some of its leaves above and others below the water surface. Despite identical genetic information in the leaves above and beneath the water, the phenotypes differ. Leaves found above the water are broad, lobed, and flat, while those found below the water are thin and finely divided.

Figure 20.29

The sun may lighten hair and darken freckles.

Figure 20.30

The water buttercup is an example of how variation can be caused by the external environment. The submerged leaves are finely divided, compared with the ones growing in air. This variation can even occur within the same leaf.

SOCIAL ISSUE:
Genetic Screening

Before the discovery of insulin, many people who had the recessive gene for diabetes in the homozygous condition died before passing on their gene to their offspring. Today genetic screening can tell potential parents if they carry the gene. This information could help a couple decide whether or not they want to have children.

Huntington's chorea is a dominant neurological disorder that only begins to establish itself later in life. The disease is characterized by the rapid deterioration of nerve control, eventually leading to death. Early detection of this disease by genetic screening is possible.

Statement:

Compulsory genetic screening would improve the level of health in our society.

Point

- People are living longer thanks to new medicines. As a result, there are also more people who carry recessive genes for genetic disorders. Screening individuals who plan to have children could cut down on the incidence of these disorders.
- Genetic screening could provide valuable information about worker health and safety. Individuals who are susceptible to industrial chemicals, air pollutants, or nuclear radiation could be identified and then transferred to alternate types of employment.
- Identifying and analyzing human genes could enable doctors to begin treating patients long before the symptoms of disease become serious. This might limit the devastation caused by genetic diseases. The screening could also provide important information used for the diagnosis of diseases related to genetic disorders.

Counterpoint

- Screening individuals who plan to have children is an invasion of privacy. Neither doctors nor the state has the right to interfere in our personal lives to this extent.

- Identification of a genetic disease in an employee could cause early dismissal or end a promising career. All individuals are susceptible to the harmful effects of chemicals, pollution, and nuclear radiation; companies are obligated to provide a safe workplace for all their employees.
- Once genetic screening is done, who gets access to the results? Banks and other financial institutions might refuse to extend loans or give insurance coverage. Companies might only hire workers who can tolerate higher levels of toxic emissions rather than take measures to reduce such emissions.

Research the issue.
Reflect on your findings.
Discuss the various viewpoints with others.
Prepare for the class debate.

CHAPTER HIGHLIGHTS

- Genetics is the study of heredity.
- Gregor Mendel established the principles of genetics by observing the phenotypes arising from the crossing of successive generations of peas.
- The dominant trait will mask the effect of the recessive trait in heterozygous individuals.
- Incomplete dominance occurs when neither gene exerts a dominant influence.
- Hybrids combine characteristics present in both parents.
- The law of segregation states that the two alleles of a gene move to opposite poles during meiosis.

- The principle of independent assortment states that genes for different traits, which are located on separate chromosomes, are organized independently during gamete formation.
- The rules of probability can be applied to genetics.
- Some genetic traits are regulated by more than two possible alleles.
- Dihybrid crosses involve two distinct traits.
- Genes located on one chromosome can affect the expression of other genes that are located on different loci of the same chromosome or on independent chromosomes.

APPLYING THE CONCEPTS

1   Explain why Mendel's choice of the garden pea was especially appropriate.

2   Long stems are dominant over short stems for pea plants. Determine the phenotypic and genotypic ratios of the F_1 offspring from the cross-pollination of a heterozygous long-stem plant with a short-stem plant.

3   Cystic fibrosis is regulated by a recessive allele, c. Explain how two normal parents can produce a child with this disorder.

4   In horses, the trotter characteristic is dominant to the pacer characteristic. A male trotter mates with three different females, and each female produces a foal. The first female, a pacer, gives birth to a foal that is a pacer. The second female, also a pacer, gives birth to a foal that is a trotter. The third female, a trotter, gives birth to a foal that is a pacer. Determine the genotypes of the male, all three females, and the three foals sired.

5   For shorthorn cattle, the mating of a red bull and a white cow produces a calf that is described as roan. Roan is intermingled red and white hair. Many matings between roan bulls and roan cows produce cattle in the following ratio: 1 red, 2 roan, 1 white. Is this a problem of codominance or multiple alleles? Explain your answer.

6   For ABO blood groups, the A and B genes are codominant, but both A and B are dominant over type O. Indicate the blood types possible from the mating of a male who is blood type O with a female of blood type AB. Could a female with blood type AB ever produce a child with blood type AB? Could she ever have a child with blood type O?

7   Thalassemia is a serious human genetic disorder that causes severe anemia. The homozygous condition ($T^m T^m$) leads to severe anemia. People with thalassemia die before sexual maturity. The heterozygous condition ($T^m T^n$) causes a less serious form of anemia. The genotype $T^n T^n$ causes no symptoms of the disease. Indicate the possible genotypes and phenotypes of the offspring if a male with the genotype $T^m T^n$ marries a female of the same genotype.

8   For guinea pigs, black fur is dominant to white fur color. Short hair is dominant to long hair. A guinea pig that is homozygous for white and homozygous for short hair is mated with a guinea pig that is homozygous for black and homozygous for long hair. Indicate the phenotype(s) of the F_1 generation. If two hybrids from the F_1 generation are mated, determine the phenotypic ratio of the F_2 generation.

9 For chickens, the gene for rose comb (*R*) is dominant to that for single comb (*r*). The gene for feather-legged (*F*) is dominant to that for clean-legged (*f*). Four feather-legged, rose-combed birds mate. Rooster A and hen C produce offspring that are all feather-legged and mostly rose-combed. Rooster A and hen D produce offspring that are feathered and clean, but all have rose combs. Rooster B and hen C produce birds that are feathered and clean. Most of the offspring have rose combs, but some have single combs. Determine the genotypes of the parents.

10 For mice, the allele *C* produces color. The allele *c* is an albino. Another allele, *B*, causes the activation of the pigment and produces black color. The recessive allele, *b*, causes the incomplete activation of pigment and produces brown color. The alleles *C* and *B* are located on separate chromosomes and segregate independently. Determine the F_1 generation from the cross *CcBb* × *CcBb*.

CRITICAL-THINKING QUESTIONS

1 Baldness (H^B) is dominant in males but recessive in females. The normal gene (H^n) is dominant in females, but recessive in males. Explain how a bald offspring can be produced from the mating of a normal female with a normal male. Could these parents ever produce a bald girl? Explain your answer.

2 Use the phenotype chart at the right to answer the following questions.
   a) How many children do the parents A and B have?
   b) Indicate the genotypes of the parents.
   c) Give the genotypes of M and N.

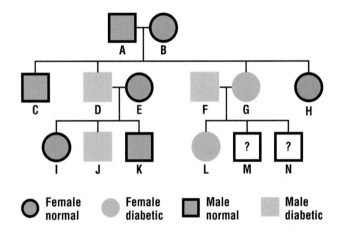

Female normal   Female diabetic   Male normal   Male diabetic

3 In Canada, it is illegal to marry your immediate relatives. Using the principles of genetics, explain why inbreeding in humans is discouraged.

ENRICHMENT ACTIVITIES

1 The ability to curl the tongue in a U shape is controlled by the dominant allele *T*. The recessive allele *t* results in the inability to curl the tongue. Find an individual who is unable to curl his or her tongue. Test other family members and construct a pedigree chart.

2 Suggested reading:
   • Diamond, Jared. "Curse and Blessing of the Ghetto." *Discover*, March 1991, p. 60. An excellent case study of Tay-Sachs disease.
   • Grady, Denis. "The Ticking of a Time Bomb in Your Genes." *Discover*, June 1987, p. 26. An interesting article on the identification of the gene for Huntington's chorea.

*T*he Source of Heredity

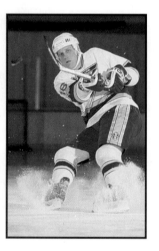

IMPORTANCE OF THE CHROMOSOMAL THEORY

Early scientists believed that hereditary traits were located in the blood. The term "pure bloodline" is still used today and is a reminder of this early misconception. Genes, which are located along the thread like chromosomes found in the nucleus of each cell, are responsible for producing or influencing specific traits in offspring. A description of the structure and location of the gene is essential to an understanding of how it works. Structure provides important clues to function.

During the Dark Ages, strict laws and social pressures prohibited the dissection of corpses; the dead were to be left alone. Yet, in spite of the laws, early physicians and scientists performed dissections secretly in caves. They began to sketch and label different parts of the body, compiling a guide to anatomy in the process. As a composite structure of organs began to appear, theories about function also emerged.

Early geneticists were driven by the same desire for knowledge, but, unlike anatomists, they were not able to see much of the structure they were studying. Science

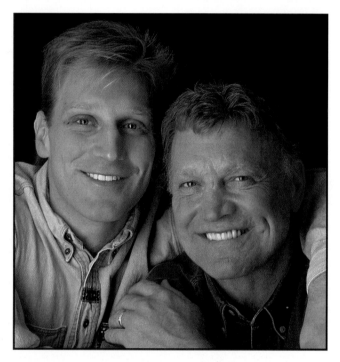

Figure 21.1

Bobby Hull and his son Brett have starred in the National Hockey League, becoming the first father and son to win the Hart trophy. However, genetic heritage is only part of the equation. Brett Hull has had to work very hard for many years to fulfill his potential as a hockey player.

had to wait for technology to catch up. The development of the light microscope allowed genetic investigations to progress. In this chapter and the next, you will discover how science and technology are intertwined. The light microscope, electron microscope, and biochemical analysis techniques, such as X-ray diffraction and gel electrophoresis, have provided a more complete picture of the mechanisms of gene action.

CYTOLOGY AND GENETICS

Over 2000 years ago, the Greek philosopher Aristotle suggested that heredity could be traced to the power of the male's semen. He believed that hereditary factors from the male outweighed those from the female. Other scientists speculated that the female determined the characteristics of the offspring and that the male gamete merely set events in motion. By 1865, the year in which Mendel published his papers, many of these misconceptions had been cleared up. Nineteenth-century biologists knew that the egg and sperm unite to form a new individual, and it was generally accepted that factors from the egg and sperm were blended in developing the characteristics of the offspring. However, Mendel knew nothing about the structure or location of the hereditary material. He did not know about meiosis, nor did he understand how the genetic code worked. Yet, in spite of the limited knowledge of the day, he set geneticists on the right path. Unfortunately, the lack of complementary information and a comprehensive theory of gene action meant that Mendel's work was interpreted as a mere experiment with garden peas.

About the same time Mendel was conducting his experiments with garden peas,

new techniques in lens grinding were providing better microscopes. A new branch of biology, called cytology, began to flourish. The nucleus was discovered in 1831, just 34 years before Mendel published his results. In 1882, Walter Fleming described the separation of threads within the nucleus during cell division. He called the process *mitosis*. In the same year, Edouard van Benden noticed that sperm and egg cells of roundworms had two chromosomes, but the fertilized egg had four chromosomes. By 1887, August Weisman offered the theory that a special division took place in sex cells. The reduction division is now known as *meiosis*. Weisman had added an important piece to the puzzle of heredity, and in doing so, provided a framework in which Mendel's work could be understood. When scientists rediscovered Mendel's experiments in 1900, the true significance of his work became apparent.

DEVELOPMENT OF THE CHROMOSOMAL THEORY

In 1902, an American biologist, Walter S. Sutton, was studying gamete formation in grasshoppers. At the same time, Theodor Boveri, a German scientist, was studying gamete formation independently of Sutton. Both scientists observed that chromosomes came in pairs, and that they segregated during meiosis. The chromosomes then formed new pairs when the egg and sperm united. The concept of paired, or homologous, chromosomes supported Mendel's two-factor explanation of inheritance. Today, these factors are referred to as genes. One factor, or gene, for each characteristic must come from each sex cell.

The union of two different factors in offspring and the formation of new com-

binations of factors in succeeding generations could be explained and supported by cellular evidence. The behavior of chromosomes during gamete formation could help explain Mendel's law of segregation and law of independent assortment.

Sutton and Boveri knew that the egg was much larger than the sperm, but that the expression of a trait was not tied to just the male or just the female sex cell. Some structure in both the sperm cell and the egg cell must determine heredity. Sutton and Boveri deduced that Mendel's factors (genes) must be located on the chromosomes. The fact that humans have 46 chromosomes, but thousands of different traits, led Sutton to hypothesize that each chromosome contains many different genes.

CHROMOSOMAL THEORY

The development and refinement of the microscope led to advances in cytology and the union of two previously unrelated fields of study: cell biology and genetics. As you continue reading and exploring genetics, you will discover how other branches of science, such as biochemistry and nuclear physics, have also become integrated with genetics.

The chromosomal theory of inheritance can be summarized as follows:
- Chromosomes carry genes, the units of hereditary structure.
- Paired chromosomes segregate during meiosis. Each sex cell or gamete has half the number of chromosomes found in a **somatic cell.**
- Chromosomes assort independently during meiosis. This means that each gamete receives one of the pairs and that one chromosome has no influence on the movement of a member of another pair.

Somatic cells *are all the cells of an organism except the sex cells.*

- Each chromosome contains many different genes.

MORGAN'S EXPERIMENTS

Few people have difficulty distinguishing gender in humans. Mature females look very different from mature males. (Even immature females can be identified on the basis of anatomy.) But can you distinguish whether a blood cell or cheek cell comes from a female or a male? The work of the American geneticist Thomas Hunt Morgan (1866–1945) provided a deeper understanding of gender and the inheritance of some characteristics.

Morgan was one of many geneticists who used the tiny fruit fly, *Drosophila melanogaster*, to study Mendel's principles of inheritance. There are several reasons why the fruit fly is an ideal subject for study. First, the fruit fly reproduces rapidly. Females lay over 100 eggs after mating, and each offspring is capable of mating shortly after leaving the egg. The large number of offspring is ideally suited for genetics, which is based on probability. Because *Drosophila*'s life cycle tends to be only 10 to 15 days, it is possible to study many generations in a short period of time. A second benefit arises from the small size of the *Drosophila*. Many can be housed in a single test tube. A small, solid nutrient at the bottom of the test tube can maintain an entire community. The third and most important quality of *Drosophila* is that males can easily be distinguished from females.

Morgan discovered various mutations in *Drosophila*, noting that some of the mutations seemed to be linked to other traits. Morgan's observations added support to the theory that the genes responsible for the traits were located on the chromosomes.

While examining the eye color of a large number of *Drosophila*, Morgan noted the appearance of a white-eyed male among many red-eyed offspring. He concluded that the white-eyed trait must be a mutation. Morgan was interested in tracing the action of the white-eyed gene, so he mated the male with a red-eyed female. All members of the F_1 generation had red eyes. Normal Mendelian genetics indicated that the red gene was dominant. Most researchers might have stopped at that point, but Morgan did not. Pursuing further crosses and possibilities, he decided to mate two hybrids from the F_1 generation. An F_2 generation produced 3/4 red and 1/4 white, a ratio that could again be explained by Mendelian genetics. But further examination revealed that all the females had red eyes. Only the males had white eyes. Half of the males had red eyes and half had white eyes. Did this mean that the phenotype white only appeared in males?

With subsequent crosses, Morgan was able to prove that females can express the white mutation. However, when he crossed a white male with a red female, all members of the F_1 generation had red eyes, and only males of the F_2 generation had white eyes. Why could males express the white-eyed trait but not females? How did the pattern of inheritance differ between males and females? To find an answer, Morgan turned to cytology.

Morgan selected the giant cells of the salivary glands of *Drosophila*. The chromosomes were stained and viewed under the microscope. *Drosophila* have eight chromosomes. Morgan observed that the females had four homologous pairs. The males, however, had only three homologous pairs; the final pair, the sex chromosomes, were not completely homologous. A small, hooked-shaped chromosome, the Y chromosome, was unique to males. Females have two X chromosomes, while males have a single X and a Y. Morgan concluded that because the X and Y chromosomes were not completely homologous, they must contain different genes.

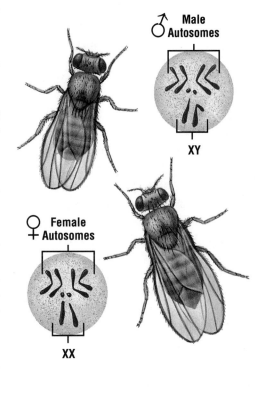

Figure 21.3

Drosophila contain three paired chromosomes called autosomes, and a single set of sex chromosomes. In females, the sex chromosomes are paired X chromosomes, but in males, the sex chromosomes are not completely homologous. Males have a single X and a single Y chromosome.

Morgan reasoned that the traits governing eye color in *Drosophila* must be located on the X chromosomes. Traits located on sex chromosomes are called **sex-linked traits.** The Y chromosome, because it is not completely homologous, does not carry the trait. The initial problem can now be re-examined. The red-eyed female can be indicated by the genotype $X^R X^R$ and the white-eyed male by the genotype $X^r Y$. The symbol X^R indicates that the red gene is dominant

Figure 21.2

In *Drosophila*, the gene that codes for white eyes is recessive to the gene that codes for red eyes.

Sex-linked traits *are controlled by genes located on the sex chromosomes.*

and located on the X chromosome. Notice no gene for eye color is expressed on the hook-shaped Y chromosome. A Punnett square, as shown in Figure 21.4, can be used to determine the genotypes and the phenotypes of the offspring. All members of the F_1 generation have red eyes. The females have the genotype $X^R X^r$, and the males have the genotype $X^R Y$.

The F_2 generation is determined by a cross between a male from the F_1 generation and a female from the F_1 generation. Examine the males of the F_1 and F_2 generations. Do the males inherit the trait for eye color from the mother or father? The male offspring always inherit a sex-linked trait from the mother. The father supplies the Y chromosome, which makes them a male.

The F_2 male *Drosophila* are $X^R Y$ and $X^r Y$. The females are either homozygous for red eye color, $X^R X^R$, or het-

erozygous for red eye color, $X^R X^r$. Although Morgan did not find any white-eyed females from his initial cross, white-eyed females do occur in nature. To get a white-eyed female, a female with at least one white gene must be crossed with a white-eyed male. The $X^r X^r$ genotype is possible. Notice that females have three possible genotypes, but males have only two. Males cannot be heterozygous for a sex-linked trait.

Females	Males
$X^R X^R$	$X^r Y$
$X^R X^r$	$X^R Y$
$X^r X^r$	$X^r Y$

SEX DETERMINATION

Sex-linked genes are also found in humans. Red-green color blindness is

Figure 21.4

Punnett squares showing F_1 and F_2 generations for a cross between a homozygous red-eyed female and a white-eyed male.

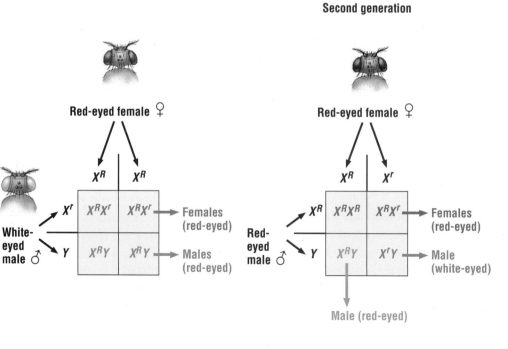

Second generation

Unit Six:
Heredity

determined by a recessive gene located on the X chromosome. The reason males are color-blind more often than females is that females require two recessive genes to exhibit color blindness. On the other hand, because males have only one X chromosome, the incidence of color blindness is greater for them.

Does the fact that females have two X chromosomes make them more likely to survive than males? To an extent, yes. More male than female babies die in infancy, for example. However, the ratio of males to females in the human population is still very close.

Sex chromosomes

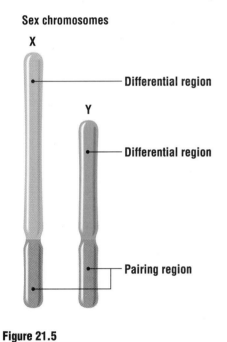

Figure 21.5

Sections of the X and Y chromosomes are homologous; however, few genes are common to both chromosomes.

The difference between a male and female lies within the X and Y chromosomes. Do female cells differ from male cells? A dark spot, called a **Barr body,** can be seen in the somatic cells of female mammals during interphase. Male cells do not have a Barr body. In 1961, the British geneticist, Mary Lyon, proposed that the dark spot was actually a dormant X chromosome. Lyon hypothesized that

one of the X chromosomes becomes inactive, but that the inactive chromosome varies, depending on the cell. This means that all female cells are not identical, depending on which of the two X chromosomes is active. Lyon's hypothesis has since been tested and confirmed. In many ways, the female body is a mosaic, made of segments of genetically different cells. Heterozygous females who carry a lethal gene may escape the disease because the gene is only active in approximately 50% of their cells.

The example of calico cats supports Lyon's explanation. Male cats tend to be black (X^BY) or orange (X^OY). Female cats, by contrast, can be black (X^BX^B), orange (X^OX^O), or calico (X^BX^O). (Calico is a patchwork blending of the black and orange genes.) Areas in which the orange genes are active appear orange, while those areas in which the black genes are active appear black. Males cannot be heterozygous, but occasionally a male calico appears. These rare males are believed to contain an extra X chromosome. The explanation is supported by the fact that these males are always sterile.

You have 46 chromosomes. If you are a female, you have 23 pairs of homologous

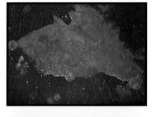

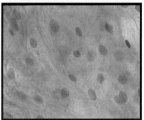

Figure 21.6

Barr bodies are found in the cells of female mammals.

Barr bodies *are small, dark spots of chromatin, located in the nuclei of female mammalian cells.*

Figure 21.7

The calico coat color of the female cat is the result of a blending of the genes for black and those for orange.

chromosomes: 22 autosomes and two X sex chromosomes. If you are a male, you have only 22 pairs of homologous chromosomes and one X and one Y sex chromosome. It has been estimated that the human X chromosome carries between 100 and 200 different genetic traits. The Y chromosome, much smaller than the X chromosome, carries the information that determines gender. In 1987, geneticists pinpointed the location of the gene that specifies gender. This gene was named the testes determining factor (TDF).

A male fetus does not differ from a female fetus until the sixth or seventh week of pregnancy. At this point, if the TDF gene is activated, it initiates production of a protein that stimulates the testes to begin secreting male hormones. These hormones cause the development of male characteristics. It is interesting to note that the testes develop inside the body cavity in the same location as the ovaries in the female fetus, and descend only when activated. It would appear that the absence of the TDF gene causes the development of the female characteristics. However, things may not be quite this simple. Researchers have speculated that the activation of the female system might be even more complicated than that of the male system. Although some genetic differences for gender have been identified, many questions remain unanswered. For example, are gender-determining genes located on the X chromosome? Does the Y chromosome receive any information from the X chromosome in males?

▮ REVIEW QUESTIONS ▮ ?

1 A recessive sex-linked gene (h) located on the X chromosome increases blood-clotting time. This causes the genetic disease, hemophilia.

a) Explain how a hemophilic offspring can be born to two normal parents.

b) Can any of the female offspring develop hemophilia?

2 A mutant sex-linked trait called "notched" (N) is deadly in *Drosophila* when homozygous in females. Males who have a single (N) allele will also die. The heterozygous condition (Nn) causes small notches on the wing. The normal condition in both males and females is represented by the allele n.

a) Indicate the phenotypes of the F_1 generation from the following cross: $X^n X^N \times X^n Y$.

b) Explain why dead females are never found in the F_1 generation, no matter which parents are crossed.

c) Explain why the mating of female $X^n X^N$ and a male $X^N Y$ is unlikely.

3 In *Drosophila*, eye color is determined by two different genes located on different chromosomes. A recessive gene (b), found on the second chromosome, produces brown eye color. The second chromosome is an autosomal chromosome. A recessive sex-linked gene (v) causes vermilion eye color to be produced. The presence of the dominant genes (B) and (V) results in a wild-type color. The presence of both the brown and vermilion alleles in the homozygous condition results in an individual with white eyes. Indicate the genotypes and phenotypes produced from the following crosses:

a) BB,VV (wild-type female) $\times$ bb,vY (white-eyed male)

b) Bb,Vv (wild-type female) $\times$ Bb,VY (wild-type male)

c) bb,VV (brown-eyed female) $\times$ Bb,vY (vermilion-eyed male)

GENDER VERIFICATION AT THE OLYMPICS

The need for gender verification at sports competitions might seem unnecessary in light of the obvious differences between male and female anatomy. Although most physical exams can readily identify gender, there are situations in which a true distinction between male and female is not as easy as it might appear.

Although the sex chromosomes help define gender, not all of the genes that code for sex are located on the sex chromosomes. The autosomes carry some traits that control the production of sex hormones. Males carry genes that code for female sex hormones. Most organisms contain genes for both sexes.

Gender is determined by both the genetic makeup of the person and the development of the sex organs. Females with Turner's syndrome, which is characterized by a single X chromosome, do not have Barr bodies. (It is believed that the Barr body results from a second X chromosome, which is inactive.) If the test for gender were based on the presence or absence of Barr bodies, individuals with Turner's syndrome would not be classified as females. However, a karyotype would prove that they were females.

An extreme example of difficulty in gender verification arises from testicular feminization syndrome. These XY individuals appear to be females, but show a positive test for male Y chromatin. The syndrome is caused by a gene mutation on the X chromosome that acts only in XY zygotes. These individuals do not react to injections of male sex hormones, and, therefore, do not display the muscle development associated with males.

The present method of testing for gender at the Olympics is done by a buccal mucosa smear, which uses scrapings from the inside of the mouth. The smear is examined for evidence of X and Y chromatin. This testing program has received considerable criticism. Opponents claim that the method has technical pitfalls and increases the risk of disease transmission. A much less expensive and more reliable screening method would be a simple physical inspection by a physician.

Figure 21.8

Gender verification at the Calgary Winter Olympics in 1988 was done by a buccal mucosa smear.

LABORATORY
HUMAN SEX-LINKED GENES

Objective

To investigate the inheritance of color blindness.

Materials

color-blindness chart

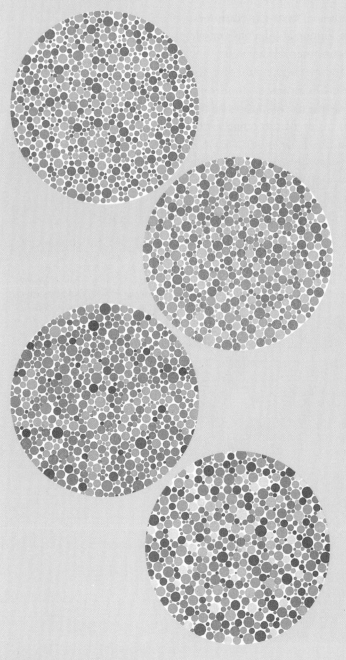

Procedure

1   Look at the color-blindness chart. The chart is composed of a number of colored circles. Some of the circles in each of the charts combine to form numbers.
2   Copy the table in your data book and complete.

a)

Plate	Number identified	Actual number
1		
2		
3		
4		

3   Combine class results, and copy and complete the second data table. (This laboratory works best if two or three class results can be combined.)

b)

Gender	Total number of individuals	Number identified	Number not identified
Females			
Males			

Laboratory Application Questions

1   How would your laboratory results differ if color blindness were not sex-linked?
2   Explain why a woman who is not color-blind, but whose father was color-blind, can give birth to a son who is color-blind.
3   Diabetes is caused by a recessive gene located on an autosomal chromosome. You already know that color blindness is caused by a recessive sex-linked trait. Explain why the ratio of women to men who have diabetes is much closer than the ratio of women to men who have color blindness.
4   Hemophilia A, another sex-linked disorder, is very rare in females, yet color blindness is fairly common. Explain why the color-blindness gene is more common. ■

GENE LINKAGE AND CROSSING-OVER

It is often said that great science occurs because good questions are asked. Thomas Hunt Morgan, like Mendel, asked great questions. Such questions were raised when Morgan obtained a few odd gene combinations while performing dihybrid crosses with *Drosophila*. The appearance of combinations of recessive traits in very small numbers challenged the principles of Mendelian genetics. The mating of fruit flies that were homozygous for wild-type body coloring and straight wings (*AABB*) with those that had black body color and curved wings (*aabb*) produced offspring that were heterozygous for both traits (*AaBb*). Is it possible to get a member of the F_2 generation that demonstrates a recessive phenotype such as curved wings (*bb*) along with the dominant gene for wild-type body color? How would you explain it if this combination occurred in only 9 of 300 offspring, as Morgan found? The phenotypic ratios of the offspring could not be explained in terms of a normal 9:3:3:1 ratio. The phenotypes of the F_2 generation could not be explained by normal Mendelian genetics.

By studying the frequencies of other genes located on similar chromosomes, Morgan was able to provide a tentative explanation to these questions. His work with sex-linked traits established that specific genes are located on specific chromosomes. For example, eye color for *Drosophila* is located on an X chromosome. Morgan knew that two or more genes located on nonhomologous chromosomes segregate independently during meiosis. But what about genes located on the same chromosome? Should these genes not segregate together? Figure 21.9 shows what should happen during normal segregation.

Morgan concluded that the gene for wing shape and the gene for body color were located on the same chromosome. The two genes did not segregate independently. Genes located on the same chromosome tend to be transmitted together. These are referred to as **linked genes.**

However, an event called *crossing-over* can provide new combinations. Crossing-over occurs during synapsis of meiosis. This means that a single chromosome can change as it passes from generation to generation. In Figure 21.10, consider the blue-colored chromosome to be the one inherited from the father and the red chromosome to be the one inherited from the mother. Those gametes with chromosomes that have recombined have sections of the chromosome that are maternal (coming from the mother) and other sections that are paternal (coming from the father).

Linked genes *are located on the same chromosome.*

Figure 21.9

Complete linkage. During normal meiosis, the homologous blue and red chromosomes move to opposite poles. One sex cell inherits the *AB* alleles, while the other inherits the *ab* alleles.

Figure 21.10

Incomplete linkage. Gametes with the gene combination *Ab* and *aB* would not be possible without crossing-over.

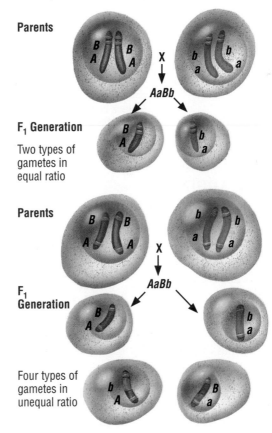

Armed with the knowledge of recombination of genes due to crossing-over, we can re-examine the problem that faced Morgan. Consider the mating between fruit flies with wild-type body color and straight wings (*AABB*) and those with black body color and curved wings (*aabb*). If the genes are located on the same chromosome we would expect all members of the F1 generation to be AaBb. In the F2 generation, AABB, AaBb, and aabb are possible. However, Morgan found a few F2 individuals that were Aabb or aaBb. The new combinations can be explained in terms of crossing-over.

The frequency of crossing-over can be stated as a percentage:

$$\text{Crossover \%} = \frac{\text{number of recombinations}}{\text{total number of offspring}} \times 100\%$$

$$= \frac{18}{300} \times 100\% = 6\%$$

Figure 21.11

New combinations can be explained in terms of crossing-over.

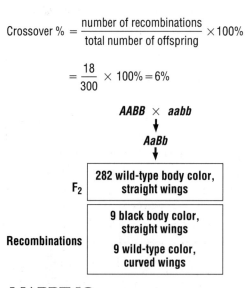

AABB × *aabb*

↓

AaBb

↓

F₂ — 282 wild-type body color, straight wings

Recombinations —
9 black body color, straight wings
9 wild-type color, curved wings

MAPPING CHROMOSOMES

Genes located on the same chromosome segregate with each other. A recessive characteristic like white eyes for *Drosophila* can often be used as a **gene marker.** White eyes can be easily observed in the offspring. However, some genes do not code for traits that are readily observable. These genes can be tracked by the use of marker genes located on the same chromosome. The appearance of white eyes in offspring may

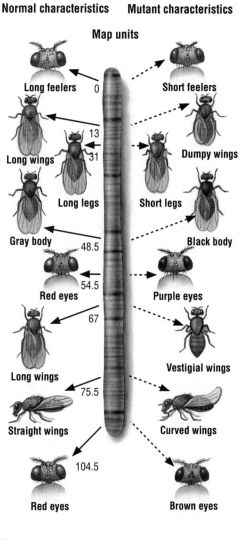

Normal characteristics **Mutant characteristics**

Map units

Long feelers — 0 — Short feelers

— 13 —
Long wings — 31 — Dumpy wings

Long legs — Short legs

Gray body — 48.5 — Black body

— 54.5 —
Red eyes — Purple eyes

— 67 —

Vestigial wings

Long wings — 75.5 —

Straight wings — Curved wings

— 104.5 —

Red eyes — Brown eyes

Figure 21.12

Gene mapping of chromosome number 2 for *Drosophila melanogaster*. Note that many genes are located on one chromosome.

also provide a key to many other genes located on the same chromosome.

Unfortunately, crossing-over can alter gene linkages along a chromosome. If new segments of DNA are exchanged at the site of the marker, the mapping becomes impossible. By following the offspring from a variety of crossovers, geneticists were able to determine that genes located near each other almost always ended up on the same chromosome. In contrast, genes separated from each other by greater distances were more likely to be affected by crossing-over. This means that a crossover value of 1% indicates that the genes are close to each other, while a value of 12% indicates that the genes are much farther apart. The crossover frequency can help scientists determine the relative position of the genes along the chromosome. The greater the frequency of crossover, the greater the **map distance.** A crossover frequency of 5% means that the two genes are 5 map units apart. A crossover frequency of 16% means that the genes are much farther apart: 16 map units.

The frequency of crossovers can be used to determine gene maps. Consider the following problem. Assume the crossover frequency between genes *A* and *B* is 12%, between *B* and *C* is 7%, and between *A* and *C* is 5%. The fact that map distances are additive allows you to determine the proper sequence of genes. If you assume that gene *A* is in the middle, then the sum of the distances between *B* and *A*, and *A* and *C* must equal the distance between *B* and *C*. These distances are not equal. Therefore, gene *A* is not in the middle.

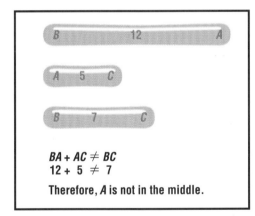

BA + AC ≠ BC
12 + 5 ≠ 7
Therefore, *A* is not in the middle.

If you assume that gene *B* is in the middle, then the sum of the distances between *A* and *B*, and *B* and *C* must equal the distance between *A* and *C*. These distances are not equal. Therefore, gene *B* is not in the middle.

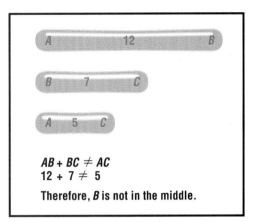

AB + BC ≠ AC
12 + 7 ≠ 5
Therefore, *B* is not in the middle.

If you assume that *C* is the middle gene, then the sum of the distances between *A* and *C*, and *C* and *B* must equal the distance between *A* and *B*. These distances are equal. Therefore, gene *C* is in the middle.

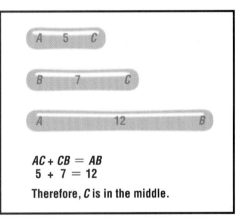

AC + CB = AB
5 + 7 = 12
Therefore, *C* is in the middle.

The development of special stains for chromosomes has also advanced gene mapping. These stains give the chromosomes distinctive light- and dark-colored bands, which can be used to identify the position of distinctive genes along a particular chromosome.

Map distance *refers to the distance between two genes along the same chromosome.*

B I O L O G Y C L I P
To date, only 0.1% of all human genes have been mapped.

CASE STUDY

MAPPING CHROMOSOMES

Objective

To use crossover frequencies to construct a gene map.

Background Information

A.H. Sturtevant, a student who worked with Thomas Morgan, made the following hypotheses:

Genes are located in a linear series along a chromosome, much like beads on a string. Genes that are closer together will be separated less frequently than those that are farther apart. Crossover frequencies can be used to construct gene maps.

Sturtevant's work with *Drosophila* helped establish techniques for mapping chromosomes.

Procedure

1 Examine the diagram of the chromosome shown. Crossing-over takes place when breaks occur in the chromatids of homologous chromosomes during meiosis. The chromatids break and join with the chromatids of their homologous chromosomes. This causes an exchange of alleles between chromosomes.

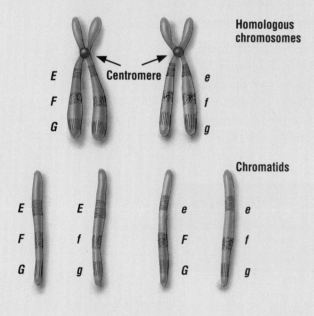

a) Indicate the areas of the chromatids that show crossing-over.

b) According to the diagram, which genes appear farthest apart? (Choose from *EF*, *FG*, or *EG*.)

c) Which alleles have been exchanged?

2 In 1913, Sturtevant used crossover frequencies of *Drosophila* to construct chromosome maps. To determine map distances, he arbitrarily assigned one recombination for every 100 fertilized eggs. For example, genes that had a crossover frequency of 15% were said to be 15 units apart. Genes that had 5% recombinations were much closer: 5 units apart.

d) Using the following data, determine the distance between genes *E* and *F*.

Cross	*EF* × *ef*
Offspring	*EF* + *ef* (from parent) = 94%
Frequency	*Ef* + *eF* (recombination) = 6%

3 Use the data table below to construct a complete gene map.

Cross	Offspring	Frequency
EF × *ef*	*EF* + *ef* (from parent)	94%
	Ef + *eF* (recombination)	6%
EG × *eg*	*EG* + *eg* (from parent)	90%
	Eg + *eG* (recombination)	10%
FG × *fg*	*FG* + *fg* (from parent)	96%
	Fg + *fG* (recombination)	4%

e) What is the distance between genes *E* and *G*?

f) What is the distance between genes *F* and *G*?

Case-Study Application Questions

1 What mathematical evidence indicates that gene *F* must be found between genes *E* and *G*?

2 Draw the gene map to scale. (1 cm = 1 unit)

3 For a series of breeding experiments, a linkage group composed of genes *W, X, Y,* and *Z* was found to show the following gene combinations. (All recombinations are expressed per 100 fertilized eggs.)

Genes	*W*	*X*	*Y*	*Z*
W	–	5	7	8
X	5	–	2	1
Y	7	2	–	1
Z	8	3	1	–

Construct a gene map. Show the relative positions of each of the genes along the chromosome and indicate distances in gene units.

4 For a series of breeding experiments, a linkage group composed of genes *A, B, C,* and *D* was found to show the following gene combinations. (All recombinations are expressed per 100 fertilized eggs.)

Genes	*A*	*B*	*C*	*D*
A	–	12	15	4
B	12	–	3	8
C	15	3	–	11
D	4	8	11	–

Construct a gene map. Show the relative positions of each of the genes along the chromosome and indicate distances in gene units. ■

GENE RECOMBINATIONS IN NATURE

Scientists once believed that, with the exception of a few new combinations that might occur because of crossing-over, chromosome structure was fixed. However, in the late 1940s, Barbara McClintock, an American biologist, interpreted her results of experiments with Indian corn and came to a conclusion that would shatter the traditional view of gene arrangement on chromosomes. McClintock suggested that genes can move to a new position. Her theory was dubbed the "jumping gene theory."

Color variation in the kernels of Indian corn led McClintock to hypothesize the existence of transposable elements, called **transposons.** The insertion of some genes into a new position along a chromosome inactivated the genes that affect the production of pigment. Despite McClintock's well-documented experiments, the vast majority of the scientific community ignored her work or dismissed her results. The idea of jumping genes challenged the widely accepted theory that genes are fixed in position along the chromosomes. Undaunted by her critics, McClintock continued to gather data on jumping genes. Her contributions to science were finally recognized in 1983, when McClintock was awarded the Nobel Prize for physiology and medicine.

Transposons *are specific segments of DNA that can move along the chromosome.*

Figure 21.13

In the late 1940s, Barbara McClintock's theory of "jumping genes" was greeted with tremendous skepticism by the traditional scientific community.

Transposable genes have since been discovered in other organisms. Bacteria can insert genes randomly along a circular chromosome. The transposon can also move from the chromosome in one bacterium to a chromosome in another bacterium. The inserted genes can even be integrated with a secondary structure of DNA, called a **plasmid,** located in a bacterium. The plasmid is a small ring of genetic material, and can be considered extra DNA.

Some bacteria are capable of a form of sexual reproduction called **conjugation.** During bacterial conjugation, two or more cells fuse, and plasmids are passed from one cell to another. This special property has been exploited by geneticists using revolutionary new technology. But bacterial conjugation has also created some serious consequences for humans. Some disease-causing bacteria have evolved genes that are resistant to certain types of antibiotics, such as penicillin. Under normal conditions, antibiotics interfere with chemical reactions that occur within these harmful microbes. Genes that provide resistance to an antibiotic permit the disease-causing bacteria to continue living and wreaking havoc for the cells of the body.

Plasmids are small rings of genetic material.

Conjugation is a form of sexual reproduction in which genetic material is exchanged between two cells.

GENE SPLICING TECHNIQUES AND GENE MAPPING

Prior to the 1970s, geneticists studied the inheritance of traits within successive generations of families. Earlier in the chapter, you learned that genes located on the same chromosome are usually inherited together. Recessive genes can be used as markers to follow the inheritance of the linked trait that is being studied; however, linked traits are often disrupted by crossing-over. As described earlier, geneticists have been able to turn the anomaly of crossing-over into an advantage. The closer two genes are on a specific chromosome, the less likely they are to separate during crossing-over. The frequencies of crossing-over can be used to determine gene position. However, the process of following genes during crossover events is tedious and time-consuming. Two recent innovations have increased the speed at which genes can be mapped.

Scientists have been able to fuse the cell of a human with the cell of a mouse. Although the hybrid cell tends to lose most of the human chromosomes during cell replication, all of the rodent chromosomes are maintained. In most cells, a single human chromosome remains, but in each case it appears to be a different chromosome. The production of a human protein by the hybrid cell led geneticists to realize that the genes controlling the production of that protein must exist somewhere on the chromosome that remains in the hybrid cell.

The other major breakthrough in gene mapping technique came on the heels of recombinant DNA research. You will read more about this technology in the next chapter. Special enzymes are used to chop strands of DNA into small segments. Remember that each enzyme

Figure 21.14

Bacteria exchange genetic material during conjugation. The long appendage between the bacteria is called a *pilus*. The pilus will draw the cells closer together and act as a bridge for the exchange of genetic material.

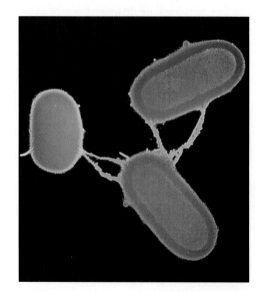

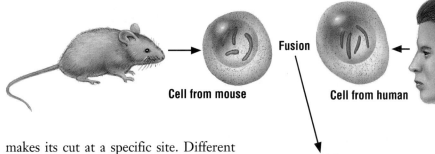

Cell from mouse   Fusion   Cell from human

Figure 21.15

The fusion of a human cell with a mouse cell.

Single remaining human chromosome directs the cell's manufacture of protein.

makes its cut at a specific site. Different pieces of DNA are referred to as restriction-fragment-length-polymorphisms, or RFLPs for short. The word polymorphism refers to the fact that each segment has a different length. Because each of us has a distinctive length of RFLP segments, these segments can be used to identify individual people.

Like early cartographers who mapped the seas by using the stars, gene mappers have used RFLP markers to determine the position of genes on a specific site, or locus, on chromosomes. By combining two other existing technologies— radioactive labelling and gel electrophoresis— with the use of restriction enzymes, scientists have been able to significantly increase the rate at which gene maps are made. Radioactive labels can be attached to one end of the DNA. Strands of DNA are then added to four separate test tubes. Each test tube contains specific enzymes that cleave specific bond sites.

Gel electrophoresis was originally developed by Linus Pauling to identify proteins. However, the same principles also apply to strands of DNA. The DNA segments are removed from the test tubes and applied to a thin layer of gel. An electrical field pulls the pieces of DNA through the gel. The smaller pieces of DNA move faster and farther than the larger ones. The final positions of the

BIOLOGY CLIP

Using biochemical analysis techniques, a team of British scientists has discovered a gene that causes obesity. This gene increases the speed at which food is converted into fat.

DNA segments can be identified by the radioactive label that reacts with the gel to form a distinctive banding pattern. This pattern of banding can be used to determine the sequence of the DNA.

Although technological improvements have greatly increased the speed at which gene maps can be constructed, scientists are far from understanding the complete human **genome.** Of the estimated 100 000 genes that reside on your chromosomes, only about 4500 have been identified. The number mapped changes every day. Even with accelerated techniques, an estimated 1000 person years of effort are still required to map all the genes on your chromosomes. However, some progress has been made. Scientists have already found genes for Huntington's chorea, cystic fibrosis, and Dutchenne's muscular dystrophy. They have also located the genes that make people susceptible to Alzheimer's disease, some forms of schizophrenia, and cancer.

Figure 21.16

A scientist works under ultraviolet light during the preparation of an electrophoresis gel used in DNA separation techniques.

*The **genome** is the complete set of instructions contained within the DNA of an individual.*

RESEARCH IN CANADA

The Cystic Fibrosis Gene

Cystic fibrosis, one of the most serious respiratory disorders, may well be eliminated because of the work of a group of Toronto doctors. Working with a research group from Michigan, Drs. Lap-Chee Tsui, Frank Collins, and Jack Riordin have located the gene responsible for cystic fibrosis. Cystic fibrosis causes secretions of sticky mucus that can block the airways and create digestive problems. It is the most common of all hereditary disorders in Canada, affecting about 1 in 2000 children. One estimate indicates that 1 in every 20 people carry the defect in a heterozygous condition. The recent scientific breakthrough means that carriers can now be identified by gene mapping. Counselling can be provided for couples who might have a child with cystic fibrosis. Although the discovery of the gene's location does not mean a cure, it does pinpoint the source of the problem. As new techniques become available, it may one day be possible to turn off the defective gene or remove it from the chromosome.

**DR. JACK RIORDIN
DR. FRANK COLLINS
DR. LAP-CHEE TSUI**

FRONTIERS OF TECHNOLOGY: GENE THERAPY

Over 3500 different genetic diseases have so far been linked to single defective genes. Cystic fibrosis, diabetes, hemophilia, Huntington's chorea, and sickle-cell anemia are but a few of the hereditary diseases known in Canada. Although treatments, such as insulin injections for diabetes, control the disease, no true cure has been found. Medical advances, coupled with modified diet and environmental adjustments, have enabled people with such diseases to continue living productive lives.

Imagine the potential of transforming a defective gene! Although **gene therapy** is in the early stages of development, the prospect of gene correction provides exciting possibilities in the quest to conquer genetic diseases. There are three possible strategies for gene therapy. The first strategy involves *gene insertion*. The normal gene is inserted into position on the chromosome of a diseased cell. Because not every cell uses a particular gene, the insertion can be restricted to those cells in which the gene is active. For example, diabetes occurs when the cells of the pancreas do not produce sufficient amounts of insulin. The normal gene for insulin production need only be inserted into specialized cells within the pancreas. The gene would not have to be inserted into a muscle cell as muscle cells do not produce insulin. A second method involves *gene modification*. The defective gene is modified chemically in an effort to recode the genetic message. This method is much more delicate and requires greater knowledge about the chemical composition of the normal and defective genes. The third technique, *gene surgery*, is the most ambitious: the defective gene is extracted and replaced with a normal gene.

The first attempt at gene therapy dates back to 1980, when Dr. Martin Cline attempted to help two patients who had the genetic disorder called thalassemia (Cooley's anemia). The disorder is characterized by a defective hemoglobin molecule. Hemoglobin, the molecule of red blood cells, is responsible for transporting oxygen to the tissues of the body.

> **BIOLOGY CLIP**
> It has been estimated that 8% of the human population will show some signs of genetic disorder by age 25.

Cline knew that red blood cells were produced in the bone marrow of the long bones. Bone marrow therefore was extracted from the thigh bones, and the defective cells were incubated with normal hemoglobin. It was hoped that the genes from the normal hemoglobin would become embodied within the defective cells. Unfortunately, the procedure failed to produce any significant improvement.

Two years later, the technique was revived. Once again, researchers turned to genetic problems associated with the hemoglobin molecule. Sickle-cell anemia, like thalassemia, is characterized by poor oxygen delivery to the tissues. Sickle-cell anemia affects 1 in 500 people of African heritage in the United States. Red blood cells carrying oxygen appear normal, but once oxygen is released, the hemoglobin turns and folds in an unusual pattern, bending the cell into a sickle shape. Unfortunately, the sickle-shaped cells become lodged in the tiniest blood vessels of the body, the capillaries. The clogged red blood cells impair the transport of oxygen to the tissues and bring about painful swelling.

Gene therapy *is a procedure by which defective genes are replaced with normal genes in order to cure genetic diseases.*

Two genes control the production of hemoglobin during fetal development. As the fetus approaches term, one of the genes is turned off. The gene that remains "on" produces adult hemoglobin. Unfortunately, the gene that is turned on is the very gene that harbours the mutation in people with sickle-cell anemia. If the gene that produces fetal hemoglobin is switched back on and the defective gene turned off, the disorder can be cured. A drug called 5-azacytidine turns the fetal hemoglobin back on; however, this drug is highly toxic, and it is still unclear whether or not the drug turns on other genes as well.

The success of gene therapy has been modest, but its boundaries are extended almost daily. So far, the spotlight of gene therapy has focused on somatic cells in an attempt to cure disease. But what will happen when gene therapy is directed at reproductive cells? Will the ability to select genes for our offspring be far behind? The ethical and moral questions of manipulating life are complex. Is the development of our morality keeping pace with technological advances?

◼ REVIEW QUESTIONS ▪ ?

4 List three difficulties that arise when genes are studied in human populations.

5 What are gene crossovers? How does this process affect segregation?

6 What are linked genes?

7 How can gene crossover frequencies be used to construct gene maps?

8 What are some possible benefits of gene therapy?

Figure 21.17

Retroviruses can be used for gene therapy. (a) Genes from the virus are replaced with therapeutic genes. (b) Viral genes are placed in the cell. (c) Genetic information from the virus directs production of proteins which make up the viral shell. The virus assembles (d) and enters the target cell. Genes carried by the virus are spliced into chromosomes of the target cell. Transplanted genes direct the synthesis of needed proteins.

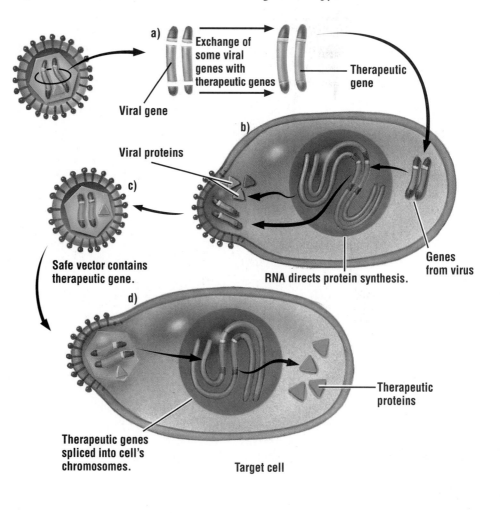

SOCIAL ISSUE:
The Potential of Gene Therapy

Scientists have located the genes for such hereditary diseases as cystic fibrosis and Huntington's chorea. As new techniques in gene therapy become available, it may soon be possible to switch off defective genes and replace them with normal genes, thereby providing a cure for many genetic diseases. As with most new medical technologies, however, important ethical questions must first be addressed.

Statement:

As a technology, gene therapy is beneficial.

Point

- Gene therapy will eventually provide a cure for most forms of genetic disease, and the cure will be permanent.

- People who once died because of hereditary diseases are being kept alive by modern medicine and their genes are being passed on. Therefore, more and more people will carry genetic disorders unless gene therapy is used. Ultimately, gene therapy will provide a better gene pool.

Counterpoint

- The question of which genes are "good" or "bad" and which genes should be changed is difficult to decide. Also, gene therapy could be used to do more than just cure diseases. What happens if people decide to use gene therapy to select "desirable" traits for their children?

- The human genome is infinitely complicated. Most characteristics are controlled not by a single gene, but by the interaction of many genes. With such a complicated intertwining of factors, the attempt to control the gene pool would only lead to disaster.

> *Research the issue.*
> *Reflect on your findings.*
> *Discuss the various viewpoints with others.*
> *Prepare for the class debate.*

CHAPTER HIGHLIGHTS

- Chromosomes carry genes, the units of hereditary structure.
- Paired chromosomes segregate during meiosis. Each sex cell or gamete has half the number of chromosomes found in a somatic cell.
- Chromosomes assort independently during meiosis. This means that each gamete receives one of the pairs and that one chromosome has no influence on the movement of a non-paired chromosome.
- Each chromosome contains many different genes.
- Sex-linked traits are located on either the X or Y chromosome.

- Barr bodies are small, dark spots of chromatin located in the nuclei of female mammalian cells.
- Genetics can be used to study human populations.
- Crossing-over occurs between homologous chromosomes during meiosis. Small segments of DNA are exchanged, thereby creating new combinations of linked genes.
- Crossover frequencies can be used to determine gene maps.
- Genes do not remain in fixed positions along the chromosomes. Transposons are specific segments of DNA that can move along a chromosome.
- Gene therapy is directed at curing hereditary disorders.

APPLYING THE CONCEPTS

1   In what ways was the development of the chromosomal theory linked with improvements in microscopy?

2   Discuss the contributions made by Walter Sutton, Theodor Boveri, Thomas Morgan, and Barbara McClintock to the development of the modern chromosomal theory of genetics.

3   The gene for wild-type eye color is dominant and sex-linked in *Drosophila*. White eyes are recessive. The mating of a male with wild-type eye color with a female of the same phenotype produces offspring that are 3/4 wild-type eye color and 1/4 white-eyed. Indicate the genotypes of the P_1 and F_1 generations.

4   Use the information from the pedigree chart to answer the following questions.

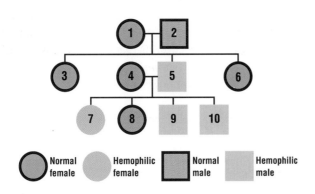

a) State the phenotypes of the P_1 generation.

b) If parents 1 and 2 were to have a fourth child, indicate the probability that the child would have hemophilia.

c) If parents 1 and 2 were to have a second male child, indicate the probability that the boy would have hemophilia.

d) State the genotypes of 4 and 5.

5   The autosomal recessive gene *tra* transforms a female into a phenotypic male when it occurs in the homozygous condition. The females who are transformed into males are sterile. The *tra* gene has no effect in XY males. Determine the F_1 and F_2 generations from the following cross: XX, +/*tra* crossed with XY, *tra*/*tra*. (Note: the + indicates the normal dominant gene.)

6   Edward Lambert, an Englishman, was born in 1717. Lambert had a skin disorder characterized by very thick skin that was shed periodically. The hairs on his skin were very coarse, like quills, giving him the name "porcupine man." Lambert had six sons, all of whom exhibited the same traits. The trait never appeared in his daughters. In fact, the trait has never been recorded in females. Provide an explanation for the inheritance of the "porcupine trait."

7   A science student postulates that dominant genes occur with greater frequency in human populations than recessive genes. Using the information that you have gathered in this chapter, either support or refute the hypothesis.

8   In 1911, Thomas Morgan collected the following crossover gene frequencies while studying *Drosophila*. Bar-shaped eyes are indicated by the *B* allele, and carnation eyes are indicated by the allele *C*. Fused veins on wings (*FV*) and scalloped wings (*S*) are located on the same chromosome.

Gene combinations	Frequencies of recombinations
FV/B	2.5%
FV/C	3.0%
B/C	5.5%
B/S	5.5%
FV/S	8.0%
C/S	11.0%

Use the crossover frequencies to plot a gene map.

9   Huntington's chorea is a dominant neurological disorder that usually appears when a person is between 35 and 45 years of age. Many people with Huntington's chorea, however, do not show symptoms until they are well into their sixties. Explain why the slow development of the disease has led to increased frequencies in the population.

10   Explain the significance of locating the cystic fibrosis gene.

CRITICAL-THINKING QUESTIONS

1 Aristotle suggested that heredity could be linked to male semen. Other early scientists suggested that the female determined the traits of the offspring. Based on the knowledge that you have gathered about genetics, provide evidence that would refute both of these theories.

2 Despite meeting all the criteria for good scientific research, the work of Barbara McClintock was ignored for almost 20 years. The significance of her work was only acknowledged when other scientists, working with bacteria, confirmed the fact that genes move along chromosomes. Why do you think her work was not readily accepted?

3 Discuss ways in which gene mapping technology can be used for beneficial purposes.

4 Discuss ways in which gene mapping technology might be applied for harmful purposes.

5 The gene for Huntington's chorea has been located near another gene that can be used as a genetic marker. The marker travels with the gene for Huntington's chorea. When segments of DNA are chopped up with restriction enzymes, scientists test for the Huntington's gene by locating the marker. If the marker is present, there is a 96% chance that the patient also carries the Huntington's gene. However, despite the success of the test, some people who are at risk have decided not to take the test. Why would some people be reluctant to take the test?

6 The ability to fix genes of people who have genetic diseases lies on the horizon of genetic research. Gene therapy has been employed for some blood diseases, such as sickle-cell anemia. This disease is characterized by a single defective gene that forms abnormal hemoglobin. As you read earlier in the chapter, the hemoglobin is unable to carry oxygen efficiently. In fetal development, two different genes control the production of hemoglobin. In infancy, one gene is turned off, and the other, called the beta gene, is turned on. Unfortunately, the defective beta gene produces the abnormal hemoglobin, and sickle-shaped cells result.

One technique of gene therapy involves manufacturing a drug that turns off the defective beta gene and switches the infant gene back on. Although results have so far proved inconclusive, the technique shows promise. Should scientists attempt to develop drugs that turn genes on and off? Do scientists have the right to alter human genes? Support your opinion.

ENRICHMENT ACTIVITIES

1 Research different genetic diseases. Identify symptoms of the diseases and explain any treatment.

2 Investigate the Human Genome project, which represents 17 nations and has a budget of over $1 billion. This endeavor, which involves hundreds of scientists, will take at least 10 years to complete. Why do you think this project has been compared to the last great scientific undertaking, the Manhattan Project (the development of the nuclear bomb)?

3 Are there any biotechnology companies working in your area? You may want to organize a class visit. List some of the products that these companies manufacture.

DNA: The Molecule of Life

IMPORTANCE OF DNA

The nucleus of every cell in your body contains **deoxyribonucleic acid,** or **DNA.** This molecule is found in the cells of all organisms, from mushrooms to trees, from sponges to mammals. Scientists' fascination with DNA arises from the fact that it is the only molecule known that is capable of replicating itself. Sugar molecules, protein molecules, and fat molecules cannot build duplicates of themselves inside your body. DNA can duplicate itself, thereby permitting cell division.

DNA provides the directions that guide the repair of worn cell parts and the construction of new ones. This information is provided by way of chemical messages carried between the nucleus, the control center of the cell, and the cytoplasm, the part of the cell that contains the functioning organelles. Sometimes referred to as the language of life, the genetic code is contained in 46 separate chromosomes in your body. One estimate indicates that there are approximately three billion base pairs of DNA and 100 000 **genes** located on the chromosomes. Characteristics such as your hair color, skin

Deoxyribonucleic acid (DNA) *is the carrier of genetic information in cells.*

Genes *are the units of heredity in the chromosome.*

Figure 22.1

DNA contains the information that ensures that pea plants produce seeds that grow into other pea plants.

color, and nose length are all coded within the chemical messages of DNA. Packed within the DNA are all the instructions that make you unique. Unless you are an identical twin, your DNA code is distinctively one of a kind.

DNA contains instructions that ensure **continuity of life.** Bean plants produce seeds that grow into other bean plants because the DNA holds the chemical messages for the roots, stems, leaves, and seed pods of a bean. In a similar way, guinea pigs give birth to other guinea pigs, and humans procreate other humans. However, you have learned that not all offspring are identical to their parents. The uniqueness of descendants can often be explained by new combinations of genes and by **mutations.** In order to understand how genes affect the expression of an organism's traits, you will have to learn how DNA regulates the production of cell protein. Proteins are the structural components of cells. DNA, therefore, not only provides for a continuity of life, but also accounts for the diversity of life forms.

SEARCHING FOR THE CHEMICAL OF HEREDITY

By the early 1940s biologists began to accept the hypothesis that hereditary material resided within **chromosomes.** Chromosomes are long threads of genetic material found in the nucleus of cells. Chemical analysis indicated that the chromosome is made up of roughly equal portions of proteins and nucleic acids. Proteins are composed of 20 different amino acids, which can be organized into an almost infinite variety of proteins. By altering the sequence of a single amino acid in the protein chain, new proteins can be formed. Nucleic acids, like pro-

teins, are very large molecules. However, all nucleic acids share many similarities. The basic unit of nucleic acids is the **nucleotide,** which is arranged in long chains. Nucleotides are made of phosphates, sugar molecules, and one of four different nitrogen bases.

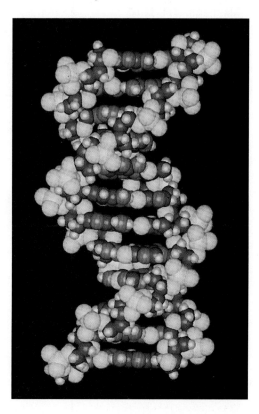

Continuity of life *is a succession of offspring that share structural similarities with those of their parents.*

Mutations *arise when the DNA within a chromosome is altered. Most mutations change the appearance of the organism.*

Chromosomes *are long threads of genetic material found in the nucleus of cells. Chromosomes are composed of many nucleic acids and proteins.*

Nucleotides *are the building blocks of nucleic acids. Nucleotides are composed of a ribose sugar, phosphate, and a nitrogen base.*

Scientists soon began to speculate that the protein component of the chromosome complex contained the chemical message of life. The protein molecule within the nucleus was thought to act as a master molecule, directing the arrangement of amino acids in the cytoplasm. Nucleic acids were thought to be monotonous repetitions of a single sugar molecule, identical phosphates, and four nitrogen bases. The same sugar, phosphate, and nitrogen bases were found in all living things. How could a language be constructed upon such a limited alphabet? The key to the genetic code was believed to lie in the proteins. The hypothesis was logical, but incorrect.

Background Information

In 1944, Oswald Avery, Colin MacLeod, and MacLyn McCarty at the Rockefeller Institute in New York City conducted a series of experiments to determine the origin of inheritance. The scientists worked with a bacterium called *pneumococcus.* This bacterium, which causes pneumonia, exists in two forms. The first form is surrounded by a sugar-like compound called a capsule. The other variety has no capsule. When cells with a capsule divide, they form other cells with a capsule. Cells without capsules form new cells without capsules.

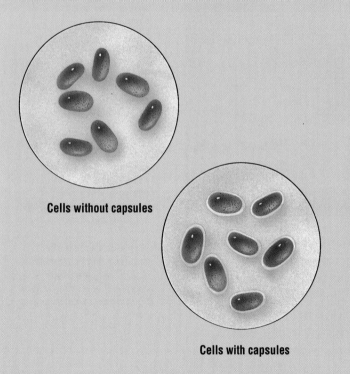

Cells without capsules

Cells with capsules

The following is an abbreviated summary of Avery, MacLeod, and McCarty's procedures and results.

Objective

To analyze experimental data.

Material

pencil

Procedure

1 Mouse A was injected with capsuled cells, while mouse B was injected with noncapsuled cells.

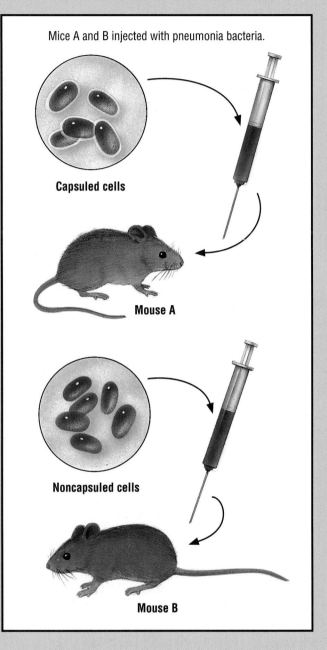

Mice A and B injected with pneumonia bacteria.

Capsuled cells

Mouse A

Noncapsuled cells

Mouse B

Observation: Mouse A contracted pneumonia and died, while mouse B continued to live.

a) What conclusion can you derive from the experimental results?

b) Why might a scientist decide to repeat this experimental procedure on other mice?

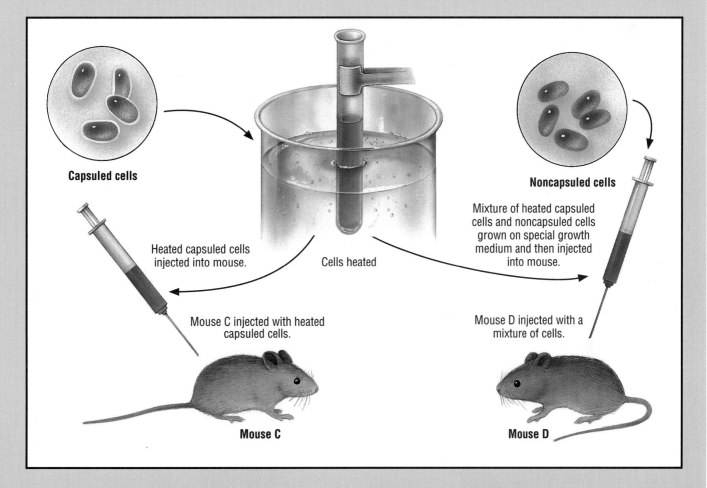

Capsuled cells

Noncapsuled cells

Heated capsuled cells injected into mouse.

Cells heated

Mixture of heated capsuled cells and noncapsuled cells grown on special growth medium and then injected into mouse.

Mouse C injected with heated capsuled cells.

Mouse D injected with a mixture of cells.

Mouse C

Mouse D

2 The capsuled pneumonia cells were heated and killed, and then injected into mouse C.

Observation: Mouse C continued to live.

c) Explain these experimental results.

d) Predict what would have happened to the mouse if the noncapsuled cells had been heated and then injected.

3 The heated capsuled cells were mixed with noncapsuled cells. The mixture was grown on a special growth medium. Cells from the culture medium were injected into mouse D.

Observation: Mouse D died. An autopsy indicated that the mouse had died of pneumonia, and capsuled bacteria were isolated from the mouse.

e) Would you have predicted this observation? Explain why or why not.

Case-Study Application Questions

1 A microscopic examination of the dead and live cell mixture revealed cells with and without capsules. What influence did the heat-destroyed cells have on the noncapsuled cells?

2 Avery, MacLeod, and McCarty hypothesized that a chemical in the dead, heat-treated, capsuled cells must have altered the genetic material of the living noncapsuled cells. In similar previous experiments, scientists had dubbed this chemical the *transforming principle.* Avery and his associates believed that the transforming principle was DNA. What must have happened to the DNA when the cells divided?

3 To test whether or not DNA was the transforming principle, Avery and his associates crushed capsuled cells to release their contents. The DNA was extracted. Next they added the DNA extract to a cell culture that contained noncapsuled pneumonia cells. These cells were later found to contain some cells with capsules. Did this confirm or disprove their hypothesis? Give your reasons. ■

HISTORICAL PROFILE: JAMES WATSON AND FRANCIS CRICK

When scientists confirmed that DNA was the material of heredity, their focus shifted to understanding how it works. Part of that understanding would come from knowing its structure because in biology, like other subjects, structure provides many clues about function. In the race to be the "first" to discover the structure of DNA, scientists around the world employed emerging technologies to help them gain new insights into this mysterious "molecule of life." In the end, the honor would go to a pair of scientists, James Watson and Francis Crick.

James Watson was considered a child prodigy when he entered the University of Chicago at the age of 15. He began studying ornithology, but eventually turned his attention to genetics and molecular biology. In 1951, he began studies at England's Cambridge University, where he met Francis Crick, a physicist who had served with the British army during World War II. Each would bring to bear his experience from a different area of science to interpret and synthesize the experimental data that were rapidly mounting.

One source of important data came from the Cambridge laboratory of Maurice Wilkins, where researcher Rosalind Franklin used a technique called X-ray diffraction to help determine the structure of the DNA molecule.

Figure 22.3

X rays are short, high-frequency electromagnetic waves. If an X ray is passed through material, it will bend (be diffracted) a certain way. The amount of bending depends on the material. If X rays are passed through a fragment of DNA and the diffraction patterns are captured on film, the resulting picture looks like this. Just as a shadow of your hand gives some information about its shape, the X-ray diffraction shadow of the DNA molecule provides a glimpse of the molecule's shape.

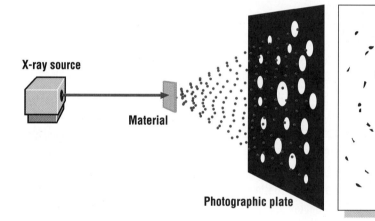

X-ray source

Material

Photographic plate

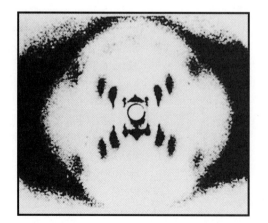

Another source of data involved the comparison of the chemical structure of DNA molecules in different organisms. Scientists already knew that molecules of DNA were made up of sugars (deoxyribose), phosphate, and four different nitrogen bases: adenine, guanine, cytosine, and thymine. What they did not know was the way in which these bases were arranged. New research revealed that, even though the proportion of the nitrogen bases varies from species to

520 ■

species, the proportion stays the same in the DNA of all of a given species' cells. That is, the number of adenine molecules is the same as the number of thymine molecules, and the number of guanine molecules is the same as those of cytosine. These observations suggested that the nitrogen bases were arranged in pairs.

Watson's background enabled him to understand the significance of the emerging chemical data, while Crick was better able to appreciate the significance of the X-ray diffraction results. With such data, Watson and Crick developed a three-dimensional model of the DNA molecule, which they presented to the scientific community in 1953. This model, which was visually confirmed in 1969, is still used by scientists today. Some additional information gathered since Watson and Crick's time has been incorporated into it.

Models are very useful tools for scientists. A model airplane looks like the real thing, except that it is much smaller. By studying the replica, one can learn more about how the real plane works. However, the model of a molecule cannot be scaled down for detailed study; molecules are already too small to see. Instead, molecules are made larger to show how the different atoms interact. X-ray diffraction techniques provide a picture that indicates how different chemical bonds interact with each other. Scientists use models as visual devices that help them understand the relationship and interactions of different parts.

Politics and Science

Watson and Crick might not have been credited as the co-discoverers of DNA were it not for politics. The X-ray diffraction technique developed in England had been used by Maurice Wilkins and Rosalind Franklin to view the DNA molecule. At that time, the American scientist Linus Pauling, a leading investigator in the field, was refused a visa to England to study the X-ray photographs. Pauling, along with others, had been identified by Senator Joseph McCarthy as a communist sympathizer for his support of the anti-nuclear movement. Many scientists believe that the United States passport office may have unknowingly determined the winners in the race for the discovery of the double-helix model of DNA.

The McCarthy era of the early 1950s is considered by many historians as a time of paranoia and repression. Many creative people had their careers stifled or destroyed because of their perceived association with communism. In most cases the charges were unfounded. It is perhaps ironic that, in 1962, Linus Pauling was awarded a Nobel Prize, this time for his dedication to world peace.

> **BIOLOGY CLIP**
> If all the DNA in your body were uncoiled and stretched out, it would reach to the sun and back approximately 3000 times.

Figure 22.4

James Watson and Francis Crick were awarded the Nobel Prize for physiology/medicine in 1962 for what many people believe to be the most significant discovery of the twentieth century.

STRUCTURE OF DNA

DNA is most often described as a double helix. DNA closely resembles a twisted ladder. The sugar and phosphate molecules form the backbone of the ladder, while the nitrogen bases form the rungs. Nitrogen bases from one spine of the ladder are connected with nitrogen bases from the other spine of the ladder by means of hydrogen bonds. A hydrogen bond is a weak bond that forms between the positive charge on the end of one molecule and the negative charge on the end of another molecule. The backbone

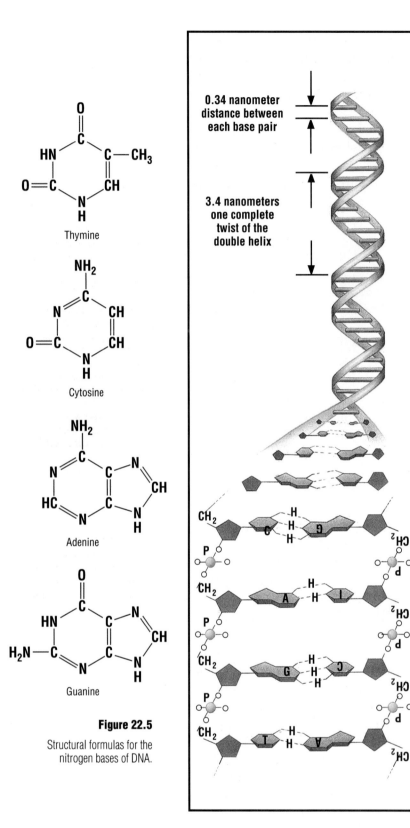

O
‖
C
HN   C—CH₃
     ‖
O=C   CH
    \   /
     N
     |
     H

Thymine

NH₂
|
C
N   CH
‖   ‖
O=C   CH
    \   /
     N
     |
     H

Cytosine

NH₂
|
C
N   C   N
‖   \   / \
HC   C   CH
    \   \ /
     N   N
         |
         H

Adenine

O
‖
C
HN   C   N
     \   / \
H₂N—C   C   CH
    ‖   \ /
     N   N
           |
           H

Guanine

Figure 22.5

Structural formulas for the nitrogen bases of DNA.

0.34 nanometer distance between each base pair

3.4 nanometers one complete twist of the double helix

Figure 22.6

Representation of DNA molecule. Note the complementary hydrogen bonds.

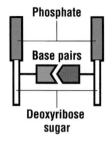

Phosphate

Base pairs

Deoxyribose sugar

of the DNA molecule becomes twisted, which makes the molecule look like a winding, spiral staircase. The DNA molecule is made of individual units composed of deoxyribose sugars, phosphates, and nitrogen bases. Each unit is referred to as a nucleotide. The model to the right shows two complementary nucleotides. Notice how the nitrogen base fits into its complementary pair. An adenine molecule always pairs with a thymine molecule, while a guanine molecule always pairs with a cytosine molecule.

■ REVIEW QUESTIONS ■   ?

1 Name two ways in which the DNA molecule is important in the life of a cell.
2 What chemicals make up chromosomes?
3 What are nucleotides?
4 What chemicals are found in a nucleotide?
5 On what basis did some scientists conclude that proteins provided the key to the genetic code?
6 Who were the co-discoverers of the double-helix model of DNA?
7 How did the X-ray diffraction technique provide a clue to the structure of DNA?
8 Which nitrogen base pairs with guanine? Which base pairs with adenine?

REPLICATION OF DNA

DNA is the only molecule known that is capable of duplicating itself. This is accomplished through a process known as **replication.** Replication helps explain how one cell can divide into two identical cells. Each of the daughter cells requires a complete set of genetic information that can only be obtained if the DNA molecule makes an exact duplicate of itself.

During replication, the weak hydrogen bonds that hold the complementary nitrogen bases together are broken. The two edges of the ladder seem to "unzip," as shown in Figure 22.7. The parent strands are conserved (remain intact). Therefore, this process is also known as **semiconservative replication**—semiconservative because only the parent strands are conserved; the original, inte-

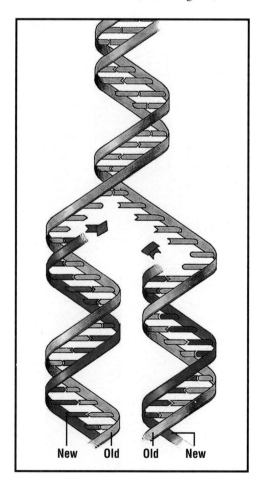

New   Old   Old   New

grated molecule is not. Semiconservative replication produces two "half-old, half-new" strands of DNA. Each parent strand acts as a template, or mold, to which free-floating nucleotides in the cell can attach. The nucleotides attach themselves at their complementary bases: adenine with thymine, and cytosine with guanine. A series of enzymes, called **polymerases,** fuse the free nucleotides together in the complementary chain of DNA.

The free-floating nucleotides in your cells (and those in the cells of other organisms) are derived from the food you eat. A steak supplies you with the muscle cells from a cow. Each of these cells contains the nucleus and the genetic information from the cow. However, the intake of cow DNA does not mean that you will begin to look like a cow! Specialized enzymes in your digestive tract break apart the cow DNA into nucleotides, which you use to make human DNA.

The duplication of DNA in humans is a marvelous feat. It has been estimated that the human genome contains approximately three billion bases. An error rate as infrequent as one per million bases would produce 3000 mistakes, or mutations, in a single division. Imagine what would happen if each cell had 3000 genetic mistakes. Although genetic mistakes do occur during the duplication of genetic information, they are very infrequent.

Environmental factors such as hazardous chemicals or radiation are one source of the genetic mistakes that do occur. These factors can cause uncomplementary nitrogen bases to become paired. Permanent damage is prevented by enzymes that act as proofreaders. These enzymes run along the strands of DNA looking for mismatched pairs. Once a damaged section is detected, it can be repaired. Another enzyme snips the error from the chain and replaces it with the correct nucleotide sequence.

Replication *is the process in which a single strand of nucleotides acts as a template for the formation of a complementary strand. A single strand of DNA can make a complementary strand.*

Semiconservative replication *is the process in which the original strands of DNA remain intact and act as templates for the synthesis of duplicate strands of DNA.*

Polymerases *are enzymes that join individual nucleotides together in complementary strands of DNA.*

Figure 22.7

The DNA molecule unzips, and each strand serves as a blueprint for the synthesis of a complementary strand of DNA.

RESEARCH IN CANADA

With recent advances in reproductive technology and applied genetics, couples who were once unable to have children can now become parents. However, these new techniques raise some very important ethical and legal questions. For example, will frozen embryos be considered property or a life form?

Reproductive Technologies

In 1987, the United States embarked on a massive $3 billion project that will attempt to decipher the entire human genome. It has been predicted that doctors will be able to screen for defective genes by the end of the century. How will this knowledge be applied in Canada? Will couples be allowed to select the genes of their offspring? Who will decide how genetics is to be used? These questions cannot be ignored.

Dr. Patricia Baird, a medical geneticist from the University of British Columbia, heads the Royal Commission on New Reproductive Technologies. The mandate of the commission is to examine how new information is applied to emerging technologies concerned with human reproduction. Who will be able to apply these procedures? Who will be responsible for the expense? How much public money will be spent? Although the commission's mandate extends beyond the confines of science, a knowledge of science and technology is critical to understanding the issues.

DR. PATRICIA BAIRD

LABORATORY

EXPLORING DNA REPLICATION

Objective

To investigate how the double helix of DNA replicates.

Background Information

The DNA molecule is made up of nucleotides, each comprising a deoxyribose sugar, a phosphate, and a nitrogen base. During the replication process, the double strands of DNA separate along the bonds between the nitrogen bases. Each parent strand serves as a template for the arrangement of new nucleotides. An enzyme joins the nucleotides into a complementary strand of DNA. A second enzyme checks the ordering of bases for errors.

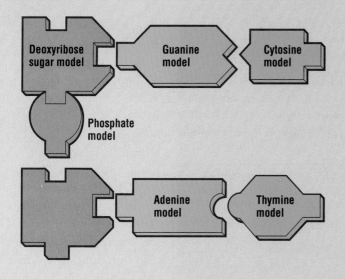

Materials

scissors
toothpicks

transparent tape or glue
blank sheet of paper

Procedure

1 You will be supplied with a page of symbols representing the molecules that make up DNA. Cut out the individual molecules. The toothpicks will be used to represent bonds between the molecules.

a) Why are the adenine and guanine molecules represented by larger shapes than the other two nitrogen bases?

2 Using a toothpick, bond the adenine molecule to a deoxyribose sugar molecule. Then use another toothpick to bond the phosphate molecule to the sugar molecule. Place the phosphate molecule along the left margin of a sheet of paper.

b) What is this structure called?

3 Assemble four different nucleotides, as described in the procedure above. Keeping the phosphate molecules along the left margin of the page, attach each of the nitrogen bases to a different sugar molecule.

4 The phosphate molecules of the DNA bond with two different sugar molecules. Place a second toothpick on the phosphate molecule, and attach a second sugar molecule to the phosphate molecule.

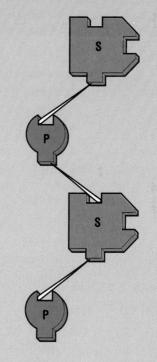

5 Using toothpicks to represent bonds, attach a line of four nucleotides together.

c) Record the genetic code by indicating the letters of the nitrogen bases, beginning from the top of the page.

6 Make a complementary strand of DNA by matching nitrogen bases. The complementary strand should have the phosphate molecules aligned along the right of the page.

d) Record the genetic code of the complementary strand.

7 The hydrogen bonds between the nitrogen bases are easily broken. Separate the two strands so that each is alongside each margin. Then make complementary strands for each of the original strands.

8 Glue each of the strands to a blank sheet of paper.

e) What do you notice about the two strands of DNA?

Laboratory Application Questions

1 Thymine always bonds with adenine. Explain why the thymine used in your model does not bind with guanine.

2 What determines the nitrogen base sequence of DNA in a new strand of DNA?

3 Adenine is a double-chain purine molecule. Purines always bond with pyrimidines. What special problems would be created if the purine adenine joined with guanine? with another purine?

4 A special enzyme scans the DNA helix in search of improper nitrogen base pairings. Using your model, indicate how the enzyme would be able to detect incorrect nitrogen base pairings.

5 Radiation can cleave the bonds between phosphate and sugar molecules, thereby cutting a section of the DNA out of the ladder. Special enzymes have the responsibility of gluing the spliced segment of DNA back into place; however, the segment is not always glued back in the correct place. Explain why it is important to place the broken section of DNA back in its original position. ■

FRONTIERS OF TECHNOLOGY: DNA FINGERPRINTING

In 1986, DNA matching was used to identify a rapist-murderer in Leicester, England. More than 1000 men were tested in three villages near Leicester. The technique, called DNA fingerprinting, was used to free one suspect, and led to the arrest and subsequent confession of another. Later the same year, a rapist in Orange County, Florida, was convicted on the basis of genetic evidence.

The DNA fingerprinting test used in Britain was developed by Alec Jeffreys, a geneticist from the University of Leicester. Jeffreys found that, although long stretches of the DNA molecule are similar from one person to another, particular segments contain unique arrange-

ments of nitrogen bases. Only identical twins share the same nitrogen base arrangements in these sections. These short sequences of DNA appear to have no function. Geneticists believe that the codes are nonsense codes, which seem to repeat in almost a chemical type of stuttering.

In the DNA fingerprinting test, a DNA segment is taken from the semen found in the rape victim and compared with the DNA segment taken from a blood sample of the suspect. The DNA samples are transferred to a nylon sheet, where they are tagged with a radioactive probe that identifies the unique segments of the DNA chain. The nylon sheet is then placed against an X-ray film. Black bands appear where the probes have attached to the segments used to establish identity. A print is then made from the film and used to compare samples.

> **BIOLOGY CLIP**
> Fisheries Canada officials are considering using DNA fingerprinting to identify persons who catch protected species of fish.

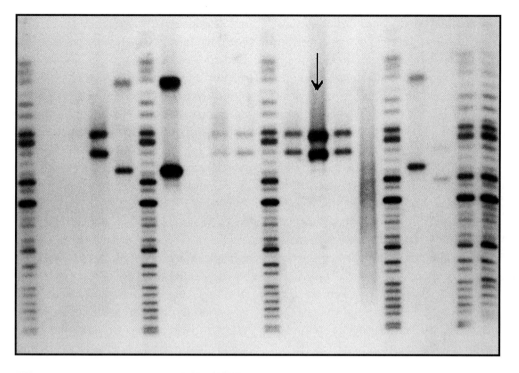

Figure 22.8

DNA profiling can be used to identify criminals.

REVIEW QUESTIONS ?

9 What is the significance of DNA replication for your body?

10 Name the nitrogen bases in the DNA molecule, and state which normally pairs with which.

11 Briefly describe the events of DNA replication.

12 Proofreading enzymes scan the strands of DNA to check the nitrogen base pairings. Explain why these enzymes are important.

13 When you eat fish, you take in fish protein and fish DNA; however, you do not assume the characteristics of a fish. Explain why the nucleic acids of a fish do not change your appearance.

14 What is DNA fingerprinting and how does it work?

GENE RECOMBINATIONS IN THE LABORATORY

All cells have enzymes called *endo-nucleases*, which can cut strands of DNA.

Bacteria have special endonucleases called **restriction enzymes,** which cut apart foreign DNA that has invaded the cell. Restriction enzymes act like biological scissors because they snip sites along the DNA molecule. There are over 200 different restriction enzymes, each of which cuts a different segment along the chromosome.

The discovery of the scissor enzyme was preceded by the discovery of another equally important enzyme, the **ligase enzyme,** or the DNA glue. The ligase enzyme is part of the normal DNA repair system. Many things in your environment can damage your DNA. Even sunlight can split chromosomes apart. The ligase enzyme helps restore the fragmented chromosome. Molecular biologists use both enzymes for gene splicing. The restriction enzymes can also be used to snip a recipient chromosome apart. The ligase enzyme permits the DNA snipped from the donor chromosome to be glued into the recipient.

Biologists have recently discovered that restriction enzymes from bacteria can be used to cut genes from many different living things. Herbert Boyer and Stanley Cohen conducted experiments using a common bacterium of the gut, *Escherichia coli* (or *E. coli*). First, they used restriction enzymes to extract a gene from a toad. Next, they used another restriction enzyme to cut the small circular plasmid of an *E. coli* bacterium. The gene from the toad was then inserted into the opening of the plasmid and glued in place. A gene from a toad had thus been incorporated into the genetic material of a bacterium. Boyer and Cohen watched in amazement as the toad DNA was duplicated along the DNA in the plasmid. Even more incredible, the DNA from the toad began providing direction to the cytoplasm of the bacterial cell as if it had always been there. Two organisms that would never exchange genetic material in nature had been joined in a most unusual union. This technique has been called **recombinant DNA**.

The genes from the toad duplicate right along with the genetic information that is native to the bacteria. As the *E. coli* cells conjugate, the toad DNA located in the plasmid is passed to other bacteria. The bacteria that receive the spliced DNA express traits coded for by the new genes just like their conjugate partners. Therefore, the new gene combination can soon be found in many cells as the bacteria reproduce by conjugation.

Imagine blending bits of your own DNA into bacteria cells. These primitive microbes would have segments of your characteristics. Does this sound like science fiction? It is not. Scientists have already placed a number of human genes into bacteria. The gene that produces human insulin, for example, has been spliced into *E. coli* bacteria. The human DNA directs the bacteria to produce human insulin, a hormone essential for regulating blood sugar. People who do not produce adequate amounts of this hormone have the disease sugar diabetes. Diabetics can regulate their blood sugar by taking an insulin injection. Traditionally, insulin has been extracted from the pancreases of cows and pigs. But cow and pig insulin differs from human insulin, and some patients have allergic reactions to it. Human insulin, produced by *E. coli* bacteria, was first marketed in Canada in 1983.

DNA AND INDUSTRY

Nobody knows when the first human accidentally discovered the intoxicating union of yeast cells with sugar. But once people began to take full advantage of this discovery, biotechnology was born. Biotechnology involves the use of living things—usually but not exclusively microbes—for making products. Today, microbes are a multimillion-dollar business that extends far beyond the traditional boundaries of beer and wines. Modified yeast cells have been employed to produce feed for pigs, fuel for cars, and vaccines for humans. Bacteria have been altered to produce human insulin and growth hormones, as well as glues and artificial sweeteners.

Biotechnology has become a tremendous growth industry. It has been estimated that, by 1995, 40% of Canada's $4 billion pharmaceutical industry will be derived from biotechnology. That figure is expected to increase to 70% by the turn of the century. Canadian biotechnology companies compete with similar companies in Japan, the United Kingdom, Germany, Russia, Australia, and the United States for their share of this promising industry. The Science Council of Canada estimated that biotechnology will be a $186 billion industry by the year 2000. Some economists have predicted

Figure 22.9

Giant fermenter at Alberta Research Park.

Recombinant DNA *is an application of genetic engineering in which genetic information from one organism is spliced into the chromosome of another organism.*

528 ■

that biotechnology will have a greater impact on the 1990s than the computer chip had on the 1980s.

Table 22.1 Cloned Human Gene Products

Protein	Used in treating
Atrial natriuretic factor	High blood pressure
Erythropoietin	Anemia
Factor VIII	Hemophilia
Factor IX	Hemophilia
Insulin	Diabetes
Interferons	Some cancers, viral infections
Interleukin-2	Cancer
Monoclonal antibodies	Infectious disease
Somatotropin (growth hormone)	Pituitary dwarfism
Tissue plasminogen factor	Heart attack, stroke
Tumor necrosis factor	Cancer

U.S. biotechnology firms like Genetech Inc. and Cetus Corp. have become economic giants virtually overnight. In 1981, Wall Street stockbrokers watched in amazement as stock for a little company, Genetech, skyrocketed from $39 to $89 a share within minutes of trading. Herbert Boyer, who had borrowed $500 to begin the company in 1976, found it had an estimated worth of $80 million only five years later. The trading was spurred by the announcement that interferon, a natural product of the body, had been produced by genetically engineered bacteria. In 1981, there was evidence that interferon could lead to a cure for cancer. Although the drug could be collected by conventional methods, it was very expensive. The promise of mass-producing the drug sent speculators into a frenzy.

NEW, IMPROVED MOUSE

Biotechnology involving recombinant DNA was first directed at placing genes from plants and animals into lower life forms. More recent research has concerned itself with placing genes in higher organisms. In April 1988, researchers used a mouse to combine specific genes customized for cancer research. The mouse contains an oncogene that makes it susceptible to breast cancer. This allows researchers to study which chemicals cause breast cancer in a living model. The mouse can also be used to test cancer-preventing drugs.

E.I. du Pont de Nemours, the company that engineered the new mouse, applied for and received a patent that would prevent other companies from copying their technique. Although other companies have applied for and received patents on recombinant DNA bacteria, no company has received a patent on a mammal. Another group of researchers have placed the genes that carry HIV (the virus that causes AIDS) into a mouse. Like the mouse that contains the oncogenes, this mouse will prove invaluable in determining how HIV destroys the immune response.

This kind of research raises many concerns. Opponents of animal patenting fear that this practice will allow a company to own a life form. Animal-rights supporters are concerned that patents on genetically engineered animals will promote their suffering. Some researchers fear that, by placing patents on mice, companies can control research; only those groups that have enough money or prestige to buy their mice will have an opportunity to study oncogenes. Should companies receive profits for technology that has worldwide benefit? What would happen if the laboratory mice escaped and began to breed with normal mice?

Figure 22.10

Both mice are the same age, but the mouse on the right had the gene for human growth hormone inserted into it.

SOCIAL ISSUE:
Nonhuman Life Forms as Property

Statement:

Companies and research institutes should be forbidden to apply for patents on or restrict access to living organisms.

Point

- A company should not be allowed to own a life form such as an oil-digesting microbe or to gain exclusive profits from its use. Environmental disasters such as oil spills have such far-reaching effects on water quality, wildlife, native people, and other groups that any helpful technology should be available to all.

- Research institutes should not be able to impede research by other researchers. If they or private companies could have a patent on a mouse, it would not be long before similar patents were sought on more complex life forms.

Counterpoint

- Developing an oil-ingesting microbe is a very expensive process. The patent would protect a company by enabling it to get a return on its investment and encouraging further research. Since the microbe may reduce the dangers imposed by crude-oil spills, the company may be acting for the benefit of both humanity and the environment.

- Profit has motivated many discoveries. An example is the development of antibiotics by drug companies. Few people would choose to do without penicillin or tetracycline, just because a drug company is making a profit on them.

CHAPTER HIGHLIGHTS

- The experiments of Avery and associates demonstrated that DNA was the chemical of heredity.
- Nucleotides of DNA are composed of deoxyribose sugar, phosphates, and four different nitrogen bases.
- The Watson and Crick model of DNA is described as a double helix.
- The Watson and Crick model is based on the fact that specific nitrogen base pairs are constant. Adenine always forms a hydrogen bond with thymine, while guanine always bonds with cytosine.

- DNA, unlike other molecules, is capable of replicating. The double strand of DNA separates, and each of the single strands acts as a template for the alignment of nucleotides and subsequent synthesis of a complementary strand of DNA.
- DNA fingerprinting has many applications, one of which is the identification of individuals.
- Biotechnology is an emerging industrial force, but along with the new technology come moral and ethical questions.

APPLYING THE CONCEPTS

1 What led scientists to speculate that proteins were the hereditary material?
2 Explain how Avery, MacLeod, and McCarty's experiment pointed scientists toward nucleic acids as the chemical of heredity.
3 Science and technology are often referred to as *synergists*. This means that not only do scientific breakthroughs provide information for technologi-

cal applications, but technological advances also spur scientific progress. Explain how X-ray diffraction techniques led to the discovery of DNA.
4 Using DNA as an example, explain why scientists use models.
5 Compare the amount of DNA found inside one of your muscle cells with the DNA found in one of your brain cells.

6 Why are organ transplants more successful between identical twins than between other individuals?

7 A drug holds complementary nitrogen bases with such strength that the DNA molecule is permanently fused in the shape of a double helix. Predict whether or not this drug might prove harmful. Provide your reasons.

8 DNA fingerprinting has been used to identify rapists. Suggest at least two other applications for the DNA matching technique.

9 What follows is a hypothetical situation. Genes that produce chlorophyll in plants are inserted into the chromosomes of cattle. Indicate some of the possible advantages of this procedure.

10 Analysis of chloroplasts and mitochondria reveals that DNA is located within these organelles. Explain how this discovery helps support the theory that these organelles might actually be descendants of individual living creatures?

11 Explain the significance of the following statement to the search for the structure of the genetic material: Structure provides many clues about function.

CRITICAL-THINKING QUESTIONS

1 The European Economic Community (EEC) has ruled that organisms created by biotechnology can be patented. Such patents have been awarded in the United States since 1984. The Supreme Court of Canada ruled that a patent application for a hybrid soybean could not be granted. Although this does not prevent another company from applying for patents in Canada, it does raise an interesting question: Will successful biotechnology companies choose not to locate in Canada unless they are given the same protection as companies in Europe and the United States? Give your opinions on this matter.

2 DNA fingerprinting has had a tremendous impact on law enforcement. Speculate about how DNA fingerprinting will affect criminal trials. Should DNA fingerprinting ever be used to convict a criminal? Why or why not?

3 In December 1988, researchers at Toronto's Hospital for Sick Children transplanted human bone marrow into mice. The mice allow scientists to study human blood cells, immune responses, and genetic blood disorders in a living system. Previously, bone marrow could only be studied in a living culture, not in a functional body. Many supporters of animal rights are appalled by this procedure. They believe that the mice are being exploited. Should animal modeling experiments, like the ones performed at the Hospital for Sick Children, be continued? Give reasons for your answer.

4 Recombinant DNA technology has been described as the 20th century's most powerful technique since the splitting of the atom. Do you agree or disagree with this comparison?

ENRICHMENT ACTIVITIES

1 Isolate some germinating seeds and ask your dentist to treat the seeds with various levels of radiation. Plant the seeds and compare the growth rates of seeds treated with radiation with those of a control group.

2 Using the squash technique, prepare cells from the meristematic region of the germinating roots. Use a stain to make the chromosomes more prominent. Compare the chromosomes of plants treated with radiation with those of a control group.

Protein Synthesis

Amino acids *are organic chemicals that contain nitrogen. Amino acids can be linked together to form proteins.*

Proteins *are the structural components of cells.*

IMPORTANCE OF PROTEINS

Why do organ transplants present medical problems? Why do identical twins look alike? Why does our hair turn white as we age? What causes a baby's blue eyes to turn to hazel as the child ages? To answer these questions we must begin investigating the structural components of cells—the proteins.

The proteins within your body are composed of 20 different **amino acids.** The amino acids can be strung together in an almost endless variety to make different **proteins.** Proteins are the chemicals that make up the structure of cells. Cell membranes and cellular organelles are composed of proteins. Proteins make up the muscle filaments that enable the body to move. Hair and hair color are also produced by proteins. However, proteins are responsible for much more than just appearance. They also produce the enzymes that regulate the speed of chemical reactions within cells. Antibodies—disease-controlling agents—and hormones—chemical messengers produced by certain cells of the body—are all made up of proteins.

Figure 23.1

The spectacular colors of the kingfisher are produced by proteins, the structural components of cells.

The production of each protein is controlled by one gene. A simple protein may comprise as few as eight amino acids, while complicated proteins may contain in excess of 50 000 amino acids. The sequencing of the amino acids is regulated by the DNA. The replacement of a single amino acid can change the protein. Since the protein structure within each of your tissues is unique, your body is capable of recognizing foreign proteins. That is why some heart patients reject a transplanted heart—the proteins found in the recipient's heart are different from proteins found in the donor's heart.

How does the nucleus instruct the organelles within the cytoplasm to make proteins? How does the cell know which proteins to make? These are but a few of the questions that you will explore in this chapter. An investigation of the control mechanisms of DNA and their function as the directors of protein synthesis will provide you with a better understanding of the continuity and diversity of life.

ONE GENE, ONE PROTEIN

Advancements in one field of biology often lead to advancements in other fields, even when the two fields are not closely linked. But no one could have predicted that one scientist's hypothesis concerning symptoms of hereditary diseases in humans would receive support from another's research on bread mold.

In the early 1900s, Archibald Garrod, a physician, was studying metabolic disorders. Garrod was aware that most chemical reactants in the body are changed to products through a series of metabolic reactions. Each step in the reaction is controlled by an **enzyme,** a catalyst that speeds up the rate of the reaction. Each step of the conversion of reactants to products is regulated by a specific enzyme. The chemicals that are formed in the steps between reactants and products are referred to as **intermediary metabolites.**

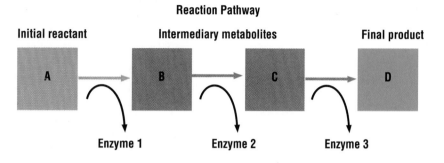

Reaction Pathway

Initial reactant · Intermediary metabolites · Final product

Enzyme 1 · Enzyme 2 · Enzyme 3

Figure 23.2

Each step of the reaction pathway is controlled by a specific enzyme.

Enzymes *are protein catalysts that speed up reactions.*

Intermediary metabolites *are the chemicals that form as reactants are converted to products during a series of chemical reactions.*

Garrod analyzed blood and urine samples from patients with hereditary disorders and noted higher concentrations of intermediary metabolites than were found in the normal population. Could one of the steps in the reaction pathway, shown in Figure 23.2, be blocked? Garrod concluded that one of the enzymes must be defective. According to Garrod, the accumulation of substance B must indicate that enzyme 2 was defective. An accumulation of substance C would indicate that enzyme 3 was defective. But why did the enzyme not work? Garrod hypothesized that the enzyme is synthesized under the control of hereditary material, since genes regulate the production of the protein enzymes. Could the gene be defective?

Garrod's hypothesis remained an educated guess for 33 years. A hypothesis must be tested by experiment to achieve acceptance. In the early 1940s George Beadle and Edward Tatum at Stanford University provided the experiment. Working with a variety of bread mold called *Neurospora crasa*, Beadle and Tatum examined metabolic pathways. *Neurospora* will grow on a minimal-nutrient medium containing only sugar, salts, and one of the B vitamins. All other materials can be synthesized by the bread mold. If a gene synthesized one of the enzymes, what would happen to the enzyme should the gene mutate?

Beadle and Tatum found that a mutated strain of *Neurospora* would only grow when vitamin B_6 was added to the minimal-nutrient medium. Other strains only grew when vitamin B_1 was added. Mutated strains of *Neurospora* that could not synthesize a specific vitamin had to be supplied with the vitamin in order to grow. But why could the *Neurospora* not synthesize the vitamins? Analysis of cell extracts showed that each gene mutation could be associated with alterations of a specific enzyme. Beadle and Tatum concluded that defective enzymes must be linked to genetic mutations. Since genes regulate the production of proteins, Beadle and Tatum's explanation has been referred to as the "one gene, one protein hypothesis."

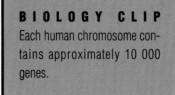

Codons are three-base codes for amino acids.

Ribonucleic acid (RNA) is a single-stranded nucleic acid used to translate the genetic information of DNA into protein structure.

THE ROLE OF DNA IN PROTEIN SYNTHESIS

One of the most baffling questions that confronted biologists was how the DNA molecule provides directions for the construction of proteins. Most researchers were aware of the similarities of DNA in all living things. Although the length of the DNA molecule might vary from organism to organism, the chemical seemed identical. Each chromosome had a deoxyribose sugar backbone linked with a phosphate molecule, and only four different nitrogen bases. How could something this simple create a complex chemical code?

Researchers began to notice that the sequence of base pairs from the top of the ladder to the bottom made one DNA molecule different from another. It seemed reasonable to suggest that the

sequencing of nitrogen bases provided the chemical code. Proteins can be composed of 20 different amino acids, but there are only 4 different nitrogen bases.

Clearly, one nitrogen base cannot code for a single amino acid. Even a combination of two different nitrogen bases will not provide enough information. Four nitrogen bases can be arranged into 16 different combinations of two (4^2 = 16 combinations). However, a triplet code of 3 nitrogen bases provides 64 different combinations (4^3 = 64 combinations). The 20 different amino acids can be accommodated by a triplet code. Experiments confirm that the three-base code, called a **codon,** identifies the 20 different amino acids.

The DNA does not leave the nucleus during protein synthesis. A carrier molecule called messenger **RNA,** often referred to as mRNA, is charged with the responsibility of reading the DNA code and carrying it to the ribosomes. Although mRNA is very similar in structure to DNA, it has some striking differences, most notably in the sugar molecule. RNA contains ribose sugar in place of the deoxyribose sugar, characteristic of the DNA molecule. RNA has no thymine; another nitrogen base, called uracil, works in its place. Figure 23.3 shows the similarities between the nitrogen bases thymine and uracil.

Figure 23.3

The similarities between the nitrogen bases thymine and uracil.

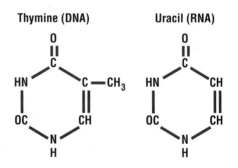

The mRNA reads the code from the DNA molecule. The RNA molecule codes are listed in Table 23.1. Notice that some of the codes provided by the mRNA are not for amino acids. These codes, called *terminators*, end protein synthesis in much the same way that a period indicates the end of a sentence. Other codes are thought to be *initiators*, and are responsible for turning protein synthesis on.

strand of DNA as a blueprint. DNA cytosine binds with mRNA guanine, while DNA adenine pairs with mRNA uracil. The mRNA nucleotides are soon joined into a long chain. Once the chain has been fused, the mRNA molecule moves away from the parent DNA strand. The two strands of the original DNA then rejoin. The process of **transcription** has been completed.

Transcription *is the process by which the genetic code is transferred from the DNA molecule to the messenger RNA molecule.*

Table 23.1 Dictionary of Messenger RNA Code Words

Amino acids	Codons
Alanine	GCU GCC GCA GCG
Arginine	CGA CGU CGC CGG AGA AGG
Asparagine	AAU AAC
Aspartate	GAU GAC
Cysteine	UGU UGC
Glutamate	GAA GAG
Glutamine	CAG CAA
Glycine	GGA GGC GGU GGG
Histidine	CAU CAC
Isoleucine	AUC AUU AUA
Leucine	UUA UUG CUU CUC CUA CUG
Lysine	AAA AAG
Methionine	AUG
Phenylalanine	UUU UUC
Proline	CCA CCC CCU CCG
Serine	UCG UCU UCA UCC AGU AGC
Threonine	ACU ACC ACG ACA
Tryptophan	UGG
Tyrosine	UAU UAC
Valine	GUU GUC GUA GUG
Terminator	UAA UAG UGA
Initiator	AUG

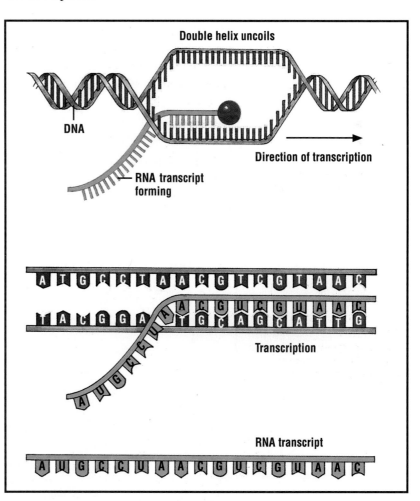

Figure 23.4

The DNA molecule uncoils and acts as a template for the synthesis of mRNA.

PROTEIN SYNTHESIS: TRANSCRIPTION

Protein synthesis begins once the double-stranded DNA molecule in the nucleus "unzips." As the double helix uncoils, nucleotides from the mRNA find the appropriate pair by using the single

The single-stranded mRNA molecule moves through the nuclear membrane and carries the nitrogen base code to the ribosomes in the cytoplasm. The fact that the DNA remains protected within the nucleus helps prevent environmental agents from changing the genetic code. Alterations of DNA would cause mutations.

PROTEIN SYNTHESIS: TRANSLATION

Anticodons *are the three-base codes found in tRNA that pair with the codons of mRNA.*

Translation *is the process by which proteins are synthesized using the DNA instructions encoded in the mRNA.*

The mRNA attaches itself to the ribosome much like a ribbon. An initiator codon turns on the protein synthesis. Another type of RNA, called transfer RNA (tRNA), picks up amino acids that are circulating within the cytoplasm and shuttles them to the mRNA. Specific tRNA molecules use energy to pick up a single amino acid. The amino acid is held in place against one end of the cloverleaf-shaped molecule. The other end of the

> **BIOLOGY CLIP**
> Each human cell has about three billion nitrogen bases that code for genetic characteristics.

tRNA molecule has a plug-shaped structure with three nitrogen bases exposed. The exposed bases are called the **anticodon**. Each kind of tRNA molecule has a specific anticodon. Glutamate acid will be carried by a tRNA molecule that carries either the CUU or the CUC anticodon, while the amino acid valine will only be carried by a tRNA molecule that has the CAA anticodon.

The mRNA strand attached to the ribosome will select the plug-shaped tRNA molecule if the codon and anticodon base pairs match. A UUU codon of messenger RNA will only

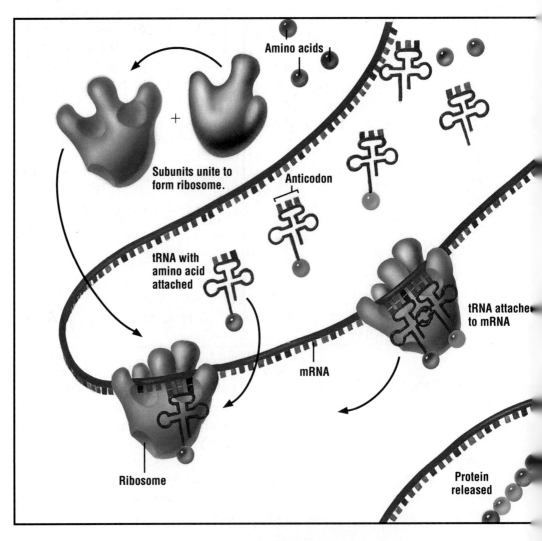

Figure 23.5

A simplified picture of the translation stage of protein synthesis.

Amino acids

Subunits unite to form ribosome.

Anticodon

tRNA with amino acid attached

mRNA

tRNA attached to mRNA

Ribosome

Protein released

receive an AAA anticodon of tRNA. The tRNA will place its amino acid in the ribosome and move away from the ribosome in search of another amino acid that fits into its structure. The mRNA then moves along the ribosome much like a typewriter ribbon. A new codon of mRNA is then exposed for fitting with another tRNA anticodon. The amino acids carried by the tRNA molecules are fused into long-chain proteins on the ribosome. The sequencing of the amino acids is determined by the message carried from the nucleus by the mRNA molecule. Once the protein molecule has been built, a terminator codon turns the synthesis off and the process of **translation** is complete.

■ REVIEW QUESTIONS ?

1 In what ways is the structure of mRNA similar to DNA? How does mRNA differ from DNA?
2 Provide an example of a type of message that mRNA might carry.
3 What is the function of tRNA?
4 What is a codon? an anticodon?
5 Differentiate between transcription and translation.
6 What anticodon on the tRNA molecule would gain access to the codon UUG?
7 Why is protein synthesis essential for life?

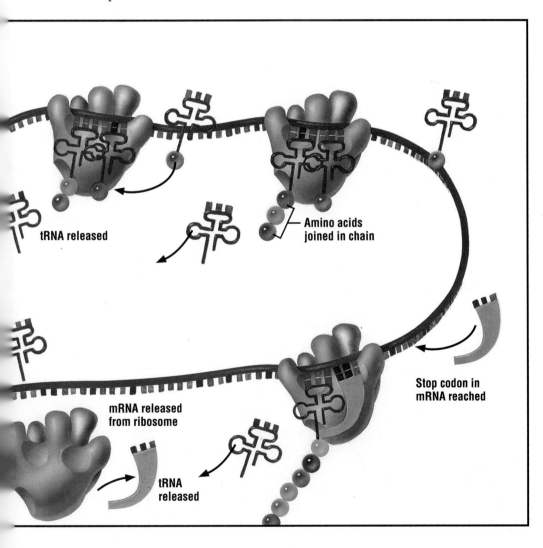

tRNA released

Amino acids joined in chain

Stop codon in mRNA reached

mRNA released from ribosome

tRNA released

LABORATORY

ANTIBIOTICS AND PROTEIN SYNTHESIS

Objective

To study the effect of antibiotics on bacteria.

Background Information

A number of different antibiotics prevent protein synthesis by bacteria. Deprived of the ability to produce new proteins, bacteria are unable to reproduce.

A group of antibiotics, referred to as tetracyclines, block the site where amino acids attach to the tRNA. Another antibiotic, called puromycin, binds to the ribosome and becomes incorporated into the protein chain forming on the ribosome. The linking of amino acids on the ribosome is stopped once the puromycin becomes attached. Streptomycin works in much the same way.

Materials

wax marker
petri dish with nutrient agar
sterile cotton swab
3 antibiotic disks
masking tape
ruler (millimeters)

Procedure

1 Using a wax marker, divide the bottom (small plate) of the petri dish into three sections, and number each section as shown in the diagram below.

2 Remove a sterile swab from its package and run it along your forehead to collect bacteria.

3 Remove the upper lid of the petri dish and streak the entire plate with the swab, running it along the surface of the agar first in horizontal strokes and then in vertical strokes, as shown in the diagram below. Make sure that you streak the entire plate.

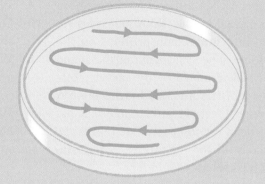

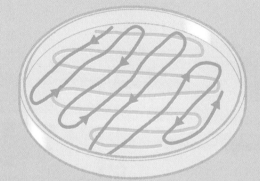

4 Place three different antibiotic disks in the center of each section of the petri dish and close the lid. Seal the petri dish with adhesive tape. Initial the petri dish and indicate the date.

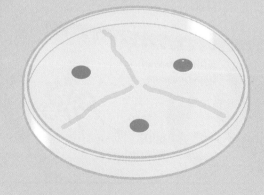

a) Record which antibiotic was placed in each section of the petri dish.

5 Invert the petri dish and place it in a bacteriological incubator set at 37°C.

6 Check the petri dishes after 48 h. Measure the growth ring around each disk. Four separate measurements should be taken, as indicated in the diagram below.

Antibiotic	Measurement (mm)				
	#1	#2	#3	#4	Average
1					
2					
3					

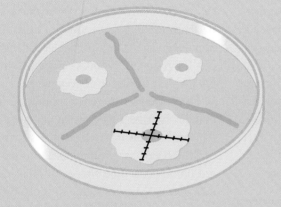

b) Record your data in tabular form as shown.

Laboratory Application Questions

1 On the basis of the experimental evidence, indicate which antibiotic was the most effective.

2 Would the antibiotic that best controlled the bacteria found on your face necessarily be the most effective at controlling the bacteria that causes strep throat? How would you go about testing your hypothesis?

3 What potential problems might be created by placing low dosages of antibiotics in skin creams?

4 Some antibiotics prevent protein synthesis in more advanced cells (eukaryotic cells). Indicate some of the advantages of these antibiotics, and explain why their dosages must be carefully administered. ■

DNA AND MUTATIONS

Mutations are inheritable changes in the genetic material. They can arise from mistakes in DNA replication when one nitrogen base is substituted for another. Cosmic rays, X rays, ultraviolet radiation, and chemicals that alter DNA are referred to as **mutagenic agents.** By changing the arrangement of the nucleotides in the double helix, the mutagen changes the genetic code. The ribosome will read the new code and assemble amino acids according to the new instructions provided. The shift of a single amino acid will, unfortunately, produce a new protein. The new protein has a different chemical structure and, in most cases, is incapable of carrying out the function of the required protein. Without the required protein, cell function is impaired, if not completely destroyed. Although some mutations can, by chance, improve the functioning of the cell, the vast majority of mutations produce adverse effects.

Sometimes the error may arise because of a shortage of a particular type of nucleotide during the replication process. A more plentiful nitrogen base is substituted for the scarce one but, in the process, the genetic code is altered. One of the most common types of errors is created when one nitrogen base is substituted for another. For example, a chemical mutagen called hydroxylamine can modify cytosine so that it pairs with thymine. Normally, adenine forms hydro-

Mutagenic agents *are things that cause changes in the DNA.*

gen bonds with thymine. The hydroxy-lamine removes a nitrogen group attached to the adenine molecule, making it appear much like the guanine molecule. The guanine then bonds with a thymine-containing nucleotide. Occasionally, X rays will break the backbone of the DNA molecule. Special enzymes will repair the break. Unfortunately, the spliced segment of the DNA ladder is not always placed in the correct section. The misplaced segment of DNA may alter the entire library of genetic information. The impact would be similar to reading the first sentence of this paragraph if the word breaks were moved two letters to the left. "Sometimes the error may arise" would read "Sometim est heerr orm ayar ise."

Figure 23.6

Mutations sometimes occur when spliced sections of DNA are returned in the incorrect position.

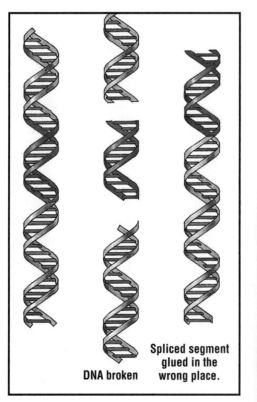

DNA broken | **Spliced segment glued in the wrong place.**

When mutations occur, they are repeated each time the cell divides. A mutation in an egg or sperm cell will lead to permanent change in the characteristics of an offspring. In the adult, many genes code for characteristics that have already developed. For example, your hands, feet, and toes are already in place. If the gene that controls the production of hair color were altered in a single cell, it would have little effect on your body. However, if this same gene were altered in a fertilized egg, the mutation would be coded into the DNA of every succeeding cell of the body. The single mistake can be repeated billions of times. This is why mutagenic agents are particularly dangerous for pregnant women, especially during the first trimester of pregnancy.

One well-known genetic mutation is a human disorder called sickle-cell anemia. This genetic disorder affects the structure of the oxygen-carrying molecule found in red blood cells. The alteration of a single nitrogen base causes valine to replace glutamate as the sixth amino acid in one of the protein chains. Unfortunately, even this slight change has devastating consequences. The red blood cell assumes a sickle shape and is unable to carry an adequate amount of oxygen. To make matters worse, the sickle-shaped cells clog the small capillaries, starving the body's tissues of oxygen.

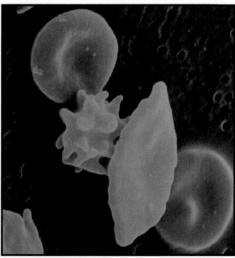

Figure 23.7

Sickle-shaped red blood cells characteristic of sickle-cell anemia. This disease is a hereditary, genetically determined anemia.

CASE STUDY

HUMAN IMMUNODEFICIENCY VIRUS

Objective

To investigate the genetic properties of HIV.

Background Information

Acquired immune deficiency syndrome, or AIDS, describes a number of disorders associated with the infection of the human immunodeficiency virus, or HIV. Two different types of HIV have been identified. HIV-1 was discovered in 1981, HIV-2 in 1985. HIV invades the very cells whose function is to protect the body from **pathogens,** or disease-causing agents. HIV progressively damages the immune system, predisposing a person to a number of opportunistic infections and malignancies. In 1992, there is no cure for AIDS. However, advances in microbiology, genetics, and molecular biology, along with improvements in the microscope are a cause for cautious optimism. For the time being, an educated public may prove to be the best defense against the virus.

HIV must be directly transmitted. Unlike the chicken pox and flu viruses, which can be transmitted through the air, HIV must enter the bloodstream. HIV has been found in human body fluids. It is spread primarily through sexual intercourse and by the introduction of blood or blood components into the bloodstream through the sharing of needles or syringes for injection drug use. HIV can also be transmitted to infants during pregnancy or at the time of birth. In rare cases, HIV has been transmitted through the breast milk of an infected mother.

Although HIV is very tiny, the damage it causes can be devastating. The virus contains a control center of RNA, a sister molecule of the more familiar DNA, and an envelope of protein. What makes HIV even more insidious than other viruses is that it attacks the immune system directly. The helper T cells (sometimes referred to as T4 lymphocytes), the cells that act as guards against invading pathogens, are the targets of HIV. Thus, HIV destroys the body's own defenses, rendering it incapable of defeating other invading substances.

The Case Study

HIV attacks the helper T cells. The shape of the protein coat of the virus permits binding to the cell membrane of the helper T cell much like a lock and key. The unique binding site of the T cell is not designed for HIV, but as a port for hormones or other needed chemicals. HIV takes advantage of the contours of the port of the T cell to launch its invasion. The outer membrane of the HIV is compatible with the outer membrane of the T cell. This explains why certain viruses only infect certain cells.

1 View the diagram of the outer membranes of the helper T cell, the skin cell, and the muscle cell. The drawings are models of the cell membranes, not the actual structures.

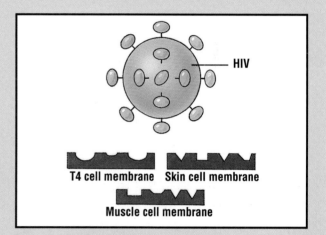

a) Can HIV attach itself to a muscle cell or a cell from the skin?

b) Explain why you cannot get AIDS by shaking hands. (Use the information that you have gained about binding sites.)

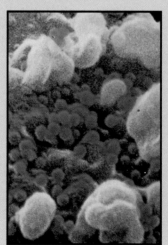

Color-enhanced scanning electron micrograph of the T cell being attacked by HIV particles. The HIV particles appear green in color.

The helper T cell mistakes the virus for a needed substance and engulfs it in a process known as **phagocytosis**. Cells normally engulf large molecules by phagocytosis. The entire virus enters the cell. Once inside, the virus sheds its coat, and the RNA core is set free.

2   View the diagram that shows how HIV enters the helper T cell. The genetic material from HIV is incorporated into the genetic material of the host cell.

   c) Most viruses leave their coat on the membrane of the infected cell. Indicate why these viruses are much more easily identified than HIV.

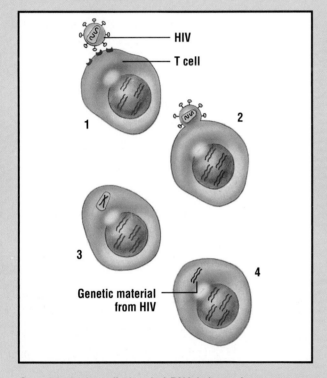

HIV

T cell

1

2

3

4

Genetic material
from HIV

Once inside the cell, the viral RNA behaves in a very special way. Normally, the DNA acts as a blueprint for instructions, and the mRNA molecule reads the hereditary message from it. The process is referred to as *transcription*. Under normal circumstances, the RNA acts as a messenger by carrying the genetic information from the DNA in the nucleus to the cytoplasm.

Once in the cytoplasm, the mRNA provides the ribosome with the correct instruction for the assembly of a protein. Another molecule of RNA, tRNA, carries amino acids to the ribosome. Here the code provided by the mRNA dictates the position of the amino acids. The ribosome links individual amino acids together in a process known as *translation*. Proteins are made from amino acid building blocks.

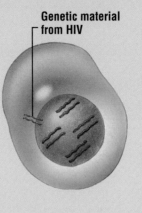

Genetic material
from HIV

Most viruses contain DNA, but HIV is different; that is, its genetic material is RNA. The RNA molecule of HIV is much more than just a messenger. A special enzyme called **reverse transcriptase** allows the genetic message contained in the RNA of the virus to be printed along a strand of DNA.

   d) Why is the enzyme referred to as reverse transcriptase?

The newly constructed viral DNA now slips into the nucleus of the infected cell. Here it can splice into healthy DNA. The instructions of HIV are now part of the helper T cell. The virus can remain dormant for many years. The viral DNA becomes part of the human DNA.

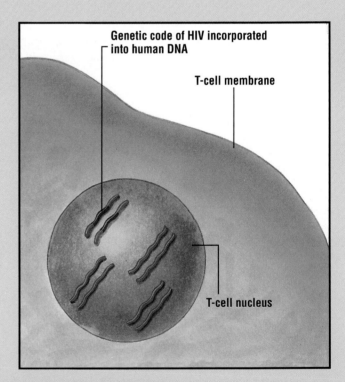

Genetic code of HIV incorporated
into human DNA

T-cell membrane

T-cell nucleus

3   Use the previous diagrams for reference to answer the following two questions.

   e) What happens to the viral DNA if the T cell divides?

   f) Explain why it is possible for a human to be infected with HIV and not exhibit any of the symptoms of AIDS?

542   ■

No one knows what activates HIV. Some scientists speculate that it could be other co-infections. Others believe the virus has a gene that acts like a ticking time bomb. Whatever the reason, once activated, the virus reproduces at a furious pace.

Viral DNA stimulates the production of viral RNA, which carries the virus message into the cytoplasm. The transcribed RNA attaches itself to the ribosome and directs it to produce viral proteins. The proteins will serve as coats for the newly released RNA.

The once-healthy helper T cell has been transformed into an HIV factory. Eventually, the overworked and exhausted T cell bursts, and the virus particles are released to infect other T cells.

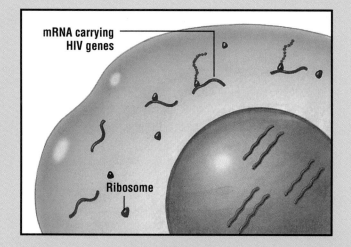

mRNA carrying HIV genes

Ribosome

g) Normally, the killer T cells would destroy a cell infected with a virus long before it could become a virus factory. Why are the infected helper T cells not held in check?

HIV creates yet another problem for the helper T cells. Once infected, the outer membrane of the helper T cells changes, causing them to fuse together. This allows the virus to pass from one cell to another without entering the bloodstream. Because antibodies move freely through the fluid portions, HIV avoids them by slipping from one cell to another. The T cell not only produces other HIV viruses, but also protects them from wandering antibodies.

h) Indicate why people infected with HIV most often die of another infection such as pneumonia?

i) David, "the boy in the plastic bubble," suffered from a disorder called "severe combined immunodeficiency syndrome." How does this disorder differ from acquired immune deficiency syndrome?

The challenge presented in finding a cure for AIDS stems from its variable protein coat. Each mutation produces a distinctive coat.

j) Why is it so difficult to destroy a virus that changes shape?

(a) Electron micrograph showing HIV escaping from the cell membrane of a T cell.

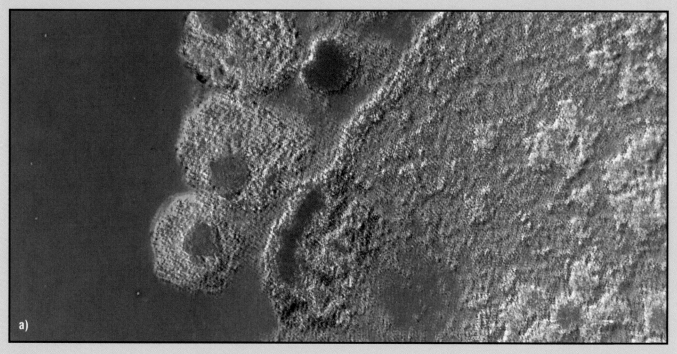

a)

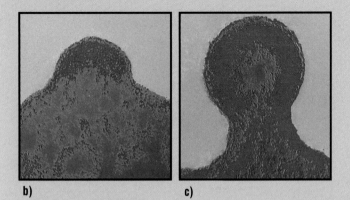

b)   c)

Pictures (b) and (c) show a closer view of the infected T cell as the virus escapes through the cell membrane.

A drug known as Zidovudine (formerly known as AZT) hampers the replication of HIV. Zidovudine has proved somewhat effective if the virus lies dormant. However, for those who already exhibit the symptoms of AIDS, it has limited results. Although Zidovudine may help increase the life span of people infected with HIV, it cannot be viewed as a cure. At $10 000 a year per patient, the cost of the drug is also a major concern.

The testing of vaccines has been underway in both France and the United States. Daniel Zagury of the Université de Pierre et Marie Curie announced that he had injected himself and others with a vaccine developed by Bernard Moss. To date, the vaccine has produced no observable side effects, but hope remains tempered with caution.

Major breakthroughs have come in the form of tests for detecting HIV. Blood collected by the Red Cross in Canada has been screened for the presence of HIV since 1985.

People receiving transfusions or transplants no longer have to worry about acquiring HIV through donated blood. In June 1987, the Du Pont company announced a new test that can detect HIV antibodies. The new test is much cheaper and easier to run. It takes only about five minutes to complete and does not require refrigeration. This breakthrough is especially significant for central Africa, the area with the highest incidence of AIDS cases. Unfortunately, it is also one of the poorest areas in the world. Currently, the United States spends more money testing for AIDS than central African states allocate for their entire health budget.

Case-Study Application Questions

1 Why does the Canadian Red Cross inquire about a person's travel before they accept blood donations?
2 How does transcription for HIV differ from normal cell transcription?
3 Can AIDS be transmitted through either food or beverages? Explain your answer.
4 Can AIDS be contracted by casual contact such as shaking hands, or using the same telephone or toilet seat? Explain your answer.
5 Do tattoos and pierced ears pose any potential risks for infection with AIDS?
6 How is it possible to catch HIV from a person who shows no symptoms associated with AIDS?
7 Should people with AIDS be quarantined? Justify your answer.
8 Should health-care workers such as doctors, dentists, and nurses be screened for HIV? Justify your answer. ■

ONCOGENES: GENE REGULATION AND CANCER

Cancer is characterized by uncontrolled cell division. Cancer cells are often described as cells that are "too alive." But what causes normal cells to become cancerous? Two lines of evidence indicate that cancer results from changes in the genetic code. First, cancer cells often display nitrogen base substitution, or the movement of genetic material from one part of the chromosome to another. A second factor that supports the idea that cancer arises because of alterations of DNA is the fact that many known mutagens are also known to cause cancer. X rays, ultraviolet radiation, and mutagenic chemicals can also induce cancer.

In 1982, molecular biologists were able to provide additional evidence to support the hypothesis that cancer could be traced to genetic mutations. Segments of chromosomes extracted from cancerous mice transformed normal mouse cells growing in tissue cultures into cancerous cells. The cancer-causing genes, called **oncogenes,** seemed to turn on cell division.

Further studies indicate that cancer-causing oncogenes are present in normal strands of DNA. But if oncogenes are found in normal cells, why do normal cells not become cancerous? One of the current theories that has gained acceptance from the scientific community suggests that the cancer gene has been transposed to another gene site. Such transpositions may have been brought about by environmental factors or mutagenic chemicals. The movement of the oncogene away from its regulator gene may have caused the problem.

You have read how genes direct the ribosomes to assemble amino acids into needed proteins. These genes are referred to as **structural genes** because they direct the protein building of individual cells. The DNA signals the production of hemoglobin proteins in immature red blood cells and of keratin protein in skin cells. But what turns on the production of required proteins? The genes are not active all the time. Another important question is, "What causes the skin cells to produce keratin, a skin protein, but prevents them from making the protein component of hemoglobin?" The skin cell contains all the genetic instruction to produce all of the proteins made by every cell of your body. How does a skin cell turn "on" certain genes, while turning "off" others? A gene called the **regulator gene** acts like a switch that turns "off" segments of the DNA molecule.

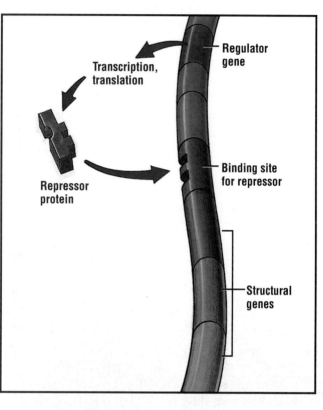

Figure 23.8

The attachment of the repressor to the binding site turns off the structural genes.

Cancer genes leave the switch regulating cell division open. In the past 10 years, a great number of oncogenes have been uncovered. In 1982, researchers exposed mouse cells to a cancer-producing chemical, methlycholanthrene. By extracting DNA from the treated cells and injecting it into normal cells, researchers discovered that some of the normal cells became cancerous. Evidently, the DNA from the treated cells had been incorporated into the normal DNA. The DNA contained oncogenes, which are mutated forms of the normal genes.

Oncogenes *are cancer-causing genes.*

Structural genes *are genes that direct the synthesis of proteins.*

Regulator genes *control the production of repressor proteins, which switch off structural genes.*

The most common oncogene, *ras*, is found in 50% of colon cancers and 30% of lung cancers. Present in normal cells, *ras* makes a protein that acts as an "on" switch for cell division. *Ras* ensures that cells divide to replace damaged or dead cells. After a sufficient number of cells have been produced, the *ras* gene should be turned off. But the cancer-causing oncogene produces a protein that blocks the "off" switch. With the switch left on, cell division continues at an accelerated rate. Some researchers speculate that the mutation separates the regulator and the structural genes.

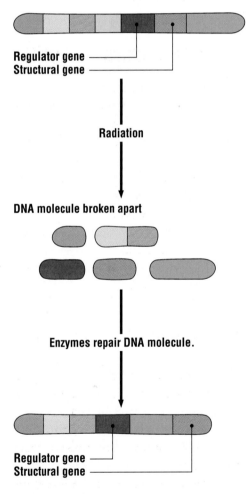

DNA molecule

Regulator gene
Structural gene

Radiation

DNA molecule broken apart

Enzymes repair DNA molecule.

Regulator gene
Structural gene

Figure 23.9

Mutagenic agents may cause the separation of the regulator and structural genes.

FRONTIERS OF TECHNOLOGY: THE AMES TEST

The relationship between mutagens and carcinogens is becoming more evident as research progresses. The ultraviolet radiation that comes to earth with sunlight has long been associated with skin cancer. Benzene, a chemical once used as a solvent, is a known carcinogen. Certain food preservatives, some artificial sweeteners, and types of food dyes have been identified as mutagenic and carcinogenic. Pesticides, industrial vinyl chlorides, and a host of other chemicals have also been shown to be cancer-causing. With new products bombarding consumers daily, how can so many products be adequately tested before a segment of the population has been exposed to them?

A new procedure has been developed that can test cancer-causing agents in drugs, foods, and even in drinking water. Named after its founder, Dr. Bruce Ames, a professor at the University of California, Berkeley, this method is faster and less expensive than conventional methods that employ test animals. Most important, this revolutionary procedure is also extremely accurate. The technique involves testing the potential of a chemical to alter the DNA of an organism. It is generally accepted that cells can become cancerous if their genetic information is altered. Although all gene mutations do not lead to cancer, it is assumed that a chemical's ability to cause a mutation is a measure of its potential to cause cancer. Any chemical that rapidly alters the DNA molecule could produce a mutation that affects a cell's ability to regulate its own rate of reproduction.

The Ames test, shown in Figure 23.10, is performed on microorganisms. The bacterium *Salmonella typhimurium* is most commonly used. The bacteria have undergone a mutation that has made them unable to produce an essential chemical called histidine. In order for the bacteria to grow, histidine must be supplied. A salmonella culture is placed on a petri dish, and the test chemical is added. Because the growth medium in the petri dish does not contain histidine, no growth would be expected. If bacteria colonies are found, the conclusion must be that the microbe has mutated and is now producing its own histidine. The more colonies there are, the more mutagenic is the test chemical. The chemical is often added to liver enzymes. Although the chemical itself may be harmless, it may be broken down in the liver into toxic metabolites.

Like most scientific models, the mutation of a microbe does not signify that the substance will cause cancer in all humans. It is possible that a toxic chemical could cause many genetic changes in a person, yet the changes might never manifest themselves as cancer. On the other hand, this same chemical could cause cancer in another individual. Although our DNA is composed of essentially the same chemicals, the arrangement of the nitrogen bases is unique. Some sequences seem more susceptible to change by chemical agents than others. The Ames test not only allows carcinogens to be identified in a less costly manner, but it also provides researchers with information on how chemicals alter DNA.

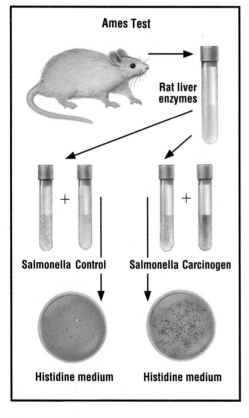

Figure 23.10

The appearance of colonies of bacteria indicate that some mutation has occurred. Should the experimental plate have more colonies than the control, scientists conclude that the tested chemical is causing the mutations. Chemicals that cause mutations are often carcinogens.

■ REVIEW QUESTIONS ?

8 What are mutations?

9 Indicate three factors that can produce gene mutations.

10 Explain why a food dye that has been identified as a chemical mutagen poses greater dangers for a developing fetus than for an adult.

11 What are oncogenes?

12 How does the Ames test identify cancer-causing agents?

13 How do regulator genes turn off structural genes?

BIOLOGICAL WARFARE

The history of biological warfare goes back to about 600 B.C. when Solon, an Athenian legislator, poisoned the water supply of the city of Kirrha. He had roots from the *Helleborous* plant, which contain a particularly potent toxin, placed in the drinking water. During the 14th century, the Tartar army hurled the disease-ridden dead bodies of plague victims over the walls of the city of Kaffa.

RESEARCH IN CANADA

Detecting Carcinogens

More than one thousand different chemicals are introduced into North American markets each year. How can we be sure these new chemicals are safe? What happens when these chemicals combine with others in the environment? Although a particular chemical may be harmless by itself, it could prove to be lethal in combination with another chemical. How can we be sure the combinations are adequately tested for toxicity? Past experience indicates that some toxic chemicals have escaped detection. Traditional animal testing can take from two to four years to complete. The longer the testing period, the more expensive the program.

Molecular biology provides at least a partial solution to the testing dilemma. Dr. Ram Mehta is a genetic toxicologist and president of Prairie Biological Research, a biotechnology company in Alberta, concerned with identifying carcinogenic materials in our environment. Using the same principle as Dr. Bruce Ames, Dr. Mehta has placed gene markers in yeast, a primitive organism that is capable of rapid reproduction. However, unlike the Ames test, which relies on a prokaryote modeling experiment to determine mutagenic capacity, Dr. Mehta's method employs a eukaryotic yeast cell. The mechanisms of cell reproduction in the yeast cell more closely mirror those in human cells than do those in bacteria.

DR. RAM MEHTA

American soldiers took advantage of the North American native peoples' susceptibility to smallpox to drive them from their lands. Blankets, contaminated with the deadly virus, were used in trade.

During World War II, British, American, and Canadian armies were involved in the development of biological weapons. Porton Down in England, Camp Detrick in Maryland, and Suffield in Alberta were designated as research and testing stations. The preferred microbe was the deadly anthrax bacterium, which affects both cattle and humans. The deadly spores of the rod-shaped bacterium live for long periods of time, are highly contagious, and are resistant to many environmental factors. American and British armies had planned to make thousands of anthrax bombs that would disperse the microbes on impact. The war ended before the plan was implemented.

The Japanese army was also engaged in the development of biological weapons during World War II. Pingfan, a small village near Harbin in China, housed over 3000 researchers, technicians, and soldiers who were dedicated to exploiting the disease-causing properties of typhoid fever, anthrax, and cholera.

Almost any disease-causing agent can be exploited for biological weaponry. The microbe, or toxin produced by the microbe, need not be harmful to humans. An enemy could target the destruction of livestock, cereal grains, or bacteria found in the soil to create food shortages and cause economic ruin. Fortunately, few organisms are well-suited for mass destruction. HIV, the virus that causes AIDS, cannot be transmitted through the air or by casual contact. It is transmitted in body fluids such as blood. Therefore, releasing the virus into a city's drinking water would not create an epidemic. The bacterium *Clostridium botulinum* produces one of the most powerful poisons known to humans. It has been estimated that one kilogram of the toxin placed in a typical water reservoir could kill 50 000 people. Sixty percent of the population would die in less than 24 h. However, this microbe would have little effect if released into the air. *Clostridium* cannot live in environments where oxygen is present. The microbe must be cultured in oxygen-free environments, where it is capable of producing the harmful toxin. Even the most dangerous of microbes have some natural controls that have evolved over a period of time. No disease-causing agents can survive if their hosts are eliminated.

Combining genes for weaponry is a frightening prospect. Merging genes that permit rapid reproduction with those that demonstrate a resistance to environmental factors could create a "superbug." Bacteria that carry drug-resistant genes in their plasmids already create problems. The genetic information can be duplicated and passed between microbes during sexual reproduction. Consider the possibility of a microbe that is resistant to a range of antibiotics being introduced to an enemy population. The attacking army might be able to construct a secret drug that is capable of protecting their allies while selectively removing the enemy. By the time the enemy finds an antibiotic for the disease, a significant number would have died and their resistance weakened. The hybrid microbe might provide yet another advantage. It is quite likely that disease-causing microbes could escape detection by both the body's immune system and physicians. For example, *E. coli* bacteria are a normal fauna of the human gut. The body will not mobilize an immune response against this organism. By splicing disease-causing genes into the *E. coli*, the source of the infection would be masked. Most physicians would have little difficulty identifying an anthrax infection, but associating the symptoms of a disease with the *E. coli* microbe would be much more difficult.

CAREER INVESTIGATION

The applications and uses of genetic technology are increasing every day. Biotechnology can be applied to almost everything in our lives. The scope of this field ranges from agriculture to pharmacology, from oil spill cleanup to outer space exploration.

Forensic scientist

Police are using more sophisticated techniques to determine the identity of criminals. DNA fingerprinting, for example, is a powerful tool in the hands of a trained technician.

- Identify a career associated with biotechnology.
- Investigate and list the features that appeal to you about this career. Make another list of features that you find less attractive about this career.
- Which high-school subjects are required for this career? Is a post-secondary degree required?
- Survey the newspapers in your area for job opportunities in this career.

Astronaut

As well as being physically and mentally fit, an astronaut must possess a wide-ranging scientific knowledge. Astronauts are called on to perform many different experiments in space, some of which deal with biology and biotechnology.

Fertility specialist

Many couples who are unable to have children are turning to fertility specialists. A fertility specialist must determine the nature of the problem and provide a solution. The solution may involve a number of techniques such as hormone treatments or *in vitro* fertilization.

Stockbroker

Many biotechnology companies have started up over the past few years. Biotechnology stocks are a billion-dollar business. Knowledge about which stocks to buy requires scientific knowledge. A stockbroker must be able to predict how well the company will do in order to properly advise clients.

SOCIAL ISSUE:

Biological Warfare

The development of chemical weapons during World War I shows the extent to which science and technology can be exploited for destructive purposes. During that war, chlorine gas, phosgene gas, and mustard gas disabled over one million soldiers.

Advancements in genetics may have made nature an unwilling accomplice in war. Recent biological techniques, responsible for the development of antibiotics, vaccines, and genetically created industrial products, have opened the door for more harmful applications.

Statement:

Research that could be used for biological warfare must be banned.

Point

- In 1985, the United States government reported spending $39 million on defensive biological weapons programs. The programs were directed at developing vaccines for genetically-created microbes. The money would be much better spent on research to develop vaccines for naturally occurring diseases.
- Biological weapons present a real threat to world peace. Unlike nuclear weapons, biological weapons are cheap to produce, portable, and have a built-in time delay. Therefore many smaller countries with militaristic governments are able to build and use them.

Counterpoint

- The money spent on the development of this type of defense cannot be construed as money spent on weapons. The development of vaccines and antibiotics has historically been associated with the growing and culturing of disease-causing organisms.

- The threat posed by biological weapons is exaggerated. It is not likely that humans can construct microbes that are more dangerous than those that nature has taken years to perfect.

Research the issue.
Reflect on your findings.
Discuss the various viewpoints with others.
Prepare for the class debate.

CHAPTER HIGHLIGHTS

- The genetic code is determined by the arrangement of nitrogen bases within the strands of DNA.
- Each gene codes for the production of a specific protein.
- Messenger RNA reads the chemical message inscribed on the DNA and carries the information to the ribosomes located in the cytoplasm. The transfer of the genetic message from the DNA to the tRNA is known as transcription.

- Ribosomes are the site of protein synthesis.
- Transfer RNA molecules carry amino acids to the ribosome for protein synthesis. The tRNA molecules place the amino acids along a strand of mRNA, which is on the ribosome. The mRNA determines the sequencing of the amino acids. The anticodons of tRNA must combine with complementary codons from the mRNA. The process is referred to as translation.

- Gene mutations occur when the nitrogen base sequence in DNA is altered.
- Gene control mechanisms regulate gene expression.
- The operon consists of operator and structural genes.
- Cancer genes, called oncogenes, leave the molecular switches for cell division on.

- The Ames test can be used to identify carcinogens.
- Biological warfare is the exploitation of microbes or the products of living things for military ends. Many ethical questions must be explored as our knowledge of gene action is extended and applied.

APPLYING THE CONCEPTS

1 Why are somatic cell mutations less harmful than germ cell mutations?
2 Explain how Beadle and Tatum's experiments with the bread mold *Neurospora* helped explain protein synthesis.
3 In what ways does mRNA differ from DNA?
4 Suppose that during protein synthesis, nucleotides containing uracil are in poor supply. The uracil is substituted with another nitrogen base to complete the genetic code. How will the protein be affected by the substitution of nitrogen bases?
5 In what ways do the codons and anticodons differ?
6 A scientist discovers a drug that ties up the site at which mRNA attaches to the ribosome. Although mRNA is transcribed in the nucleus, it has no way of attaching itself to the ribosome. Speculate about how the drug will affect the functioning of a cell. If the drug is introduced into a single brain cell, will the organism be destroyed?
7 Outline the advantages of Dr. Ram Mehta's genetic tests using yeast cells over those that use animal testing.
8 Compare protein synthesis in cells with an automobile assembly plant. Match the parts of the auto assembly plant with the correct part of the cell. Provide reasons for each of the matches.

Auto assembly plant:	Cell parts:
1) Corporate headquarters	a) tRNA
2) Master blueprint for the car	b) cytoplasm
3) Entire shop area	c) nucleus
4) Supervisor who carries blueprints	d) DNA
5) Stockperson who brings parts to the assembly worker	e) mRNA
6) Assembly worker	f) amino acid
7) Parts of the automobile	g) ribosome

9 Thalidomide was a drug given to pregnant women in the early 1960s to reduce morning sickness. Unfortunately, the drug caused irreparable damage to the developing fetus. Thalidomide inhibited proper limb formation. Many children were born without arms and legs. Explain how the Ames test could have prevented this tragedy.
10 Recombinant DNA provides a valuable technique for those engaged in biological warfare. The gene that produces the botulism toxin, a deadly food poison, can be extracted from its resident bacteria and placed into a harmless bacterium, called *Escherichia coli*, which is a natural inhabitant of the large intestine of humans. Why would the transfer of the botulism gene to the bacteria found in your gut be so dangerous?

CRITICAL-THINKING QUESTIONS

1 Recombinant DNA has produced human insulin in bacteria. Because millions of people suffer from diabetes, the market for human insulin is enormous, and so are the profits. Because of economic pressures, the nature of scientific research has changed from one of sharing information through publishing to one of patents and secretiveness. Companies are unwilling to release any breakthrough for fear that their ideas might be stolen. Scientific research, at least in some fields, is no longer controlled by researchers, but by investors, who seek the advice of accountants and lawyers.

Should government remove biotechnology from private enterprise? Support your opinion.

2   John Moore suffered from a rare form of leukemia. A team of doctors from UCLA removed his spleen, and he went on to live a fruitful life. However, years after his operation, Moore discovered that the surgeons were cloning cells from his extracted spleen for cancer research. The cells produced significant amounts of a substance called interferon, which has tremendous commercial value. Moore now wants what he believes is his fair share of the money, and he is suing the people who are cloning the cells. Should Moore be entitled to any money? Is the spleen Moore's property? Support your conclusions.

3   *Pseudomonas syringae* is a bacterium found in raindrops and in most ice crystals. Researchers have been able to snip the frost gene from its genetic code, thereby preventing the bacteria from forming ice crystals. The *Pseudomonas* has been aptly named "frost negative." By spraying the bacteria on tomato plants, scientists have been able to reduce frost damage. The bacteria can extend growing seasons, thus increasing crop yields, especially in cold climates. A second version, called "frost positive," has also been developed. The frost-positive strain promotes the development of ice and, not surprisingly, has been eagerly accepted by some ski resorts. When this microbe is sprayed on ski slopes, a longer season and better snow base can be assured. However, environmental groups have raised serious concerns about releasing genetically engineered bacteria into the environment. Could these new microbes gain an unfair advantage over the naturally occurring species? What might happen if the genetically engineered microbes mutate? Could the mutated microbe become a super microbe? Do you think genetically engineered microbes should be introduced into the environment? Support your conclusions.

4   The gene for the growth hormone has been extracted from human chromosomes and implanted into bacteria. The bacteria produce human growth hormone, which can be harvested in relatively large quantities. The production of human growth hormone is invaluable to people with dwarfism. Prior to the development of this hormone, people with dwarfism relied on costly pituitary extracts. Although the prospect of curing dwarfism has met with approval from a majority of the scientific community, some concerns about the potentially vast supply of growth hormone have been raised. How can scientists ensure that the growth hormone produced by these genetically engineered bacteria will not be used by normal individuals who wish to grow a few more centimeters? Do people have the right to choose their own height? Give your opinion and support your conclusions.

ENRICHMENT ACTIVITIES

Suggested reading:

- Barnes, James. "The Ancient Quest." *Canada and the World*, January 1990, pp. 14–16.
- Beardsely, Tim. "Smart Genes." *Scientific American* 265(2) (August 1991).
- DeDuve, Christian. *A Guided Tour of the Living Cell*. New York: Freedman and Company, 1984.
- Eberlee, John. "Biology's Holy Grail." *Canada and the World*, January 1990, pp. 24–25.
- ——. "The Life Molecule." *Canada and the World*, January 1990, pp. 17–19.
- Grady, Denise. "The Ticking Time Bomb." *Discover* 6 (6) (June 1987).
- Mackenzie, R.C., "Designer Genes." *Canada and the World*, January 1990, pp. 20–24.
- Maxson, Linda R., and Charles H. Daughtery. *Genetics: A Human Perspective*. Dubuque, Iowa: Brown, 1986.
- Petruz, Max. "The Birth of Protein Engineering." *New Scientist* 1460 (June 1985).
- Suzuki, David, and Peter Knudson. *Genethics*. Toronto: Stoddart, 1988.
- Suzuki, David, Eileen Thalenburg, and Robert Sinshiner. *Let's Talk about Aids*. Toronto: General Paperbacks, 1987.

Change in Populations and Communities

■

U N I T **7**

CHAPTER 24

Population Genetics

IMPORTANCE OF VARIATION

Variation among living organisms is not restricted solely to physical appearance. It can also be expressed in an organism's metabolism, fertility, mode of reproduction, behavior, or other measurable characteristics. Although Darwin knew from observation of nature that the majority of species possessed variable phenotypes, he was unable to explain the source of variation or how it is passed on. Ironically, in 1859, the same year in which Darwin published *On the Origin of Species*, Mendel began his genetic research with experiments involving garden peas. This work would eventually lay the foundation for genetics, and demonstrate conclusively that the source of genetic variation among individuals is sexual reproduction.

The modern theory of evolution recognizes that the main source of variation in a population lies in the differences in the genes carried by the chromosomes. Genes determine an organism's appearance, and mutations (permanent genetic changes) can cause new variations to arise. These variations can be passed on from generation to generation.

Certain genotypes may be better equipped than others for survival. Organisms with these genes might be better able to obtain necessary resources such as food and water, or to protect themselves against predators, or

Figure 24.1

Birds like the albatross have developed elaborate courtship behavior.

they may have higher reproductive potentials. Through sexual reproduction, the genes for these variations would be transmitted to the offspring. Given that such offspring are also more likely to survive, subsequent generations would include an increased frequency of the variant genes. Consequently, there would be natural selection, within the group, of individuals better adapted to prevailing conditions.

Over billions of years the change in genetic makeup of the individual within the species has been a major contributing factor to earth's most valuable resource: biological diversity, or biodiversity.

GENES IN HUMAN POPULATIONS

The principles of genetics, established by studies on plants and fruit flies, can be applied to humans. Human chromosomes, like those of other organisms, are composed of DNA, and undergo mitosis and meiosis. However, the study of human genetics presents some unique problems. Unlike garden peas and *Drosophila*, humans produce few offspring, which makes it difficult to determine the genotypes of both the parents and offspring for any particular trait. Another problem with human genetics is that observing successive generations requires time. *Drosophila*, in contrast, can reproduce every 14 days, and many different generations can be studied within a few months. A third problem is that many human traits, including body size, weight, or even intelligence, are affected by environment as well as by genes.

One of the most common techniques used to study human populations is **population sampling.** In the sampling technique a representative group of individuals within the population is selected, and the trends or frequencies displayed by the selected group are used as indicators for the entire population.

One example of a trait that can be studied by means of sampling is tongue rolling. The ability to roll the tongue is controlled by a dominant gene. People with two recessive genes cannot roll their tongues. Approximately 65% of the population carry the dominant gene.

Blood type is another example of genes that can easily be studied within a human population. The I^A, I^B, and I^O alleles combine to make different blood types. However, unlike the tongue-rolling trait, in which the dominant allele occurs in greater frequency, in blood types the recessive allele is more common. Type O blood is the most common among North American whites and blacks. An estimated 45% of whites and 49% of blacks contain two recessive I^O alleles. Despite the fact that the I^A and I^B alleles are both dominant to I^O, only 4.0% of the white population and 3.5% of the black population contain both the I^A and I^B alleles. Blood type AB is considered rare.

The recessive Rh negative alleles are found only in 15% of Canadians. However, the frequency of Rh– alleles is much higher along a valley region that borders France and Germany. Some population geneticists have postulated that the origin of the Rh– allele can be traced to that valley.

All of the genes that occur in a population are referred to as the **gene pool.** The gene pool maintains continuity of traits from generation to generation. Although some gene frequencies remain the same over many generations, others change quickly. Geneticists have used gene frequencies to study changes in the human population. Certain gene frequencies have been associated with a particular population of people. For example, red hair is often associated with people of Irish or Scottish ancestry.

Population sampling *is a technique in which gene frequencies for a particular genetic trait are determined in a small sample of the population, and results are applied to the whole population.*

Gene pools *are all of the genes that occur within a specific population.*

CASE STUDY
TRACING THE HEMOPHILIA GENE

Objective

To use pedigree charts to trace the hemophilia gene from Queen Victoria.

Background Information

A pedigree chart provides a means of tracing the inheritance of a particular trait from parents through successive generations of offspring. Hemophilia A is a blood clotting disorder that occurs in about one in 7000 males. The disorder is associated with a recessive gene located on the X chromosome. The fact that a female must inherit one of the mutated genes from her mother and another of the mutated genes from her father helps explain why this disorder is very rare in females.

Procedure

1 Study the pedigree chart of Queen Victoria and Prince Albert. Note the legend. Males are designated by a square, while females are designated by a circle.

a) Who was Queen Victoria's father?

b) How many children did Queen Victoria and Prince Albert have?

2 Locate Alice of Hesse and Leopold, Duke of Albany, on the pedigree chart.

c) Using the legend, provide the genotypes of both Alice of Hesse and Leopold.

3 Locate the royal family of Russia on the pedigree chart. Alexandra, a descendant of Queen Victoria, married Nikolas II, Czar of Russia. Nikolas and Alexandra had four girls (only Anastasia is shown), and one son, Alexis.

d) Explain why Alexis was the only child with hemophilia.

Case-Study Application Questions

1 Is it possible for a female to be hemophilic? If not, explain why not. If so, identify a male and female from the pedigree chart who would be capable of producing a hemophilic, female offspring.

2 On the basis of probability, calculate the number of Victoria's and Albert's children who would be carriers of the hemophilic trait. ■

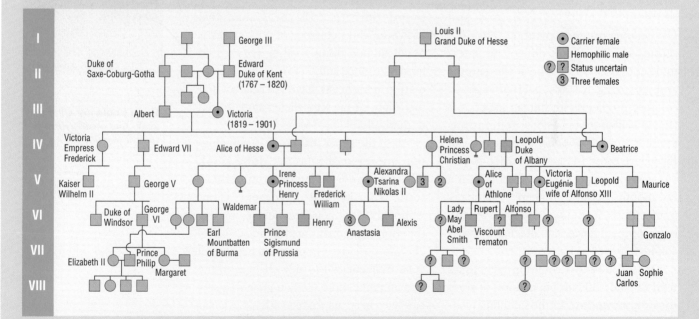

RESEARCH IN CANADA

Dr. Norma Ford Walker (1893–1968) was one of Canada's leading medical geneticists. Working in the Toronto area, Dr. Walker became an authority on genetics in children. She came to prominence because of her work with multiple births, following the birth of the celebrated Dionne quintuplets. By tracing the inheritance of genetic defects through pedigree charts, Dr. Walker provided valuable data that helped establish the genetic basis for many disorders.

Genetic Research

Dr. Frank Clarke Fraser was born in Norwich, Connecticut, in 1920. He received his B.Sc. from Acadia University and his M.Sc., Ph.D., and M.D. from McGill. Dr. Fraser's work, showing how genes and environmental factors interact to cause malformations in mice, provided a basis for investigations in human genetics. Under his guidance, the first medical genetics clinic was established in Canada in Montreal. Dr. Fraser has published more than 250 works on medical genetics and in 1985 was made a member of the Order of Canada.

DR. F. C. FRASER
DR. NORMA FORD WALKER

POPULATION EQUILIBRIUM

Populations are characterized by two competing factors: the tendency to remain stable and the tendency toward variability. In 1908, an English mathematician, G.H. Hardy, and a German physician, W. Weinberg, independently derived the basic principle of population genetics. Known as the **Hardy-Weinberg principle**, it predicts that, if all other factors remain constant, the gene pool will have the same composition generation after generation. This stability is called *genetic equilibrium*; only if that equilibrium is upset can the population evolve.

The Hardy-Weinberg principle is expressed by the following mathematical equation:

$$p^2 + 2pq + q^2 = 1$$

where p = the frequency of allele A and
q = the frequency of allele a in a population

If the values for p and q are known, this equation can be used to calculate the frequency of the three genotypes AA, Aa, and aa. On the other hand, if the frequencies of the three genotypes are known, the frequencies of the alleles can be calculated. The equation, which is based on probabilities, applies to an idealized population of sexually reproducing organisms that is not evolving. The conditions under which no change will occur in a gene pool are:

- Large populations—This condition is necessary to ensure that changes in gene frequencies are not the result of chance alone.
- Random mating.
- No mutations.
- No migration—No new genes enter or leave the population.

The Hardy-Weinberg principle indicates conditions under which allele and gene frequencies will remain constant from generation to generation.

- Equal viability, fertility, and mating ability of all genotypes (i.e., no selection advantage).

Today, scientists know that evolution, as a change in the genetic composition of a population, is a change in allele frequencies (and hence genotype frequencies) within gene pools. It is possible to determine what factors cause evolution by determining what factors cause a shift in the allele frequencies.

Application of the Hardy-Weinberg Principle

Consider the simplest situation: one gene with two alleles, A and a. The genotypes that might be found in a large population will be AA, Aa, and aa. In mathematical terms, the frequencies with which the alleles will occur must add up to 1 (and so must the frequencies of the genotypes). Hence, if the dominant allele A has a frequency of 0.7 (i.e., is found in 70% of the genes), then the recessive allele a will have a frequency of $1.0 - 0.7 = 0.3$. The expected frequencies of the three possible genotypes can be calculated by means of a Punnett square. (Note that, unlike the genetic Punnett square used for determining individual traits, the eggs and sperm in this Punnett square represent the genes for the entire population.)

	Sperm	
	A (0.7)	a (0.3)
A (0.7)	*AA* (0.49)	*Aa* (0.21)
a (0.3)	*aA* (0.21)	*aa* (0.09)

(Eggs)

The equation predicts that the frequencies of the three genotypes possible in the next generation will be:

$$p^2 + 2pq + q^2 = (0.7)^2 + 2(0.7 \times 0.3) + (0.3)^2 = 1$$
Genotypes: 49% - *AA*; 42% - *Aa*; 9% - *aa*

Given this distribution of genotypes, it is possible to predict the frequency of the A and a alleles in the population as follows:

F$_1$ generation: 0.49AA 0.42Aa 0.09aa

Potential gametes: Ⓐ Ⓐ Ⓐ ⓐ ⓐ ⓐ

A 0.49 + 0.21 = 0.7

a 0.21 + 0.09 = 0.3

Because the allele frequencies are the same as in the original gametes, identical results will be obtained generation after generation. In nature, however, a population rarely meets the ideal conditions for genetic equilibrium. Why, then, bother with the principle? The principle makes it possible to make predictions about populations that are *not* evolving. From this reference point, it is possible to consider the special conditions that serve as a measure of the rates of evolutionary change.

In the above example, allele frequencies were assumed so that the corresponding genotype frequencies could be calculated. But the Hardy-Weinberg principle can be applied to other situations, as in the following examples:

Example 1

Suppose a recessive genetic disorder occurs in 9% of the population. How is it possible to determine what percentage of the population is heterozygous, or "carries" the allele for the disorder, but does not have the disorder? Since the disease occurs only in homozygous recessive individuals, the frequency of the aa genotype is 0.09. By the Hardy-Weinberg formula,

$$p^2 + 2pq + q^2 = 1$$

Let q^2 stand for the frequency of aa; then,

the frequency of $a = 0.3$, and the frequency of A must be $1.0 - 0.3 = 0.7$. By substituting into the formula it is possible to calculate the frequencies of the genotypes:

$$p^2 + 2pq + q^2 = 1$$
$$(0.7)^2 + 2(0.7 \times 0.3) + q^2 = 1$$
$$(0.49) + (0.42) + (0.09) = 1$$

The term $2pq$ stands for the frequency of the heterozygous genotype. Therefore, the percentage of the population who are **carriers** is 42%.

Example 2

The Hardy-Weinberg formula can also be used to calculate changes in allele frequencies when only the phenotype frequencies are known.

For a hypothetical moth population that is freely interbreeding, suppose that 60% of the moths are white-colored and 40% are dark-colored. White color (allele W) is dominant. In three years, the observed color percentages change to 65% white and 35% black. What does this shift say about the dark phenotype?

Since w is recessive, dark moths must be ww, and the frequency of the ww genotype is 0.40. The value of w can now be determined. By setting $q^2 = 0.40$, it can be calculated that $w = \sqrt{0.40} = 0.63$. Since $w + W = 1.0$, the frequency of W must be 0.37. By the same procedure, the frequency of the alleles in year three is $w = 0.59$ and $W = 0.41$.

The frequency of W has therefore changed from 0.37 to 0.41, and the frequency of w from 0.63 to 0.59. This change in frequencies indicates that evolution might be occurring in the moth population and that the allele W appears to have the selective advantage.

> **BIOLOGY CLIP**
> Population geneticists speculate that blood type A arose as a mutation of blood type O. Blood type B is considered a more recent mutation because it is less frequent.

Carriers *are individuals that are heterozygous.*

EVOLUTIONARY CHANGE

A population gene pool is very unstable. It is constantly influenced by external factors—factors that were intentionally ignored by Hardy and Weinberg. These factors change a population's genetic makeup, upset the tendency toward genetic stability, and lead to evolutionary change.

Factors that bring about evolutionary change are mutation, genetic drift, and migration (gene flow). While each of these factors may not be acting equally at all times, each is important in certain circumstances.

Mutation

Mutations are changes in the genetic makeup of an organism and occur in a cell as it undergoes meiosis to form an egg or sperm. Mutations can be one of two types: chromosome mutation or gene mutation. In *chromosome mutation*, the resulting gamete will either lack a particular chromosome or gain an extra one. Down syndrome, for example, is caused by an extra copy of chromosome 21. The second type of mutation, *gene mutation*, results from a chemical change inside an individual gene. This mutation type occurs when there is rearrangement of or damage to bases in the DNA molecule. Gene mutations may cause such serious illnesses as sickle-cell anemia, cystic fibrosis, and Tay-Sachs disease.

Mutations, or genotypic variations, are the raw material for evolution. There is general agreement among scientists that mutations are the original source of variation and that they have been accumulating for billions of years. Yet studies on gene mutations in diverse organisms such as bacteria, corn, and humans indicate that mutations are rare events for any given gene. Estimates suggest that mutation rates may be on the order of one per half-million genes per generation. Some researchers believe that a higher mutation rate might have the effect of losing the variation before natural selection has the opportunity to act on it. Nevertheless, because each individual contains many genes, and because there are so many individuals in a population, it is safe to assume that new mutations occur in each generation.

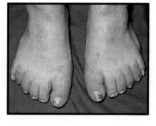

Figure 24.2

What is unusual about these feet?

As genes are paired, mutations resulting in recessive genes may remain "masked" in the population for long periods of time without affecting an organism's phenotype. In this way, a gene that might be harmful can remain in the gene pool. If environmental change occurs, the harmful gene might later turn out to have a selective advantage.

Mutations, in and of themselves, are neither good nor bad. A mutation considered beneficial in one environment may be detrimental in another environment. Many mutations are neutral, and have no effect. Other mutations, with long-ranging effects, can be harmful and often deadly.

Genetic Drift

Genetic drift, also called "random genetic drift," explains the disruption of the genetic equilibrium in small populations. (Remember, the Hardy-Weinberg principle applies only to large populations and is based on the laws of probability rather than natural selection.) For example, consider a population of 10 guinea pigs, in which only one member displays an allele designated *B* for black coat color. If, by chance, the black individual does not mate, the black allele will disappear from the population (assuming that genetic drift is the only evolutionary factor at work). This situation would change the gene frequencies in successive generations of the population.

Migration (Gene Flow)

Migration is another way in which variations originating within a population become distributed. Movement of members of a species into (immigration) or out of (emigration) a population alters its equilibrium. In immigration, new genes are added to the existing gene pool; in emigration, genes are removed.

Consider a population of squirrels with red fur color. The population has its own characteristic gene pool. If a group of brown-colored squirrels of the same species is introduced into the same area, the frequencies of the genes will change because a new set of genes is added from outside the population. The genetic equilibrium is upset, and the gene pool of the original population changes.

■ REVIEW QUESTIONS   ■   ?

1 What is the main source of variation in living organisms?

2 How would you define a gene pool?

3 Why is population sampling a useful technique in studying population genetics?

4 What purpose does a pedigree chart serve in tracing particular genetic traits? Give an example.

5 What prediction does the Hardy-Weinberg principle make with respect to populations?

6 Why does the Hardy-Weinberg principle apply mainly to large populations?

7 What are three factors that may bring about evolutionary change?

8 Describe the two main categories of mutation.

CHI-SQUARE TEST

Phrases such as "the luck of the draw" and "those are the odds" attempt to describe the outcomes of chance events such as coin tosses, dice throws, and lottery draws. But what are "chance" events? "Chance" describes any situation in which the outcome cannot be known ahead of time. If you toss a coin, for example, you know it will eventually come down. You do not know, however, which side will be up when it lands. When you do laboratory experiments or make observations in the real world, it is important to know whether your results are due to chance, or have been affected by other factors. Suppose, for example, a new drug is tested on 1000 people. If all 1000 show improvement, it is reasonable to assume that the drug is effective. But what if only 240 show improvement? Is the drug effective, or could 240 persons have improved just by chance? The examples in this section reflect efforts to explain the probability of an event happening by chance alone.

The mathematics of probability, originally developed by students of gambling, determine the likelihood of an event taking place by chance. Probability predicts only what *should* happen—it cannot predict with certainty what *will* happen in a given situation. For instance, most people would predict a 50% probability (chance) of obtaining heads with the flip of a coin. This hypothesis would be immediately challenged if, out of 10 tosses, the results turned out to be 3 heads and 7 tails. However, when the number of trials is extended to, say, 100 or 1000, the chances of obtaining the expected outcome increase. Hence, sample size affects the chances of achieving the expected results.

Figure 24.3
Guinea pigs display a wide variety of coat colors.

The following example further illustrates random (chance) behavior. A probability chart, like the one in Table 24.1, gives all possible outcomes for a particular event, in this case throwing a pair of dice. By using the chart, one can calculate the chances of throwing totals from 2 to 12. According to the chart, for a large number of throws, there should be five times more 6's than 2's. What are some other probabilities that can be predicted from the chart?

Table 24.1 Probability Chart

Die 1 / Die 2	1	2	3	4	5	6
1	2	3	4	5	6	7
2	3	4	5	6	7	8
3	4	5	6	7	8	9
4	5	6	7	8	9	10
5	6	7	8	9	10	11
6	7	8	9	10	11	12

To obtain the odds of throwing a particular total, divide the number of times that total appears in the chart by the total number of combinations (36). Hence the odds for a total of 2 are 1 in 36, for a total of 3 are 2 in 36, and so on.

Mathematical techniques and statistical analysis are particularly important to the study of genetics and evolution. Many believe Mendel's greatest contribution to science was to apply the concept of probability to biology. Probability enters into genetic concepts such as the Punnett square and the Hardy-Weinberg principle, studies of cross-breeding, predicting phenotypes of offspring, and determining the frequency and distribution of genes in a population.

It is important to realize that populations, not individuals, evolve. Any change in the relative frequencies of alleles in a population gene pool constitutes evolution, and hence the process of generating new genes, new alleles, and new combinations produces evolutionary change.

When scientists perform experiments, the results are often not as clear-cut as in the coin-tossing example, which had only two possible outcomes (heads or tails). However, scientists cannot simply accept results as "close to" or approximating the expected results. They employ mathematical techniques to make better, more informed interpretations of their experiments. One mathematical tool, the chi-square test, is valuable in testing the discrepancy between a set of observed values (experimental) and the corresponding theoretical values (expected). The theoretical values are based on some hypothesis concerning a population.

The following two examples illustrate the use of the chi-square (χ^2) test in determining how often chance alone can be expected to account for deviations in experimental data.

The chi-square formula can be written as follows:

$$\chi^2 = \Sigma \frac{(O-E)^2}{E} = \Sigma \frac{(d^2)}{E}$$

where O = observed numbers
E = expected numbers
Σ = "sum of all"
d = deviation from expected result

Example 1: Analysis of Data for Tossing a Coin

Suppose an individual tosses a coin first 20 times (trial 1) and then 100 times (trial 2). One might predict equal numbers of heads and tails in each trial; however, trial 1 results in 5 heads and 15 tails, while trial 2 results in 45 heads and 55 tails. Are these outcomes a result of chance, or is there some other explanation?

By applying the formula to each trial,

Trial 1: $\chi^2 = \dfrac{(5-10)^2}{10} + \dfrac{(15-10)^2}{10}$

$= 2.5 + 2.5$

$= 5.0$

Trial 2: $\chi^2 = \dfrac{(45-50)^2}{50} + \dfrac{(55-50)^2}{50}$

$= 0.5 + 0.5$

$= 1.0$

The chi-square for trial 1 is 5.0 and for trial 2 is 1.0. What do these numbers mean? Each experiment involved only 2 classes (heads and tails); that is, the χ^2 was calculated on the basis of only two squared deviations.

By consulting the table of chi-square values below (derived by mathematicians to provide the basis for judging whether any χ^2 value is greater than can be expected by chance alone), the probability of each event occurring by chance can be found.

Table 24.2 Chi-square Values

No. classes	χ^2 values							
2	0.0002	0.004	0.455	1.074	1.642	2.706	3.841	6.635
3	0.020	0.103	1.386	2.408	3.219	4.605	5.991	9.210
4	0.115	0.352	2.366	3.665	4.642	6.251	7.815	11.345
Probability *	99	95	50	30	20	10	5	1

*The probability value refers to the number of times out of 100 in which chance alone could have produced the observed deviation.

Scientists now agree that if there is less than a 5% probability that an observed deviation was due to chance alone, then factors other than chance have likely affected the results. Similarly, when the probability is greater than 5%, the observed deviation can be disregarded. The change is most likely due to chance alone.

The χ^2 for trial 1 was 5.0. In the table, this value falls between 3.841 and 6.635 for two classes. This means that the probability the deviation resulted from chance is less than 5% but greater than 1%. In mathematical terms, it suggests that the difference is significant. In scientific terms, it indicates that some other factor(s) influenced the outcome. Could sample size be a factor?

In trial 2, the χ^2 of 1.0 falls between 0.455 and 1.074. There is a 30 to 50% chance the outcome resulted from chance alone. Therefore, the deviation is not statistically significant, and can be disregarded.

Example 2: Analysis of Sickle-Cell Anemia Data

Sickle-cell anemia is a common inherited disorder caused by a mutant recessive gene. Homozygous individuals who have the condition develop sickled, or C-shaped, red blood cells due to the presence of an unusual form of hemoglobin. Because sickle cells severely limit the oxygen-carrying capacity of the blood, these individuals die at a young age. On the other hand, heterozygous members ("carriers") are not associated with the disorder and, at worst, only periodically show a mild form of sickle cells.

From the above description it would seem that the mutant gene would eventually disappear as the trait is "lost" through natural selection. At the very least, one would expect the gene to be present only in a low percentage in any given population. However, in certain parts of Africa the gene frequency has remained constant (up to 40% of the population) over the years. This indicates that carriers (those who have both the gene for sickle cells and normal red blood cells) have a high survival rate.

Interestingly, areas with high levels of sickle-cell anemia also display a high incidence of the parasite responsible for malaria. A number of studies have been conducted to determine whether the presence or absence of the gene is associated with the presence of malaria parasites. The example presented here combines results from two independent studies conducted in similar locations and under similar conditions. Repeated studies have produced similar results.

The study involved 600 African children, all the same age, and all living under the same nutritional conditions in the same area. Of this population, 16.5% were sicklers and approximately 44.7% were found to have malarial parasites in their blood.

The results include only two classes: sicklers and non-sicklers. To obtain the "expected value" for each class, an additional calculation for each of the four possible combinations must be performed. To obtain this value, multiply the totals in the right-hand column by a ratio obtained from the totals in the third line. For example, for sicklers with malaria the value would be $268 \div 600 \times 99$, and for non-sicklers with malaria it would be $332 \div 600 \times 501$. The expected values arrived at by this calculation are as follows:

Class 1 (Sicklers):
    with malaria   = 44.22
    without malaria = 54.78

Class 2 (Non-sicklers):
    with malaria   = 233.78
    without malaria = 277.22

Applying the chi-square test to the two classes, the results are:

Class 1: $\chi^2 = \dfrac{(23-44.22)^2}{44.22} + \dfrac{(76-54.78)^2}{54.78}$

$= 10.2 + 8.2$

$= 18.4$

Class 2: $\chi^2 = \dfrac{(245-233.78)^2}{233.78} + \dfrac{(256-277.22)^2}{277.22}$

$= 0.54 + 1.62$

$= 2.16$

	With malaria	Without malaria	Total
Sicklers	23	76	99
Non-sicklers	245	256	501
Total	268	332	600

Table 24.2 indicates that the χ^2 for sicklers (18.4) with respect to malaria is significant. The probability that sickling has remained high in this population by chance alone is much less than 1%. Does this agree with the original hypothesis? Might an explanation for the results be that the gene is not being removed by natural selection because the sickling trait is conferring some measure of protection from the malarial parasite? If so, the heterozygotes have a selective advantage in their resistance to the disease.

The χ^2 for non-sicklers (2.16) with respect to malaria falls between 2.706 and 1.642 in Table 24.2. The probability of this deviation being due to chance is therefore between 10 and 20%. That is 2 to 4 times greater than the 5% sufficient to suggest a chance occurrence. Therefore, the deviation is not significant and can be disregarded.

Table 24.3 Human Inherited Disorders

Disorder	Main consequences
Gene mutations	
Autosomal recessive inheritance:	
Albinism	Absence of pigmentation
Sickle-cell anemia	Severe tissue, organ damage
Tay-Sachs	Loss of motor control
Autosomal dominant inheritance:	
Polydactyly	Extra fingers, toes, or both
Achondroplasia	A type of dwarfism
Huntington's disorder	Progressive, irreversible degeneration of nervous system
X-linked inheritance:	
Hemophilia A	Deficient blood-clotting
Testicular feminizing syndrome	Absence of male organs, sterility
Chromosomal mutations	
Cri-du-chat	Mental retardation, malformed larynx
Down syndrome	Mental retardation, heart defects
Turner syndrome	Sterility, abnormal development of ovaries and sexual traits
Klinefelter syndrome	Sterility, mental retardation

CASE STUDY
GENETIC DISORDERS AS MODELS FOR EVOLUTION

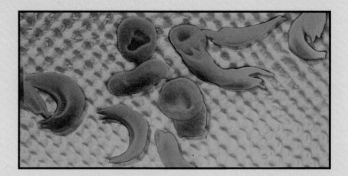

Objective

To investigate two recessive genetic disorders as models for studying evolution in human populations.

Background Information

Human genetic disorders are useful models for studying evolution. Because of their implications for human populations, there is more incentive to study genetic disorders than to study the inheritance and evolution of normal characteristics. Individuals with genetic disorders are often readily identifiable in the larger population. Many disorders are associated with specific populations, reflecting not only differences in lifestyle, but often differences in patterns of genetic inheritance as well. Furthermore, studies of genetic disorders frequently provide evidence of long periods of geographic and genetic isolation within the human population. Examples of disorders that place certain groups and their descendants at a higher risk than others are: cystic fibrosis and European whites; diabetes and Pacific Islanders; sickle-cell anemia and African blacks; Tay-Sachs and eastern European Jews.

Many of the well-known inherited disorders are classified as autosomal (i.e., not sex-linked), recessive disorders. They result from point, or gene, mutations that cause errors in the body's metabolism. The recessive alleles can persist at fairly high frequencies in populations because heterozygous members may still survive and reproduce; only one normal allele is necessary to carry out the specific function.

Two extensively studied recessive disorders are sickle-cell anemia and Tay-Sachs disorder. In both cases, the condition is expressed in homozygous, recessive individuals and is usually fatal. The heterozygotes (or "carriers") are usually symptom-free.

Sickle-Cell Anemia

Discovered by a Chicago physician in the early 1900s, this disease gets its name from the sickle-shaped red blood cells found in the blood of those who have the condition. While rare in most human populations, it is common in certain groups of African blacks and in those of African descent. Sickle-cell anemia causes general body pains, loss of appetite, yellowish eyes, a low resistance to infection, and shortness of breath. Death usually occurs in early childhood, but a few cases are known to have survived to adulthood.

It was Dr. Linus Pauling, a noted chemist and double Nobel prizewinner, who uncovered the reason for the presence of sickle cells in individuals with the disorder. By analyzing the hemoglobin molecule from patients' blood, Pauling concluded that the sickle cells occur when one amino acid (valine) is substituted for another (glutamic acid) in one of the four chains of the hemoglobin molecule. He determined that the change in the shape of red blood cells occurs when oxygen levels are low (e.g., at high altitudes, during physical exertion, and so on) and is irreversible.

Since specific amino acids can only be produced from the existing genes within a gene pool, Pauling's work demonstrates the importance of the gene to the production of variation (change) within a population. The shuffling of genes in the process of inheritance provides great variation in gene combinations. The sickle-cell trait, like other genetic traits, is inherited by a simple Mendelian pattern. The genotype of a carrier with the defective gene is written $Hb^A Hb^S$ where Hb is the symbol for hemoglobin and the superscripts A and S represent the genes for normal hemoglobin and the sickle-cell gene respectively.

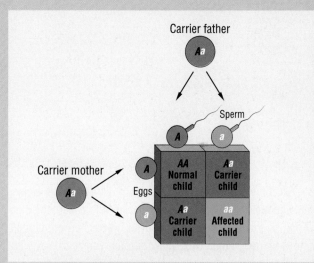

Carrier father

Aa

Sperm

A   **a**

Carrier mother

Aa

Eggs

A

a

	A	**a**
A	**AA** Normal child	**Aa** Carrier child
a	**Aa** Carrier child	**aa** Affected child

Using the diagram, answer the following questions:

a) Write the genotypes of the offspring from a cross between two carriers of the sickle-cell gene. What is the probability that the couple will have **(i)** a normal child? **(ii)** a child with the sickle-cell disorder?

b) What is the probability that a child born of a normal parent and a carrier parent would have a normal child? A child with the sickle-cell disorder?

c) Why would there be no concern about the offspring of two homozygous individuals having the sickle-cell disorder?

d) How does Pauling's discovery provide an answer to the observation that carriers experience mild "sickling" during strenuous exercise?

One biological deduction from this information might be that the recessive allele will eventually disappear from the population through natural selection. However, this has not happened. As many as 40% of the population in certain parts of Africa and 10% of Americans of African descent still carry the trait. What accounts for the frequency of the sickle-cell gene remaining relatively constant over the years?

A review of Example 2: Analysis of Sickle-Cell Anemia Data will shed light on this question and provide additional information for answering the following questions:

e) What selective advantage is afforded a heterozygous individual ($Hb^A Hb^S$) in certain African populations?

f) What might happen to the frequency of the sickle-cell gene if malaria were eliminated in Africa? Why?

g) Estimates suggest that the frequency of the sickle-cell gene in North American blacks has decreased from 22% in the early slavery period to a current value of 10% or less. What might explain this difference?

Tay-Sachs Disorder

Tay-Sachs disorder is also helping scientists understand how genetic disorders evolve, as well as providing insight into precisely how genes (even lethal ones) can persist and spread over the centuries. Tay-Sachs, which occurs in the general population in one out of every 400 000 births, was co-discovered by W. Tay, a British ophthalmologist, and B. Sachs in the 1880s. A notable exception to its occurrence is found in the eastern European Jewish population known as the Ashkenazim. In this group, the Tay-Sachs disorder is 100 times more frequent—about one in 3600 births.

The condition causes a deterioration of the central nervous system and becomes noticeable in infants at about six months after birth. Babies with the disorder lose much of their motor control, have convulsions, and usually die between the ages of two and four years. As with sickle-cell anemia, the genotypes of individuals with Tay-Sachs are homozygous recessive. Carriers of the trait are generally unaffected by the presence of the defective gene.

The cause of the disorder was not determined until 1962, when researchers discovered that people with Tay-Sachs lacked a gene that produces the enzyme ß-N-hexosaminidase A (Hex A). Hex A is required to control excessive accumulation of a fatty substance in body cells, especially nerve cells. Normally, fat is present only in modest levels in cell membranes because it is constantly being broken down by Hex A. A deficiency of the enzyme alters the metabolism of the fat, which then accumulates around the nerve sheath and in time destroys the nerve cells.

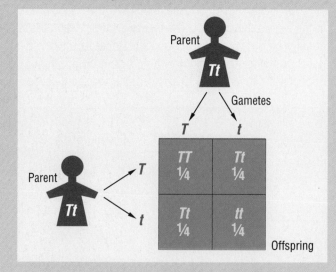

Parent

Tt

Gametes

T   **t**

Parent

Tt

T

t

	T	**t**
T	**TT** 1/4	**Tt** 1/4
t	**Tt** 1/4	**tt** 1/4

Offspring

Using the diagram and the preceding text, answer the following questions.

h) Are the odds in this example like those of Mendel's F_2 generation? Under what circumstances would the next generation yield the same odds?

i) Why can scientists be certain that each parent must have had at least one good copy of the gene along with the defective copy?

j) What similarities exist between the biochemical abnormalities that cause sickle-cell anemia and those that cause Tay-Sachs disorder?

k) Explain how scientists recognize that the selection pressure against Tay-Sachs (tt) is high?

To understand why there is an excessively high risk of Tay-Sachs disorder among the Ashkenazim, two unique features associated with the group must be examined. One concerns history and lifestyle, and the other relates to a potential selective advantage bestowed by the gene on carriers of the disease.

The lifestyle of the Ashkenazim has been unique among the world's ethnic groups. As a result of certain events over the past several thousand years, including war and various forms of persecution, the population has remained isolated from the general European population. Some reports suggest that this lifestyle has kept intermarriage with other groups down to a mere 15%. Another manifestation of the population's isolation is their susceptibility to 10 other genetic disorders that do not occur to the same extent in other Jews or eastern European non-Jews.

The most plausible hypothesis put forward to explain the high frequency of the Tay-Sachs gene suggests a potential selective advantage conferred upon carriers of the disorder. In 1972, a questionnaire administered to parents of Tay-Sachs children in the United States produced a surprising piece of valuable information. The majority of the children's grandparents who did not emigrate from the old country died of common causes such as stroke or heart attack. Only one of 306 grandparents died of tuberculosis, even though TB was a common killer during their youth. A follow-up study, a decade later, on the distribution of TB and the Tay-Sachs gene within Europe revealed that the Tay-Sachs gene was three times more frequent among the eastern European Jews (9 to 10% of the population were heterozygous) than among other populations of European Jews. If the recessive gene did lend some measure of protection against TB, then the gene should occur more often in areas with a high incidence of TB.

The ghetto conditions in which the eastern European Jews were forced to live is also an environment in which TB thrives. With little or no intermarriage, some investigators believe that the ghetto-bound population was under the strongest pressure to evolve genetic resistance to TB.

l) What are the two selection pressures acting on the eastern European Jewish population?

m) The Tay-Sachs recessive gene has survival value in areas where there is a high incidence of TB. How does natural selection operate in its favor in these areas?

n) What significance can be attached to the fact that eastern European Jews are susceptible to 10 other genetic disorders that are not found in other Jewish and non-Jewish populations?

o) How does this case illustrate that evolution results from the interaction between an organism's genetic makeup and its environment?

Case-Study Application Questions

1 How are the factors in the evolutionary process illustrated in this case study? In your answer consider the ideas of mutation, natural selection, and survival value.

2 What advice should a genetics counsellor give to carriers who are contemplating giving birth to a child?

3 What is the meaning of the statement, "Recessive genetic disorders can be both a blessing and a curse?" ■

FRONTIERS OF TECHNOLOGY: MITOCHONDRIAL DNA AND EVOLUTION

Much is already known about the mitochondrion (the so-called "power plant" of the cell) and its vital importance in cellular respiration. Tens of thousands of these sausage-shaped organelles (roughly the size of bacteria) are present in every eukaryotic cell, from microbes up to the most complex living organisms, including humans. They are particularly abundant in muscle cells and parts of nerve cells, and near the surface of cells that specialize in the transport of nutrients. The role of the mitochondrion (plural: "mitochondria") in the production of ATP, the fuel that provides the energy to power all the body cells' activities, is also well documented.

However, the discovery in the mid-1960s that mitochondria contain their own genetic material shook the scientific community. Scientists had previously believed that all of a cell's structure and activities were under the control and direction of DNA inside the nucleus. This new revelation suggested that mitochondria, by carrying their own genes, maintain a measure of control of their own destiny. Furthermore, because mitochondria are found in all eukaryotic cells, these organelles must somehow figure directly in the evolutionary events that led to the diversity of organisms on earth.

Several other features of the mitochondrion made it a prime candidate for the study of evolutionary relationships. Unlike the nuclear DNA of eukaryotes, mitochondrial DNA (mtDNA) forms small, looping chains, resembling the DNA of viruses and bacteria. The mtDNA is relatively tiny, containing only 17 000 nucleotide pairs (compared with the nearly three billion in nuclear DNA). However, while a cell has only one copy of the nuclear DNA, it can have numerous mitochondria, each with its own mtDNA. The mtDNA also divides and reproduces independently of its host cell.

Recent electron microscope studies and biochemical analyses show that mtDNA contains enough nucleotide pairs to carry the genetic code for 10 to 20 proteins associated with ATP synthesis, the vital energy-transforming function associated with mitochondria. Does this support the idea that the mitochondrion is a "power plant" for the cell?

This recent research and other lines of evidence have led to speculation that mitochondria originally had sufficient genetic material to exist on their own. Lynn Margulis and other evolutionary biologists have expanded this idea into the *symbiotic* or *endosymbiotic hypothesis*. This hypothesis states that at least two organelles, mitochondria and chloroplasts, are the descendants of prokaryotic organisms. The first living cells are thought to have been primitive, anaerobic, one-celled ancestors of some of today's bacteria. These primitive organisms probably obtained their energy from nucleotides such as ATP, likely in a process similar to *fermentation*.

A major event occurred about 1.5 billion years ago. An aerobic organism was captured (either invaded or engulfed) by an anaerobic form. This arrangement was mutually beneficial. The guest (aerobe) was provided protection and increased access to nutrients, while the host (anaerobe) gained an efficient means of respira-

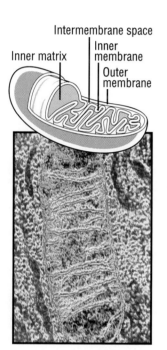

Figure 24.4

Transmission electron micrograph of a mitochondrion.

Intermembrane space
Inner membrane
Inner matrix
Outer membrane

> **BIOLOGY CLIP**
> The mitochondria from the male sperm cell do not enter the egg during fertilization. You have mtDNA from your mother but not from your father.

tion. Later, by a now-familiar genetic mechanism, certain guest genes (mtDNA) were transferred to the host nucleus. The host then took over much, but not all, of the control from the guest. Since then, the aerobes have continued to process oxygen both for themselves and their hosts. From this initial symbiotic relationship, or partnership, all other life forms are believed to have evolved. Indeed, the mtDNA in human cells is directly related to that in the ancestral organisms.

Current research on human mtDNA links faulty mitochondrial genes to a growing list of human genetic disorders such as Leber's disease, and less obvious familial illnesses such as Parkinson's disease. Research at the University of California at Berkeley has resulted in the successful cloning of mtDNA fragments from a zebra-like animal that lived a hundred years ago on the steppes of South Africa. The fragments used in the experiment were taken from tissues of a preserved skin in a German museum, and have been used to determine the relationships between and changes in zebras and horses. Some scientists speculate that one day DNA may be retrieved from other animal parts or even mummies to create banks of genes from extinct species. Given that only fragments of mtDNA and no whole intact nuclear DNA has been found, one can only imagine that this pioneering work could one day result in the recreation of extinct species.

SPECIATION

The process by which species originate is called **speciation**. However, it is important to recognize that the origin of species and evolution are not necessarily the same thing. Similarly, natural selection, while the major cause of disruption in genetic equilibrium, does not always lead to speciation. In the peppered moth case discussed in the chapter Adaptation and Change, evolution occurred without the creation of a new species. In the modern sense of the term, a species is a group of similar organisms that can interbreed and produce fertile offspring in their natural environment.

Scientists agree that the number of species today is much greater than it was in the past, even though many species have become extinct. Furthermore, since a species can only arise from existing species, there must be some process or mechanism in which a single species can develop into one or more descendant species.

Two ways in which a new species may arise are through *geographic isolation* and *reproductive (genetic) isolation*. In both instances, populations or parts of a population become isolated and must adapt to conditions in a new environment. Since all environments differ to varying degrees, the "selective" pressures on the populations also vary. It is important to understand that species are not created instantly, but usually evolve over a long period of time.

Geographic Isolation
In geographic isolation, the separation is caused by physical obstacles or barriers such as mountain ranges and bodies of water, or even barriers created by humans. When this happens, gene flow between the isolated group and the main population ceases. Eventually, the groups

Speciation *refers to the formation of a new species.*

Figure 24.5

Mummies may provide scientists with DNA of extinct lineages.

become so different that individuals of one population can no longer interbreed with those of another. Reasons for the differences include different adaptations of populations in the separate environments, the development of different gene frequencies within the separate populations, and different mutations within the populations.

Geographic isolation is used to explain the existence of the 14 species of finches found by Darwin in the Galápagos Islands. The species likely descended from individuals that reached the islands from the mainland of South America. The finches probably arrived by being blown off course in a storm or by getting lost. Finches do not normally fly over great distances. When the finches reached Galápagos, the water that separates the islands acted as a barrier to pre-

vent interaction among the separate populations. Over time, this isolation would have resulted in the finches adapting to new conditions of vegetation, food, and so on, characteristic of the different islands. These changed conditions could have caused the populations to evolve in different directions.

Many other examples support the idea of geographic isolation leading to speciation. Among these are: the spread of the house sparrow (*Passer domesticus*) in North America; populations of certain gulls of the genus *Larus*, which circle the North Pole and overlap in Great Britain; and variation among turtles on the different islands in the Galápagos.

The question of the "exact" moment at which speciation occurs, or whether, indeed, speciation *has* occurred, cannot be answered precisely. For instance, while

Figure 24.6

Masses of marine iguana at Espinosa Point, Galápagos. These lizards eat marine algae from rocks.

lions and tigers are markedly different in appearance, and do not interbreed in the wild, they are known to reproduce in captivity (producing "ligers"). A similar situation occurs between the leopard frog (*Rana pipiens*) and the wood frog (*Rana sylvatica*). The interpretations of whether or not speciation has taken place in these cases will require further study.

While the rate of evolution of a new species remains a contentious issue, the widely held view is that speciation is a gradual and lengthy process. There is also ample evidence that in some instances, such as in **polyploidy**, a new species may arise suddenly. Although an unusual event in the animal kingdom, polyploidy is common in plants. It often produces hardier and larger varieties of grains, fruits, and other useful plants.

Polyploids can mate with each other, but not with members of the parent generation, because of different chromosome numbers. In spite of the benefits associated with polyploids, it is not uncommon for them to possess undesirable traits or genes of each parent, which may be passed on to successive generations.

Polyploidy *is a condition in which an organism possesses more than two complete sets of chromosomes.*

Figure 24.7

A liger, also known as a tiglon. The father of this animal was a tiger and the mother was a lion.

Reproductive Isolation

Geographic isolation may also lead to reproductive, or genetic, isolation. Reproductive isolation occurs when organisms in a population can no longer mate and produce offspring, even following the removal of the geographic barriers. Factors that contribute to reproductive isolation include differences in mating habits and courtship patterns, seasonal differences in mating, and the inability of the sperm to fertilize eggs. In some cases, even where fertilization has taken place, the genes and chromosomes are so different that the zygote does not develop or the embryo is prevented from developing normally.

REVIEW QUESTIONS ?

9 What does the concept of probability enable scientists to predict?

10 What do scientists mean when they say a chi-square value is "significant"?

11 Why are human genetic disorders useful models for studying evolution?

12 What common features are shared by the two genetic disorders in the case study?

13 Why did the discovery that mitochondria contain their own DNA surprise many scientists?

14 What features of the mitochondrion make it useful in the study of evolutionary relationships?

15 What is speciation?

16 Explain two ways in which new species may arise.

SOCIAL ISSUE:
Interbreeding of Plains and Woodland Bison

The North American bison are considered by some biologists to be two different subspecies: the plains bison and the woodland bison. Under natural conditions, the plains bison grazed the prairie regions and the woodland bison occupied lowland meadows and delta regions several hundred kilometers away from the plains bison. In the early 1800s, Canada's plains bison totalled around 60 million. By 1885, they were almost extinct. In 1909, the Canadian government bought plains bison and established Bison Recovery Park in Wainwright, Alberta. In 1922, Canada's largest national park, Wood Buffalo National Park, was established to provide a sanctuary for the woodland bison. To relieve overcrowding in Alberta's Elk Island National Park, plains bison were transported to Wood Buffalo National Park. The resulting hybridization almost destroyed the woodland bison as a separate subspecies.

Statement:

Human beings should not interfere with the natural evolutionary process.

Point

- Distinctive environmental pressures select different genotypes in woodland and plains bison. By placing the subgroups together, the gene pool has been altered. This means that many genes not suited for Wood Buffalo National Park have been reintroduced.
- If left alone, the changes in the two subspecies could become so pronounced that they may develop into two distinct species. Humans should not interfere with such diversification.

Counterpoint

- Recent DNA studies do not support the idea that the two groups of bison are different subspecies. The differences in appearance can be accounted for by a difference in their environments. The heavy fur coat of the more northern bison may be a response to a colder environment.
- The assumption that humans should not interfere with the natural evolutionary process can also be disputed. If we consider humans part of nature, then human interference is but another selective pressure.

Research the issue.
Reflect on your findings.
Discuss the various viewpoints with others.
Prepare for the class debate.

CHAPTER HIGHLIGHTS

- Genetic variation (mutation) is the raw material for evolution.
- All of the genes that occur within a population make up its gene pool.
- The measure of the relative occurrence of genes in a population is called the gene frequency. Evolution occurs when there is a change in the genetic makeup (frequency) of a population.

- The Hardy-Weinberg principle states that the frequency of genes (alleles) in a population stays the same when a population is in genetic equilibrium.
- In nature, populations rarely, if ever, meet the conditions required for genetic equilibrium. Therefore, evolution is recognized as a continuous and ongoing process.

- Three factors that bring about evolutionary change are mutation, genetic drift, and migration. Genetic drift is governed by the laws of probability.

- Two ways in which speciation occurs are through geographic isolation and reproductive isolation.

APPLYING THE CONCEPTS

1 Would it be more correct to say "an organism evolves" or "a species evolves"? Explain.

2 The five conditions of the Hardy-Weinberg principle are rarely met in nature. Yet the theory is still useful for studying "real" populations. How can you account for this apparent contradiction?

3 In a given population of organisms, the dominant allele (p) has a frequency of 0.7, and the recessive allele (q) has a frequency of 0.3. Use the Hardy-Weinberg formula to determine the genotype frequencies within the population.

4 In Tanzania, 4% (0.04) of the population are homozygous sickle-cell anemics (ss) and 32% (0.32) are heterozygotes (Ss). From these data, calculate the proportion of alleles that are s or S.

5 Mutation rates are usually quite low in sexually reproducing organisms, yet mutations are known to be the raw material for evolution. Explain how this is so.

6 A cross between two pea plants, in which tall (T) is dominant to short (t), yields 1000 seeds. Of this number, 550 produce plants that are tall, while 450 produce short plants. Use the chi-square test to determine whether the deviation is the result of chance or some other complicating factor.

7 Describe why long periods of geographic isolation of a small group from other members of a population favor speciation.

CRITICAL-THINKING QUESTIONS

1 How do the genetic disorders discussed in this chapter illustrate the point that evolution involves interactions between an organism's genetic makeup and its environment?

2 The Hardy-Weinberg formula, $p^2 + 2pq + q^2 = 1$, is said to represent all possible genotypes in a population. Verify this statement by determining (a) the frequencies of dominant and recessive alleles, and (b) the number of heterozygotes in a population of 200 pigs in which 72 have the recessive trait. If natural selection removed all of the individuals with the recessive trait, what would be the gene frequencies in the next generation?

ENRICHMENT ACTIVITIES

1 Do library research to learn how mitochondrial DNA investigators are providing insight into the causes of certain illnesses such as Leber's disease and Parkinson's disease.

2 Review the case studies in the chapter Adaptation and Change and in this chapter and develop an argument for the following viewpoint: "Natural selection is no longer in the realm of pure theory—it is now considered an operating principle of biology."

*P*opulations and Communities

IMPORTANCE OF POPULATIONS AND COMMUNITIES

*A **population** is a group of individuals of the same species occupying the same area at a given time.*

Figure 25.1

Snow geese migration, Cap-Tourmente Wildlife Sanctuary.

Twice a year the sounds of the greater snow geese can be heard near Cap-Tourmente, Quebec. Cap-Tourmente is one of the most important staging areas for the geese during their annual migration to their breeding grounds in the high arctic and their wintering areas in the coastal marshes of the eastern United States. The flats along the shoreline turn white as thousands of these birds spend days probing for the roots of aquatic plants. Up to 80 000 geese can be seen feeding at any one time, and visitors from around the world come to view this spectacular scene.

Several questions probably come to mind when you look at this picture of snow geese. Why do they return to Cap-Tourmente? Are the geese endangered? If so, do they require special protection? Is there enough food to support so many geese and the other species of water-fowl living in the same area? To answer these questions and others you must examine the biosphere in terms of populations and communities.

A **population** refers to all of the individuals of the same species living in the same place at a certain time. Although individuals from a population can be studied to gain information about life span, food preferences, and reproductive cycle, this tells little about the entire population. Information about competition among members of the same species for food, territory, and reproduction requires a study of populations. Since changes in a population are closely linked with predators, parasites, and food sources, population studies also provide information about relationships among different organisms.

When you investigate how two or more populations interact, you are studying a **community.** A community includes all the species that occupy a given area. The study of a community involves only the organisms, whereas the study of an ecosystem includes both the biotic (living) and abiotic (nonliving) components of a specific area. In spite of this difference between community and ecosystem, it is virtually impossible to examine the structure and activities of any community without some reference to the abiotic factors that may influence its populations.

By comparing population data from a wide variety of organisms, ecologists are able to make a number of generalizations. They have discovered that certain principles govern the growth and stability of populations over time. Analysis of the data allows them to predict what might happen to a species if the environment is altered or a new species is introduced. Ecologists can also use the data to describe how a population interacts with another (interspecific competition) and how individual members of one species relate to each other (intraspecific competition).

In this chapter you will examine the ecological factors that influence the distribution, size, density, and growth of populations.

HABITATS, GEOGRAPHIC RANGE, AND THE ECOLOGICAL NICHE

To understand some of the terms used in population studies, consider the northern flying squirrel (*Glaucomys sabrinus*). The flying squirrel has a flap of loose skin extending between the ankles of its front and hind legs on each side of its body. It is able to glide from tree to tree by extending its front and back legs so that the loose skin forms a kind of parachute. The flying squirrel can make gliding turns and spirals by moving its front legs and thereby changing the shape of the skin flaps.

To learn more about this unique animal, an ecologist would have to know the flying squirrel's **geographic range.** This is a region, usually outlined on a map, where sightings of the animal have occurred. While helpful, the map does not tell you exactly where individual populations are found. The **habitat** of the squirrel is the place where it lives. It is usually determined by the environmental conditions under which the squirrel population has the best chance of survival. For example, a flying squirrel would more likely be spotted in a forest ecosystem than a grassland ecosystem in the same locale. The squirrel's habitat is limited by factors such as vegetation, soil conditions, and climate. In western Canada, its preferred habitat is the northern boreal forest, particularly where the trees are spaced about two to three meters apart. In eastern and Atlantic Canada, its preferred habitat is a mixed-woodland environment dominated by hemlock and yellow birch trees.

Ecologists also need to know the size of the squirrel population and to what extent the population is distributed throughout its range. Changes in the distribution of flying squirrels from season to season would also provide key infor-

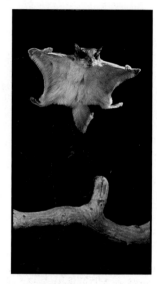

Figure 25.2
A flying squirrel leaps from a branch.

*A **community** is made up of the populations of all organisms that occupy an area.*

*A **geographic range** is a region where a given organism is sighted.*

*A **habitat** is the physical area where a species lives.*

Figure 25.3

Geographic range map of the northern flying squirrel.

Ecological niche *refers to the overall role of a species in its environment.*

Clumped distribution *occurs in aggregates. The distribution of organisms is affected by abiotic factors.*

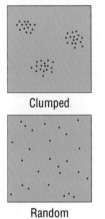

Clumped

Random

Uniform

Figure 25.4

Population patterns.

Sloughs *are depressions often filled with stagnant water.*

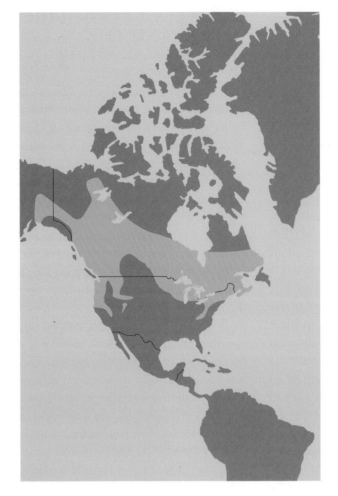

Since habitat preference plays an important role in population distribution patterns, ecologists must take great care when determining the population size of any species in a region. Population patterns can be divided into three patterns: clumped, random, and uniform.

A **clumped distribution** occurs when individuals are grouped in patches or aggregations. Organisms are distributed according to a certain environmental factor. For example, in river valleys trees often grow only on the south slopes and grasses dominate the north slopes. The north slope receives direct sunlight, which evaporates moisture from the soil. The drier soils of the north slope cannot support trees. By contrast, the south slope of the valley receives less direct sunlight and the soil tends to hold more water. Figure 25.5 illustrates grassland and woodland communities on opposite sides of the valley. Another example of clumped distribution can be seen in the arrangement of trees around **sloughs** in grassland ecosystems. Some sloughs have trees growing at the water's edge, while others do not. Permanent sloughs have water throughout the summer, whereas temporary sloughs dry up during the summer and do not provide enough water for trees during late fall. As shown in Figure 25.6, the most important limiting factor that affects the distribution of trees along river banks and the edges of sloughs is the presence of water.

mation. Within a habitat, every population occupies an **ecological niche.** This term refers to the population's role in the community, including all the biotic and abiotic factors under which the species can successfully survive and reproduce. Not surprisingly, the niche is not something you can actually see. For example, the niche of a flying squirrel includes its specific feeding habits, its ability to produce a large number of offspring, and its capacity to serve as food for a variety of parasites and predators. The flying squirrel's body wastes enrich the soil, while its use of seeds as a primary food source helps in the distribution of many grasses. The flying squirrel interacts with a number of other organisms and influences several biotic factors in its community.

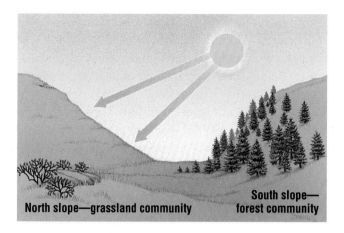

Figure 25.5

North slopes receive direct sunlight. This increases evaporation and decreases the water needed to support trees.

Random distributions of individuals, although not very common, occur when there is neither an attraction nor repulsion among the members. Biotic and abiotic factors have little effect on random distributions.

Uniform distributions occur when there is competition among individuals for factors such as moisture, nutrients, light, and space. In desert environments, moisture and soil nutrients are so limited that competition among roots results in regular spacing of plants. Coastal nesting sites for sea birds are often so crowded that uniform distribution occurs.

CHAOS THEORY AND BIOLOGICAL SYSTEMS

Is there some order to apparent random distributions of animals? A new way of looking at many complex systems, referred to as *chaos theory*, is relevant in many fields of science. Classical opinion has always held that to understand the behavior of an entire system it is sufficient just to study each of its individual components. By identifying generalizations or tendencies through the study of each of these parts, scientists attempted to predict events. Predictions were based on identifying repeating, ordered events. Conversely, the chaos theory assumes that since randomness is a basic feature of many complex systems, long-term predictions may well be extremely difficult, even impossible. Chaos theory suggests that small uncertainties in short-term prediction of individual events may be magnified to such an extent over the long term that the expected behaviors in complex systems become quite unpredictable. Even if the events appear to be random and without connections, there may be an orderly system producing the phenomena. Chaos theory may actually allow us to discover order in some situations.

There are indications that chaos is a normal feature in biological systems. Medical research is using chaos theory to diagnose possible illnesses, particularly heart disease. Biological development, evolution, and the complex ecological data associated with populations, communities, and ecosystems are also being looked at in light of chaos theory. A major

a)

b)

Figure 25.6

(a) A temporary slough.
(b) A permanent slough.

Figure 25.7

Atlantic puffins in front of their burrows.

Random distributions *are arbitrary and appear to be unaffected by biotic factors.*

Uniform distributions *are orderly and appear to be affected by competition.*

advantage of this new way of looking at nature is that chaos can account for the ability of complex systems to respond with flexibility to the environment.

Much of the chaos theory is based on mathematical models. As with any new hypothesis, the definition of a complex system and even the term chaos itself are still under debate. The acceptance or rejection of this novel idea will be determined by further computer simulations and much debate.

SIZE AND DENSITY OF POPULATIONS

Ecologists use a number of measurements to describe populations. A statement of the size of a population must include the numbers of the named organisms, the location of the population, and the time when the numbers were determined. Statements such as "There were 27 642 northern pike in Sylvan Lake, Alberta, in 1981" or "There were 295 990 people in the Halifax, Nova Scotia, metropolitan area in 1986" are statements of **population size.** The numbers themselves may have resulted from an exact count or, as is usually the case, an estimate of the total population size using appropriate sampling methods.

For the data to be meaningful, the population size is usually recorded in terms of some convenient unit of measure, such as a unit of area. For example, the snow goose population at Cap-Tourmente can be described as the number of birds per hectare. For populations of aquatic organisms, the number of individuals per unit volume is used. **Population density** describes the number of organisms in a defined area. The density (D) of any population is calculated by dividing the total numbers counted (N) by the space (S) occupied by the population. A simple formula for determining population density is as follows:

$$D = \frac{N}{S}$$

If 200 lemmings were living in a 25 ha (hectare) area of tundra near Churchill, Manitoba, in 1980, their density could be calculated as follows:

$$D = \frac{200 \text{ lemmings}}{25 \text{ ha}}$$
$$= 8 \text{ lemmings/ha}$$

Therefore, there were eight lemmings per hectare near Churchill in 1980.

Population size *is the number of organisms of the same species sharing the same habitat at a certain time.*

Population density *is the number of organisms per unit of space.*

Figure 25.8

Population of snow geese in St. Lawrence Valley, Quebec. The population of snow geese in any given year, by itself, provides you with limited information. However, when you compare the data obtained between 1969 and 1984, a trend is observed.

Table 25.1 Numbers of Greater Snow Geese Counted in the St. Lawrence Valley, Quebec

Year	Number of geese	Year	Number of geese
1969	68 800	1977	160 000
1970	89 600	1978	192 600
1971	123 300	1979	170 100
1972	134 800	1980	180 000
1973	143 000	1981	170 800
1974	165 000	1982	163 000
1975	153 000	1983	185 000
1976	165 600	1984	225 000

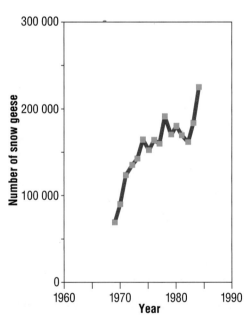

The population density by itself does not provide you with much information. Is eight lemmings per hectare normal or abnormal? The data can even be misleading. Imagine a visitor to Canada, observing the Toronto metropolitan region from the CN Tower and being told that the population density of Canada is two persons per square kilometer. If Toronto were used to calculate Canadian population density, an inappropriate answer would result.

DETERMINING CHANGES IN POPULATION DENSITY

One way of comparing population densities is to see if there have been changes within the same population over a certain time period. This is referred to as the **rate of change** (R) in a population. (Rate refers to some factor that changes over time. This example deals with density.) The rate of change can be expressed as a simple mathematical formula:

$$\text{Rate of density change} = \frac{\text{change in density}}{\text{change in time}}$$

$$R = \frac{\Delta D}{\Delta t}$$

The Greek symbol Δ (delta) stands for "change in." Therefore, the rate of change is equal to the change in density divided by the change in time (time elapsed). By convention, the ΔD calculation must begin with the density that has the most recent date minus the density that has the earlier date. This will indicate whether there has been an increase ($+$) or decrease ($-$) in the population density. The two densities must have the same area units.

Referring back to the Churchill lemming population of eight animals per hectare in 1980, assume that an ecologist determined that the population density in 1990 was 22 lemmings per hectare. The rate of change can be calculated using the above formula.

$$D_{1980} = 8 \text{ lemmings/ha}$$
$$D_{1990} = 22 \text{ lemmings/ha}$$
$$R = \frac{\Delta D}{\Delta t}$$
$$= \frac{D_{1990} - D_{1980}}{1990 - 1980}$$
$$R = \frac{22 - 8}{1990 - 1980}$$
$$R = \frac{14}{10}$$
$$= +1.4 \text{ lemmings/ha/yr}$$

Since 1.4 is a positive number, the lemming population was increasing.

This calculation raises a number of questions. Does it indicate that something unusual is occurring that is worthy of further investigation? Shouldn't populations remain constant? What factors are responsible for this increase? How long is this growth likely to continue? How does this growth compare with other lemming populations in similar habitats? How will this growth affect other organisms, such as those on which the lemmings feed, and those that feed on the lemmings? Should further population measurements be made to see if a long-term trend is being established? Is the habitat undergoing changes beneficial to the lemming? To answer these questions, some of the factors that regulate population growth must be examined.

Rate of change *refers to the change in a population over a period of time.*

■ REVIEW QUESTIONS ■ ?

1 List three characteristics of a population.

2 Distinguish between a population and a community.

3 How does an animal's habitat differ from its geographic range?

4 Explain the meaning of the term ecological niche.

5 Distinguish among clumped, random, and uniform population distribution patterns and describe the factors that may be responsible for each pattern.

6 Identify examples of clumped and uniform distributions around your home or school. Explain the environmental factors responsible for each type of distribution.

7 Calculate the density of a meadow vole population if 78 animals were observed in a 20 ha area.

8 Define rate of change in density of a population.

9 The following are data on the density of lemmings in a defined area of the tundra:

September 1981: 15 animals per hectare
September 1991: 3 animals per hectare

Calculate the rate of change in density of the lemming population. Is the population increasing, decreasing, or remaining stable? Explain.

POPULATION GROWTH PATTERNS

Changes in population size occur when individuals are added to or removed from a population. Four factors determine population size:

1. *Natality*—the number of offspring of a species born in one year.
2. *Mortality*—the number of individuals of a species that die in one year.
3. *Immigration*—the number of individuals of a species moving into an existing population.
4. *Emigration*—the number of individuals of a species moving out of an existing population.

If all the factors remain the same except for an increase in birth rate, the population will increase. The same holds true if immigration increases. The reverse effect occurs if there is an increase in mortality or emigration. In populations, all four factors interact, with natality and mortality generally having the greatest impact. Applying these factors to the lemming population of Churchill, one can surmise that the combined natality and immigration rates must have exceeded the combined mortality and emigration rates.

The concept of population growth can be generalized in the formula below:

$$PG = \frac{[\text{births} + \text{immigration}] - [\text{deaths} + \text{emigration}]}{\text{initial number of organisms}} \times 100$$

$$= \frac{[b + i] - [d + e]}{n} \times 100$$

Using the formula above, calculate the percentage growth of the sandhill cranes from Table 25.2:

$$PG = \frac{40 - 55}{200} \times 100$$

$$= \frac{-15}{200} \times 100$$

$$= -7.5\%$$

Table 25.2 Sandhill Cranes, Banks Island Breeding Site, Canadian Arctic 1991

Births	Immigration
40	0

Deaths	Emigration
55	0

Note: The colony contained 200 sandhill cranes in spring.

In Table 25.2, immigration and emigration are recorded as zero for the early summer months, the time when the study was conducted. By fall, all sandhill cranes migrate from the area, to return next spring. Because the study was conducted during the breeding season, the percentage growth can be stated as -7.5% per breeding season. (Note: Rate refers to a change over time, $R = \Delta PG/\Delta T$.)

Does it seem unusual that the population decreased during the breeding season? A decrease of 7.5%, especially at this time of year, signals that something is wrong. The population is decreasing at a time when it should be increasing. Has the population been exposed to a new disease, to predators, or to adverse weather conditions? In many cases human intervention may be responsible for rapid changes in the size of a population. The use of pesticides, hunting and fishing regulations, creation of wilderness areas, flooding of valleys, and draining of wetlands are some of the factors that can account for fluctuations in population.

In mature ecosystems, populations tend to remain relatively stable over the long term. This "balance" is referred to as **dynamic equilibrium,** or steady state. In many ways dynamic equilibrium, a term used by ecologists, can be compared with homeostasis, the term used by physiologists. Dynamic equilibrium describes how populations adjust to changes in the environment to maintain equilibrium, while homeostasis describes how an organism tends to maintain a relatively constant internal environment despite a changing external environment.

Biologists generally classify populations as either "open" or "closed." In natural or **open populations,** all four factors (natality, mortality, immigration, and emigration) are functioning. However, when populations exist in laboratory settings and in some game pre-

serves, immigration and emigration do not occur. Consequently, any changes in the size of the population will result only from natality and mortality. Populations living under these conditions are referred to as **closed populations.**

To record population changes, numbers and density information can be tabulated and graphed, with time used as the manipulative (dependent) variable and numbers or density of organisms as the responding (independent) variable. These **growth curves** are then plotted so that population fluctuations can be examined and analyzed. Rather than connect each coordinate, scientists represent growth curves as *best-fit graphs* in which the line is positioned to pass through as many coordinates as possible to show a tendency.

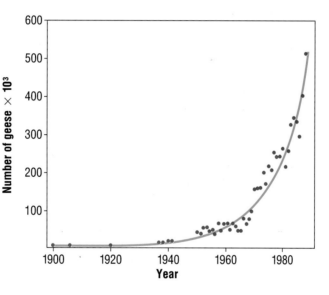

Growth Curves for Closed Populations
Biologists study simple closed populations to determine and analyze changes in populations. Most of this work is done using bacteria, yeasts, and other microorganisms, which are easy to handle and maintain. In addition, their relatively short life spans allow many generations to be grown in just a few days. In all cases the population curves are similar. The

Dynamic equilibrium *refers to any condition within the biosphere that remains stable within fluctuating limits.*

An **open population** *is one in which density changes result from the interaction of natality, mortality, immigration, and emigration.*

A **closed population** *is one in which density changes are the result of natality and mortality with neither food nor wastes being allowed to enter or leave the given environment.*

A **growth curve** *is a graph used to show the changes in a population over a specific length of time.*

Figure 25.9

Greater snow goose population.

experiments begin with the introduction of a few individuals as a breeding stock into an environment that is ideal in terms of nutrients, oxygen, space, and temperature. A growth curve like the one shown in Figure 25.10 usually results.

Four definite phases can be recognized in this idealized growth curve. The first is the **lag phase,** a delay that occurs before the population enters a phase of active reproduction. The precise reasons for the

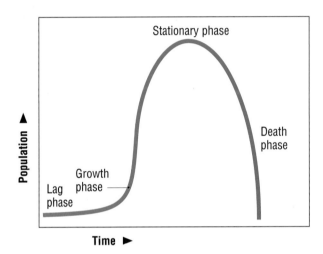

delay in population growth are unknown; however, it has been suggested that the cells must adjust to their new environment before cell division can begin, and that time may be required for the synthesis of new enzymes for cell growth and reproduction.

During the **growth phase,** the population is increasing at its fastest rate. For example, in binary fission of microorganisms the number of individuals in the population increases at a constant rate, doubling at each division. This is referred to as logarithmic, or exponential, growth (2-4-8-16-32, etc.). The expected population increase in a given time (I) can be calculated from the following formula:

$$I = R \times N$$

where R = growth rate
N = current population

Consider what would happen if paramecia in a closed population increased at a rate of 7.5% per day. If the initial population is 200, then

$$I = 7.5\% \times 200$$
$$= 15,$$

an increase of 15 paramecia.

If the growth rate continues at 7.5%, the colony will contain 215 paramecia one day later. For day 2, $I \times 7.5\% \times 215 = 16.13$. By day 3 the population would be 231.13. Although the decimal makes no "sense" in actual numbers of organisms, it does make statistical sense. The number 16.13 indicates that the increase will most often be 16 paramecia, but occasionally it may be 17 paramecia. The expected increase is determined as a probability: biologists have no way of counting organisms that do not yet exist. Can you determine on what day the population will have tripled? This should be on about day 16.

Table 25.3 Predictions of Population Increase Based on 7.5% per Day Growth Rate

Day	N	I
1	200	15
2	215	16.125
3	231.13	17.33
4	248.46	18.63
5	267.09	21.53
6	287.12	21.53
7	308.65	23.15
8	331.80	24.89

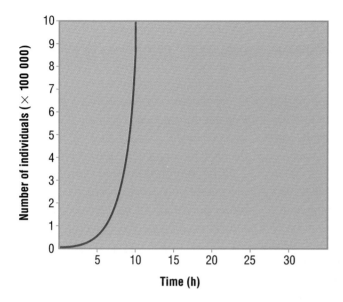

Figure 25.11

Exponential growth of a bacterial population.

Eventually, the paramecium population would no longer increase. A lack of space, a shortage of nutrients, and an accumulation of toxic metabolic wastes cause a reduction in the rate of increase. Population growth slows and eventually ceases. This is the maximum **stationary phase,** when the rates of natality equal mortality. This phase may continue for some time as new cells are produced and old cells die.

The last phase is the **death phase,** when mortality exceeds natality as nutrients run out and wastes accumulate. The number of individuals decreases at a constant rate; the death rate, like the logarithmic growth rate, reaches a constant value, and often the entire population will die out.

Growth Curves for Open Populations

When some limiting factor, such as a nutrient, is introduced into a closed popula- tion, the plotted points on the graph usually result in an S-shaped curve (Figure 25.12). The S-shaped curve is typical of an organism placed in a new environ- ment. In this case an addi- tional nutrient is added to the ecosystem. The popula- tion responds to increased food, and natality increases in relation to mortality. Equilibrium is established once again. At the point where the curve levels off, the maximum number of individuals the environment can support has been reached. This number is called the **carrying capacity.**

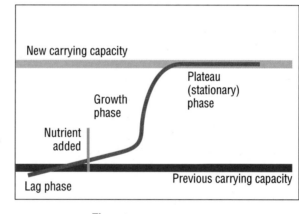

Figure 25.12

The introduction of extra nutrients will increase the carrying capacity of the environment.

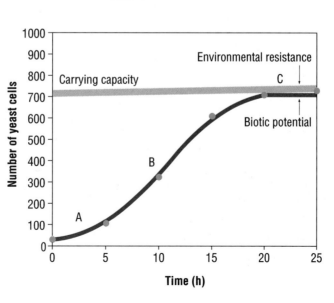

Figure 25.13

Population growth curve for a yeast population.

In the natural environment, all species have the ability to reproduce at a rate typical for that species. For example, most mature female black bears give birth to one or two cubs after a gestation period of 7.5 months. Generally, female bears take at least two years to mature. In contrast, flying squirrels have litters of between three to six young after a gestation period of only about 40 days. Field mice can have as many as 10 young with a gestation period of 22 to 35 days and usually more than one brood in a single year. Depending on the species, most rodents mature and begin reproducing during their birth year. The term **biotic potential** (R_{max}) is used to refer to this theoretical maximum birth rate. Biotic potential is regulated by the following factors:

1. *Offspring*—the maximum number of offspring per birth.
2. *Capacity for survival*—the chances the organism's offspring will reach reproductive age.
3. *Procreation*—the number of times per year the organism reproduces.
4. *Maturity*—the age at which reproduction begins.

Since it is typical of populations to remain in dynamic equilibrium, there must be factors that tend to repress excess population growth. The general term **environmental resistance** describes all factors that limit population growth. These limiting factors can be both biotic and abiotic, and are continually changing, thus affecting the carrying capacity of an environment. For example, a climatic change may result in a greater-than-normal production of the vegetation that supports a deermouse population, causing the mouse population to increase. A return to normal amounts of vegetation would lower the carrying capacity as well as the number of mice. This is fairly typical of the small fluctuations that occur in most natural populations. In addition to food, some of the more common factors of environmental resistance are predation, competition for space, and disease. Populations fluctuate regularly due to the interaction of factors that increase or decrease the number of individuals.

The dependence of population increase (I) on carrying capacity (K) and biotic potential can be expressed by the following mathematical equation:

Population = Biotic × Number of × Relationship
increase   potential   individuals   between
             (reproduction   carrying
             per individual)   capacity and
                                         resources
                                         available

$$I = R_{max} \frac{N(K-N)}{K}$$

where R_{max} = biotic potential
N = number of individuals in the population
K = carrying capacity
I = population change

When the population is low, the I value is close to $N \times R$; however, as the population increases and resources become depleted, I decreases sharply to zero when $N = K$.

J-Shaped Population Curve

A J-shaped curve results when rapid population growth is followed by a sharp decline in the population. J-shaped curves are typical of a number of insect populations as well as some phytoplankton (algae) species. The usual growth phase occurs, and the population increases rapidly. However, the quick increase in population can suddenly exceed the carrying capacity of its habitat, and the population declines. The period of rapid growth, exceeding the carrying capacity, and the subsequent drop in population may lead to population fluctuations of a cyclic

Biotic potential *is the maximum number of offspring that can be produced by a species under ideal conditions.*

Environmental resistance *includes all the factors that tend to reduce population numbers.*

nature. In the case of an introduced organism, the initial crash is often followed by growth then a relatively stable stationary phase.

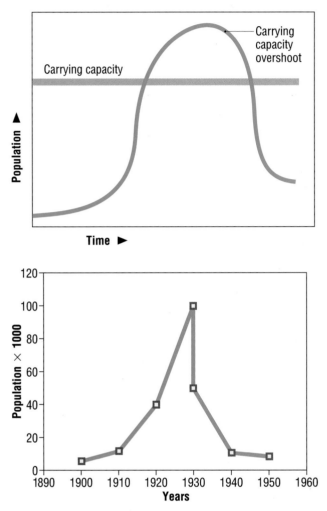

Food is usually the most important limiting factor for animal populations. As the population reaches its carrying capacity, ill health becomes more common. This can result in higher predation rates. The inevitable result is a decrease in the population. However, death is not the only method by which animal populations adjust to changes in carrying capacity. Overcrowding, often resulting from rapid growth, affects the reproductive ability of many animals. For example, some bird species may reduce the number of eggs laid from the normal eight to ten down to

one or two and sometimes none. Thus, by reducing the reproductive rate, a population may limit its growth and match the environment's carrying capacity.

A decrease in the number of predators can result in the population growth of a prey species beyond the original carrying capacity. This has been well documented in a study of a deer population on the Kaibob Plateau in Arizona. To build up the deer herd in this natural preserve, all its natural predators, such as cougars, wolves, coyotes, and bobcats, were killed during the period from 1906 to 1924. With plenty of food and no predators to control their numbers, the deer population rose dramatically. By 1918, the population began to exceed the region's carrying capacity, reaching approximately 70 000 animals in 1924. Without enough food, the deer began to starve, and nearly 80% of the population died between 1928 and 1930. Although biologists had warned of the consequences of predator elimination, public pressure forced the program to its conclusion.

Figure 25.14

A representative J-shaped population curve.

Figure 25.15

Population curve for the Kaibob Plateau deer, 1900–1950.

Figure 25.16

The introduction of an organism into an environment can lead to a rapid growth exceeding the carrying capacity. This is followed by a drop in population.

LIMITING FACTORS IN POPULATIONS

The survival and reproduction of an organism depends on adequate supplies of nutrients and the ability of the organism to withstand many of the abiotic factors in the environment. In the mid-1800s, the German chemist Justus von Liebig noted that plants required essential substances for growth and development. If any one of these substances is reduced below the required concentration, the development of the plant will be affected regardless of the amounts of the other substances. This idea later became known as the **law of the minimum.** In 1913, Victor Shelford added to Liebig's work by noting that an excess of nutrients can also be detrimental to an organism's survival. Subsequently, it was determined that an organism will grow only within a range between the upper and lower limits of abiotic factors. The greater the range of tolerance, the greater the survival ability of an organism. This principle is often called **Shelford's law of tolerance.** As seen in Figure 25.17, there is an optimum range of conditions for maximum population size. If one could plot tolerance curves for each environmental factor on a single graph, few would fit exactly on

each other. Therefore, the overall optimum range for a population is restricted. The organisms may respond to changes in any one abiotic factor by having their population reduced or increased as the case may be.

Limiting factors such as daily and seasonal temperature extremes may extend beyond the limits of tolerance for a particular organism. In Canada, temperature extremes are not unexpected. If a tree reaches its lowest limit of tolerance to cold at $-30\,^{\circ}$C, and a week of $-40\,^{\circ}$C temperatures occur, there is a good chance that the tree will not survive. To overcome situations such as these, many organisms have developed adaptations that help them tolerate and survive such extremes. Trees can produce a type of antifreeze to keep the water in their cells from expanding. Hibernation and estivation in animals are adaptations for avoiding prolonged cold or heat and dryness. Any abiotic factors such as those mentioned above are said to be **density independent**—they will affect a population regardless of its size. However, because of the variations among individuals within a population, there is a high probability that some members will survive some extremes of limiting factors and thus continue to perpetuate the species.

When the density of a population increases, other factors brought on by the population size may limit further growth and/or reduce population numbers. These factors are said to be **density dependent.** An excellent example would be limits to the food supply in the Kaibob deer population. Density-dependent factors, such as increased incidence of disease, starvation, and predation, can lead to

The **law of the minimum** *states that, of the number of essential substances required for growth, the one with the minimum concentration is the controlling factor.*

Shelford's law of tolerance *states that too little or too much of an essential factor can be harmful to an organism.*

Density-independent *factors affect members of a population regardless of population density (e.g., flood, fire).*

Density-dependent *factors are factors arising from population density (e.g., food supply) that affect members of a population.*

Figure 25.17

The distribution of a population through its tolerance range for an abiotic factor.

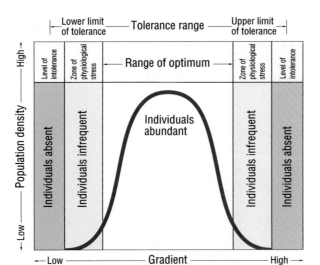

higher mortality rates. Reducing the birth rate and increasing emigration can also decrease population stress. When the population density returns to lower levels, the problems are reduced and an increase in numbers can occur.

r AND K POPULATION STRATEGIES

Most natural populations fall between two general extremes called r-selected and K-selected populations. These populations are associated with specific habitat conditions. **K-selected populations** are found where the environmental conditions are fairly stable and fluctuations are few. Even the changes of the seasons are regular and predictable. K-selected populations become crowded, and competition between members of the same species soon becomes intense. As indicated previously, crowding has detrimental effects on survival rate and the number of adults able to breed. The best-adapted adults are selected to reproduce and the overall population maintains itself close to the carrying capacity (*K*) of the habitat. Members of the K-selected populations are usually large in size and produce young that are slow-growing and require parental care. Their overall reproduction rate is low both in terms of the group and the individual. If an environmental change does occur, the young have the poorest survival rate. Large mammals such as elk, bears, and humans are classified as K-selected animals.

By contrast, **r-selected populations** are found where environments undergo many changes, some of which are unpredictable. These individuals are usually small in size, have a short life span, and reproduce (*r*) at a high rate. The offspring grow rapidly, with little, if any, parental care provided. If conditions are favorable, the population can grow quickly and competition is not a significant factor. A sudden environmental change can result in a massive number of deaths, particularly of the adults. Nevertheless, the rapid birth rate and short life span of these populations seem to favor their ability to adapt to a changing habitat. Insects are examples of r-selected populations.

■ REVIEW QUESTIONS ■ ?

10 Puffins are small marine birds found off the coast of Atlantic Canada. Calculate the population growth rate of a puffin colony based on the following population for 1990.

Original population:	200 000
Natality:	15 000
Mortality:	10 000
Immigration:	175 000
Emigration:	160 000

11 Define dynamic equilibrium.

12 How does an open population differ from a closed population?

13 For growth curves of a closed population, describe the characteristic features of the lag, growth, stationary, and death phases.

14 Define carrying capacity, biotic potential, and environmental resistance.

15 Name four factors that affect environmental resistance.

16 Differentiate between S-shaped and J-shaped growth curves.

17 Differentiate between r and K population strategies. Give at least two examples of each.

18 Define the law of the minimum and Shelford's law of tolerance.

19 Distinguish between density-dependent and density-independent factors. Provide examples of each.

K-selected populations are found where environmental conditions are stable. These populations are characterized by intense intraspecific competition.

An r-selected population undergoes many changes, many of which cannot be predicted. These populations are characterized by a high birth rate and a short life span.

LIFE-HISTORY PATTERNS

Some organisms exhibit regular population fluctuations of growth and decline known as population cycles. A number of small rodents, including mice, voles, and lemmings, have cycles that are about four years in length (see Figure 25.18).

The growth curves for rodents appear to be variations on the basic J-shape, with a characteristic major population decrease on a regular basis, in this case every fourth year. The decrease can be attributed to a depletion of food supply as a result of overfeeding. Interestingly enough, the further north you go, the greater the extremes in the cycles. Considerable research is still going on in an attempt to provide more information about these natural cycles.

One of the more accurate long-term records of population cycles comes from the Hudson's Bay Company's pelt counts for the snowshoe hare and lynx. The snowshoe hare cycle is about 11 years in length. The lynx pelt records reveal a similar cycle. When the population cycles for both animals are plotted on the same graph, the lynx population is seen to cycle along with the snowshoe hare but about a year behind. Because the lynx is a predator of the hare, this set of data has often been used to illustrate how predators can help to control a population. As tempting as this explanation may be, the situation is not quite that simple. Keep in mind that the data could have been influenced by many factors. For example, the demand for pelts in a specific year varies due to changes in fashion. Other natural predators, fluctuations in the number of parasites, or dwindling food supplies due to competition by other primary consumers must also be considered.

The puzzle of the hare-lynx relationship becomes even more confusing when hare populations are observed on arctic islands where there are no lynx. These populations also exhibit cyclic changes. Today it is generally believed that the hares were probably limited more by overgrazing their habitat than by any predators. Therefore, it is quite likely that the lynx numbers were affected by the hares, and not the reverse.

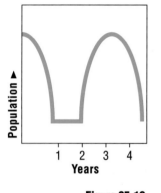

Figure 25.18

Representative population cycle for rodents.

Figure 25.19

Population fluctuations of lynx and snowshoe hare based on Hudson's Bay Company records.

> **BIOLOGY CLIP**
> The Norwegian lemming, famous for its four-year population cycle, changes habitats each spring and fall by migrating. The animals may move several hundred kilometers from their usual range. They are easily seen and are quite aggressive, having been observed attacking vehicles. Contrary to popular belief, lemmings do not commit mass suicide either by leaping off cliffs or swimming out into bodies of water and drowning.

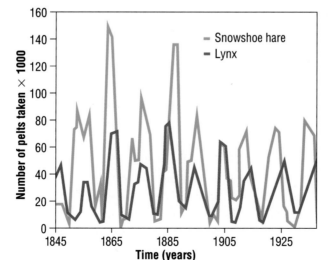

CASE STUDY

CALCULATING THE SIZE OF A SMALL MAMMAL POPULATION

Objective

To use field study data on shrew populations to examine the characteristics of a natural population.

Materials

graph paper   notebook paper
ruler   calculator (optional)

Procedure

The masked shrew (*Sorex cinereus*) is one of the smallest mammals in North America. As a predator of many types of insects, this tiny animal is beneficial to humans. It can significantly reduce the population of sawflies, which attack both larch and spruce trees. It was successfully introduced into Newfoundland in 1958 for this purpose.

1 The data provided have been simulated to represent a typical population study that could have occurred in Newfoundland a few years after the introduction of the shrews. Information on shrews can be found in the reference book *Mammals of Canada* as well as other sources.

  a) Could a population of masked shrews have evolved in Newfoundland?

  b) Would you expect there to be natural predators of the shrew living in Newfoundland? Explain.

2 Biologists released 100 breeding pairs of the shrew into a defined area of Newfoundland during the spring. Each year thereafter for a period of 10 years, the population size was determined by live trapping in late August and the resulting data were recorded.

3 The release area was extremely large, measuring 10.0 km^2. Samples were taken from the area and the overall population was estimated.

4 Four trapping stations were randomly selected. Quadrats measuring 20 × 20 m were marked off and live traps were placed in pairs at 10 m intervals around the quadrat. In addition, one pair of traps was placed at the center, making a total of 18 traps in all. The traps were baited and left for a three-day period. Any animals caught were

marked and then released. This procedure was repeated at the same locations for the duration of the study.

A 20 × 20 m trapping quadrat.

The symbols "xx" in the diagram represent the locations of each pair of traps.

  c) What is the advantage of sampling the population rather than trapping the entire area?

  d) Why were the quadrats the same size, and the traps set at the same time and location each year?

  e) Why did the biologist mark each trapped shrew?

5 Each year the trapping results from the four quadrats were combined and then averaged.

  f) Why were four trapping stations used rather than one?

6 The results of the 10 years of trapping are recorded in the table below.

Trapping Results for the Masked Shrew, from 4 Quadrats over a 10-year Period

Year	Quadrat #1	Quadrat #2	Quadrat #3	Quadrat #4
1	1	1	1	2
2	1	1	0	3
3	2	1	0	4
4	2	1	0	4
5	3	2	2	6
6	5	3	2	8
7	8	4	2	9
8	10	5	4	12
9	9	5	3	12
10	12	7	4	15

  g) Record the total number of shrews trapped each year, then calculate the average number caught in each quadrat over the ten years. Tabulate your results.

7 Draw a graph of the population changes by plotting the annual number of shrews caught (responding variable) against the time in years (manipulated variable). Leave sufficient room on your graph to permit extrapolation of your data.

8 Examine your table of data and your graph carefully.

h) Are there any indications of a preferred habitat for the shrews? What factors might account for these differences?

i) Does the graph show continuous growth in the population? Explain any trends.

j) Compare your graph with the theoretical population growth curves described in the text.

k) Predict what might happen to the population over the next two years.

l) What may have happened to the population in year 9?

m) If other species of shrew were already living in the area, how might this have affected the population being studied? Explain.

Case-Study Application Questions

(For each of the following questions use the appropriate formula when necessary and show your work.)

1 Calculate the following:
a) How many square meters are there in a trapping quadrat?
b) What is the size of a quadrat in hectares?
c) How many hectares are there in the total study area?

2 Using the average number of shrews per quadrat for year 10, calculate the population density in shrews per hectare.

3 From your answer to question 2, calculate the number of shrews for the total study area.

4 Explain why the answer to question 3 is usually called a population "estimate."

5 Calculate the rate of change in shrew density between years 1 and 5 and years 6 and 10. Which four-year period shows the greater population change?

6 Although there is a wide range of shrews and other small mammals all across Canada, only the meadow vole, a mouse-like animal, naturally inhabits Newfoundland. Suggest reasons why so few rodents live in Newfoundland. ■

POPULATION HISTOGRAMS

Although population growth curves show how populations change over time and allow predictions, they do not tell you about the age distribution of the members. The *population histogram* is particularly useful when studying populations of animals in which individuals have a life span of more than a couple of years. The pyramids in a histogram allow you to examine the population of an organism in terms of its age structure and the proportions of males and females at a specific instant in time. Using the data from this type of graph, it is possible to predict, with some accuracy, whether the popula-

tion will grow, stabilize, or decline. Figure 25.20 shows demographic data from three different human populations.

These histograms indicate the percentages of males and females plotted separately on the X (horizontal) axis. The Y (vertical) axis shows age groups of the males and females in five-year intervals. For organisms with shorter life spans, year-by-year or other age intervals may be used. The population's potential for future growth is determined by the proportion of individuals in the different age groups. However, natality and mortality are only two of the four factors that account for changes in a population's size. Large-scale immigration and emigration can significantly alter the demographics

592 ■

Unit Seven:
Change in Populations and Communities

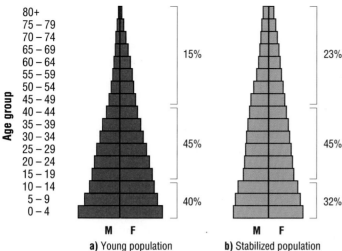

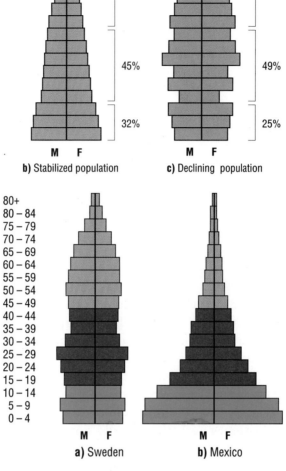

Figure 25.20

Age pyramids representing
(a) growing, (b) stable, and
(c) declining population.

of a population, and therefore must be considered in any long-range projections of population growth.

An age pyramid with a wide base is characteristic of a rapidly growing population (see Figure 25.20a). Not only does it indicate a high percentage of young offspring but it also shows that the number of animals capable of reproduction, usually those in the central portion of the pyramid, is also high. Figure 25.20(b) represents a fairly stable population. There are slightly more young being produced than adults producing them, but with infant mortality higher than the mortality of most other age groups, the population is growing very slowly. This population is approaching what is often described as **zero population growth.** When the base of the pyramid is narrower than the middle section, as shown in Figure 25.20(c), this indicates that fewer offspring are being produced and that, over time, the population will decline. These pyramids have been very useful in the study of human population growth, particularly in comparisons of one country or region of the world with another.

Human Population Growth

The first significant change in the growth in human population occurred following the invention and practice of agriculture about 10 000 years ago. With the change from hunting and gathering to farming, humans were able to harvest and store their food for use during the winter months or periods of drought. A variety of agriculture-related occupations arose that allowed individuals to exchange their services for food. How do you think this affected where people lived? During the years that followed, there was a steady rise in population but no explosion of numbers. By the late 1600s the human population reached about 500 million.

Figure 25.21

Sweden (a) has zero population growth, while Mexico (b) has a rapidly growing population.

Zero population growth *occurs when the population of a species shows no increase or decrease in size over time.*

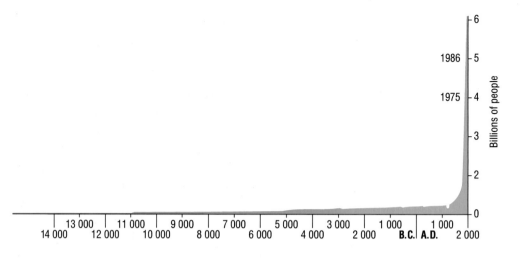

Figure 25.22

Human population growth.

At least three major factors led to a dramatic increase in the human population in the 19th century. The industrial revolution had a major impact on agriculture, resulting in the production of more food. The development of transport systems allowed food to be distributed, particularly to regions where even subsistence agriculture could not be practiced. The reduction of infant mortality due to improved water purification, sewage treatment, and advances in medical science increased the number of children surviving to reproductive age.

During the last three centuries the global population has risen at an exponential rate. Between 1650 and 1850, the population doubled to about one billion. Over the next 80 years it doubled to two billion. From 1930 to 1975 the population doubled to four billion, this time taking only 45 years. At the current annual growth rate of 1.8%, which adds about 80 million people per year, the next doubling will take only 39 years from 1975. This rapid growth of the human population is not without consequence. About one out of five persons is malnourished and lacks clean drinking water and adequate housing.

INTRASPECIES AND INTERSPECIES COMPETITION

Every population will grow to the limits of its ability to obtain resources. The more similar the ecological niches of different species, the greater the competition and the more difficult it becomes to maintain a variety of different populations. G. F. Gause, a Russian biologist, attempted to investigate competition by growing cultures of two closely related species of paramecium separately and together. In all cases the food source was the same. As Figure 25.23 illustrates, each population could survive when isolated, but when grown together, following an initial population growth for each group, one population eventually dominated. This has led to the idea, sometimes called **Gause's principle,** that if two populations of organisms occupy the same ecological niche, one of the populations will be eliminated. This phenomenon would represent the ultimate in **interspecific competition.** Under most natural conditions there is enough separation of niches even within similar groups of animals (e.g., mouse and vole) that many populations can coexist in the same habitat. In these cases there is more **intraspecific competition** in a population, and the principles of natural selection and the survival of the fitter become significant. Food, space, and suitable breeding areas are factors that can cause competition within a single population.

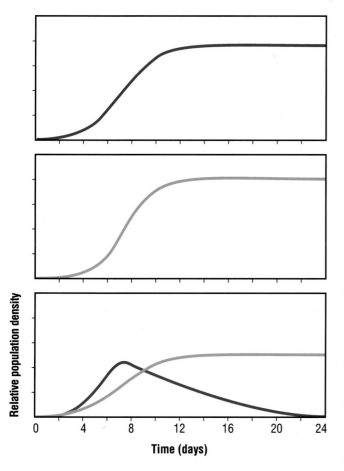

Figure 25.23

Both species of paramecium thrived when grown separately, but one died when they were grown together.

PREDATION

Predation is the most commonly described interaction between populations. The example of the lynx-hare cyclic populations, described earlier, provides a classic model of how the fate of two organisms is closely tied. Fluctuations in the hare (prey) population have a delayed effect on the lynx (predator) population. But the hare itself can be considered a consumer. The hare feeds on the buds and shoots from shrubs and trees. As crowding occurs and winter food becomes scarce, the hare turns to plants such as alder, poplar, and birch, which produce toxins in their tissues. Ecologists have speculated that the toxins eaten by the hares can ultimately be linked with a drop in the population.

All too often the prey in predator-prey relationships is described as a predetermined victim. Nothing could be further from the truth. The fact that varying hares turn white during winter to blend in with the background emphasizes the evolutionary pressures at work in predator-prey relationships. **Camouflage** is used by a great number of organisms to avoid predators. Other organisms avoid predation by physiological adaptations; for example, the production of toxins by the alder, birch, and poplar trees deters hares and other grazers from consuming their tissues. Butterflies use a similar mechanism. Slow-flying, brightly-colored butterflies are no match to out-maneuver most birds, but, once caught, the butterfly tastes extremely bitter. After this experience, the bird will avoid butterflies with that coloration. Although this does not save the individual butterfly, it does ensure species survival. The bright colors of the butterfly thus serve as warning coloration. So successful is this form of defense that other organisms, in a process called **mimicry,** often evolve a similar pattern.

Figure 25.24

The fur of hares turns white during the winter months to provide camouflage.

Camouflage *is an adaptation in form, shape, or behavior that better enables an organism to avoid predators.*

Mimicry *involves developing a similar color pattern, shape, or behavior that has provided another organism with some survival advantage.*

Coevolution *occurs when two different species exert selective pressures on each other.*

Symbiosis *is a relationship in which two different organisms live in a close association.*

Predators and prey exert continuous pressure on each other to gain some survival advantage. Nowhere is this **coevolution** more evident than in the predator-prey relationship of moths and bats. Biologists have long been aware of how bats use sound waves to locate moths, but were surprised to discover that some species of moths have evolved thicker powderlike scales on their wings. Biologists have speculated that the scales work as soundproofing. Because the sound waves are partially absorbed by the wings, the bats have greater difficulty locating their position. To counter this strategy, many species of bat have developed irregular flight patterns and use sound waves from two different positions to locate the moth.

SYMBIOTIC RELATIONSHIPS

At one time, biologists interpreted all relationships between organisms in terms of competition. The idea of competition for survival may have been borrowed from the theory of evolution, but the interactions of most organisms extend beyond the boundaries of predator-prey relationships. **Symbiosis** refers to the relationships between two individuals of a variety of different species. Terms used to describe symbiotic relationships are *parasitism*, in which one organism benefits and the other is harmed; *commensalism*, in which one organism benefits, but the other neither benefits nor is harmed; and *mutualism*, in which both organisms benefit from the relationship (see Table 25.4).

Table 25.4 Symbiotic Relationships

Term	Species #1	Species #2
Parasitism	+	−
Commensalism	+	0
Mutualism	+	+

Note: + = benefits from relationship; − = harmed by relationship; 0 = no effect.

Parasitism

Parasites obtain nourishment from their **hosts,** but do not usually kill the host. Since killing the host would create a major problem for the parasite—finding a new food source—it is in the parasite's best interests to live with the host as long as possible.

Dutch elm disease is a parasitic fungus that has infected hundreds of thousands of North American elm trees. The fungus is believed to be spread by a beetle. The dwarf mistletoe, a parasitic shrub, has infected the lodgepole pine and jack pine. The sea lamprey, a primitive, jawless fish, has migrated from the St. Lawrence through locks and canals into the Great Lakes. The lampreys attach their mouth to fish and, using a file-like tongue, rasp through the body wall. The blood of the fish is ingested by the lampreys.

Some of the most unusual examples of parasitism can be classified as social parasitism. One species of ant from the Amazon area will invade other colonies and steal their cocoons. When the unsuspecting ant emerges from the cocoon, it is conscripted into a life of slavery by its captors. The cowbird displays a form of social parasitism by abdicating parental responsibility. The cowbird will remove an egg of a warbler and replace it with one of its own, leaving the adoptive parents to care for the fledgling along with their own. Warblers will usually feed the cowbird fledgling as one of their own, but the larger cowbird will throw the adopted siblings from the nest, assuring itself of an ample supply of food.

Commensalism

There are many naturally occurring examples of **commensalism,** an association in which one of the partners appears to benefit while the other shows neither positive nor negative effects. A dramatic example comes from Canada's arctic. Explorers noticed that the arctic fox often followed migrating caribou. Could the fox be looking for sick or injured caribou? The prospect of a 4 kg arctic fox attacking a sick caribou that could weigh between 100 and 250 kg is highly unlikely. Observers noted that the caribou's shovel-like feet are ideal for kicking away snow that covers lichens, their main food source. As the snow is moved, the subnivean environment of the many small mammals is disturbed, exposing the animals to watchful foxes. While the caribou is unaffected by the arctic fox's behavior, the fox benefits greatly from the actions of the caribou.

> **BIOLOGY CLIP**
> Some years ago, a caribou wearing a radio-tracking collar was followed across 7 km of open ocean to the island of Ramea, off the south coast of Newfoundland. Provincial wildlife officials, curious about how the caribou managed the trip, took a closer look and located the caribou's remains in a garage on Ramea. An unsuspecting poacher had carried it across the water in a small boat—but neglected to remove the collar!

Figure 25.26

Parasitism. (top) The tapeworm reaches maturity in the intestine. (bottom) Cocoons of a wasp are carried by a host caterpillar.

Parasites *live in or on another organism, from which they obtain their food.*

Hosts *are living organisms from which a parasite obtains its food supply.*

Commensalism *is an association between two organisms in which one benefits and the other is unaffected.*

Figure 25.27

Commensalism. A shark with remora attached.

Another well-known example of commensalism exists between sharks and the remora. The remora is a small fish that uses a suction disk located on the back of its head to attach to the shark. Not only does the remora save energy by being carried about by the shark, but it is also able to feed on pieces of the shark's prey. The shark is not harmed in any way by the remora.

Mutualism

Mutualism is a condition in which two different organisms live together and both benefit from the relationship. An excellent example is the relationship between nitrogen-fixing bacteria and the legume plants such as clover and alfalfa in the nitrogen cycle. The bacteria are provided with sugar by the plant while the plant receives nitrates, the usable form of nitrogen. Bacteria living in your large intestine also provide a dramatic example of mutualism. Bacteria are provided with food, and you are provided with needed vitamins from the microbes.

Pollination is an example of mutualism and coevolution: flowers provide the pollinators with nectar, while the pollinators provide a mechanism for the exchange of genetic material, thereby increasing the diversity of the species. Plants with red flowers attract hummingbirds. Red is a color the hummingbird seems to prefer. Trumpet-shaped flowers are ideal for the long beak of the hummingbird: as the bird hovers beside the flower, it places its head near the opening of the flower and inserts its beak. Bees are most often attracted to yellow flowers, which tend to have many petals to provide footing for the bees. By crawling into a trumpet-shaped flower, a bee risks damaging its delicate wings; however, the many-petaled yellow flowers tend to be open. Flowers that attract moths tend to be white in color. Color serves no advantage when light intensity is limited, since moths tend to be most active at dusk. However, white flowers usually have a strong fragrance, which helps the moth locate it in the dark. The long tongue of the moth is adapted for getting nectar from the flower.

Figure 25.28

Mutualism. Both the honeybee and the flowering plant profit from the relationship.

■ REVIEW QUESTIONS ■ ?

20 What information is provided by a population histogram?

21 Name the basic factors that have resulted in the rapid growth in the human population during the past 150 years. Explain.

22 What is the difference between interspecific and intraspecific competition?

23 Why is the parasite-host relationship considered to be symbiotic?

24 Distinguish between mutualism and commensalism.

25 How does a predator differ from a parasite?

26 A symbiotic relationship between two organisms is described as +/0. Classify the relationship. Support your answer.

RESEARCH IN CANADA

A senior researcher for the Canadian Wildlife Service, Dr. Austin Reed has conducted ongoing studies on the significance of habitat types on populations of waterfowl. Currently he is involved with a research team at Bylot Island in the eastern high arctic, a major breeding area for snow geese. They are investigating the greater snow goose population using techniques such as banding and radio telemetry to track migration routes, and to locate stop-off points such as Cap-Tourmente, Quebec. These researchers have obtained critical information on the survival rates of the snow geese.

Population Ecology

In 1988 Dr. Reed received the Atlantic Center for the Environment Conservation Award in recognition of his work on the conservation of migratory birds. The studies of Dr. Reed and others associated with this research have provided ecologists with a better picture of the total population dynamics of the snow goose.

DR. AUSTIN REED

FRONTIERS OF TECHNOLOGY: RADIO TELEMETRY

Have you ever wondered how ecologists trace the migration route of mountain caribou, or determine the home range of the kit fox? Radio transmitters attached to specific animals have been used to track the movement of larger mammals and birds. Today, scientists take advantage of smaller and lighter electronic transmitters that utilize integrated circuits and memory chips. The miniature transmitters, like the new hearing aids and heart pacemakers, are a product of the space program, where smaller and lighter equipment is essential.

Because of this new technology smaller animals can now be monitored in their natural environment. New telemetry devices provide information about the location of an animal, and some, fitted with special receptors, can also monitor body temperature and heart rate. More compact and longer-lasting batteries have enabled ecologists to follow animals for a greater duration of time, while more sensitive receiving equipment enables them to follow the animals over a much wider range.

Using the latest technology, a major study on the effects of low-level jet flights on caribou herds in Labrador has recently been completed. Satellite-tracked radio collars were attached to a number of animals in the herds. The signals were received by polar-orbiter satellites and transmitted to the researchers, who could determine whether the animals were walking or running. This was one of the first studies of its kind ever attempted. In the future, ecologists may routinely track an animal, monitor blood pressure, determine hormone levels, and check its general health from satellites that relay information from the study site back to computers in their office.

Figure 25.29

New life begins after a devastating forest fire.

SUCCESSION IN COMMUNITIES

Few things appear as devastating as the destruction of a mature forest by a severe fire. All that remains is a blackened landscape with a few solitary tree trunks and stems starkly pointing to the sky. Within a few weeks, however, the ground will slowly turn green as annual and perennial plants, tolerant of the sunlight and the resulting high soil temperatures, take root, grow, and reproduce in a soil made fertile by the mineral content of the ash. Within two or three years shrubs and young trees are quite evident and growing rapidly. A few years later, an untrained observer would probably never know that the area had once been burned out. Over the long term, the forest will again reach maturity. Once mature, the forest will remain in a steady state until another disturbance, natural or man-made, once again alters the abiotic environment and vegetation. Along with the changing vegetation is a corresponding progression in the variety of animals (birds, mammals, insects) present. Populations enlarge and then decline as the habitat slowly but surely changes.

The pattern described is not limited to forest communities. Other terrestrial regions of the biosphere, such as prairie and tundra, also show regular regrowth following environmental change. This universal process is referred to as **succession.** Succession describes the gradual changes in the vegetation of an area as it develops toward a final stable community, called a **climax community.** At all times, succession is influenced by the abiotic factors characteristic of the region in which the process is occurring.

There are two possible types of succession. **Primary succession** occurs in an area in which no community existed previously. This could be exemplified by the invasion of plant life on a newly formed volcanic island in the ocean or on land released from a retreating glacier. Soil is lacking and must be developed to a stage where vegetation can take hold. Through the death of the plants or plant parts, the various processes of decay further contribute to soil development. Soil building is a long, slow process and can be modified, depending on abiotic factors such as temperature, slope, and the composition of its ground materials.

Secondary succession occurs following the partial or complete destruction of a community. The regrowth of an area after a forest fire is an example of secondary succession. Since soil is already present, the lengthy process of soil formation found in primary succession is not necessary.

BIOLOGY CLIP

Following a forest fire, the rapid regrowth of vegetation as secondary succession commences, usually resulting in a wide variety of animal populations such as deer and moose. In cold-weather regions the reverse is possible. Due to the cold climate of the northern boreal forest and tundra areas, lichens, a major food source for the woodland caribou, are extremely slow growing. A fire in these regions may result in a massive decline in the caribou population.

The first plant community to appear, along with its associated animal species is referred to as the **pioneer community.** Lichens, moss, and some insects are common pioneer species. These plants must be able to exist in intense solar radiation and fluctuating soil temperatures. Moisture loss, both from the soils and the organisms, dictates which plants survive. The development of vegetation, however, now sets up new ground-level conditions. The plants provide a measure of shade, resulting in a **microclimate** at the base of the plants that differs from that above them. The temperature becomes somewhat lower and evaporation is reduced. Decay processes increase the thickness and fertility of the soil cover. As soils develop, their ability to hold water also improves.

The development of shade gradually sets up conditions that are more hospitable to the spores and seeds of other plants. Plants that are more tolerant of shade begin to grow, and a new community of plants begins to take over. These plants tend to be taller than the pioneer plants and effectively block out much of the solar radiation, contributing even more to a changing microclimate and soil conditions.

As the plant communities are changing, so are the ecological niches available to the animals of these communities. As a result, there is a parallel succession of animal species. Their activities and wastes contribute to the community develop-

Succession *is the slow, orderly, progressive replacement of one community by another during the development of vegetation in any area.*

A **climax community** *is the final, relatively stable community reached during successional stages.*

Primary succession *refers to the occupation, by plant life, of an area not previously covered by vegetation.*

Secondary succession *occurs in an area that was previously covered by vegetation and still has some soil.*

Pioneer communities *are the first species to appear during succession.*

Microclimate *refers to the climate in a small part of a habitat.*

ment. These communities, or **seral stages,** follow one another until a final community is reached that can self-perpetuate. This is the climax community.

Throughout these stages the communities overlap, and species from all stages can be present at any time. Disturbances can cause isolated patches to regress to an earlier seral stage, at which time succession continues. A climax community is therefore in a state of dynamic equilibrium, dominated by the climax vegetation but containing extensive areas of vegetation representing every seral stage.

Seral stages *are specific stages in succession identified by the dominant species present.*

Figure 25.30

Generalized pattern for succession.

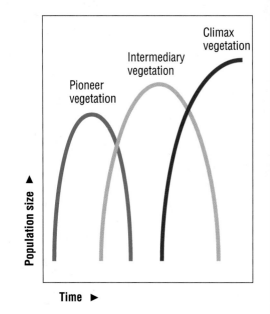

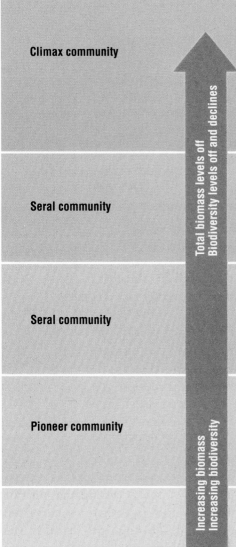

If all the conditions for forest development are met in an area undergoing primary succession, the sequence of seral stages is generally as follows: lichens; mosses; annual and biennial plants; grasses and perennial herbaceous plants; shrubs and tree seedlings; pioneer tree species; one or more intermediate tree stages; climax forest tree species. All of the stages overlap and the species from any one stage may still appear in any subsequent stage. It should also be noted that individual species within a family of plants may reach their optimum development in different seral stages. For example, many

species of moss are characteristic of the early, dry, poor-soil stage, while others can only survive under the shade provided by a heavy canopy of trees where the ground is very moist. Stages can also be skipped, depending both on local conditions and the availability of seeds and spores from specific plants.

Generalizations about Succession

- Species composition changes more rapidly during the earlier stages of succession.

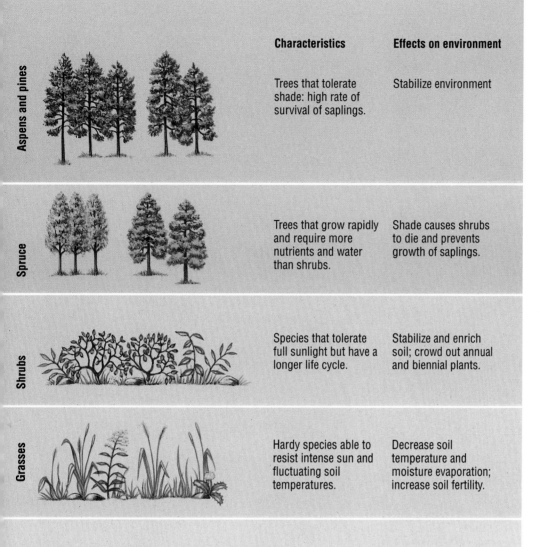

	Characteristics	Effects on environment
Aspens and pines	Trees that tolerate shade: high rate of survival of saplings.	Stabilize environment
Spruce	Trees that grow rapidly and require more nutrients and water than shrubs.	Shade causes shrubs to die and prevents growth of saplings.
Shrubs	Species that tolerate full sunlight but have a longer life cycle.	Stabilize and enrich soil; crowd out annual and biennial plants.
Grasses	Hardy species able to resist intense sun and fluctuating soil temperatures.	Decrease soil temperature and moisture evaporation; increase soil fertility.

Bare land

Figure 25.31

Secondary succession.

- The total number of species increases dramatically during the early stages of succession, begins to level off during intermediary phases, and usually declines as the climax community becomes established.
- Food webs become more complex and the relationships more clearly defined as succession proceeds.
- Both the total biomass and nonliving organic matter increase during succession and begin to level off during the establishment of the climax community.

REVIEW QUESTIONS ?

27 Define succession.

28 Distinguish between primary and secondary succession.

29 What is meant by a climax community? How would you recognize a climax forest community?

30 Name two human activities that can result in secondary succession.

31 Why does succession proceed in a series of stages?

Objective

To study colonization by plants following the retreat of a glacier.

Procedure

1 Observe diagrams (i) and (ii) below. As a glacier retreats, it leaves a harsh, stony landscape behind.

(i)

(ii)

a) Algae are often found in the snow and ice. Indicate what function they would serve in this ecosystem.

2 As the ice melts, the newly exposed minerals are carried away from the topsoil by streams. Some minerals collect in temporary ponds. Ice immediately below the surface soils also greatly reduces nitrogen fixation. Examine diagram (iii). This shows the first pioneer plants, the mountain avens.

(iii)

b) Provide a possible explanation for the uneven distribution of the pioneer plants seen in diagram (iii).

c) The mountain avens contain nitrogen-fixing bacteria in their roots. What special advantage do the mountain avens gain from this mutualistic relationship?

3 The next pioneer species, shown in diagrams (iv) and (v), are deciduous shrubs called alders. The alders also benefit from a relationship with nitrogen-fixing bacteria.

d) Describe how the landscape has changed between diagrams (iv) and (v).

4 Eventually, hemlock and cottonwood begin to grow. Examine diagrams (vi) and (vii). Diagram (vi) shows an intermediary phase of succession dominated by cottonwood and hemlock, while (vii) shows a mature forest dominated by spruce trees. The climax community establishes itself approximately 80 years after a glacier has completely retreated.

e) Referring to diagram (vi), speculate why the cottonwood shrubs only begin to colonize the area after the mountain avens are established.

f) Compare diagram (iii) with diagram (vii). Which plant provides the greatest biomass?

Case-Study Application Questions

1 Compare the pioneer vegetation shown in diagram (iii), the intermediary vegetation shown in diagram (iv), and the vegetation of the climax community shown in diagram (vii).

a) Which community would support the greatest number of organisms? Give your reasons.

b) Which community would support the greatest diversity of organisms? Give your reasons.

2 Provide sample food webs for diagrams (iii) and (vii).

3 Briefly explain how the abiotic factors within the community change because of the succession of vegetation. ■

B I O L O G Y   C L I P

Increasing acidity in lakes of the Laurentian Shield region can result in "succession in reverse." Fish species lose their ability to reproduce. In an Ontario study all five species found in one lake ceased to reproduce between 1979 (pH = 5.6) and 1982 (pH = 5.1). They did so in sequence from the least to the most acid-tolerant species.

Chapter Twenty-five:
Populations and Communities

CAREER INVESTIGATION

No living organism is completely independent. We all interact with other organisms and with the abiotic environment. If we are to continue to live on this planet, we must understand and preserve the organisms we share it with.

Ski resort operator

Ski hills must be designed not only to provide an exciting ski run for skiers but also to have a minimum impact on the area's ecology. The interactions between the local organisms must be studied and evaluated before construction begins.

Fishing boat operator

Fish are an important source of food for people around the world. Overfishing can result in a severe depletion of fish stocks, causing loss of employment and production. Scientific management of fish stocks is necessary if we are to continue to enjoy this valuable resource.

Farm manager

On a large farm, a farm manager may be required to supervise up to 100 full-time farm workers. The manager must coordinate production, purchasing, and marketing. Farm managers must keep up to date on new technological developments affecting crops, growing conditions, and plant and animal diseases.

- Identify a career associated with change in populations and communities.
- Investigate and list the features that appeal to you about this career. Make another list of features that you find less attractive about this career.
- Which high-school subjects are required for this career? Is a postsecondary degree required?
- Survey the newspapers in your area for job opportunities in this career.

Community planner

Community planners analyze the demographic, economic, and social characteristics of municipalities or cities. They are responsible for setting zoning restrictions, designing transportation routes, and preventing land erosion and pollution. A community planner must attempt to preserve the environment while providing a safe, clean, comfortable area for people to live in.

Unit Seven:
Change in Populations and Communities

SOCIAL ISSUE:
Forest Fires and Ecology

Over 9000 forest fires occur in Canada every year. About 65% of these fires are caused by humans and the remainder by lightning. Fire is an important component of ecosystems. Fire keeps forests from taking over grasslands. Trees such as the black spruce, jack pine, and white birch are all adapted to regenerate after fire. Other species, including the white pine and Douglas fir, require ground that has been prepared by fire in order to reproduce properly. Some trees, such as the jack pine, require fire to reproduce.

Statement:

Naturally occurring forest fires should not be fought unless they endanger lives.

Point

- Each year, more than $250 million is spent fighting forest fires. If these fires were allowed to run their course, the dollar savings would be considerable.

- Some trees, such as the jack pine, require fire to reproduce. Without fire, these species would eventually disappear.

Counterpoint

- Forest fires annually account for millions of dollars in damage to commercial forests. In addition, many animals and birds are destroyed or displaced by forest fires.
- Although fire causes succession, many other factors in nature ensure change. By stopping fires, succession will not be prevented, only redirected.

Research the issue.
Reflect on your findings.
Discuss the various viewpoints with others.
Prepare for the class debate.

CHAPTER HIGHLIGHTS

- A population is a group of individuals of the same species that occupy the same area.
- Community refers to the populations of all organisms that occupy an area.
- An ecological niche is the overall role of a species in its environment.
- Dynamic equilibrium refers to any condition within the biosphere that remains stable within fluctuating limits.
- Carrying capacity refers to the supply of resources (including nutrients, energy, and space) that can sustain a population in an ecosystem.
- Biotic potential is the number of offspring that could be produced by a species under ideal conditions.
- Environmental resistance includes all the factors that tend to reduce population numbers.

- Gause's principle states that no two species can occupy the same ecological niche without one being reduced.
- Interspecific competition involves competition among different species; intraspecific competition involves competition within the ecological niche among members of the same species.
- Symbiosis is a relationship in which two different organisms live in a close association.
- Succession is the slow, orderly, progressive replacement of one community by another during the development of vegetation in any area. A climax community is the final, relatively stable community reached during successional stages.

APPLYING THE CONCEPTS

1 Calculate the rate of change in a moose population using the following data:

1978: 25 moose; area surveyed = 40 ha

1984: 11 moose; area surveyed = 30 ha

2 Plot a population curve from the following data, which were obtained after the introduction of deer mice on an isolated hillside.

a) Is the curve characteristic of an open or closed population? Why?

b) Using a line graph, show changes in the population between 1979 and 1990.

c) Use a red line to indicate the probable position of the carrying capacity of the ecosystem. Why have you placed it where you did?

Date	Numbers
1979	20
1980	20
1981	22
1982	26
1983	25
1984	28
1985	40
1986	80
1987	130
1988	128
1989	133
1990	132

3 Using the data provided above, calculate the growth rate between 1986 and 1987. Identify factors that might account for the accelerated growth rate.

4 Explain how the carrying capacity of an ecosystem is related to the biotic potential of a species and its environmental resistance.

5 Predict what might happen to a population of moose if their numbers exceeded the carrying capacity of their environment. Provide the reasons for your prediction.

6 Design a research study that would supply you with data on the effect of latitude on the population cycle of the common deer mouse.

7 In what ways is Shelford's law a more appropriate principle in population studies than the law of the minimum?

8 Draw a population histogram from the following data on the white-tailed deer.

Age	Males	Females
1	72	75
2	35	33
3	24	25
4	17	15
5	14	11
6	8	9
7	7	6
8	5	5
9	4	3
10	2	3

What information is provided by this histogram? (Note: The few animals over 10 years of age are not included.)

CRITICAL-THINKING QUESTIONS

1 How could habitat preference influence the results of a population density study? Is it possible to account for this factor when designing a population study? Explain clearly.

2 Compare the concept of dynamic equilibrium in an ecosystem with homeostasis in the body of an advanced animal.

ENRICHMENT ACTIVITIES

Suggested reading:
- Gordon, Anita, and David Suzuki. *It's a Matter of Survival*. Toronto: Stoddart Publishing, 1990.

- Mungall, Constance, and Digby McLaren. *Planet under Stress*. Toronto: Oxford University Press, 1991.

Unit Seven:
Change in Populations and Communities

abdominal cavity: Extends from the diaphragm to the upper part of the pelvis. The stomach, liver, pancreas, small intestine, large intestine, and kidneys are contained in the abdominal cavity.

abiogenesis: A theory that states that nonliving things can be transformed into living things.

abiotic components: The nonliving components of the biosphere. They include chemical and physical factors.

absorption: The movement of fluids in the direction of a diffusion, or osmotic, gradient.

accommodation reflexes: Adjustments made by the lens and pupil for near and distant objects.

acetylcholine: A transmitter chemical released from vesicles in the end plates of neurons. Acetylcholine makes the postsynaptic membranes more permeable to Na+ ions.

acids: Substances that release hydrogen ions in solution.

acrosome: The cap found on sperm cells. It contains packets of enzymes that permit the sperm cell to penetrate the gelatinous layers surrounding the egg.

action potentials: Nerve impulses. The reversal of charge across a nerve cell membrane initiates an action potential.

activation energy: The energy required to initiate a chemical reaction.

active site: The area of the enzyme that combines with the substrate.

active transport: The use of cell energy to move materials across a cell membrane against the concentration gradient.

adaptation: An inherited trait or set of traits that improve the chances of survival and reproduction of organisms.

adrenal cortex: The outer region of the adrenal gland. It produces glucocorticoids and mineralocorticoids.

adrenal medulla: Found at the core of the adrenal gland. The adrenal medulla produces epinephrine and norepinephrine.

aerobic respiration: The complete oxidation of glucose in the presence of oxygen.

afferent arteriole: Carries blood to the glomerulus.

agglutination: The clumping of blood cells caused by antigens and antibodies.

albedo: A term used to describe the extent to which a surface can reflect light that strikes it. An albedo of 0.08 means that 8% of the light is reflected.

aldosterone: A hormone produced by the adrenal cortex. It helps regulate water balance in the body by increasing sodium and water reabsorption by the kidneys.

allantois: An embryonic membrane.

alleles: Two or more alternate forms of a gene.

all-or-none response: The all-or-none response of a nerve or muscle fiber means that the nerve or muscle responds completely or not at all to a stimulus.

allosteric activity: The change in the protein enzyme caused by the binding of a molecule to the regulatory site of the enzyme.

alveoli: Blind-ended sacs of the lung. The exchange of gases between the atmosphere and the blood occurs in the alveoli.

amino acids: Organic chemicals that contain nitrogen. Amino acids can be linked together to form proteins.

amniocentesis: A technique used to identify certain genetic defects in a fetus or embryo.

amnion: A fluid-filled embryonic membrane.

amylase: An enzyme that hydrolyzes complex carbohydrates.

amyloplasts: Colorless plastids that store starch.

anabolic steroids: Strength-enhancing drugs.

anabolism: Chemical reactions in which simple chemical substances are combined to form complex chemical structures.

anaerobic bacteria: Bacteria that grow best in environments that have little oxygen.

anaerobic respiration: Takes place in the absence of oxygen.

analogous structures: Similar in function and appearance but not in origin. The wing of an insect and the wing of a bird are analogous structures.

anemia: The reduction in blood oxygen due to low levels of hemoglobin or poor red blood cell production.

aneurysm: A fluid-filled bulge found in the weakened wall of an artery.

angina: Literally means to suffocate. It is often used to describe the chest pain produced by heart attack.

antibodies: Proteins formed within the blood that react with antigens.

anticodons: The three-base codes found in tRNA that pair with the codons of mRNA.

antidiuretic hormone (ADH): Acts on the kidneys to increase water reabsorption.

antigen: A substance, usually protein in nature, that stimulates the formation of antibodies.

aorta: Carries oxygenated blood to the tissues of the body.

aqueous humor: Supplies the cornea with nutrients and refracts light.

arteries: High-pressure blood vessels that carry blood away from the heart.

arterioles: Fine branches from arteries.

artificial involution: A process by which egg cells are extracted from a donor and placed in a nongenetic mother.

astigmatism: A vision defect caused by the abnormal curvature of the surface of the lens or cornea.

atherosclerosis: A disorder of the blood vessels characterized by the accumulation of cholesterol and other fats along the inside lining.

atmosphere: The air that encircles the earth.

atoms: The smallest particles of matter. They are composed of smaller subatomic particles: neutrons, protons, and electrons.

ATP (adenosine triphosphate): A compound that stores chemical energy.

atria: Thin-walled heart chambers that receive blood from veins.

atrioventricular (AV) valves: Prevent the backflow of blood from the ventricles into the atria.

auditory canal: Carries sound waves to the eardrum.

autonomic nerves: Motor nerves designed to maintain homeostasis. Autonomic nerves are not under conscious control.

autosomes: Chromosomes not involved with sex determination.

autotrophs: Organisms capable of obtaining their energy from the physical environment and building their required organic molecules.

axon: An extension of cytoplasm that carries nerve impulses away from the dendrites.

Barr bodies: Small, dark spots of chromatin located in the nuclei of female mammalian cells.

bases: Substances that release hydroxide ions in solution.

basilar membrane: Anchors the receptor hair cells in the organ of Corti.

B cells: Make antibodies.

benthos: The bottom of any body of water.

bile salts: The components of bile that emulsify fats.

binary fission: A form of asexual reproduction in which one cell divides into two equal cells.

biogeochemical cycle: The complex, cyclical transfer of nutrients from the environment to an organism and back to the environment.

biological amplification: The buildup of toxic chemicals in organisms as tissues containing the chemical move through the food chain.

biomes: Large-scale ecosystems such as tundra, boreal forest, or grassland.

biosphere: The narrow zone around the earth that harbors life.

biotic components: The biological or living components of the biosphere.

biotic potential: The maximum number of offspring that can be produced by a species under ideal conditions.

blastocyst: An early stage of embryo development.

blastula: The early stage of development in which the cells of the dividing embryo form a hollow ball of cells.

blind spot: The area in which the optic nerve attaches to the retina.

blood pressure receptors: Specialized nerve cells that are activated by high blood pressure.

boreal forest: A worldwide forest region found in the upper latitudes that is characterized by coniferous trees.

Bowman's capsule: A cuplike structure that surrounds the glomerulus. The capsule receives filtered fluids from the glomerulus.

brackish: A mixture of fresh and salt water.

bronchial asthma: Characterized by a reversible narrowing of the bronchial passage.

bronchioles: The smallest passageways of the respiratory tract.

bronchitis: An inflammation of the bronchioles.

Brownian motion: The random movement of molecules.

browsing: Feeding on leaves, twigs, buds, bark, and similar vegetation.

buffers: Absorb excess acid or base, thereby preventing any significant fluctuation in pH values.

camouflage: An adaptation in form, shape, or behavior that better enables an organism to avoid predators.

capillaries: Tiny blood vessels that connect arteries and veins. The site of fluid and gas exchange.

carbonic anhydrase: An enzyme found in red blood cells. The enzyme speeds the conversion of CO_2 and H_2O to carbonic acid.

cardiac output: The amount of blood pumped from the heart each minute.

carnivore: An animal that eats other animals in order to obtain food.

carriers: Individuals that are heterozygous.

carrying capacity: The maximum population that can be sustained by a given supply of resources (nutrients, energy, and space).

catabolism: Reactions in which complex chemical structures are broken down into simpler molecules.

catalysts: Chemicals that regulate the rate of chemical reactions without themselves being altered.

cataracts: Occur when the lens or cornea become clouded.

cell fractionation: The process by which cell fragments are separated by centrifugation.

cellular respiration: The process by which living things convert the chemical energy in sugars into the energy used to fuel cellular activities.

cellulose: A plant polysaccharide that makes up plant cell walls.

centrioles: Small protein bodies that are found in the cytoplasm of animal cells.

centromeres: Structures that hold chromatids together.

cerebellum: The region of the brain that coordinates muscle movement.

cerebral cortex: The outer lining of the cerebral hemispheres.

cerebrospinal fluid: Fluid that circulates between the innermost and middle membranes of the brain and spinal cord.

cerebrum: The largest and most highly developed part of the human brain. The cerebrum stores sensory information and initiates voluntary motor activities.

cervix: A muscular band that prevents the fetus from prematurely entering the birth canal.

chemical compounds: Formed when two or more elements are joined by chemical bonds.

chemoreceptors: Specialized nerve receptors that are sensitive to specific chemicals.

chemosynthesis: The formation of carbohydrates from energy resulting from the breakdown of inorganic substances rather than from light.

chioneuphores: Animals that can withstand winters with snow and cold temperatures.

chionophiles: Animals whose ranges lie within regions of long, cold winters ("snow lovers").

chionophobes: Animals that avoid snow-covered regions ("snow haters").

chlorophyll: The pigment that makes plants green. Chlorophyll traps sunlight energy for photosynthesis.

chloroplasts: Organelles that specialize in photosynthesis, and contain the green pigment chlorophyll found in plant cells.

cholinesterase: An enzyme released from vesicles in the end plates of neurons shortly after acetylcholine. Cholinesterase breaks down acetylcholine.

chordae tendinae: Support the AV valves.

chorion: The outer membrane of a developing embryo.

chorionic villus sampling (CVS): A prenatal diagnosis technique that secures cells from the outer membrane of the embryo for analysis.

choroid layer: The middle layer of the eye. Pigments prevent scattering of light in the eye by absorbing stray light. Many blood vessels are found in this layer.

chromatids: Single strands of a chromosome that remain joined by a centromere.

chromatin: The material found in the nucleus. It is composed of protein and DNA.

chromoplasts: Store orange and yellow pigments.

chromosomes: Long threads of genetic material found in the nucleus of cells. Chromosomes are composed of many nucleic acids and proteins.

cilia: Tiny hairlike protein structures found in eukaryotic cells. Cilia sweep foreign debris from the respiratory tract.

cirrhosis of the liver: A chronic inflammation of liver tissue characterized by an increase of nonfunctioning fibrous tissue and fat.

climax community: The final, relatively stable community reached during successional stages.

climax vegetation: The long-enduring steady-state plant community.

closed population: A population in which density changes are the result of natality and mortality with neither food nor wastes being allowed to enter or leave the given environment.

clumped distribution: Occurs in aggregates. The distribution of organisms is affected by abiotic factors.

coagulation: Occurs when the bonds holding a protein molecule are disrupted, causing a permanent change in shape.

cochlea: The coiled structure of the inner ear that identifies various sound waves.

codons: Three-base codes for amino acids.

coenzymes: Organic molecules synthesized from vitamins that help enzymes combine with substrate molecules.

coevolution: Occurs when two different species exert selective pressures on each other.

cofactors: Inorganic molecules that help enzymes combine with substrate molecules.

collecting duct: Receives urine from a number of nephrons and carries urine to the pelvis.

colon: The largest segment of the large intestine. Water reabsorption occurs in the colon.

commensalism: An association between two organisms in which one benefits and the other is unaffected.

community: Includes the populations of all organisms that occupy an area.

competitive inhibitor: A molecule that has a shape complementary to a specific enzyme, thereby permitting it access to the active site of the enzyme. Inhibitors block chemical reactions.

complementary proteins: Help phagocytotic cells engulf foreign cells.

cones: Photoreceptors that identify color.

conjugation: A form of sexual reproduction in which genetic material is exchanged between two cells.

connective tissue: Provides support and holds various parts of the body together.

consumers: Heterotrophic organisms.

continuity of life: A succession of offspring that share structural similarities with those of their parents.

controls: Standards used to verify a scientific experiment. Controls are often conducted as parallel experiments.

convergent evolution: The development of similar forms from unrelated species due to adaptation to similar environments.

cornea: A transparent tissue that refracts light toward the pupil.

coronary arteries: Supply the cardiac muscle with oxygen and other nutrients.

corpus callosum: A nerve tract that joins the two cerebral hemispheres.

corpus luteum: Made up of the follicle cells of the ovary following ovulation. The corpus luteum secretes estrogen and progesterone.

cortex: The outer layer of the kidney.

cortisol: A hormone that stimulates the conversion of amino acids to glucose by the liver.

counter-current exchange: An anatomical variation of the circulatory system designed to reduce heat loss.

covalent bonds: Formed when electrons are shared between two or more atoms.

Cowper's (bulbourethral) gland: Contributes a mucus-rich fluid to the seminal fluid (semen).

cranial cavity: Surrounded by the skull, which protects the brain, eyes, and inner ear.

crossing-over: The exchange of genetic material between two homologous chromosomes.

cyclic AMP: A secondary chemical messenger that directs the synthesis of protein hormones by ribosomes.

cytokinesis: The division of cytoplasm.

cytoplasm: The area of the protoplasm outside of the nucleus.

Dalton's law of partial pressure: States that each gas in a mixture exerts its own pressure, which is proportional to the total volume.

deamination: The removal of an amino group from an organic compound.

death phase: Phase that marks a constant decline in the population. Mortality exceeds natality.

decomposer food chains: Usually bacteria and fungi; consume wastes and dead tissue from organisms.

decomposers: Bacteria and fungi that break down the remains or wastes of other organisms in the process of obtaining their organic nutrients.

dehydrolysis synthesis: The process by which larger molecules are formed by the removal of water from two smaller molecules.

dendrites: Projections of cytoplasm that carry impulses toward the cell body.

denitrifying bacteria: Soil bacteria that reduce nitrates or nitrites to gaseous nitrogen and some nitrous oxide.

density-dependent factors: Factors arising from population density (e.g., food supply) that affect members of a population.

density-independent factors: Factors that affect members of a population regardless of population density (e.g., flood, fire).

deoxyribonucleic acid (DNA): The carrier of genetic information in cells.

depolarization: Caused by the diffusion of sodium ions into the nerve cell. Excess positive ions are found inside the nerve cell.

detoxify: To remove the effects of a poison.

detritus: Any organic waste from animals and plants.

diabetes mellitus: A genetic disorder characterized by high blood sugar levels.

diapedesis: The process by which white blood cells squeeze through clefts between capillary cells.

diaphragm: A sheet of muscle that separates the organs of the chest cavity from those of the abdominal cavity.

diastole: Refers to heart relaxation.

diffusion: The movement of molecules from an area of higher concentration to an area of lower concentration.

diploid chromosome number: The full complement of chromosomes. Every cell of the body, with the exception of sex cells, contains a diploid chromosome number.

disaccharides: Formed by the joining of two monosaccharide subunits.

distal tubule: Conducts urine from the loop of Henle to the collecting duct.

dominant genes: Determine the expression of the genetic trait in offspring.

Down syndrome: A trisomic disorder in which a zygote receives three homologous chromosomes for chromosome pair number 21.

duodenum: The first segment of the small intestine.

dynamic equilibrium: Any condition within the biosphere that remains stable within fluctuating limits.

ecological niche: Refers to the overall role of a species in its environment.

ecosystem: A community and its physical and chemical environment.

edema: Tissue swelling caused by decreased osmotic pressure in the capillaries.

efferent arteriole: Carries blood away from the glomerulus to a capillary net.

electrocardiograph: An instrument that monitors the electrical activity of the heart.

electron transport system: A series of progressively stronger electron acceptors. Each time an electron is transferred, energy is released.

elements: Pure substances that cannot be broken down into simpler substances. There are 109 different elements.

embolus: A blood clot that dislodges and is carried by the circulatory system to vital organs.

embryo: Refers to the early stages of an animal's development. In humans, the embryo stage lasts until the ninth week of pregnancy.

emphysema: An overinflation of the alveoli. Continued overinflation can lead to the rupture of the alveoli.

endergonic reaction: Reaction that requires the continuous addition of energy. Low-energy reactants are converted into high-energy products.

endocrine hormones: Chemicals secreted by glands directly into the blood.

endocytosis: The process by which particles too large to pass through cell membranes are transported within a cell.

endometrium: The glandular lining of the uterus that prepares the uterus for the embryo.

endorphins: A group of chemicals classified as neuropeptides. Containing between 16 and 31 amino acids, endorphins are believed to reduce pain.

endoscope: An instrument that views the interior of the body.

energy systems: Involve energy input, energy conversion, and energy output.

enkephalins: A group of chemicals classified as neuropeptides. Containing five amino acids, the enkaphalins are produced by the splitting of larger endorphin chains.

enterokinase: An enzyme of the small intestine that converts trypsinogen to trypsin.

entropy: A measure of non-usable energy within a system.

enucleated cells: Do not contain a nucleus.

environmental resistance: All the factors that tend to reduce population numbers.

enzymes: Special protein catalysts that permit chemical reactions within the body to proceed at low temperatures.

epididymis: A compact, coiled tube located along the posterior border of the testis. Consists of coiled tubules that store sperm cells.

epiglottis: A structure that covers the opening of the trachea (glottis) during swallowing.

epilimnion: The upper layer of water in a lake; it heats up in the summer.

epinephrine: A hormone produced by the adrenal medulla that initiates the flight-or-fight response.

epistatic genes: Mask the expression of other genes.

epithelial tissue: A covering tissue that protects organs, lines body cavities, and covers the surface of the body.

equilibrium: A condition in which all acting influences are balanced, resulting in a stable condition.

erepsins: Enzymes that complete protein digestion by converting small-chain peptones to amino acids.

erythroblastosis fetalis ("blue baby"): Occurs when the mother's antibodies against Rh+ blood enter the Rh+ blood of her fetus.

erythropoiesis: The process by which red blood cells are made.

esophagus: A tube that carries food from the mouth to the stomach.

estrogen: A female sex hormone.

estuary: The place where rivers enter the ocean.

eukaryotic cells: Cells that have a true nucleus. The nuclear membrane surrounds a well-defined nucleus.

eustachian tube: An air-filled tube of the middle ear that equalizes pressure between the external and internal ear.

eutrophic lakes: Lakes that are shallow and warm, and are rapidly becoming filled in.

eutrophication: The filling in of a lake by organic matter and silt.

evolution: The cumulative changes in characteristics of populations of organisms in successive generations.

exergonic reaction: A reaction that releases energy.

exocytosis: The passage of large molecules through the cell membrane to the outside of the cell.

experimental variables: Designed to test a hypothesis. Experimental groups test a single variable at a time.

external intercostal muscles: Muscles that raise the rib cage, decreasing pleural pressure.

extracellular fluids (ECF): Occupy the spaces between cells and tissues.

Fallopian tube: See "oviduct."

farsightedness: Occurs when the image is focused behind the retina.

fats: Animal lipids composed of glycerol and saturated fatty acids.

feedback inhibition: The inhibition of an enzyme in a metabolic pathway by the final product of that pathway.

fertilization: Occurs when a male and a female sex cell fuse.

fetus: Refers to the later stages of an unborn offspring's development. In humans, the embryo is called a fetus after the ninth week of development.

filtration: The selective movement of materials through capillary walls by a pressure gradient.

first law of thermodynamics: States that energy can be changed in form but cannot be created or destroyed. It is often referred to as the law of conservation of energy.

first trimester: Extends from conception until the third month of pregnancy.

flow phase: Phase of the menstrual cycle marked by the shedding of the endometrium.

follicles: Structures in the ovary that contain the egg and secrete estrogen.

follicle-stimulating hormone (FSH): A gonadotropin that increases sperm production in males; promotes the development of the follicles in the ovary of the female.

follicular phase: Phase of the menstrual cycle marked by the development of the ovarian follicles prior to ovulation.

food chains: Illustrate a step-by-step sequence of who eats whom in the biosphere.

food web: A series of interlocking food chains representing the transfer of energy through various trophic levels in an ecosystem.

fovea centralis: The most sensitive area of the retina. It contains only cones.

gallstones: Crystals of bile salts that form in the gallbladder.

gametes: Sex cells. They have a haploid chromosome number.

gametogenesis: The formation of sex cells in animals.

Gause's principle: No two species can occupy the same ecological niche without one being reduced in numbers or being eliminated.

gene markers: Are often recessive traits that are expressed in the recessive phenotype of an organism. The markers can be used to identify other genes found on the same chromosome.

gene pools: All of the genes that occur within a specific population.

genes: Units of instruction located on chromosomes that produce or influence a specific trait in the offspring.

gene therapy: A procedure by which defective genes are replaced with normal genes in order to cure genetic diseases.

genome: The complete set of instructions contained within the DNA of an individual.

genotype: The genes an organism contains.

geographic range: A region where a given organism is sighted.

glaucoma: A disorder of the eye caused by the build-up of fluid in the anterior chamber to the lens.

glomerulus: A high-pressure capillary bed that is surrounded by Bowman's capsule. The glomerulus is the site of filtration.

glucagon: A hormone produced by the pancreas. When blood sugar levels are low, glucagon promotes the conversion of glycogen to glucose.

glycogen: The form of carbohydrate storage in animals. Glycogen is often called the animal starch.

glycolipids: Compounds consisting of specialized sugar molecules attached to lipids.

glycolysis: The process in which ATP is formed by the conversion of glucose into pyruvic acid.

glycoproteins: Compounds consisting of specialized sugar molecules attached to the proteins of the cell membrane. Many distinctive sugar molecules act as signatures that identify specialized cells.

goiter: An enlargement of the thyroid gland.

Golgi apparatus: A protein-packaging organelle.

gonadotropic hormones: Produced by the pituitary gland. Regulate the functions of the testes and ovaries.

gonadotropin-releasing hormone (GnRH): A chemical messenger from the hypothalamus that stimulates secretion of gonadotropins (FSH and LH) from the pituitary.

grana: Green disks stacked together. The disks are part of the thylakoid membrane.

grazer food chain: Originate with plants that are consumed by herbivores (grazers).

grazing: Feeding on grass or grasslike vegetation.

growth curve: A graph used to show the changes to a population over a specific length of time.

growth hormone: Produced by the cells of the anterior pituitary. Prior to puberty, the hormone promotes growth of the long bones.

growth phase: Phase marked by accelerated reproduction by the population. Natality exceeds mortality.

habitat: The physical area where a species lives.

haploid chromosome number: One-half of the full complement of chromosomes. Sex cells have haploid chromosome numbers.

Hardy-Weinberg principle: Indicates conditions under which allele and gene frequencies will remain constant from generation to generation.

HCG: A placental hormone that maintains the corpus luteum.

helper T cells: Identify antigens.

hemoglobin: The pigment found in red blood cells.

herbivore: An animal that obtains its food exclusively from plant tissue.

heredity: The passing of traits from parents to offspring.

heterotrophs: Organisms that obtain food and energy from autotrophs or other heterotrophs; they are unable to synthesize organic food molecules from inorganic molecules.

heterozygous: A genotype in which the gene pairs are different.

hibernation: A dormant (sleep) state in which body temperature and functions are reduced well below normal.

homeostasis: A process by which a constant internal environment is maintained despite changes in the external environment.

homologous chromosomes: Chromosomes that are similar in shape, size, and gene arrangement.

homologous structures: Have similar origins but different uses in different species. The front flipper of a dolphin and the forelimb of a dog are homologous structures.

homozygous: A genotype in which both genes of a pair are identical.

hormones: Chemicals released by cells that affect cells in other parts of the body. Only a small amount of a hormone is required to alter cell metabolism.

hosts: Living organisms from which a parasite obtains its food supply.

hybrids: Offspring that differ from their parents in one or more traits. Interspecies hybrids result from the union of two different species.

hydrogen bonds: Formed between a hydrogen proton and the negative end of another molecule.

hydrologic or water cycle: The movement of water through the environment—from the atmosphere to earth and back.

hyperpolarized membranes: Membranes that are much more permeable to potassium than usual. The inside of the nerve cell membrane becomes even more negative.

hypertonic solutions: Solutions in which the concentration of solutes outside the cell is greater than that found inside the cell.

hypolimnion: The lower layer of a lake; it maintains a constant low temperature.

hypothermia: A condition in which an animal's body core temperature drops to a level that will eventually cause death.

hypothesis: A possible solution to a problem or an explanation of an observed phenomenon.

hypotonic solutions: Solutions in which the concentration of solutes outside the cell is lower than that found inside the cell.

implantation: The attachment of the embryo to the endometrium.

inbreeding: The process whereby breeding stock is drawn from a limited number of individuals possessing desirable phenotypes.

insulin: A hormone produced by the islets of Langerhans in the pancreas. Insulin is secreted when blood sugar levels are high.

intermediary metabolites: The chemicals that form as reactants are converted to products during a series of chemical reactions.

internal intercostal muscles: Muscles that pull the rib cage downward, increasing pleural pressure.

interneurons: Carry impulses within the central nervous system.

interspecific competition: Competition among similar species for a limited resource (e.g., food or space).

interstitial spaces: The spaces between the cells.

intraspecific competition: Competition within an ecological niche between members of the same species.

in vitro fertilization: Occurs outside of the female's body. *In vitro* is Latin for "in glass."

ionic bonds: Formed when electrons are transferred between two atoms.

ionosphere: A region of the upper atmosphere consisting of layers of ionized gases that produce the northern lights and reflect radio waves.

ions: Atoms that have either lost electrons or gained electrons to become positively or negatively charged.

iris: Regulates the amount of light entering the eye.

isomers: Chemicals that have the same chemical formula but a different arrangement of molecules.

isotonic solutions: Solutions in which the concentration of solute molecules outside the cell is equal to the concentration of solute molecules inside the cell.

isotopes: Elements with the same number of protons but different numbers of neutrons.

jaundice: The yellowish discoloration of the skin and other tissues brought about by the collection of bile salts in the blood.

karyotypes: Pictures of chromosomes arranged in homologous pairs.

killer T cells: Puncture the cell membranes of cells infected with foreign invaders, thereby killing the cell and the invader.

Klinefelter syndrome: A trisomic disorder in which a male carries an XXY condition.

K-selected populations: Populations found where environmental conditions are stable. These populations are characterized by intense intraspecific competition.

lag phase: The adjustment period prior to accelerated reproduction by the population.

larynx: The voice box.

law of the minimum: Of the number of essential substances required for growth, the one with the minimum concentration is the controlling factor.

ligase enzyme: A biological glue that permits one section of DNA to be fused to another.

limnetic zone: The open water area of a lake.

linked genes: Genes that are located on the same chromosome.

lipases: Lipid-digesting enzymes.

littoral zone: The edge around a lake or pond where the water is shallow enough to permit the growth of aquatic vegetation.

luteal phase: Phase of the menstrual cycle characterized by the formation of the corpus luteum following ovulation.

luteinizing hormone (LH): A gonadotropin that promotes ovulation and the formation of the corpus luteum in females; regulates the production of testosterone in males.

lymph: The fluid found outside capillaries. Most often, the lymph contains some small proteins that have leaked through capillary walls.

lymph nodes: Contain white blood cells that filter lymph.

lymphocytes: Antibody-producing white blood cells.

lymphokine: A protein produced by the T cells that acts as a chemical messenger between cells.

macrophages: Phagocytotic white blood cells found in lymph nodes or in the blood (in bone marrow, spleen, and liver).

magnetosphere: A region found above the outer atmosphere consisting of magnetic bands caused by the earth's magnetic field.

map distance: The distance between two genes along the same chromosome.

matrix: The noncellular material secreted by the cells of connective tissue.

medulla: The area inside of the cortex in the kidney.

medulla oblongata: The region of the hindbrain that joins the spinal cord to the cerebellum. The medulla is the site of autonomic nerve control.

meiosis: The two-stage cell division in which the chromosome number of the parental cell is reduced by half. The process by which sex cells are formed.

memory T cells: Retain information about the geometry of antigens.

meninges: Protective membranes that surround the brain and spinal cord.

menopause: Marks the termination of the female reproductive years.

menstruation: The shedding of the endometrium.

mesosphere: The region of the atmosphere found between the stratosphere and the upper atmosphere.

metabolism: The sum of all chemical reactions that occur within the cells.

metastasis: Occurs when a cancer cell breaks free from the tumor and moves into another tissue.

microclimate: The temperature and moisture that occur at the ground level in a plant community.

micropipette: A thin glass rod that can be used to extract urine from the nephron.

microvilli: Infoldings of the cell membrane.

migration: The movement of organisms between two distant geographic regions.

mimicry: A form of camouflage that involves developing a similar color pattern, shape, or behavior that has provided another organism with some survival advantage.

mitochondria (singular: mitochondrion): Organelles that specialize in aerobic respiration.

mitosis: A type of cell division in which daughter cells receive the same number of chromosomes as the parent cell.

molecules: Units of matter composed of one or more different atoms.

monoculture: Involves growing a single species of plant to the exclusion of others.

monohybrid cross: Involves one gene pair of contrasting traits.

monosaccharides: Single sugar units. All monosaccharides include carbon, hydrogen, and oxygen in a ratio of 1:2:1.

monosomy: The presence of a single chromosome in place of a homologous pair.

motor neurons: Carry impulses from the central nervous system to effectors.

murmurs: Caused by faulty heart valves, which permit the backflow of blood into one of the heart chambers.

mutagenic agents: Things that cause changes in the DNA.

mutations: Arise when the DNA within a chromosome is altered. Most mutations change the appearance of the organism.

mutualism: A relationship in which two different organisms living together both benefit from each other.

myelin sheath: A fatty covering over the axon of a nerve cell.

myogenic muscle tissue: Contracts without external nerve stimulation.

NAD+: A strong electron acceptor important for electron transport systems in cellular respiration. NADH is the reduced form of NAD⁺.

NADP+: A strong electron acceptor important for electron transport systems in photosynthesis. NADPH is the reduced form of NADP⁺.

nearsightedness: Occurs when the image is focused in front of the retina.

negative feedback system: A control system designed to prevent chemical imbalances in the body. The body responds to changes in the external or internal environment. Once the effect is detected, receptors are activated and the response is inhibited, thereby maintaining homeostasis.

nephrons: The functional units of the kidneys.

neurilemma: The delicate membrane that surrounds the axon of nerve cells.

neurons: Cells that conduct nerve impulses.

neutralization: Occurs when the pH is brought to 7, i.e., the $[H^+]$ = $[OH^-]$.

nitrogen cycle: The cycling of nitrogen between organisms and the environment.

nitrogen fixation: The conversion of nitrogen gas (N_2) into nitrates (NO_3) and ammonium ions (NH_4^+), which can then be used by plants.

nitrogen-fixing bacteria: Bacteria that convert atmospheric nitrogen to nitrogen compounds such as ammonia and nitrate.

nodes of Ranvier: The regularly occurring gaps between sections of myelin sheath along the axon.

norepinephrine: A hormone produced by the adrenal medulla that initiates the flight-or-fight response.

nuclear imaging: Technique that uses radioisotopes to view organs and tissues of the body.

nuclear magnetic resonance (NMR): Technique that employs magnetic fields and radio waves to determine the behavior of molecules in soft tissue.

nucleolus: A small spherical structure located inside the nucleus.

nucleotides: The building blocks of nucleic acids. Nucleotides are composed of a ribose sugar, phosphate, and a nitrogen base.

nucleus: The control center for the cell. Contains hereditary information.

nutrients: Chemicals that provide nourishment. Nutrients provide energy or are assimilated to form protoplasmic structures.

oils: Plant lipids composed of glycerol and unsaturated fatty acids.

olfactory lobes: Areas of the brain that detect smell.

oligotrophic lakes: Lakes that are cold and deep, and have only begun the process of eutrophication.

omnivore: An organism that eats both animals and plants.

oncogenes: Cancer-causing genes.

ootids: Unfertilized egg cells.

open population: A population in which density changes result from the interaction of natality, mortality, immigration, and emigration.

organic molecules: Compounds that contain carbon.

organ of Corti: The primary sound receptor in the cochlea.

organs: Structures composed of different tissues specialized to carry out a specific function.

organ systems: Groups of organs that have related functions. Organ systems often interact.

osmotic pressure: The pressure exerted on the wall of a semi-permeable membrane resulting from differences in solute concentration. (In this case, the more concentrated the plasma proteins, the greater is the osmotic pressure.)

ossicles: Tiny bones that amplify and carry sound in the middle ear.

otoliths: Tiny stones of calcium carbonate found within the saccule and utricle. The tiny stones are embedded in a gelatinous coating. Gravity causes the otoliths to slide downward as the head is lowered or raised.

oval window: Receives sound waves from the ossicles.

ovaries: The female gonads, or reproductive organs. Female sex hormones and egg cells are produced in the ovaries.

oviduct (Fallopian tube): The passageway through which an ovum moves from the ovary to the uterus, or womb.

ovulation: The release of the egg from the follicle held within the ovary.

oxidation: Occurs when an atom or molecule loses electrons.

oxytocin: A hormone from the posterior pituitary gland that causes strong uterine contractions.

ozone: An inorganic molecule. A layer of ozone found in the stratosphere helps to screen out ultraviolet radiation.

ozone hole: A region in the ozone layer in which the ozone levels have been considerably reduced and the layer has become very thin.

Pangaea: A large supercontinent that existed approximately 225 million years ago.

parasite: An organism that lives in or on another organism, from which it obtains its food.

parasympathetic nerves: A division of the autonomic nervous system. These nerves are designed to return the body to normal resting levels following adjustments to stress.

pathogens: Disease-causing agents.

pelvis: The area in which the kidney joins the ureters.

pepsin: A protein-digesting enzyme produced by the cells of the stomach.

peptide bonds: Bonds that join amino acids.

pericardium: A saclike membrane that protects the heart.

peristalsis: The rhythmic, wavelike contraction of smooth muscle that moves food along the gastrointestinal tract.

phagocytosis: A form of endocytosis in which cells engulf large molecules and incorporate them into the cytoplasm.

pharynx: The passage for both food and air.

phenotype: The observable traits of an organism that arise because of the interaction between genes and the environment.

pheromone: A chemical substance produced by an organism that serves as a stimulus to another organism of the same species.

phospholipids: The major components of cell membranes in plants and animals. Phospholipids have a phosphate molecule attached to the glycerol backbone, making the molecule polar.

phosphorylation: The addition of one or more phosphate groups to a molecule.

photolysis: The splitting of water by means of light energy.

photosynthesis: The process by which plants and some bacteria use chlorophyll, a green pigment, to trap sunlight energy. The energy is used to synthesize carbohydrates.

photosystems: Light-trapping units composed of pigments within the thylakoid membranes.

pH scale: Used to measure the concentrations of acids and bases.

pinna: The outer part of the ear. The pinna acts like a funnel, taking sound from a large area and channeling it into a small canal.

pinocytosis: A form of endocytosis in which liquid droplets are engulfed by cells.

pioneer communities: The first species to appear during succession.

placenta: An organ made from the cells of the baby and the cells of the mother. It is the site of nutrient and waste exchange between mother and baby.

plasma: The fluid portion of the blood.

plasmids: Small rings of genetic material.

plastids: Organelles that function as factories for the production of sugars or as storehouses for starch and some pigments.

pleiotropic genes: Affect many characteristics.

polar bodies: Formed during meiosis. These cells contain all the genetic information of a haploid egg cell but lack sufficient cytoplasm to survive.

polarized membranes: Charged membranes. Polarization is caused by the unequal distribution of positively charged ions.

polar molecules: Molecules that have positive and negative ends.

polygenic traits: Inherited characteristics that are affected by more than one gene.

polymerases: Enzymes that join individual nucleotides together in complementary strands of DNA.

polymers: Molecules composed of from three to several million subunits. Many polymers contain repeating subunits.

polypeptides: Proteins composed of amino acids and joined by peptide bonds

polyploidy: A condition in which an organism possesses more than two complete sets of chromosomes.

polysaccharides: Composed of many single sugar subunits.

pons: The region of the brain that acts as a relay station by sending nerve messages between the cerebellum and the medulla.

population: A group of individuals of the same species occupying a given area at a certain time.

population density: The number of organisms per unit of space.

population sampling: A technique in which gene frequencies for a particular genetic trait are determined in a small sample of the population, and results are applied to the whole population.

population size: The number of organisms at a certain time.

postsynaptic neurons: The neurons that carry impulses away from the synapse.

precursor activity: The activation of the last enzyme in a metabolic pathway by the initial reactant.

presynaptic neurons: The neurons that carry impulses to the synapse.

primary succession: The occupation, by plant life, of an area not previously covered by vegetation.

producers: Autotrophic organisms.

profundal zone: The region of a lake to which light cannot penetrate.

progesterone: A female sex hormone.

prokaryotic cells: Primitive cells that do not have a true nucleus or a nuclear membrane.

prolactin: A hormone produced by the pituitary that is associated with milk production.

prostaglandins: Hormones that have a pronounced effect in a small localized area.

prostate gland: Contributes to the seminal fluid (semen), a secretion containing buffers that protect sperm cells from the acidic environment of the vagina.

protein denaturation: Occurs when the bonds holding a protein molecule are disrupted by physical or chemical means, causing a temporary change in shape.

protein hormones: Composed of chains of amino acids. This group includes insulin, growth hormone, and epinephrine.

proteins: The structural components of cells.

protoplasm: All the material within a cell. The protoplasm is composed of the nucleus and the cytoplasm.

proximal tubule: A section of the nephron joining Bowman's capsule with the loop of Henle. The proximal tubule is found within the cortex of the kidney.

pukak: A layer of open space containing ice crystals at the base of a snow pack.

pulmonary artery: Carries deoxygenated blood from the heart to the lungs.

pulmonary circulatory system: Carries deoxygenated blood to the lungs and oxygenated blood back to the heart.

pulmonary veins: Carry oxygenated blood from the lung to the heart.

pulse: Caused by blood being pumped through an artery.

Punnett square: A chart used by geneticists to show the possible combinations of alleles in offspring.

pus: Substance formed when white blood cells engulf and destroy invading microbes. The white blood cell is also destroyed in the process. The remaining protein fragments are known as pus.

pyramid of biomass: An energy pyramid based on the dry mass of the tissue of organisms at each trophic level.

pyramid of energy: A pyramid drawn on the basis of the energy produced (as heat) at each trophic level.

pyramid of numbers: An energy pyramid based on the numbers of organisms at each trophic level.

qali: Snow that collects on the branches of trees.

qamaniq ("snow shadow"): The depression in the snow found around the base of trees, particularly conifers.

radioisotopes: Unstable chemicals that emit bursts of energy as they break down.

random distributions: Are arbitrary and appear to be unaffected by biotic factors.

rate of change: The change in a population over a period of time.

receptor sites: Act as ports along cell membranes. Nutrients and other needed materials fit into specialized areas along cell membranes.

recessive genes: Genes that are overruled by dominant genes, which determine the genetic trait.

recombinant DNA: An application of genetic engineering in which genetic information from one organism is spliced into the chromosome of another organism.

reduction: Occurs when an atom or molecule gains electrons.

refractory period: The recovery time required before a neuron can produce another action potential.

regulator genes: Genes that control the production of repressor proteins, which switch off structural genes.

relaxin: A hormone produced by the placenta prior to labor that causes the ligaments within the pelvis to loosen.

rennin: An enzyme that coagulates milk proteins.

replication: The process in which a single strand of nucleotides acts as a template for the formation of a complementary strand. A single strand of DNA can make a complementary strand.

repolarization: A process in which the original polarity of the nerve membrane is restored. Excess positive ions are found outside the nerve membrane.

respiration: The chemical process in which nutrients are broken down to provide energy.

resting membranes: Maintain a steady charge difference across the cell membrane. These membranes are not being stimulated.

restriction enzymes: Enzymes that cut strands of DNA at specific sites.

retina: The innermost layer of the eye.

reverse transcriptase: An enzyme that allows the genetic message from the RNA of a virus to be transcribed into the DNA.

rhodopsin (visual purple): The pigment found in the rods of the eye.

ribonucleic acid (RNA): A single-stranded nucleic acid used to translate the information of DNA into protein structure.

rods: Photoreceptors used for viewing in dim light.

r-selected population: Population that undergoes many changes, many of which cannot be predicted. These populations are characterized by a high birth rate and a short life span.

saccule: Responsible, with the utricle, for static equilibrium.

sclera: The outer covering of the eye that supports and protects the eye's inner layers.

scrotum: The sac that contains the testes.

second law of thermodynamics: States that, during an energy transformation, some of the energy produced, usually in the form of heat, is lost from the system.

second trimester: Extends from the third month to the sixth month of pregnancy.

secondary succession: Occurs in an area that was previously covered by vegetation and still has some soil.

secretin: A hormone that stimulates pancreatic and bile secretions.

segregation: The separation of paired genes during meiosis.

selective breeding: The crossing of desired traits from plants or animals to produce offspring with both characteristics.

selectively permeable membranes: Membranes that allow some molecules to pass through the membrane, but prevent other molecules from penetrating the barrier.

semen (seminal fluid): A secretion of the male reproductive organs that is composed of sperm and fluids.

semicircular canals: Fluid-filled structures within the inner ear that provide information about dynamic equilibrium.

semiconservative replication: The process in which the original strands of DNA remain intact and act as templates for the synthesis of duplicate strands of DNA.

semilunar valves: Prevent the backflow of blood from arteries into the ventricles. (Semilunar means "half-moon.")

seminal vesicles: A gland located along the vas deferens that contributes to the seminal fluid (semen), a secretion that contains fructose and prostaglandins.

seminiferous tubules: Coiled ducts found within the testes, where immature sperm cells divide and differentiate.

sensory adaptation: Occurs once you have adjusted to a change in the environment.

sensory neurons: Carry impulses to the central nervous system.

sensory receptors: Modified ends of sensory neurons that are activated by specific stimuli.

seral stages: Specific stages in succession identified by the dominant species present.

Sertoli cells: Nourish sperm cells.

sex chromosomes: Pairs of chromosomes that determine the sex of an individual.

sex-linked traits: Traits that are controlled by genes located on the sex chromosomes.

Shelford's law of tolerance: Too little or too much of an essential factor can be harmful to an organism.

sinoatrial node: The heart's pacemaker.

sloughs: Depressions often filled with stagnant water.

solutes: Molecules that are dissolved in water. Salt and sugars are common solutes.

somatic cells: All the cells of an organism except the sex cells.

somatic nerves: Nerves that lead to skeletal muscle. Somatic nerves are under conscious control.

speciation: The formation of a new species.

sphincter: A ring of smooth muscle in the wall of a tubular organ. Contraction of the sphincter closes the opening.

sphygmomanometer: A device used to measure blood pressure.

spindle fibers: Protein structures that guide chromosomes during cell division.

starch: A plant storage carbohydrate.

stationary phase: Phase that marks equilibrium between natality and mortality.

steroid hormones: Made from cholesterol. This group includes male and female sex hormones and cortisol.

stratosphere: The region of the earth's atmosphere found above the troposphere. The ozone layer is found in this region.

stroke volume: The quantity of blood pumped with each beat of the heart.

stroma: The gel-like substance containing proteins that surrounds the grana.

structural genes: Genes that direct the synthesis of proteins.

subnivean: Means "beneath a snow cover."

substrate molecules: Molecules that attach to enzymes.

succession: The slow, orderly, progressive replacement of one community by another during the development of vegetation in any area.

summation: The effect produced by the accumulation of transmitter chemicals from two or more neurons.

supercooling: Occurs when a water solution is chilled well below the point at which the solution crystallizes spontaneously into ice.

suppressor T cells: Turn off the immune system.

surrogate: Means "substitute." A nongenetic mother is a surrogate mother.

symbiosis: A relationship in which two different organisms live in a close association.

sympathetic nerves: A division of the autonomic nervous system. These nerves prepare the body for stress.

synapses: The regions between neurons or between neurons and effectors.

synapsis: The pairing of homologous chromosomes.

systemic circulatory system: Carries oxygenated blood to the tissues of the body and deoxygenated blood back to the heart.

systole: Refers to heart contraction.

target tissues: Tissues that have specific receptor sites that bind with the hormones.

T cells: Produced in the lymph nodes. Different types of T cells regulate an immune response.

testes: The male gonads, or primary reproductive organs. Male sex hormones and sperm are produced in the testes.

testosterone: The male sex hormone produced by the interstitial cells of the testes.

tetrads: Contain four chromatids.

thermocline: The zone separating the epilimnion from the hypolimnion.

third trimester: Extends from the seventh month of pregnancy until birth.

thoracic cavity: The chest cavity. Heart and lungs are contained in the thoracic cavity.

threshold level: The minimum level of a stimulus required to produce a response.

threshold level: In terms of reabsorption, the maximum amount of material that can be moved across the nephron.

thrombus: A blood clot that forms within a blood vessel.

thylakoid membranes: Specialized cell membranes found in chloroplasts.

thyroxine: A hormone secreted by the thyroid gland that regulates the rate of body metabolism.

tissues: Groups of similarly shaped cells that work together to carry out a similar function.

totipotent cell: A cell that has the ability to support the development of an organism from egg to adult.

trachea: The windpipe.

transcriptase: An enzyme that allows the genetic message from the RNA of a virus to be transcribed into the DNA.

transcription: The process by which the genetic code is transferred from the DNA molecule to the messenger RNA molecule.

translation: The process by which proteins are synthesized using the DNA instructions encoded in the mRNA.

transpiration: The loss of water through the leaves of a plant.

transposons: Specific segments of DNA that can move along the chromosome.

triglycerides: Lipids composed of glycerol and three fatty acids.

trisomy: The presence of three homologous chromosomes in every cell of an organism.

trophic level: The number of energy transfers an organism is from the original solar energy entering an ecosystem.

troposphere: The lowest region of the earth's atmosphere. It extends upward about 12 km. Most weather occurs here.

trypsin: A protein-digesting enzyme.

tundra: The most northerly major life zone. Characterized by low precipitation, low temperatures, permafrost, and a lack of trees.

turgor pressure: The pressure exerted by water against the cell membrane and the nonliving cell walls of plant cells.

Turner's syndrome: A monosomic disorder in which a female has a single X chromosome.

tympanic membrane: The eardrum.

ulcer: A lesion along the surface of an organ.

ultraviolet radiation: The electromagnetic radiation from the sun. It can cause burning of the skin (sunburn) and cellular mutations.

umbilical cord: Connects the fetus to the placenta.

uniform distributions: Are orderly and appear to be affected by competition.

urea: A nitrogen waste formed from two molecules of ammonia and one molecule of carbon dioxide.

ureters: Tubes that conduct urine from the kidneys to the bladder.

urethra: A tube that carries urine from the bladder to the exterior of the body.

uric acid: A waste product formed from the breakdown of nucleic acids.

uterus (womb): The female organ in which the fertilized ovum normally becomes embedded and in which the embryo and fetus develop.

utricle: Responsible, with the saccule, for static equilibrium.

vagus nerve: A major cranial nerve that is part of the parasympathetic nervous system.

varicose veins: Distended veins.

vas deferens: Tubes that conduct sperm toward the urethra.

vasoconstriction: The narrowing of a blood vessel. Less blood goes to the tissues when the arterioles constrict.

vasodilation: The widening of the diameter of the blood vessel. More blood moves to tissues when arterioles dilate.

veins: Carry blood back to the heart.

venae cavae: Carry deoxygenated blood back to the heart.

ventricles: Muscular, thick-walled heart chambers that pump blood through the arteries.

venules: Small veins.

vestibule: A chamber found at the base of the semicircular canals that provides information about static equilibrium.

villi: Small fingerlike projections that extend into the small intestine. Villi increase surface area for absorption.

waxes: Long-chain lipids that are insoluble in water.

womb: See "uterus."

zero population growth: Occurs when the population of a species shows no increase or decrease in size over time.

zygote: The cell resulting from the union of a male and female sex cell, until it divides.

Hummingbird, 598
Hunting, 58
Huntington's chorea, 509, 514, 515
Huygens, Christiaan, 388
Hybridization, 485
Hybridoma, 149, 290–91
Hybrids, 469, 472
Hydra, 437
Hydrochloric acid, 158, 159, 230–31, 233
Hydrogen, 154
    and aerobic respiration, 200–201
    and biogeochemical cycle, 30
    and carbon-fixation reaction, 194
    and oxidation, 197
    and photolysis, 193
Hydrogen bond, defined, 157
Hydrogen ion
    in acids and bases, 157–59
    and pH scale, 158
Hydrogen peroxide, 186
Hydrologic cycle. *See* Water cycle
Hydrosphere, 21
Hydroxide ion
    in acids and bases, 157–59
Hyperglycemia, 341, 343
Hyperopia (farsightedness), 392
Hypertension, 142
Hypertonic solution, 145–46
Hypolimnion, 73
Hypothalamus, 320, 321, 333, 336–37, 339, 370, 411, 417
Hypothermia, 83, 90
Hypothesis, defined, 123
Hypothyroidism, 345
Hypotonic solution, 145
Hysterectomy, 415

Identical twins, 440
Igloo, 91
Immigration, 582ff.
Immune response, 280ff.
Immune system, 142
    and allergies, 286
    cells of (summary), 286
    memory of, 285–86
Immunosuppressant, 285, 286
Implantation, 420
Impotency, 408
Incomplete dominance, 476–77
Incus, 396, 397
Independent assortment, law of, 481
Independent events, rule of, 482
Indoor environment, 62

Induced-fit model, 180
Infertility, 425
Influenza virus, 282
Inheritance, single-trait, 473–74
Initiator, 535
Inner ear, 395ff.
Inorganic compound, defined, 159
Insecticide. *See* Pesticides
Insects, and winter, 80
Insulin, 101, 169, 332, 333, 340, 491
    discovery of, 342
    human, 528
Intercostal muscles, 300
Interferon, 529
Interkinesis, 451
Intermediate inheritance, 476
Interneuron, 351ff., 368
Interphase, 432
Interstitial cell, 410
Interstitial space, 249
Intestine
    large, 213, 235–36
    small, 232–33, 237–38
Inuit
    clothing, 29
    eyeglasses, 390
    snow shelters, 91
    terms for snow, 78
In vitro fertilization, 425, 461, 462
Iodine, 163–64, 345
Ion
    defined, 156
    gate, 358
    and nerve impulse, 357ff.
Ionic bond, defined, 156
Ionosphere, 23
Iris, 385, 386
Irish setter, and inbreeding, 485
Iron, 272, 273
Island biogeography, 103–4
Islets of Langerhans, 333, 340, 341, 344
Isomer, defined, 160
Isotonic solution, 145
Isotope, 154–55

James Bay Project, 94
Jaundice, 235
Jeffreys, Alec, 526
Jenner, Edward, 287–88
Johnson, Ben, 173
J-shaped population curve, 586
Juxtaglomerular apparatus, 323

Kaibob Plateau, deer population of, 587, 588
Kangaroo, 103
Karyotype, 455, 459
Kidney, 213
    and blood pressure, 322
    disease, 323–24, 326, 328
    and filtration, 317–18
    and nephron, 316–17
    and reabsorption, 318–19
    and REF, 273
    and secretion, 319
    stone, 324
    structure of, 315–16, 317
    and water balance, 315, 320–22, 325
King, Thomas, 438
Klinefelter syndrome, 456
Kobuk Valley, 78
Komatik, 78
Krause, Margarida, 446
Krebs cycle, 201–2
K-selected population, 589
Kuhne, Wilhelm, 388, 389

Labor, 423
Labrador retriever, 474
Lactation, 424–25
Lacteal, 266
Lactic acid, 198, 199, 261
Lactobacillus bulgaricus, 123
Lactose, 160, 233
Laennec, Rene, 258
Lag phase, 584
Lake Baykal, 70
Lake Erie, 75
Lake(s)
    bed, 75
    Canadian, 75
    classification of, 69
    eutrophic, 70
    eutrophication of, 70
    oligotrophic, 70
    oxygen level in, 69ff.
    and photosynthesis, 70, 71
    pollution of, 75
    and spring, 73
    structure of, 71ff.
    and summer, 73
    thermal pollution of, 75
    and winter, 72
Lake Superior, 70

Stearic acid, 165
Stefansson, Baldur, 486
Steptoe, Patrick, 425
Stereoscopic vision, 394
Sterility, male, 407
Steroids, 172, 173
Stethoscope, 258
Stewart, Frederick, 437
Stigma, 470
Stomach, 230–31, 280
Stratosphere, 23
Strawberry, 437
Stream, 68
Streptococcus, 123, 290
Stress response, 339
Stretching, 250
Stroke, 372
Stroke volume, 259
Stroma, 191
Sturtevant, A.H., 506
Style, 470
Subnivean environment, 85, 87, 88, 90–91
Substantia gelatinosa, 375
Substrate, defined, 180
Succession, 604–5
    defined, 69, 601
    in forests, 600ff.
    generalizations about, 602–3
    primary, 601
Sucrose, 160, 194
Sudan IV solution, 171
Sugar. *See also* Fructose, Glucose
    aldehyde, 161
    composition of, 160
    ketone, 161
Sulfanilamide, 289
Sulfur, 46
Sulfur dioxide
    and acid deposition, 32–33
    properties of, 157
Summation, 363
Summer
    and lakes, 73
Sun, 26–27
Sundew plant, 39, 65
Sunlight, 26–27, 176. *See also* Energy;
    Light
    and food chains, 57
    and photosynthesis, 30–31, 45, 46, 176–77, 189ff.
Supercooling, 80–81, 89
Surrogate, 461, 462, 464
Sutton, Walter S., 495
Suzuki, David, 62
Sweating, 223
Symbiosis, 596–98
Symbiotic hypothesis, 571
Sympathetic nerve, 351, 366, 367

Synapse, 362ff.
Synapsis, 450
Syndrome, 456
Syphilis, 289
Systole, 258
Systolic blood pressure, 260

Tachycardia, 256
Taiga, 78
Target tissue, 332
Taste bud, 229, 383
Taste receptor, 383
Tatum, Edward, 533–34
Tay-Sachs disease, 138, 562, 568–69
T-cell lymphocyte, 274
    helper, 284–85, 541ff.
    killer, 284–85
    memory, 286
    suppressor, 285, 286
Tear gland, 298
Telophase
    meiosis, 451
    mitosis, 433
Temperature, as limiting factor, 588
Temporal lobe, 370
Terminator, 535
Test cross, 475
Testes determining factor (TDF), 500
Testis, 406–7, 409ff., 500
Testosterone, 172, 183, 335, 410–11
"Test-tube baby," 425, 464
Tetrad, 450
Thalamus, 370
Thalassemia, 492, 511
Thalidomide, 429
Theory, scientific, 108
Thermocline, 73
Thermodynamics, laws of, 52–53, 177–78
Thirst, 321
Thompson, J.J., 133
Thoracic cavity, 216
Threshold level
    and electrical stimulus, 361
    and reabsorption, 319
Thrombin, 277
Thromboplastin, 276–77
Thrombosis, 277
Thrombus, 277
Thylakoid membrane, 190–91
Thymine, 534
Thyroid gland, 344–45
Thyroid-releasing factor (TRF), 345

Thyroid-stimulating hormone (TSH), 335, 337, 345
Thyroxine, 344–45
Tiger, 47
Tin, 155
Tissue, 211–12. *See also* specific types
Tissue cloning, 446
Toboggan, 78
Tongue, 383
Tongue rolling, 557
Totipotent cell, 438
Tra, 514
Trachea, 297
Transcription, 535, 542
Transforming principle, 519
Translation, 536–37, 542
Translucence test, 171
Transmission electron microscope, 133
Transmitter chemical, 363–64
Transpiration, and the water cycle, 32
Transposons, 507–8
Transverse plane, 216
Trees, and adaptation to winter, 89, 588
Triglyceride, 164
Triiodothyronine, 344
Trisomy, 455, 456
Tritium, 154
Trophic level, defined, 47
Tropism, 112
Troposphere, 23
Trout, 74
Trypsin, 181, 182, 233
Trypsinogen, 233
Tsui, Lap-Chee, 510
Tundra
    biome, 68
    and global warming, 36
Turgor pressure, 146
Turner's syndrome, 456, 501
Turtle, 81
Twins, 440
Tylenol, 379
Tympanic membrane (eardrum), 396, 397

Uchida, Irene, 457
Ulcer, 231, 232
Ultraviolet (UV) radiation, 23, 24, 382
Umbilical cord, 422
Uniform distribution, 579
Uracil, 534
Uranium, 155
Urea, 315
Ureter, 315

Urethra, 315, 408
Uric acid, 315
Urinary system
    male vs. female, 315
    structure of, 316
Urine, 315ff., 341
Urine formation, 317–19
Ussher, James, 105
Uterine contraction, 423–24
Uterus, 412ff.
Utricle, 396, 398

Vaccine, 287–88
Vacuole, 138
Vagina, 410, 413
Vagus nerve, 366
Valium, 185, 364, 375
Van Allen belts, 23
Van Benden, Edouard, 495
Van Helmont, Jean, 123–24
Van Leeuwenhoek, Anton, 124–25, 130
Variation, 556ff.
Varicose vein, 250
Vas deferens, 408
Vasectomy, 408
Vasoconstriction, 247
Vasodilation, 247–48
Vavilov, Nikolai, 480
Vein, 249–50, 252
Vena cava, 252
Ventricle, 252, 258–59
Venule, 249
Venus, 43
Venus flytrap, 46, 111
Vesicle, 137
    and exocytosis, 148
Vestibule, 396
Villi, 237–38
Virus, 283, 284, 285, 541ff.
Viscera, 216
Vision, 384ff.
Vision defects, 392–93
Visual acuity, 394
Vitalists, 152
Vitamin A, 390, 391
Vitreous humor, 385
Vocal cords, 297
Voice box. See Larynx
Volcanoes
    and albedo, 27
    and the carbon cycle, 34
Von Baer, K.E., 100
Von Liebig, Justus, 588
Von Merring, Joseph, 333

Vulva, 315

Waders, 81
Walker, Norma Ford, 560
Wallace, Alfred, 102
War of the Worlds, 231
Waste, human
    in lakes, 72, 75
Water
    and aerobic respiration, 200–201
    balance, 315, 320–22, 325
    and body composition, 66, 157
    and capillary fluid exchange, 264–65
    dissociation of, 158
    importance of, 31, 66
    molecule, 67, 157
    and osmosis, 144–45
    pH of, 158
    and photosynthesis, 190, 192ff.
    properties of, 67, 157
    solubility of oxygen in, 73
Water buttercup (Ranunculus aquatilis),
    490
Water cycle, 31–33
    and acid deposition, 32–33
Watson, James, 520–21
Wax, 165
Weasel, 88
Weinberg, W., 560
Weisman, August, 495
Wells, H.G., 231
Wenman, Wanda, 292
Wheat, 480, 485
    Marquis, 486
    Red Fife, 486
White blood cell (leukocyte). 148,
    273–74, 275
    and leukemia, 441
    and immune response, 281
White matter, 353, 368
Wilkins, Maurice, 520–21
Wilson, Edward, 59
Wilson, John (Tuzo), 102
Windpipe. See Trachea
Winter
    adaptations of animals to, 78ff.
    energy available in, 58
    heat loss in, 83–85
    and lakes, 72
Woese, Carl, 116
Wolf, 47, 48, 49, 53
Womb. See Uterus

X chromosome, 500
    and Barr body, 499
X-ray diffraction, 520–21
X-ray technology, 214
XYY male, 460

Y chromosome, 500
Yeast, 198, 199, 200

Zero population growth, 593
Zidovudine. See AZT
Zygote, 407, 420

UNIT OPENING PHOTOS: Unit 1 Main photo: Tom Van Sant/Geosphere Project, Santa Monica/Science Photo Library/Masterfile; Sidebar: Bill Brooks/Masterfile; Unit 2 Main photo: Freeman Patterson/Masterfile; Sidebar: Dr. Bryan Eyden/Science Photo Library; Unit 3 Main photo: CNRI/Science Photo Library; Sidebar: Manfred Kage/Science Photo Library; Unit 4 Main photo: Tom Sanders/Masterfile; Sidebar: Mel Di Giacomo, Bob Masini/The Image Bank; Unit 5: Main photo: Michael Melford/The Image Bank; Sidebar: Steve Satushek/The Image Bank; Unit 6: Main photo: Oxford Scientific Films/Animals Animals/Earth Scenes; Sidebar: Science Photo Library/Photo Researchers; Unit 7: Main photo: Comstock/Miller Comstock; Sidebar: Telegraph Colour Library/Masterfile; **ONE:** **1.1** NASA/Science Photo Library; **20** NASA/Masterfile; **21** J.A. Kraulis/Masterfile; **1.2** Washnik Studio/Masterfile; **1.3** Martyn Lengden after C. Starr and R. Taggart, *Biology 5ed.*, Wadsworth Publishing Company, 1989; **1.4** Jack Finch/Science Photo Library; **1.5** NASA; **1.6A** Philippe Plailly/Science Photo Library; **1.6B** Government of Canada, Department of Communication; **1.7** Martyn Lengden; **1.8** Martyn Lengden; **1.9** Martyn Lengden; **1.10** Prof. Stewart Lowther/Science Photo Library; **28** Richard Siemens; **29** Left, Bob Rose Productions/Miller Comstock; Right, Rick Riewe; **1.11** Martyn Lengden; **1.12** Stephen Mader; **1.13** Margo Stahl; **1.14A** FourByFive Inc.; **1.14B** Will McIntyre/Photo Researchers; **1.15** Stephen Mader; **1.16** Martyn Lengden; **1.20** Stephen Mader; **1.21** Adam Hart-Davis/Science Photo Library; **1.22** Stephen Mader; **1.23** W. Griebeling/Miller Comstock; **41** Kam Yu; **TWO:** **2.1** Comstock/Miller Comstock; **44** W. Metzen/Miller Comstock; **2.2** Martyn Lengden; **2.3** Kerry T. Givens/Tom Stack & Associates; **2.4** Margo Stahl; **2.5A** Dr. E.R. Degginger; **2.5B** Biological Photo Service; **2.5C** J.R. Carmichael, Jr./The Image Bank; **2.5D** Johnny Johnson/Animals Animals/Earth Scenes; **2.6** Margo Stahl; **2.7** Martyn Lengden; **2.8** Margo Stahl; **51** Carolina Biological Supply Co.; John Gerlach/Tom Stack & Associates; **2.9** Martyn Lengden; **2.10** Margo Stahl; **55** Right, Bill Corbett; Left, Lowell Georgia, The National Audubon Society Collection/Photo Researchers; **2.11** Martyn Lengden; **2.12** Martyn Lengden; **2.13** Martyn Lengden; **2.14** Margo Stahl; **2.15** Martyn Lengden; **2.16** Martyn Lengden; **2.17** Martyn Lengden; **2.19** Dr. Morley Read/Science Photo Library; **2.20** Thomas Kitchin/First Light; **2.21** Martyn Lengden; **62** Larry Fisher/Masterfile; **THREE:** **3.1** Johnny Johnson/Animals Animals/Earth Scenes; **66 and 67** Walt Anderson/Tom Stack & Associates; **3.2** Martyn Lengden; **3.3** Margo Stahl; **70** Top, Steve Short/First Light; Bottom, W. Griebeling/Miller Comstock; **3.4** Margo Stahl; **3.5** Margo Stahl; **3.6** Stephen Mader; **3.8** Stephen Mader; **3.9** J.A. Kraulis/Masterfile; **76** Stephen Mader; **3.10** Martyn Lengden; **3.11A** Alan Marsh/First Light; **3.11B** Joe Devenney/The Image Bank; **3.12** Martyn Lengden; **3.13** Lowell Georgia, The National Audubon Society Collection/Photo Researchers; **3.14** Ken Highfill, The National Audubon Society Collection/Photo Researchers; **3.15** David M. Dennis/Tom Stack & Associates; **3.16** Malak/Miller Comstock; **3.17** Bruce Drysdale; **3.18A** Gregory K. Stott/The National Audubon Society Collection/Photo Researchers; **3.18B** Jerry Kobalenko/First Light; **3.18C** Dieter & Mary Plage/Survival Anglia; **3.18D** Tom & Pat Leeson, The National Audubon Society Collection/Photo Researchers; **3.20** Dave Mazierski; **3.21** Margo Stahl; **3.22** Rod Planck/Tom Stack & Associates; **3.23** Martyn Lengden; **3.24** Martyn Lengden; **3.25** Martyn Lengden; **3.26** Bruce Drysdale; **88** Joe Van Os/The Image Bank; **3.28** Kathleen Norris Cook/The Image Bank; **3.29** Bob Rose Productions/Miller Comstock; **3.30** Superstock/FourByFive Inc.; **3.31** Margo Stahl; **3.32** W. Belsey/Miller Comstock; **92** Left, W. Heck/University of Manitoba; Right, Jim Brandenburg/Minden Pictures; **93** Martyn Lengden; **FOUR:** **4.1** Stephen Dalton. The National Audubon Society Collection/Photo Researchers; **4.2** George Calef/Masterfile; **4.3A** John Reader/Science Photo Library; **4.3B** Sinclair Stammers/Science Photo Library; **4.3C** Kevin Schafer/Tom Stack & Associates; **4.6** Martyn Lengden; **4.7** Steve Gilbert; **4.8A** Yogi/Peter Arnold Inc.; **4.8B** Dr. Morley Read/Science Photo Library; **4.8C** Stephen Dalton/Animals Animals/Earth Scenes; **4.8D** Stephen Dalton/Animals Animals/Earth Scenes; **4.9** Martyn Lengden; **4.10** Martyn Lengden; **4.11A** Greg Stott/Masterfile; **4.11B** Horus - ZEFA/Masterfile; **4.12A** National Library of Medicine/Science Photo Library/Science Source/Photo Researchers; **4.12B** Dr. Jeremy Burgess/Science Photo Library; **4.13** The Bettmann Archive; **4.14A** John Cancalosi/Tom Stack & Associates; **4.14B** Stephen Mader after C. Starr and R. Taggart, *Biology 5ed.*, Wadsworth Publishing Company, 1989; **4.15** Stephen Mader; **109** Michael Tweedie, The National Audubon Society Collection/Photo Researchers; **4.16** Right, Nuridsany et Perennou, The National Audubon Society Collection/Photo Researchers; **4.16** Left, Dr. Paul A. Zahl, The National Audubon Society Collection/Photo Researchers; **4.17** Stephen Mader; **4.18** David Parker/Science Photo Library; **114** Martyn Lengden; **116** Right, Dr. Ford Doolittle/Dalhousie University; Left, Dr. Tony Brain & David Parker/Masterfile; **117** Left, E. Otto/Miller Comstock; Bottom, Burton McNeely/The Image Bank; Top right, SuperStock Inc.; Bottom right, Michael Melford/The Image Bank; **FIVE:** **5.1** CNRI/Science Photo Library; **122** M.I. Walker/Photo Researchers; **123** Allen Haslinger/Masterfile; **5.2A** Burndy Library; **5.2B** Dr. Jeremy Burgess/Science Photo Library; **5.2C** Burndy Library; **5.3** Martyn Lengden; **5.4** Martyn Lengden; **127** Top, Kam Yu; Bottom left, Martyn Lengden; Bottom right, Kam Yu; **128** Kam Yu; **5.5** Dave Mazierski; **5.6** Kam Yu; **5.7** Dr. Jeremy Burgess/Science Photo Library; **5.8** Martyn Lengden after C. Starr and R. Taggart, *Biology 5ed.*, Wadsworth Publishing Company, 1989; **5.9** Dave Mazierski; **5.10** Martyn Lengden; **5.11** Secchi-Lecaque-Roussel-Uclaf/CNRI/Science Photo Library; **5.12** Comstock/Miller Comstock; **5.13** Dr. Jeremy Burgess/Science Photo Library; **134** Top left, Canapress Photo Service; Bottom left, Dr. Thomas Chang/McGill University; Right, Dr. Brian Eyden/Science Photo Library; **5.14** Dave Mazierski; **5.15** Dave Mazierski; **5.16A** Dr. Gopal Murti/Science Photo Library; **5.16B** Dave Mazierski; **5.17A** Dave Mazierski; **5.17B** CNRI/Science Photo Library; **5.18** Dave Mazierski; **5.19** Dr. B.H. Satir/Albert Einstein College of Medicine; **5.20A** W.P. Wergin, courtesy of E.H. Newcomb, University of Wisconsin /Biological Photo Service; **5.20B** Dave Mazierski; **5.21** Dave Mazierski; **140** Kam Yu; **5.22** Kam Yu; **142** Kam Yu; **5.23** Dave Mazierski; **5.24** Dr. E.R. Degginger; **144** Kam Yu; **5.25A** Kam Yu after C. Starr and R. Taggart, *Biology 5ed.*, Wadsworth Publishing Company, 1989; **5.25B** Dennis Kunkel/Phototake/First Light; **5.26** Photographs courtesy of Thomas Eisner; **5.27A** Dave Mazierski; **5.27B** Leanard Lessin; **5.28** Kam Yu; **5.29** Photographs courtesy of M.M. Perry from M.M. Perry and A.B. Gilbert (1979), *Journal of Cell Science*, Vol. 39, pp. 257-272; **SIX:** **6.1** Oxford Molecular Biophysics Laboratory/Science Photo Library/Masterfile; **152** Science Photo Library/Masterfile; **6.2** Bob Chambers/Miller Comstock; **6.3** Kam Yu; **6.4** Kam Yu; **6.5** Kam Yu; **6.6** Kam Yu; **6.7** Kam Yu; **6.8** Kam Yu; **6.9** Martyn Lengden after C. Starr and R. Taggart, *Biology 5ed.*, Wadsworth Publishing Company, 1989; **6.10** Martyn Lengden; **6.11** Martyn Lengden; **6.12** Martyn Lengden; **6.13** Martyn Lengden; **6.14A** Carolina Biological Supply Co. **6.14B** Martyn Lengden after C. Starr and R. Taggart, *Biology 5ed.*, Wadsworth Publishing Company, 1989; **163** Kam Yu; **6.15** Martyn Lengden; **6.16** Martyn Lengden; **167** Left, U.S. National Cancer Institute/Science Photo Library; Bottom right, Dr. Leslie Levin; Top right, Dr. Dianne Goettler; **168** Martyn Lengden; **6.17** Martyn Lengden; **6.18** Martyn Lengden; **6.19** Martyn Lengden; **6.20** Laboratory of Molecular Biology, MRC/Science Photo Library; **6.21** Laboratory of Molecular Biology, MRC/Science Photo Library; **172** Top, Martyn Lengden; **6.22** Martyn Lengden; **175** Martyn Lengden; **SEVEN:** **7.1** Adam Hart-Davis/Science Photo Library; **176** Mark Tomalty/Masterfile; **7.2** Martyn Lengden; **7.3** Martyn Lengden; **7.4** Margo Stahl; **7.5** Martyn Lengden; **7.6** Martyn Lengden; **7.8** Martyn Lengden; **7.9** Martyn Lengden; **7.11** Martyn Lengden after art by Victor Royer in C. Starr and R. Taggart, *Biology 5ed.*, Wadsworth Publishing Company, 1989; **7.13** Photograph courtesy of Mike Connolly; **7.14** Martyn Lengden; **7.15** Martyn Lengden; **7.16** Martyn Lengden; **185** Left, Dr. Arthur Lesk, MRC Laboratory of Molecular Biology/Science Photo Library; Right, Tom Walker, Photo Journalist; **7.17** Martyn Lengden; **7.18** Martyn Lengden; **7.19** Martyn Lengden; **7.20** Martyn Lengden; **7.22** Margo Stahl; **7.23A** D. Cavagnaro/Peter Arnold. Inc.; **7.23B** Dave Mazierski; **7.23C** Dr. Jeremy Burgess/Science Photo Library; **7.23D** Dave Mazierski; **7.23E** Dr. Kenneth R. Miller/Science Photo Library; **7.24** Martyn Lengden after art by Victor Loyer in C. Starr and R. Taggart, *Biology 5ed.*, Wadsworth Publishing Company, 1989; **7.25** Kam Yu; **7.26** Dave Mazierski; **7.27** Martyn Lengden after art by L. Calver in C. Starr and R. Taggart, *Biology 5ed.*, Wadsworth Publishing Company, 1989; **7.28** Martyn Lengden after art by L. Calver in C. Starr and R. Taggart, *Biology 5ed.*, Wadsworth Publishing Company, 1989; **7.29** Martyn Lengden; **7.30** Martyn Lengden; **7.31** Martyn Lengden; **7.32** Martyn Lengden; **196** Kam Yu; **7.33A** Kam Yu; **7.33B** Richard Siemens; **7.34** Martyn Lengden; **7.35** Martyn Lengden; **7.36** Left, John Kelly/The Image Bank; Right, First Light; **7.37** Canapress Photo Services/Pictor Uniphoto; **7.38** Martyn Lengden; **7.39** Martyn Lengden; **7.40** Martyn Lengden; **7.41** Martyn Lengden; **203** Left, Grant Johnson/Miller Comstock; Center, Harvey Pincis/Science Photo Library; Top right, Lena Rooraid/Photo Edit; Bottom right, Schmid-Langsfeld/The Image Bank; **205** Martyn Lengden; **206** Margo Stahl; **207** Margo Stahl; **EIGHT: 8.1** CNRI/Science Photo Library; **210** CNRI/Science Photo Library; **8.2** Manfred Kage/Science Photo Library; **8.3** Dave Mazierski; **8.4** CNRI/Science Photo Library; **8.5** Ohio-Nuclear Corporation/Science Photo Library; **8.6** CNRI/Science Photo Library; **8.7** Mehau Kulyk/Science Photo Library; **8.8** Kam Yu; **8.9** Kam Yu; **217-222** Steve Gilbert; **8.10** Martyn Lengden; **224** Peter Menzel; **225** Left, CNRI/Science Photo Library; Right, Dr. Arthur Prochazka/University of Alberta; **NINE: 9.1** Larry Lefever from Grant Heilman/Miller Comstock; **228** Greg Biss/Masterfile; **9.2** Dave Mazierski; **9.3** Kam Yu; **9.4** Martyn Lengden; **9.5** Martyn Lengden; **9.6** Martyn Lengden; **9.7A/B** IMS Creative Communications **9.7C** Manfred Kage/Science Photo Library; **9.7D** Dr. R.F.R. Schiller/Science Photo Library; **9.8** Kam Yu; **9.9** Dave Mazierski; **9.10A** Dave Mazierski; **9.10B** Manfred Kage/Science Photo Library; **9.11** From TISSUES AND ORGANS: A Text-Atlas of Scanning Electron Microscopy. by Richard G. Kessel and Randy H. Kardon. Copyright ©1979 by W.H. Freeman and Company. Reprinted by permission. **240** Top left, Service de Polycopie et de Photographie Université de Montréal; Bottom left, Dr. Gilles Caillé/Université de Montréal; Right, CNRI/Science Photo Library; **242** Martyn Lengden; **243** Martyn Lengden; **TEN: 10.1** Lennart Nilsson from *Behold Man*, © 1974 by Albert Bonniers Forlag and Little, Brown and Company, Boston; **244** Dr. E.R. Degginger; **245** Phototake/First Light; **10.2** Burndy Library; **10.3** Dave Mazierski; **10.4** Dave Mazierski; **10.5** Kam Yu; **10.6** Dave Mazierski; **10.7** Dr. Jeremy Burgess/Science Photo Library; **10.8** Dave Mazierski; **251** Carolina Biological Supply Company; **10.9** Dave Mazierski; **10.10** M. Huberland/Science Source/Photo Researchers; **10.11** Dave Mazierski; **10.12** Dave Mazierski; **10.13** Cardio-Thoracic Centre, Freeman Hospital, Newcastle-Upon-Tyne/Science Photo Library; **10.14** Dave Mazierski; **10.15** Dave Mazierski; **10.16** Martyn Lengden; **10.17** Martyn Lengden; **257** Left, Lennart Nilsson from *Behold Man*, © 1974 by Albert Bonniers Forlag and Little, Brown and Company, Boston; Right, Dr. Alexandra Lucas/University of Alberta; **10.18** Dave Mazierski; **10.19** Sheila Terry/Science Photo Library; **10.20** Martyn Lengden; **10.21** Dave Mazierski; **10.22** Martyn Lengden; **10.23** Martyn Lengden; **10.24** Kam Yu; **10.25** Kam Yu; **10.26** Dave Mazierski; **10.27** Dave Mazierski; **ELEVEN: 11.1** Baylor College of Medicine/Peter Arnold, Inc.; **270** Lennart Nilsson/Bonnier Fakta; **11.2** Kam Yu; **11.3** CNRI/Science Photo Library; **11.4** Dave Mazierski; **11.5** Martyn Lengden and Dave Mazierski; **11.6** Dave Mazierski; **276** Martyn Lengden; **11.7** Dave Mazierski; **11.8** Kam Yu; **11.9A** Martyn Lengden; **11.9B** Kam Yu; **280** Richard Ustinich/The Image Bank; **11.10** Kam Yu; **11.11** Kam Yu; **11.12** Kam Yu; **11.13** Kam Yu; **11.14** Kam Yu; **11.15** Kam Yu; **11.16** Kam Yu; **11.17** Photographs courtesy of Dr. Gilla Kaplan, Cohen Simon Lab, Rockefeller University, New York, NY.; **11.18** Dave Mazierski; **11.19** Martyn Lengden; **11.20** Martin Bond/Science Photo Library; **11.21** Dave Mazierski; **292** Left, Dr. Wanda Wenman/University of Alberta; Right, CNRI/Science Photo Library; **294** Martyn Lengden; **295** Martyn Lengden; **TWELVE: 12.1** NASA **296** Martyn Lengden; **12.2** Dave Mazierski; **12.3** Dave Mazierski; **12.4** Martyn Lengden; **12.5** Dave Mazierski; **12.6** Dave Mazierski; **12.7** Martyn Lengden; **12.8A** Science Photo Library; **12.8B** Lennart Nilsson from *Behold Man*, ©1974 by Albert Bonniers Forlag and Little, Brown and Company; **12.9A** Dave Mazierski; **12.9B** James Stevenson/Science Photo Library; **305** Top, Custom Medical Stock Photo; Bottom, Steinmark/Custom Medical Stock Photo; **12.10** Martyn Lengden; **12.11** Martyn Lengden; **12.12** Martyn Lengden; **12.13** Martyn Lengden; **309** Left, Manfred Kage/Peter Arnold, Inc.; Right, John Pym; **310** Dave Mazierski; **312** Martyn Lengden; **313** Stephen Mader; **THIRTEEN: 13.1** Geoffrey Gove/The Image Bank; **314** Brad Nelson/Custom Medical Stock Photo; **13.2** Martyn Lengden; **13.3** Dave Mazierski; **13.4** Dave Mazierski; **13.5** Kam Yu; **13.6** Kam Yu; **13.7** Dave Mazierski; **13.8** Martyn Lengden; **13.9** Martyn Lengden; **13.10** Kam Yu; **13.11** Martyn Lengden; **325** Left, CNRI/Science Photo Library; Right, Dr. Linda Peterson/University of Ottawa; **327** Left, FPG/Masterfile; Bottom, Stephen Derr/The Image Bank; Top right, Comstock/Miller Comstock; Bottom right, Simon Fraser, Hexham General/Science Photo Library; **FOURTEEN: 14.1** St. Bartholemew's Hospital/Science Photo Library; **332** Ellen Schuster/The Image Bank; **333** Lars Ternblad/The Image Bank; **14.2** Kam Yu; **14.3** Kam Yu; **14.4** Stephen Mader and Martyn Lengden; **14.5** Stephen Mader; **14.6** Stephen Mader; **14.7** Stephen Mader after C. Starr and R. Taggart, *Biology 5ed.*, Wadsworth Publishing Company, 1989; **14.8A** Manny Millan/Sports Illustrated; **14.8B** The Bettman Archive; **14.9** Photographs courtesy of Dr. William H. Daughaday, Washington University School of Medicine. From A.I. Mendelhoff and D.E. Smith, eds., *American Journal of Medicine*, 20:133 (1956); **14.10** Stephen Mader and Martyn Lengden; **14.11A** Secchi-Lecaque/Roussel-Uclaf/CNRI/Science Photo Library **14.11B** Dave Mazierski; **14.12** Martyn Lengden; **342** Left, Canapress Photo Service/Wide World Photo; Right, Biophoto Associates/Photo Researchers; **14.13** Dave Mazierski; **14.14** Martyn Lengden and Stephen Mader; **346** Martyn Lengden; **FIFTEEN: 15.1** Obremski/The Image Bank; **350** Hank deLespinasse/The Image Bank; **15.2** Martyn Lengden; **15.3** Stephen Mader; **15.4A** Dave Mazierski; **15.4B** Ed Rescuke/Peter Arnold Inc.; **15.5** Dave Mazierski; **355** Kam Yu; **356** Top right, Stephen Mader; **15.6** A. Boccaccio/The Image Bank; **15.7A** Stephen Mader after C. Starr and R. Taggart, *Biology 5ed.*, Wadsworth Publishing Company, 1989; **15.7B** Martyn Lengden after C. Starr and R. Taggart, *Biology 5ed.*, Wadsworth Publishing Company, 1989; **15.8** Martyn Lengden; **15.9A** Stephen Mader; **15.9B** Martyn Lengden; **15.10** Martyn Lengden; **15.11** Martyn Lengden; **15.12** Stephen Mader; **15.13** Martyn Lengden; **15.14A** CBC/Phototake NYC/First Light; **15.14B** Kam Yu; **15.15** Martyn Lengden; **15.16** Stephen Mader and Martyn Lengden; **364** Stephen Mader; **365** Left, Custom Medical Stock Photo Inc.; Top right, Dr. John McClean/Memorial University of Newfoundland; Bottom right, Dr. Sergey Federoff/University of Saskatchewan; **15.17** Dave Mazierski; **15.18** Dave Mazierski; **15.19** Steve Gilbert; **15.20A** Steve Gilbert; **15.20B** Dr. Colin Chumbley/Science Photo Library; **15.21** Dave Mazierski; **15.22** Steve Gilbert; **373** Left, Joseph Drivas/The Image Bank; Right, Canapress Photo Service/Montreal Gazette; **374** Top, Photo courtesy of the Warren Anatomical Museum/Harvard Medical School; Bottom, Alexander Tsiaras/Science Photo Library; **15.23** Stephen Mader; **378** Left, Stephen Mader; Right, Martyn Lengden; **SIXTEEN: 16.1** Geoffrey Gove/The Image Bank; **380** Douglas J. Fisher/The Image Bank; **381** A. Gesar/The Image Bank; **16.2** Richard Siemens; **16.3** Dave Mazierski; **16.4A** Dave Mazierski; **16.4B** Sinclair Stammers/Science Photo Library; **16.5** Dave Mazierski; **384** Bottom right, Martyn Lengden; **16.6A** Dave Mazierski and Martyn Lengden; **16.6B** Omikron/Science Source/Photo Researchers; **387** Left, Kelly Dobler, Rufus Day and Rob Hale/Cross Cancer Institute; Right, Frank LoCicero/Cross Cancer Institute; **16.7** Martyn Lengden; **16.8** Martyn Lengden; **16.9** Martyn Lengden; **390** Fred Bruemmer; **16.10** Kam Yu; **16.11** Martyn Lengden; **16.12** Martyn Lengden; **392** Martyn Lengden; **393** Martyn Lengden; **394** Right, Kam Yu; **16.13** Steve Gilbert; **397** Frink/Waterhouse/H. Armstrong Roberts/Miller Comstock; **16.14** Steve Gilbert; **16.15** Steve Gilbert; **16.16** Martyn Lengden; **16.17** Dave Mazierski; **16.18** Stephen Mader; **400** Kam Yu; **401** Left, Snowdon/Canapress Photo Service; Center, Comstock/Miller Comstock; Top right, Victor Fisher/Canapress Photo Service; Bottom right, Robert J. Herko/The Image Bank **402** Top, Dr. Goran Bredberg/Science Photo Library; Center, Secchi, Lecaque, Roussel, Uclaf, CNRI/Science Photo Library; Bottom, Scanning electron micrograph prepared by Robert Preston, Courtesy of Prof. J.E. Hawkins, Kresge Hearing Research Institute, University of Michigan; **403** Martyn Lengden; **SEVENTEEN: 17.1** Carl Haycock/First Light; **406** A.B. Dowsett/Science Photo Library/Masterfile; **17.2A** Dave Mazierski; **17.2B** Secchi, Lecaque, Roussel, Uclaf, CNRI/Science Photo Library; **17.3** Dave Mazierski; **17.4**

Dave Mazierski; **17.5** Martyn Lengden; **17.6** Dave Mazierski; **17.7** Dave Mazierski; **17.8** Kam Yu; **17.9** Martyn Lengden; **419** Left, Kam Yu; Right, Martyn Lengden; **420** Martyn Lengden; **17.10** From Lennart Nilsson, *A Child is Born*, ©1966,1977, Dell Publishing Company Inc. **17.11** Andy Walker, Midland Fertility Services/Science Photo Library; **17.12** Dave Mazierski; **17.13** From Lennart Nilsson, *A Child is Born*, ©1966,1977, Dell Publishing Company Inc.; **17.14** Dave Mazierski; **17.15** Martyn Lengden; **17.16** Kam Yu; **426** Left, Dr. Renee Martin. Alberta Children's Hospital, Child Health Centre; Right, Howard Sochurek/Masterfile; **428** Left, Dave Mazierski; Right, Martyn Lengden; **429** Martyn Lengden; **EIGHTEEN: 18.1** Dave Mazierski; **18.2** Science Photo Library/Photo Researchers; **18.3** Dave Mazierski; **18.4** Biophoto Associates/Science Source/Photo Researchers; **18.5** Andrew Bajer/University of Oregon; **18.6** Dave Mazierski; **18.7** Carolina Biological Supply Co.; **18.8** Runk/Schoenberger from Grant Heilman/Miller Comstock; **18.9** Kam Yu; **18.10** Dave Mazierski; **18.11** Canapress Photo Service/AP/Wide World Photos; **18.12** Dave Mazierski; **18.13A** John Kelly/The Image Bank; **18.13B** André Gallant/The Image Bank; **18.14** Kam Yu; **18.15** Martyn Lengden; **18.16** National Cancer Institute - NCI Laboratory of Tumor Virus Biology; **18.17** Canapress Photo Service; **445** Left, Dr. Tony Brain/Science Photo Library; Right, Dr. Margarida Krause/Dept. of Biology/University of New Brunswick; Dr. Grant McFadden/Dept. of Biochemistry/University of Alberta; Dr. Nancy Simpson/Queen's University; **NINETEEN: 19.1** Dr. Jeremy Burgess/Science Photo Library; **448** Dr. Jeremy Burgess/Science Photo Library; **19.2** Martyn Lengden; **19.3** Kam Yu; **19.4** Kam Yu; **19.5** Dave Mazierski; **19.6** Dave Mazierski; **19.7** Kam Yu; **452** Andy Walker, Midland Fertility Services/Science Photo Library; **19.8** Kam Yu; **454** Martyn Lengden; **19.9A** Photo courtesy of Oshawa General Hospital; **19.9B** Photo courtesy of McMaster University; **456** Photographs courtesy of the Ontario Special Olympics; **19.10** Martyn Lengden; **19.11** Kam Yu; **457** Left, Colin Milkins Oxford Scientific Films/Animals Animals/Earth Scenes; Right, Dr. Irene Uchida/National Film Board of Canada; **19.12** A. Tsiaras/Photo Researchers; **459** Martyn Lengden; **19.13** UPI/The Bettman Archive; **19.14** Kam Yu; **19.15** Kam Yu; **19.16** Kam Yu; **463** Left, David Young-Wolff/Photo Edit; Center, Grant Heilman/Miller Comstock; Top right, Custom Medical Stock Photo; Bottom right, Steve Dunwell/The Image Bank; **TWENTY: 20.1** Konrad Wothe/Survival Anglia; **468** Konrad Wothe/Survival Anglia; **20.2** Science Photo Library **20.3** Kam Yu; **20.4** Kam Yu; **20.5** Kam Yu; **20.6** Kam Yu and Martyn Lengden; **20.7** Kam Yu and Martyn Lengden; **20.8** Kam Yu and Martyn Lengden; **20.9** Kam Yu and Martyn Lengden; **20.10** Kam Yu and Martyn Lengden; **20.11** Valda Glennie and Martyn Lengden; **20.12A** Dave Mazierski; **20.12B** Dr. Jeremy Burgess/Science Photo Library; **20.13** Dave Mazierski and Martyn Lengden; **20.14A** Dr. E.R. Degginger; **20.14B** Valda Glennie and Martyn Lengden; **20.15** Kam Yu and Martyn Lengden; **479** Martyn Lengden; **20.16** Novosti/Science Photo Library; **20.17** Kam Yu and Martyn Lengden; **20.18** Kam Yu and Martyn Lengden; **20.19** Dave Mazierski; **20.20** Kam Yu and Martyn Lengden; **20.21** Richard Siemens; **20.22** Martyn Lengden; **484** Carolina Biological Supply Co. **20.23** Kam Yu; **486** Left, Agriculture Canada; Centre, Dr. Baldur Stefansson; Right, Chris Alan Wilton/The Image Bank; **20.24** Martyn Lengden; **20.25** Martyn Lengden; Photos courtesy of Tedd Somes from C. Starr and R. Taggart, *Biology 5ed.*, Wadsworth Publishing Company, 1989 **20.26** Martyn Lengden; **20.27** Martyn Lengden; **20.28** The Bettman Archive; **20.29** Derek Caron/Masterfile; **20.30** Dave Mazierski; **493** Martyn Lengden; **TWENTY-ONE: 21.1** Mark Tucker; **494** David Klutho/Sports Illustrated; **495** Dave Mazierski; **21.2** Dr. E.R. Degginger; **21.3** Dave Mazierski; **21.4** Dave Mazierski and Martyn Lengden; **21.5** Martyn Lengden; **21.6** Dr. E.R. Degginger; **21.7** Norva Behling/Animals Animals; **21.8** Larry J. MacDougal/First Light; **502** The color blindness charts were reproduced from *Ishihara's Test for Colour Blindness* published by KANEHARA & Co., LTD. Tokyo, Japan. Tests for color blindness cannot be conducted with this material. For accurate testing, the original plates should be used. **21.9** Dave Mazierski; **21.10** Dave Mazierski; **21.11** Martyn Lengden; **21.12** Dave Mazierski; **505** Martyn Lengden; **506** Dave Mazierski; **21.13** UPI/The Bettman Archive; **21.14** Dr. L. Caro/Science Photo Library; **21.15** Dave Mazierski; **21.16** Philippe Plailly/Science Photo Library; **510** Left, Reuters/The Bettman Archive; Right, Dr. Chris Burton/Science Photo Library; **21.17** Dave Mazierski; **514** Martyn Lengden; **TWENTY-TWO: 22.1** R. Scherer/Miller Comstock; **516** Shoji Yoshida/The Image Bank; **22.2** Philippe Plailly/Science Photo Library; **518** Dave Mazierski; **519** Dave Mazierski; **22.3A** Martyn Lengden; **22.3B** Omikron/Photo Researchers; **22.4** Omikron/Photo Researchers; **22.6** Valda Glennie after art by D. & V. Hennings in C. Starr and R. Taggart, *Biology 5ed.*, Wadsworth Publishing Company, 1989; **522** Top right, Martyn Lengden; **22.7** Kam Yu; **524** Left, Ralph Bower/Pacific Press Ltd./Vancouver Sun; Right, Science Source/Photo Researchers; **525** Martyn Lengden; **22.8** Pamela Newell, Head of DNA Unit, Biology Section, Centre of Forensic Sciences, Toronto; **22.9** Alberta Research Council; **22.10** R.L. Brinster and R.E. Hammer, School of Veterinary Medicine, University of Pennsylvania; **TWENTY-THREE: 23.1** Joe Van Os/The Image Bank; **532** Gary S. Chapman/The Image Bank; **23.2** Martyn Lengden; **23.3** Martyn Lengden; **23.4** Kam Yu; **23.5** Dave Mazierski; **538** Martyn Lengden; **539** Martyn Lengden; **23.6** Kam Yu; **23.7** Jackie Lewin, Royal Free Hospital/Science Photo Library; **541** Top, Kam Yu; Bottom, Science Photo Library/Photo Researchers; **542** Kam Yu; **543** Top, Kam Yu; Bottom, Prof. Luc Montagnier, Institut Pasteur/CNRI/Science Photo Library; **544** NIBSC/Science Photo Library; **23.8** Kam Yu; **23.9** Martyn Lengden; **23.10** Dave Mazierski; **548** Left, Dr. Ram Mehta/Prairie Biological Research Ltd.; Right, Dr. Jeremy Burgess/Science Photo Library; **549** Reuters/The Bettman Archive; **550** Left, N.R. Rowan/Custom Medical Stock Photo; Center, Bill Becker/Canapress Photo Service; Top right, Martin Dohrn/IVF Unit, Cromwell Hospital/Science Photo Library; Bottom right, Bill Becker/Canapress Photo Service; **TWENTY-FOUR: 24.1** The Image Bank Canada; **556** Grant Heilman/Miller Comstock; **557** Scott Camazine/Science Source/Photo Researchers; **558** Martyn Lengden; **559** Left, Eric Grave/Phototake NYC/First Light; Right, Dr. Frank Fraser, Van Dyck & Meyers Ltd.; Dr. Norma Walker, University of Toronto Archives; **24.2** Biophoto Assoc./Science Source/Photo Researchers; Jim Stevenson/Science Photo Library/Photo Researchers; **24.3** Fritz Prenzel Photo/Animals Animals/Earth Scenes; **567** Science Source/Photo Researchers; **568** Martyn Lengden; **24.4A** Martyn Lengden; **24.4B** CNRI/Science Photo Library/Photo Researchers; **24.5** Larry Fisher/Masterfile; **24.6** Asa C. Thorensen, The National Audubon Society Collection/Photo Researchers; **24.7** Zig Leszczynski/Animals Animals/Earth Scenes; **TWENTY-FIVE: 25.1** Jessie Parker/First Light; **576** K. Wothe/The Image Bank; **25.2** Nick Bergkessel, The National Audubon Society Collection/Photo Researchers; **25.3** Martyn Lengden; **25.4** Martyn Lengden; **25.5** Margo Stahl; **25.6A** J.R. Wicinsons/Animals Animals/Earth Scenes; **25.6B** Greg Vaughn/Tom Stack & Assoc.; **25.7** Leef Snyder, The National Audubon Society Collection/Photo Researchers; **25.8** Martyn Lengden; **25.9** Martyn Lengden; **25.10** Martyn Lengden; **25.11** Martyn Lengden; **25.12** Martyn Lengden; **25.13** Martyn Lengden; **25.14** Martyn Lengden; **25.15** Martyn Lengden; **25.16** G.C. Kelley, The National Audubon Society Collection/Photo Researchers; **25.17** Martyn Lengden; **25.18** Martyn Lengden; **25.19A** Alan Carey, The National Audubon Society Collection/Photo Researchers; **25.19B** Martyn Lengden; **25.20** Martyn Lengden; **25.21** Martyn Lengden; **25.22** Martyn Lengden; **25.23** Martyn Lengden; **25.24** Marty Stouffer/Animals Animals/Earth Scenes; **25.25** Hans Pfletschinger/Peter Arnold Inc.; **25.26** Top, Cath Ellis, Dept. of Zoology, University of Hull; Bottom, Robert Hermes, The National Audubon Society Collection/Photo Researcher; **25.27** Henry Ausloos/Animals Animals/Earth Scenes; **25.28** Dr. E.R. Degginger; **597** Left, P. Wallick/Miller Comstock; Right, Dr. Austin Reed; **25.29** Barbara Von Hoffman/Tom Stack & Associates; **25.30** Martyn Lengden; **25.31** Margo Stahl; **604** Top left, Victor Last; Bottom left, Richard Kolar/Animals Animals/Earth Scenes; Right, Brian Milne/Animals Animals/Earth Scenes; **605** Top to Bottom, A.E. Sirulnikoff/First Light; Doug Wechsler/Animals Animals/Earth Scenes; Doug Wechsler/Animals Animals/Earth Scenes; Dr. E.R. Degginger; **606** Left, LaFoto/H. Armstrong Roberts/Miller Comstock; Top center, Eric Hayes/Miller Comstock; Bottom center, J.D.Taylor/Miller Comstock; Top right, Barbara K. Deans/Canapress Photo Service; Bottom right, Elena Rooraid/Photo Edit.